Green Energy and Technology

Climate change, environmental impact and the limited natural resources urge scientific research and novel technical solutions. The monograph series Green Energy and Technology serves as a publishing platform for scientific and technological approaches to "green"—i.e. environmentally friendly and sustainable—technologies. While a focus lies on energy and power supply, it also covers "green" solutions in industrial engineering and engineering design. Green Energy and Technology addresses researchers, advanced students, technical consultants as well as decision makers in industries and politics. Hence, the level of presentation spans from instructional to highly technical.

Indexed in Scopus.

More information about this series at http://www.springer.com/series/8059

Saad Motahhir · Ali M. Eltamaly
Editors

Advanced Technologies for Solar Photovoltaics Energy Systems

Springer

Editors
Saad Motahhir
ENSA
Sidi Mohamed Ben Abdellah University
Fes, Morocco

Ali M. Eltamaly
Electrical Engineering Department
Mansoura University
Mansoura, Egypt

ISSN 1865-3529 ISSN 1865-3537 (electronic)
Green Energy and Technology
ISBN 978-3-030-64564-9 ISBN 978-3-030-64565-6 (eBook)
https://doi.org/10.1007/978-3-030-64565-6

This Springer imprint is published by the registered company Springer Nature Switzerland AG
The registered company address is: Gewerbestrasse 11, 6330 Cham, Switzerland

Preface

With the steady increase in the consumption of electric energy worldwide and with the increase of technological industries, dependence on fossil fuels has led to depletion and high prices in addition to the harmful effects it causes to the surrounding environment. All these effects made an urgent need for more energy sources to meet these requirements and remedy the effects of the generation from fossil fuels. That is why, renewable energy sources have become a solution to this dilemma because they are inexhaustible and do not pollute the environment, and there is no way to prevent societies from using these sources of energy. Solar photovoltaic energy was one of the most important of these sources and now it competes with even some traditional energies, and it has many uses that have increased its importance. With the steady decline in the prices of solar panels, as they appeared in the year 1958 with about $1000 to generate only 1 watt, and now in 2020, the solar panels capable of generating 1000 watts are less than $1000. With the scientific and technical progress, scientists have added many improvements to the generation system from solar cells, and these studies focused on two basic methods. The first is to improve the efficiency of the photovoltaic cells by using materials that can generate electricity with high efficiency and the other way is to improve the efficiency of the electrical energy extraction circuits from cells such as sun-tracking systems and maximum power point tracking systems, cell cleaning systems from dust, etc. All of these methods to improve the efficiency of the generation system from solar cells were highlighted in this book.

The book introduced a study in improving the efficiency of the photovoltaic modules using new materials and improved manufacturing technology. Moreover, mathematical models used to create simulations of these cells were introduced. The book also dealt with improving sun tracking systems and using inexpensive systems that can track the sun throughout the year. Studies of automatic cleaning systems that use robots to clean cells from dust were also presented to increase the efficiency of a photovoltaic power generation system.

Many studies have been presented that explains the issue of partial shading on solar cells and its harmful effects and ways to get rid of these effects, as well as explaining, reviewing, and comparing systems that track the highest generation power of cells, clarifying the best ones, and adding many modifications to these systems. The book also contains hybrid renewable energy generation systems and the possibility of using them in feeding remote communities that are far from the electrical grid. The study of the hybrid systems also includes the design and sizing of the components to build these hybrid systems. Moreover, this book covers the use of new advanced technologies as embedded system, Internet of Thing (IoT), and blockchain technologies for PV systems through different applications such as MPPT controller, solar tracker, cleaning system, and monitoring system.

The book will be very useful for all undergraduate and postgraduate students, as well as those interested in the field of generating electricity from renewable energy sources, especially solar energy. Moreover, this book will be very interesting for the readers who are looking for using solar modules to feed loads in isolated areas as well as on the utility scale. It will also help them to know the photovoltaic energy systems' characteristics, modeling, operation, challenges, maximum power tracking, and practical implementation. This book will help the researchers, designers, and operators, as well as undergraduate/postgraduate students, to be familiar with the new trends in the field of photovoltaic energy systems.

Acknowledgments
The editors of this book would like to thank the authors and reviewers for their contributions and efforts. Moreover, we would like to thank all colleagues from K. A. CARE Energy Research and Innovation Center, Riyadh, Saudi Arabia for their help and efforts.

Fes, Morocco Saad Motahhir
Mansoura, Egypt Ali M. Eltamaly

Contents

About the Editors

Saad Motahhir (Eng., Ph.D., IEEE member) has previous expertise acting in industry as Embedded System Engineer at Zodiac Aerospace morocco from 2014 to 2019, and more recently became a professor at ENSA, SMBA University, Fez, Morocco, since 2019. He received the engineer degree in Embedded System from ENSA Fez in 2014. He received his Ph.D. degree in Electrical Engineering from SMBA University in 2018. He has published a good number of papers in journals and conferences in the last few years, most of which are related to Photovoltaic (PV) solar energy and embedded systems. He published a number of patents in the Morocco Patent Office. He acted as a guest editor for different special issues and topical collections. He is a reviewer and in the editorial board of different journals. He was associated with more than 30 international conferences as a Program Committee/Advisory Board/Review Board member.

Ali M. Eltamaly (Ph.D—2000) is a full professor at Mansoura University, Egypt, and King Saud University, Saudi Arabia. He received his B.Sc. and M.Sc. degrees in electrical engineering from Al-Minia University, Egypt, in 1992 and 1996, respectively. He received his Ph.D. degree in Electrical Engineering from Texas A&M University in 2000. His current research interests include renewable energy, smart grid, power electronics, motor drives, power quality, artificial intelligence, evolutionary and heuristic optimization techniques, and distributed generation. He published 20 book and book chapters and he has authored or coauthored more than 200 refereed journal and conference papers.

He published several patents in the USA Patent Office. He has supervised several M.S. and Ph.D. theses worked on several National/International technical projects. He got a distinguish professor award for scientific excellence, Egyptian Supreme Council of Universities, Egypt, June 2017, and he has been awarded many prizes in different universities in Egypt and Saudi Arabia. He is participating as an editor and associate editors in many international journals and chaired many international conferences' sessions. He is the chair professor of Saudi Electricity Company Chair in power system reliability and security, King Saud University, Riyadh, Saudi Arabia.

Abbreviations

ABC	Artificial Bee Colony Algorithm
ACO	Ant Colony Optimization
AI	Artificial Intelligent
ANN	Artificial Neural Network
APPSO	adaptive perceptive PSO
BA	Bat Algorithm
BNIA	Bio/Natural Inspired Algorithms
BOA	Beta Optimization Algorithm
BST	Bisection Search Technique
CDC	counts of dimension to change
CFA	Curve-fitting Algorithm
CFA	Cuttlefish Algorithm
COA	Chaos Optimization Algorithm
COA	Chaos optimization algorithms
CS	Cuckoo Search
CSO	Cat Swarm Optimization
DCCS	Dual Carrier Chaotic Search
DCLCDC	DC-Link Capacitor Droop Control
DE	Differential Evolution
DEM	Direct estimated methodology
EA	Evolutionary Algorithms
EOA	Earthquake Optimization Algorithm
ESC	Extremum Seeking Control
FFA	Fireflies Algorithm
FLC	Fuzzy Logic Controller
FOCV	Fractional Short-circuit Voltage
FPA	Flower Pollination Algorithm
FSA	Fibonacci Search Algorithm
FSCC	Fractional Short-circuit Current
GA	Genetic Algorithm
GHO	Grass Hopper Optimization
GP	Global peak

GWO	Grey Wolf Optimizer
HC	Hill-climbing
HMT	Hybrid MPPT Techniques
HPO	Human Psychology Optimization
IMKE	Intelligent Monkey King Evolution
InCond	Incremental Conductance
JOA	Jaya Optimization Algorithm
LCLVM	Load Current or Load Voltage Maximization
LPs	Local peaks
LuT	Look-up Table
MA	Metaheuristic Algorithms
MBA	Mathematical Based Algorithms
MEE	MPPT energy efficiency
MFO	Moth-Flame Optimization
MML	Mismatch loss
MPP	Maximum power point
MPPT	Maximum power point tracker
MPT	Maximum Power Trapezium
P&O	Perturb & observe
PSC	Partial shading condition
PSO	Particle Swarm Optimization
PV	Photovoltaic
PWM	Pulse Width Modulation
RCC	Ripple Correlation Control
SA	Simulated Annealing
SCO	Stepped-up Chaos Optimization
SC-SC	Soft-computing with soft-computing
SIA	Swarm Intelligence Algorithms
SMA	Skipping Mechanism Algorithm
SMC	Slide Mode Control
SMP	Seeking memory pool
SPC	Self-position considering
SRD	Seeking range of the selected dimension
SSA	Salp Swarm Algorizm
SSJ	Search-Skip-Judge
TBM	Transient Based MPPT
TLA	Teaching Learning Algorithm
TPBP	Three-point Bidirectional Perturbation
TSA	Tabu Search Algorithm
T-SC	Traditional with soft-computing
T-T	Traditional with traditional MPPT Techniques
VWS	Voltage Window Search
WCA	Water Cycle Algorithm
WOA	Whale Optimization Algorithm

Nomenclature

Forecast Model Implementation

$r_{\alpha\beta}$	Pearson correlation coefficient of variables α and β
$\hat{h}$	A positive integer reflecting $\hat{h}$ time steps ahead
$G(\cdot)$	Nonlinear mapping relationship between input and output variables
K	Data sample size for the forecast model
u	Input data array for the forecast model
u_{norm}	Normalized input data array for the forecast model
$u_1(t)$	Discrete PV power time series obtained from phase space reconstruction
$u_2(t)$	Weather data array obtained from correlation analysis
u_{min}, u_{max}	Minimum and maximum values of the input data array for the forecast model
$y_{act}\left(t+\hat{h}\right)$	Observed value of the PV output power at time $\left(t+\hat{h}\right)$
$y_p\left(t+\hat{h}\right)$	Forecast value of the PV output power at time $\left(t+\hat{h}\right)$

Short-Term Memory Network

$\circ$	Multiply symbol by element
$\sigma(\cdot)$	Sigmoid activation function
$\tanh(\cdot)$	tanh activation function
C_t	Candidate value of memory cell state at time t
b_f, b_i, b_c, b_o	Bias vector for the forget gate, input gate, cell memory state, output gate
C_t	Cell memory state at time t
f_t	Output value of the memory cell's forget gate at time t
h_t	Hidden state at time t

i_t, O_t	Output value of the memory cell's input gate and output gate at time t
W_{cx}, W_{ch}	Weight matrix of the cell memory state corresponding to x_t and h_{t-1}
W_{fx}, W_{fh}	Weight matrix of the forget gate corresponding to x_t and h_{t-1}
W_{ix}, W_{ih}	Weight matrix of the input gate corresponding to x_t and h_{t-1}
W_{ox}, W_{oh}	Weight matrix of the output gate corresponding to x_t and h_{t-1}
x_t	Input to the memory cell at time t

Nonlinear Reconstruction Technique

$\bar{S}\ (\lambda)$	Average of the statistical metric $S\left(m, N, r_j, \lambda\right)$ of all sub-sequences
$\Delta\ \bar{S}\ (\lambda)$	Average of the statistical metric $\Delta S(m, N, \lambda)$ of all sub-sequences
Δt	Sample interval for the time series
λ	Number of disjoint sub-sequences from time series with length N
$\varepsilon(m, N, r, \tau)$	Correlation integral of the time series
ε_η	Correlation integral of the η-th sub-sequence
τ	Time delay of the phase space
$A(i)$	The i-th phase point in an m-dimension phase space
$a(n)$	The n-th point of scalar time series in phase reconstruction
$a_{\mathrm{PV}}(t)$	The t-th point of PV power time series in phase reconstruction
$H(\cdot)$	Heaviside function
$int(\cdot)$	Taking an integer operation
M	Number of phase points in the phase space
m	Embedding dimension of the vector $A(i)$
N	Length of the time series
q	Number of non-zero elements in $d_i(\mu)$
r	Threshold distance
$y(\mu)$	Average value of $\ln d_i(\mu)$ for i
$d_i(0)$	Initial distance between the i-th point $A(i)$ and its nearest neighbor $A(\hat{i})$
$d_i(\mu)$	Distance between the i-th pair of the nearest neighbors after μ discrete time steps
$S\ (m, N, r_j, \lambda)$	Statistical metric for each sub-sequence with neighbor radius r_j

Acronyms

AC	Alternating Current
D/R	Demand/Response
DC	Direct Current
EEBLab	Energy Efficient Building laboratory
EM	Energy Management
HVAC	Heating, Ventilation, and Air-Conditioning
ICT	Information and Communication Technologies
IoE	Internet of Energy
IoS	Internet of Service
IoT	Internet of Things
MG	Micro-Grid
PV	Photovoltaic
RES	Renewable Energy Source
SG	Smart Grid
SoC	State-of-Charge
TEG	Traditional Electric Grid

Advanced Materials for Solar Cell Applications: Case of Simple and Composite Oxides

Abderrahman Abbassi

Abstract The study aims to investigate the electronic and optical properties of two types of oxides in their stable phases (the co-doped ZnO and $BiYO_3$, the simple and composite compound) in order to build a new sufficient solar cell transparent electrodes. This work may contribute to the development of solar cell electrodes by the exploitation of the optoelectronic properties of these compounds. This study is made with the numerical techniques ab initio based on the DFT that we will describe in more details in this chapter. The result was found to show that the transmittance of $BiYO_3$ and the simple oxide treated is significant (about 90% for the simple oxide), bandgap varies and the behavior of the conductivity is ensured by the presence of an important concentration of electrons.

Keywords Oxides · Semiconductors · Ab initio · DFT · Optical properties · Electronic properties, DOS

1 Introduction

The manufacturing of solar electrodes has grown significantly in recent years; scientists are working to increase the performance of solar panels, improve their yields and other parameters related to their constructions. The use of technologies based on Silicon is starting to have a huge change and especially by the use of other materials, which are considered as high-performance materials compared to the old generation of solar electrodes. This is where our idea comes from, the development of window layers, active layers in PV applications. We focus on two types of oxides, simple and composite. In this work, we study only the first layer in PV cells, which is called the window or transparent layer with the capacity to collect a maximum of solar radiation. We calculate this transmittance rate by formalisms and computer code, which are reliable and known in materials science. A second calculation is made to

A. Abbassi (✉)
Polydisciplinary Faculty of Beni Mellal, Sultan Moulay Slimane University, Mghila BP: 592, Beni Mellal, Morocco
e-mail: abbassi.abder@gmail.com

S. Motahhir and A. M. Eltamaly (eds.), *Advanced Technologies for Solar Photovoltaics Energy Systems*, Green Energy and Technology,
https://doi.org/10.1007/978-3-030-64565-6_1

estimate the conductivity of these materials. The first type of materials treated in this chapter is the simple oxide, we will focus on the investigation of different properties of zinc oxide doped and co-doped with various amounts of dopants as aluminum and silicon. Several types of research have been published by many researchers in which they investigate the properties of ZnO [1–8]. We will briefly discuss the properties that make zinc oxide material of great importance according to his application. The zinc oxide ZnO is a powder in the solid state; it has some advantages such as being noncombustible, it is abundant and non-toxic. The crystallization of this compound can be made in two different structures, hexagonal/wurtzite (Fig. 1) and cubic zinc blend. The wurtzite form is thermo-dynamically stable more than other structures at ambient conditions. The structural parameters of the wurtzite used in this study are: $a = 0.32495$ nm and $c = 0.52069$ nm, with $c/a \sim 1.60$ which is close to the ideal ratio $c/a = 1.633$. The unit cell positions are (0; 0; 0), and (2/3; 1/3; 1/2). Each atom of zinc in tetrahedral site is encircling by four oxygen atoms and vice versa. The oxygen and c bond has an important ionic character due to the high electronegativity of the oxygen atom.

The point group of wurtzite zinc oxide is 6 mm according to the Hermann–Mauguin notation, this point group can also be found with Schoenflies notation C_{6v}, $P6_3mc$ and C_{6v}^4 mentioned the space group of the studied structure. ZnO bonding is ionic (Zn^{2+}–O^{2-}), as in several compounds classified in groups II–VI. The radii used is 0.140 nm (O^{2-}) and 0.074 nm (Zn^{2+}). This characteristic is important concerning the formation of hexagonal structure comparing with the blend structure. ZnO has also a significant piezoelectricity, which can be used in other different applications, especially the manufacturing of sensors. The polar Zn–O bonds make zinc and oxygen planes electrically charged.

The second type to study in this chapter is the multiferroics compound as composite oxide with as single phase, which present firstly two or more primary ferroic properties. These composite materials were discovered many years ago and are still not extensively investigated. These composite oxides present different aspects: the magnetic and ferroelectric, the optical one is still not investigated widely.

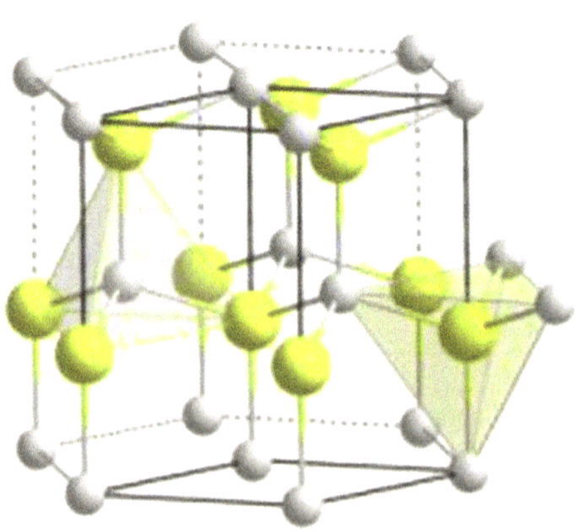

Fig. 1 Wurtzite structure of ZnO https://commons.wikimedia.org/wiki/File:Wurtzite_polyhedra.png

The ferroelectric behavior is mainly due essentially to the hybridization d and p states of its atoms. The electronic transition was observed in $BaTiO_3$. The multiferroic term was developed to take into consideration several properties such as ferroelasticity, spontaneous electric polarization, and magnetic. Therefore, the multiferroic materials can present simultaneously various properties of ferroelasticity, ferromagnetism, ferroelectricity and can take antiferromagnetic or ferromagnetic aspects. The 3+ cations of atom A contribute to show the ferroelectricity aspect, however, 3d of atom B (ABO_3) contributes to show magnetic effect. Now, we focus in this chapter on the study of the ability to use these materials in solar applications. Some recent works report the interest of simple perovskite/composite oxides and simple co-doped oxides example of ZnO [9–13] in these applications.

The ABO_3 present in some case a coupling between the polarization and magnetic effect. The study will take consideration of tetragonal (Fig. 2) as a stable structure in the calculation, we estimate that the structural parameters are $a = b = 3.729A^{\circ}$, $c = 4.72A^{\circ}$ and $\alpha = \beta = \gamma = 90$ [14]. This family of composite oxide can be the aim of several types of research in order to investigate their new applications and manufacturing, especially in solar cell electrodes. BiYO3 will be treated in this chapter, Y is the position that can take various atoms. The position Y is chosen in order to stabilize the structure and to increase the required properties, Y will not cause any distortion of the tetragonal structure, Y = Fe, Zn, V, Co have approximately the same atomic radius. The optical result of this work will be based mainly on the transmittance rate and absorption coefficient found made by the DFT calculation.

Fig. 2 Tetragonal structure of $BiYO_3$ (made by Wien2K with XCrySDen interface)

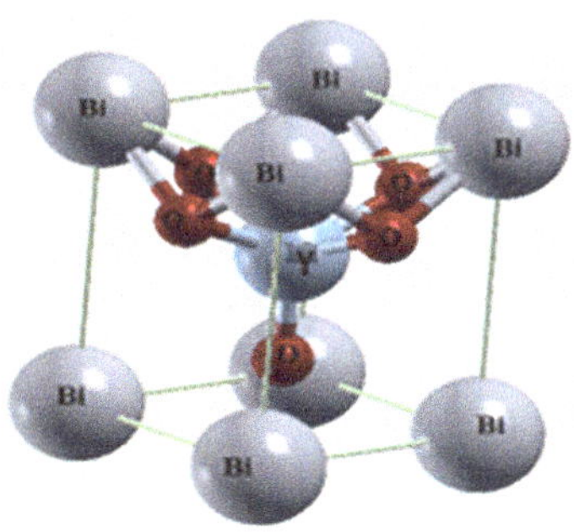

2 Modeling Method

The Wien is a code that works on Linux, UNIX, it was developed at the Institute of Materials Chemistry—Austria, published by Blaha et al. in 1990. In order to improve it and make it adaptable to different calculations and to treat different physical properties of materials, Wien code has undergone several changes and every change takes a published notation, which depends on the year of publication, e.g. Wien93, Wien95, Wien97, WIEN2k. This work was developed and made with the Wien2k_13.1 version that presents the best advantage in terms of calculation time and formalism used for the calculation of the physical properties of systems, efficiency, and reliability. This version has also a graphic interface "w2web" to access the web. w2web means the Wien to the web. The WIEN2k code has shown efficiency in the field of quantum chemistry and the physics of condensed matter. It is based on the augmented plane wave method implemented within the DFT. This package is a set of independent programs written with Fortran and calculates several physical properties such as electronic properties of materials for the study of the band structure, the total/partial electronic density (TDOS and PDOS), and the optical properties. This code can calculate the optical parameters such as absorption, reflectivity, refractive index, optical conductivity, etc. It may also treat the total energy of the system and the optimization of structures, structural properties, thermodynamic (enthalpy, etc.), and magnetic properties (calculation of ferromagnetic and anti-ferromagnetic state to study the magnetic stability of the systems), polarization, electric field gradients, and hyperfine fields, etc.

The electrical and optical properties of the studied simple and composite oxides were calculated with the first principles using DFT theory with different approaches and approximation of the correction: GGA and modified Beak Johnson [15], using the wien2k package.

The imaginary part of the dielectric function tensor can be written by the following expression:

$$Im\,\varepsilon_{\alpha\beta}(\omega) = \frac{4\pi e^2}{m^2\omega^2}\sum_{c,v}\int dk c_k\left|P^{\alpha}|V_k\rangle\langle V_k|P^{\beta}\right|c_k\delta(\varepsilon_{C_K} - \varepsilon_{V_K} - \omega) \quad (1)$$

And the optical conductivity can be written also by

$$Re\,\sigma_{\alpha\beta}(\omega) = \frac{\omega}{4\pi} Im\,\varepsilon_{\alpha\beta}(\omega) \quad (2)$$

V_k and C_K mentioned the wave functions, they mention respectively also VB and BC bands, and k is the vector of the concerned waves.

In Eq. (2), $\sigma_{\alpha\beta}$ is the conductivity parameter, which linked directly to the current density J_α between VB and CB bands following α direction. This density generate sE_β as an electric field, which follows the β (direction). The dielectric function is written by the following expression:

$$\varepsilon(\omega) = \varepsilon_1(\omega) + i\varepsilon_2(\omega) \tag{3}$$

The parameter $\varepsilon_1(\omega)$ (real part) can be found using $\varepsilon_2(\omega)$.

$$\varepsilon_1 = Re(\varepsilon(\omega)) = 1 + \frac{2}{\pi} L \int_0^\infty \frac{\omega\prime \varepsilon_2(\omega\prime)}{\omega\prime^2} d\omega\prime \tag{4}$$

With

L(P) is the main value of the integral.

And

$\varepsilon_2(\omega)$ is equal to:

$$\varepsilon_2 = \left(\frac{4\pi e^2}{\omega^2 m^2}\right) \sum_{a,b} \int a|M|b^2 f_a(1 - f_b) * \delta(E_b - E_a - \omega) d^3k \tag{5}$$

2.1 *Electronic Properties of the Simple and Composite Oxides*

For the simple oxide treated, Figs. 3 and 4 show the obtained band structure of three different situations of ZnO, pure, doped, and the co-doped with Al and Si. An important difference in energy levels gap energy is shown for all these cases studied. The gap energy of the pure structure is direct and to 1.0 eV at the Γ point calculated with Wien2k. This result will be improved with the introduction of approximation into the calculation made, some other similar results show less than the obtained

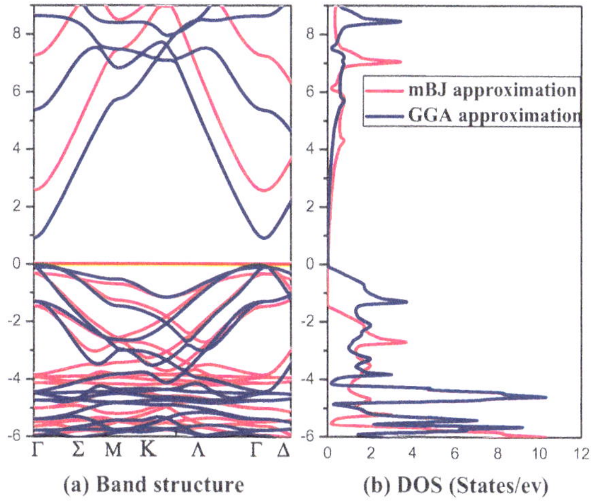

Fig. 3 Band structure of pure structure (**a**) and total DOS (**b**)

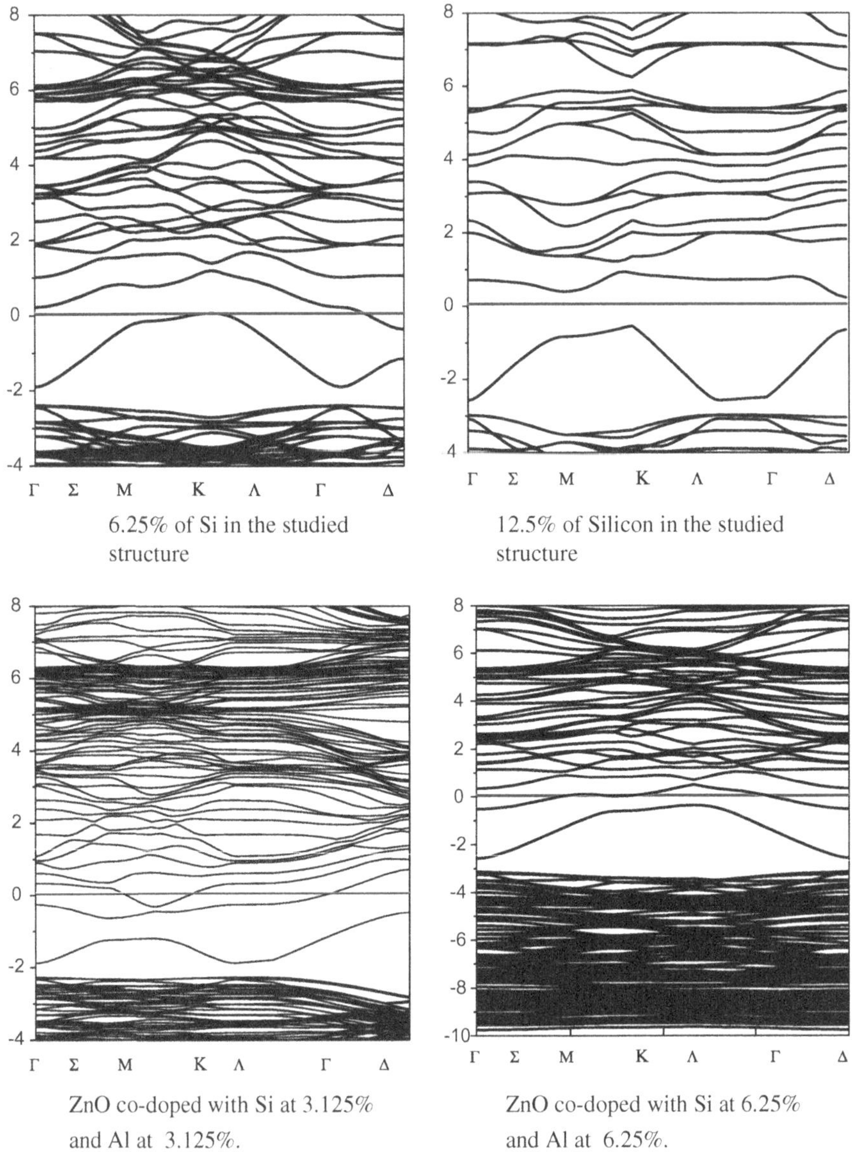

Fig. 4 Band structure of the doped and co-doped structure

result with our Code [15]. The difference between this result and the experimental one is due to the limitation of the DFT calculation. For the result realized by the GGA approach. Therefore, the same structure of pure ZnO has been investigated with another approach (i.e. mBJ). The bandgap is found equal to 2.6 eV, which is approximately in agreement with experimental works. The conduction and valence bands move away from each other for showing the bandgap. The doped and co-doped ZnO with the reported elements Si and Al are shown with the band structure in Fig. 4. The presence of Si increases the energy levels and the E_F change position in the CB band, which present totally occupied and half-occupied states. The VB of Si outdated Zn one, so the n-type occurs as a conduction aspect. This aspect leads to electronic transitions between bands or intra-band under excitation. It can be stated that an increase in the concentration of Si may contribute to an increase in the probability of free electrons entering the conduction band. The behavior of the conductivity is still present even with the co-doping with aluminum, the structure of the ZnO doped with 6.25% of Al and 6.25% of Si acquired a metallic behavior, CB and VB bands are very close and almost no gap and the conductivity is ensured. It is as noticed that the co-doping with aluminum present also energetic levels, which occur on the E_F level. This metallic behavior will arise in the curve of the density of state by a peak that explains the electrical conductivity of the ZnO structure doped and co-doped with both elements (i.e. Al and Si). The total density of states of pure structure is shown in Fig. 3. The silicon gives an additional intense peak around Fermi energy, E_F, which leads to an increase of the ZnO conductivity.

The second type of oxides is also investigated, with the variation of the Y atom by V, Mn, Zn, and Fe, we discuss in this part, the electronic properties of the composite oxide are reported in this chapter. When Y = V, the CB contains essentially an important electrical zone region contributed by orbital p of Bi and d of V, the gap energy is estimated equal to zero because of the electrical peak around the E_F level. Around −11.34 eV and −12.9 eV, a band energy level occurs by s orbital of Bi (see Fig. 5).

When Y = Zn, the VB contains two important electrical zones. The first one from 0.74 to −0.4 eV contributed by d of Zn and orbital p of oxygen. This behavior gives a metallic aspect of $BiZnO_3$. We not also a very strong hybridization in this electrical zone. The second one is reported around −3 eV contributed by p of oxygen and zinc. An energy level appears around −9 eV due to s of Bi. Finally, the CB of $BiZnO_3$ consists of orbital p of Bi atom.

For Y = Mn, the CB is due mainly to orbital p of Bi atom in spin/up and spin/down of the density of states and orbital d of Mn in spin/down specifically in the electrical zone of 1.6–4.2 eV.

For this case, a metallic aspect occurs around E_F, the VB consists of p-oxygen and d-Mn, and around 5 eV, a low hybridization is shown. Y = Fe, the VB is between −6.4 eV and 0 eV, due to spin/up of d orbital of iron and in spin/down is due to the p-oxygen. Y = Co, many zones occur contributed by p-Bi atom. Around 4 eV, a peak appears in the spin/down mainly contributed by d-cobalt.

Figure 6 presents the band structure of $BiYO_3$ with Y = Z, V, Mn, Fe, and Co [14], the illustrations are in good agreement with the analyses made for the total density of states. When Y is substituted with V, Zn, and Mn, the gap energy is close to zero showing a metallic aspect, which is due to d and p orbitals. For cobalt and iron, gap energy occurs. Finally, the simulation made with Wien2k shows that with Y = Fe and Co the composite oxide can be considered as an excellent semiconductor material suitable for the aimed application (solar cell electrode).

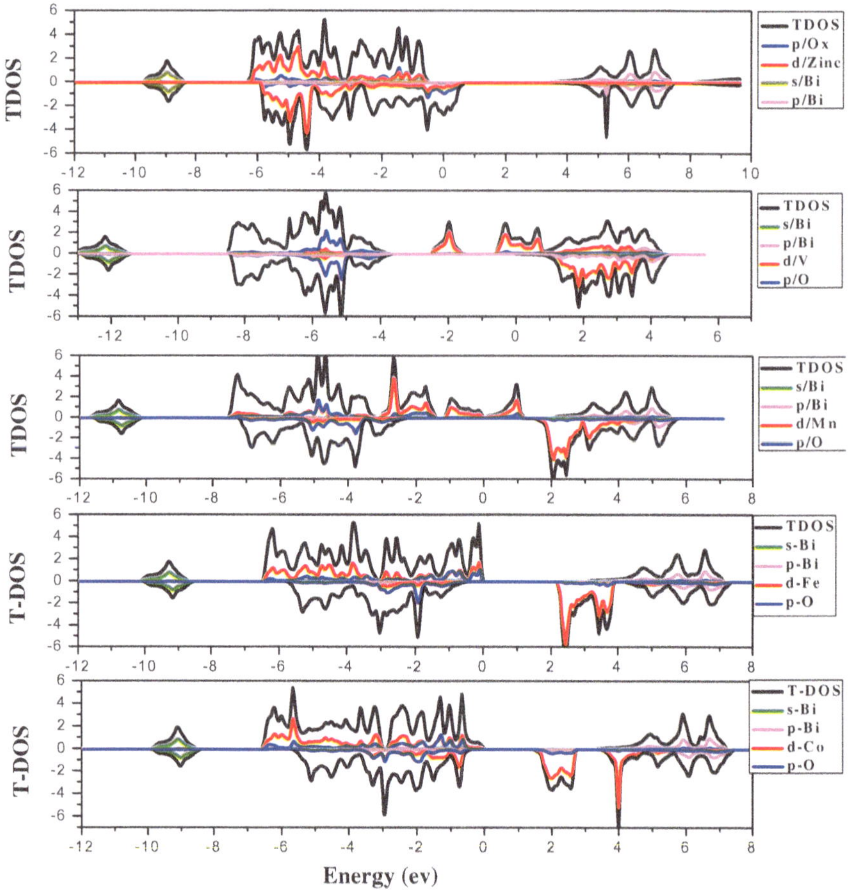

Fig. 5 Total and partial density of state of $BiYO_3/Y$ = Zn, V, Mn, Fe, Co [14]

2.2 *Optical Properties of Both Simple and Composite Oxides*

As shown in Fig. 7, for the pure ZnO, the average transparency is about 86% in UV (200–400 nm) and visible (400–800 nm) regions. Thus, for wavelengths $\lambda >$ 400 nm, the light is observed without absorption and consequently, the pure structure of simple oxide treated becomes transparent. Moreover, with the incorporation of Si, the transmittance increase until 94% and 96% for doping at 6.25% and 12.5%, respectively, which is higher than that of pure structure in the visible zone (i.e. 400–600 nm). Doping with 6.5 and 12.5% of Si present an unstable behavior although the transmittance reaches its maximum value. The transparency at 6.25% amounts of Si is stable at a wide range as shown in Fig. 5 compared with 12.5% of the same impurity. Thus, the transparency with 12.5% is not so significant due to the absorption caused by Si atoms. The co-doping of the pure structure changes and increases the

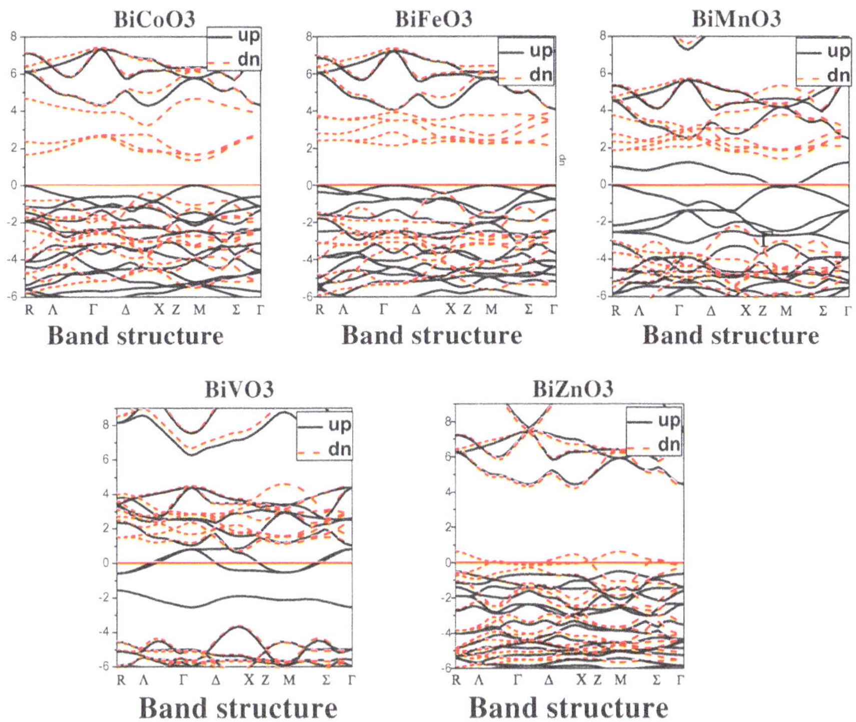

Fig. 6 Band structures of $BiYO_3$

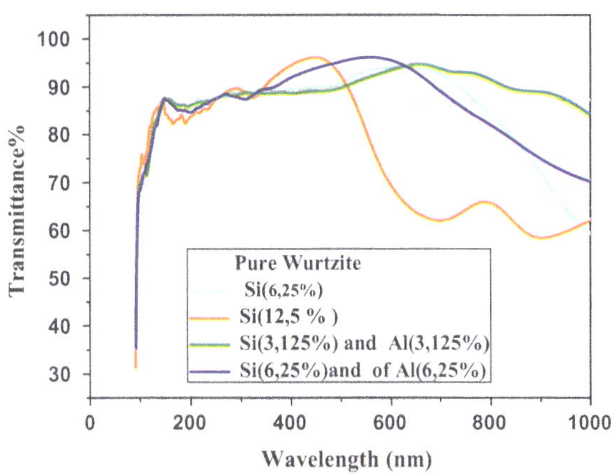

Fig. 7 Transparency of the studied oxide (Simple oxide)

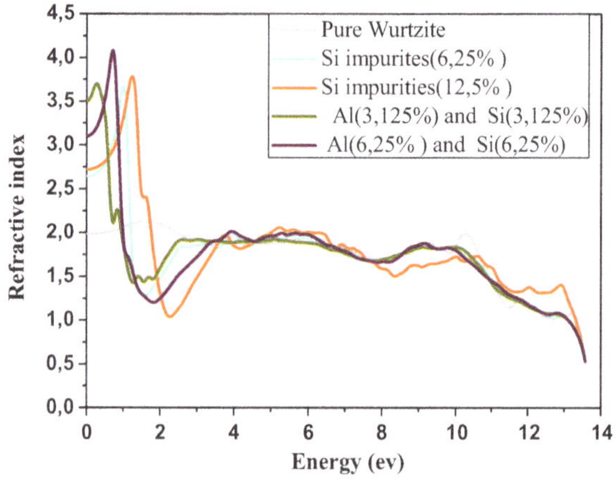

Fig. 8 Refractive index of the pure structure of simple oxide

transparency, the curve clearly shows that in the UV, the transparency is important for Si–Al co-doping, the instability behavior of T presented by the Si at 12.5% amount between 470 nm and 800 nm is corrected with 6.5% Al and 6.5% Si co-doping. For example, at 600 nm, the substitution with 6.25% Si leads to a transparency of 69%, in the case of Al–Si co-doped ZnO, the transparency is about 96%. The co-doping with aluminum increases the transmittance compared to the doping of ZnO with only 6.5% of Si, practically in the 363–631 nm range. The Si–Al co-doped pure structure shows an important improvement in its properties.

$\varepsilon_1(0)$ is called the static dielectric constant calculated based on the dielectric function, this constant leads us to have an idea of how the refractive index behaves, it is directly linked by the following expression $n(0) = \sqrt{\varepsilon_1(0)}$. The refractive index of pure structure, substituted with Si and Al is illustrated in Fig. 8, which is extracted using Eqs. (1) and (2). In the pure structure, the refractive index of solar radiations (2.4 eV) is about 1.9, which is in agreement with some experimental values; n = 2.00. n(hv) increases with increasing doping concentration. This substitution imposes a change in the variation of n(hv). The low value of n(hv) corresponds to the high transparency, and for radiation of 1.62 eV, the transmission is about 90%.

The second step consists of the discussion for the second type of oxides treated, the composite one, $BiYO_3$. In this paragraph, we focus on the variation of the absorption and reflectivity of different elements Y in different zones throughout the wavelengths. This calculation may help us to deduce the interest of integrating this composite oxide in solar electrodes.

According to the absorption illustrate in the figure above, the substitution of Y by iron and cobalt leads to absorption which is not negligible, the increase of radiation leads to an increase if the absorption. From 476 nm, Y = iron is considered as a transparent material in the VL zone, an electrode TCO based on the composite oxide. Beyond 652 nm, Y = cobalt transmits are also transparent in the VL zone, however, this transmittance does not cover a wide zone in the visible light range. Y =

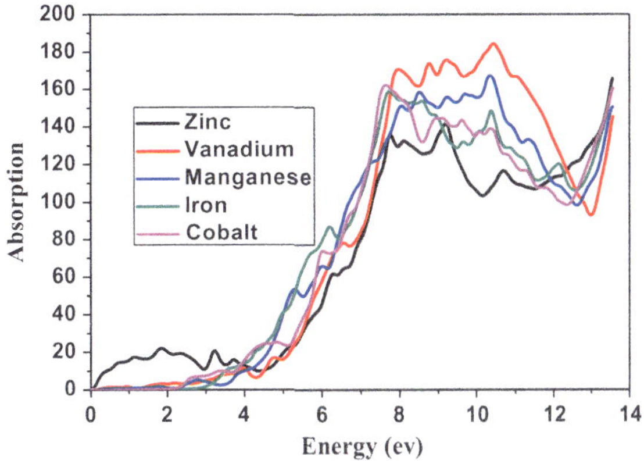

Fig. 9 Absorption (E in xx direction of radiation)

Z, Mn, and V absorption of energy (hν) start from small energies, this characteristic makes these composite materials a candidate to absorb radiation hν in VL zone.

$BiYO_3$ is a highly and sufficient absorbent compound between 100 nm and 250 nm, in particular for Y = V. The substitution of Y atoms by Vanadium improves the absorption, for Y substituted by zinc, the absorption becomes low comparing with Co, Mn, and Fe.

According to the result of Fig. 10, the reflectivity is very important for Y = Mn, V, Zn, Co in the UV range where $\lambda < 200$ nm. Y = Fe and Co, the reflectivity stay stable, the substitution with this element improves the optical characteristic of this oxide. Figure 10 confirms the interpretation made for the previous result of absorption in Fig. 9.

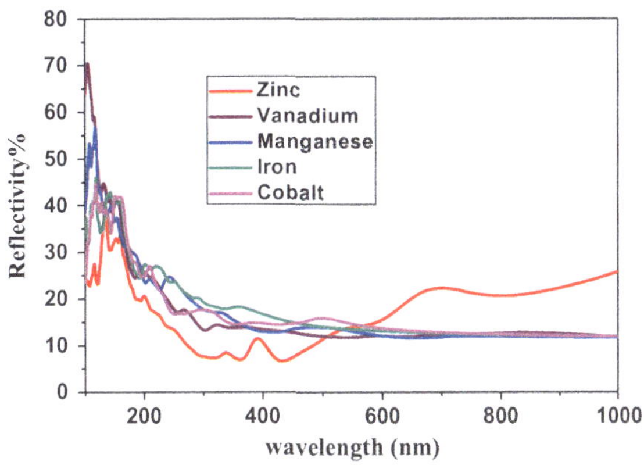

Fig. 10 Reflectivity of the composite oxide

3 Conclusion

We studied the optical electronic properties of two types of oxides simple and composite. The simple oxide that is ZnO doped and co-doped with various amounts of Si and Al shows that the electronic results found and the transparency behavior can be exploited to manufacture a new generation of transparent electrodes as a window layers in solar cells (transparency rate is around 90%). The same concept can be applied for composite oxide such as $BiYO_3$ in a stable structure. In both cases, conductivity and transparency are ensured. But $BiYO_3$, in some cases, can act as active thin layers with an important absorption. Both types of oxide are suitable and alternative candidates to exceed and replace old generations of TCO electrodes.

Special Acknowledgements I address with sadness my feelings of thanks to my father who left us recently. Thanks for all encouragement and efforts that you gave me. May your soul rest in peace.

References

1. Ozgur U, Alivov Y, Liu C, Teke A, Reshchikov M, Dogan S, Avrutin V, Cho S, Morkoc H (2005) A comprehensive review of ZnO materials and devices. J Appl Phys 041301 (cf. p.15, 22–23, 26, 181, 204)
2. Jagadish C, Pearton SJ (2006) Zinc oxide bulk, thin films and nanostructures: processing, properties and applications. Elsevier (cf. p 15, 23, 43)
3. Allen MW, Miller P, Reeves RJ, Durbin SM (2007) Influence of spontaneous polarization on the electrical and optical properties of bulk, single crystal ZnO. Appl Phys Lett 90(6):062104–3 (cf. p 15)
4. Yang X, Xu C, Giles NC (2008) Intrinsic electron mobilities in CdSe, CdS, ZnO, and ZnS and their use in analysis of temperature-dependent Hall measurements. J Appl Phys 104(7):073727 (cf. p 15, 72–73, 75, 124)
5. Yang X, Giles NC (2009) Hall effect analysis of bulk ZnO comparing different crystal growth techniques. J Appl Phys 105(6):063709 (cf. p 15, 124, 147)
6. Humphreys CJ (2008) Solid-State Lighting. MRS Bull 33(04):459–470 (cf. p 21)
7. Choi Y-S, Kang J-W, Hwang D-K, Park S-J (2010) Recent advances in ZnO based light-emitting diodes. English. IEEE Trans Electron Dev 57(1):26–41 (cf. p 22)
8. Klingshirn C, Fallert J, Zhou H, Sartor J, Thiele C, Maier-Flaig F, Schneider D, Kalt H (2010) 65 years of ZnO research—old and very recent results. physica status solidi (b) 247(6):1424–1447 (cf. p 22, 27)
9. Alexandrov A, Zvaigzne M, Lypenko D et al (2020) Al-, Ga-, Mg-, or Li-doped zinc oxide nanoparticles as electron transport layers for quantum dot light-emitting diodes. Sci Rep 10:7496. https://doi.org/10.1038/s41598-020-64263-2
10. Abbassi A, Nainaa F, Arejdal M, Toma M, El Amrani A, Ez-Zahraouy H, Iacomi F (2020) Structural and optical properties of Zn1 − x−yAlx SiyO wurtzite hetero structure thin film for photovoltaic applications. Mater Sci Eng B 260:114614. ISSN 09215107, https://doi.org/10.1016/j.mseb.2020
11. Huang LK, Zhang DL, Bu SX, Peng RX, Wei Q, Ge ZY (2020) Synergistic interface energy band alignment optimization and defect passivation toward efficient and simple-structured Perovskite Solar Cell. Adv Sci 7:1902656. https://doi.org/10.1002/advs.201902656
12. Yu Z, Yang Z, Ni Z et al (2020) Simplified interconnection structure based on C_{60}/SnO_{2-x} for all-perovskite tandem solar cells. Nat Energy. https://doi.org/10.1038/s41560-020-0657-y

13. Li W, Zheng J, Hu B et al (2020) High-performance solar flow battery powered by a perovskite/silicon tandem solar cell. Nat Mater. https://doi.org/10.1038/s41563-020-0720-x
14. Abbassi A, Arejdal M, Ez-Zahraouy H et al (2016) Investigation of optoelectronic properties of $BiMO_3/M = TM$, within the full potential-linearized augmented plane wave method. Opt Quant Electron 48:38. https://doi.org/10.1007/s11082-015-0331-y
15. Abbassi A, Ez-Zahraouy H, Benyoussef A (2015) First principles study on the electronic and optical properties of Al- and Si-doped ZnO with GGA and mBJ approximations. Opt Quant Electron 47:1869–1880. https://doi.org/10.1007/s11082-014-0052-7

A Fractional-Order Dynamic Photovoltaic Model Parameters Estimation Based on Chaotic Meta-Heuristic Optimization Algorithms

Dalia Yousri, Dalia Allam, and M. B. Eteiba

Abstract Modeling of Photovoltaic (PV) solar modules is an essential target in achieving an efficient emulation for the PV system. Recently, the dynamic PV models were considered to recognize the influence of the switching circuits and load change proprieties. The fractional-order dynamic PV model was the currently proposed one to boost the reliability, accuracy, and efficiency of the classical dynamic PV model. The optimal parameters of this model should be identified; therefore, in this chapter, several Chaotic biologically-inspiring Optimization techniques are proposed to demonstrate the most efficient one for this non-linear optimization problem. The introduced techniques are the chaotic variants of Grasshopper Optimizer, Moth-Flame Optimizer, and Flower Pollination Algorithm in addition to their original versions. To assess the efficiency of the endorsed algorithm, its results are compared to the non-linear least-squares method based on the accuracy, the convergence speed, and the fitting of the experimental dataset. Additionally, another comparison is carried out between the recent fractional-order dynamic PV model and its integer version based on the same algorithm to evaluate the efficiency of using the fractional calculus in the modeling of the PV modules. The overall results show that Chaotic Flower Pollination Algorithm with Chebyshev and Singer chaotic maps in the case of the fractional dynamic PV model offers the best fitting on the load current-time curve. Moreover, the fractional-order dynamic model that can emulate the physical behavior of the real system is efficient than the integer-order dynamic model.

Keywords Dynamic PV equivalent circuit · Fractional-order dynamic PV equivalent circuit · Parameters estimation · Chaos maps · Chaotic flower pollination algorithm

D. Yousri (✉) · D. Allam · M. B. Eteiba
Faculty of Engineering, Electrical Engineering Department, Fayoum University, Fayoum, Egypt
e-mail: day01@fayoum.edu.eg

S. Motahhir and A. M. Eltamaly (eds.), *Advanced Technologies for Solar Photovoltaics Energy Systems*, Green Energy and Technology,
https://doi.org/10.1007/978-3-030-64565-6_2

1 Introduction

Recently, there is a persistent need to search for alternative energy resources because of the high price of fossil fuels and their rising rate of pollution that may threaten the creatures' lives. Solar energy is a dominant renewable energy resource because of its availability and free-pollution [1]. The accurate prediction of PV system behavior and the optimal capturing of the available energy need a robust, efficient, and attested representation for the designed PV systems before and after the installation [2].

Several equivalent circuits have been published to present the characteristics of the PV solar modules like the static PV models and the dynamic PV equivalent circuits. The static PV equivalent circuits have been evolved from the ideal models to the Single Diode Model (SDM) [3] up to the more complicated ones named the Double Diode Model (DDM), the Modified Double Diode Model (MDDM), and the Three Diode Model (TDM)[4]. Unfortunately, the switching operation of the inverter and DC/DC converter, as well as the load variation, have not been taken into account using these models [5]. Therefore, developing a novel circuit for the PV models as the dynamic PV models has been attracted to cover the aforementioned shortcoming of the static models [6, 7].

Lately, the fractional calculus has been utilized to increase the flexibility and reliability of the classical dynamic PV model (Integer-Order Model (IOM)) in [6]. Thereby, a new generation of the dynamic models named the Fractional-Order dynamic PV Model (FOM) has been created as in [7]. In [7], the FOM has been formulated mathematically via fractionating the inductor and capacitor elements of the integral-order model, and the effect of this fractionalization on the model efficiency and flexibility have been investigated as well. Therefore, it's expected that a deeper vision into the physical processes of the PV modules underlying a long-range memory behavior may be achieved using this newly developed model. Consequently, this may reflect in turn on design, control, and operation of PV system near MPP [7].

The accuracy of the identified PV models parameters has a large influence on the accuracy of the reported PV models. These parameters are usually unknown by manufacturer data sheets [6, 7]. Therefore, reliable and efficient techniques should be proposed to estimate these parameters accurately which may achieve a better fit on the experimental load current-time (I-T) curves of the dynamic PV model.

Several methods have been published to estimate the parameters of the static PV models. These methods can be categorized as deterministic methods and stochastic algorithms. Deterministic methods are as reported in [8, 9]. We have examples for the stochastic algorithms like the Flower Pollination Algorithm [10] (FPA), Moth-flame Optimization Algorithm (MFO) [4], Time-Varying Acceleration Coefficients Particle Swarm Optimization (TVACPSO) [11], Differential Evolution [12], and Particle Swarm Optimization are proposed previously as in [13]. The marine predators algorithm, Slime mould algorithm, atom search optimization, Political Optimizer, Parasitism Predation algorithm as well as harris hawk optimizer and salp swarm algorithm have been implemented in [14] to identify the parameters of simple and detailed static PV models. The artificial ecosystem-based optimization approach has been proposed in [15] and self-adaptive ensemble-based differential evolution has been recently introduced in [16], and Ibrahim et al. [17] used an adaptive wind-driven algorithm for SDM parameters estimation process. According to the literature, the stochastic optimization algorithms prove their efficiency and superiority in extracting the parameters of the PV static models compared to the deterministic techniques. Up till now, there is no wide prevalence of using meta-heuristics to extract the parameters of the dynamic PV models although they are powerful tools for simultaneous estimation of the optimal parameters of several complicated models. Where the conventional linear Least Squares (LS) and the Non-linear Least Squares (NLS) methods have been reported to extract the parameters of the IOM and the FOM, respectively [6, 7]. In the circumstance of meta-heuristic techniques, in [18], Chaotic Heterogeneous Comprehensive Learning Particle Swarm Optimizer variants (C-HCLPSO) was applied to identify the dynamic parameters based on Root Mean Square Error (RMSE) between the current-time measured and estimated curves as the first technique for this optimization problem. This motivates the authors of this work to employ several meta-heuristic optimization techniques with utilizing another objective function on parameters extraction of the integer and the FOMs and endorse the most suitable one for this optimization problem.

Grasshopper Optimizer (GOA), Moth-flame Optimization Algorithm (MFO), and Flower Pollination Algorithms (FPA) are recently introduced and applied on several non-linear complicated optimization problems [19–21]. Where these algorithms have significant features that may improve diversification and intensification during searching for optimal solutions. Therefore, they are selected to be used in identification of the dynamic order PV models. According to no-free-lunch theorem, there is no super technique that can be applied to solve all the optimization problems [22]. That's why different algorithms should be tested to prove their validity and suitability in a specific application. As the performances of GOA, FPA, and MFO are influenced by their random parameters, it is necessary to introduce newly developed approaches to modify the algorithms' performances by controlling their random parameters to balance between the exploration and intensification phases [23–25].

Developing stochastic optimization techniques by integrating the chaotic maps with the original techniques to adaptively tune some of their factors is considered as a new approach [23–25]. It's worth mentioning that this combination improves the balance between diversification and/or exploitation capability of the basic algorithms depending on the randomization effectiveness of the chaos theory which may take a turn for better quality and decaying of convergence rate of the basic techniques [26–28].

In this chapter, the IOM and FOM models' parameters are estimated using different meta-heuristic algorithms. Three optimization techniques known as GOA, MFO, and FPA as well as their chaotic variants are tested. The chaotic variants are obtained by merging ten chaos maps with the original algorithms to adjust their parameters which may improve their accuracy, consistency, and the decaying convergence rate. The results of the three original algorithms and their chaotic variants are compared together via an intensive statistical analysis to select the most suitable variants. subsequently, the recommended variants for each model are compared to the previously reported techniques in literature known as the least-squares method for the integral-order model and the non-linear least-squares method for the fractional-order model. Moreover, another round of comparison is accomplished among the recommended variants of the integer-order model and the fractional-order one to highlight the more robust, reliable, and efficient model. The final outcome is that the fractional-order PV model and its optimal parameters estimated by CFPA algorithm with Chebyshev and Singer maps are the final recommended variant that provides the best emulation of the physical behavior of the dynamic data. Where they achieve the least deviation between the measured and the estimated load current curve with faster speed of convergence to the optimal parameters. It is worth mentioning that more accuracy of the selected model and its optimal parameters estimation may achieve in turn more accurate design, control, and operation of the PV system.

The rest of the chapter is ordered as follows: the equivalent circuit of the PV dynamic models are presented in Sect. 2. The problem definition is documented in Sect. 3. The basic background of the applied algorithms as well as the chaos maps equations furthermore the algorithms' chaotic variants are explained in detail in Sect. 4. Simulation and results are discussed in Sect. 5. At the last Sect. 6, the main outcomes are listed.

2 PV Dynamic Equivalent Circuit

Dynamic PV models are accounted as recent and endorse trend dut to in these models the switching operation of the converters and inverters as well as the load variation are recognized [6, 7]. The dynamic models have two versions integer-order dynamic model and fractional-order one. The two dynamic models are reported in this work as follows.

2.1 Integer-Order Dynamic PV Equivalent Circuit (IOM)

The integer-order dynamic equivalent circuit of a PV module and its connected load has been investigated in [6] as shown in Fig. 1a where the influence of the variation in the load and the converters/ inverters switching circuits are taken into account. This model consists of two parts, the first one is the static part of the PV module that is shortened to a constant voltage source V_{oc} and a series resistance R_s as illustrated in Fig. 1b. While the other one is the dynamic part of the model that consists of a capacitor (C) representing the junction capacitance and (R_c) to account for the conductance as well as a series inductance (L) to take the cabling inductance and connection into consideration.

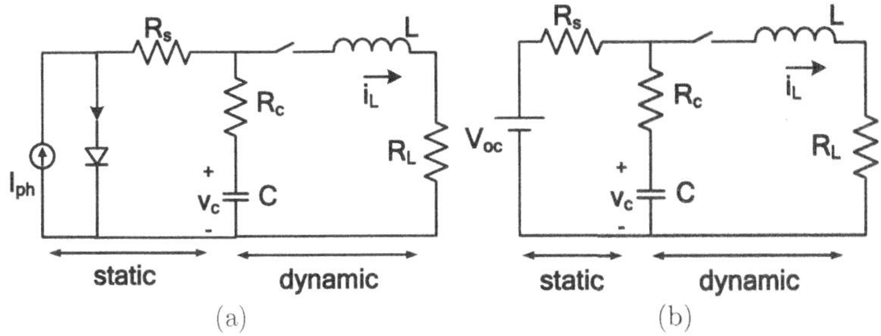

Fig. 1 The integral-order dynamic PV model [7] of **a** complete circuit , and **b** equivalent circuit

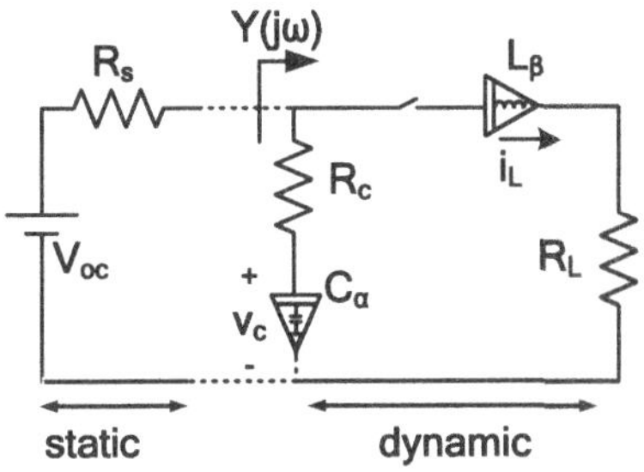

Fig. 2 The fractional-order dynamic PV model [7]

To analyse the PV model in Fig. 2, the load current–voltage relationship can be written in s-domain as in the following Eq. 1 [6].

$$i_L(s) = \frac{V_{OC}}{s}\frac{a_{21}(s+b_1)+b_2(s-a_{11})}{(s-a_{22})(s-a_{11})-a_{12}a_{21}}, \tag{1}$$

where

$$\begin{pmatrix} a_{11} & a_{12} \\ a_{21} & a_{22} \end{pmatrix} = \begin{pmatrix} \frac{-1}{C(R_c+R_s)} & \frac{-R_s}{C(R_c+R_s)} \\ \frac{R_s}{L(R_c+R_s)} & \frac{-[R_L R_c+R_s R_c+R_L R_s]}{L(R_c+R_s)} \end{pmatrix}, \tag{2}$$

$$\begin{pmatrix} b_1 \\ b_2 \end{pmatrix} = \begin{pmatrix} \frac{1}{C(R_s+R_c)} \\ \frac{R_c}{L(R_c+R_s)} \end{pmatrix}. \tag{3}$$

Based on Eqs. 1 and 3, it's obvious that the unknown parameters are (R_C, C, and L) with knowing the parameters of the static part.

2.2 Fractional-Order Dynamic PV Equivalent Circuit (FOM)

Lately, in [7], fractional calculus is utilized to introduce the fractional-order dynamic PV equivalent circuit in 2018 where the capacitor and the inductor of the integral-order PV equivalent circuit are interchanged by the fractional counterparts of orders α and β, respectively, as illustrated in Fig. 2. Where a better mathematical model of a system leads to a better fitting on the measured data, a better description of the real system response, and subsequently a better emulation of the physical behavior of the real system. This is accomplished by a novel compact fractional model of the PV module that has been published recently in [7] as two more parameters are added to the differential equation describing the transient (I-T) response of the PV module that are the derivative orders added by fractionating the capacitor and the inductor of the integral-order mode. These extra parameters provide in turn a better accuracy for the fractional mathematical PV dynamic equivalent circuit over the integer one. It should be noticed that the value of the resistor R_c will be lower than that of the integral-order equivalent circuit due to the effect of the real frequency dependent part of the fractional capacitor impedance that has partially replaced the series resistor connected to it [29].

To implement the fractional-order dynamic PV equivalent circuit, the load current–voltage relationship has been modeled in s-domain as described in the following Eq. 4 [30]

$$i_L(s) = \frac{V_{oc}}{s}\,\frac{a_{21}(s^{\alpha}+b_1)+b_2(s^{\alpha}-a_{11})}{(s^{\beta}-a_{22})(s^{\alpha}-a_{11})-a_{12}a_{21}}, \tag{4}$$

where

$$\begin{pmatrix} a_{11} & a_{12} \\ a_{21} & a_{22} \end{pmatrix} = \begin{pmatrix} \frac{-1}{C_{\alpha}(R_c+R_s)} & \frac{-R_s}{C_{\alpha}(R_c+R_s)} \\ \frac{R_s}{L_{\beta}(R_c+R_s)} & \frac{-[R_L R_c+R_s R_c+R_L R_s]}{L_{\beta}(R_c+R_s)} \end{pmatrix}, \tag{5}$$

$$\begin{pmatrix} b_1 \\ b_2 \end{pmatrix} = \begin{pmatrix} \frac{1}{C_{\alpha}(R_s+R_c)} \\ \frac{R_c}{L_{\beta}(R_c+R_s)} \end{pmatrix}, \tag{6}$$

C_{α} and L_{β} illustrate the fractional capacitance and fractional inductance, respectively. α and β indicate the derivative orders, they have values that are less or greater than 1 in the fractional-order model.

Based on the equivalent circuit of the fractional-order dynamic PV equivalent circuit and Eqs. 4–6, there are five unknown parameters should be estimated which are (R_C, C_{α}, L_{β}, α, and β).

3 Problem Formulation

Identification of the unknown variables of the dynamic PV models can be described as an optimization problem where the difference between the experimental dynamic data of the load current and the extracted one via using the extracted parameters is required to be minimized. The global unknown parameters are obtained by using a fitness function known as the sum of the absolute errors between the experimental and estimated current curves.

$$SAE = \sum_{i=0}^{m} \mid I_m(t_i) - I_e(z, t_i) \mid, \tag{7}$$

where m is the number of the experimental points. z is the vector of the variables (R_C, C, L) for the integral-order equivalent circuit and $(R_C, C_\alpha, L_\beta, \alpha, \beta)$ for the fractional-order dynamic PV equivalent circuit. I_e and I_m show the estimated and the measured current as functions of time (t_i).

The major target of the applied optimization algorithms is to estimate the optimal values of the integral-order and the fractional-order dynamic PV equivalent circuits parameters z that attain less deviation between the experimental and predicted dynamic datasets.

4 An Overview of the Used Optimization Algorithms

In this part, the basic background of the implemented techniques in identification of the integral-order and the FOMs parameters is presented.

4.1 Grasshopper Optimization Algorithm (GOA)

Saremi et al. [21] used the features of the grasshoppers in nature as the basic idea in the Grasshopper Optimization Algorithm (GOA). GOA is published lately in 2017 [21]. The main concept of grasshoppers' behavior is as follows:

1. There are two phases of life of the grasshoppers, the first one is the nymph phase while the other one is the adulthood phase. During the nymph stage, the motion of grasshopper is in slow profile with small steps while during the adulthood phase, the long-periodic stage and the abrupt motion are the main features of the flock [21].
2. Seeking for the food sources is the other main aim of the grasshopper flocks [21].

Those two aforementioned features have been mathematically formulated to give the GOA [21].

The behavior of the grasshoppers flock can be modeled mathematically as below in (8) [21].

$$Z_i = S_i + G_i + A_i, \tag{8}$$

where Z_i is the i-th grasshopper location, S_i is the social interaction between grasshoppers, G_i is the gravity force, and A_i is the wind advection [21].

The social interaction between the grasshopper can be calculated as below in Eq. 9 [21].

$$S_i = \sum_{j=1, j \neq i}^{N} s(d_{ij})\widehat{d_{ij}}, \tag{9}$$

where d_{ij} is the distance between i-th and j-th grasshopper and $\widehat{d_{ij}}$ is the unit vector of the distance. d_{ij} and $\widehat{d_{ij}}$ are calculated as in Eq. 10. The s function defines the social forces between grasshoppers and it can be modeled as in Eq. 11 [21].

$$\begin{aligned} d_{ij} &= |z_j - z_i| \\ \widehat{d_{ij}} &= \frac{|z_j - z_i|}{d_{ij}}, \end{aligned} \tag{10}$$

$$s(r) = f e^{\frac{-r}{l}} - e^{-r}, \tag{11}$$

where f shows the intensity of attraction and l indicates the attractive length scale [21].

Furthermore, the gravity force G and the wind advection A are written as below in Eqs. 12 and 13 [21].

$$G_i = -g\widehat{e_g} \tag{12}$$

$$A_i = u\widehat{e_w}, \tag{13}$$

where g in Eq. 12 is the gravitational constant and $\widehat{e_g}$ in Eq. 12 is a unity vector toward the earth center. In Eq. 13, u is a constant drift and $\widehat{e_w}$ is a unity vector in the wind direction.

Equation 8 is accounted to be a main equation of the GOA. However, this mathematical formulation cannot be directly utilized to solve the optimization problems due to the grasshoppers' quick stuck in the local zone. Thus, the main equation Eq. 8 of GOA has been upgraded as in the following Eq. 14 [21].

$$Z_i^d = c\left(\sum_{j=1, j \neq i}^{N} c\frac{ub_d - lb_d}{2} s\left(|z_j - z_i|\right)\frac{|z_j - z_i|}{d_{ij}}\right) + \widehat{T_d}, \tag{14}$$

where ub_d and lb_d are the upper and lower limits in the D_{th} dimension. T_d is the value of the best solution attained so far and c is a decreasing coefficient from 1 to zero across the iterations number, this relation $\left(c_{max} - t\left(\frac{c_{max}-c_{min}}{T}\right)\right)$, t is the current iteration and T is the maximum number of iterations. c_{max} , c_{min} are 1 and 0.00001, respectively [21].

The first term of Eq. 14, $c\left(\sum_{j=1, j\neq i}^{N} c\frac{ub_d-lb_d}{2}s\left(|z_j - z_i|\right)\frac{|z_j-z_i|}{d_{ij}}\right)$ describes the relationship between the grasshoppers in nature while the second term, T_d, modulates their tendency to move toward the source of food [21].

4.2 *Moth-Flame Optimizer (MFO)*

MFO algorithm is based on the Moths behavior during their flying at night known as transverse orientation [31]. Where they keep a certain angle with a very far source of light which helping them to fly in a straight path. As the light source becomes nearer, the moths fly around it spirally until reach to it. MFO consists of three parts. The first one is the initialization where the random positions of moths are generated. The second part is the main part of the algorithm in which the transverse orientation of moths are formulated mathematically and last part is the termination part that concern with the algorithm stop.

In MFO algorithm, both the moths and the flames are solutions, whereas they have different features while they are modified. The moths are the agents that fly around the search landscape while flames are the personal best location of each candidate attained so far [31]. Moreover, The logarithmic spiral function is utilized to describe mechanism of transverse orientation of moths around the light source. The mathematical formula of this function is modeled as follows [31]:

$$S(M_i, F_j) = D_i e^{cl} cos(2\pi t) + F_j, \tag{15}$$

where M_i is the i-th moth, F_j illustrates the j-th flame, and S indicates the spiral function. D_i is the distance of the i-th moth for the j-th flame ($|F_j - M_i|$). c is a constant for describing the shape of the logarithmic spiral. l is a random number in $[d, 1]$ and d denotes the adaptive convergence constant that is linearly decreased from -1 to -2 to accelerate the convergence speed.

To enhance the response of the MFO technique, the number of flames reduces over the iterations as in Eq. 16. This feature helps to balance between the diversification and exploitation stages of MFO while searching for the global solution. Accordingly, moths modified their locations using only the location of the best flame across the last steps of the iterations [31]:

$$flame\ Number = round\left(M - t \times \frac{M-1}{T}\right), \tag{16}$$

where t is the current iteration, T is the iterations maximum number, and the M is the maximum number of flames.

4.3 Flower Pollination Algorithm (FPA)

The main concept of the pollination in the planets is the usage idea in the FPA technique [19]. The mating in the plants occurs during the pollen transferred through biotic and/or abiotic pollination processes. The biotic process is known as a cross-pollination as the bees and butterflies transport the pollen grains across plants of different species. For abiotic process, it is considered as a self-pollination as the wind transports the pollen across flowers of the same species.

Yang et al. [19] used the main concept of the biotic process to implement the global search stage of FPA as below

$$Z_i^{t+1} = Z_i^t + \gamma L(\lambda)(g_* - Z_i^t), \tag{17}$$

where Z_i^t is the pollen i (solution vector) Z_i at iteration t. The global solution can be denoted by g_*. The symbol of γ is a scaling coefficient to adjust the step size. The symbol of $L(\lambda)$ is the levy factor that is responsible for the transfer of pollens between several species of flowers. It can be calculated as follows [19]:

$$L \sim \frac{\lambda\Gamma(\lambda)\sin(\frac{\pi\lambda}{2})}{\pi}\frac{1}{m^{1+\lambda}} \qquad (m \gg m_o > 0), \tag{18}$$

where $\Gamma(\lambda)$ is the gamma function, and this distribution is valid for large steps m > 0. h is step size and the value of m_o can be as small as 0.1.

For the local search capitiy of FPA, Yang et al. [19] considered the pollination during the abiotic process as follows:

$$Z_i^{t+1} = Z_i^t + \epsilon(Z_j^t - Z_k^t). \tag{19}$$

where Z_j^t, Z_k^t are the different pollens of same plant species. ϵ is drawn from a uniform distribution $\in$ [0, 1].

To switch between the global and local search capabilities of FPA, a switching probability factor S is chosen in the interval of [0.2, 1] [19].

4.4 Chaotic Variants of GOA, MFO, and FPA Optimization Algorithms (CGOA, CMFO, and CFPA)

Stochastic optimization techniques are randomly based techniques. This randomization is accomplished via using the Gaussian or the uniform distribution. Currently, a novel trend has been proposed to exchange this distribution by the chaos maps to avail of the better characteristics of the chaotic maps randomization. In this approach, integration between the properties of chaotic maps and basic techniques enable them to converge to the optimal solution accurately and rapidly while optimizing the multi-modal test functions [23].

4.5 Chaos Maps

In this work, ten different one-dimensional chaotic maps are used to adjust some coefficients of the standard versions of GOA, MFO, and FPA techniques. As a result, the Chaotic Grasshopper Optimizer (CGOA), Chaotic Moth-Flame Optimizer (CMFO), and Chaotic Flower Pollination Algorithm (CFPA) are proposed.

The selected chaos maps are listed as follows:

- Chebyshev map
$$x_{i+1} = \cos(i\cos^{-1}(x_i)), \tag{20}$$
where x_i is the ith chaotic number, i is the times of iteration, x ∈ (0, 1) under the initial condition $x_0 \in (0, 1)$, thus x_0 is 0.7.
- Circle
$$x_{i+1} = mod\left(x_i + b - \left(\frac{a}{2\pi}\right)\sin(2\pi x_k, 1)\right), \tag{21}$$
where a, and b are the control parameters of chaotic behavior and equal to 0.5 and 0.2, respectively. The inital conditions x_0 is selected to be 0.7 to ensure a complete chaotic state x ∈ (0, 1).
- Gauss/mouse
$$x_{i+1} = \begin{cases} 1 & x_i = 0 \\ \frac{1}{mod(x_i, 1)} & \text{Otherwise,} \end{cases} \tag{22}$$
where x_0 is 0.7.
- Iterative
$$x_{i+1} = \sin\left(\frac{a\pi}{x_i}\right), \tag{23}$$
where a = 4 and x_0 is 0.7.
- Logistic
$$x_{i+1} = ax_i(1 - x_i), \tag{24}$$
where a = 4 and x_0 is 0.7.
- Piecewise
$$x_{i+1} = \begin{cases} \frac{x_i}{P} & 0 <= x_i < P \\ \frac{x_i - P}{0.5 - P} & P <= x_i < 0.5 \\ \frac{1 - P - x_i}{0.5 - P} & 0.5 <= x_i < 1 - P \\ \frac{1 - x_i}{P} & 1 - P <= x_i < 1 \end{cases}, \tag{25}$$
where P = 0.4 and x_0 is 0.7.
- Sine
$$x_{i+1} = \frac{a}{4} sin(\pi x_i), \tag{26}$$
where a = 4 and x_0 is 0.7.

- Singer

$$x_{i+1} = \mu(7.86x_i - 23.31x_i^2 + 28.75x_i^3 - 13.302875x_i^4), \tag{27}$$

where $\mu = 1.07$ and x_0 is 0.7.

- Sinusoidal

$$x_{i+1} = ax_i^2 \sin(\pi x_i), \tag{28}$$

where a =2.3 and x_0 is 0.7.

- Tent

$$x_{i+1} = \begin{cases} \frac{x_i}{0.7} & x_i < 0.7 \\ \frac{10}{3}(1 - x_i) & x_i >= 0.7 \end{cases}, \tag{29}$$

where x_0 is 0.7.

4.5.1 Chaotic Grasshopper Optimizer (CGOA) Technique

Depending on Eq. 14, the GOA performance and its convergence speed for the global optimal solution are controlled by three main coefficients. These variables are c, f, and l. Variable c is a main coefficient that supports the balance between the diversification and exploitation processes. The c factor follows a linearly decreasing function from 1 to 0 across the iteration numbers. For the other coefficients f and l, they used to manage the social interaction among the grasshoppers to avoid trapping in the local minima. The f and l have values of 0.5 and 1.5, respectively [28]. Whereas, in the introduced CGOA variant, the values of c, f, and l are changed according to the patterns of the employed chaotic maps as clarified in the following Eqs. (30) and (31)

$$c = \left(c_i - t.\frac{(c_f - c_i)}{T}\right).Chaos(t); \tag{30}$$

$$\begin{gathered} Norm\vec{C}haos = \frac{\vec{Chaos}.(d - e)}{b - a} + e, \\ f = Norm\vec{C}haos, \quad l = Norm\vec{C}haos, \end{gathered} \tag{31}$$

where c_i, c_f are the initial and final values of c, they are tuned as 1 and 0.00001, respectively. The symbols of t, and T are the current iteration and the maximum number of iterations, respectively. $Norm\vec{C}haos$ is the normalized chaotic map. The $[a \quad b]$ is the interval of chaotic maps. The $[e \ d]$ is the normalization range that can be selected as [0.3 0.7] and [1.3 1.7] for f and l, respectively.

4.6 Chaotic Moth-Flame Technique (CMFO)

In the standard MFO technique, there is an essential parameter named t that is controlling the locations of the moths with respect to the flames position where t defines how much the next position of the moth should be near to the flame. This parameter trades between the diversification and the exploitation phases. This parameter has random distribution values between [r 1] where r is linearly decreased from -1 to -2. While in CMFO, the parameter t is modified by the chaos maps to enhance the convergence speed of the MFO algorithm such as follows (32):

$$\begin{aligned} r &= \left(r_i + t.\frac{(r_f - r_i)}{T}\right).Chaos_k; \\ l &= (r-1).rand + 1 \end{aligned} \tag{32}$$

,

where r_i, r_f are the initial and final values of the deceasing function. They are adjusted as -1, -2, respectively. The symbols of t, and T are the current iteration and the maximum number of iterations, respectively. $Chaos_k$ is the chaotic map of index k.

4.7 Chaotic Flower Pollination Algorithm

The FPA has three important factors that govern its response while searching for the global solution that are S, L, and ϵ. The symbol of S is the switching probability between the global and local pollination. It starts from 0.2 until reaches 1. While in CFPA, S changes from 0.2 to 1 chaotically using chaos maps as in Eq. 33. The parameter L is considered as the strength of the pollination, it has values greater than 0 ($L > 0$) from the Levy distribution. In CFPA, chaos maps are merged within the Levy distribution as in Eq. 34. Moreover, in FPA, ϵ is drawn from the uniform distribution in the interval of [01] while in CFPA the ϵ is drawn from the chaos maps in the same interval as in Eq. 35.

$$s = \left(s_i + t.\frac{(s_f - s_i)}{T}\right).Chaos_k; \tag{33}$$

$$L \sim \left(\frac{\lambda\Gamma(\lambda)\sin\left(\frac{\pi\lambda}{2}\right)}{\pi}\frac{1}{m^{1+\lambda}}\right).Chaos_k; \tag{34}$$

$$\epsilon = Chaos(t); \tag{35}$$

where s_i, s_f are the initial and final values of a that are selected as 0.2 and 1.2, respectively. $Chaos_k$ is a chaos map of index K.

5 Simulation and Results

In this section, the quality of the proposed chaotic variants of the GOA, MFO, and FPA (CGOA, CMFO, CFPA) as well as their original versions are tested, evaluated and demonstrated for extracting the IOM and the FOM models' parameters based on an experimental dynamic dataset of the load current for a connected PV module with resistive load (R_L= 23.1 Ω). The PV module sybjected to an irradiance and temperature level of 655 w/m^2 and 25 °C [6]. The module is fixed tilted at 50° and the electric specification of the module are V_{oc} = 19.6 V, I_{sc} = 0.96 A, V_{mp} = 14.96 V, and I_{mp} = 0.92 A at the irradiance and the temperature levels.

The structure of this section can be composed of two parts.

- **The first part** is accomplished by applying the introduced algorithms on the IOM and FOM. Subsequently, the results of the chaotic variants are compared with that of the corresponding standard versions to evaluate and demonstrate the performance of these techniques and endorse the better ones. The comparison is based on the best, mean ± and the Standard Deviation (STD) of the extracted parameters and the corresponding fitness function as well as the mean convergence curves to reach these values of the fitness function. Moreover, to evaluate the accuracy of the extracted parameters with respect to that of the reported techniques, the estimated load current curves by the proposed meta-heuristic algorithms are compared with that of the previously published methods LS and NLS for the IOM and the FOMs, respectively [6, 7]. Furthermore, the Absolute Error (AE) curves between the experimental load current curves and the identified ones by all variants are calculated and compared to that of the previous methods. For extra validation, non-parametric statistical analysis called Wilcoxon Rank-Sum Test is carried out among the recommend variants to endorse the most significant one.
- **In the second part**, the more appropriate variants' results for the integral-order dynamic PV model are compared with that of the fractional-order one to discuss the importance of employing the fractional calculus for the PV modeling as well as to focus on the additional features of utilizing the fractional PV dynamic models. The winner algorithm for this application is highlighted often in these successive stages of comparisons.

For accurate comparison between the introduced techniques, each one is run with 500 iterations and 30 population size for 20 independent runs. The upper and lower limits of the unknown variables of integral-order dynamic PV equivalent circuit (IOM) and fractional-order one (FOM) are set as in Table 1.

Table 1 The upper and lower ranges of models' parameters

IOM variables	Lower limit	Upper limit	FOM variables	Lower limit	Upper bound
$R_c(\Omega)$	0	20	$R_c(\Omega)$	0	20
$C(F)$	$20e^{-9}$	$600e^{-7}$	$C_\alpha(F)$	$20e^{-9}$	$600e^{-7}$
$L(H)$	$5e^{-6}$	$100e^{-6}$	$L_\beta(H)$	$5e^{-6}$	$100e^{-6}$
			α	0.8	1.1
			β	0.8	1.1

5.1 Discussion of the Integral-Order Dynamic PV Model (IOM) Results

The best, the mean ± the STD values of the identified parameters of the IOM and their corresponding fitness function (SAE) over 20 independent run times are listed in Table 2. The reported results show that the chaotic variants of the three algorithms under test provide more accurate and consistent results than their original versions. Where cooperating the logistic map with GOA improves the best, mean ± STD of SAE from 5.217, 10.08 ± 9.260 × 10^0 to 4.629, 4.925 ± 5.304 × 10^{-1} , respectively. Chebyshev map has a remarkable influence on the accuracy and reliability of the MFO results as CMFO1 provides best, mean, and STD of SAE equal to 4.629, 4.629, and 1.204 × 10^{-7} whereas MFO offers 4.629, 5.713, and 2.648 × 10^0 , respectively. Likewise, integration of sinusoidal map with FPA improves the consistency of its results where STD value are updated from 1.065 × 10^{-5} to 4.039 × 10^{-7}, respectively. Additionally, the tabulated results indicate that CFPA1 to CFPA10 especially CFPA9 and CMFO1 have the lower best, mean, and STD values of SAE than CGOA, CMFO2 to CMFO10 variants as well as their basic versions (MFO and GOA). Therefore, CFPA9 and CMFO1 variants are recognized as the most endorsed variants for identifying the parameters of the IOM accurately and with highest consistent rate.

To discuss the impact of the chaotic maps on the convergence speed of the recommended chaotic variants (CFPA9 and CMFO1), their mean convergence curves versus the standard versions of FPA and MFO are plotted in Fig. 3. The Figures of the convergence curves in Fig. 3a and d of CMFO1 and CFPA9, respectively, exhibit that integrating the Chebyshev map into CMFO has noteworthy effect not only on MFO convergence speed but also on the accuracy of the obtained fitness function over the selected number of iterations. For the decaying rate of convergence of CFPA9 becomes lower than FPA for the first 150 iterations then basic FPA try to nearly achieve the same mean value of the SAE at the end of the whole iterations.

To validate the accuracy of the identified parameters by the proposed chaotic algorithms, the estimated load current curves by these variants and that of LS method [6] versus the experimental one are plotted as in Fig. 3b–e. Additionally, the AE curves between the measured load current curve and the identified ones are drawn in Fig. 3c–f for the studied algorithms. It is illustrated from the fitting and absolute

Table 2 The best, mean ± STD of the extracted parameters as well as the fitness function (SAE) for the IOM by the introduced algorithms over 20 runs

Algorithms		$R_c(\Omega)_{best,mean \pm STD}$	$C(F)_{best,mean \pm STD}$	$L(H)_{best,mean \pm STD}$	$SAE_{best,mean \pm STD}$
CGOA	GOA	$5.477, 9.241 \pm 5.841$	$3.442e-07, 5.030e-06 \pm 1.377e-05$	$7.833e-06, 8.010e-06 \pm 2.092e-06$	$5.217, 1.008e+01 \pm 9.260$
	CGOA1	$5.835, 7.731 \pm 4.681$	$3.570e-07, 3.997e-07 \pm 1.821e-07$	$7.518e-06, 7.178e-06 \pm 7.219e-07$	$5.125, 7.656 \pm 2.092$
	CGOA2	$5.063, 5.881 \pm 2.527$	$4.057e-07, 3.589e-07 \pm 8.386e-08$	$8.016e-06, 8.097e-06 \pm 1.932e-06$	$4.629, 6.800 \pm 3.834$
	CGOA3	$1.065e+01, 1.509e+01 \pm 4.212$	$5.951e-07, 2.757e-05 \pm 2.020e-05$	$7.791e-06, 8.735e-06 \pm 2.216e-06$	$9.701, 2.475e+01 \pm 7.775$
	CGOA4	$4.346, 5.835 \pm 5.288$	$4.225e-07, 3.460e-07 \pm 1.155e-07$	$8.389e-06, 9.794e-06 \pm 4.095e-06$	$4.955, 1.125e+01 \pm 9.388$
	CGOA5	$5.063, 5.361 \pm 6.885e-01$	$4.065e-07, 3.857e-07 \pm 4.758e-08$	$8.015e-06, 7.919e-06 \pm 1.916e-07$	$4.629, 4.925 \pm 5.304e-01$
	CGOA6	$5.025, 5.565 \pm 1.247$	$4.077e-07, 3.899e-07 \pm 4.989e-08$	$8.060e-06, 7.649e-06 \pm 3.260e-07$	$4.631, 4.997 \pm 7.206e-01$
	CGOA7	$5.065, 5.800 \pm 1.800$	$4.067e-07, 3.812e-07 \pm 6.301e-08$	$8.015e-06, 7.794e-06 \pm 3.194e-07$	$4.629, 5.037 \pm 9.523e-01$
	CGOA8	$5.062, 4.664 \pm 1.541$	$4.057e-07, 4.077e-07 \pm 3.394e-08$	$8.016e-06, 8.472e-06 \pm 1.853e-06$	$4.629, 6.046 \pm 4.126$
	CGOA9	$5.113, 6.435 \pm 2.415$	$3.998e-07, 3.457e-07 \pm 7.147e-08$	$7.977e-06, 7.620e-06 \pm 3.134e-07$	$4.643, 5.471 \pm 1.119$
	CGOA10	$5.147, 8.717 \pm 4.812$	$4.003e-07, 5.012e-06 \pm 1.482e-05$	$7.916e-06, 7.009e-06 \pm 9.321e-07$	$4.652, 7.899 \pm 4.249$
	MFO	$5.477, 9.241 \pm 5.841$	$4.066e-07, 3.486e-07 \pm 1.416e-07$	$8.014e-06, 7.889e-06 \pm 3.072e-07$	$4.629, 5.713 \pm 2.648$
	CMFO1	**5.067**, **5.066** ± **6.513e-04**	**4.066e-07**, **4.066e-07** ± **1.867e-11**	**8.014e-06**, **8.014e-06** ± **1.030e-10**	**4.629**, **4.629** ± **1.204e-07**

(continued)

Table 2 (continued)

Algorithms		$R_c(\Omega)_{best,mean\pm STD}$	$C(F)_{best,mean\pm STD}$	$L(H)_{best,mean\pm STD}$	$SAE_{best,mean\pm STD}$
CMFO	CMFO2	$5.067, 6.559 \pm 4.597$	$4.066e-07, 3.680e-07 \pm 1.190e-07$	$8.014e-06, 7.930e-06 \pm 2.581e-07$	$4.629, 5.352 \pm 2.225$
	CMFO3	$5.067, 7.306 \pm 5.471$	$4.066e-07, 3.594e-07 \pm 1.222e-07$	$8.014e-06, 7.889e-06 \pm 3.055e-07$	$4.629, 5.532 \pm 2.308$
	CMFO4	$5.066, 7.306 \pm 5.471$	$4.066e-07, 3.487e-07 \pm 1.416e-07$	$8.014e-06, 7.889e-06 \pm 3.072e-07$	$4.629, 5.713 \pm 2.648$
	CMFO5	$5.066, 5.812 \pm 3.339$	$4.066e-07, 3.981e-07 \pm 3.835e-08$	$8.014e-06, 7.973e-06 \pm 1.843e-07$	$4.629, 4.810 \pm 8.082e-01$
	CMFO6	$5.066, 8.053 \pm 6.129$	$4.066e-07, 3.293e-07 \pm 1.587e-07$	$8.014e-06, 7.847e-06 \pm 3.442e-07$	$4.629, 6.075 \pm 2.967$
	CMFO7	$5.067, 5.813 \pm 3.339$	$4.066e-07, 3.873e-07 \pm 8.646e-08$	$8.014e-06, 7.972e-06 \pm 1.875e-07$	$4.629, 4.990 \pm 1.616$
	CMFO8	$5.067, 6.559 \pm 4.597$	$4.066e-07, 3.680e-07 \pm 1.190e-07$	$8.014e-06, 7.931e-06 \pm 2.581e-07$	$4.629, 5.352 \pm 2.225$
	CMFO9	$5.067, 6.560 \pm 4.597$	$4.066e-07, 3.680e-07 \pm 1.190e-07$	$8.014e-06, 7.930e-06 \pm 2.581e-07$	$4.629, 5.352 \pm 2.225$
	CMFO10	$5.067, 6.559 \pm 4.597$	$4.066e-07, 3.680e-07 \pm 1.190e-07$	$8.014e-06, 7.931e-06 \pm 2.581e-07$	$4.629, 5.352 \pm 2.225$
	FPA	$5.065, 5.066 \pm 2.809e-03$	$4.067e-07, 4.066e-07 \pm 1.082e-10$	$8.015e-06, 8.015e-06 \pm 7.721e-10$	$4.629, 4.629 \pm 1.065e-05$
	CFPA1	$5.066, 5.066 \pm 6.186e-04$	$4.066e-07, 4.067e-07 \pm 1.707e-11$	$8.014e-06, 8.014e-06 \pm 1.207e-10$	$4.629, 4.629 \pm 5.186e-07$
	CFPA2	$5.066, 5.065 \pm 1.246e-03$	$4.067e-07, 4.067e-07 \pm 2.994e-11$	$8.014e-06, 8.015e-06 \pm 3.471e-10$	$4.629, 4.629 \pm 2.456e-06$

(continued)

Table 2 (continued)

Algorithms		$R_c(\Omega)_{best,mean\pm STD}$	$C(F)_{best,mean\pm STD}$	$L(H)_{best,mean\pm STD}$	$SAE_{best,mean\pm STD}$
CFPA	CFPA3	$5.086, 5.069 \pm 1.830e-01$	$4.065e-07, 4.085e-07 \pm 4.733e-09$	$8.021e-06, 7.956e-06 \pm 1.330e-07$	$4.629, 4.656 \pm 3.180e-02$
	CFPA4	$5.066, 5.066 \pm 1.033e-03$	$4.067e-07, 4.067e-07 \pm 2.373e-11$	$8.014e-06, 8.014e-06 \pm 2.084e-10$	$4.629, 4.629 \pm 7.696e-07$
	CFPA5	$5.066, 5.066 \pm 9.612e-04$	$4.066e-07, 4.067e-07 \pm 1.916e-11$	$8.014e-06, 8.014e-06 \pm 1.716e-10$	$4.629, 4.629 \pm 6.440e-07$
	CFPA6	$5.066, 5.066 \pm 1.072e-03$	$4.067e-07, 4.066e-07 \pm 3.071e-11$	$8.014e-06, 8.015e-06 \pm 3.358e-10$	$4.629, 4.629 \pm 2.620e-06$
	CFPA7	$5.067, 5.065 \pm 1.214e-03$	$4.066e-07, 4.067e-07 \pm 2.691e-11$	$8.014e-06, 8.015e-06 \pm 2.520e-10$	$4.629, 4.629 \pm 1.174e-06$
	CFPA8	$5.066, 5.066 \pm 9.115e-04$	$4.066e-07, 4.066e-07 \pm 1.742e-11$	$8.014e-06, 8.014e-06 \pm 1.403e-10$	$4.629, 4.629 \pm 6.696e-07$
	CFPA9	**5.066, 5.066 ± 5.814e-04**	**4.066e-07, 4.066e-07 ± 1.931e-11**	**8.014e-06, 8.014e-06 ± 1.170e-10**	**4.629, 4.629 ± 4.039e-07**
	CFPA10	$5.066, 5.065 \pm 1.038e-03$	$4.066e-07, 4.066e-07 \pm 3.153e-11$	$8.014e-06, 8.015e-06 \pm 4.564e-10$	$4.629, 4.629 \pm 2.449e-06$

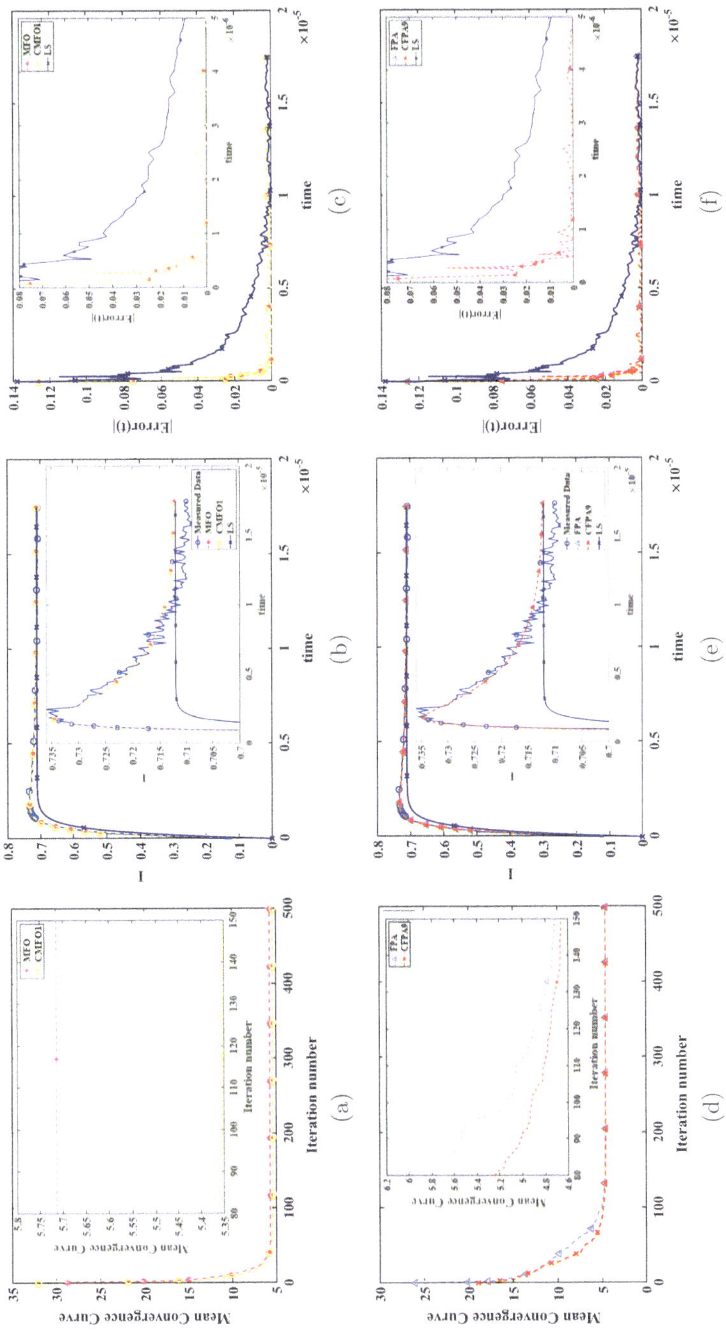

Fig. 3 Convergence curves, I-T fitting and absolute error curves by the recommended variants of CMFO and CFPA and basic FPA and MFO versions versus LS, **a** CMFO convergence curve, **b** CMFO fitting, **c** CMFO Absolute error, **d** CFPA convergence curve, **e** CFPA fitting, and **f** CFPA Absolute error

error curves that the meta-heuristic optimization techniques provide the better fitting on the experimental dynamic datasets with lower AE than LS method in most of the points on the curve especially in the transient part at the knee of the current curve. Moreover, the Chaotic variants of the proposed algorithms achieve lower AE than their standard variants at different points on the load current curve.

According to the results in Table 2 as well as the depicted curves in Fig. 3, it's concluded that CMFO and CFPA with Chebyshev, and sinusoidal maps, respectively, provide the most accurate and consistent results with fastest convergence speed than the other variants, respectively. Additionally, CMFO1 and CFPA9 have the first rank in consistency of the results followed by CGOA as shown in Table 2. To endorse the best one of these variants for IOM part statistically, the Wilcoxon Rank-Sum Test is carried out between CMFO1 and CFPA9 variants as observed from Table 2. Based on the Wilcoxon Rank-Sum analysis in Table 3, it's obvious that the CFPA9 is significantly different from the CMFO1 where the obtained *p-values* are less than 0.05. Therefore, depending on the obtained best, mean ± and STD values of the computed objective function by CFPA9 as well as the Wilcoxon Rank-Sum analysis results shown in Table 3, CFPA9 is selected as the recommended algorithms for identifying the IOM parameters with better fitting on the experimental current-time curve, less error and faster convergence speed.

5.2 Discussion of the Fractional-Order Dynamic PV Model (FOM)

In this part, the proposed algorithms search for five unknown parameters (R_C, C_α, L_β, α and β), the best mean ± STD of these parameters and the corresponding fitness function are listed in Table 4. The tabulated data clarifies that the chaotic variants offer lower best, mean ± STD of SAE values than their original versions especially, CGOA with Sine map, CMFO with Logistic map and CFPA with Chebyshev and singer maps. Where CGOA7 modifies the best, mean ± STD of SAE (by GOA) from $3.821, 6.931 \pm 3.233 \times 10^0$ to $2.743, 3.705 \pm 4.962 \times 10^{-1}$, respectively. Similarly, CMFO5 provides best, mean, and STD of SAE 2.740, 3.583 and 7.580×10^{-1} while MFO exhibits 2.935, 3.745, and 1.153×10^0, respectively. Likewise, CFPA1 and CFPA8 improves the accuracy and consistency of the FPA results where best, mean, and STD values are updated from 2.881, 3.527, and 4.624×10^{-1} to (2.738, 3.128 and 3.579×10^{-1}) and (2.817, 3.131 and 2.602×10^{-1}), respectively. It is obvious that CFPA1 and CFPA8 have the first rank in the accuracy and consistency followed by CMFO5 and at last CGOA7.

From the obtained mean convergence curves by the endorsed variants8 (CGOA7, CMFO5, CFPA1, and CFPA8) in Fig. 4a, d, g, respectively, it's concluded that the chaos maps have a remarkable influence on their convergence rate especially in case of GOA, as merging the sine map (CGOA7) modifies the convergence curve decaying rate nearly by 50% compared to te GOA convergence curve. CMFO5 achieves SAE

Table 3 Wilcoxon Rank-Sum test for the recommended variants for IOM

Algorithms		*ranksum*	*Zval*	*p-value*	h_0	Winner	Recommended
IOM	CFPA9 versus CMFO1	$5.710e+02$	4.342	$1.415e-05$	✗	CFPA9	CFPA9

equals to 4 at 110 iterations however MFO arrives for 4.4 at the same number of iterations. Similarly, CFPA1 and CFPA8 provide values of SAE lower than FPA at the same number of iterations.

To discuss the accuracy of the extracted parameters by the algorithms, both of the estimated current curves versus the measured one as well as the AE curves between these current curves are plotted in Fig. 4b, c, e, f, h, i, respectively. The obtained curves by the only previously published method NLS in [7] is also included on the fitting curves and error curves for comparison. It's observed from the figures that the proposed variants can provide better fitting on the experimental dataset especially at the transient part of the load current curve than NLS with lower AE values in several points as in the drawn error curves. It is noticed that the meta-heuristic optimization algorithms provide better performance than the conventional method (NLS), whereas CFPA1 and CFPA8 come on the top of these algorithms followed by CMFO5 and after that the CGOA7.

As a part of the successive comparisons among the winner variants, Wilcoxon Rank-Sum Test is carried out between the two last selected ones and the other recommended algorithms (CMFO5, and CGOA7). To determine which algorithm is the most appropriate one for the parameters identification of FOM, the results of CFPA1 and CFPA8 are compared statistically with (CMFO5, and CGOA7) as in Table 5. The listed results in this Table indicate that the CFPA1 and CFPA8 are the most suitable techniques for this problem where they have a significant difference with respect to the other algorithms as well as they achieve the least best mean ± STD values of the fitness function in addition to the best fitting with the measured data.

5.3 Comparison Between IOM and FOM

In the current subsection, the comparison is carried out between the IOM and FOM in fitting on the experimental datasets and the accuracy of the results. It's observed from Tables 2 and 4 that the best and mean values of SAE function between the identified and measured load current curves are lower in case of FOM than IOM. For more detailed investigation, the recommended algorithms for each model are compared together. For IOM, CFPA9 is the most suitable algorithm while in case of FOM, CFPA1, and CFPA8 are the most adequate ones. The best, mean ± STD values by these algorithms are reported in Table 6, their convergence curves and their fitting on the experimental load current curve are plotted in Fig. 5.

Table 4 The best, mean ± STD of the extracted parameters as well as the fitness function (SAE) for the FOM by the introduced algorithms over 20 runs

Algorithms		$R_c(\Omega)_{best,mean \pm STD}$	$C_\alpha(F)_{best,mean \pm STD}$	$L_\beta(H)_{best,mean \pm STD}$	$\alpha_{best,mean \pm STD}$	$\beta_{best,mean \pm STD}$	$SAE_{best,mean \pm STD}$
CGOA	GOA	$2.382, 1.965 \pm 1.848$	$8.654e-07, 3.301e-06 \pm 3.130e-06$	$1.443e-05, 2.708e-05 \pm 1.977e-05$	$9.592e-01, 8.895e-01 \pm 7.195e-02$	$9.570e-01, 9.222e-01 \pm 5.942e-02$	$3.821, 6.931 \pm 3.233$
	CGOA1	$1.443, 6.148e-01 \pm 1.094$	$5.341e-07, 2.523e-06 \pm 9.538e-07$	$3.358e-05, 4.290e-05 \pm 2.179e-05$	$1.027, 9.063e-01 \pm 5.481e-02$	$8.973e-01, 8.929e-01 \pm 4.588e-02$	$3.068, 3.909 \pm 1.142$
	CGOA2	$2.800e-01, 1.335 \pm 2.570$	$1.627e-06, 1.791e-06 \pm 5.585e-07$	$4.606e-05, 5.020e-05 \pm 2.788e-05$	$9.425e-01, 9.331e-01 \pm 3.369e-02$	$8.775e-01, 8.925e-01 \pm 7.374e-02$	$3.003, 3.888 \pm 1.411$
	CGOA3	$3.098, 1.392e+01 \pm 5.907$	$3.230e-06, 1.564e-05 \pm 1.909e-05$	$1.024e-05, 2.938e-05 \pm 1.678e-05$	$8.233e-01, 9.096e-01 \pm 8.870e-02$	$9.824e-01, 8.808e-01 \pm 6.107e-02$	$6.922, 1.959e+01 \pm 5.233$
	CGOA4	$9.174e-03, 2.366 \pm 6.258$	$4.535e-05, 3.955e-05 \pm 2.528e-05$	$2.293e-06, 5.818e-06 \pm 4.410e-06$	$9.136e-01, 8.468e-01 \pm 3.307e-02$	$8.797e-01, 9.011e-01 \pm 4.871e-02$	$3.211, 6.030 \pm 3.430$
	CGOA5	$7.315e-01, 8.067e-01 \pm 7.626e-01$	$1.241e-06, 2.129e-06 \pm 9.386e-07$	$3.399e-05, 4.253e-05 \pm 2.933e-05$	$9.565e-01, 9.199e-01 \pm 5.512e-02$	$8.981e-01, 8.971e-01 \pm 4.947e-02$	$2.975, 3.937 \pm 1.012$
	CGOA6	$3.484e-01, 8.534e-01 \pm 1.714$	$1.344e-06, 1.720e-06 \pm 7.928e-07$	$4.863e-05, 4.731e-05 \pm 1.836e-05$	$9.601e-01, 9.424e-01 \pm 4.004e-02$	$8.733e-01, 8.850e-01 \pm 5.015e-02$	$2.997, 3.596 \pm 9.888e-01$
	CGOA7	**3.529, 1.236 ± 2.528**	**1.829e-07, 1.885e-06 ± 1.203e-06**	**1.520e-05, 6.029e-05 ± 3.018e-05**	**1.093, 9.637e-01 ± 7.817e-02**	**9.512e-01, 8.728e-01 ± 6.205e-02**	**2.743, 3.705 ± 4.962e-01**
	CGOA8	$3.786e-01, 1.038 \pm 1.281$	$1.646e-06, 1.767e-06 \pm 5.279e-07$	$3.899e-05, 4.021e-05 \pm 2.358e-05$	$9.370e-01, 9.295e-01 \pm 4.245e-02$	$8.893e-01, 9.009e-01 \pm 5.335e-02$	$3.064, 3.844 \pm 9.870e-01$
	CGOA9	$5.283e-01, 1.004 \pm 1.921$	$1.188e-06, 1.656e-06 \pm 5.002e-07$	$3.737e-05, 4.459e-05 \pm 2.506e-05$	$9.633e-01, 9.373e-01 \pm 1.326e-02$	$8.927e-01, 8.949e-01 \pm 5.737e-02$	$3.033, 3.584 \pm 9.605e-01$
	CGOA10	$2.654, 7.262e-01 \pm 1.014$	$2.707e-07, 3.819e-06 \pm 2.636e-06$	$2.280e-05, 4.322e-05 \pm 3.199e-05$	$1.072, 8.873e-01 \pm 7.971e-02$	$9.233e-01, 8.964e-01 \pm 5.074e-02$	$3.004, 4.812 \pm 1.314$
	MFO	$6.647e-01, 3.088 \pm 5.143$	$1.107e-06, 1.516e-06 \pm 1.577e-06$	$3.967e-05, 3.780e-05 \pm 2.566e-05$	$9.706e-01, 9.867e-01 \pm 8.918e-02$	$8.872e-01, 9.186e-01 \pm 7.409e-02$	$2.935, 3.745 \pm 1.153$

(continued)

Table 4 (continued)

Algorithms		$R_c(\Omega)_{best,mean\pm STD}$	$C_\alpha(F)_{best,mean\pm STD}$	$L_\beta(H)_{best,mean\pm STD}$	$\alpha_{best,mean\pm STD}$	$\beta_{best,mean\pm STD}$	$SAE_{best,mean\pm STD}$
CMFO	CMFO1	1.856, 3.044 ± 3.223	4.883e − 07, 1.239e − 06 ± 1.274e − 06	2.418e − 05, 2.968e − 05 ± 2.628e − 05	1.025, 1.001 ± 8.779e − 02	9.203e − 01, 9.359e − 01 ± 6.879e − 02	2.800, 3.667 ± 8.002e − 01
	CMFO2	3.643, 2.605 ± 3.099	1.670e − 07, 1.380e − 06 ± 1.157e − 06	1.491e − 05, 3.430e − 05 ± 3.048e − 05	1.100, 9.873e − 01 ± 8.944e − 02	9.525e − 01, 9.260e − 01 ± 6.936e − 02	2.759, 3.695 ± 7.137e − 01
	CMFO3	2.023, 1.679 ± 2.310	5.334e − 07, 2.384e − 06 ± 2.065e − 06	2.095e − 05, 3.670e − 05 ± 2.858e − 05	1.013, 9.448e − 01 ± 1.031e − 01	9.302e − 01, 9.151e − 01 ± 6.004e − 02	2.831, 3.943 ± 1.209
	CMFO4	3.661, 3.548 ± 6.169	1.814e − 07, 1.474e − 06 ± 1.192e − 06	1.423e − 05, 4.099e − 05 ± 3.537e − 05	1.091, 9.825e − 01 ± 8.642e − 02	9.558e − 01, 9.156e − 01 ± 7.385e − 02	2.726, 4.061 ± 1.694
	CMFO5	**2.829, 2.814 ± 3.126**	**2.814e-07, 1.202e-06 ± 1.132e-06**	**1.772e-05, 2.785e-05 ± 2.363e-05**	**1.061, 9.993e-01 ± 8.945e-02**	**9.412e-01, 9.364e-01 ± 6.293e-02**	**2.740, 3.583 ± 7.580e-01**
	CMFO6	4.402, 1.473 ± 2.373	1.495e − 07, 1.829e − 06 ± 1.738e − 06	1.154e − 05, 4.263e − 05 ± 2.603e − 05	1.100, 9.688e − 01 ± 8.728e − 02	9.700e − 01, 8.998e − 01 ± 5.660e − 02	2.814, 3.365 ± 7.844e − 01
	CMFO7	3.866, 2.550 ± 3.102	1.645e − 07, 1.281e − 06 ± 1.044e − 06	1.364e − 05, 3.568e − 05 ± 2.837e − 05	1.098, 9.898e − 01 ± 8.130e − 02	9.586e − 01, 9.216e − 01 ± 7.238e − 02	2.730, 3.732 ± 1.258
	CMFO8	4.648, 3.577 ± 5.152	1.460e − 07, 1.497e − 06 ± 1.725e − 06	1.081e − 05, 3.836e − 05 ± 3.395e − 05	1.100, 9.942e − 01 ± 9.415e − 02	9.744e − 01, 9.258e − 01 ± 8.081e − 02	2.881, 4.079 ± 1.617
	CMFO9	3.820, 3.351 ± 5.047	1.622e − 07, 1.668e − 06 ± 1.986e − 06	1.396e − 05, 3.658e − 05 ± 3.105e − 05	1.100, 9.878e − 01 ± 9.915e − 02	9.570e − 01, 9.245e − 01 ± 7.501e − 02	2.734, 4.086 ± 1.663
	CMFO10	3.799, 2.335 ± 2.910	2.038e − 07, 1.557e − 06 ± 1.701e − 06	1.302e − 05, 3.650e − 05 ± 2.830e − 05	1.079, 9.863e − 01 ± 9.202e − 02	9.617e − 01, 9.189e − 01 ± 6.766e − 02	2.791, 3.547 ± 9.173e − 01
	FPA	3.210, 1.280 ± 1.740	3.052e − 07, 1.401e − 06 ± 9.218e − 07	1.757e − 05, 4.275e − 05 ± 2.091e − 05	1.052, 9.726e − 01 ± 5.977e − 02	9.411e − 01, 8.917e − 01 ± 4.346e − 02	2.881, 3.527 ± 4.624e − 01

(continued)

Table 4 (continued)

Algorithms		$R_c(\Omega)_{best,mean\pm STD}$	$C_\alpha(F)_{best,mean\pm STD}$	$L_\beta(H)_{best,mean\pm STD}$	$\alpha_{best,mean\pm STD}$	$\beta_{best,mean\pm STD}$	$SAE_{best,mean\pm STD}$
CFPA	**CFPA1**	**3.676, 1.800 ± 1.675**	**1.684e-07, 1.133e-06 ± 1.060e-06**	**1.406e-05, 3.460e-05 ± 2.149e-05**	**1.097, 1.002 ± 7.698e-02**	**9.566e-01, 9.088e-01 ± 4.342e-02**	**2.738, 3.128 ± 3.579e-01**
	CFPA2	3.673, 1.972 ± 1.762	$1.947e-07, 1.057e-06 \pm 9.146e-07$	$1.316e-05, 3.135e-05 \pm 2.226e-05$	$1.082, 9.979e-01 \pm 6.925e-02$	$9.606e-01, 9.197e-01 \pm 4.956e-02$	$2.814, 3.193 \pm 2.959e-01$
	CFPA3	3.123, 1.382 ± 2.111	$2.628e-07, 1.867e-06 \pm 1.477e-06$	$1.524e-05, 4.424e-05 \pm 2.558e-05$	$1.060, 9.625e-01 \pm 8.084e-02$	$9.514e-01, 8.932e-01 \pm 5.339e-02$	$2.806, 3.759 \pm 6.759e-01$
	CFPA4	$2.492, 8.496e-01 \pm 1.339$	$3.427e-07, 1.645e-06 \pm 8.725e-07$	$2.104e-05, 4.571e-05 \pm 1.859e-05$	$1.051, 9.585e-01 \pm 5.863e-02$	$9.292e-01, 8.844e-01 \pm 3.523e-02$	$2.802, 3.273 \pm 3.832e-01$
	CFPA5	$4.575, 8.991e-01 \pm 1.550$	$1.618e-07, 1.476e-06 \pm 7.611e-07$	$1.272e-05, 4.826e-05 \pm 2.240e-05$	$1.095, 9.690e-01 \pm 6.008e-02$	$9.630e-01, 8.833e-01 \pm 4.239e-02$	$2.855, 3.204 \pm 2.721e-01$
	CFPA6	$4.274e-01, 9.112e-01 \pm 1.475$	$1.287e-06, 1.588e-06 \pm 7.777e-07$	$4.154e-05, 4.556e-05 \pm 2.193e-05$	$9.598e-01, 9.589e-01 \pm 5.916e-02$	$8.847e-01, 8.875e-01 \pm 4.325e-02$	$2.977, 3.324 \pm 3.436e-01$
	CFPA7	3.506, 1.667 ± 1.903	$1.694e-07, 1.239e-06 \pm 9.261e-07$	$1.524e-05, 3.761e-05 \pm 2.263e-05$	$1.100, 9.892e-01 \pm 7.418e-02$	$9.505e-01, 9.035e-01 \pm 4.630e-02$	$2.799, 3.211 \pm 3.267e-01$
	CFPA8	**3.592, 1.427 ± 1.507**	**2.071e-07, 1.105e-06 ± 7.515e-07**	**1.661e-05, 3.741e-05 ± 2.061e-05**	**1.083, 9.926e-01 ± 6.227e-02**	**9.445e-01, 9.013e-01 ± 3.907e-02**	**2.817, 3.131 ± 2.602e-01**
	CFPA9	3.860, 1.328 ± 1.907	$1.753e-07, 1.444e-06 \pm 8.602e-07$	$1.338e-05, 4.040e-05 \pm 2.324e-05$	$1.090, 9.697e-01 \pm 6.532e-02$	$9.598e-01, 8.994e-01 \pm 4.779e-02$	$2.777, 3.246 \pm 2.677e-01$
	CFPA10	1.815, 1.120 ± 1.475	$6.128e-07, 1.644e-06 \pm 1.105e-06$	$2.282e-05, 4.173e-05 \pm 2.066e-05$	$1.003, 9.657e-01 \pm 7.234e-02$	$9.240e-01, 8.940e-01 \pm 4.346e-02$	$2.861, 3.341 \pm 2.987e-01$

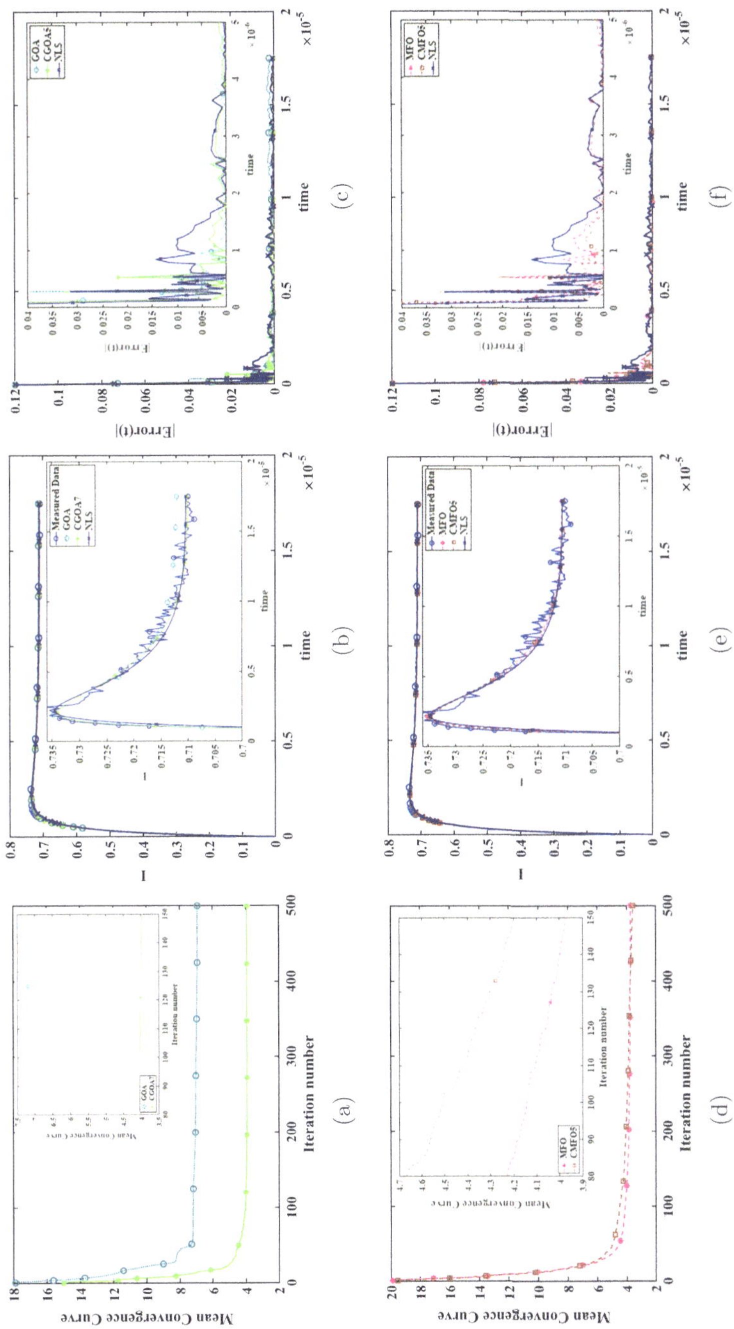

Fig. 4 Convergence curves, I-T fitting and absolute error curves by the recommended variants of CMFO and CFPA and basic FPA and MFO versions versus NLS, **a** CGOA convergence curve, **b** CGOA fitting, **c** CGOA Absolute error, **d** CMFO convergence curve, **e** CMFO fitting, **f** CMFO Absolute error, **g** CFPA convergence curve, **h** CFPA fitting, and **i** CFPA Absolute error

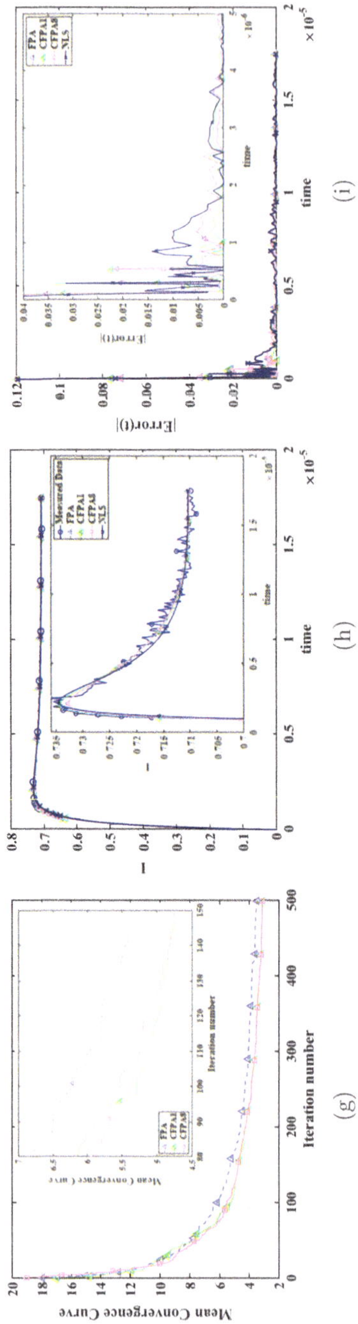

Fig. 4 (continued)

Table 5 Wilcoxon Rank-Sum test for the recommended variants for FOM

Algorithms		*ranksum*	*Zval*	$p-value$	h_0	Winner	Recommended
FOM	CFPA1 versus CGOA7	$2.470e+02$	-2.750	$5.966e-03$	✗	CFPA1	
	CFPA1 versus CMFO5	$3.430e+02$	-1.799	$7.205e-03$	✗	CFPA1	
	CFPA1 versus CFPA8	$3.810e+02$	$-7.709e-01$	$4.407e-01$	✓	–	CFPA1
	CFPA8 versus CGOA7	$2.420e+02$	-2.970	$2.982e-03$	✗	CFPA8	CFPA8
	CFPA8 versus CMFO5	$3.630e+02$	-1.258	$2.085e-02$	✗	CFPA8	
	CMFO5 versus CGOA7	$2.880e+02$	$-9.459e-01$	$3.442e-01$	✓	–	

(–) No significant difference

It's obvious from Table 6 that the best and mean values of SAE are lower in case of FOM than IOM due to the more flexibility and extra degree of freedom that fractional calculus appended for the model. Where in FOM, techniques searching for five unknown parameters rather than the three parameters of IOM is increasing the search space which make the fractional model more descriptive for the real system response. The algorithms become capable for capturing more accurate combination between the parameters that reflected in turn on the accuracy of the SAE value than the fitting of the measured dataset. Furthermore, Fig. 5 enhances the previous observation where the fractional-order model with the recommended variants provides a better fitting on the experimental datasets of the current curve not only on the transient section at the knee of the curve but also on the steady-state part as CFPA1 and CFPA8 achieve lower error values between the measured and estimated load current curves. Moreover, from the convergence curve in Fig. 5, it's concluded that the algorithms in case of FOM have the flexibility and capability to converge to lower values of the objective function where at 150 iteration CFPA1 and CFPA8 have better combination between the unknown five parameters that may affect in turn on the accuracy of the objective function. These results prove the superiority of the FOM in modeling of the experimental load current curves because of its extra degrees of freedom that is represented in the non integer derivative orders added to the model and a deep vision into the physical processes of the PV modules underlying a long-range memory behavior.

Table 6 The recommended variants for IOM and FOM

Models/algorithms		$SAE_{best,mean}$
IOM	CFPA9	4.629*e* + 00, 4.629*e* + 00
FOM	CFPA1	2.738*e* + 00, 3.128*e* + 00
	CFPA 8	2.817*e* + 00, 3.131*e* + 00

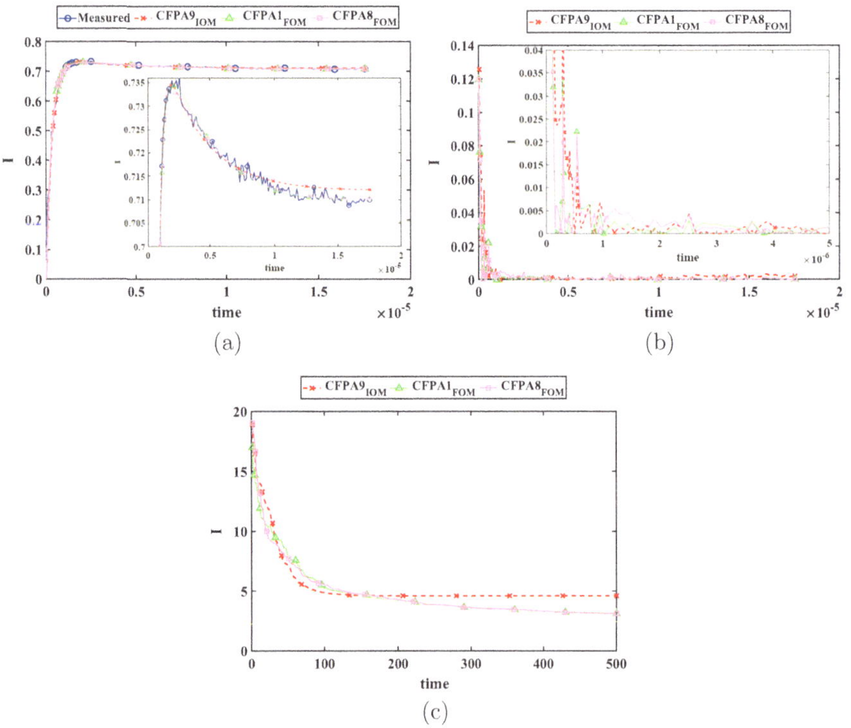

Fig. 5 Comparison between IOM and FOM based on the **a** Load Current curve fitting, **b** Absolute error, and **c** Convergence curves

6 Conclusion

This chapter proposes new optimization techniques to extract the parameters of the dynamic PV models accurately and quickly based on the experimental dynamic dataset of the load current curve. The introduced dynamic PV models are classified as the integral-order dynamic model and a newly developed fractional-order dynamic model that may provide a better fitting on the real system response. The proposed techniques for parameters estimation of the models are the Chaotic Grasshopper Optimizer, the Chaotic Moth flame Optimizer, and the Chaotic Flower Pollination Algorithm where their original versions are combined with ten different chaos maps for adaptive tuning of their parameters. Successive comparisons are established to compare The chaotic variants' results to that of the standard versions of the proposed algorithms as well as the conventional methods published in literature (Least squares and non-linear Least squares) to recommend the best variant that achieves more accuracy, consistency, higher speed of convergence, and less deviation from experimental data. For further investigation, Wilcoxon Rank-Sum Test is carried out between the chaotic and basic versions of the utilized algorithms as well as between the recommended algorithms. Based on these intensive comparisons among the results of this work and that of the previously published algorithms, it is clear that the metaheuristic algorithms are more efficient tools in providing more accurate results than the conventional ones even with the lately developed more complicated models. Moreover, the chaotic variants offer better convergence speed and consistent results than the original algorithms. For IOM, CFPA with Sinusoidal is the most suitable one, whereas CFPA merged with the Chebyshev and Singer chaos maps are the most appropriate variants in case of fractional-order PV models as they achieve better fitting on the utilized datasets with faster decaying rate of convergence. From the comparison between the recommended algorithms in both models, it's concluded that combination between the CFPA (Chebyshev and Singer chaos maps) with FOM achieves an accurate representation for the physical behavior of the dynamic PV system compared to the IOM one where it exhibits a better description of the real system response not only on the transient part but also on the steady-state part as well. For the future work, the FOM will be employed for different PV modules types, moreover, novel algorithms with innovative objective function will be examined for optimizing this model.

References

1. Yang B, Yu T, Shu H, Zhu D, Zeng F, Sang Y, Jiang L (2018) Perturbation observer based fractional-order pid control of photovoltaics inverters for solar energy harvesting via yin-yang-pair optimization. Energy Conver Manage 171:170–187
2. Chen Y, Sun Y, Meng Z (2018a) An improved explicit double-diode model of solar cells: fitness verification and parameter extraction. Energy Convers Manag 169:345–358
3. Chen X, Xu B, Mei C, Ding Y, Li K (2018b) Teaching-learning-based artificial bee colony for solar photovoltaic parameter estimation. Appl Energy 212:1578–1588
4. Allam D, Yousri D, Eteiba M (2016) Parameters extraction of the three diode model for the multi-crystalline solar cell/module using moth-flame optimization algorithm. Energy Convers Manag 123:535–548
5. Hsieh Y-C, Yu L-R, Chang T-C, Liu W-C, Wu T-H, Moo C-S (2019) Parameter identification of one-diode dynamic equivalent circuit model for photovoltaic panel. IEEE J Photovolt 10:219–225
6. Piazza MCD, Luna M, Vitale G (2013) Dynamic PV model parameter identification by least-squares regression. IEEE J Photovolt 3:799–806

7. AbdelAty AM, Radwan AG, Elwakil AS, Psychalinos C (2018) Transient and steady-state response of a fractional-order dynamic PV model under different loads. J Circ Syst Comput 27:1850023
8. Boutana N, Mellit A, Haddad S, Rabhi A, Pavan AM (2017) An explicit i–v model for photovoltaic module technologies. Energy Convers Manag 138:400–412
9. Elbaset AA, Ali H, Sattar MA-E (2014) Novel seven-parameter model for photovoltaic modules. Solar Energy Mater Solar Cells 130:442–455
10. Alam D, Yousri D, Eteiba M (2015) Flower pollination algorithm based solar PV parameter estimation. Energy Convers Manag 101:410–422
11. Jordehi AR (2016) Parameter estimation of solar photovoltaic (PV) cells: a review. Renew Sustain Energy Rev 61:354–371
12. Chin VJ, Salam Z, Ishaque K (2016) An accurate modelling of the two-diode model of PV module using a hybrid solution based on differential evolution. Energy Convers Manag 124:42–50
13. Bana S, Saini R (2017) Identification of unknown parameters of a single diode photovoltaic model using particle swarm optimization with binary constraints. Renew Energy 101:1299–1310
14. Yousri D, Elaziz MA, Oliva D, Abualigah L, Al-qaness MA, Ewees AA (2020) Reliable applied objective for identifying simple and detailed photovoltaic models using modern metaheuristics: Comparative study. Energy Convers Manag 223:113279
15. Yousri D, Rezk H, Fathy A Identifying the parameters of different configurations of photovoltaic models based on recent artificial ecosystem-based optimization approach. Int J Energy Res
16. Liang J, Qiao K, Yu K, Ge S, Qu B, Xu R, Li K (2020) Parameters estimation of solar photovoltaic models via a self-adaptive ensemble-based differential evolution. Solar Energy 207:336–346
17. Ibrahim IA, Hossain M, Duck BC, Fell CJ (2019) An adaptive wind-driven optimization algorithm for extracting the parameters of a single-diode pv cell model. IEEE Trans Sustain Energy 11:1054–1066
18. Yousri D, Allam D, Eteiba M, Suganthan PN (2019) Static and dynamic photovoltaic models parameters identification using chaotic heterogeneous comprehensive learning particle swarm optimizer variants. Energy Convers Manag 182:546–563
19. Yang X-S (2012) Flower pollination algorithm for global optimization. In: Unconventional computation and natural computation, Springer, pp 240–249
20. Mirjalili S (2015) Moth-flame optimization algorithm: a novel nature-inspired heuristic paradigm. Knowledge-Based Syst
21. Saremi S, Mirjalili S, Lewis A (2017) Grasshopper optimisation algorithm: theory and application. Adv Eng Softw 105:30–47
22. Wolpert DH, Macready WG (1997) No free lunch theorems for optimization. IEEE Trans Evol Comput 1:67–82
23. Mirjalili S, Gandomi AH (2017) Chaotic gravitational constants for the gravitational search algorithm. Appl Soft Comput
24. Rizk-Allah RM, Hassanien AE, Bhattacharyya S (2018) Chaotic crow search algorithm for fractional optimization problems. Appl Soft Comput
25. Ibrahim RA, Elaziz MA, Lu S (2018) Chaotic opposition-based grey-wolf optimization algorithm based on differential evolution and disruption operator for global optimization. Expert Syst Appl 108:1–27
26. Yang D, Li G, Cheng G (2007) On the efficiency of chaos optimization algorithms for global optimization. Chaos Solitons Fract 34:1366–1375
27. Sayed GI, Khoriba G, Haggag MH (2018) A novel chaotic salp swarm algorithm for global optimization and feature selection. Appl Intell 1–20
28. Yousri D, Allam D, Eteiba M (2018) Parameters identification of fractional order permanent magnet synchronous motor models using chaotic meta-heuristic algorithms. In: Mathematical techniques of fractional order systems. Elsevier, pp 529–558

29. Radwan AG, Salama KN (2012) Fractional-order RC and RL circuits. Circ Syst Signal Process 31:1901–1915
30. AbdelAty AM, Radwan AG, Elwakil A, Psychalinos C (2016) A fractional-order dynamic PV model In: 2016 39th International conference on telecommunications and signal processing (TSP). IEEE
31. Mirjalili S (2015) Moth-flame optimization algorithm: a novel nature-inspired heuristic paradigm. Knowledge-Based Syst 89:228–249

Dust Accumulation and Photovoltaic Performance in Semi-Arid Climate: Experimental Investigation and Design of Cleaning Robot

Alae Azouzoute, Massaab El Ydrissi, Houssain Zitouni, Charaf Hajjaj, and Mohammed Garoum

Abstract Photovoltaic technology is still under development in many countries around the world. However, the desert regions are still the most attractive zones in terms of solar radiation and land use availability. On the other hand, the semi-arid and arid climates predominate in those regions where dust concentration is relatively high, which influence drastically the efficiency of the PV system. This study investigated the effect of dust accumulation on the transmittance of the glass samples and the overall electrical efficiency of the PV module for different cleaning frequencies. It was found that the broadband transmittance loss of a glass sample decreases by up to 52% after 3 months of exposure, whereas it was in the range of 6.5% after 7 days during the dry period. The dust accumulation rate was about 5.6 g/m^2 and 0.4 g/m^2 after 90 days and 7 days of exposure, respectively. For the PV output performance, dust accumulation has significantly influenced the soiling ratio of the PV system with no cleaning as it reached 0.77 after 68 days of exposure and no significant impact has been seen for the weekly cleaned PV module with a soiling ratio in the range of 0.97 during all the period of investigation. Thereafter, a new self-guided cleaning

A. Azouzoute (✉) · M. El Ydrissi · H. Zitouni
Institut de Recherche en Energie Solaire et Energies Nouvelles–IRESEN, 16 Avenue S.A.R. Sidi Mohamed, 10090, Rabat, Morocco
e-mail: azouzoute@iresen.org

M. El Ydrissi
e-mail: elydrissi@iresen.org

H. Zitouni
e-mail: zitouni@iresen.org

A. Azouzoute · M. Garoum
MEAT, EST Salé, University Mohamed V, Avenue Prince Héritier, BP 227, Salé Medina, Salé, Morocco
e-mail: garoum1@yahoo.fr

C. Hajjaj
Laboratory of Applied Sciences for the Environment and Sustainable Development, Higher School of Technology of Essaouira, Cadi Ayyad University, BP. 383 Essaouira, Morocco
e-mail: hajjaj.charaf@gmail.com

S. Motahhir and A. M. Eltamaly (eds.), *Advanced Technologies for Solar Photovoltaics Energy Systems*, Green Energy and Technology,
https://doi.org/10.1007/978-3-030-64565-6_3

robot system has been designed to optimize the cleaning schedule under the local weather conditions.

Keywords Photovoltaic system · Dust accumulation · Transmittance · Density of dust deposition · Cleaning robot · Semi-arid climate

1 Introduction

The photovoltaic module is characterized by the electrical power that it delivers when it is conventionally subjected to the sunshine of 1000 W/m^2 and a cell temperature of 25 °C. This electrical power is determined just after the modules have been manufactured. However, these data are not constant over time; when the module undergoes degradations, the electrical power decreases, as well as the other magnitudes short-circuit current and voltage.

Today, there is a lack of information on the different modes of degradation of photovoltaic modules in terms of frequency, rate of evolution, and degree of impact on the life and reliability of PV modules. In recent years, research on photovoltaic modules has rather focused on the race to develop new technologies without having enough feedback on those that are already operational.

The meteorological parameters and the characteristics of the installation site remain one of the drastically most influencing parameters on the performance of photovoltaic solar panels. They can significantly reduce both the efficiency of electrical performance and the capacity of the panels, knowing that most panels are designed for proper operation for more than two decades, but this time is reduced due to the hard environmental conditions. However, the accumulation of dust on the photovoltaic panels remains one of the most influencing parameters on the performance of the panels as well as their lifespan.

a. Soiling phenomena

Solar radiation is one of the most abundant clean resources existing on the planet especially in regions with desert land. MENA region and other countries in the Sunbelt enjoy the high potential of incoming solar irradiation and the availability of area for the development and deployment of solar power plants [1–3]. However, those regions are known for their harsh weather conditions; hot and dry in the summer season, and wet climate with rainy days in winter. Soiling is the main challenge that impacts directly the efficiency of a solar power plant (concentrated solar power or photovoltaic systems) [4–9]. Soiling on the front glass of PV modules results in optical losses due to the area shaded by dust particles [10–14]. On the other hand, the deposition of contaminants such as mineral dust deposit, biofilms of bacteria, algae, lichen, mosses or fungi, plant debris or pollen, bird droppings, engine exhausts or agricultural emissions, and industry emissions [15–19] onto the PV module surface leads to an excessive reduction of power generation. This can be quantified by more than 1% power loss per day [10, 15, 20, 21]. It often makes the installation economically unreliable. To this end, the MENA region was reported to be the worst region that exhibits dust accumulation regarding the other regions [22, 23].

Dust accumulation was found to be the most significant factor affecting the efficiency of a PV module in interaction with environmental parameters as the relative humidity, temperature, wind speed, and wind direction [8, 9, 24, 25]. On the other hand, the accumulation rate of dust on the surface of the PV module increases over the time of exposition [26]. For example, in Minia region, Egypt, the mass accumulation rate of 150–300 mg/m^2/day [27] has been reported depending on the tilting angle, as well in Dhahran, Saudi Arabia, the dust surface density has reached 6.2 g/m^2 for the period from February to December (10 months) [28]. In another study, the accumulation rate of 132 mg/m^2/day in Mesa, Arizona [29], and 1–50 mg/m^2/day in Colorado [30], and from 0.01 to 0.02 mg/cm^2/day has been recorded in Lahore, Pakistan for panels with a fixed tilt angle of 30 ° [31]. In the Mediterranean climate, in Athens, Greece, the dust loading of 0.1–1 g/m^2 was recorded for a period of exposition of 2–8 weeks [32]. This accumulation rate is strongly dependent on dust particles concentration at the upper of the atmosphere and weather conditions of the local site [33].

In reading the literature, several methods have been used to evaluate the effect of soiling and especially dust deposition on the performance of PV panels. In term of optical efficiency, the effect of dust on the transmittance of the front glass of the PV module has been widely used to evaluate the impact of dust deposition on the transmittance of light radiation with regard to the density of dust deposition [9, 27, 34–36], thus, will consequently reduce the incoming light to the PV cell and decrease the electrical output. However, the most reliable method is to assess the direct impact of dust deposition on the surface of the PV module in terms of the electrical parameters as the maximal power, short-circuit current, and the total energy production (in the case of a PV system). As dust deposits tend to attenuate short wavelengths with regards to the selective aspect of the spectral response of the PV cell, which explains the difference between light transmission loss and power loss [37].

Therefore, understanding the soiling mechanisms is highly crucial in order to develop optimized cleaning scenarios for the dusty region. The next section in this work will describe the process of the dust life cycle and the interference of different parameters on it.

b. Dust life cycle

Recently, many research studies have been done on the processes governing the transport and deposition of dust in solar power plants [38, 39]. All these studies have concluded that the impact of dust on PV modules follows a cycle, which is called the life cycle of dust. This cycle contains four main stages (generation, deposition, adhesion, and finally removal or elimination at the level of the panel) as indicated in Fig. 1.

Generation

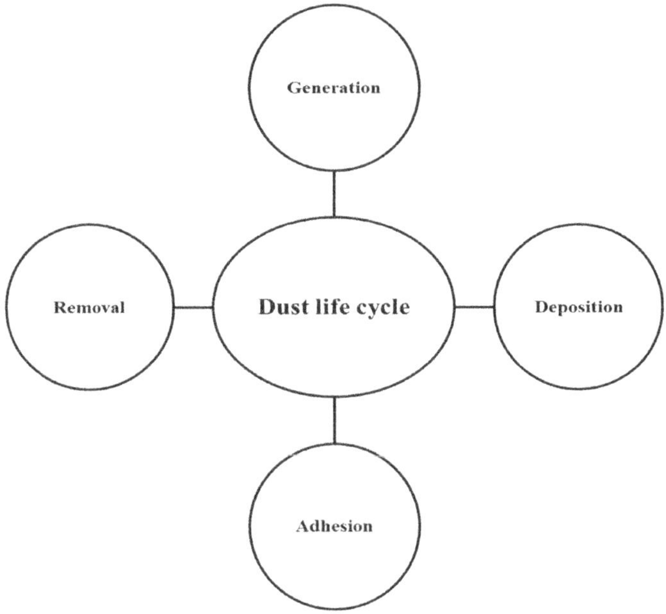

Fig. 1 The dust life cycle

The charging of the atmosphere with dust particles is mainly linked to soil erosion due to the wind [40, 41]. The wind emission produces when the wind has enough power to move the granular crushed material [42]; therefore the wind is the main phenomenon that allows raising dust to the atmosphere. We can distinguish three different modes of transport of particles by the wind speed, presented in Fig. 2 [41].

Suspension: In general, fine dust can only be carried away if it has been thrown into the air by the impact of larger grains. Once in the turbulent layer, they can be lifted to great heights by ascending air currents and form dust clouds reaching altitudes of 3–4,000 meters. Their appearance can be impressive; the essential mechanism of wind erosion remains saltation because without it such clouds could not occur.

Saltation: The initial movement of soil particles is a series of jumps. The diameter of the saltation particles is between 0.5 and 1.1 mm. After having jumped, the particles fall back under the action of gravity. The descending part of the trajectory is very inclined towards the ground and practically straight. Few particles reach an altitude greater than 1 m and about 90% of them jump less than 30 cm. The horizontal amplitude of a jump is generally between 0.5 m and 1 m. The saltation phenomenon is essential to initiate wind erosion. It is the cause of two other modes of transport of soil elements by the wind: surface crawling and air suspension.

Crawling: Larger particles roll or slide across the soil surface. Too heavy to be lifted, their movement is triggered by the impact of saltation particles rather than by the action of the wind. The particles that move in this way to a diameter of between 0.5 mm and 2 mm depending on their density and the wind speed.

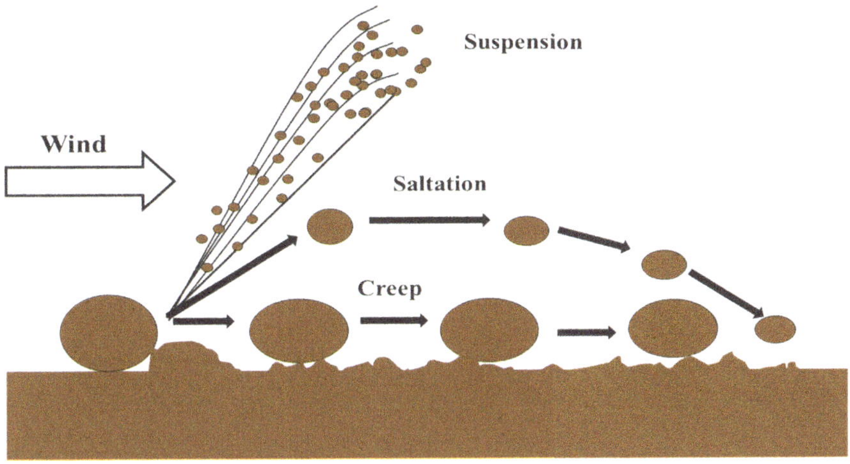

Fig. 2 Modes of transport of particles by wind speed

Deposition

Once the dust particles have been assessed and suspended in the atmosphere. Many factors influence the deposition of dust on photovoltaic panels, two main types of deposit can be mentioned:

- Dry deposition: The desert aerosol cycle ends with the deposition of particles on the PV module surfaces, under dry or wet atmospheric conditions. Dry deposition is mainly controlled by gravitational forces [43], which cause the particles to sediment. The particle dry deposition depends on several variables such as wind speed, friction speed, turbulence intensity, and atmospheric stability [44].

The sedimentation rate depends on the size of the particle. Sedimentation is a function of particle size, with the larger particles falling first and the smaller ones last. As a result, the larger and heavier particles will be deposited near the region of origin, while the smaller ones will be deposited farther away. Turbulence can also play a role, disrupting the flow of particles [45]. If the turbulence occurs near the surface, then the particles are deposited faster.

The transfer of the dust particles to the surface of the panels is done through different mechanical processes as shown in Fig. 3; sedimentation, Brownian diffusion, turbulent impaction, and interception processes [46].

- Wet deposition: Wet deposition occurs when dusty air masses are mixed by convection with moist or cloudy air masses. Mineral particles can be captured directly by precipitation or by droplets within the cloud.

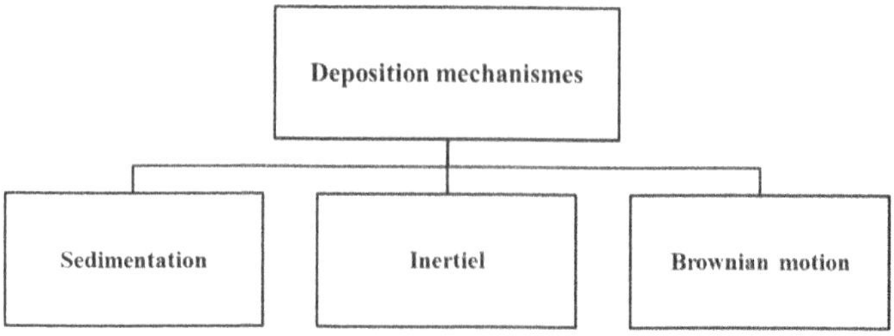

Fig. 3 The deposition mechanisms

The rate of wet deposition depends on the rate of precipitation and the rate of droplet fall. Particles between 0.1 and a few μm in size have the lowest sedimentation velocities, so they will be mostly deposited by wet deposition [38].

Adhesion

Dust is initially deposited on the surface of the photovoltaic panels, the adhesion forces such as van der Waals forces, electrostatic forces, and capillary forces are the active forces that cause the particles to bind to the surface. The adhesion processes of these forces are described in Fig. 4 [47].

- Due to the strength of van der Waals, small dry dust particles stick to a dry surface. This force is considered to be the dominant force between a solid platform and a dry particle under dry ambient conditions. The van der Waals forces are always present between the surface and the particles and act over a short distance since they originate from two surfaces that are in contact with interacting dipoles.
- The capillary force depends on both the air's humidity content (RH; relative humidity) and the surface. Capillary forces act when two moist bodies meet. The

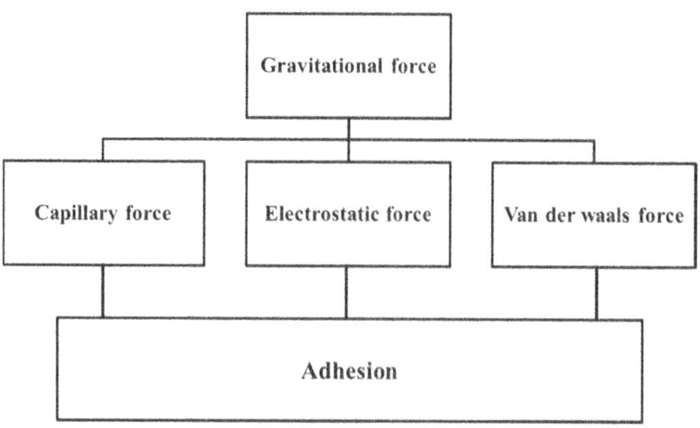

Fig. 4 The adhesion mechanism

water vapor condenses into fine particles, thus enabling the gap between the dust particles and the surface of a PV module to be bridged leading to adhesion.
- The electrostatic force causes adhesion in the presence of charges. Dust particles tend to acquire electric charges in the atmosphere by colliding with each other, and these charged particles tend to attract a positive charge to the surface, inducing a coulomb force.

All adhesion forces are active almost everywhere, but the extent of their adhesion is determined by the environmental condition and the property of the dust particles.

Removal

The elimination phase is the process of cleaning the particles from the surface. Basically, there are two types: Natural elimination and artificial elimination. The natural removal phase is when particles are removed by natural causes such as wind and rain. It consists of two phases: rebound and resuspension. However, artificial removal is done using specific tools to remove dust particles [48].

- Natural elimination: Two factors influence the natural cleaning process: the particle properties (composition, number, and size range) and the local weather conditions (frequency, intensity, rain, and wind duration).

The conditions taken into consideration for this process are the alteration conditions, duration, and the angular orientation of the surfaces (angle of inclination).

Consequently, it has been found that particles with a diameter of less than 10 μm which are the most present on the surface of the panels are eliminated by the natural cleaning forces.

- Artificial elimination: As regards artificial cleaning, they depend mainly on the type of contact device (brush and soft cloth), the quality of the water used (demineralized and tap water), the water pressure, additives, and the state of the water (liquid and vapor).

The main objective of this chapter is devised into two. The first one is to highlight the impact of soiling on photovoltaic performance by calculating the transmittance drop and the dust mass density by using photovoltaic glass and the soiling ratio using photovoltaic panels, over a period of 6 months of measurement at ground level.

The second objective is to present a developed self-guided cleaning robot in order to maintain the performance of photovoltaic solar panels and to optimize the cleaning cost.

2 Soiling Measurements

In this section, the results of an experimental study conducted at Green Energy Park research facility in the mid-south of Morocco (in Benguerir city; 32.12 °N, −7.94

Fig. 5 Glass samples were exposed in the fixed rack at the Green Energy Park research facility

°E) are presented. Glass samples with dimensions of 10 cm × 10 cm and thickness of 3 mm were used in order to evaluate the dust effect on their transmittance values. The samples were exposed for different periods: weekly, monthly, and 3 months in a fixed rack tilted by 32° and 1 m above the ground facing south over the dry period of the year from May to August 2018 (see Fig. 5). Before the exposure, reference mass of all the samples was determined, and the amount of dust on the surface of samples after weekly, monthly, and 3 months of exposition was measured as well as the transmission values for each sample was measured in 3 points to solve the non-uniformity of dust deposition on the sample surfaces. To measure the amount of dust accumulated on the sample surface, the samples were weighted under restrained conditions using the Kern ABT analytical balance with an accuracy of 0.1 mg. Then the density of dust deposition was determined by dividing the difference between the sample reference mass value (clean) and the mass of dusted sample (dirty) by the sample surface area.

In addition, the transmission values were measured using PerkinElmer Lambda 1050 UV-Vis-NiR Spectrophotometer. The transmission loss was determined by comparing the initial transmittance value of the clean state of the glass samples and dirty state due to dust settlement.

Moreover, in order to assess the effect of dust deposition on the electrical performance of the PV system, three identical PV modules were exposed for the same period as the glass samples. The cleaning schedule was defined as daily for the reference module, weekly cleaning for the second, and the last was kept uncleaned all the period of investigation. The drop on the electrical output was quantified by the short-circuit current and the daily soiling ratio coefficient was obtained by dividing the daily short-circuit currents of dirty modules by the clean ones (reference module).

2.1 Light Transmission

As mentioned above that the transmittance of PV glass samples was measured with the spectrophotometer at day 0 (reference day before exposure), after 7 days (weekly exposure), 30 days (monthly exposure), and 90 days (3 months of exposure). The results of these measurements are illustrated in Fig. 6. As can be noticed from the figure, dust settlement on glass sample tends to impact the transmittance at shorter wavelength (below 350 nm) region as reported in other studies [49–51]. Although, the transmittance response decreases over the period of exposure from day 0 until day 90 after 3 months of exposure. During the period of investigation, we noticed days with different amounts of precipitation that range from 0.1 mm/day to 11.3 mm/day (see Fig. 7). The relative humidity was varying between 63% in May and 48.6% in August 2018. All the weather data have been collected from a meteorological station installed next to the exposure site.

In order to describe transmittance loss due to the effect of dust accumulation on the glass surface for the three different periods of exposition, the losses as defined by Eq. 1.

$$\tau_{loss} = \left[1 - \frac{\tau_{dirty}}{\tau_{clean}}\right] \times 100\% \quad (1)$$

where τ_{loss} is the broadband transmittance loss, τ_{dirty}, and τ_{clean} are respectively the broadband transmittance average over the 380–1100 nm spectral range.

The broadband transmittance loss was 6.5% by day 7 (the first week of exposure). The daily transmittance loss was in the range of 0.93%/day after 7 days of exposure, which describes the high accumulation rate of dust during the first days of exposure, as the glass surface intensity is high. However, the transmittance loss was in the range of 13% after 30 days, which presents a daily transmittance loss of 0.43%/day.

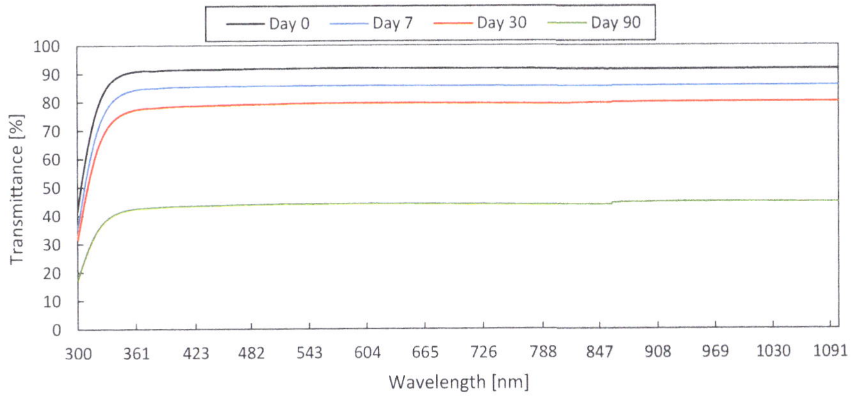

Fig. 6 Transmittance variation for a wavelength range 300–1100 nm before exposition (day 0) and for three different exposure periods

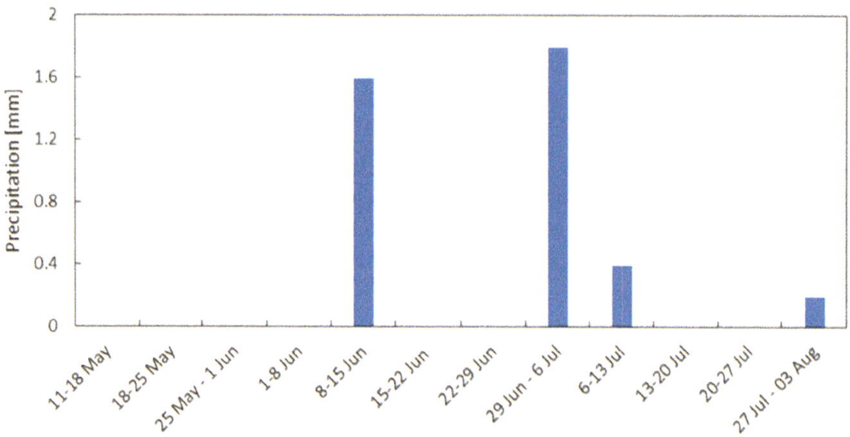

Fig. 7 Weekly accumulated precipitation all over the period of investigation from May 11th to August 3rd, 2018

This could be explained by high relative humidity and low dust concentration (not shown in this study) over the first 30 days. Whereas by the end of 90 days (3 months of exposure), the loss has reached 52% with the increase of dust concentration and the decrease of relative humidity by the end of June (after 30 days of exposure). Despite, the rainy days noticed during this period the concentration of dust and the cumulative dust deposition have increased the losses of the broadband transmittance after 3 months of exposure.

To summarize, the accumulation rate of dust is high during the first days of exposure with daily transmittance loss of 0.93%/day (after 7 days), whereas it decreases to reach 0.43%/day after 30 days. In fact, the climate during the first 30 days (first month) was with no rainfall and an average relative humidity in the range of 61% and a daily temperature average of 18.5 °C. Thereafter, the daily transmittance loss has increased under cumulative soiling conditions after 90 days of exposure with 0.57%/day. This increase could be attributed to a high concentration of dust from day 30 over the period of investigation (the last 2 months) and relatively moderate relative humidity with a daily average of 54.5% and rainy days with a maximum amount of 1.6 mm/day. In fact, high dust concentration accompanied with light rainfall will increase dust settlement on the surface of the glass and therefore increase the soiling transmittance loss.

2.2 *Soil Mass*

The amount of dust settled on the surface of a PV panel is strongly dependent on the weather conditions, the aerosol concentration in the up of the atmosphere, and the period of exposure. Table 1 presents the surface density of dust deposition ρ (g/m^2)

Table 1 The density of dust deposition for a different period of exposition

Time [days]	7	30	90
ρ [g/m^2]	0.4	1.1	5.6
τ_{loss} [%]	6.5	13	52

on the glass sample after 7, 30, and 90 days. Once it can be noticed that the density of dust deposition increases with the increase in the period of exposure. By day 7, the amount of dust settled was about 0.4 g/m^2; however, during the experiment, the samples exposed for the weekly period have shown a significant amount of dust settled on their surfaces that reached 1.6 g/m^2 in the period from June 22 to June 29 (not shown in this work). Whereas by day 30, and after 90 days, the dust surface density was 1.1 g/m^2 and 5.6 g/m^2, respectively. The accumulation rate of dust particles is higher during the first days of exposition and it decreases with time until saturation, where the daily accumulation rate was ~ 0.06 g/m^2/day for the 3 months of exposure. This dust accumulation rate tends to be low in comparison to other regions with arid climate in the Middle East and Pakistan. The authors have found a daily dust accumulation rate of 0.14 g/m^2/day for the winter season and 0.098 g/m^2/day for the summer season after 2 months of exposition in Doha, Qatar [52]. Besides, in Islamabad, Pakistan, the authors reclaim a dust density deposition of 4.6 g/m^2 after 1 month of exposure [53].

On the other hand, the surface dust density can be correlated to the broadband transmittance loss as the amount of dust increase on the surface of the glass sample the transmittance decrease progressively as presented in Table 1. However, as reported in other studies [34, 35], the relationship between dust deposition and transmittance was linearly dependent until reaching its upper limit. Thereafter, the dust deposition load will no longer influence the transmittance coefficient.

2.3 PV Electrical Output

The accumulation of soiling and especially dust on the surface of the PV module reduces the incoming sunlight transmitted through the glass cover to the PV cells as described in the last two sections and, thus, the electrical output produced by the module. In the present chapter as described above, we will present the results of an experimental investigation for a CdTe Thin-film solar module with different cleaning scenarios for the period from May 29 until October 5, 2018. The main electrical parameters of the PV module used in this experiment are presented in Table 2.

The electrical outputs of the clean and dirty modules have been analyzed and quality checked, only data recorded between 12:00 pm and 2:00 pm have been used in order to remove any occurring disruption as cloudy days (irradiance below 500 W/m^2) or missing data. The soiling impact has been evaluated by the daily soiling ratio factor [54, 55], as described in Eq. 2.

Table 2 Electrical characteristics of the PV module

Parameter	Value
Nominal power, P_{max}	77.5 [W]
Short circuit current, I_{sc}	1.98 [A]
Open circuit voltage, V_{oc}	62.5 [V]
Current at maximum power, I_{mp}	1.68 [A]
Voltage at maximum power, V_{mp}	46.7 [V]
Standard test conditions, STC	1000 W/m^2, 25 °C, AM 1.5

$$SR_{daily} = \frac{I_{sc,dirty}}{I_{sc,clean}} \tag{2}$$

where $I_{sc,dirty}$ and $I_{sc,clean}$ are respectively the daily average short-circuit current of the dirty modules (weekly and no cleaning ones) and the cleaned module (daily cleaned as reference).

The analysis of the results from the variation of daily soiling ratio leads to some interesting outcomes. Once it can be noticed from Fig. 8, that the soiling pattern can drastically influence the electrical power of the PV module if no cleaning was occurring especially for the dry period. For the no cleaned PV module, the soiling ratio decreases progressively from ~0.99 (clean state) in day 0 to 0.77 in day 66 (from May 29 to August 3, 2018), which explains the 23% losses in the electrical output. The weekly cleaned PV module was less affected and the soiling ratio has reached a limit decrease of ~0.97 during the same period as the last one (no cleaned one). This founding affirms the influence of the frequency of the cleaning schedule even if during the high soiling period (dry period).

In addition, for the period from August 3 to September 1, 2018, no cleaned PV module shows an increase in term of soiling ratio of ~+12% in day 69 (August 5, 2018). This gain in electrical output was due to the rainy days (day 66 and day 68) with total precipitation of 0.3 mm accompanied with a high average relative humidity of 68% from day 64. Thereafter, the soiling ratio decreases progressively from day 69 until day 96 (September 1, 2018), where it reaches ~0.81. Once it can be noticed that the daily soiling loss for the first period (from day 0 to day 69) was in the range

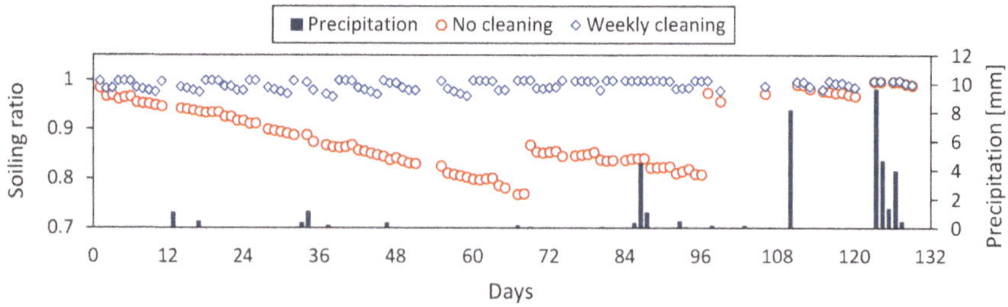

Fig. 8 Soiling ratio for the PV module with weekly cleaning and PV module with no cleaning for the period from May 29 to October 5, 2018

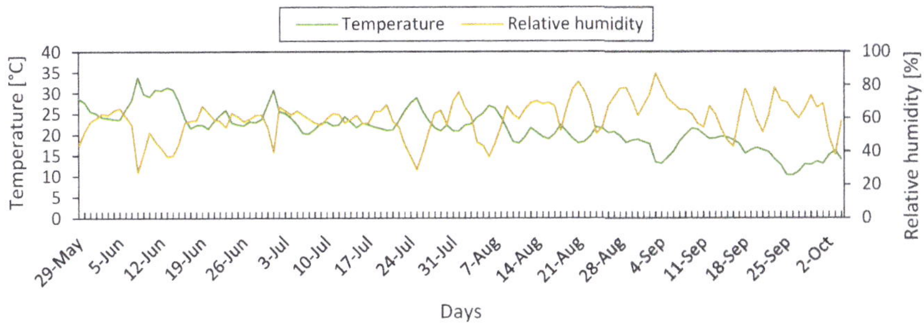

Fig. 9 Daily average temperature and relative humidity for the investigation period

of 0.34%/day, whereas, it was decreasing with a daily average of 0.24%/day in the second period (from day 70 to day 96). The difference between the first period and the second period could be explained by rainy days with a total amount of 7.8 mm in the second period compared to 4 mm in the first period. Another reason that emphasizes this difference was the high dust concentration (not shown in this work) during the first period accompanied by low relative humidity (see Fig. 9), which increase the soiling drop. Although the second period was enjoying days with low dust concentration and high relative humidity, which increase the natural self-cleaning and reduce the settlement of dust particles.

Moreover, in day 97, the no cleaned PV module has shown a significant increase with an average daily soiling ratio that reached ~0.97, which means a gain of ~+17% on the electrical output, which could be explained by the precipitation recorded at the night of day 96 (0.2 mm). However, there is no conclusion about defining a rain threshold for which a cleaning event will occur for a PV module [55], whereas in a recently published study in Evora, Portugal, the authors have reported that a threshold of ~2 mm has a probability of 50% to reduce the soiling ratio [56]. Therefore, it can be concluded that the soiling ratio recovery can be highly dependent on the duration and the intensity of the precipitation.

For the remaining investigation period from day 97 until day 129, the soiling ratio was in the range of ~0.96, which is due to the rainy days with a total accumulation of precipitation of 29 mm and high level of relative humidity and low temperature (see Fig. 8 and Fig. 9).

3 Cleaning Techniques

To avoid the soiling problem, the use of cleaning systems is mandatory in order to remove the dust accumulation on the front surface of PV panels and increase its efficiencies. In this, since several cleaning systems have been developed like a natural method, manual cleaning, electrostatic method, self-cleaning nanofilm, and automatic cleaning system.

3.1 Natural Cleaning

The natural cleaning process is a method that uses nature as its cleaning tool. Among these natural tools, we find wind speed, rainwater as well as the earth gravitation. The high wind speed or a water droplet can remove or roll-off the dust on the PV panel surfaces. However, less gravitational occurs at night and early morning can float the dust naturally onto the surface of PV panels.

Gair et al. [57] reported in their study that dust removal can be done easily if we turn the PV panels to a vertical position during a rainy day, evening night, and early morning. This method has advantages, low cost, and clean energy for the environment. However, the problem of this technique is the tilting rotation of PV plants on a large scale and not operational for high humidity climate.

3.2 Manual Cleaning

This method is the same as the one used in the cleaning of high-rise building windows. Dust particles are removed by special brushes, which are equipped with bristles to avert any scratches on the glass of the PV module. These brushes are connected with a water supply that enables washing. The uneven movement of brushes over the PV module surface leads to a risk of abrasive effect. This latter can be minimized soft cleaning cloth or brush with soft bristles [58, 59]. The cleaning brush is used with a water pressure of 100–160 bar, which allows easy manual cleaning. For high dirty surfaces, a cleaning product is added to the water to remove easily dirt. Then the surfaces are rinsed with demineralized water. This cleaning system is recommended for small areas of 150–300 m^2.

To clean the solar panels, we need a bucket of warm soapy water, a dry cloth, and a rubber squeegee. The surface of the panel is washed with soap water and a cloth to remove dust, debris, and water spots. A squeegee is used to remove excess water. This method has the advantage of being simple and less expensive. On the other hand, it has many disadvantages: it is not suitable for large areas, it uses large quantities of water, which is not compatible with desert areas, it leads to a degradation of the PV module surface (appearance of micro-scratches under the effects of friction), and finally, it has a direct impact on the damping time of the installation.

Manual cleaning is usually used for cleaning small facilities or domestic solar panels [60, 61]. This cleaning method is expensive and requires skilled labor to clean off soiling onto the PV plants. However, fully automated systems can bring flexible cleaning. At the level of our research facility [62], we use manual cleaning since we dispose of small scale PV plants that are dedicated to research and development.

3.3 Electrostatic Cleaning

The electrostatic method is based on an electrostatic charge (from the electric curtain on the PV panel) to remove dust on the surface of the PV panel. This latter has been developed at NASA in 1967 [63]. The action of electrostatic and dielectrophoretic forces to remove the dust has been studied by Calle et al. [64]. In order to generate electrostatic and dielectrophoretic force, the electrodes have been used to transport charged and uncharged particles on the PV module surface. Stable electrostatic force occurs, which makes to unbalance between charged particles and surface, these phenomena deal with the particles to create their force. This kind of force generates a movement of a dust particle on the PV module surface, which are called dielectrophoretic forces.

3.4 Self-Cleaning Nanofilm

The self-cleaning nanofilm method is the method that uses the coating process to add a nanofilm layer on the PV panel surface. Several coating processes have been innovated by researchers such as chemical vapor deposition (molecular beam epitaxy, electrostatic spray assisted vapor deposition, vapor deposition, chemical and electrochemical technique, physical vapor deposition, plasma spraying, roll-to-roll coating process, spin-coating, and dip coating. This method deals to modify the normal PV module surface by turning it to superhydrophobic surfaces and superhydrophobic surfaces using a special nanofilm coating characterized by superhydrophobic and superhydrophilic proprieties [65–70].

3.5 Automatic Cleaning System

Robotic systems have emerged as an attractive solution for cleaning the dirty surfaces of the photovoltaic module [71]. Besides, the geographical land and the area of application are highly important, where the existing solutions can be further compared based on capital costs and performance ratio.

PV module cleaning robot comprises a mobile robot that carries the cleaning payload and cleaning tool, which performs the cleaning work [71].

Serbot Swiss Innovations has developed a robotic cleaning system called 'Gekko Solar' and 'Gekko Solar Farm' in order to be used for mobile deployment onto PV plants [72]. This latter uses the rotating brush and demineralized water to clean the PV module's surface. The movement of this system is based on feet with vacuum technology enabling the robot to astonishing flexible movement in every chosen direction.

Raybot is an eco-friendly robot designed to clean solar panels [73]. It can move on surfaces with a slope of up to 65 degrees using suction cups without risk of damaging them. It can clean around five solar panels per hour and in most cases, it is enough to use it at the end of two months. The robot safely sweeps surfaces that are often characterized by difficult access. Raybot is an eco-friendly robot designed to clean solar panels. It can move on surfaces with a slope of up to 65 degrees using suction cups without risk of damaging them. It can clean around five solar panels per hour and in most cases, it is enough to use it at the end of two months. The robot safely sweeps surfaces that are often characterized by difficult access. By sweeping, blowing, and vacuuming, it removes the dust and dirt that naturally settles on the solar panels. Using several sensors, Raybot mobilizes without any risk of falling. In addition, to carry out its task, it is equipped with an interchangeable battery that allows it to resume work without the need for a charging station.

The HYCLEANER black SOLAR allows simple, quick, and economical cleaning [74]. This robot allows optimal cleaning with its mechanical power and low water consumption. It works with a radio remote control, so the user does not need to walk, whether on the roof or the solar surface. Lithium batteries guarantee electric drive, so a power outlet near the work area is not mandatory.

Ecoppia's E4 is a robot that operates during the night to maintain maximum energy production at all times [75]. It operates at a cleaning rate of 54 square feet (approximately 5 m) per 30 s. The robot moves along a rigid aluminum structure, its wheels are covered with polyurethane to ensure that the movements are light without carrying a load on the solar panel surface. This robot is fully autonomous and independent of energy, it has its own solar panel for self-loading and its self-cleaning mechanism. The robot recovers energy during its descent along with the solar panel, which will be reused in the next cleaning cycle, which optimizes the robot's performance.

SOL BRIGHT developed a cleaning robot that eliminates 99% of dirt and pollution on photovoltaic panels. Tested in large solar power plants, the robots improve their electricity production rate by 7–15%. The robot operates at night using its own photovoltaic module as an energy source, without interrupting the conversion of solar energy during the day. The cleaning robot is equipped with a roller brush, which is not supplied with water, which helps to save energy and protect the environment.

Table 3 gives a summary of the different robotic cleaning technologies described above.

4 New Cleaning System

4.1 Robot Design and Functionality

Based on the investigations and the obtained results discussed in the previous sections, and in order to increase the PV plant efficiency by continuous cleaning activities,

Table 3 Comparison of the presented robotic cleaning systems

Robot	Gekko	Raybot	Hycleaner	Eccopia	Solbright
Tilt	45°	65°	35°	Adaptable to any inclination	Adaptable to any inclination
Cleaning technology	By heated demineralized water and pressurized	Dry	Wet	Dry	Dry
Cleaning equipment	Polyamide12 nylon brush with a rotation speed of 350 rpm	Brushes, vacuum cleaners and fan	Brush	2 microfiber brushes	Roller brush
Displacement mechanism	By suction cups	• By suction cups • By lengthening and shrinking his body	By strap	By rail and polyurethane wheels	Roll along with the chassis of the PV panels
Moving speeds	4 m/min	–	25 m/min	–	10–20 m/min
Course control system	By remote control	By position sensors	By radio control	Remote control by masterE4 application	Using intelligent control software.
Yield	300–400 m^2/h	5 panels/h	400–800 m^2/h	600 m^2/h	1800–3600 m^2/h

an autonomous cleaning robot is proposed. The proposed system can maintain the high efficiency of the solar panels by ensuring continuous cleaning without the need for any guide or human intervention (self-guided robot). In addition, the robot can be monitored and controlled in real-time through a web link application. Users can easily monitor the robot status (ON or OFF), battery charge, water level, and can also set a schedule with a specific time for cleaning. The proposed robot is designed to be mounted and adapted to all PV systems and technologies by adjusting only the vertical supports (over the PV system width). In addition, system flexibility is the main advantage of the proposed system. The robot can be controlled by three modes: a manual mode that the operator can immediately turn on/off the robot, the robot can also move to a specific position via this mode; following, users can control the time or frequency of cleaning using the web link application; the third mode is used to communicate with other electronic devices by receiving a digital signal to start the cleaning process (the user sets a default time to start cleaning when a signal is received, 6 pm. for example). For instance, an electronic device calculates the soiling ratio or PV efficiency. In addition, the robot is designed to be lightweight and small to facilitate installation and maintenance and also to reduce the cost of the system

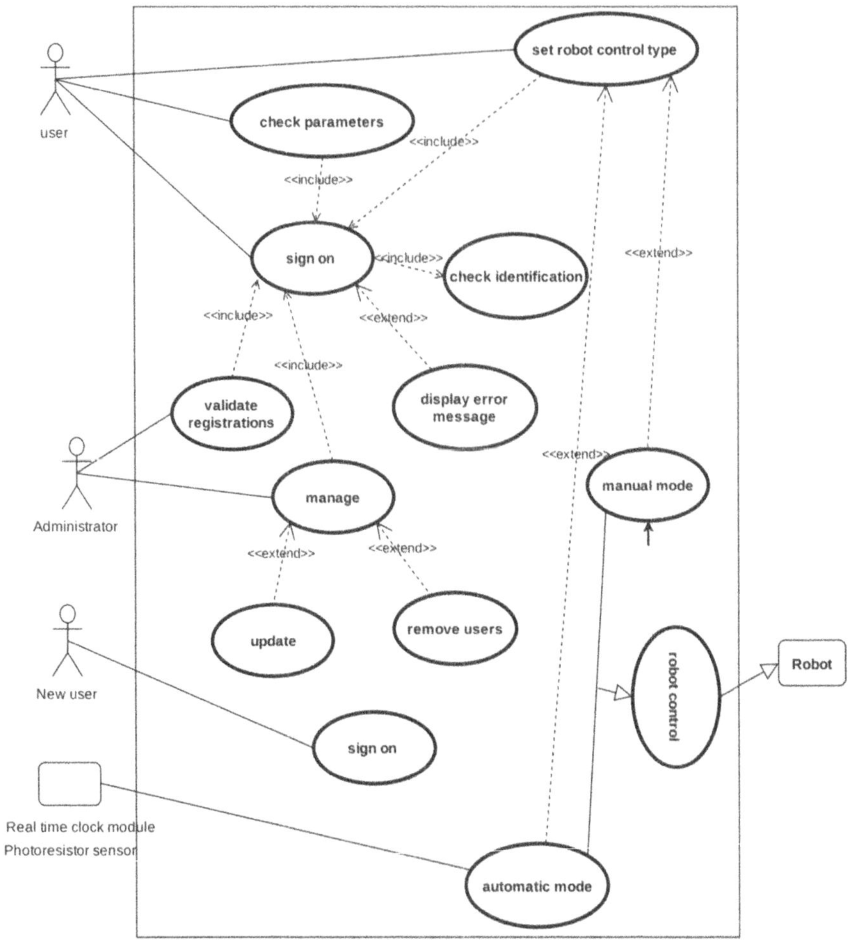

Fig. 10 Use case diagram of the proposed robot cleaning

by minimizing mechanical structures and motor torques. The proposed solution uses also a low number of electronic sensors and motors to minimize even more the robot cost and reduce its maintenance (details can be seen in Fig. 10).

4.2 Mechanical Robot Design

The mechanical design and simulation were done using CATIA V5. All the robot components were designed and simulated in order to establish an optimal design and to study all the mechanical aspects and uncertainties that may occur in real working conditions. For this reason, mechanical design and modeling address to the following technical issues: to choose the optimal material and size for each block; to better distribute the robot and supports mass on the solar panels; to size the electrical motors for brush rotation and robot movement; to minimize the possibility of robot slipping or sliding.

The proposed robot is divided into two main systems. The first system performs a horizontal movement along the PV system length (see Fig. 11a). This system is made up of mechanical supports and wheels grip the frame of the panels in order to avoid any scratch on the PVP surfaces. Furthermore, the battery and water reservoir are mounted in this frame, in which the weight is shared between the top and the bottom. Two windshield wipers are used at the front and rear of the system to remove any water or sand that can propagate or re-deposition on previously cleaned panels. The main advantage of this system is its flexibility, it can be adapted to any length or technology. The second system represents the cleaning unit and ensures the vertical movement over the entire PV system width (see Fig. 11b). It is composed of a mechanical

(a)

(b)

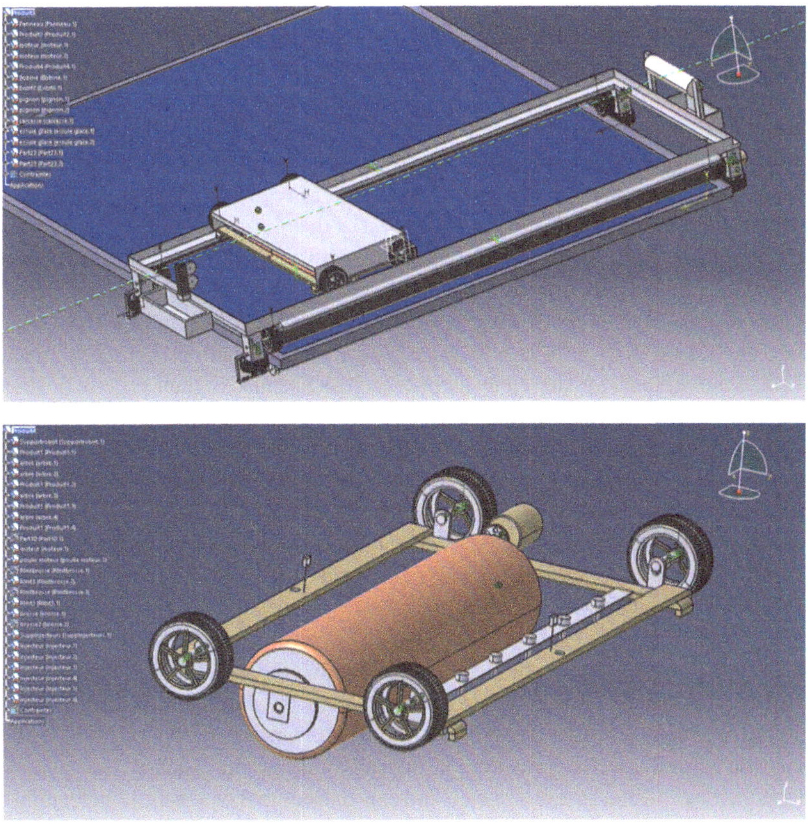

Fig. 11 Robot design **a** mainframe **b** second frame

frame, which moves vertically using four wheels and contains a cylindrical brush in the middle. The vertical movement is ensured by a belt pulley system.

To summarize, the proposed robot is made up of two systems or frames: the mainframe used to move horizontally using wheels, which grip on the frame of the panel. The second frame composed of one cleaning and rotating brush, moves up and down along each PV system column. The material used for supports and frames design is the Aluminum 7075 due to its high resistivity and good density. The weight of the mainframe is 21 kg including wipers, motors, battery, water reservoir, and others. The weight of the cleaning frame is 4 kg including brush, DC motor, wheels, and belt pulley system. Therefore, the total weight of the robot is 25 kg.

4.3 Electronic Control and Powering System

The overall electronic devices are mounted inside the robot and powered by a small PV panel. The electronic circuits developed are designed and simulated using Isis Proteus and the Arduino software (IDE). The on-board control system uses Arduino Mega 2560 based on ATmega2560 microcontroller. After establishing the total energy consumption of the robot including all the electronic devices, the robot is powered by 12 V DC, 12 Ah. For this, the appropriate PV panel is dimensioned and fixed to a 25 W monocrystalline panel with a PWM regulator. In addition, the motors' torque is determined to ensure the movement of the robot and the rotation of the brush as a function of the weight of each corresponding frame. For the mainframe (horizontal movement), the Nema 11 (MS14HS1P4024) is used. Regarding the second frame (vertical movement), the Nema 24 (MS24HS5L4420) is employed. In addition, the SPOMHNK3054 motor is used for rotating the brush. Both horizontal and vertical movements are driven by stepper motors because of their precise positioning and controlled by PWM based signals given by Arduino Mega 2560 board, while a DC motor is used for rotating the brush with high speed and simply activated or deactivated by a digital signal generated by the board. Two servo motors are used to lift the wipers because of their ability to hold the position (DS04-NFC). Furthermore, four high-resolution ultrasonic sensors (HC-SR04) are used to detect the real-time current position of the robot as well as to adjust its direction and speed. These sensors are mounted on the four corners of the robot, powered by +5 V DC, and connected to Arduino Mega 2560 inputs. The water reservoir is also measured in real time by the ST045 sensor. Finally, the ESP8266 integrated circuit allows the connection via WIFI and is widely used to control devices over the Internet (a GSM module can also be used in this sense). This integrated circuit communicates and transfers data in real time with the web application. Users can monitor and visualize the robot status, battery charge, water reservoir level, and can also control or schedule a cleaning time. Figure 12 illustrates the internal block diagram of the proposed robot and describes the interactions between blocs.

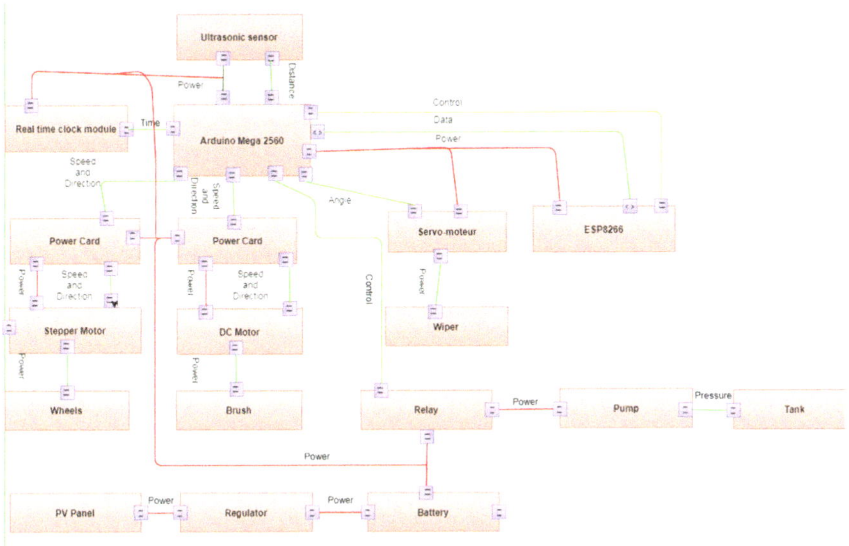

Fig. 12 The internet block diagram of the proposed robot

4.4 *Robot Movement and Cleaning Strategy*

As stated above, three modes can be used to control the robot and start cleaning: manual control; automatic control by communicating with other electronic devices installed in the field; automatic control via weblink application allowing users to set a scheduled time for cleaning. Figure 13 illustrates the robot movement control algorithm and the cleaning strategy. It is essential to mention that the robot is designed to ensure PVP cleaning in both directions (forward and backward). Therefore, the front and rear sections of the robot are indistinguishable and the cleaning process and the direction of movement are the same. Before starting the cleaning, the robot is located each on the right or on the left of the PV system and mounted on metal support appropriate to its shape and away by 0.5 m from the edges of the PVP. The system is always waiting for control signals each from users or electronic devices. Then, the robot checks its current position to move each in the right or left directions with a distance of 0.5 m, while the appropriate wiper is powered. Following, the water injection is done for 2 s and the robot again checks its current positions, but this time its vertical position, in order to move each up or down while the brush is rotating. When the edges of the PV system column are detected, the vertical movement ends. Then, the robot is horizontally moved with 0.5 m and the cleaning frame goes up or down (when the robot moves up, the rotating brush direction is changed). This process is repeated until the horizontal ends of the edges of the PV system are detected. The robot prototyping is being finalized and validated at the Green Energy Park research facility (GEP), taking into account several PV technologies and different PV systems geometries. The proposed cleaning robot is low cost and lightweight equipment due

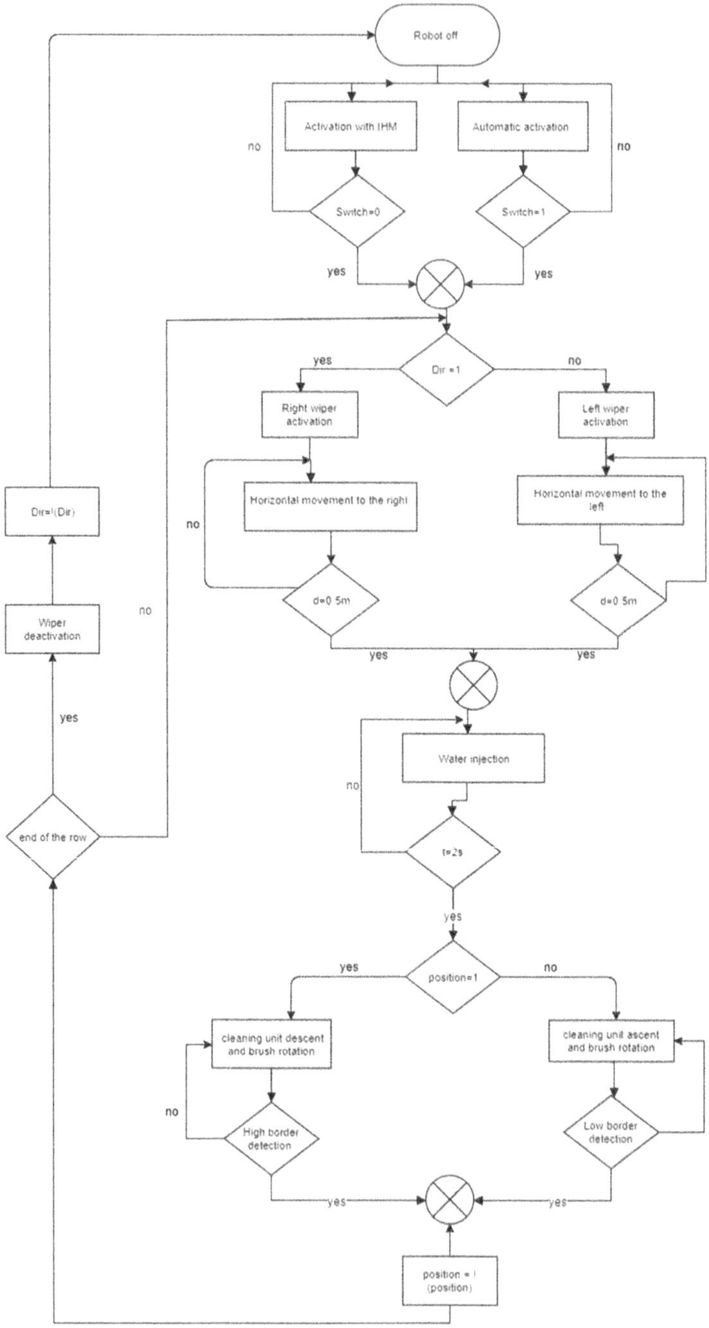

Fig. 13 Flowchart illustrating the robot movement control algorithm and the cleaning strategy

to its mechanical design, which divides the entire system into two mainframes. This aspect allows the robot to be more efficient and stable, in particular, its mass is well distributed on the PVP surface. Furthermore, the flexibility of the proposed solution is the main key allowing the robot to be easily adapted to any PV system or technology. The front and rear sides of the robot are the same, allowing the cleaning process in two directions since the robot locates exactly its current position before starting cleaning. In addition, users can easily monitor the robot status and control it using the web application. This functionality is very advantageous if several cleaning robots are installed in the same field, which facilitates monitoring and maintenance and considerably reduces human intervention. The proposed robot architecture and features seem to be very promising, robust, and low-cost technology.

4.5 Cleaning Cost

The manual cleaning cost at Green Energy Park is calculated with Eq. 3:

$$C_m = (Pw\ .Q\ .Nb\ .Nn) + (P) + (L\ .Nn) \tag{3}$$

with:

C_m: The cleaning cost per month with manual cleaning (€/month).
Pw: The price of a cubic meter of water (€/m^3).
Q: The amount of water needed to clean one module per day (m^3/day).
Nb: Number of modules in the string.
Nn: Number of cleaning days per month (day/month).
P: The price of cleaning equipment per month (€/month).
L: The labor for cleaning a day (€/day).

For example, the cleaning of a string composed with 23 modules (Nb = 23), the amount of water used for cleaning a module is (Q = 0.002 m^3/day) with a frequency of eight times per month (Nn = 8 days/month), using cleaning equipment of (P = 3€/month). It is assumed that a person can clean the entire string with a salary of (L = 5€/day).

From the previous Eq. 3, the cleaning cost of PV strings is calculated as:

$$C_m = (1.23\ \times 0.0005\ \times 23\ \times 8) + 3 + (5 \times 8) = 43.1\ €/\mathrm{month}$$

Based on our first assumptions and Eq. 3, the cleaning cost C_r using the new self-cleaning robot is calculated as:

Where, the assumed price of the self-guided cleaning robot is 500 € per unit with a total replacement cost of 10% per year and a lifetime of 10 years.

$$C_r = (1.23\ \times 0.0001\ \times 23\ \times 8) + 8.33\ + 0 = 8.35\ €/\mathrm{month}$$

As noticed from the economical point of view, the amount of water used for the cleaning and the labor cost can drastically increase the cleaning cost, and therefore, using a cleaning robot system will reduce the cleaning cost of ≈35 € per month.

5 Conclusion

The abundant solar radiation especially in regions with semi-arid and arid climates promotes the development and deployment of photovoltaic technology. Although the weather conditions in those regions as soiling cause a significant degradation on the power production and performance of the PV systems.

In this chapter, we assess the effect of dust accumulation on the optical efficiency in terms of transmittance of glass samples and on PV modules electrical output. The results have illustrated that the soiling phenomenon is highly significant in a region with semi-arid climate especially for the dry period of the year. Dust accumulation is strongly dependent on seasonal conditions, the electrical loss of the PV module is important in periods with high dust concentration and low relative humidity. This electrical loss can be worst if the particles of dust on the atmosphere coincide with light rainfall, which will significantly deteriorate the PV output performance.

By investigating dust accumulation on glass samples, the broadband transmittance loss was respectively 6.5%, 13%, and 52%, for a period of exposure of 7, 30, and 90 days. This founding was explained by the dust accumulation rate of 0.4 g/m^2, 1.1 g/m^2, and 5.6 g/m^2 after 7, 30, and 90 days of exposure, respectively. In fact, high dust concentration and rainy days with low amounts have increased the impact of dust accumulation, which explains the transmittance loss of 52% after 3 months of exposure. On the other hand, the same pattern has been illustrated for the PV modules exposed for a weekly cleaning period and no cleaning all over the period of investigation. The results have shown that the weekly cleaning was highly efficient and keep the soiling ratio in the range of 0.97 all over the period of the experiment. For the no cleaned PV module, the soiling ratio has decreased to reach 0.77 after 68 days of exposure. Thereafter, the no cleaned PV module has recorded an electrical gain of 12% and 17% respectively in day 69 and day 97, which has been explained by the rainfall event noticed during the period of exposure.

To this end and based on our previous soiling studies at the local climate conditions, we have presented a self-guided cleaning robot that will be highly efficient and flexible for different technologies and structures. Thereby, the cleaning cost will be reduced by five times using this system instead of manual cleaning.

References

1. Trieb F, Schillings C, Pregger T, O'Sullivan M (2012) Solar electricity imports from the Middle East and North Africa to Europe. Energy Policy 42:341–353
2. Kahia M, Aïssa MSB, Lanouar C (2017) Renewable and non-renewable energy use-economic growth nexus: the case of MENA net oil importing countries. Renew Sustain Energy Rev 71:127–140
3. A. Azouzoute, A. A. Merrouni, et S. Touili, « Overview of the integration of CSP as an alternative energy source in the MENA region », Energy Strategy Rev., vol. 29, p. 100493, 2020.
4. Azouzoute A, Merrouni AA, Garoum M, Bennouna EG, Ghennioui A, Ydrissi ME (2019) The impact of optical soiling losses on the electrical production of CSP power plant. AIP Conf Proc 2123(1):020090. https://doi.org/10.1063/1.5117017
5. Azouzoute A, Chouitar M, Garoum M, Bennouna EG, Ghennioui A (2019) A new PV soiling monitoring device for optimized cleaning strategy, vol 2190, p 020068
6. Azouzoute A, Merrouni AA, Garoum M, Bennouna EG (2019) Comparison of soiling effect of two different solar mirrors in mid-south of Morocco. AIP Conf Proc 2126(1):190002. https://doi.org/10.1063/1.5117699
7. Abderrezek M, Fathi M (2017) Experimental study of the dust effect on photovoltaic panels' energy yield. Sol Energy 142:308–320
8. Saidan M, Albaali AG, Alasis E, Kaldellis JK (2016) Experimental study on the effect of dust deposition on solar photovoltaic panels in desert environment. Renew Energy 92:499–505
9. Azouzoute A, Merrouni AA, Garoum M (2020) Soiling loss of solar glass and mirror samples in the region with arid climate. Energy Rep 6:693–698
10. Sarver T, Al-Qaraghuli A, Kazmerski LL (2013) A comprehensive review of the impact of dust on the use of solar energy: History, investigations, results, literature, and mitigation approaches. Renew Sustain Energy Rev 22:698–733. https://doi.org/10.1016/j.rser.2012.12.065
11. Bergin MH, Ghoroi C, Dixit D, Schauer JJ, Shindell DT (2017) Large reductions in solar energy production due to dust and particulate air pollution. Environ Sci Technol Lett 4(8):339–344
12. Piedra P, Moosmüller H (2017) Optical losses of photovoltaic cells due to aerosol deposition: role of particle refractive index and size. Sol Energy 155:637–646
13. Ilse KK et al (2018) Comprehensive analysis of soiling and cementation processes on PV modules in Qatar. Sol Energy Mater Sol Cells 186:309–323
14. Zitouni H, Azouzoute A, Hajjaj C, El Ydrissi M, Regragui M, Polo J, Oufadel A, Bouaichi A, Ghennioui A (2021) Experimental investigation and modeling of photovoltaic soiling loss as a function of environmental variables: a case study of semi-arid climate. Solar Energy Mat Solar Cells 221:110874
15. Ilse KK, Figgis BW, Naumann V, Hagendorf C, Bagdahn J (2018) Fundamentals of soiling processes on photovoltaic modules. Renew Sustain Energy Rev 98:239–254
16. Martin-Sanchez PM et al (2018) Monitoring microbial soiling in photovoltaic systems: a qPCR-based approach. Int Biodeterior Biodegrad 129:13–22
17. Einhorn A et al (2018) Evaluation of soiling and potential mitigation approaches on photovoltaic glass. IEEE J Photovolt 9(1):233–239
18. Shirakawa MA et al (2015) Microbial colonization affects the efficiency of photovoltaic panels in a tropical environment. J Environ Manage 157:160–167
19. Conceição R, Silva HG, Mirão J, Collares-Pereira M (2018) Organic soiling: the role of pollen in PV module performance degradation. Energies 11(2):294
20. Costa SC, Diniz ASA, Kazmerski LL (2018) Solar energy dust and soiling R&D progress: literature review update for 2016. Renew Sustain Energy Rev 82:2504–2536
21. Costa SC, Diniz ASA, Kazmerski LL (2016) Dust and soiling issues and impacts relating to solar energy systems: literature review update for 2012–2015. Renew Sustain Energy Rev 63:33–61
22. Ghazi S, Sayigh A, Ip K (2014) Dust effect on flat surfaces–A review paper. Renew Sustain Energy Rev 33:742–751

23. Al-Addous M, Dalala Z, Alawneh F, Class CB (2019) Modeling and quantifying dust accumulation impact on PV module performance. Sol Energy 194:86–102
24. Said SA, Hassan G, Walwil HM, Al-Aqeeli N (2018) The effect of environmental factors and dust accumulation on photovoltaic modules and dust-accumulation mitigation strategies. Renew Sustain Energy Rev 82:743–760
25. Figgis B et al (2018) Investigation of factors affecting condensation on soiled PV modules. Sol Energy 159:488–500
26. Mastekbayeva G, Kumar S (2000) Effect of dust on the transmittance of low density polyethylene glazing in a tropical climate. Sol Energy 68(2):135–141
27. Hegazy AA (2001) Effect of dust accumulation on solar transmittance through glass covers of plate-type collectors. Renew Energy 22(4):525–540
28. Adinoyi MJ, Said SA (2013) Effect of dust accumulation on the power outputs of solar photovoltaic modules. Renew Energy 60:633–636
29. Boppana S (2015) Outdoor soiling loss characterization and statistical risk analysis of photovoltaic power plants. Arizona State University
30. Boyle L, Flinchpaugh L, Hannigan M (2015) Natural soiling of photovoltaic cover plates and the impact on transmission. Renew Energy 77:166–173
31. Ullah A, Amin A, Haider T, Saleem M, Butt NZ (2020) Investigation of soiling effects, dust chemistry and optimum cleaning schedule for PV modules in Lahore, Pakistan. Renew Energy 150:456–468
32. Kaldellis J, Kokala A (2010) Quantifying the decrease of the photovoltaic panels' energy yield due to phenomena of natural air pollution disposal. Energy 35(12):4862–4869
33. Said SA, Walwil HM (2014) Fundamental studies on dust fouling effects on PV module performance. Sol Energy 107:328–337
34. Elminir HK, Ghitas AE, Hamid RH, El-Hussainy F, Beheary MM, Abdel-Moneim KM (2006) Effect of dust on the transparent cover of solar collectors. Energy Convers Manag 47(18–19):3192–3203
35. Gholami A, Saboonchi A, Alemrajabi AA (2017) Experimental study of factors affecting dust accumulation and their effects on the transmission coefficient of glass for solar applications. Renew Energy 112:466–473
36. Azouzoute A, Garoum M, Jeffali F, Ghennioui A (2020) Experimental study of dust effect on the transmission of a glass PV panel for a fixed and tracking system. Mater Today Proc
37. Qasem H, Betts TR, Müllejans H, AlBusairi H, Gottschalg R (2014) Dust-induced shading on photovoltaic modules. Prog Photovolt Res Appl 22(2):218–226. https://doi.org/10.1002/pip.2230
38. Picotti G, Borghesani P, Cholette M, Manzolini G (2018) Soiling of solar collectors–Modelling approaches for airborne dust and its interactions with surfaces. Renew Sustain Energy Rev 81:2343–2357
39. Chanchangi YN, Ghosh A, Sundaram S, Mallick TK (2020) Dust and PV performance in Nigeria: a review. Renew Sustain Energy Rev 121:109704
40. Shao Y (2001) A model for mineral dust emission. J. Geophys Res Atmosp 106(D17):20239–20254
41. Shao Y, Raupach MR, Leys JF (1996) A model for predicting aeolian sand drift and dust entrainment on scales from paddock to region. Soil Res 34(3):309–342
42. Kok JF, Parteli EJ, Michaels TI, Karam DB (2012) The physics of wind-blown sand and dust. Rep Prog Phys 75(10):106901
43. Kim E, Kalman D, Larson T (2000) Dry deposition of large, airborne particles onto a surrogate surface. Atmos Environ 34(15):2387–2397
44. Aluko O, Noll KE (2006) Deposition and suspension of large, airborne particles. Aerosol Sci Technol 40(7):503–513
45. Friedlander S (1977) Smoke, dust haze-fundam. Aerosol Behav
46. Hinds WC (1999) Aerosol technology: properties, behavior, and measurement of airborne particles. John Wiley & Sons

47. Isaifan RJ, Johnson D, Ackermann L, Figgis B, Ayoub M (2019) Evaluation of the adhesion forces between dust particles and photovoltaic module surfaces. Sol Energy Mater Sol Cells 191:413–421
48. Figgis B, Ennaoui A, Ahzi S, Rémond Y (2017) Review of PV soiling particle mechanics in desert environments. Renew Sustain Energy Rev 76(C):872–881. https://doi.org/10.1016/j.rser.2017.03.100
49. Piedra PG, Llanza LR, Moosmüller H (2018) Optical losses of photovoltaic modules due to mineral dust deposition: experimental measurements and theoretical modeling. Sol Energy 164:160–173. https://doi.org/10.1016/j.solener.2018.02.030
50. Smestad GP et al (2020) Modelling photovoltaic soiling losses through optical characterization. Sci Rep 10(1):1–13
51. Micheli L et al (2019) Correlating photovoltaic soiling losses to waveband and single-value transmittance measurements. Energy 180:376–386. https://doi.org/10.1016/j.energy.2019.05.097
52. Javed W, Wubulikasimu Y, Figgis B, Guo B (2017) Characterization of dust accumulated on photovoltaic panels in Doha, Qatar. Sol Energy 142:123–135. https://doi.org/10.1016/j.solener.2016.11.053
53. Majeed R, Waqas A, Sami H, Ali M, Shahzad N (2020) Experimental investigation of soiling losses and a novel cost-effective cleaning system for PV modules. Sol Energy 201:298–306
54. Micheli L, Muller M, Kurtz S (2016) Determining the effects of environment and atmospheric parameters on PV field performance. In: 2016 IEEE 43rd photovoltaic specialists conference (PVSC), pp 1724–1729. https://doi.org/10.1109/pvsc.2016.7749919
55. Micheli L, Muller MT, Deceglie MG, Ruth D (2017) Time series analysis of photovoltaic soiling station data: version 1.0. Nat Renew Energy Lab (NREL) Golden CO (United States)
56. Conceição R, Vázquez I, Fialho L, García D (2020) Soiling and rainfall effect on PV technology in rural Southern Europe. Renew Energy
57. Gaier JR, Perez-Davis ME (1990) Aeolian removal of dust types from photovoltaic surfaces on Mars
58. China mini clean brush from jinjiang manufacturer: Jinjiang Jiaxing Company. https://jiaxingbest.manufacturer.globalsources.com/si/6008845131807/pdtl/Carwash/1159952867/Mini-Clean-Brush.htm
59. 3 M Commercial solutions division. technical data sheet scotch-brite^{TM} high performance cloth. tech data sheet 2017:1. https://multimedia.3m.com/mws/media/1459376O/scotch-brite-high-performance-cloth-2010-technical-data-sheet.pdf
60. Kegeleers M (2015) The development of a cleaning robot for PV panels. Fac Eng Technol
61. Gheitasi A, Almaliky A, Albaqawi N (2015) Development of an automatic cleaning system for photovoltaic plants, pp 1–4
62. Green energy park research facility (2014). http://www.greenenergypark.ma/ (consulté le mars 07, 2018)
63. Syafiq A, Pandey A, Rahim N (2016) Photovoltaic glass cleaning methods: an overview
64. Calle C et al (2008) Dust particle removal by electrostatic and dielectrophoretic forces with applications to NASA exploration missions vol 2008
65. Gurav AB et al (2015) Superhydrophobic coatings prepared from methyl-modified silica particles using simple dip-coating method. Ceram Int 41(2):3017–3023
66. Mahadik SA et al (2013) Superhydrophobic silica coating by dip coating method. Appl Surf Sci 277:67–72
67. Liu S, Latthe SS, Yang H, Liu B, Xing R (2015) Raspberry-like superhydrophobic silica coatings with self-cleaning properties. Ceram Int 41(9):11719–11725
68. Choi T, Kim J-S, Kim JH (2016) Transparent nitrogen doped TiO2/WO3 composite films for self-cleaning glass applications with improved photodegradation activity. Adv Powder Technol 27(2):347–353
69. Lee WH, Lai CW, Hamid SBA (2015) One-step formation of WO3-loaded TiO2 nanotubes composite film for high photocatalytic performance. Materials 8(5):2139–2153

70. Jumeri F, Lim H, Zainal Z, Huang N, Pandikumar A (2014) Titanium dioxide-reduced graphene oxide thin film for photoelectrochemical water splitting. Ceram Int 40(9):15159–15165
71. Marques L (2008) Advances in mobile robotics. In: proceedings of the Eleventh International Conference on Climbing and Walking Robots and the support technologies for mobile machines, Coimbra, Portugal, 8–10 September 2008. World Scientific
72. Innovations SS Gekko solar
73. Koebler J Here's an autonomous, solar panel-cleaning robot. https://www.vice.com/en_us/article/wnj4kx/heres-an-autonomous-solar-panel-cleaning-robot
74. Fuchs-Barbana A, Schlittler S (2016) Einsatz von Reinigungs-Robotern in FM-Services, p 71
75. Grando MT, Maletz ER, Martins D, Simas H, Simoni R et al (2019) Robots for cleaning photovoltaic panels: state of the art and future prospects Rev Tecnol Cienc 35:137–150

Internet of Things-Based Solar Tracker System

Aboubakr El Hammoumi, Saad Motahhir, Abdelaziz El Ghzizal, and Aziz Derouich

Abstract Internet of Things (IoT) technologies, along with economies of scale and advances in hardware, software, and network technologies, have accelerated the explosion of connected objects across the Internet. A connected object can be controlled online from an IoT platform and can send, receive, and process various and varied data. In this chapter, we leverage some of the IoT technologies to propose a simple and low-cost IoT solution to monitor and control a smart dual-axis solar tracker system for performance evaluation. The solution also includes alert notifications to inform a remote user through phone or mail (or both) when a sensor has reached a certain predefined event. The solution is designed based on low-cost and easy-to-use hardware and software and an online open-source IoT platform. The design aspects of the IoT-based solar tracker are extensively described in this chapter. Moreover, a prototype of the IoT-based solar tracker has been manufactured and tested. Test results demonstrate that solar tracker data can be sent easily and properly and can be directly monitored online, as well as the solar tracker, can take commands from the IoT monitoring application.

Keywords Internet of things · IoT · Monitoring · Arduino · Solar tracker

---

A. El Hammoumi (✉) · A. El Ghzizal
InnovativeTechnologies Laboratory, EST, SMBA University, 30000 Fez, Morocco
e-mail: aboubakr.elhammoumi@usmba.ac.ma

A. El Ghzizal
e-mail: abdelaziz.elghzizal@usmba.ac.ma

S. Motahhir
Engineering, Systems, and ApplicationsLaboratory, ENSA, SMBA University, 30080 Fez, Morocco
e-mail: saad.motahhir@usmba.ac.ma

A. Derouich
Industrial Technologies and Services Laboratory, EST, SMBA University, 30000 Fez, Morocco
e-mail: aziz.derouich@usmba.ac.ma

S. Motahhir and A. M. Eltamaly (eds.), *Advanced Technologies for Solar Photovoltaics Energy Systems*, Green Energy and Technology,
https://doi.org/10.1007/978-3-030-64565-6_4

1 Introduction

The International Telecommunication Union (ITU) has defined the IoT as a global infrastructure for the information society that enables the provision of advanced services by connecting (physical and virtual) things, based on existing and evolving interoperable information and communication technologies [1]. The IoT, or as it is called the Internet of Everything (IoE), includes all devices that can communicate with the Internet and that can collect, send and process the data they capture from their surrounding environment using embedded sensors and processors in addition to the communication networks [2, 3]. IoT applications are expected to equip billions of objects with connectivity and intelligence [4]. It is already being deployed extensively, in various fields, namely: wearables [5], smart buildings [6], smart cities applications [7], health care [8], agriculture [9], industrial automation [10], solar monitoring systems [11], etc. In this chapter, we leverage some of the IoT technologies to design and build an IoT-based solar tracker system, where an IoT application is proposed to control and monitor this system.

To maximize the absorption of sunlight and thereby increasing energy production, it is necessary to integrate solar tracker systems into conventional solar energy systems, where the solar panels can be fixed on a structure that moves according to the sun's path. 10–50% additional output energy can be obtained by using solar tracker systems that track the sunlight instead of conventional systems that attach at a fixed angle [12]. Depending on the mechanisms used to orient the solar panels, solar tracker devices can be divided into single or double axis devices. Single-axis devices can only track sunlight by rotating around a horizontal or vertical axis, i.e. they track the sun's movement in one direction (toward East and West or toward South and North). While dual-axis solar tracker devices can rotate vertically and horizontally to ensure solar panels are always perpendicular to the sunlight [13]. Various solar tracker systems have been reported in the literature and they differ according to employed tracking methods [14]. For instance, a sensor-based solar tracker has been proposed in our previous works, it uses light sensors to predicts the sun's position (intensity of light) to track the sun for maximum power generation [15, 16]. The solar tracker system detects the sun position with the help of Light Dependent Resistor (LDR) sensors and sends the data to the controller. This latter then processes these data to command two servomotors that rotate a photovoltaic (PV) panel, in the optimal directions, to move toward the sunlight. For more details, authors in [17] have categorized solar tracker systems based on five tracking methods: sensor-based tracker method, geometric and astronomical equation-based method, open- or closed loop-based method, artificial intelligent-based method, and a combination of two or more of these methods. Indisputable that solar tracker systems have manifested a high ability for increasing the efficiency of solar panels to produce more energy. Besides, making a solar tracker device as a connected object using IoT technologies can be more profitable and advantageous, where the user can remotely control the device and access its data, including the electrical and environmental parameters linked to the solar panels, from an IoT platform. These data can be used to evaluate the solar tracker operation, as well as to assess the PV energy potential, early detection and diagnosis of electrical faults, evaluate the weather variations, and preventive maintenance.

However, to the authors' knowledge, there are only a few attempts available in the literature that deal with this subject. Authors in [18] have developed a dual-axis solar tracker with IoT monitoring using the Ubidots IoT platform. A WiFi ESP8266 board has been employed to connect their tracker device with the internet to communicate with the IoT monitoring application, where its data, including voltage, current, and power are displayed. The same WiFi board has been used by the Authors in [19], who have elaborated a single axis solar tracker, to send the same mentioned data to the cloud server of the ThingSpeak IoT platform so that they can be visualized in a dashboard that preconfigured on ThingSpeak. Furthermore, a monitoring solution of a solar tracker using Raspberry Pi3 (RPi3) board and a personal developed cloud server has been established in [20]. It uses socket programming using Python language to communicate between the "client" that runs on a remote laptop and "server" that runs on RPi3. There are two ways to design an IoT monitoring platform, either we design it ourselves, or using one of the available IoT platforms, which most of them are open source. By using the second way, the development process of an IoT project can be done easily and as early as possible. Because IoT platforms are designed to reduce an IoT project development time by enabling ready-made, reusable technology stack and are compatible with and support various hardware platforms (such as Arduino *and* Raspberry) [21]. To this end, IoT platforms are widely used by engineers and researchers in their IoT projects [22–25].

This chapter aims to present a simple and low-cost IoT solution to monitor and control a dual-axis solar tracker system. A low-cost and popular embedded board (Arduino) is used along with LDR sensors, servomotors, and associated circuits to control a PV panel to track the sunlight for maximum power generation. Different sensors are employed to measure electrical output parameters (voltage, current, and power) and environmental parameters (temperature, humidity) linked to the solar tracker system. An Ethernet shield is used to connect the system over the Internet and to exchange data between hardware and the cloud server using Message Queuing Telemetry Transport (MQTT) protocol. Data processing and activities that occur can be monitored online through an IoT monitoring application developed on Cayenne IoT platform. The solar tracker can also take commands from the monitoring platform. In addition, the application includes an alert system to notify the user when a sensor has reached a certain predefined event. The hardware and software used have been chosen to be simple and inexpensive. Arduino board is used due to its low-cost and its easy-to-use hardware and software [26, 27]. Likewise, Cayenne IoT platform is used due to its easy-to-use interface and protocols. It is an open-source IoT platform that has a simple Application Programming Interface (API) to store and retrieve data from things using the MQQT protocol over the Internet or via a Local Area Network [28].

The rest of this chapter is structured around three sections. Section "Research methodology" describes the architecture of the proposed IoT-based solar tracker system and presents the hardware and software used to develop it. Section "Results and discussion" lists and discusses the experimental results. Finally, the main conclusions of this chapter are drawn in Section "Conclusion".

2 Research Methodology

2.1 System Description

The proposed IoT-based solar tracker system is depicted in Fig. 1. It is a dual-axis solar tracker that can rotate automatically to track the sun position using LDR sensors, or manually by the user through the dashboard of an IoT application. The system starts with detects the sun position (intensity of light) by LDR sensors and sends the data to the controller (Arduino Mega board). This latter then processes these data to command servomotors (SM1 and SM2) that hold the PV panel to rotate toward the sun. The values of the generated PV voltage and current, temperature, and humidity are also sent to the Arduino through associated sensors. Next, the Ethernet shield, which is mounted with Arduino and allows it to be connected to the Internet, will send the data that has been taking and/or processed by Arduino to the cloud (webserver). Lastly, the solar tracker data, including LDR sensors, PV power, temperature, and humidity, are displayed in real time in the IoT monitoring application via pre-created Widgets. The IoT monitoring application is designed using Cayenne myDevices platform. Once the user is connected to the internet from his computer or smartphone, he can visualize, in the dashboard of the IoT application, all solar tracker data in their associated widgets. Therefore, the user has the necessary data linked to the environment and performance of the PV panel. In addition, in the manual mode, the servomotors will take angle directions from their associated widgets in the dashboard. Hence, the user can control his system to seek the best

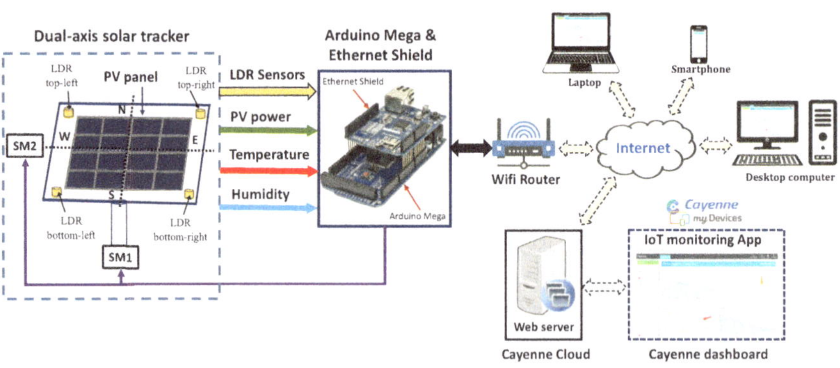

Fig. 1 Schematic of the IoT-based solar tracker system

environmental conditions and extract the maximum energy from the PV panel. The IoT application is also programmed to send notification alerts (SMS or Email) when a senor reaches a predefined threshold value.

2.2 *Hardware Design*

As shown in Fig. 2, the IoT solar tracker system consists of the PV panel, two servomotors, four LDR sensors, a voltage divider circuit, temperature and humidity sensor, a Led and the Arduino Mega board.

The used PV panel is 115 by 85 mm in size with a 1.6 W output and can generate a voltage up to 6 V [29]. Two 180° servomotors are used to motorize the solar tracker and they are controlled by the Arduino board through PWM pins 5 and 6. The left-right (L-R) servomotor (MG996R) rotates the solar tracker on the vertical axis (East/West), while the Up-down (U-D) servomotor (SG90) rotates the solar tracker on the horizontal axis (South/North).

Four LDRs (Cds GL5528) are used to sense the sun's position and which have been fixed in the four corners of the PV panel. The LDR sensors are connected to the Arduino through analog pins from A0 to A3. The LDR is a resistor whose value decreases with increasing light intensity incident on its surface. The LDR sensor is designed as a voltage divider circuit as can be seen in Fig. 2. The output of the voltage divider is connected to an analog input (A0 for instance) of the Arduino. Then, the

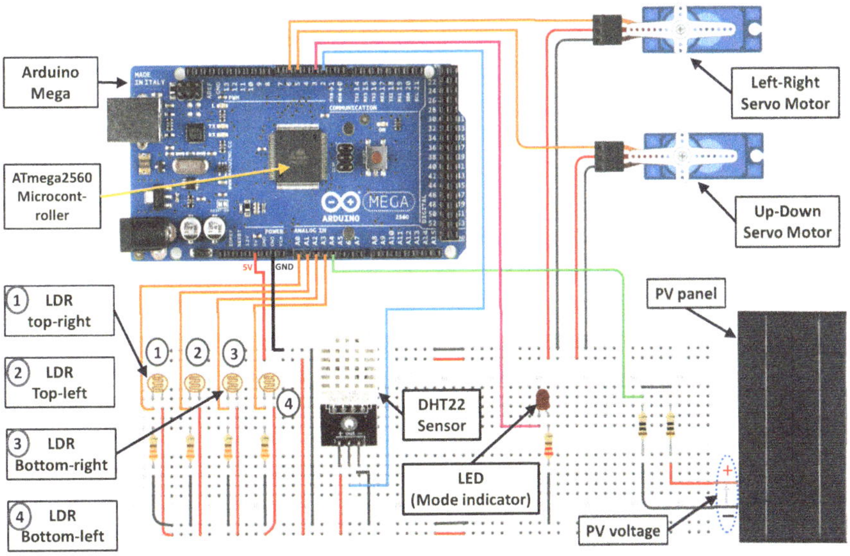

Fig. 2 Electronic circuit of IoT-based solar tracker system

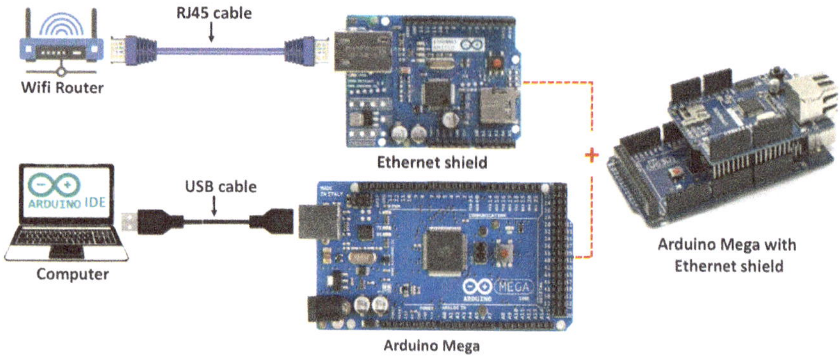

Fig. 3 Hardware interface between Arduino and Ethernet shield

Analog to Digital Converter (ADC) of the microcontroller converts the analog value read by A0 into a digital value between 0 and 1023 because the ADC is coded in 10 bits. The value of the series resistor in the LDR sensor circuit is 330 Ω.

The temperature and humidity are measured through the DHT22 sensor, which is an ultra-low-cost sensor that is widely used in embedded projects. DHT22 has a thermistor and a capacitive humidity sensor embedded in it to measure temperature and relative humidity. Its temperature range is from −40 to 80 °C with < ± 0.5 °C of accuracy, and its humidity range is from 0 to 100% with ± 2% (Max ± 5%) of accuracy [30]. This sensor uses one signal wire to transmit data to Arduino (digital pin 2), and two wires for power supply.

The PV voltage and current are measured through a voltage divider circuit that acts also as a load and which consists of two series resistors of 10 Ohms. The divider circuit output is connected to the Arduino's analog pin A4. Furthermore, a LED, which is connected to digital pin 3, reflects in the system circuit the mode state of solar tracker (manual or automatic).

The Arduino Mega with ATmega2560 microcontroller is used as the embedded controller that interacts with the Arduino Ethernet shield along with the monitoring platform. The Ethernet shield, which is mounted above the Arduino board, must be connected with a Wi-Fi router (or PC) through an RJ45 cable as shown in Fig. 3. The Ethernet Shield is based on the Wiznet W5100 Ethernet chip that provides a network (IP) stack for TCP and UDP protocols [31].

2.3 Prototype

Figure 4 presents the solar tracker prototype in its detached and assembled state. It consists of the PV panel, the L-R, and U-D servomotors and LDR sensors. The panel is attached to the U-D servomotor on one side and with a bearing on the other side to ensure better flexibility when the solar tracker rotates around the horizontal axis. The

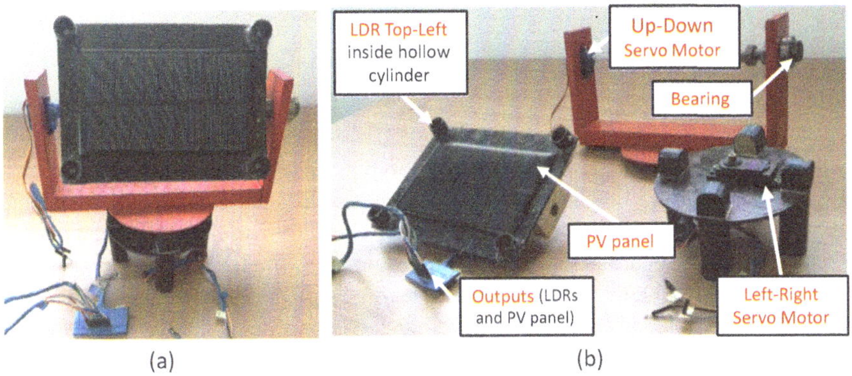

Fig. 4 Solar tracker prototype in its detached and assembled state

assembly is attached to the L-R servomotor. The LDR sensors are fixed in the four corners of the panel inside hollow cylinders. If the panel is not perpendicular to the sun, at least one LDR will be covered by shadow caused by the surrounding cylinder. Hence, there will be a difference in light intensity. The best orientation is when the light intensities are equal in all LDR sensors. Figure 5 shows the entire prototype of the IoT-based solar tracker system, and it is clear that all reported components in the hardware part have been used to build it.

Fig. 5 IoT-based solar tracker prototype

2.4 Software Design

a. Arduino IDE

Arduino is an open-source electronics prototyping platform with easy-to-use hardware and software [32]. The Arduino platform provides an integrated development environment (IDE), which includes support for C and C++ programming languages. The used Arduino board in this work is programmed by the IDE that serves as a code editor and from which the program code can be uploaded to the microcontroller through USB cable, as can be shown in Fig. 3. The Arduino Megaboard is utilized to implement all software requirements of the IoT-based solar tracker.

b. MyDevices Cayenne

myDevices is a company that offers IoT solutions. It offers an end-to-end platform for the IoT. In our project, we will focus on Cayenne, one of the solutions from myDevices. This tool allows developers, designers, and engineers to build prototypes of the IoT. Cayenne uses the Message Queuing Telemetry Transport (MQTT) protocol to connect any device with the Cayenne cloud. Once connected, the user can send and receive data from the device to the Cayenne dashboard via the Widgets created. MQTT is a publish–subscribe messaging protocol based on the TCP/IP protocol. The publish–subscribe methodology uses a message agent that is responsible for delivering messages to the client. The MQTT is the API for sending information to the Cayenne cloud, or devices controlled by Cayenne. The messaging agent in this connection is the cloud, it manages the different clients (sensors and actuators) that send and receive the data.

To use MQTT with Cayenne, we need to use the Cayenne libraries. For Arduino, the CayenneMQTT library can be installed from the IDE's Library Manager. To program our Cayenne IoT platform-based IoT application, we will take advantage of the predefined functions. For example, to establish the connection between Cayenne cloud and Arduino Mega equipped with the Ethernet module, we call the CayenneMQTT Ethernet library where we declare our authentication information (the username, password and the ClientID) which should be obtained from the Cayenne Dashboard. Then, in the setup part of the program, we call *Cayenne.begin ()* function to establish the connection with Cayenne dashboard. For each actuator, we create a function with an integer parameter between 0 and 31 imperatively called *CAYENNE IN (VIRTUAL CHANNEL)*. For each sensor, we create a function with an integer parameter between 0 and 31 imperatively called *CAYENNE_OUT(VIRTUAL_CHANNEL)*. In the loop part of the program, we call the predefined function *Cayenne.loop ()*, this function itself calls the functions *CAYENNE_OUT* and *CAYENNE_IN*. The virtual channel as its name suggests is a channel that does not physically exist, it characterizes visualization or command widgets. It allows them to be linked with the corresponding sensor or actuator.

c. The Embedded Software Design

The embedded software is the piece that will be embedded in the Arduino Megato interact between the Ethernet module and Cayenne cloud (see Appendix). It is designed as follow:

(i) The IoT-based solar tracker has two function modes: manual and automatic. A button created in the Cayenne dashboard has a role to switch between the two modes. When it is inactive, the manual mode is selected, otherwise automatic mode. Besides, a function is established in the Arduino code that allows recovering the state of the button. The LED in the system circuit reflects the state of this switch.

Therefore, for the controller to know the selected operating mode, we just need to test the state of the pin in which the LED is connected. For example, if the LED state is low, the controller will call the manual mode function to execute, otherwise, it will call the automatic function.

(ii) If the manual mode is selected, the user can directly control the positions of the servomotors to orient the PV panel from east to west by L-R servomotor or from south to north by the U-D servomotor. The control is made from the associated widgets of servomotors in the dashboard of the IoT application.

In this mode, the controller calls *Cayenne.loop ()* function which itself calls all the functions *CAYENNE_IN*, including those related to servomotors, to execute. The Cayenne.loop () function will also call all the functions CAYENNE_OUT, linked to the sensors, to execute. Where the data related to LDR sensors, PV current, voltage and power, temperature and humidity would be sent to the server so that they can be visualized in their associated widgets in the IoT application.

(iii) If the automatic mode is selected, the algorithm shown in Fig. 6 will be executed. The algorithm starts by reading the analog values returned by LDR sensors. Then, it processes these data to command servomotors that move the PV panel toward the sun position. Considering the vertical axis-based solar tracker movement, the average values of the two LDRs on the left and the two LDRs on the right are compared and if the lefts receive more light, the PV panel will move in that direction (clockwise) through the L-R servomotor. The latter will stop when the difference result is between -10 and 10. This range is used to stabilize the controller and to reduce the power consumption of servomotors. Otherwise, if the right set of LDRs receives more light, the PV panel will move in that direction (Counterclockwise) through the L-R servomotor and will continue to rotate until the difference result is in the range $[-10, 10]$. The same approach is used for the horizontal axis-based solar tracker movement where the average values of the two LDRs on the top and the two LDRs on the bottom are compared.

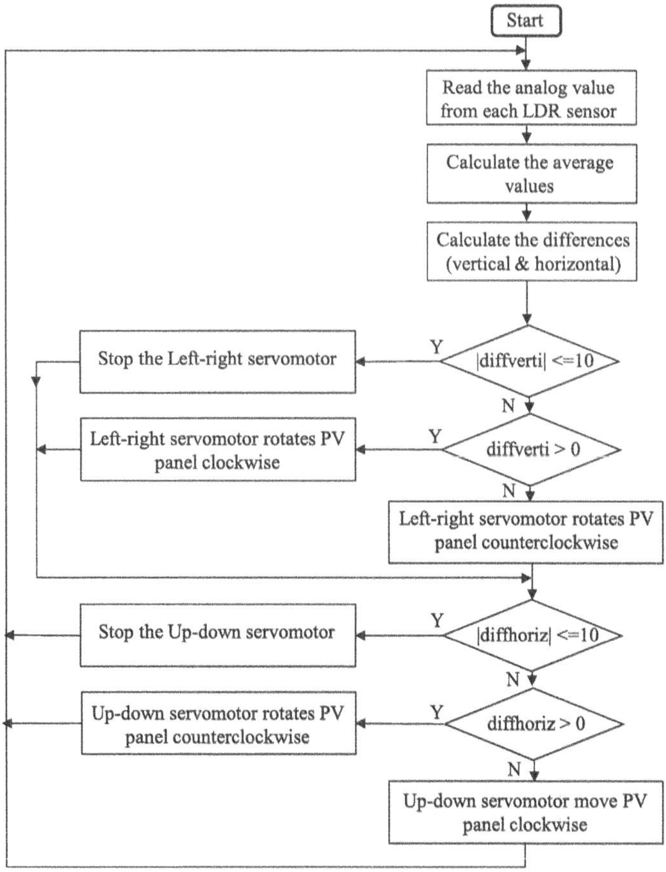

Fig. 6 The flowchart for the automatic mode of the solar tracker

As well as in the automatic mode, the controller will also call the *Cayenne.loop ()* function to send the solar tracker data to the IoT application.

For

d. Development of the IoT Monitoring Application

(i) *Hardware interfacing with Cayenne IoT platform*

To interface the hardware, including sensors and actuators, with the IoT platform, we need to follow the next steps:

- Log in on Cayenne myDevice website after creating an account (Fig. 7a).
- Then, click on "Bring Your Own Things" from Cayenne API (Fig. 7b).
- Copy the MQTT credentials (username, password and client ID) from Create App (Fig. 8), and paste them in Arduino source code as described previously. After successfully compiling and uploading the entire code to Arduino Mega, open

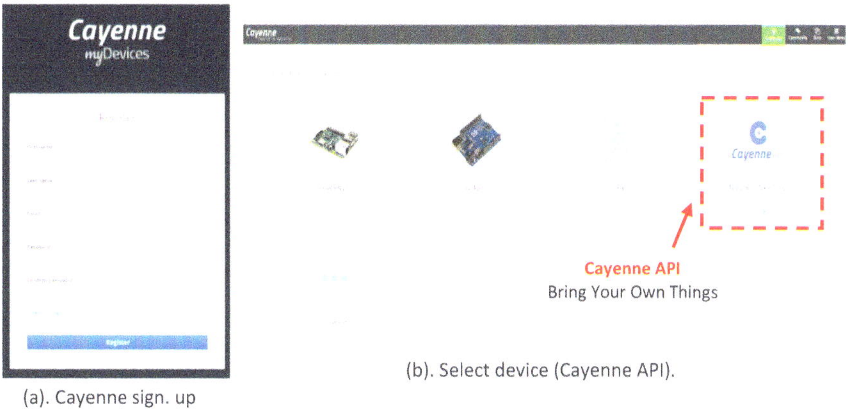

Fig. 7 Cayenne IoT Platform sign up (**a**). Cayenne API (**b**)

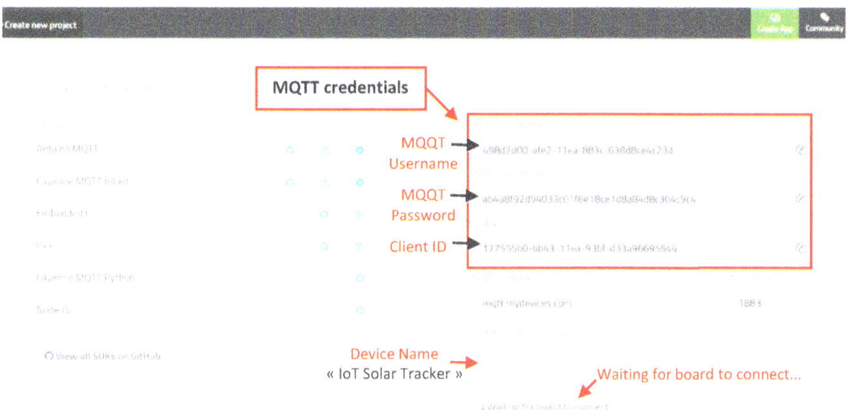

Fig. 8 MQTT credentials and device connection to Cayenne

Serial Monitor in Arduino IDE to get the Cayenne log prints (Fig. 9). As soon as our device comes online and connects to Cayenne, the previous page (Fig. 8) is automatically updated and we will see our device in the online dashboard as can be seen in Fig. 10.

Fig. 9 Cayenne log prints on Serial Monitor

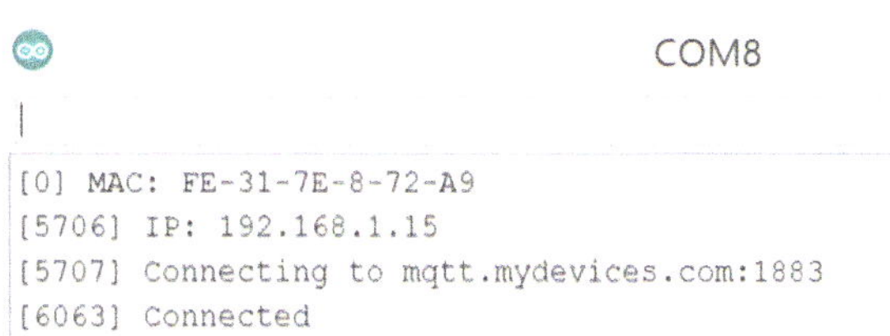

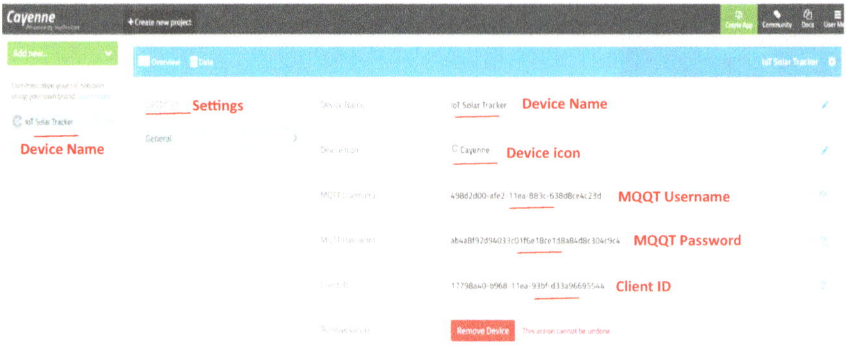

Fig. 10 Device settings

- Then, to interface sensors and actuators, i.e. create their widgets, click on "Add new…", select "Device/Widget" and click on "Custom Widgets" (Fig. 11). Then, select a widget and populate all its associated settings (the channel number must be the same as in code), and finally click on "Add Widget" to add it to the dashboard of your device. For us, we chose the "value" widget for all sensors, "Button" widget for mode switch and the "Slider" widget for servomotors.

Finally, Fig. 12 illustrates the designed IoT application for monitoring solar tracker data. Once the connection with the solar tracker system is established, sensor data can be visualized on their associated widgets, the tracking mode (automatic or manual) can be selected from the switch button, as well as controlling servomotors' angles through their widgets. Sensor data can also be obtained in graphical form by modifying the representation type in their settings, or just by clicking on the graph icon above the widget.

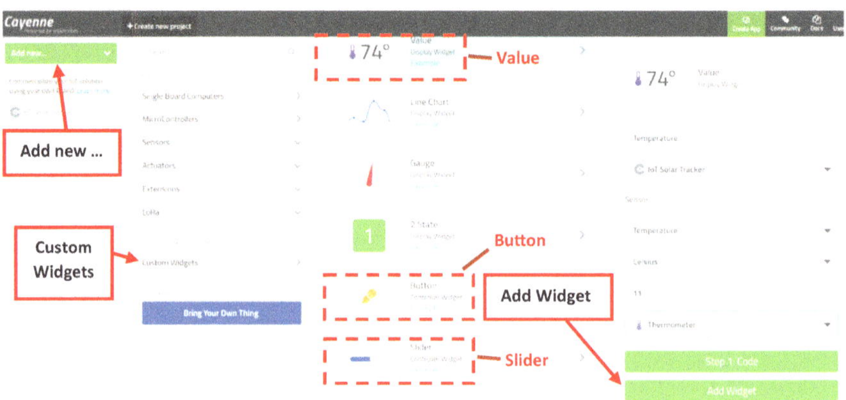

Fig. 11 Cayenne custom widgets

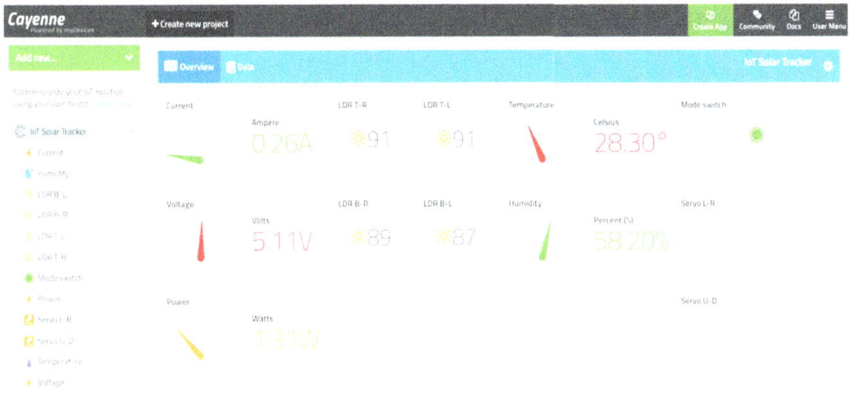

Fig. 12 IoT monitoring application of solar tracker system

(ii) *Alerts creation*

One of the most important criteria in a monitoring system is its ability to send notification alerts to inform users when an event, related to their monitored devices, occurs. To this end, we take advantage of one of Cayenne's features [33] to add alerts to our IoT application, where we can preprogram our application to send a notification alert (SMS, Email, or both) or to perform a specified action. For example, a temperature alert is created to send an email notification to the user (or recipients) when the monitored temperature is reached a threshold value, as can be shown in Fig. 13. To create an alert, click on "Add new…" and select "Trigger", then set the event and its action and finally click on "save" to add it to the dashboard.

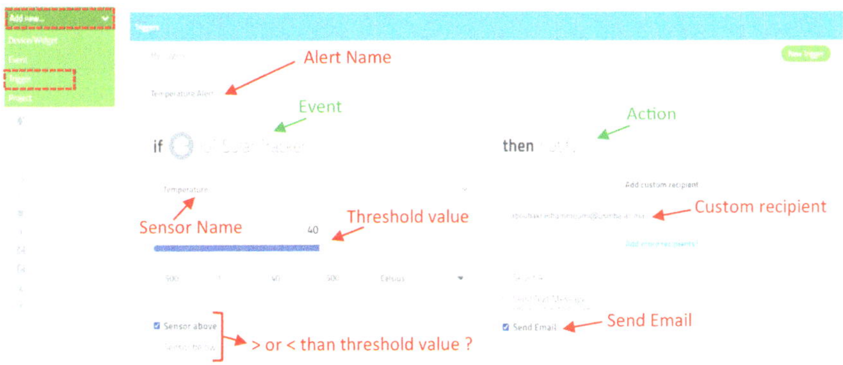

Fig. 13 Temperature alert configuration

3 Results and Discussion

Different tests have been carried out to examine the developed IoT-based solar tracker prototype. The experimental setup is illustrated in Fig. 14. The Arduino board is powered with the computer through a USB cable, which is also used to display, in the Serial Monitor of Arduino IDE, the measured parameters and data received from the IoT application. This will allow us to verify whether the captured data from Arduino are correctly and in real-time sent to the IoT application or not. Whereas, Arduino can be powered with an external DC power supply. The Ethernet shield connects the Arduino board to the internet via RJ45 cable. Once the connection with the IoT application is established, the data of the solar tracker system are sent to the monitoring application, where we can view these data live and send commands to the controller.

The system is programmed to send all data from the device regardless of the tracker mode (manual or automatic). First, the automatic mode has been tested, which is activated when the switch mode is in the high state; the LED (mode indicator) in the circuit lights up. The servomotors were automatically controlled according to intensities captured by LDR sensors. Figure 15 presents the samples of data recorded in real time from the solar tracker system in the IoT application during the test period. Figure 15a displays the recorded electrical measurements, namely the current, voltage and power. While Fig. 15b displays the recorded environmental measurements, namely the temperature, humidity and the intensity of light (captured by the top-right LDR) with the accurate time and date. It has been verified, by comparing the data sent from Arduino and those received on the dashboard of the

Fig. 14 The experimental setup of the prototype

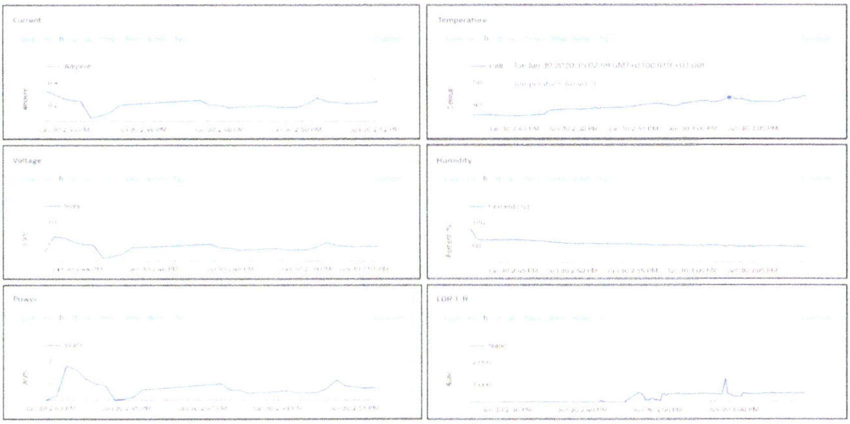

(a) Electrical measurements. (b). Environmental measurements.

Fig. 15 Samples of data recorded in real time in the IoT application

IoT application, that all electrical and environmental measurements are sent properly and in real-time.

To check the reliability of the monitoring application to notify the user when an event occurs, it has been programmed to send an alert. For example, when the monitored temperature is higher than 40 °C. Figure 16 shows the alert notification received in our mailbox at the same time when the temperature exceeds 40° as can be seen in Fig. 15b. Other alerts can be added to the application, such as a malfunction of one of the sensors and/or actuators and a rapid decrease in PV power.

Moreover, the proposed IoT prototype has been tested in manual mode, which is activated when the switch mode is in the low state; the LED in the circuit turns off. The servomotors were controlled through their associated widgets in the dashboard. For instance, we have set the slider linked to the L-R servomotor at the center (i.e. at a value of 0.5) and the other slider of U-D servomotor at 0.3, which means that the L-R and U-D motors will rotate by 90° and 30°, respectively. Also, tests have shown that solar tracker properly and rapidly executes commands from the monitoring platform with a time not exceeding 2 s. In the manual mode, the user can remotely position his device in an optimal direction according to the surrounding environment and device location. Moreover, in this mode, the power consumption of motors can be too minimized or, where the user can intervene to position the solar tracker for example in only one direction according to each month or season (winter, spring, summer, and autumn) in the year. However, an amount of PV energy can be lost due to a limitation of the solar tracker operation according to the daily movement of the sun. The servomotor commands can be programmed beforehand without user intervention by creating events and associated actions in the IoT application.

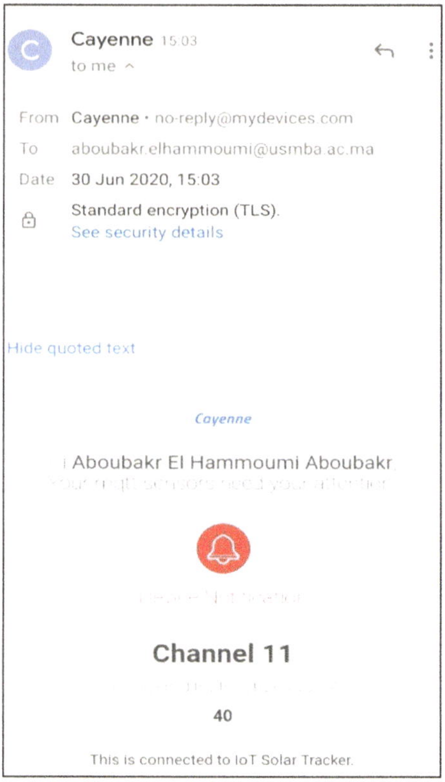

Fig. 16 The received notification alerts in the mailbox

4 Conclusion

In this chapter, a smart prototype has been designed to monitor and control a dual-axis solar tracker system using a simple and efficient IoT solution. The prototype has been tested experimentally. Test results demonstrate that the developed IoT-based solar tracker provides users with a simple monitoring application, in which users can easily and in real-time monitor electrical and environmental parameters of the solar tracker system for further processing and management. Other sensors could be added, for example, solar irradiation and wind sensors to help understand more about the PV power output as well as to test the solar tracker system on its flexibility during high wind. Due to its simplicity, the proposed IoT solution can be employed in various fields to connect devices or things to the internet as well as for research or educational purposes.

Appendix

The embedded code of the IoT-based solar tracker system

```
#define CAYENNE_PRINT Serial
#include <CayenneMQTTEthernet.h>    //CayenneMQTT library
#include <Servo.h>                  //Servo motor library
#include <DHT.h>                    //DHT library
#define DHTTYPE DHT22
#define DHTPIN 2
DHT dht(DHTPIN,DHTTYPE);

//MQTT credentials
char username[]="498d2d00-afe2-11ea-883c-638d8ce4c23d";
char password[]="ab4a8f92d94033c01f6e18ce1d8a84d8c304c9c4";
char clientID[]="17798a40-b968-11ea-93bf-d33a96695544";

Servo servo_x;                    //up-down servomotor
int servoh = 0;
int servohLimitHigh = 170;
int servohLimitLow = 10;

Servo servo_z;                    //left-right servomotor
int servov = 0;
int servovLimitHigh = 170;
int servovLimitLow = 10;

int topl,topr,botl,botr;
int threshold_value=10;
float Vout;

void setup()
{ Serial.begin(9600);
  Cayenne.begin(username, password, clientID);
  servo_x.attach(5);
  servo_z.attach(6);
  dht.begin();
  pinMode(3,OUTPUT);
  digitalWrite(3,LOW);
}

void loop()
{ topr= analogRead(A2);
  topl= analogRead(A3);
  botl= analogRead(A4);
  botr= analogRead(A5);
  Vout=(analogRead(A1) * 5.0) / 1023;
  Serial.println(" Manual-mode");
  Cayenne.loop();

  if (digitalRead(3)==HIGH){
    Serial.println(" Automatic-mode");
    servoh = servo_x.read();
    servov = servo_z.read();
    int avgtop = (topr + topl) / 2;
    int avgbot = (botr + botl) / 2;
    int avgright = (topr + botr) / 2;
    int avgleft = (topl + botl) / 2;
    int diffhori= avgtop - avgbot;
    int diffverti= avgleft - avgright;

    if (abs(diffhori) <= threshold_value)
    {
     servo_x.write(servoh);           //stop the servo up-down
    }else {
```

```
        if (diffhori > threshold_value)
           { Serial.println(" x - 2 ");
           servo_x.write(servoh -2);     //Clockwise rotation CW
           if (servoh > servohLimitHigh)
           {
            servoh = servohLimitHigh;
           }
           delay(10);
           }else {
            servo_x.write(servoh +2);    //CCW
            if (servoh < servohLimitLow)
            {
            servoh = servohLimitLow;
            }
            delay(10);
            }
       }

     if (abs(diffverti) <= threshold_value)
     {
      servo_z.write(servov);        //stop the servo left-right
     }else{
        if (diffverti > threshold_value)
        {
        servo_z.write(servov -2);  //CW
        if (servov > servovLimitHigh)
        {
        servov = servovLimitHigh;
        }
        delay(10);
        }else{
         servo_z.write(servov +2);  //CCW
         if (servov < servovLimitLow)
         {
         servov = servovLimitLow;
         }
         delay(10);
         }
      }
   }
}
// Cayenne Functions
CAYENNE_IN(8){
  int value = getValue.asInt();
  CAYENNE_LOG("Channel %d, pin %d, value %d", 8, 3, value);
  digitalWrite(3,value);
}
CAYENNE_IN(7){ //up-down servo motor
  if (digitalRead(3)==HIGH){ //Automatic_mode
  }
  else{ //Manual_mode
  servo_x.write(getValue.asDouble() * 180);
  }
}
CAYENNE_IN(6){ //left-right servo motor
  if (digitalRead(3)==HIGH){
  }
  else{
  servo_z.write(getValue.asDouble() * 180);
  }
}
```

```
CAYENNE_OUT(0) { //Current
  float current = Vout/10;
  Cayenne.virtualWrite(0, current);
  Serial.print("Current: ");
  Serial.println(current);
}
CAYENNE_OUT(1) { //Voltage
  float voltage = Vout * 2;
  Cayenne.virtualWrite(1, voltage);
  Serial.print("Voltage: ");
  Serial.println(voltage);
}
CAYENNE_OUT(2){ //LDR Top-right
  Cayenne.virtualWrite(2, topr);
}
CAYENNE_OUT(3){ //LDR Top-left
  Cayenne.virtualWrite(3,topl);
}
CAYENNE_OUT(4){ //LDR Bot-left
  Cayenne.virtualWrite(4,botl);
}
CAYENNE_OUT(5){ //LDR Bot-right
  Cayenne.virtualWrite(5,botr);
}
CAYENNE_OUT(10) { //Power
  float power = (Vout * 2 * Vout)/10 ;
  Cayenne.virtualWrite(10, power);
  Serial.print("Power: ");
  Serial.println(power);
}
CAYENNE_OUT(11){ //Temperature
  float t = dht.readTemperature();
  //int chk = dht.read(DHT11PIN);
  Cayenne.virtualWrite(11, t, TYPE_TEMPERATURE, UNIT_CELSIUS);
  Serial.print("temperature: ");
  Serial.println(t);
}
CAYENNE_OUT(12){ //Huidity
  float h = dht.readHumidity();
  //int chk = dht.read(DHT11PIN);
  Cayenne.virtualWrite(12, h);
  Serial.print("  humidity: ");
  Serial.println(h);
}
```

References

1. International Telecommunications Union (2012) Overview of the internet of things
2. Kang B, Kim D, Choo H (2017) Internet of everything: a large-scale autonomic iot gateway. IEEE Trans Multi Scale Comput Syst 3(3):206–214
3. Ahmed MA, Eltamaly AM, Alotaibi MA, Alolah AI, Kim YC (2020) Wireless network architecture for cyber physical wind energy system. IEEE Access 8:40180–40197
4. Business Insider Intelligence (2020) The internet of things report
5. Haghi M, Thurow K, Stoll R (2017) Wearable devices in medical internet of things: Scientific research and commercially available devices. Healthc Inform Res 23(1):4–15
6. Jia M, Komeily A, Wang Y, Srinivasan RS (2019) Adopting internet of things for the development of smart buildings: a review of enabling technologies and applications. Autom Constr 101:111–126

7. Khajenasiri I, Estebsari A, Verhelst M, Gielen G (2017) A review on internet of things solutions for intelligent energy control in buildings for smart city applications. Energy Procedia 111:770–779
8. YIN Y, Zeng Y, Chen X, Fan Y (2016) The internet of things in healthcare: an overview. J Indust Inform Integr 1(Elsevier B.V):3–13
9. Tzounis A, Katsoulas N, Bartzanas T, Kittas C (2017) Internet of things in agriculture, recent advances and future challenges. Biosyst Eng 164:31–48. (Academic Press)
10. Li JQ, Yu FR, Deng G, Luo C, Ming Z, Yan Q (2017) Industrial internet: a survey on the enabling technologies, applications, and challenges. IEEE Commun Surv Tutorials 19(3):1504–1526
11. Rahman MM, Selvaraj J, Rahim NA, Hasanuzzaman M (2018) Global modern monitoring systems for PV based power generation: a review. Renew Sustain Energy Rev 82:4142–4158. (Elsevier Ltd)
12. Nsengiyumva W, Chen SG, Hu L, Chen X (2018) Recent advancements and challenges in Solar Tracking Systems (STS): a review. Renew Sustain Energy Rev 81:250–279. (Elsevier Ltd)
13. Hafez AZ, Yousef AM, Harag NM (2018) Solar tracking systems: Technologies and trackers drive types – a review. Renew Sustain Energy Rev 91:754–782. (Elsevier Ltd)
14. Sumathi V, Jayapragash R, Bakshi A, Kumar Akella P (2017) Solar tracking methods to maximize PV system output – a review of the methods adopted in recent decade. Renew Sustain Energy Rev 74:130–138. (Elsevier Ltd)
15. El Hammoumi A, Motahhir S, El Ghzizal A, Chalh A, Derouich A (2018) A simple and low-cost active dual-axis solar tracker. Energy Sci Eng 6(5):607–620
16. Motahhir S, EL Hammoumi A, EL Ghzizal A, Derouich A (2019) Open hardware/software test bench for solar tracker with virtual instrumentation. Sustain Energy Technol Assessm 31:9–16
17. AL-Rousan N, Isa NAM, Desa MKM (2018) Advances in solar photovoltaic tracking systems: a review. Renew Sustain Energy Rev 82:2548–2569. (Elsevier Ltd)
18. Nur M, Mohd A, Jumaat A, Rimong C, Jawa A, Jumaat SA (2020) Dual axis solar tracker with IoT monitoring system using arduino. Int J Power Electron Drive Syst 11(1):451–458
19. Pulungan AB, Risfendra R, Purwanto W, Maksum H, Setiawan O (2020) Design and development of real time monitoring single axis solar tracker by using internet of things
20. Williams K, Qouneh A (2017) Internet of things: solar array tracker. Midwest Symp Circuits Syst 2017:1057–1060
21. Singh KJ, Kapoor DS (2017) Create your own internet of things: a survey of IoT platforms. IEEE Cons Electron Mag Inst Electr Electron Eng Inc 6(2):57–68
22. Lopez-Vargas A, Fuentes M, Vivar M (2019) IoT application for real-time monitoring of solar home systems based on arduinotm with 3G connectivity. IEEE Sens J 19(2):679–691
23. Mekala MS, Viswanathan P (2019) CLAY-MIST: IoT-cloud enabled CMM index for smart agriculture monitoring system. Meas J Int Meas Conf 134:236–244
24. Amir Alavi S, Rahimian A, Mehran K, Alaleddin Mehr Ardestani J (2018) An IoT-based data collection platform for situational awareness-centric microgrids. In: Canadian Conference on Electrical and Computer Engineering, vol 2018
25. Motahhir S et al (2020) Optimal energy harvesting from a multistrings PV generator based on artificial bee colony algorithm. IEEE Syst J 1–8
26. El Hammoumi A, Motahhir S, Chalh A, El Ghzizal A, Derouich A (2018) Real-time virtual instrumentation of Arduino and LabVIEW based PV panel characteristics. IOP Conf Ser Earth Environ Sci 161(1):012019
27. El Hammoumi A, Motahhir S, Chalh A, El Ghzizal A, Derouich A (2018) Low-cost virtual instrumentation of PV panel characteristics using Excel and Arduino in comparison with traditional instrumentation. Renew Wind Water Sol 5(1):1–16
28. Cayenne-myDevices (2020). https://mydevices.com/. (Accessed 23 June 2020)
29. V/1.6 W Small Solar Panel (2020). https://fr.enfsolar.com/pv/panel-datasheet/crystalline/34434. (Accessed 23 June 2020)

30. DHT22 Sensor Datasheet (2020). https://datasheetspdf.com/pdf/792211/Aosong/DHT22/1. (Accessed 23 June 2020)
31. W5100 | WIZnet Co., Ltd. (2020). https://www.wiznet.io/product-item/w5100/. (Accessed 23 June 2020)
32. Zlatanov N (2020) Arduino and open source computer hardware and software
33. myDevices, Cayenne Docs (2020). https://developers.mydevices.com/cayenne/docs/features/#features-alerts. (Accessed 10 June 2020)

Impact on the Performance of Solar Photovoltaic System with the Innovative Cooling Techniques

N. Beemkumar, S. Dinesh Kumar, A. D. Dhass, D. Yuvarajan, and T. S. Krishna Kumar

Abstract The photovoltaic panel converts only some parts of solar radiation energy into electrical energy and the rest of energy will remain as heat energy, which results in raising the panel temperature and decreases electrical proficiency. The decrease in PV module efficiency depends on the assortment of limitations including temperature; the yield power diminishes by 0.2–0.5% per 1 K differs in the hotness of the photovoltaic module. The PV system efficiency could be improved by providing efficient cooling techniques like active and passing cooling system with the cooling medium air, water, phase change material (PCM), etc. This chapter summarizes the recent trends in PV cooling techniques and also discusses the impact of the innovative cooling technique on solar PV module performance by combining PCM and thermoelectric generators (TEG).

N. Beemkumar (✉)
Department of Mechanical Engineering, Faculty of Engineering and Technology, Jain (Deemed-to-be University), Bengaluru 562112, India
e-mail: beem4u@gmail.com

S. Dinesh Kumar
Department of Mechanical Engineering, Karpaga Vinayaga College of Engineering and Technology, Kanchipuram, India
e-mail: softdin@gmail.com

A. D. Dhass
Department of Mechanical Engineering, Indus Institute of Technology and Engineering, Indus University, Gujarat 382115, India
e-mail: dasaradhan.ad@gmail.com

D. Yuvarajan
Department of Mechanical Engineering, Saveetha School of Engineering, SIMATS, Chennai 602105, India
e-mail: dyuvarajan2@gmail.com

T. S. Krishna Kumar
Department of Automobile Engineering, VNR Vignana Jyothi Institute of Engineering and Technology, Hyderabad, India
e-mail: krishnakumar_ts@vnrvjiet.in

S. Motahhir and A. M. Eltamaly (eds.), *Advanced Technologies for Solar Photovoltaics Energy Systems*, Green Energy and Technology,
https://doi.org/10.1007/978-3-030-64565-6_5

Keywords Photovoltaic panel · Heat energy · Efficiency · PCM · Active cooling · Passive cooling

1 Introduction

When exposed to a hot and dry climate, the yield of a solar-powered PV panel changes from the yield of proportional solar oriented PV panels in chilly conditions. Light is an alluring piece of solar-based radiation while heat is not [1]. Adjusting the atmospheric airstream power starting is at 1–3 m/s brought down the temperature of the panel by around 10 °C [2]. Inferable from the high surrounding temperature and high panel temperature, machine execution was seen to a generous decrease in the late spring. The module showed relatively superior in winter [3]. When all is said in done, lessening the heat substance or disappointment of the solar light-based PV cell or module and expanding the electrical presentation is troublesome. Frequently, the ingestion of heat into the solar light-based PV module relies fundamentally upon the type of burden associated with the power age units. The particularly lower heat burden enthalpy, connected to the photovoltaic panel, request more power, so expel heat from the climate or heat sink utilizing a warm siphon [4].

The various m-Si PV modules, for example, Yingli-produced m-Si, m-Si, dark solar light-based PV panel, and fundamentally determined estimations of the parasitic opposition and photogenerated current of the Yingli m-Si PV module were contrasted with assessing the effect of temperature. The discoveries recommended that the m-Si 250-W module was amazingly temperature touchy to compute the shunt opposition. With an ascent in temperature, such enhancements diminished exponentially. A consistent ascent in temperature has improved the obstruction of the individual succession [5]. Indeed, the temperature decreased by about 15.13 °C since the balances installed at the base of the PV module increased the district's flow of wind current heat. Consequently, the electrical yield compared with the PV module temperature was projected to be 14.39%. Moreover, when cuts were mounted between the blades, they expanded the wind stream speed and heightened choppiness creation, in this way expanding the balances' cooling productivity [6].

As indicated by the exploratory investigation, the temperature drop in the focal fragment of the Building Integrated Concentrated Photovoltaic (BICPV) system for miniaturized scale balances finned with a PCM was found to have been diminished by around 9.6 °C. The temperature drop in a related segment was around 11.2 °C, due to the addition of nanoparticles to the PCM. The presentation of nanoparticles as stood out from typical PCM material prompted the expanded decrease in temperature [7]. When using the cooling system, the usual temperature of the PV panel drops to around 45 °C, which leads to a decrease of approximately 27% for the module temperature. This was because of the pervasive soggy circumstance on the backside of the device, while creation limit and execution improved separately by around 32.7 and 31.7% [8]. Radiative solar light-based cell cooling is a viable type of cooling to manage the temperature of solar-powered cells. However, the exemplified solar-based cells despite everything have high outflows, proposing that more changes in sun-oriented cells' radiative cooling productivity are negligible. Nonetheless, note that radiative cooling of solar-powered cells in certain conditions is without a doubt a suitable method to cool solar energy-based cells [9].

To get an economic viewpoint, the made of a cooling system is reasonable, because it is a free structure (the power deftly of the cleaning structure is done by the power set aside in the battery), and it uses modest (water or sun situated refining (nursery effect or smooth sun controlled finder)). It grants improving the electrical making of photovoltaic modules, which prompts the decrease in the force bill. It furthermore serves to the cleaning of PV modules with a water seminar on the front face [10]. The results showed that the maximum temperature drop in submerged cooling conditions about 24 °C. This prompts a rate increment of 10.06% of the open-circuit voltage, the most extreme force conveyed by the PV module additionally increments by 9.83% and this adequately expands the transformation productivity of the PV module by 9.83% [11].

The underneath conditions are analyzed with the assets of sun-based cells and the cooling necessities of the fluid chose for cooling.

(1) Great heat move execution for the fluid
(2) The ghostly reaction of sun-oriented cells ought to be coordinated by the ingestion of the daylight by the fluids
(3) Non-poisonous and great synthetic soundness must be kept up for the fluid
(4) Economical fluids are utilized for cooling [12].

It is concluded that the effect will be a 5% decrease in productivity for increasing 10 °C in cell temperature. Consequently, further attention has been orchestrated in the application of PV panel cooling strategies. The single principle technique to accomplish this cooling is by consolidating the two advancements into a half breed configuration, called photovoltaic heat (PV/T) power [13].

It was discovered that the PCM material would give positive outcomes just when the temperature of the panel surpasses the PCM dissolving point [14]. The show of three identical PV modules is the usage of module 1 without cooling, the cooling of module 2 with unadulterated vapor, and the cooling of module 3 with nanofluids (CuO–water and Al_2O_3–water with 0.1–0.7% wt).

In any case, the greatest discrepancy between the cooling module's rear temperatures and the non-cooling module was 22.13 C at 0.6% CuO centralization, with an electrical power extension of 12.57% [15]. PV cells are very vulnerable to shifts in temperature, despite the eco-friendly capacity of the photovoltaic (PV) module to generate power. This can result in a decline in skills from 0.25–0.5%/C. Scientists and scientists are committed to increasing the performance of PV cells by rising the working temperature to solve this issue. For this reason, the researchers have developed an easy and elite PV cooling system that can lower the working temperature of the module [16].

The rectangular fins reduced the temperature of the PV module by 10.6% and raised their performance by 14.5%. Nevertheless, the circular fins-based PV module will resist 112% more power than the reference PV module, with the same surface area as the rectangular fin. This indicates that the surface area of the fins is a crucial parameter from the PV module for heat dissipation and involves careful fine design [17].

A parametric analysis is carried out to evaluate the overall machine efficiency for the specific operating environments and activated capability. The developed device is also tested for two separate forms of multi-junction solar cells and three specific coolant grades: wind, ethanol, and n-pentane Water with higher concentration ratio capacity has been known to function stronger fluid due to higher latent moisture [18].

The strength of the refrigerated and non-cooled instruments was then contrasted. The temperature of the cooling modules dropped to approximately 25 °C, while the temperature of the uncooled module was 45 °C. The strongest outcomes were obtained by cooling modules with a water video, because there were no water splashes, and continuous surface cooling contributes to an improvement in power of 20%. The uncooled module obtained a maximum strength of 105.3 W/m^2, compared with 125.5 W/m^2 for its cooled equivalent during the study. Consequently, cooling the module contributed to a capacity rise of 20.2 W /m^2 [19].

The water spray cooling system used in the photovoltaic panel and its results show that the usage of this kind of cooling has a 2% improvement in the average electrical energy (EE) relative to the same level. They also observed that utilizing this sort of device would decrease the temperature of the average panel from 54 to 24 °C [20].

Some structural parameters have been examined and evaluated for impact on PV cell temperature and electrical performance. The most important parameters are summarized on specific solar irradiation, inlet air temperature, air mass flow rate, water mass flow rate, channel size, channel width, wind speed, fins length, fins tilt, fins amount, the distance between successive fins, fine thickness, container width, size of Reynolds and Prandtl numbers [21].

The purpose of this chapter, to familiarize the various types of cooling techniques were used in PV panel surface, which includes active, passive and hybrid cooling methods, effects of various methods on PV panel performance analysis are discussed in detail.

2 Types of the Cooling System in PV Panels

Various developments in cooling are studied, especially gliding using the concentration cooling method. Improving the appearance of solar-based panels is utilizing phase-changing materials; solar-based panels with water-drenching cooling methods [22]. There are two kinds of cooling strategies to boost the greatest power efficiency and PV module generation: active and passive cooling. While aloof cooling strategies include standard cooling methods for the removal of heat from the solar-powered PV floor, general air, normal water, and PCM for Fig. 1 [23].

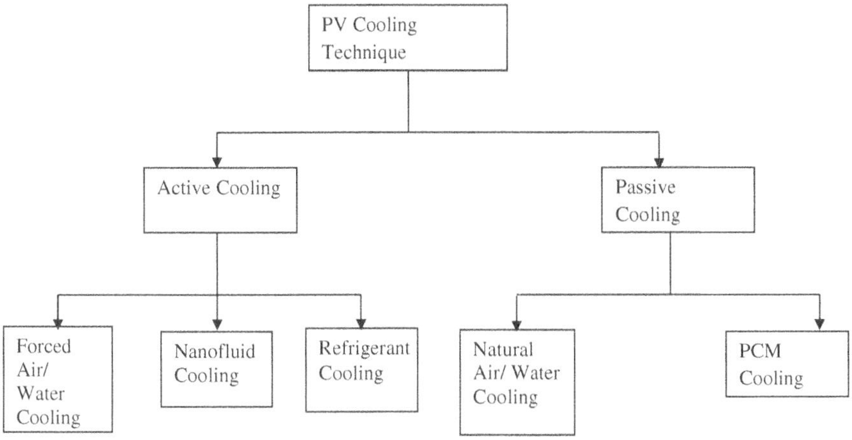

Fig. 1 Classification of PV cooling techniques based on an input source to the system [23]

The overall performance of the PV panels is greatly affected by their temperature. The temperature development affects the electrical energy created by photovoltaic cells [24]. Cooling advancements have developed towards increasingly complex methodologies, which incorporate warmth sinks or blends with different systems. The classification of cooling methods for enhancing the performance of a PV panel based on working fluid is given in Fig. 2. The epic, advanced field identifies with shaft parting (or range channel) innovation, which recognizes the frequencies utilized for PV cells from those utilized for the heat transformation of the PVT system [25].

3 Active Cooling Methods

The net proficiency yield of a solar energy-based photovoltaic cell is delicate, corresponding to temperature.

- For the most part, broadcasting a small film of water over a sun-based photovoltaic cell reduces the rate of sensation of sun-oriented radiation and cools the sun-oriented board to a comprehensive temperature.
- Theoretically, the water that is drained after passing into the channel stream set may be used as a temperature resource [26].

The revealed techniques and cooling impacts of each PV module cooling strategy with a schematic diagram are given in Fig. 3 [27]. It consists of both active and passive methods presented in the diagram.

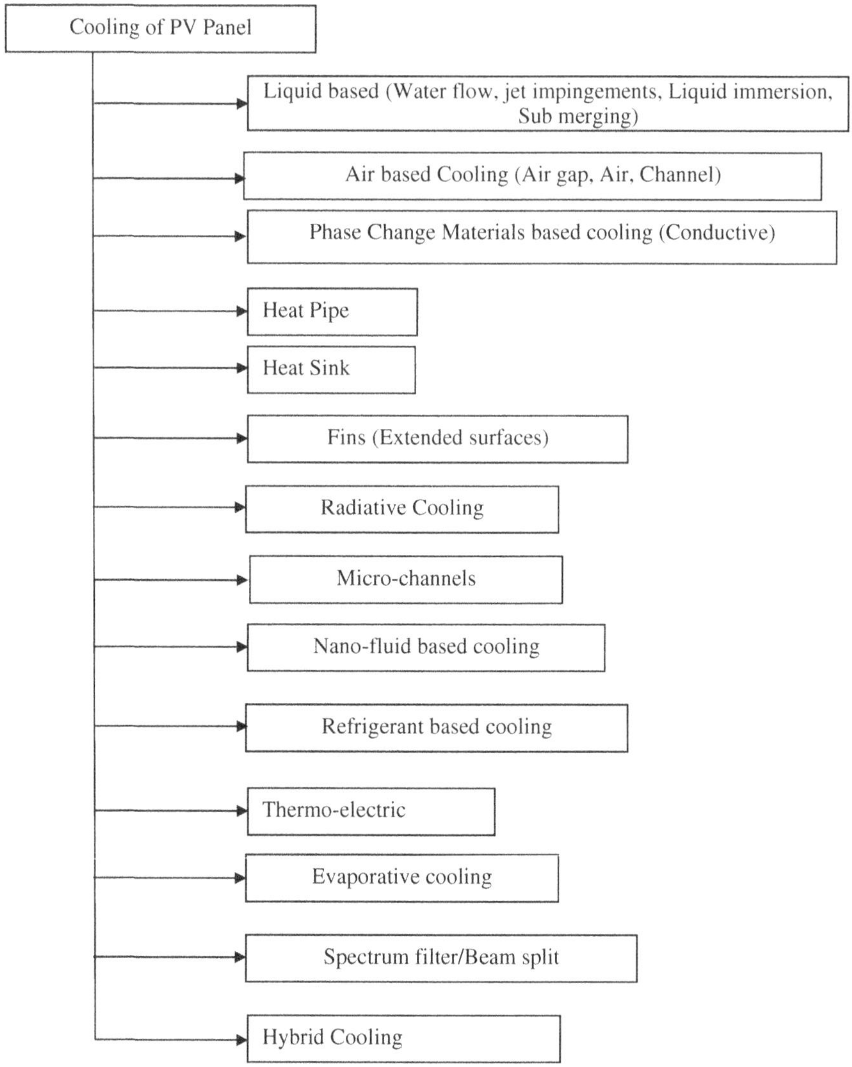

Fig. 2 Classification of cooling methods based on working fluid [25]

Heat pipe cooling

A heatpipe is a compartment tube loaded up with the working liquid. One finish of this cylinder (called evaporator area) is gotten warm contact with a hot point (PV panel back surface) to be cooled. The opposite end (called condenser segment) is associated with the dissipation point where the warmth can be dispersed (climate). A bit of the cylinder among the evaporator and condenser is called the adiabatic area.

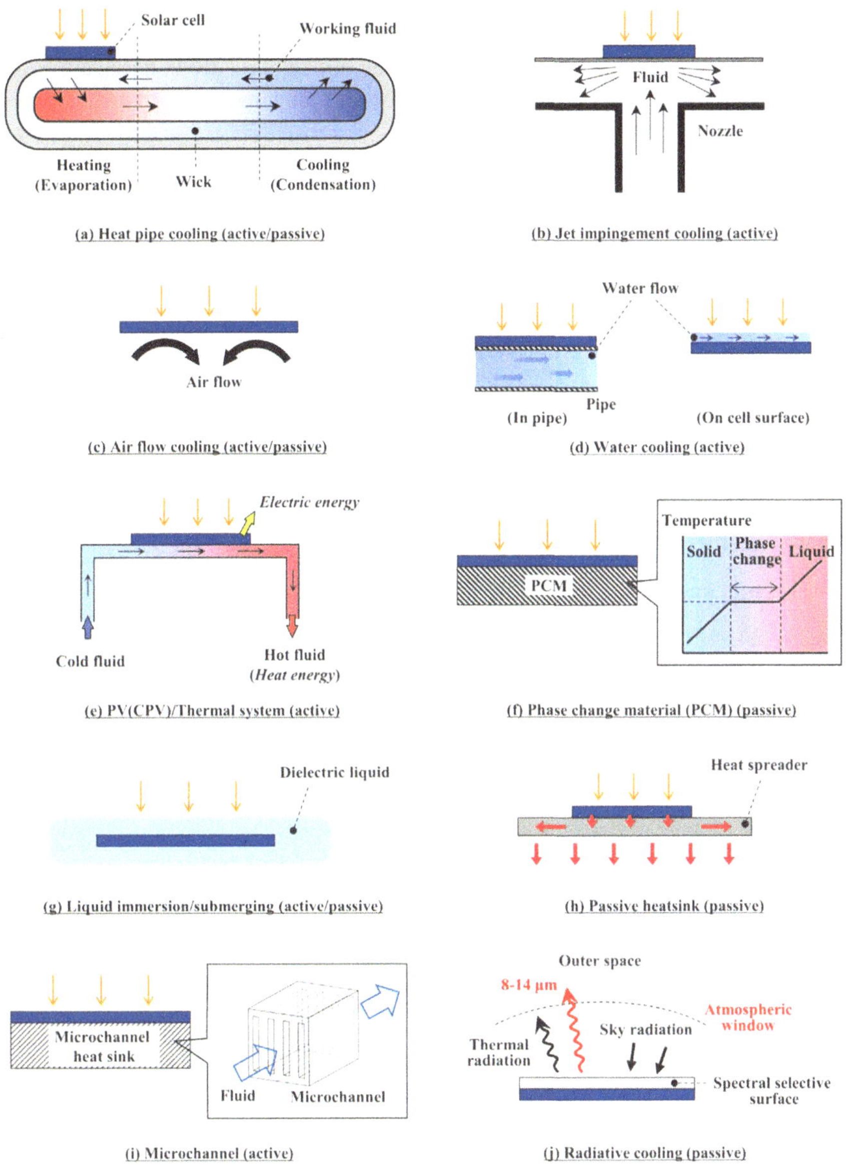

Fig. 3 Schematic diagrams of cooling methods for photovoltaic modules. The cooling type (active and/or passive) is shown in parentheses for each cooling method [27]

Jet impingement cooling

One significant cooling procedure that has been utilized broadly in hardware is the strategy for heat extraction by impinging planes. Impingement planes are equipped for removing a huge amount of heat.

Airflow cooling

Normal or constrained air dissemination is a straightforward and minimal effort mode to evacuate the warmth, yet it is less viable at low scopes, where the surrounding air temperature is over 20° C for a long time during the year. PVT/Air systems are used in pragmatic applications, predominantly as ease air-cooled Building Integrated Photovoltaics, because of the low development utilization of materials and low working expense.

Water cooling

When deep water is used specifically in cooling, the PV panel can tend to run at a virtually constant temperature all year round. That is because the temperature of the surface water will not come up against a vital variation throughout the year.

PV Thermal system

A thermoelectric cooling device allows good use of excess energy for better efficiency but has a poor rate of transition profitability and this innovation's movement is modest.

Phase change materials

Ongoing investigatation concerning uninvolved cooling of the photovoltaic panel using phase change materials has indicated to the phase change materials can accumulate a lot of warmth and while cooling the PV panel with phase change materials it continues practically consistent warmth.

Liquid Immersion/Submerging

A passive lowered water cooling technique where the PV module is lowered in the static water, and they examined the impact of the situation in the profundity course on the presentation.

Passive Heating

Detached solar-driven cooling systems operate by minimizing unwanted daytime energy benefit, providing non-mechanical airflow, exchanging warm indoor air for colder outdoor air wherever necessary, and taking away the coolness of the night to guide high daytime temperatures.

Microchannel

The microchannel heat sink shows the capacity to expel a lot of warmth from a little territory, which is an attractive component for heat move improvement. Along these lines, specialists have indicated an expanded enthusiasm for the improvement of small scale cooling innovation for different applications

Radiative cooling

An inspection into radiative cooling of sunlight-based cells has expanded as of late. An ongoing report on the issue expressed exploring the impact of upgraded radiative cooling on sun-oriented cells utilized in business PV was basic. Reenactment results uncovered that the sunlight-based cell temperature must be diminished by 1.75 K even in the perfect case.

A functioning of cooling strategy erstwhile structured in addition to displaying for PV panel utilizing TEG innovation planned to get better the PV proficiency furthermore future.

Figure 4 demonstrates the effective performance of the PV-TEC system in advance applied at 0.12 and 0.25 s for atmospheric temperature changes, holding the solar-oriented insolation steady at 1 kW /m^2. It reveals that the PV power obtained from the sensor for 25 °C surrounding temperature is 52 W without cooling activity, whereas the total PV power can be raised to 54 W with a 17 °C reduction following the cooling mechanism [28].

For illustration, nanofluid is characterized as a mixture of particles (nanometer sizes, 100 nm) and liquids such as H_2O and ethylene glycol. The liquids have improved the thermophysical properties of standard base liquids such as thermal conductivity, consistency, and overlap. This is commonly used in SPV panel cooling. So solar PV cooling is important, it diminishes the surface temperature by cooling, so ability will improve. There are various methods for refrigerating solar-based PV such as active and passive cooling. Efficient cooling utilizes air, water, nanofluids, etc. for cooling purposes, goods such as paraffin wax, eutectics, natural materials, cotton wick, etc., for passive cooling purposes [29], [30].

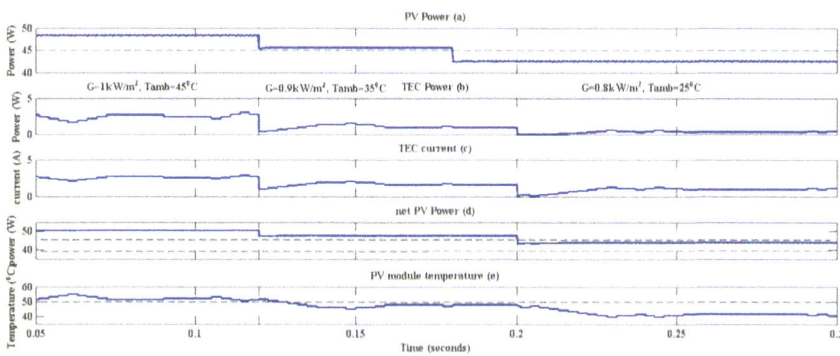

Fig. 4 Representation of photovoltaic thermoelectric generator under standard temperature condition. **a** Photovoltaic Power **b** Thermoelectric generator power **c** Thermoelectric generator Current, **d** net photovoltaic power, and **e** Photovoltaic panel temperature [28]

Table 1 Performance parameters in PV panel for different examined cooling [32]

Techniques	Power (W)	Relative increase in power output (%)	Effective increase in power output (%)	Temperature (°C)	Effective increase in electrical efficiency (%)
Without cooling	35	–	–	56	–
Back surface cooling	39.9	14.0	5.4	33.7	3.6
Front surface cooling	40.1	14.6	6.0	29.6	2.5
Simultaneous cooling	40.7	16.3	7.7	24.1	5.9

PV-PCM structure of aluminum sheet as TCE is tested by running a check under the clear daylight. The backside of the PV panel has a PCM and aluminum pocket measuring 0.0361 m^2. To improve the thermal conductivity of the PCM and heat dispersal, the aluminum sheet of zone 0.036 m^2 is conveniently mounted on the back of the PV panel. The effect of panel temperature rise is tentatively verified on Voc, Isc, and generation. The results indicate that PCM with the aluminum sheet as backplate in the solar-centered PV panel improved the performance of the transition by 24.4% on a normal [31].

A water spray cooling solution was proposed for different cooling situations (systems) and was tentatively evaluated on a monocrystalline photovoltaic stand. The best cooling solution ended up being the synchronous cooling of PV panel surfaces at the front and back. Among the various cooling, techniques are used to maximize the power output, simultaneous cooling of the back surface followed by front surface cooling has exhibited better performance than compared with the other techniques and it is given in Table 1. This produced an important 5.9% improvement in electrical output.

4 Passive Cooling Methods

Three diverse submissive lose heat situations be statistically researched, in addition to the majority encouraging one was where the photovoltaic module was furnished using cuts during the frontage photovoltaic module facade bringing about a decrease of around 4 °C. The more cases examined to end up becoming less successful with the respected drop in temperature below 1.0 °C. The concept of modern PV panel details products was seen as a realistic possibility [33, 34].

Passive cooling systems allude to innovations used to remove or potentially limit heat assimilation as of photovoltaic panel lacking extra power utilization. The component infers moving temperature anywhere it is created and dispersing it to nature.

Wide assortments of aloof cooling alternatives are accessible, most straightforward structures include the utilization of firms of soaring warm conductivity metals, for example, aluminum and copper, otherwise a variety of blades before extra expelled surfaces to upgrade heat move to the encompassing. Increasingly, unpredictable systems include the utilization of PCMs in addition to different strategies in favor of characteristic flow, notwithstanding the utilization of heat pipes that can move heat proficiently through a bubbling gathering process. A portion of the effective passive cooling systems that can limit the danger of expanding warm gains incorporates. Strategies for Passive cooling frameworks can be classified into three sorts.

- Air aloof cooling,
- Water uninvolved cooling and
- Conductive cooling.

The distinction between air detached cooling in addition to conductive cooling is to the aggregate character in the photovoltaic panel for the common heat move about components. The bigger the outside of PV cell the more it is critical to the utilization of detached cooling. Water uninvolved cooling is increasingly productive because of the warm limit of water high [35].

5 Hybrid Cooling Techniques

Joint use of PCM and nanofluid is a more feasible strategy for PV cooling than by sole use of PCM or nanofluid because additional PV panel heating is separated by two extremely heat-resistant media, for example first by PCM and then by a nanofluid. PCM fusion with nanofluid decreases PV-surface temperature as well as provides more stability in temperature due to standardized PCM interaction with the plate. The panel will gain improved warm vitality on the off chance that it blends both nanofluid and nano-PCM [36].

It has been indicated that connecting a TEG to the rear of the PV cell can produce maximum power, which can exploit solar oriented vitality. Hence, it is vital to locate the prevailing warm opposition of the partial illuminated system, and afterward locate a successful answer for control of the warm obstruction for advancement. In this area, an affectability examination is led to the warm protection of the radiation, convection, and conduction for each layer [37].

For this exploratory analysis, a number of latent cooling systems are possible utilizing a cooling tower centered on PCM and traditional water convection. Yields show that the cooling device with the rendered oil as PCM has an exceptional capacity to increase the panel's effectiveness. The findings obtained indicate that the application of Boehmite nanopowder to the composite oil decreased the hotness of the photovoltaic module in accordance with the configuration of no cooling and the usage of the composite oil as PCM [38].

Graphene-water nanofluid has obtained the most remarkable performance at a concentration of 0.1% and 40 L per minute (LPM) in general competence. Normally, performance improved by 14.1%, 12.6%, and 10.9% at 0.1% concentration for 40, 30, and 20 LPM, respectively, in water analysis. From the tests, the combined usage of nanofluid with PCM was assumed to offer favored device execution over the use of PCM alone [39]. A CPV/T-driven mixture assimilation/thermoelectric cooling system is introduced. The suggested device utilizes both warm vitality and electrical vitality that are acquired to offer cooling from the PV/T gatherer. This was guided to try to develop the PV board cooling monitor to explore the most intense level of cooling. Thus, to guide the enhancement process, TEC's COP will surpass 6.4, which means that the validity number is 70 [40].

Heat pipe helped inert heat stockpiling system are broadly utilized in warming, cooling, and waste heat recuperation applications because of their basic development and fantastic thermophysical properties, for example, high warmth stockpiling limit and capacity to move heat at a consistent temperature over significant separations. This work concentrated on applications and the extent of crossbreed framework (HP–PCM-based frameworks) in many building fields.

Following are the striking highlights of this examination:

- The half breed frameworks are exceptionally proficient and overpower the issues, for example, low warm conductivity of PCM and overheating of warmth pipes when utilized in mix.
- The crossbreed framework hinders the temperature rise and guarantees the sheltered activity of gadgets over the more drawn out periods.
- Performance of these half and half frameworks chiefly rely upon properties of PCM, sort of warmth funnel, and its direction in the framework.
- Heat pipe alongside metallic froth, foil, and blades are superior to traditional HP.
- Charging of PCM for the most part happens by convection heat move while conduction assumes a significant job during releasing procedure. The expansion of HP expands the pace of warmth move during charging and releasing [41].

In Fig. 5, the best cases in the cooling systems were tested under the appropriate solar radiation and natural conditions and in the PV panel models finned left, right and center (PV + B3), 12 number of TEGs and fins in PV panels (PV + E3), $CaCl_2H_{12}O_6$ and finned left, right and center in PV panels (PV + C2 + B3) and $CaCl_2H_{12}O_6$ and 12 number of TEGs and fins in PV panels (PV + C2 + B3) and PV panels (PV + C2 + B3) [42].

6 Effects on the Performance of PV Panels

The front-side temperature of the PV cell is greater than the rear temperature of the PV cell. Through and wide, both the front and the posterior temperatures are the same in stable conditions. In this test protocol, the influence on the execution

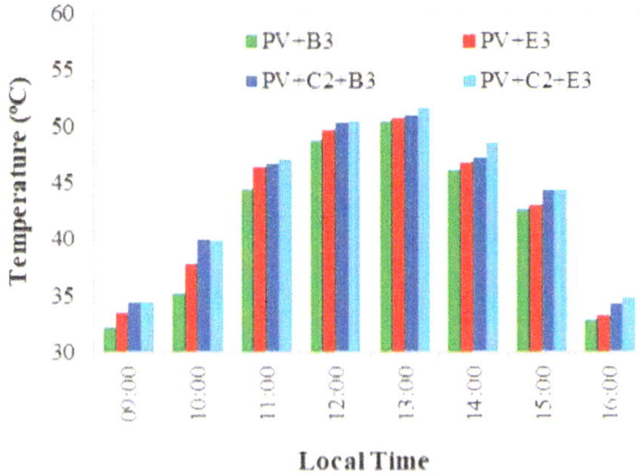

Fig. 5 Surface temperature variation in a PV panel with different cooling systems [42]

of the PV module on the front side temperature is higher than the consequence of the posterior temperature. So also, on the rear during the procedure, the impact of temperature will corrupt the fill factor esteem. The front and rear temperature of the PV module impacts on fill factor as appeared in Fig. 6a, b. The found the middle value of PV module fill factor differed from the beginning and last qualities by about 22.98, 54.16 and 26.19% with the relating variety of arrived at the midpoint of front side PV module temperature is 10.85, 11.76, and 11.42% and rear PV module temperature is 9.09, 10.76 and 12.05% separately, for the long stretch of December, January, and February. Here fill component and front side PV variance were strong in January and accompanied by months in February and December. Thus, in light of the climatic changes, the working execution boundary most likely corrupts.

On account of the backside PV module temperature, deviation levels are high in February and January and December months. The variations in temperature forecasts for December, January, and February are 9.09, 10.76, and 12.05% [43].

Figure 7 delineates the month-wise temperature pace of the front and back of the PV board at a predetermined area. The pinnacle temperature esteem 34 °C of the front side is seen in May and June and the most reduced 28 °C is seen in December. Additionally, the posterior temperature is 32 °C in May and 26 °C in December [44].

The presentation reactions of the PV board in various investigations are summed up in Table 2. The most extreme increment of 19.32% and 18.40% in electrical force yield and effectiveness were accomplished under the illumination of 900 W/m^2 by presenting the Active phase change (APC) cooling system. In the interim, the most extreme explicit force improvement picked up in this examination is 21.37 W/m^2 [45].

Behind the effect of energy generation effectiveness is introduced in Fig. 8, the outcomes demonstrate with the purpose of the photovoltaic and photovoltaic-phase change materials panel make increasingly electric vitality and encompass maximize electric effectiveness than the equivalent regular PV panel without phase change materials layer. The most noteworthy yield was estimated throughout the midyear months. The rise in electricity production for the PV-PCM panel varies from 4.3

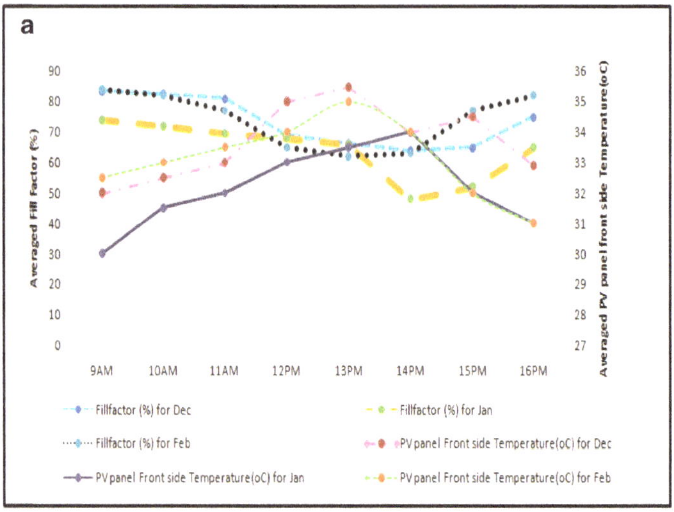

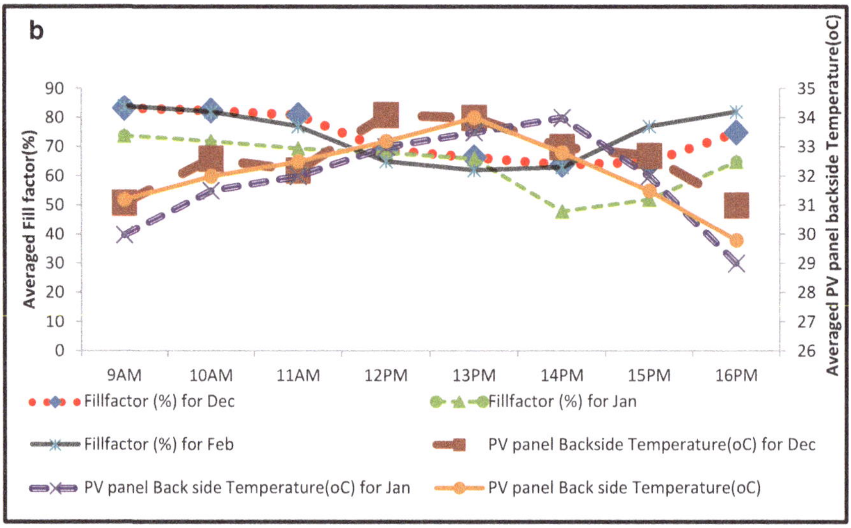

Fig. 6 **a** Performance Characteristics curve for PV system (Front side temperature) [43]. **b** Performance Characteristics curve for PV system (Back side temperature) [43]

to 8.7% and the output of vitality generation varies from 0.5 to 1%. In this way, the yearly vitality generation proficiency of the phase change materials module was 0.8% higher than the customary PV module. Through the usage of PCM in some circumstances, the energy efficiency of 12.2% and the electrical energy produced of 260.17 kWh were achieved, which dislikes an annual rise of 7.3% relative to conventional PV panels [46].

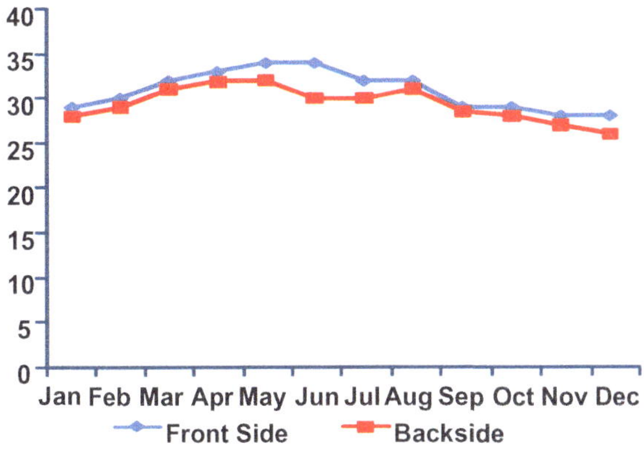

Fig. 7 The temperature profile of the front and backside of the PV panel for the entire period of the experiment [44]

Table 2 Different techniques of cooling [45]

Techniques	Temperature (°C)	Power (W)	Increase in Power Output (%)	Electrical Efficiency (%)
Uncooled condition	62.4	21.2	–	12.94
Air cooling	55.5	22.0	3.77	13.43
APC	37.4	24.6	16.03	15.02

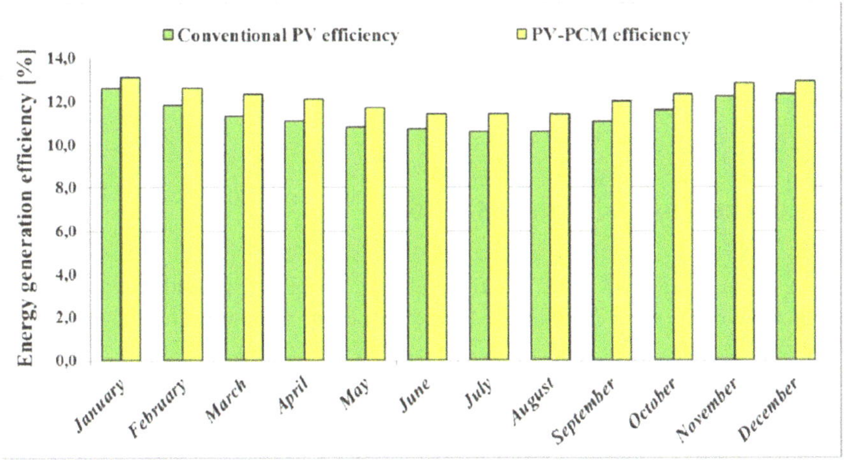

Fig. 8 The generated Energy efficiency of conservative photovoltaic and photovoltaic-phase change materials panel [46]

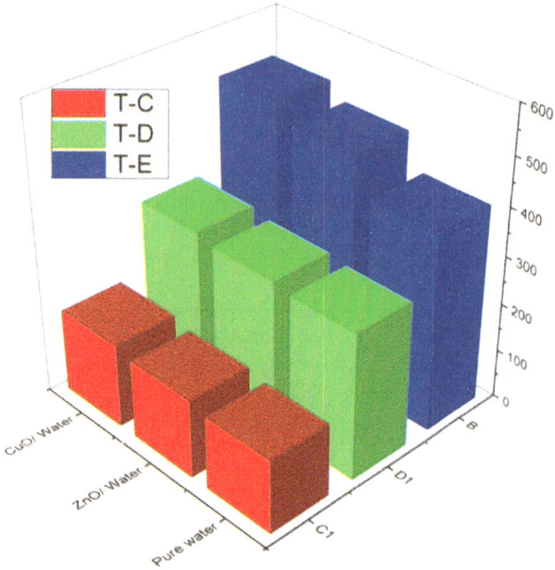

Fig. 9 Equivalent electrical output power variation of different schemes [47]

PCM records an improvement of 25% in total skills relative to the PV module shown in Fig. 9. Compared to the PV module, the normal warm yield for PCM is increased by 46%. It is further discovered that PCM can elevate the particular warmth of the warm PVT system, which quickly diminishes the warm misfortunes. Therefore, the introduction of nanoparticles to the base liquid increases the presence of the temperature exchanger regardless of the month and climate. Energy performance figures are dramatically higher inferable from the nanofluid effect [47].

7 Conclusions and Future Outlook

The behavior of a photovoltaic (PV) panel over cooling in front and rear of different advancements is examined. The PV board execution improved after it was cooled by dynamic and aloof strategies. The electrical boundaries of sun-based cells were relying upon surface temperature, which demonstrated that the cooling factor assumes a significant job in the electrical productivity upgrade. The PCM may be a great solution for cooling and homogenous diffusion of air. Nonetheless, PCM with a low liquefying point (25 °C) may lower the PV board temperature more than PCM with a high softening point (over 30 °C) for short times and trouble areas growing to develop on the surface of the PV board. The efficiency of the PV-PCM panel is capable of better by expanding the warmth move among the phase change materials and the aluminum plate. In addition, the design of the PV-TEC structure has been separately tested to include temperature advancements and solar-powered insolation variation. The extension of the nanoparticles increases power efficiency and reduces the surface temperature to obvious amounts.Reference [48] is given in the list but not cited in the text. Please cite them in text or delete them from the list.kindly delete the mentioned reference

References

1. Wani C, Gupta KK (2019) Towards improving the performance of solar photovoltaic energy system: a review. IOP Conf Ser Earth Environ Sci 227:022009
2. Rahmatmand A, Harrison SJ, Oosthuizen PH (2020) An innovative method to control surface temperature of a rooftop photovoltaic system. Sol Energy 195:581–591
3. Sajjad U, Amer M, Ali HM, Dahiya A, Abbas N (2019) Cost effective cooling of photovoltaic modules to improve efficiency. Case Stud Ther Eng 14:100420
4. Ponnusamy L, Desappan D (2014) An investigation of temperature effects on solar photovoltaic cells and modules. Int J Eng 27(11):1713–1722
5. Dhass AD, Lakshmi P, Natarajan E (2016) Investigation of performance parameters of different photovoltaic cell materials using the lambert-w function. Energy Procedia 90:566–573
6. Kim J, Nam Y (2019) Study on the cooling effect of attached fins on PV using CFD simulation. Energies 12(4):758
7. Marinić-Kragić I, Nižetić S, Grubišić-Čabo F, Čoko D (2020) Analysis and optimization of passive cooling approach for free-standing photovoltaic panel: introduction of slits. Energy Conver Manag 204:112277
8. Hasan IA, Attar D (2019) Effect of evaporative cooling combined with heat sink on pv module performance. J Univer Babylon 27(2):252–264
9. Zhao B, Hu M, Ao X, Xuan Q, Pei G (2020) Spectrally selective approaches for passive cooling of solar cells: a review. Appl Energy 262:114548
10. Fathi M, Abderrezek M, Djahli F (2019) Heating behavior of photovoltaic panels and front side water cooling efficiency. Appl Solar Energy 55(5):327–339
11. Ale TO, Rotipin KJ (2019) Cooling effects on photovoltaic module performance in the tropical region. Niger J Technol 38(3):702–706
12. Durganjali CS, Sudha R (2019) PV cell performance with varying temperature levels. In: 2019 global conference for advancement in technology (GCAT), pp 1–5. IEEE
13. Sopian K, Alwaeli AH, Al-Shamani AN, Elbreki AM (2019) Thermodynamic analysis of new concepts for enhancing cooling of PV panels for grid-connected PV systems. J Ther Anal Calorim 136(1):147–157
14. Kiwan S, Ahmad H, Alkhalidi A, Wahib WO, Al-Kouz W (2020) Photovoltaic cooling utilizing phase change materials. In: E3S Web of Conferences, vol 160, p 02004. EDP Sciences
15. Kumar R, Deshmukh V, Bharj RS (2020) Performance enhancement of photovoltaic modules by nanofluid cooling: a comprehensive review. Int J Energy Res
16. Benato A, Stoppato A (2019) An experimental investigation of a novel low-cost photovoltaic panel active cooling system. Energies 12(8):1448
17. Siecker J, Kusakana K, Numbi BP (2017) A review of solar photovoltaic systems cooling technologies. Renew Sustain Energy Rev 79:192–203
18. Amber KP, Akram W, Bashir MA, Khan MS, Kousar (2020) A experimental performance analysis of two different passive cooling techniques for solar photovoltaic installations. J Therm Anal Calorim. https://doi.org/10.1007/s10973-020-09883-6
19. Khan SA, Bicer Y, Al-Ghamdi SG, Koç M (2020) Performance evaluation of self-cooling concentrating photovoltaics systems using nucleate boiling heat transfer. Renew Energy
20. Luboń W, Pełka G, Janowski M, Pająk L, Stefaniuk M, Kotyza J, Reczek P (2020) Assessing the impact of water cooling on pv modules efficiency. Energies 13(10):2414
21. Esfe MH, Kamyab MH, Valadkhani M (2020) Application of nanofluids and fluids in photovoltaic thermal system: an updated review. Sol Energy 199:796–818

22. El Kharaz H, Khallaki K, Kadiri MS, Choukairy K (2020) Performance's improvement methods of PV solar panel by different cooling systems: A Review of Experimental and Numerical studies (no. 2774). EasyChair
23. Olawole OC, Joel ES, Olawole OF, Ikono UI, Moses C, Oyedepo SO, Fobagboye L (2019) Innovative methods of cooling solar panel: a concise review. J Phys Confer Series 1299(1):012020. IOP Publishing
24. Cotfas DT, Cotfas PA (2019) Multiconcept methods to enhance photovoltaic system efficiency. Int J Photoener
25. Dwivedi P, Sudhakar K, Soni A, Solomin E, Kirpichnikova I (2020) Advanced cooling techniques of PV modules: a state of art. Case Stud Therm Eng 100674
26. Rathour RS, Chauhan V, Agarwal K, Sharma S, Nandan G (2019) Cooling of solar photovoltaic cell: using novel technique. In: Advances in Fluid and Thermal Engineering, pp 521–529. Springer, Singapore
27. Sato D, Yamada N (2019) Review of photovoltaic module cooling methods and performance evaluation of the radiative cooling method. Renew Sustain Energy Rev 104:151–166
28. Kane A, Verma V, Singh B (2017) Optimization of thermoelectric cooling technology for an active cooling of photovoltaic panel. Renew Sustain Energy Rev 75:1295–1305
29. Yazdanifard F, Ameri M, Ebrahimnia-Bajestan E (2017) Performance of nanofluid-based photovoltaic/thermal systems: a review. Renew Sustain Energy Rev 76:323–352
30. Suresh AK, Khurana S, Nandan G, Dwivedi G, Kumar S (2018) Role on nanofluids in cooling solar photovoltaic cell to enhance overall efficiency. Mater Today Proceed 5(9):20614–20620
31. Rajvikram M, Leoponraj S, Ramkumar S, Akshaya H, Dheeraj A (2019) Experimental investigation on the abasement of operating temperature in solar photovoltaic panel using PCM and aluminium. Sol Energy 188:327–338
32. Nižetić S, Čoko D, Yadav A, Grubišić-Čabo F (2016) Water spray cooling technique applied on a photovoltaic panel: The performance response. Energy conver Manag 108:287–296
33. Nižetić S, Marinić-Kragić I, Grubišić-Čabo F, Papadopoulos AM, Xie G (2020) Analysis of novel passive cooling strategies for free-standing silicon photovoltaic panels. J Therm Anal Calorim 1–13
34. Dhass AD, Natarajan E, Ponnusamy L (2012) Influence of shunt resistance on the performance of solar photovoltaic cell. In: 2012 International conference on emerging trends in electrical engineering and energy management (ICETEEEM), pp 382–386. IEEE
35. Kalaiselvan S, Karthikeyan V, Rajesh G, Kumaran AS, Ramkiran B, Neelamegam P (2018) Solar PV active and passive cooling technologies-a review. In: 2018 International conference on computation of power, energy, information and communication (ICCPEIC), pp 166–169. IEEE
36. Senthilkumar M, Balasubramanian KR, Kottala RK, Sivapirakasam SP, Maheswari L (2020) Characterization of form-stable phase-change material for solar photovoltaic cooling. J Therm Anal Calorim 1–10
37. Ali HM (2020) Recent advancements in PV cooling and efficiency enhancement integrating phase change materials based systems–a comprehensive review. Sol Energy 197:163–198
38. Gu W, Ma T, Song A, Li M, Shen L (2019) Mathematical modelling and performance evaluation of a hybrid photovoltaic-thermoelectric system. Energy Conver Manag 198:111800
39. Abdollahi N, Rahimi M (2020) Potential of water natural circulation coupled with nano-enhanced PCM for PV module cooling. Renew Energy 147:302–309
40. Hassan A, Wahab A, Qasim MA, Janjua MM, Ali MA, Ali HM, Javaid N (2020) Thermal management and uniform temperature regulation of photovoltaic modules using hybrid phase change materials-nanofluids system. Renew Energy 145:282–293
41. Moh'd AAN, Mugdadi B (2020) A hybrid absorption/thermo-electric cooling system driven by a concentrated photovoltaic/thermal unit. Sustain Energy Technol Assessm 40:100769
42. Ali HM (2019) Applications of combined/hybrid use of heat pipe and phase change materials in energy storage and cooling systems: A recent review. J Energy Stor 26:100986
43. Bayrak F, Oztop HF, Selimefendigil F (2020) Experimental study for the application of different cooling techniques in photovoltaic (PV) panels. Energy Conver Manag 212:112789

44. Dhass AD, Lakshmi P, Natarajan E (2018) Analysis of performance degradation parameters of the photovoltaic system in Chennai. Int J Pure Appl Mathem 118(20):439–447
45. Dhass AD, Lakshmi P, Natarajan E (2016) Experimental investigation on one kw grid-connected pv system. J Electr Eng 16(4)
46. Wang Y, Gao Y, Huang Q, Hu G, Zhou L (2019) Experimental study of active phase change cooling technique based on porous media for photovoltaic thermal management and efficiency enhancement. Energy Conver Manag 199:111990
47. Stritih U (2016) Increasing the efficiency of PV panel with the use of PCM. Renew Energy 97:671–679

Photovoltaic Maximum Power Point Trackers: An Overview

Ali M. Eltamaly

Abstract The generated power from the photovoltaic (PV) array is a function in its terminal voltage. The relation between the generated power and the terminal voltage of the PV array is called the P–V curve. The point corresponding to the highest generated power in this relation is called maximum power point (MPP). This relation has only one peak in the case of uniformly distributed irradiance over the PV array. Meanwhile, it has multiple peaks in the case of partial shading conditions (PSC). The one with the highest power is called global peak (GP) and the other peaks are called local peaks (LPs). The control system should track this point to improve the efficiency of the PV system by extracting the maximum available power from the PV array. The controller used to track this point is called the maximum power point tracker (MPPT). Traditional MPPTs such as hill-climbing or incremental conductance are adequate to track the MPP in the case of uniform irradiance, but it may stick at one of the LPs in the case of PSC. For this reason an unlimited number of MPPT techniques are introduced in the literature to track this point. This chapter introduces an overview of the PV maximum power point trackers (MPPT) techniques. The classifications of MPPT of the PV system is introduced in detail in this chapter. The operating principles, advantages, and disadvantages of each technique are introduced in detail for famous and important techniques and in brief for the less famous techniques or the techniques that are not showing good performance in tracking the MPP. A comprehensive comparison between these techniques is presented in detail in this chapter. Important recommendations and conclusions are introduced at the end of this chapter to show the advantages and disadvantages of these PV MPPT techniques.

A. M. Eltamaly (✉)
Sustainable Energy Technologies Center, King Saud University, Riyadh 11421, Saudi Arabia
e-mail: eltamaly@ksu.edu.sa

Electrical Engineering Department, Mansoura University, Mansoura, Egypt

K.A. CARE Energy Research and Innovation Center, Riyadh 11451, Saudi Arabia

S. Motahhir and A. M. Eltamaly (eds.), *Advanced Technologies for Solar Photovoltaics Energy Systems*, Green Energy and Technology,
https://doi.org/10.1007/978-3-030-64565-6_6

1 Introduction

Energy is the main support for modern societies and all mankind. The excessive depletion of fossil fuels forces the researchers to explore other sources of energy that will not run out such as renewable energies. Solar energy is the most important source of renewable energy sources, where its cost is reduced over time and became mature technology. Photovoltaic (PV) energy systems are used to convert the sunlight directly into electric energy. Very rapid growth in deploying the PV energy systems where it is increased by 60% in Europe [1] and new annual installations in 2020 reached 142 GW, a 14% rise over the previous year [2]. Moreover, the total generation from solar is about 570 TWh [3]. Many efforts were introduced to increase the efficiency of the PV system which can be translated into a reduction in the cost of energy. Most of these efforts were done on improving the efficiency of the PV cells themselves via improving the materials used for their manufacturing, and the other efforts are introduced to improve the power conditioning circuit used to extract the maximum available electric power from PV systems. Moreover, much work is done in the improvement of the integration of the PV system with an electric utility or with integrating the PV system with renewable or conventional energy sources. One of the most important issues used to improve the efficiency of the PV energy system is the maximum power point tracker (MPPT) unit which will be introduced and discussed in detail in this chapter.

Numerous research works are introduced in the literature to track the maximum power point (MPP) of the PV systems. All these techniques have cons and pros which should be discussed in detail in this chapter. For this reason, many review studies were introduced to discuss these performance characteristics of these techniques. Most of the review works of MPPT are discussing certain categories of this MPPT, review a very limited number of techniques, and leave many other techniques not covered. Based on the present literature, there is no comprehensive work that covers all salient MPPT in operations, performance, implementations, and evaluation. This chapter is introduced to fill this research gap and to shed a light on the performance of different MPPT techniques. With the use of modern soft-computing in MPPT of PV systems, many new algorithms are introduced and most of the authors of these techniques claim that their technique is better than others. For this reason, a comprehensive review study for the most important MPPT techniques should be introduced to help researchers for a better understanding of different MPPT techniques. One of the most recent review works introduced a good review of the techniques that are used to mitigate the effect of partial shading [4]. This paper [4] classified the techniques that have been used to mitigate the partial shading effects into two categories, circuit-based techniques, and MPPT-based techniques. Circuit-based partial shading condition (PSC) mitigation techniques (reconfiguration techniques) will not be covered in this chapter. This paper [4] is reviewed only the MPPT in PSC as one part of the paper and leave the other part for circuit-based PSC mitigation techniques. Moreover, paper [4] classified the circuit-based MPPT techniques into four categories, namely, conventional, soft-computing, hybrid, and other techniques. This paper used all soft-computing techniques as one category as well as all hybrid techniques as one category which will be sub-classified more in this chapter. Another comprehensive review research paper evaluates 17 MPPT and gives a grade for each one [5]. This paper introduced discerptions and evaluations for 20 famous soft-computing MPPT techniques and the evaluation of hybrid between these techniques and traditional MPPT in terms of the convergence time and failure rate. A similar review paper is introduced to introduce an index to evaluate these MPPT techniques [6]. Several types of research introduced an overview of the MPPT techniques introduced in literature [7–20] each one has covered a certain point of view, but there is no one of them comprehensively covers the most important MPPT, especially in PSC.

The rest of this chapter is designed to show the modeling of PV array, and the modeling, performance of PV systems in the case of PSC, and the mismatch losses and generated efficiency calculations in the rest of Sect. 1. Section 2 introduced the classifications of MPPT techniques. Section 3 shows the traditional MPPT techniques discerptions and evaluations. Section 4 shows the different soft-computing PV MPPT and details of their performance analysis and operation. Section 5 introduces the other PV MPPT that are not classified as traditional or soft-computing techniques such as Voltage Window Search (VWS) [21], Search–Skip–Judge (SSJ) [22], and Maximum Power Trapezium (MPT) [23]. Section 6 introduces different types of hybrid PV MPPT that uses two techniques to improve the overall performance of the PV system. The lase section (Sect. 7) is introduced to summarize the conclusions, recommendations, and future work out of this review study.

1.1 Modeling of PV Arrays

The PV array is the largest building block of the PV system which consists of PV panels, then PV modules. The PV modules are consisting of several PV cells connected in series and parallel to produce the required voltage and current from the module. So, the PV cell is the basic unit of the PV systems. The PV cell is consisting of two semiconductor materials from types P and N. The PN junction absorbs the light from the Sun which adds energy to the electrons in this junction enabling it to have enough energy to cross the junction and produce voltage difference between their terminals. The voltage difference between these terminals can produce power when they are connected through an electrical load. The amount of generated power from PV cells depends on the voltage difference, temperature, and irradiance value. Different kinds of semiconductor materials have been used in the fabrication of PV cells, where crystalline silicon PV cells are the most widely used [24]. All the PV cell technologies have the same modeling with different values of parameters that will not affect the general modeling shown in this chapter.

Numerous research works have been introduced in the literature to mathematically model the PV cells [25–29]. The one-diode model is shown in Fig. 1 is widely used in the modeling of most PV cells due to its simplicity and it helps in avoidance of the redundancy that may occur in another modeling of PV cells that have a

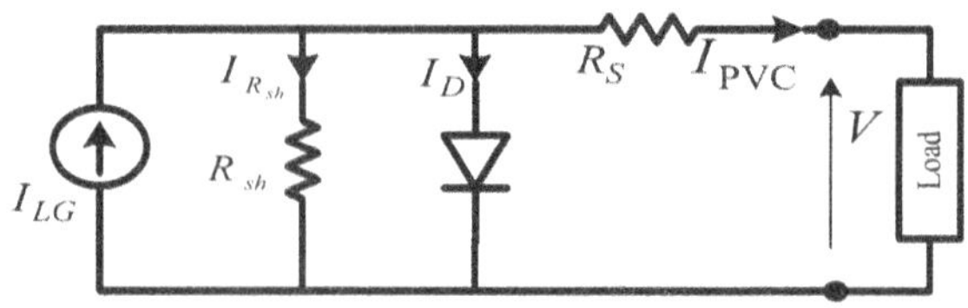

Fig. 1 Equivalent circuit of the PV cell using a one-diode model

higher number of diodes. Moreover, the one-diode model parameters can be easily determined experimentally [25, 28]. The two-diode model has also been used in the literature [26]. This model introduced one more diode to accurately model the PV cells, meanwhile, it will increase the model complexity. Some other researchers introduced a three-diode model to accurately model the PV cell [30]. The one-diode PV cell model is shown in Fig. 1 and is shown in the following equations [24, 31]. The output current generated from the PV cell is shown in (1).

$$I_{\mathrm{PVC}} = I_{\mathrm{LG}} - I_{\mathrm{sat}} * \left[e^{\left(\frac{q}{KT}(V_{\mathrm{PVC}}+R_s I_{\mathrm{PVC}})\right)} - 1 \right] - \frac{V_{\mathrm{PVC}} + R_s I_{\mathrm{PVC}}}{R_{\mathrm{sh}}} \tag{1}$$

where

I_{LG}	The light-generated current for given radiation and temperature.
I_{sat}	The reverse-saturation current.
K	Boltzmann's constant.
q	The electron charge.
V_{PVC}	Terminal voltage of PV cell.
I_{PVC}	Output current of PV cell.
T	The current surrounding temperature.
R_s, R_{sh}	Series and shunt resistors of PV model.

The light-generated current for given radiation and temperature can be obtained from (2)

$$I_{LG} = (I_{\mathrm{STC}} + K_I(T_c - T_r))\frac{G}{G_o} \tag{2}$$

where

I_{STC}	The photovoltaic current at the standard test conditions.
K_I	The short-circuit current coefficient.
G_o	The standard irradiance which is normally taken as 1000 W/m^2.
G	The current radiation in W/m^2.
T_r	The rated temperature in K$^\circ$.
Tc	The cell temperature.

The module voltage can be obtained by (3)

$$V_M = \mathrm{V}_{\mathrm{PVC}} * N_{SC} \tag{3}$$

where N_{SC} is the number of series cells within the module.

The module current can be obtained by (4)

$$I_M = \mathrm{I}_{\mathrm{PVC}} * N_{PC} \tag{4}$$

where N_{PC} is the number of parallel branches within the module.

Connecting several modules in series and parallel is forming the PV array and its voltage and current are determined from the following equations:

$$V_{PV} = V_{\mathrm{PVC}} * N_{SC} * M_S \tag{5}$$

$$I_{PV} = I_{\mathrm{PVC}} * N_{PC} * M_P \tag{6}$$

where M_S is the number of modules connected in series and M_P is the number of modules in parallel.

Multiplying the terminal voltage by the output current determines the generated power from the PV array. The relation between the terminal voltage and current in uniform condition for different irradiances, and the relation between the terminal voltage and output power are shown in Fig. 2. It is clear from Fig. 2 that the PV power is directly proportional to the voltage in the regions, where the voltage less than optimal voltage, V_{opt}, and inversely proportional to the voltage in the region of a voltage higher than V_{opt}. The maximum power, P_{max}, occurs at the value of optimal voltage, V_{opt}. The maximum power tracking techniques are used to track the MPP

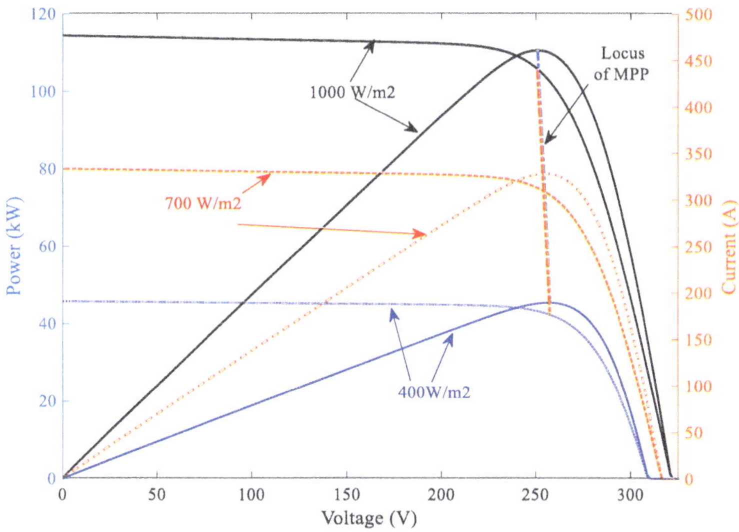

Fig. 2 The I–V and P–V characteristics of PV array

and force the PV array to work at the optimal terminal voltage, V_{opt}. Also, it is clear from the locus of MPP that the MPP lies in a narrow range of voltage.

1.2 Partial Shading Conditions

The partial shading occurred in PV array due to the shading of static and moving objects such as trees, buildings, accumulation of dust on panels, or passing clouds. The PV array characteristic is badly affected and the generated energy is considerably reduced. As has been discussed above, the PV modules should be connected in series and parallel to form the PV array. Due to static or moving objects, shading may be performed on some of these modules and it faces different irradiances than others which are called the partial shading condition (PSC). Due to different irradiances on series modules, the same current should follow through all series modules which makes some modules work as a load on the unshaded modules. Due to the current flow in the shaded PV cell higher than the generated current, the terminal voltage will become negative. Due to this negative voltage, the temperature of the shaded module will be increased especially with a high number of modules connected in series. This high temperature may destroy the shaded modules based on a phenomenon called hot-spot [32]. This condition can be dangerous where it may cause the hot-spot phenomenon on the shaded modules which can destroy the shaded modules, especially when too many modules are connected in series. For this reason, a parallel diode should be attached to each module to bypass the shaded modules when their voltage tends to be reversed to protect these modules from the hot spot phenomenon. Also, each branch should be connected in series with a blocking diode as shown in Fig. 3 to block the flow of current from another branch.

Many comprehensive types of research are introduced to the model, discuss, and to remedy the PSC [33, 34]. Due to the partial shading conditions, the P–V

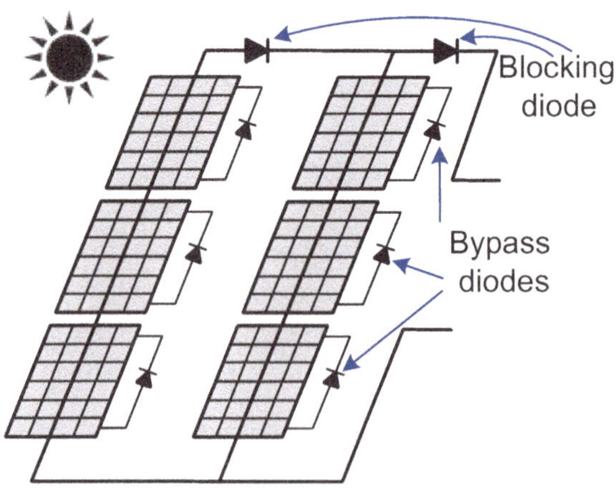

Fig. 3 PV array showing the bypass and blocking diodes connection

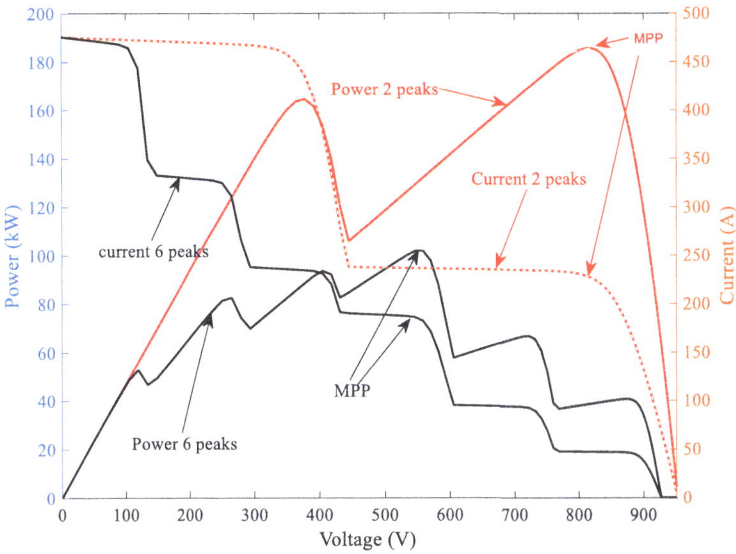

Fig. 4 The I–V and P–V characteristics of the PV array under PSC

characteristics of the PV array is having multiple peaks, the one having the highest power is called global peak (GP) and the other peaks are called the local peaks (LPs). Figure 4 shows the I–V and P–V characteristics with a different number of peaks in the case of PSCs.

It is clear from Figs. 2 and 4 that the generated power is varying with its terminal voltage which forces the designers to use a DC/DC converter at the terminal of the PV system to control this voltage and consequently control the generated power. The control system of the DC/DC converter should ensure that the PV array works at its MPP to increase the generated power and efficiency. The connection of the DC/DC converter can be connected in several configurations as shown in Fig. 5. The first configuration is done by connecting the PV array in many parallel branches and each branch is consisting of many modules in series, which is called "centralized configuration." In the centralized configuration, the PV array has a single terminal and it will be connected to a single DC/DC converter and DC/AC inverter. This configuration is using only one MPPT tracker, meanwhile, the mismatched power is the lowest among the configurations shown in Fig. 5. The other configuration is called the "*multistring configuration*" PV system. In this system, the branches of the PV array are divided among multiple DC/DC converters. This technique has higher efficiency than the centralized PV system because each string is connected to one DC/DC converter and MPPT technique which provides more freedom to each MPPT to work separately in tracking the maximum power available. The third configuration is called "string connection" in which each branch is connected to its own DC/DC converter and the MPPT technique which gives more freedom to the control system to force each branch to work at its own maximum power which increases the generated

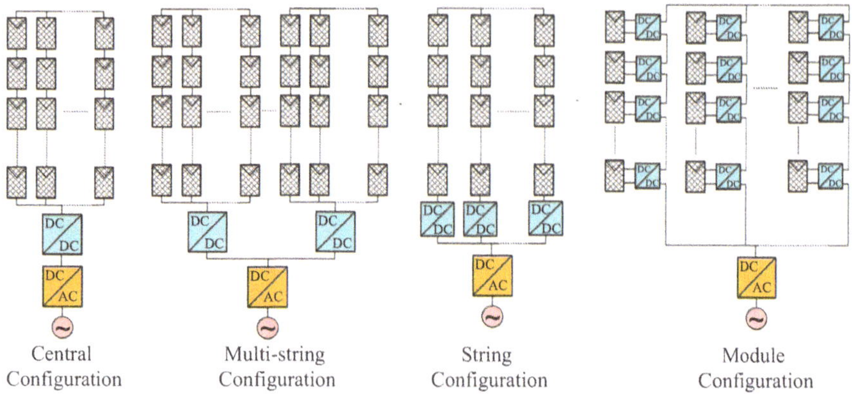

Fig. 5 Different configurations are used to interface PV energy systems to the utility grid

efficiency than the two previous techniques. The DC output can be connected to a common DC link and the inverter/inverters convert this DC power to AC or each DC converter can be connected to a separate inverter. A smart string configuration is introduced in [35, 36] using an interleaved boost converter. In this configuration, each branch is connected to one branch of the boost converter as shown in Fig. 6. In this configuration, one interleaved boost converter is used and one PSO MPPT technique is used with swarm size equal to the number of branches of the boost converter. The results obtained from this configuration is showing higher efficiency than the previous configurations discussed above. The last configuration is called "Module configuration" in which each module is connected to separate DC/DC converter and MPPT module. This configuration is complex and expensive due to the need for the DC/DC converter for each module, meanwhile, it provides the highest freedom to the control system to track the GP of each module which can increase the generation efficiency substantially. Detailed discretions of these configurations are shown in many researches [1, 37, 38].

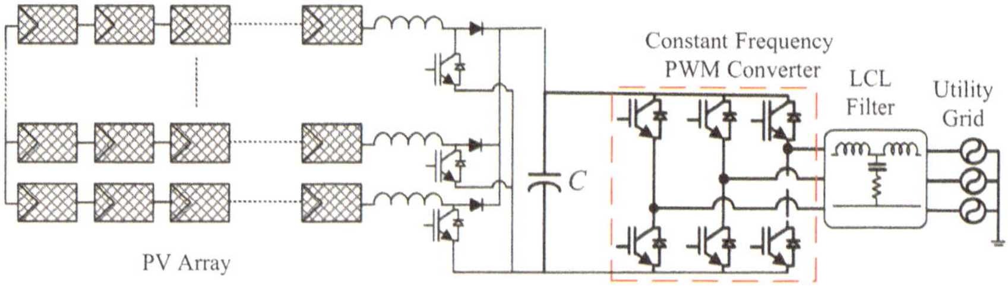

Fig. 6 String configurations used interleaved boost converter

1.3 Mismatch Power Loss

Two different kinds of mismatch occurred in the PV array, static and dynamic mismatch. The static mismatch occurs due to many reasons such as the different tolerance in the module, different aging effects, and different tilt angles of modules. The losses due to static mismatch are in the range from 0.3 to 2.5% [39].

The dynamic mismatch occurs mainly from dynamic partial shading when static or moving objects on the PV array. As discussed before the modules should be connected with bypass diodes and each branch should be connected with blocking diode as shown in Fig. 3 to avoid the hot spot and the possibility of damage to shaded modules. Due to the PSC occurrence, the generated power will not be the same in all parts of the PV array. The generated power will be lower than the sum of the available power that can be generated from a separate PV module even the PV array works at the GP. The relation between the generated power from the PV system and the sum of individual peaks from each module is called mismatch loss (MML). The formula used to determine this relationship is shown in Eq. (7). The higher values of MML mean that the generated power from the PV system is very near to the power available in the PV array and vice versa. This relation is sometimes called MPPT power efficiency (MPE) [40]. In the case of uniform irradiance and the system work at the MPP, the MML value will be 100%.

$$MML = \frac{Maximum\ power\ of\ whole\ PV\ system}{\sum_{i=1}^{N} P\max(i)} * 100 \tag{7}$$

where N is the total number of PV modules in the PV array.

Another evaluation parameter is used to evaluate the MPPT technique called MPPT energy efficiency (MEE). This parameter is used to measure the percentage of PV output energy to the maximum energy available during a certain period of time as shown in (8) [40]:

$$MPE = \frac{\int_{t=0}^{T} P(t)\,dt}{\int_{t=0}^{T} P_m(t)\,dt} * 100 \tag{8}$$

where T is the period of time

2 Classifications of MPPT Techniques

The MPPT techniques have been classified based on different methodologies. Some classifications are based on several variables used to track the MPP of the PV system [41]. Most of the classifications used are based on the use of the module parameters in the MPPT operation to model-based and non-model-based. The model-based

MPPT techniques are done using the model parameters of the PV array to determine the optimal operating model. The model-based techniques are suffering from many problems, especially the low accuracy, the high mathematical burden that can reduce the convergence time, and introduced complexity to the implementation of these techniques. Moreover, the model-based MPPT techniques need extra weather sensors to measure the radiation and temperature. These techniques are not suitable to work with systems facing PSC because it is not practical to have many weather sensors near to each PV module and it will need too much mathematical operation to get the GP in the case of PSC. These techniques are sometimes called offline techniques [41, 42]. An example of offline or model-based techniques is the fractional open-circuit voltage, fractional short-circuit current, curve fitting-based, and numerical calculation-based techniques. The other category of MPPT is the online or non-model-based are included in most of the MPPT techniques. The online-based (non-model-based) MPPT techniques can be further classified into traditional, soft-computing, hybrid, and others. The soft-computing is further classified into chaos, artificially intelligent (sometimes called brain-inspired computing), and metaheuristic techniques. These categories are further classified as shown in Fig. 7.

3 Traditional MPPT Techniques

3.1 Direct Estimated Methodology (DEM)

Directly estimated methodology (DEM) is an offline MPPT methodology that uses the module parameters and an accurate model of the PV array and determines the optimal voltage, $V_{opt,}$ based on the available weather condition (Solar irradiance and temperature) [43]. The control system used the reference value of the voltage to force the PV array to work around this value. The main shortcoming of this technique is the need for four sensors (voltage, current, radiation, and temperature sensors). Moreover, an inaccurate model of PV array parameters or sensors or the effect of degradation on the PV array can produce wrong values of the PV terminal voltage reference V_{opt} which can reduce the system efficiency. In addition to these shortcomings, this technique is not able to track the GP in the case of PSC.

3.2 Fractional Open-Circuit Voltage (FOCV)

As has been shown in Fig. 2 the terminal voltage at the MPP is located around an approximately constant voltage for all operating conditions of the uniformly distributed irradiances. Where the optimal voltage of the PV array is proportional to the open-circuit voltage as shown in (9). This technique can be classified as one of the traditional MPPT techniques and mathematical-based MPPT.

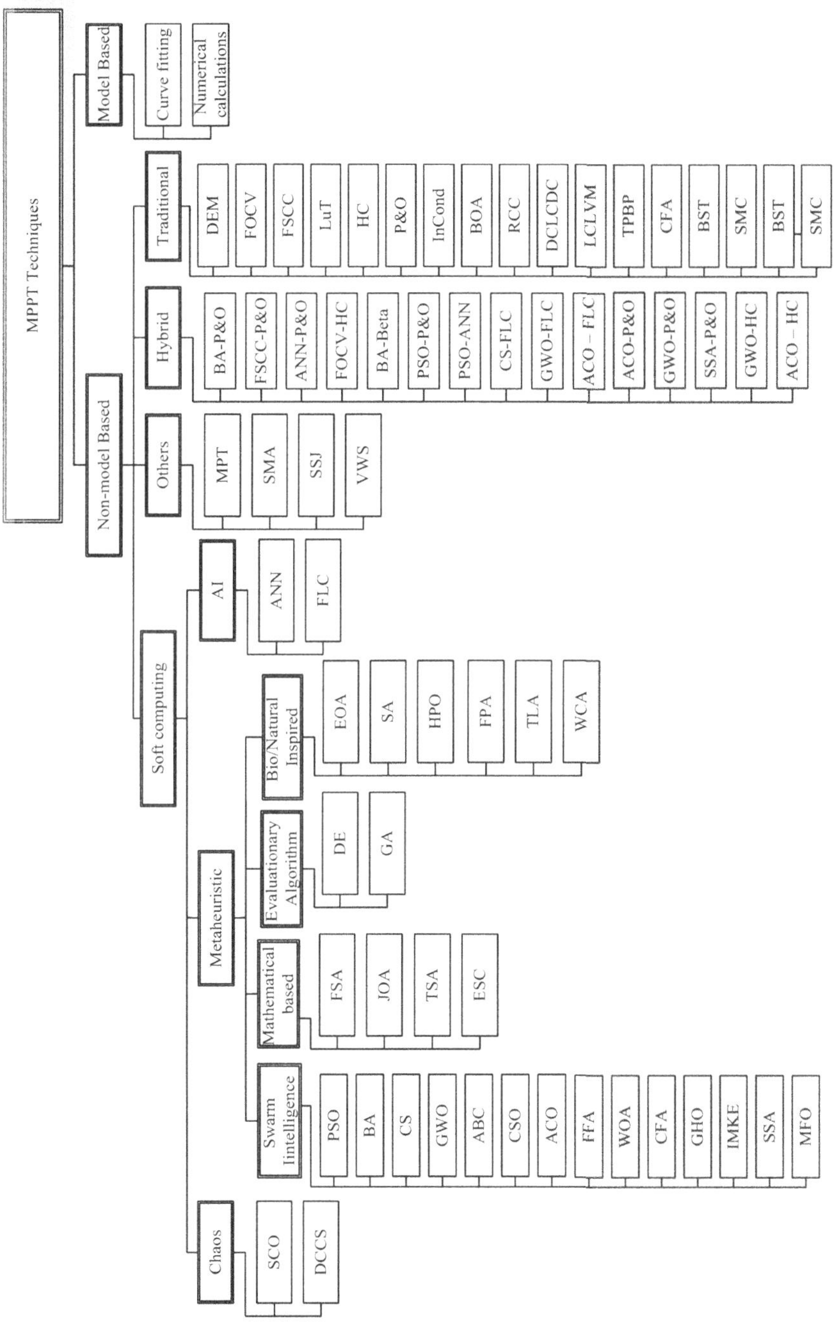

Fig. 7 Classifications of PV MPPT techniques

$$V_{\text{opt}} = k_v * V_{oc} \tag{9}$$

where k_v is a proportionality factor and it has a value between 0.71 and 0.78 [44, 45]; the accurate value of k_v is depending on the PV cell materials and this value can be determined in the lab to be used in the control system. This technique is the simplest and fastest MPPT technique. However, this technique is suffering from many problems which make its use in modern PV systems is very rare. The problems associated with this technique are the need to frequently disconnecting the PV system to measure the open-circuit voltage, the low efficiency, especially in the case of using the inaccurate value of k_v, and the inability to work with PSC.

3.3 Fractional Short-Circuit Current (FSCC)

The locus of MPP on I–V curves shown in Fig. 2 shows that the optimal current, I_{opt}, is linearly proportional to the short-circuit current. The relation between the optimal current, I_{opt}, and short circuit is shown in (10). This technique can be classified as one of the traditional MPPT techniques and mathematical-based MPPT.

$$I_{\text{opt}} = k_i * I_{SC} \tag{10}$$

where k_i is the current proportionality constant, its value is varied between 0.78 and 0.92 depending on the PV cell materials [45].

This technique is very simple and fast (as the fractional open-circuit technique) compared to other traditional MPPT techniques. The main shortcomings associated with this technique are the need to isolate the PV array from the system to perform a short-circuit on its terminals to measure the short-circuit current, the inaccurate values of current proportional constant, the inability to work with PSC. The problem of frequently short-circuit measures on the PV array with this technique can be a complex operation with a very large PV array where the short-circuit current needs special measurement tools and precautions [41].

3.4 Look-up Table (LuT)

This technique is used the module data, weather data, to calculate the voltage required for each operating condition and tabulate these data in a look-up table (LuT). This is a very fast MPPT technique compared to the other traditional MPPT techniques discussed above. The efficacy of the operation of the system is depending on the accuracy of the module parameters, sensor accuracy, and the accuracy of the model used to calculate the MPP. To overcome this shortcoming, the data of the look-up table were collected experimentally [46]. This technique is not favorite in real applications because it needs a control system with big memory size and the need for radiation and temperature sensors. This is one of the offline MPPT techniques that need accurate knowledge about the PV module parameters and characteristics. This technique cannot be used with the PSC which is one of the main shortcomings of this technique [41].

3.5 *Hill-Climbing (HC)*

The most famous traditional MPPT techniques are the hill-climbing and perturb and observe techniques. The main difference between these two techniques is the hill-climbing is using a perturbation in the duty ratio of the DC/DC converter and determines the change in duty ratio based on the change in power. Meanwhile, P&O introduces a perturbation in the terminal voltage of the PV array. This is the only difference between the operation of these two techniques, and for this reason, a detailed comparison between their operation and performance is shown in [47]. This technique needs only the voltage and current sensors. In the hill-climbing technique, when there is a positive increase in the duty ratio produces an increase in power the control system should keep an increase in duty ratio and vice versa. The flowchart of the HC MPPT technique is shown in Fig. 8. The main shortcomings of hill-climbing as most of the traditional MPPT techniques are the inability to capture the GP and the slow response to the fast change in the weather conditions. The problem of missing the GP in the case of PSC can be avoided by hybridizing the hill-climbing technique with other smart techniques to help HC to capture the GP at the beginning of tracking operation and transfer the control to HC to track the maximum power around this

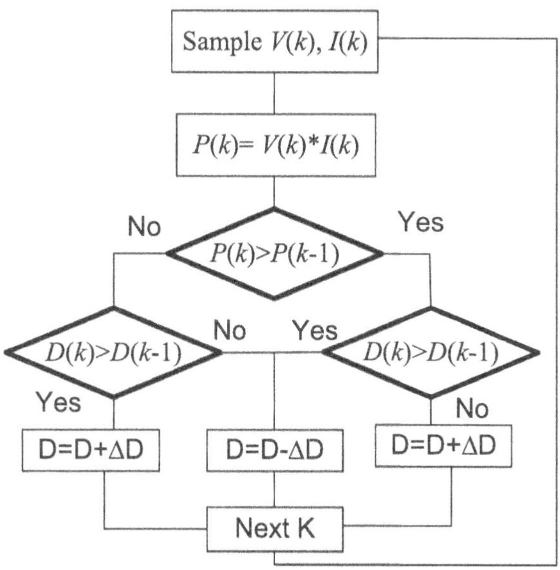

Fig. 8 The hill-climbing flowchart

GP. The problem of slow response can be avoided using the variable step size [48]. Where increased step size (ΔD) increases the convergence speed but it causes high oscillations around the MPP which reduces the generation efficiency and instability. Meanwhile, the low step size has the opposite effect. At the starting of the tracking period or acute change in radiation, the control system needs high step size value to capture the MPP swiftly but this high step size will cause oscillations around the steady state. For this reason, a variable step size technique has been introduced to avoid the sluggishness of the HC in starting and oscillations around the MPP. In this case, the HC uses a high value of step size in the starting or disturbance and low step size at steady state. An adaptive step size HC MPPT used with a boost converter is introduced to determine the optimal step size to reduce the convergence time and reduce the oscillations around the MPP [49].

3.6 Perturb and Observe (P&O)

Perturb and Observe (P&O) method has been used widely in the MPPT of the PV system due to its superior performance and simple implementation. This technique is outperforming the performance operation of HC in terms of convergence time and oscillation around the MPP. This technique needs only the voltage and current sensors. This technique perturbs the terminal voltage reference of the PV array and collects the corresponding power, if the power increased it will move in the same direction otherwise it will change the sign of the perturbation Fig. 9 shows the operation principles of the P&O technique [50]. Many modified strategies have been

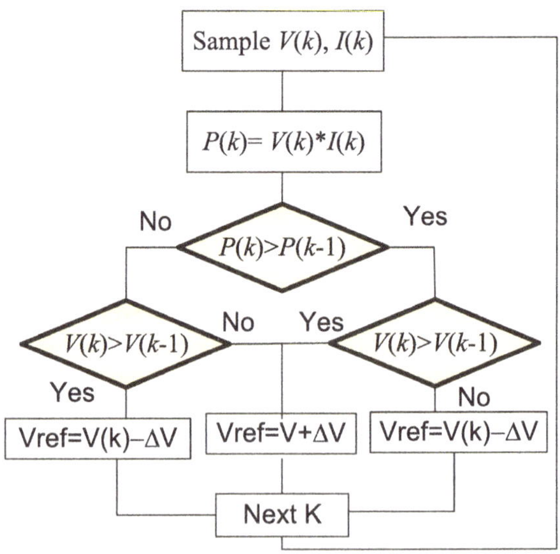

Fig. 9 The P&O flowchart

introduced to this technique to reduce the convergence time and the oscillations around the MPP. One of these modifications is done by using a variable step size [51]. This technique has good performance in uniformly distributed irradiance, meanwhile, it may stick at one of the LPs in the case of PSC. For this reason, many efforts have been introduced in the literature to improve the P&O in the case of PSC. One of these efforts used scanning values of operating voltage and force the normal P&O to work around the one having the highest power [52, 53]. This technique success with a reasonable limit to capture the place of the GP in the case of partial shading but it increases the convergence time.

$$\begin{aligned} V_{\text{ref}} &= V + \Delta V && \begin{matrix} (if\ V > V_{\text{old}}\ and\ P > P_{\text{old}}) \\ (if\ V < V_{\text{old}}\ and\ P < P_{\text{old}}) \end{matrix} \\ V_{\text{ref}} &= V - \Delta V && \begin{matrix} (if\ V > V_{\text{old}}\ and\ P < P_{\text{old}}) \\ if\ V < V_{\text{old}}\ and\ P > P_{\text{old}} \end{matrix} \end{aligned} \tag{11}$$

3.7 Incremental Conductance (InCond)

Most of the shortcomings discussed with HC and P&O techniques are now avoided by using the incremental conductance (InCond), where the convergence time associated with the IncCond is considerably reduced and the dynamic performance of the InCond with rapid change in the weather conditions is substantially improved. Moreover, the oscillation around the MPP of the PV array is substantially reduced too. The high tracking speed, accuracy, and low oscillations at steady state make the InCond is one of the most widely used traditional MPPT techniques. This technique employs the characteristics of the P–V curve of PV array to track the MPP taking into consideration that the MPP is located at zero slopes of the curve. Moreover, the slope of the curve is positive when the operating voltage is lower than the optimal voltage and negative when the operating voltage is lower than the optimal voltage. The logic used in the InCond is to determine the derivative of power concerning voltage as shown in (12) and increment the voltage based on the sign and value of this derivative. The results obtained from (12) can be written as shown in (13). Equating the left-hand side of (13) by error signal e as shown in (14) and trying to minimize this value to become zero will accelerate the convergence to the MPP. The flowchart of InCond MPPT technique is shown in Fig. 10.

The performance of InCond can be further improved in terms of convergence time and oscillations around the MPP by using variable step size as the one used with HC and P&O [54]. Regarding the high failure rate of InCond with PSC, a modified technique is employed several values of duty ratios in starting to scan the position of GP, then transfer the control to the InCond to track the MPP around this value [52, 55].

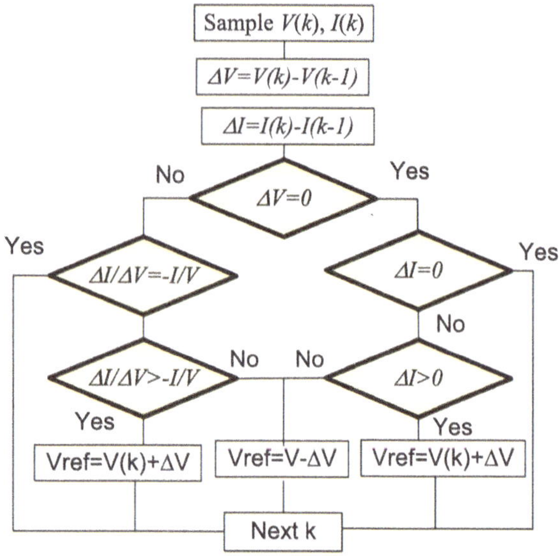

Fig. 10 Flowchart of incremental conductance MPPT technique

$$\frac{dP}{dV} = \frac{d(V \times I)}{dV} = I + V\frac{dI}{dV} = 0 \tag{12}$$

$$\frac{dI}{dV} + \frac{I}{V} = \frac{I(i) - I(i-1)}{V(i) - V(i-1)} + \frac{I(i)}{V(i)} = 0 \tag{13}$$

$$e = \frac{I(i) - I(i-1)}{V(i) - V(i-1)} + \frac{I(i)}{V(i)} \tag{14}$$

3.8 *Beta Optimization Algorithm (BOA)*

This technique uses the characteristics of the PV array to determine β factor that can capture the GP faster than most of the traditional MPPT. This technique is first introduced in 2007 by Jain and Agarwal [56]. The value of beta can be obtained from (15):

$$\beta = \ln\left(\frac{I}{V}\right) - c \times V \tag{15}$$

where c can be determined from (16)

$$c = q/(k \times T \times \eta) \tag{16}$$

where I and V are the terminal voltage and output current from the V array, respectively, q is the electronic charge, k is the Boltzmann's constant, η is the diode quality factor, T is the ambient temperature in Kelvin.

From the PV array model or the actual measurements, the two extreme values of beta, β_{min}, and β_{max} can be determined. The new value of the duty ratio of the DC/DC converter is determined from (17):

$$D(j+1) = D(j) + (\beta_g - \beta_a)\ N \tag{17}$$

where j is the iteration number, β_g is the value of β at the temperature that the PV module will work at it most of the time and it is used to determine the reference or duty ratio corrections, β_a is the actual value of β.

A comprehensive comparison between the beta algorithm and other traditional MPPT techniques is introduced in [57] showed that the beta algorithm has the highest efficiency, the fastest convergence, the lowest transient in the steady state, and has the best overall performance operation compared to the other traditional MPPT techniques.

The beta algorithm is further improved in 2016 [58] by adopting the value of N in (17) to be higher at transient than the steady-state conditions. In the case of a steady state, the control will move to the P&O to reduce the transient at steady-state conditions. This modification further improved the convergence speed and the transient at the steady-state condition which can put the beta algorithm in the best traditional MPPT techniques. The flowchart showing the modification of the beta algorithm is shown in Fig. 11. Although the superior operating performance in capturing the MPP in the case of uniformly distributed irradiances, meanwhile it will not have the ability to capture the GP in the case of PSC. Moreover, this technique needs three sensors (voltage, current, and temperature sensors) which can add a cost to the hardware implementation of this technique.

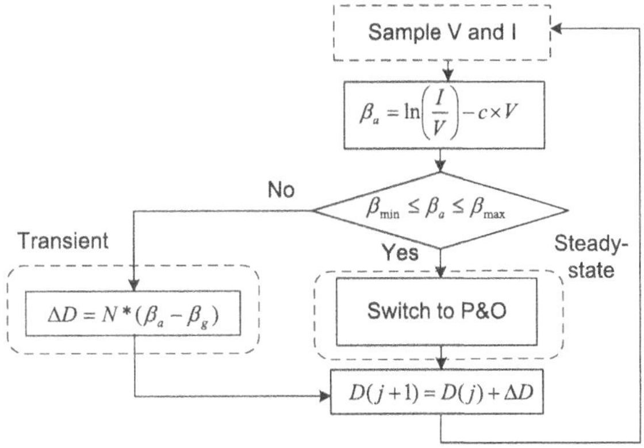

Fig. 11 The flowchart of the beta optimization algorithm for the PV MPPT technique

Fig. 12 The flowchart of RCC PV MPPT technique

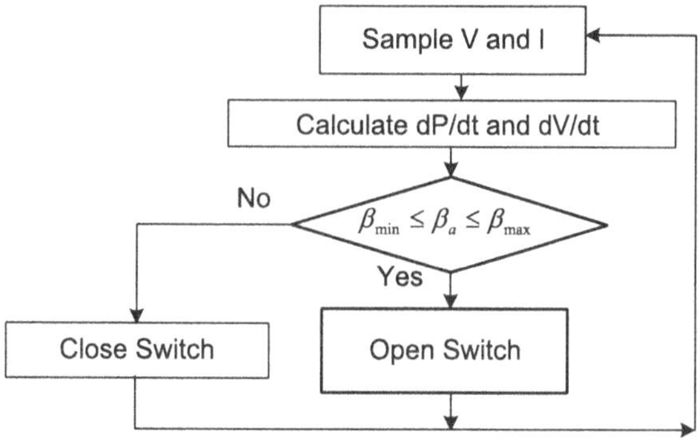

3.9 Ripple Correlation Control (RCC)

The idea used in the ripple correlation control (RCC) is to minimize the time derivative of power and current of PV array to become near to zero. The time variation of power and current near the MPP is zero, so RCC is used the ripples in the power, voltage, and current to become minimum or tend to zero to be sure the control system work at the MPP. This technique is implemented in [59] using analog circuits and is modified to reduce the convergence time in many other types of research. The advantage of this technique is it does not need prior information about the parameters of the PV array which enables it to work with any PV system with any performance characteristics. This technique will not able to capture the GP in the case of PSC. The flowchart showing the logic used in RCC is shown in Fig. 12.

3.10 DC-Link Capacitor Droop Control (DCLCDC)

This technique is designed especially for the PV systems that are integrated with the AC utility grid. This technique depends on maximizing the output power from the DC-link capacitor to the inverter without drooping the DC-link voltage. This can be accomplished by controlling the duty ratio of the DC/DC converter and the power angle and modulation index of the inverter. This technique is used with a boost converter and sine wave PWM inverter in [45, 60]. Like all the traditional MPPT techniques, the DC-link capacitor droop control is not able to capture the GP in the case of PSC.

3.11 Load Current or Load Voltage Maximization (LCLVM)

The idea behind this technique is the capturing depending on maximizing the output power connected to the DC-link of DC/DC converter. This technique has been used with a voltage source and current source converter [61, 62]. In using of voltage source converter, the control system is maximizing the output power through maximizing the output current by controlling the modulation index and power angle as well as the duty ratio of DC/DC converter. In the current source converter, the control system is maximizing the output voltage which can increase the output power. This PV MPPT will not able to capture the true GP because it is assumed that the converters are lossless. Moreover, this technique will not able to work with the PSC because it may stick at one of the LPs.

3.12 Three-Point Bidirectional Perturbation (TPBP)

Three-point bidirectional perturbations based on three-point disturbance observation are utilizing three operating points that work in different duty cycles, using two points to restore a virtual operating point which is the same PV characteristic curve as the rest of the point. In this paper, a novel three-point disturbance observation algorithm is presented based on three specially configured points continuously sampled from the PV array. The points include the current operation point, a point perturbed from the mentioned point, and another point perturbed in the opposite direction from the operation point. The proposed operation mode reduces the losses caused by the oscillation of running the MPPT algorithm [63]. The flowchart showing the three-point bidirectional perturbation (TPBP) is shown in Fig. 13.

3.13 Curve-Fitting Algorithm (CFA)

Curve-fitting MPPT technique is using the PV module parameters and weather conditions to derive third-order curve fitting polynomial as shown in (18). The first derivative of the power shown in (19) is equal to zero at the MPP of the P–V curve. The value of optimal voltage, $V_{\text{opt.}}$ can be determined from (20) [41]. This equation produces two values of optimal voltage; the real value can be easily selected. This technique can be classified as a model-based, offline, traditional, and mathematical-based MPPT technique. This technique also is not able to capture the GP in PSC.

$$P = aV^3 + bV^2 + cV + d \tag{18}$$

$$\frac{dP}{dV} = 3aV^2 + 2bV + c = 0 \tag{19}$$

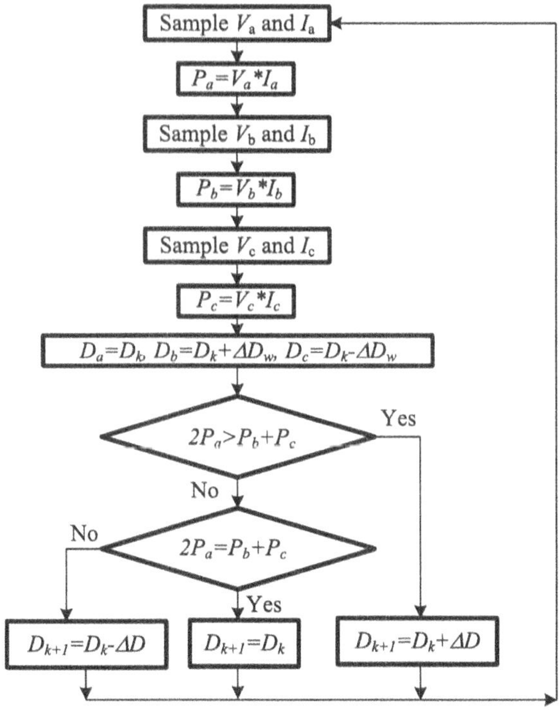

Fig. 13 The Three-point Bidirectional Perturbation (TPBP)

$$V_{\text{opt}} = \frac{-b \pm \sqrt{b^2 - 3ac}}{3a} \tag{20}$$

3.14 Bisection Search Technique (BST)

This bisection search technique is introduced in 2010 [64] to track the MPP of the PV energy system. This technique used the well-known bisection theorem to track the MPP of the PV system.

Assume $y = \Delta P/\Delta D$, it is required to get the duty ratio that has $y = 0$. Three points are selected to start the tracking process $D_a = 0$, $D_b = 0.5$, $D_c = 1$

Then determine the values of y_a, y_b, and y_c from the following Eqs. (21)–(23)

$$y_a = \frac{P(D_a + \Delta D) - P(D_a)}{\Delta D} \tag{21}$$

$$y_b = \frac{P(D_b + \Delta D) - P(D_b)}{\Delta D} \tag{22}$$

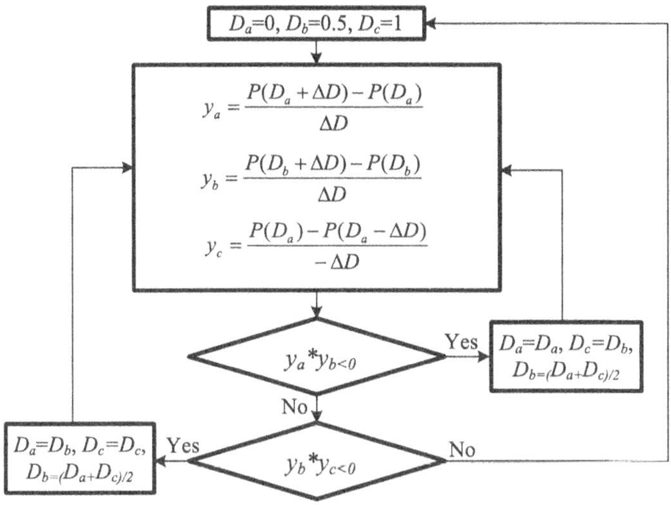

Fig. 14 The flowchart of the Bisection search technique

$$y_c = \frac{P(D_a) - P(D_a - \Delta D)}{-\Delta D} \tag{23}$$

Then check if $y_a * y_b < 0$, then $D_a = D_a$, $D_c = D_b$, and $D_b = (D_a + D_c)/2$,

Else if $y_b * y_c < 0$, then $D_a = D_b$, $D_c = D_c$, and $D_b = (D_a + D_c)/2$,

Else, (This means that all of them (y_a, y_b, y_c) have the same sign due to acute change in the radiation, and in this case, the system should start from the beginning. The flowchart of the BST is shown in Fig. 14. The value of ΔD should be chosen carefully, where a large value may capture the MPP faster but it will have oscillations around the steady state and vice versa. It is recommended to be used about $\Delta D = 0.01$ in [64].

3.15 Slide Mode Control (SMC)

Sliding mode control theory is used in the application of PV MPPT of PV systems [65]. This technique used $\Delta P/\Delta V$ to switch on and off the DC/DC converter. The value $\Delta P/\Delta V$ can be obtained from (24) [65]. The DC/DC converter used in this study is a buck converter. Based on the value of $\Delta P/\Delta V$, the DC/DC converter will be switched on and off based on the condition shown in (25).

$$\Delta P/\Delta V = I + (\Delta I/\Delta V) * V \tag{24}$$

$$S = \begin{cases} 0 & \Delta P/\Delta V \geq 0 \\ 1 & \Delta P/\Delta V < 0 \end{cases} \tag{25}$$

where I, V, P are the current, voltage, and power output from PV array.

Another research [65] is used a full-bridge single-phase PWM inverter to directly track the MPP of the PV system and to convert the DC power from the PV system directly to AC power that can be connected to the utility grid. This technique showed a very fast convergence time but it will not have the ability to trach the MPP in the case of PSC.

3.16 Transient-Based MPPT (TBM)

This PV MPPT is introduced in 2009 [15] by using a single-stage inverter single or three-phase converter. In this technique, the control system determines the maximum and minimum voltage, V_{max}, V_{min}, respectively. The control system samples the change in current and, if this change is positive, it forces the voltage to $V_{max,}$ otherwise reference voltage to V_{min}. This technique has very fast convergence, meanwhile, it suffers from many disadvantages such as the high transient around steady-state conditions and its inability to work with the PSC. A detailed discerption of this technique is shown in [15, 41].

3.17 Current Sweep MPPT (CSM)

This technique is depending on sweeping the current of PV array through the terminal capacitor and using these values of current to determine the voltage and power at MPP [66]. The mathematical modeling of this technique is performed based on that, the current function obtained from the current sweep is proportional in its derivative as shown in (26) [41].

$$f(t) = k\frac{df(t)}{dt} \tag{26}$$

where k is the constant of proportionately.

Applying the above equation to determine the time derivative of power as shown in (27).

$$\frac{dP(t)}{dt} = \left(V(t) + k\frac{dV(t)}{dt}\right)\frac{df(t)}{dt} = 0 \tag{27}$$

The solution of the above differential equation is shown in the following equation:

$$f(t) = I_{\max} e^{(kt)} \tag{28}$$

The optimal current can be determined from the above equation. The optimal voltage can be determined from the following equation:

$$\frac{dP(t)}{dt} = \left(V(t) + k\frac{dV(t)}{dt}\right)\frac{dI(t)}{dt} = 0 \tag{29}$$

3.18 Comprehensive Comparison Between Traditional MPPT Techniques

After discussing the traditional MPPT techniques in the above chapter, it has been listed in the following Table 1 for the purpose of comparison.

Table 1 A comprehensive comparison between traditional MPPT techniques

No.	MPPT technique	Convergence speed	Tracking efficiency	Oscillations at steady-state	No of sensors	Implementation complexity	PSC MPPT ability
1	DEM	High	Low	Low	3	Medium	NO
2	FOCV	High	Low	Low	3	Low	NO
3	FSCC	High	Low	Low	3	High	NO
4	LuT	High	Low	Low	3	High	NO
5	HC	Low	Low	High	2	Low	NO
6	P&O	Low	Low	High	2	Low	NO
7	InCond	High	High	Low	2	Medium	NO
8	BOA	Medium	Medium	Low	3	Medium	NO
9	RCC	Medium	Medium	Medium	2	Medium	NO
10	DCLCDC	Medium	Medium	Medium	2	High	NO
11	LCLVM	Medium	Medium	Medium	2	High	NO
12	TPBP	Medium	Medium	Low	2	Medium	NO
13	CFA	High	Low	Low	2	Medium	NO
14	BST	Low	High	Low	2	Low	NO
15	SMC	High	High	Low	2	Medium	NO
16	TBM	Medium	Medium	Medium	2	Medium	NO
17	CSM	Medium	High	Medium	2	High	NO

4 Soft-Computing MPPT Techniques

Soft-computing techniques are classified into three different categories as has been shown in Fig. 7 to four different techniques. These techniques are listed in the following points:

- Artificial Intelligent (AI)
- Metaheuristic Algorithms (MA)
- Chaos optimization algorithms (COA).

4.1 Artificial Intelligent (AI) MPPT Techniques

Two types of artificial intelligent techniques have been introduced in this chapter to work as an MPPT of PV systems. These two techniques are the fuzzy logic controller and an artificial neural network.

4.1.1 Fuzzy Logic Controller (FLC)

Fuzzy logic controller (FLC) is one of the soft-computing techniques that has been used as MPPT of PV systems [67–69], as well as in the motor drive control and renewable energy applications [69–73]. This technique is one of the most important PV MPPT techniques because it is a very fast convergence and it has very low oscillations in steady-state conditions. The fuzzy logic controller has one more advantage where it does not need accurate inputs measure or accurate PV array modeling. The operation of FLC is consisting of three parts, *fuzzification*, *Aggregation*, and *defuzzification*. In the fuzzification stage, the input variables are defined as a membership function. Moreover, linguistic relations (rules) between input and output is introduced in this part. The aggregation stage is done by combining the output fuzzy sets of each rule to perform one output fuzzy set. The defuzzification stage is done by defuzzifying the fuzzy set into crisp output. The use of FLC is introduced in many studies and it has been used separately or with other MPPT as will be discussed in the hybrid MPPT section of this chapter. The operation of PV MPPT using FLC is done by calculating the change of power divided by the change in voltage which is called the error signal as shown in (30). The value of change of error, ΔE is defined as shown in (31).

$$E(n) = \frac{P(n) - p(n-1)}{V(n) - V(n-1)} \tag{30}$$

$$\Delta E(n) = E(n) - E(n-1) \tag{31}$$

The error function and change of error that can be obtained from (30) and (31), respectively, should be expressed based on labels such as; PB (Positive Big), PM

(Positive Medium), PS (Positive Small), ZE (Zero), NS (Negative Small), NM (Negative Medium), NB (Negative Big) using a basic fuzzy subset. These linguistic variables are modeled in a mathematical membership function. The error function, E, and change of error, ΔE are two input functions in the FLC as shown in Fig. 15 [68]. In the same figure, the output will be the change in duty ratio, ΔD which will be expressed as membership in the FLC output which will be added to the old duty ratio to determine the new duty ratio to control the DC/DC converter. Many shapes of membership functions can be used to express the input and output variables, where triangle membership functions are used as shown in Fig. 17. Some researches proportionate these variables to only five fuzzy linguistic variables as shown in [74]. Table 2 shows the linguistic variables that can be translated into 7*7 fuzzy rules that can describe the logic of control as shown in the following:

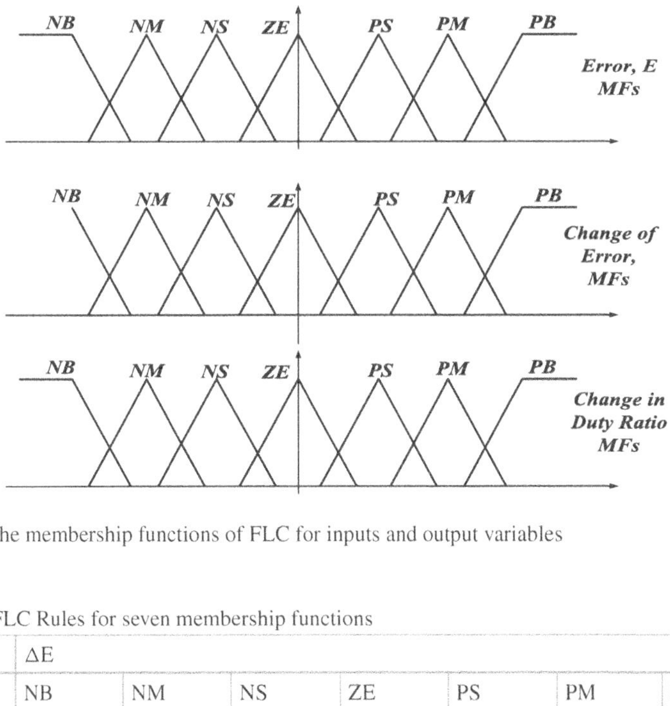

Fig. 15 The membership functions of FLC for inputs and output variables

Table 2 FLC Rules for seven membership functions

E	ΔE						
	NB	NM	NS	ZE	PS	PM	PB
NB	NB	NB	NB	NB	NM	NS	ZE
NM	NB	NB	NB	NM	NS	ZE	PS
NS	NB	NB	NM	NS	ZE	PS	PM
ZE	NB	NM	NS	ZE	PS	PM	PB
PS	NM	NS	ZE	PS	PM	PB	PB
PM	NS	ZE	PS	PM	PB	PB	PB
PB	ZE	PS	PM	PB	PB	PB	PB

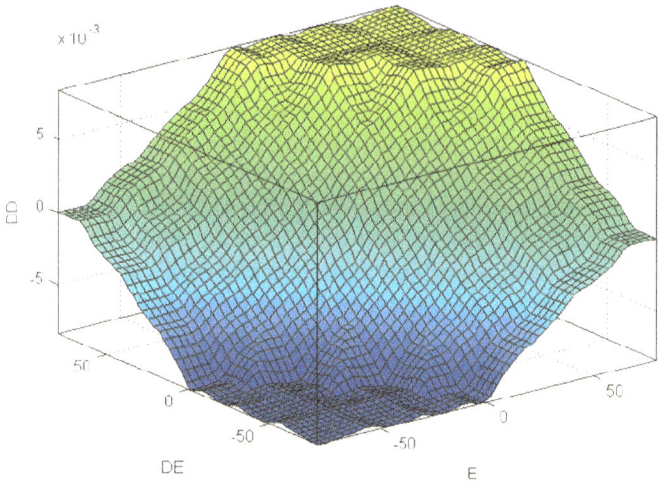

Fig. 16 FLC 3D Surface function

$$R_{25}: \; If \; E \; is \; NM \quad and \; \Delta E \; is \; PS \; then \; \Delta D \; is \; NS$$
$$R_{63}: \; If \; E \; is \; PM \quad and \; \Delta E \; is \; NS \; then \; \Delta D \; is \; PS$$
$$\ldots$$
$$R_{51}: \; If \; E \; is \; PS \quad and \; \Delta E \; is \; NB \; then \; \Delta D \; is \; NM$$

During the defuzzification stage, the output from the rules should be converted to numerical values using the output membership function. This value in the output is the change in the duty ratio, ΔD that should be added to the old duty ratio of the DC/DC converter. The height of the defuzzification can be obtained from (32) to determine the numerical value of change in duty ratio ΔD [69].

$$\Delta D = \left(\sum_{k=1}^{m} c(k) * W_k \right) / \sum_{k=1}^{n} W_K \quad (32)$$

where $c(k)$ is the peak value of each output membership function.

W_k = height of rule k, where is k = 1,2, …49.

The surface function 3-D drawing is a drawing representing the relation between the inputs and the output of the fuzzy controller is shown in Fig. 16. The surface function should be smooth to enhance the stability of the FLC.

4.1.2 Artificial Neural Network (ANN)

The artificial neural network (ANN) is a soft computing technique that has been used in many applications. This technique models the performance operations of

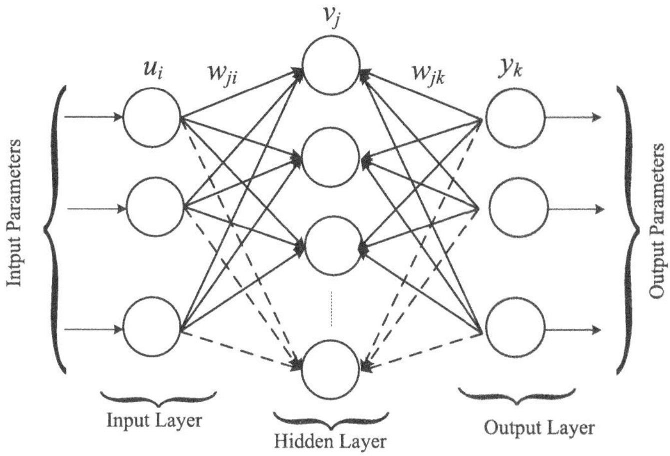

Fig. 17 The structure of the neural network

biological neural systems into a mathematical system. ANN requires so many careful training processes to enable it to learn how the system reacts to different inputs.

The ANN is used as a PV MPPT by getting accurate results including the solar radiation and temperature as input parameters and the optimal voltage or duty ratio as output parameters. The data can be collected mathematically from the model of the PV array or from the use of other MPPT in actual life to collect the input and output parameters to train the ANN and benefit from its fast response. Both data collections are not accurate because the model may be different from the actual array due to different tolerance and aging reasons. Also, the real-world data are taking an effort to collect these data and time. Despite the superiority of ANN in many applications, it is not gain the same attention when it is used as an MPPT of PV system due to many problems inherited in this application. One of these problems is the need for a higher number of good data and its inability to be used in PSCs. A lot of modifications have been introduced in the literature to improve the performance of ANN when it is used as an MPPT of the PV systems. One of these modifications is to use the results obtained from ANN (optimal voltage or optimal duty ratio) and after that, it will transfer the control to the InCond technique for accurately track the MPP [75]. The structure of the neural network is shown in Fig. 17.

4.1.3 A Comprehensive Comparison Between Artificial Intelligent (AI) MPPT Techniques

A comprehensive comparison between Artificial Intelligent (AI) MPPT Techniques is shown in Table 3.

Table 3 Comprehensive comparison between Artificial Intelligent (AI) MPPT techniques

No.	MPPT technique	Convergence speed	Tracking efficiency	Oscillations at steady-state	No of sensors	Implementation complexity	PSC MPPT ability
1	FLC	High	High	Low	2	High	NO
2	ANN	High	Medium	Low	3	High	NO

4.2 *Metaheuristic Algorithms (MA)*

Metaheuristic MPPT Techniques can be classified as shown in Fig. 7 into four different categories which are listed in the following points

- Swarm Intelligence Algorithms (SIA)
- Bio/Natural-Inspired Algorithms (BNIA)
- Evolutionary Algorithms (EA)
- Mathematical-Based Algorithms (MBA).

4.2.1 Swarm Intelligence Algorithms (SIA)

Particle Swarm Optimization (PSO)

The particle swarm optimization (PSO) is one of the best swarm optimizations that mimics the behavior of animals, birds, or fish in searching for their food. This technique is introduced in 1995 by Kennedy and Eberhart [76]. The PSO is a stochastic evolutionary optimization technique that uses several searching agents to look for optimal solutions. This technique uses the best optimal values as a social or cognitive experience and the best value for each particle as a private best experience.

The idea behind using the PSO in tracking the MPP of the P–V curve is done by sending a certain number of particles (swarm size) each one is having a certain value of duty ratio of DC/DC converter one by one to the PV system and collect the corresponding power. In many papers [77–81], the DC/DC converter used in the PV system was a boost converter but any other type of DC/DC converters can be used. The particle position, *D*, and the value, *P*, are used to determine the new position of particles using the PSO equation obtained from (33) to (34). Consecutive iterations will be used to control the movement and position to capture the GP. The new position of particles in each iteration depends on their previous position and values and social and private experiences. The movement of each particle is obtained from (33) and the new position D_{j+1}^k is equal to the previous position D_j^k plus the new movement, v_{j+1}^k.

The values of the PSO control parameters ω, c_l, and c_g substantially affect the performance of PSO in terms of convergence time, failure rate, and oscillations around the global best value. Tuning these parameters is very important to get the best performance or by using the previous experience of previous researches [79–81].

During the initialization of the PSO when it is used as an MPPT of the PV system, the particle associated with the highest power is assigned to the global best, value, PG_{best}, and position, G_{best}. Moreover, the particles' private best values, P^k_{best} and positions, D^k_{best} are equated to the particle's value and position of the initialization. The initial speed is set to zero value. The flowchart showing the logic used in the use of PSO as an MPPT of the PV system is shown in Fig. 18. The steps of the operation of PSO as an MPPT of PV systems are introduced in detail in [77–82].

$$v^k_{j+1} = \omega v^k_j + c_l r_l (D^k_{\text{best}} - D^k_j) + c_g r_g (G_{\text{best}} - D^k_j) \tag{33}$$

$$D^k_{j+1} = D^k_j + v^k_{j+1} \tag{34}$$

where j is a counter representing the iteration number that states from 1 to the maximum number of iterations, *it*. ω, c_l, and c_g are called the PSO control parameters, D^k_{best} is the personal best position of the particle k, G_{best} is the global best position, r_l and r_g are random values in between [0, 1].

Despite the superiority of using PSO as an MPPT of the PV system, it has many shortcomings and all of these shortcomings have been solved in literature. The following points are showing these shortcomings and how they are solved in literature. Most of these shortcomings in the PSO are occurring in most of other swarm optimization techniques, and for this reason, it will be discussed for PSO in detail to be as guidance for other swarm optimization techniques. The PSO also has been used in optimal sizing and allocations of hybrid renewable energy systems and distributed generation [83–93].

(a) **The problem of long convergence time and high failure rate**

There are many reasons to participate in this problem such as the random initialization of particles, this problem is solved by initializing the particles at the anticipated position of peaks [77]. The position of the anticipated peak can be determined from (35) [77]. Another technique is used by uniformly distribute the initial positions of particles within the searching space as shown in (36) [79]. Initializations of particles at positions of anticipated peaks [77] or at equal distance in the searching space [79] reduced the convergence time by more than 50% and reduced the failure rate to zero [79].

The swarm size can substantially affect the convergence time and failure rate, where the high value of swarm size can prolong the convergence time and reduces the failure rate and vice versa. This trade-off effect forces the researchers to look for the optimal value of swarm size which has been accomplished in [94] for PSO and BA when it is used as an MPPT of the PV system. This paper [94] introduced the optimal value of swarm size against the number of peaks in the P–V curve for minimum convergence time and failure rate.

$$D^k_0 = 1 - \frac{(k - 1 + k_v)}{SS} * \frac{V_{oc}}{V_{DC}} \tag{35}$$

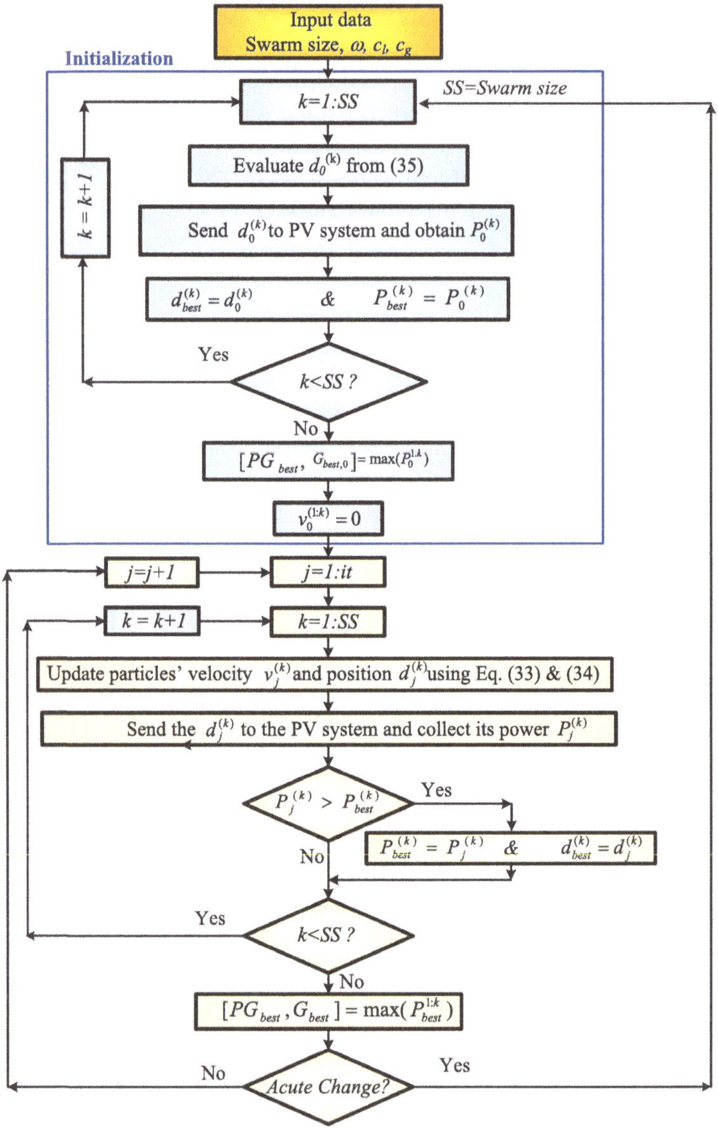

Fig. 18 The flowchart of using PSO as an MPPT of PV systems

where n is the total number of particles and i is the particle's order.

$$D_0^k = k/(SS+1) \tag{36}$$

where D_0^k is the k-th initial particle position (duty ratio), k is the counter used to represent the number of the particle in the swarm ($k = 1,2,\ldots SS$).

The unwise selection for PSO control parameters (ω, c_l, c_g) has a substantial effect on the convergence time and failure rate. Numerous researches have been introduced to improve the performance of PSO by modifying the values of control parameters. One of these efforts is done by tuning the PSO control parameters for minimum failure rate and convergence time. Another work uses linear decreasing PSO control parameters as shown in (37)–(39) [95]. Another research used a modified PSO PV MPPT control under PSC with a Gaussian particle swarm optimization method [96] to improve the performance of PSO in terms of fast and reliable convergence. Another research work used deep recurrent neural networks trained from the results obtained from PSO to improve the performance of PSO in terms of fast and reliable convergence [97]. Another work used an adaptive perceptive particle swarm optimization (APPSO) [98] technique for the same purpose. A review of different techniques used to improve the performance of PSO in terms of convergence time and failure rate when it is used as an MPPT of the PV system is introduced [99].

Another technique is introduced in [80] called scanning PSO technique, in which the control system sends a certain number of duty ratios to the PV system and collects the corresponding power. Then the duty ratio associated with the highest value of power will be selected to initialize the PSO particles to be around this optimal value. After that, the PSO will continue tracking this MPP. Actually, this is one of the fastest and highest reliable MPPT techniques where it captured the GP effectively within 0.4 s [80].

Another research paper is introduced to improve the performance of PSO when it is used in tracking the MPP of the PV system in PSC by removing the random number in the acceleration parameters of the conventional PSO velocity equation and adding a maximum allowable change in the velocity [100]. This strategy is called "The deterministic PSO (DPSO)." This strategy captured the GP with a lower number of particles in a short time. Moreover, it has only one parameter needs tuning which is the inertia weight. The only shortcoming in this technique is its need for reevaluation on different types of PSCs and systems with different numbers of peaks.

$$\omega_j = \omega_{\max} - \frac{j}{J_{\max}}(\omega_{\max} - \omega_{\min}) \tag{37}$$

$$c_{l,j} = c_{l,\max} - \frac{j}{J_{\max}}\left(c_{l,\max} - c_{l,\min}\right) \tag{38}$$

$$c_{g,j} = c_{g,\max} - \frac{j}{J_{\max}}\left(c_{g,\max} - c_{g,\min}\right) \tag{39}$$

where $J_{\max}$ is the maximum number of iterations, $\omega_{\max}$ and $\omega_{\min}$ are the highest and lowest value of inertia weight $c_{l,\max}$ and $c_{l,\min}$.

(b) **The need for reinitialization**

When all the particles are concentrated at the GP, the shading pattern may change and the GP may become in the other place. In this case, the particles will not able to capture the GP and they will continue around the previous GP. In this case, the generated power will not be the maximum available power because the GP is in another position. This problem can be avoided by reinitializing the particles when an acute change is detected. The condition that is used to detect the acute change is shown in (40). The predefined tolerance, ε is chosen between 5 and 10% [77–79]. In most of the papers, it is used with 5% where the lower value of the predefined tolerance will cause a reinitialization without a need for that and higher values of the predefined tolerance will not initialize the particles in a reasonably acute change in shading conditions.

$$\left|\frac{P_{\text{new}} - P_{\text{old}}}{P_{\text{old}}}\right| \geq \varepsilon \tag{40}$$

where P_{new} and P_{old} are the output powers captured from the PV system in the current and previous iterations, respectively. ε is the allowable power change limit that has been assumed as 5% of the old power captured.

Bat Algorithm (BA)

The bat algorithm (BA) is one of the swarm techniques that imitates the performance of bats in searching for their food. The BA is first developed in 2010 by Yang in 2010 [101].

The mechanism that the bats used in nature to track a prey are by emitting several impulses with different frequencies and amplitudes and receives the echo of these sound pulses and transfer these data to useful information to decide the next step toward the prey. The time difference between the transmitted pulse sound and the received echo represents the distance between the bat and the prey. The bats can identify the size of the prey by measuring the intensity of the echoed sound pulses. Moreover, bats can evaluate the moving speed and direction of the prey by tuning the frequency difference. In nature, the bats emit short-duration sound pulses around 10–100 times per second [101]. The searching behavior of the bats has inspired the researchers to imitate it in searching for the optimal solution for different life problems. Many generalized rules should be taken into consideration in the mathematical modeling of the BA. The following sections explain the logic of using the BA as an MPPT of the PV systems. The flowchart showing the logic of BA when it is used as an MPPT of the PV system is shown in Fig. 19.

Although the superiority of the BA compared to the PSO or any other swarm optimization techniques, it did not get its deserved weight in the MPPT of the PV systems applications where only a couple of researches have been introduced in the literature [102–107]. For this reason, the BA has been discussed deeply in this chapter with detailed performance characteristics.

BA Initializations

The initialization of bats should get their values from (35) or (36) to reduce the convergence time and failure rate compared to the random initialization of bats in

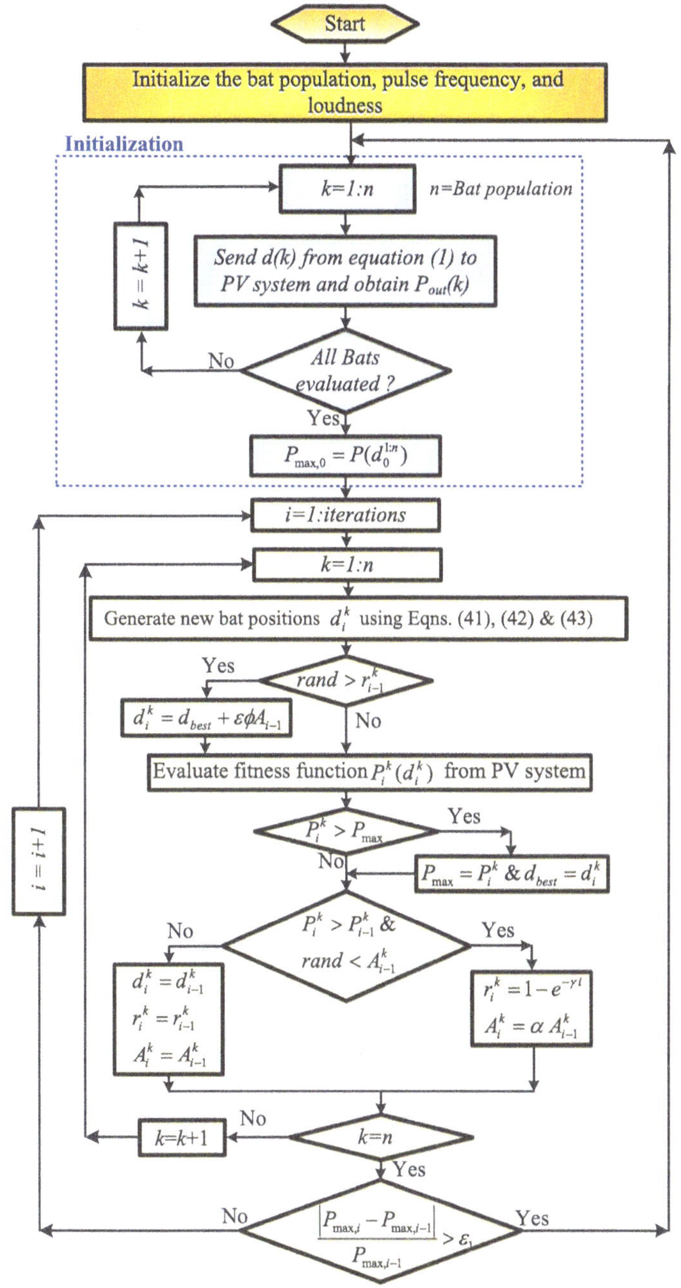

Fig. 19 The flowchart of the modified BA strategy used in tracking the GP of PV systems

the conventional BA when it is used for tracking the maximum power of PV systems [94, 108]. These results are introduced and discussed in [94, 108] When it is compared with random initialization. The BA is used as an MPPT of the PV system by giving the particles the initial values of duty ratios that can be obtained from (35) or (36) with initial frequency obtained from (41).

The initial velocity $v_0^{1:n}$ and initial frequency $f_0^{1:n}$of all bats are set to zero (where n is the swarm size). The initial values of pulse rate, r_O, loudness, A_O, and many initialization parameters are set to different values in the state-of-the-art strategies of BA which is discussed in detail in the simulation results section. The initial values of bats that can be determined from (35) or (36) will be used to start the boost converter where it will be sent to it one by one and the corresponding power $P_0^{1:n}$will be collected after waiting for the sampling time to get the steady state from the boost converter. The best value of maximum power is determined from as $P_{best} = \max\left(P_0^{1:n}\right)$ and the corresponding duty ratio d_{best}will be determined.

Global Peak Tracking using BA

The equations used to mimic the behaviors of bats are shown in Eqs. (41)–(43) where the impulse frequency is shown in (41) which will be used in (42) to determine the bats' velocities $v_i^{1:n}$. The new positions of bats $d_i^{1:n}$ can be obtained as shown in (43) by adding this velocity to the previous positions of bats.

$$f_i^{1:n} = f_{\min} + (f_{\max} - f_{\min})\beta \tag{41}$$

$$v_i^{1:n} = \omega\, v_{i-1}^{1:n} + \left(d_{\text{best}} - d_{i-1}^{1:n}\right) f_i^{1:n} \tag{42}$$

$$d_i^{1:n} = d_{i-1}^{1:n} + v_i^{1:n} \tag{43}$$

where the values of$f_{\min}$ and$f_{\max}$ are the minimum and maximum frequency ranges, respectively. The values have been chosen from [101] to be 0 and 2, respectively. β is a random value, $\beta \in [0, 1]$, as the case of PSO, the velocity of bats is multiplied by inertia weight value, ω which is used to enhance the searching stability of particles.

After determining the new position from (43), a random walk around this position should be performed to get the new position of the bats as shown in (44) [94, 108]. If the pulse emission r_i less than a random number, then the duty ration position d_i should be replaced with values shown in (44) which is a representation of a random walk around the best solution.

$$d_i^{1:n(new)} = d_{\text{best}} + \varepsilon\varphi\left\langle A_i^{1:n}\right\rangle \tag{44}$$

where ε is a random number, $\varepsilon \in [-1, 1]$, and ϕ is used to give stability or limitations to the number walk around the best solution, $\left\langle A_i^{1:n}\right\rangle$ is called the average loudness of each bat and its value equal to the average of A constant in the previous iterations, k from the beginning ($i = 1$) to the current iteration.

The value of the loudness (A_i) of the impulse should start from high-value A_0 and should be decreased as shown in (45), where it starts at 0.999 and should be decreased to 90% of its previous value. The value of r_i is called the rate of pulse transmission. The values of r_i are started at lower value $r_0 = 0$ and it increased exponentially to the end value $r_i = 1$ as shown in (46).

$$A_i^{1:n} = \alpha A_{i-1}^{1:n} \tag{45}$$

$$r_i^{1:n} = r_0^{1:n} [1 - e^{(-\gamma i)}] \tag{46}$$

where the values of α and γ have been chosen equal to 0.9 in many types of research [94, 108].

After determining the new positions of bats $d_i^{1:n}$ it will be used as a duty ratio of boost converter to control the terminal voltage. These values of duty ratio will be fed to the boost converter one by one and wait for the sampling time between each entry. The generated power for each duty ratio will be collected $P_i^{1:n}$ and the maximum value of power $P_{\max}$ and its corresponding duty ratio d_{best} can be determined.

The control system will send the new values of the duty ratios, $d_i^{1:n}$ to the PV system and will collect the corresponding power for each duty ratio, $P_i^{1:n}$. The maximum power collected from the PV system will be compared with the global best power to update the value of global best if the new power is greater than its value as shown in the following:

For $k = 1{:}\ n$; if $P_i^{1:n} > P_{\max}$ then $P_{\max} = P_i^k$ and $d_{\text{best}} = d_i^k$.

BA has been used in many types of research and it shows better performance than the PSO in terms of convergence time and failure rate. The problems of long convergence time and high failure rate shown above in PSO are inherent in BA too and it can be avoided with the same modifications as discussed in PSO, where the bats' initializations should not be random where it is better to start it with the duty ratios at the anticipated peaks which can be obtained from (35), or with equal distance between the duty ratios as obtained from (36). Moreover, the need for reinitialization discussed in PSO is also needed with the BA and has been performed with the same condition shown in (40) [94]. The performance of BA is modified considerably by using the scanning strategy discussed above in PSO [108], where, in the beginning, several values of the duty ratio will be applied to the PV system and the one associated with the highest power will be used to initialize the bats around it.

To overcome the problem of high oscillations around the GP in the steady-state operation associated with BA, a newly proposed hybrid technique is introduced [109]. In this study, the BA is used to capture the GP and once it gets it, it transfers the tracking to one of the three traditional MPPT techniques. The traditional techniques used in this study to improve the performance of BA are beta, P&O, and InCond MPPT techniques. These modifications showed improvements in the performance of BA MPPT technique in steady-state conditions, especially with the BA and beta MPPT algorithms [109].

Cuckoo Search (CS)

The cuckoo search (CS) optimization technique is introduced by Yang and Deb [110]. Three rules should be followed to use the cuckoo's brood parasitic behavior as an optimization tool.

(1) each cuckoo lays one egg at a time and places it in a randomly chosen nest,
(2) the best nest with the highest quality of eggs will carry over to the next generation, and,
(3) the number of available nests is fixed and the number of eggs that can be discovered by the host bird maintains a probability P_a, where $0 < P_a < 1$.

If the cuckoo's eggs are discovered, the host bird can abandon its nest or destroy cuckoos' eggs. Either way, a new nest will be generated with a probability of Pa for a fixed number of nests. Based on these three rules, the CS algorithm can be summarized as in the flowchart shown in Fig. 20 [111].

Cuckoo search (CS) is an optimization algorithm, inspired by the parasitic reproduction strategy of cuckoo birds [111]. It is observed that several species of cuckoos perform brood parasitism, i.e., by laying their eggs in other birds' (host birds) nests [111]. Usually, three types of brood parasitism are seen (1) intraspecific, (2) cooperative, and (3) nest takeover. Some cuckoo species such as Tapera are intelligent enough to mimic the shape and color of the host bird to increases its reproduction probability. It is also presented in [111] that cuckoos lay their eggs at some specific time so that their eggs hatch earlier than the host bird's own. After the early hatching, cuckoos destroy some of the host bird's eggs to increase the chance of their chicks getting more food. It is also a common phenomenon that the host birds discover the cuckoo's eggs and destroy these. Sometimes they abandon their nest completely and go elsewhere to build a new nest.

The first time that CS was used as an MPPT of the PV system was in 2013 [112]. Later, CS has been used extensively in these applications [110–117]. This algorithm has been also used in the optimal design of hybrid renewable energy systems in [118].

In the beginning, the initial values of eggs are selected and the corresponding power from the PV system will be sampled. Based on the values of power collected, the best nest can be selected. To enhance the private search, a random walk should be performed around each solution which can be provided by the Lévy flight model as shown in (47) [119]:

$$Levy(\lambda) \approx \mathrm{u} = 1^{-\lambda} \text{ where } (1 < \lambda < 3) \tag{47}$$

The new solution that can be determined in each iteration by the equation shown in (48) [119].

$$\mathrm{x}_\mathrm{i}^{\mathrm{t}+1} = \mathrm{x}_\mathrm{i}^\mathrm{t} + \mathrm{a} \oplus \mathrm{Levy}(\lambda) \tag{48}$$

where i is the number of eggs, t is the iteration number, the product $\oplus$ indicates entry-wise multiplication, and α is the step size. The value of α can considerably

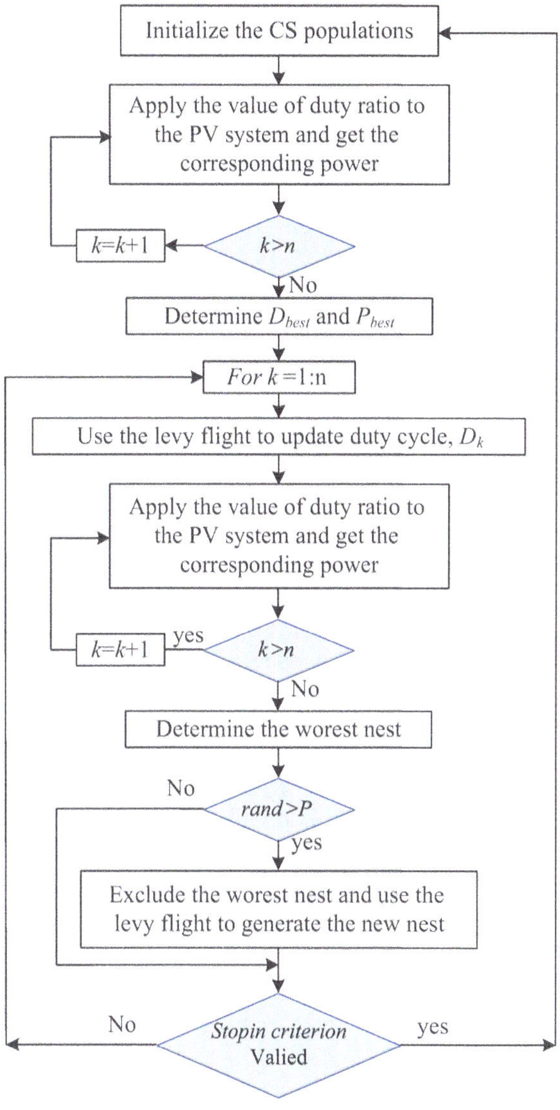

Fig. 20 The flowchart of Cuckoo search algorithm in PV MPPT

affect the performance of convergence, so careful tuning for this value should be selected. In [119], the value of α is determined by setting an initial value for it, α_0, and use the difference between two samples $\left(x_j^t - x_i^t\right)$, as shown by Eq. (49)

$$\alpha = \alpha_0 + \left(x_j^t - x_i^t\right) \tag{49}$$

Besides the value of α_0, the performance of convergence is affected also by the fraction of worse nests, parameters for Lévy distribution, and population size. The

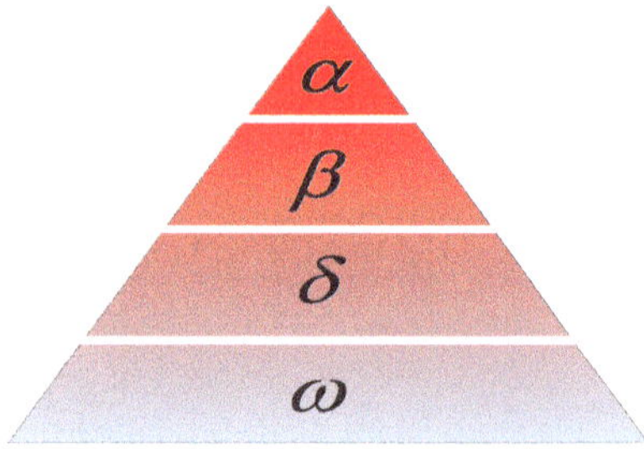

Fig. 21 Leadership pyramids with four levels of leadership (α, β, δ, and ω)

results obtained from using CS as an MPPT of the PV system in [119]. showed that the performance of this technique is having fast and reliable convergence. Meanwhile, this technique (CS) may be easily trapped in one of the LPs in the case of an unwise selection of control parameter values [119]. Figure 20 shows the flowchart of CS when it is used in the MPPT of PV systems.

Grey Wolf Optimizer (GWO)

The Grey Wolf Optimizer (GWO) is one of the best swarm optimization technique that has been used to solve several nonlinear problems like the MPPT of PV systems. This technique is inspired by the lifestyle of the gray wolves in the purse, chasing, attacking, and hunting prey in wildlife [120]. Ion nature, gray wolves like to live in a group containing 5–10 wolves with four levels of leadership. They have a pyramid leader as shown in Fig. 21 [120]. This leadership is having the high-rank leaders called alpha (α), subleaders called beta (β), as well as gamma (γ), and omega (ω), where the dominance of wolves is reduced from top to bottom. Where the strong leaders are α wolves and ω wolves are the lowest rank wolves.

As mentioned above, gray wolves encircle prey during the hunt. The mathematical model mimicking the behavior of GWO is shown in (50) and (51) [40]:

$$\vec{E} = \left|\vec{C}.\vec{D}_P(t) - \vec{D}(t)\right| \tag{50}$$

$$\vec{D}\,(t+1) = \vec{D}_p(t) - \vec{A}.\vec{E} \tag{51}$$

where t represents the current iteration, $\vec{A}$ and $\vec{C}$ are vectors based on their values the balance between the exploration and exploitation can be determined, $\vec{D}_p$ is a position vector from the wolves to the prey, and $\vec{D}$ indicates the position vector of a grey wolf. Equations (52) and (53) are used to determine the two position vectors $\vec{A}$ and $\vec{C}$, respectively [40]:

$$\vec{A} = 2\vec{a}.\vec{r}_1 - \vec{a} \tag{52}$$

$$\vec{C} = 2.\vec{r}_2 \tag{53}$$

where the coefficient a is decreasing linearly from 2 to 0 and r_1, r_2 are random vectors with a value between 0 and 1.

The value of $\vec{A}$ is considerably affecting the performance of convergence where $\left|\vec{A}\right| < 1$ is enhancing the exploitation meanwhile $\left|\vec{A}\right| > 1$ enhances the exploration.

In nature, the order of the alpha wolf ($\vec{D}_\alpha$) is the highest priority to be obeyed. Meanwhile, the rank of obeying the order is reduced in descending level for the beta wolves ($\vec{D}_\beta$) and delta ($\vec{D}_\gamma$). This leadership hierarchy can be mimicked mathematically using the following Eqs. (54), (55), and (56):

$$\vec{E}_\alpha = \left|\vec{C}_1.\vec{D}_\alpha - \vec{D}\right|, \vec{E}_\beta = \left|\vec{C}_2.\vec{D}_\beta - \vec{D}\right|, \text{and } \vec{E}_\delta = \left|\vec{C}_3.\vec{D}_\delta - \vec{D}\right| \tag{54}$$

$$\vec{D}_1 = \vec{D}_\alpha - \vec{A}_1 \cdot \vec{E}_\alpha, \vec{D}_2 = \vec{D}_\beta - \vec{A}_2 \cdot \vec{E}_\beta, \vec{D}_3 = \vec{D}_\delta - \vec{A}_3 \cdot \vec{E}_\delta \tag{55}$$

$$\vec{D}(t+1) = \frac{\vec{D}_1 + \vec{D}_2 + \vec{D}_3}{3} \tag{56}$$

The flowchart showing the use of GWO in MPPT of PV systems is shown in Fig. 22.

Artificial Bee Colony Algorithm (ABC)

The Artificial Bee Colony (ABC) algorithm proposed by Karaboga is based on the foraging behavior of honey bees [121]. In nature, artificial bees are divided into three types, employed bees, unemployed or onlooker bees, and scout bees. The employed bees function is used to search for the food and determine its place and it shares this information with other bees in the colony. The unemployed or onlooker bees' function is to watch the employed bees and help to find the place of the food. The scout bees' function is to search randomly for a new source of food. They communicate and coordinate with each other to obtain the optimal solution in a short time. In the algorithm, the location of a food source and the quantity of nectar denote a solution of the optimization problem and the fitness value of the related solution, respectively. The algorithm starts with a parameter initialization and it generates an arbitrarily initial population (P) of SS solutions, which is the population size. Each solution x_i is *an* n-dimensional vector. For the initialization process, (57) is used [122].

$$D_{i,j} = D_{\min,i} + r\left(D_{\max,i} - D_{\min,i}\right), i = 1,\ 2 \ldots SS,\ j = 1, 2, \ldots, n \tag{57}$$

where n is the number of optimization parameters ($n = 1$ in the PV MPPT because the duty ratio is the only optimization parameter), $D_{\min,i}$ and $D_{\max,i}$ are the minimum

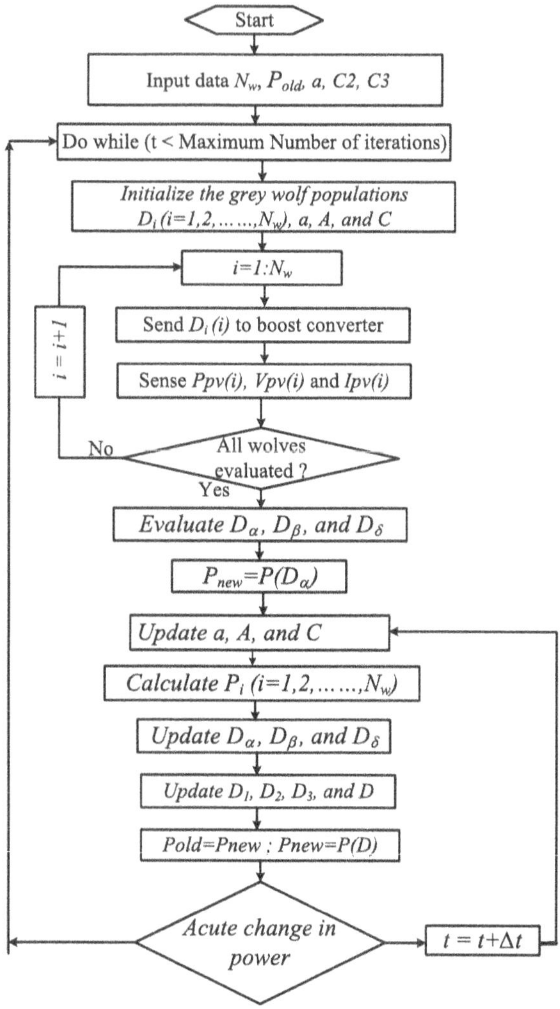

Fig. 22 The flowchart showing the use of GWO in MPPT of PV systems

and maximum allowable value of the duty ratio, respectively, r is a random number between −1 and 1. In the operation of the ABC algorithm, the employed bees evaluate the new food sources using (58) and determine the candidate food position ($v_{i,j}$) from the old value (D_i) in memory [122].

$$D_{i+1,j} = D_{i,j} + rand\,[0, 1]\left(D_{i,j} - D_{i,j}\right) \tag{58}$$

Onlooker bees that are waiting in the dancing area move closer to the position of the employed bee where the nectar quantity is the highest [123]. This movement is given as shown in (59).

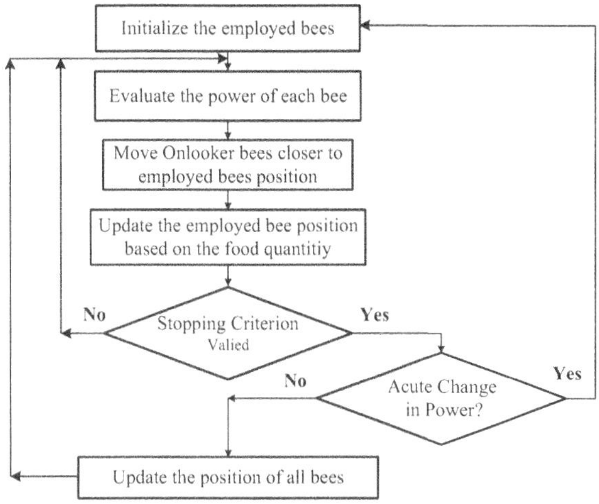

Fig. 23 The flowchart of the artificial bee colony algorithm

$$D_{i+1} = D_{h,i} + \frac{r\,(D_{\max} - D_{\min})}{\frac{SS}{2} - 1} \tag{59}$$

where D_h represents the food source position with the highest nectar amount, SS is the number of bees, r is a random number between −1, 1.

The onlooker bees select food source of the employed bee calculated on the basis of probability connected to the food source as shown in (60) [122].

$$p_i = \frac{P(D_i)}{\sum_{n=1}^{SS} P(D_i)} \tag{60}$$

where $P(D_i)$ is the fitness function of D_i.

The new value of power is compared to the old one and the new one will replace it if it is greater than the old one. This will continue until the scout bees select a new position of food based on Eq. (57). The logic shown for ABC has been used for MPPT of the PV system in [123] and it has been compared to the PSO and it is found that it has better performance than PSO in terms of convergence time and failure rate [123, 124]. Figure 23 shows the flowchart of the artificial bee colony algorithm.

Cat Swarm Optimization (CSO)

Cat swarm optimization (CSO) is one of the swarm optimization techniques that has been developed in 2006 [125]. CSO is divided into two modes of operations, namely, *seeking mode* and *tracing mode* [125]. Each cat is representing one solution and it is used in the algorithm as a searching agent. So, depending on the optimization variable, M, the cat is composed of *M* dimensions (this dimension will be only one

MPPT of PV system because only one variable will be optimized, the duty ratio of DC/DC converter or terminal voltage of PV array). So, in the case of using the CSO in MPPT of the PV system the value of M will be equal to 1 which will be used during this section.

During seeking mode, four essential factors should be defined as shown in the following points:

SMP: SMP is standing for "Seeking Memory Pool," that represents the seeking memory size for each cat.
SRD: SRD is standing for "Seeking Range Dimension," that declares the mutative ratio for the dimensions.
CDC: CDC is standing for "Counts of Dimension to Change," that discloses the dimensions will be varied.
SPC: SPC is standing for "Self-Position Considering," which is used to decides which cat will move or stand.

The logic showing the CSO performance is shown in the following steps:

Step-1: Make j copies of the present position of cat_k, where j = SMP. If the value of SPC is true, let j = (SMP-1), then retain the present position as one of the candidates.
Step-2: For each copy, according to CDC, randomly plus or minus SRD percent of the present values and replace the old ones.
Step-3: Calculate the fitness values (FS) of all candidate points.
Step-4: If all FS are not exactly equal, calculate the selecting probability of each candidate point by Eq. (61), otherwise set all the selecting probability of each candidate point to be 1.
Step-5: Randomly pick the point to move to from the candidate points, and replace the position of cat_k. If the goal of the fitness function is to find the minimum solution, $FS_b = FS_{\max}$, otherwise $FS_b = FS_{\min}$.

$$P_i = \frac{|FS_i - FS_b|}{FS_{\max} - FS_{\min}}, \; where \; 0 < i < j \tag{61}$$

Three steps are shown below that can mimic the tracing mode of cats into a mathematical form:

Step 1: Update the velocities for each cat (v_k) according to Eq. (62).
Step 2: Check the value of the velocity is within the predefined limits. If the velocity is out of the predefined limits it will be equated with the nearest limit.
Step 3: Use Eq. (63) to determine the new position of cats.

$$v_k = v_k + r_1 \; c_1 \; (x_{\text{best}} - x_k) \tag{62}$$

x_{best} is the position of the cat, who has the best fitness value; $x_{k,d}$ is the position of cat$_k$. c_1 is a constant, and r_1 is a random value in the range of [0, 1].

$$x_k = x_k + v_k \tag{63}$$

This technique (CSO) is used as an MPPT in many types of research [126]. This technique showed fast convergence to the MPP but it may stick at one of the LPs in PSC.

Ant Colony Optimization (ACO)

Ant colony optimization (ACO) is one of the swarm optimization techniques that mimic the performance of ants in their foraging behavior to be used for tracking the optimal solutions of nonlinear problems. This technique is first introduced by Dorigo and Gambardella [127]. Since then many modifications introduced in the literature to improve the performance of this technique and it has been used as MPPT of PV system [128, 129]. The use of ACO in the application of PV MPPT is done by using the voltage value of the PV array as bee location and the output power of the PV array is used as an objective function in the simulation [119]. The flowchart showing the logic of ACO when it is used as a PV MPPT is shown in Fig. 24 [128].

A new ACO pheromone updating strategy to improve the convergence performance of ACO (ACO_{NPU} MPPT) when it is used as an MPPT of the PV system is introduced in [119]. Once the ACO_{NPU} MPPT controller is developed, several tests are performed under standard test conditions to determine the ACO control parameters. The Gaussian Kernel for the *i*th dimension of the solution is shown in (64) [119].

$$\mathrm{G_i(x)} = \sum_{\mathrm{l=1}}^{\mathrm{k}} \mathrm{w_l g_l^i(x)} = \sum_{\mathrm{l=1}}^{\mathrm{k}} \mathrm{w_l} \frac{1}{\sigma_\mathrm{l}^\mathrm{i}\sqrt{2\pi}} \exp\left(-\frac{\left(\mathrm{x} - \mu_\mathrm{l}^\mathrm{i}\right)^2}{2\sigma_\mathrm{l}^{\mathrm{i}2}}\right) \tag{64}$$

where $g_i^l(x)$ is the lth sub-Gaussian function for the *i*th dimension of the solution; μ_l^i and σ_l^i are the ith-dimensional mean value and the standard deviation for the ith solution, respectively.

The formula that can be used to determine the pheromone equation, τ_{li} is shown in (65) [119]:

$$\tau_{\mathrm{li}}(\mathrm{x}) = \frac{1}{\sigma_\mathrm{l}^\mathrm{i}\sqrt{2\pi}} \exp\left(-\frac{\left(\mathrm{x} - \mu_\mathrm{l}^\mathrm{i}\right)^2}{2\sigma_\mathrm{l}^{\mathrm{i}2}}\right) \tag{65}$$

During the initial stage, the distances D_i between each x_i solution among the selected solutions are determined ($i = 1\ldots m$, *where m is the number of ants*) as shown in (66) and the best solution x_{best}.

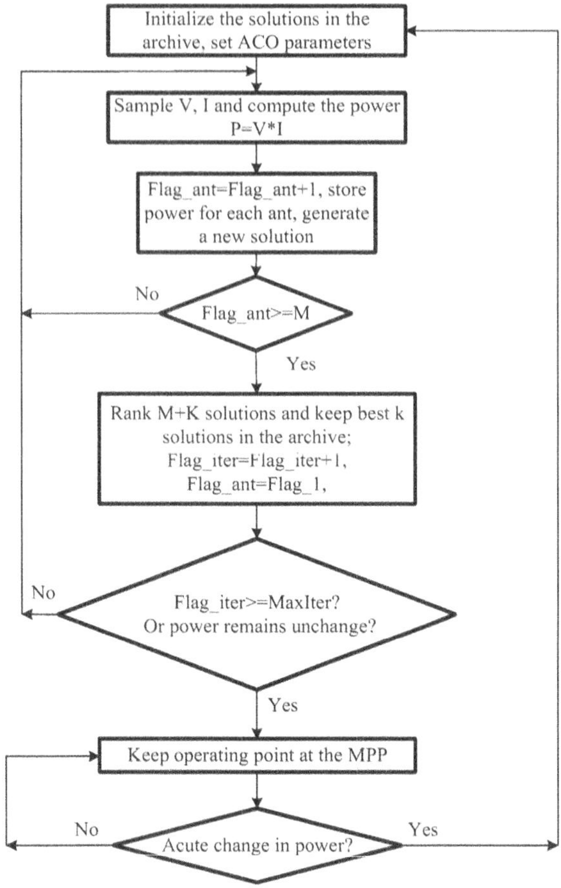

Fig. 24 Flowchart of ACO when it is used as an MPPT of the PV system [131]

$$D_i = |x_i - x_{\text{best}}| \tag{66}$$

The Gaussian, φ_i can be determined from (67).

$$\varphi_{\mathrm{i}} = \mathrm{e}^{\frac{-D_i^2}{2t}} \tag{67}$$

where t is the standard deviation of the Gaussian (usually $t = 0.05$). The pheromone's value τ_i is calculated as shown in (68):

$$\tau_i = \frac{\varphi_i}{\sum_{\mathrm{j}=1}^{\mathrm{m}} \varphi_j} \tag{68}$$

The solution vector of the i-th ant concerning the kth ant at iteration t is obtained by (69).

$$x_i = x_k(t-1) + dx \tag{69}$$

where dx is a random variable in the range of $[-\alpha, \alpha]$, the value of dx is used to determine the length of the jump. Based on the value of x_i obtained from (69), the value of di can be determined as shown in (70).

$$d_i(t) = 1 - x_i(t)/x_{\text{ref}} \tag{70}$$

The best solutions, k will be selected from all solutions $(\text{m} + K)$. After reinitializing the archive, the m best solutions will be selected and their pheromones will be updated as shown in Eqs. (66)–(68).

The corresponding generated power from the PV system can be calculated from (71) after sampling the voltage and current generated from the PV system.

$$\text{Objective function} = P = V_{PV} \times I_{PV}(G, T) \tag{71}$$

where V_{PV} and I_{PV} are the terminal voltage and current of the PV array, respectively, T is the array temperature, G is the solar radiation in W/m^2

The distance between any new solution and the best solution, V_{best} can be obtained from (72).

$$D_i = |V_i - V_{\text{best}}| \tag{72}$$

Compute a Gaussian φ_i by (73)

$$\varphi_i = e^{\frac{-D_i^2}{2t}} \tag{73}$$

where t is the standard deviation of the Gaussian. The pheromone's value i is computed as shown in (74).

$$\tau_i = \frac{\varphi_i}{\sum_{j=1}^{m} \varphi_j} \tag{74}$$

Then a perturbation of the voltage can be obtained from (75), (67).

$$V_i(t) = V_k(t-1) + d_x \tag{75}$$

The new duty ratio of each ant i is computed as shown in (76).

$$\text{d}_\text{i}(\text{t}) = 1 - \text{V}_\text{i}(\text{t})/\text{V}_{\text{ref}} \tag{76}$$

Different parameters can considerably affect the performance of ACO such as the size of the archive, balance coefficient, convergence time. Tuning these values improved the results obtained from this technique compared to the PSO and DE MPPT techniques [119].

Fireflies Algorithm (FFA)

The FFA is one of the best swarm optimization techniques which is introduced by Yang [130]. Fireflies are lightning bugs that are attracted to the light in the tropical regions. The FFA is inspired by the movement of fireflies. This light is playing an important role in attracting mating partners and preys. The rate of flashing and the amount of time form part of the signal system is responsible for brings both sexes together [131].

Let p and q be two fireflies positioned at X_p and X_q, respectively. In a single-dimensional space, the distance between these two fireflies, r_{pq} is shown in (77).

$$r_{pq} = \left\| X_p - X_q \right\| \tag{77}$$

The distance between any two fireflies p and q is a function in a factor called the degree of attractiveness, β that can be obtained from (78).

$$\beta(r) = \beta_0 e^{-\gamma \left(r_{pq}\right)^n}, \quad n \geq 1 \tag{78}$$

where, γ is called absorption coefficient which is used to controls the light intensity and its value varies between 0 and 10 and n = 2 [130], β_0 is the initial value of the absorption coefficient and its value is chosen by 1 to actively determine the position of other fireflies in its neighborhood [130]. Assuming that the brightness of firefly p is less than that of q, the new position of firefly p is given by (79).

$$X_p^{t+1} = X_p^t + \beta(r)\left(X_p - X_q\right) + \alpha(rand - 0.5) \tag{79}$$

Here, random movement factor α is constant throughout the program and falls in the range [0, 1]. The value of α enhancing the searching balance between exploitation and exploration, where the high value of α enhances exploration, meanwhile small value of α enhances exploitation [131].

The steps of the logic used with the FFA to capture the MPP of the PV system are shown in the following points:

Step 1: *Parameter Setting*: Select the values of FFA control parameters, β_o, γ, n, α, population size N, and the termination criterion. In this algorithm, the position of the firefly is taken as a duty cycle d of the DC/DC converter. The brightness of each firefly is taken as a generated power P_{PV} of the PV system, corresponding to the position of this firefly.

Step 2: *Initialization of Fireflies*: In this step, the fireflies are positioned in the allowable solution space between between $d_{\min}$ to $d_{\max}$ where $d_{\min}$ and $d_{\max}$ represent the minimum and maximum values of the duty ratio of the DC/DC converter. It is recommended to choose the starting position of the fireflies as introduced before in Eqs. (35), or (36). The swarm size is recommended to be 6 in [131].

Step 3: *Brightness Evaluation*: For each duty ratio, the corresponding PV output power, P_{pv} is taken as the brightness or light intensity of the respective firefly. This step is repeated for the position of all fireflies in the population.
Step 4: *Update the Position of Fireflies*: The firefly with maximum brightness remains in its position and the remaining fireflies update their position based on (79).
Step 5: Check if is there any acute change in the generated power which gives an indication for a big change of the shading pattern. Acute change can be detected using Eq. (40). If the acute change is detected go to step 2, otherwise go to the next step.
Step 6: Check the stopping criterion, if it is valid go to step 4 otherwise go to step 3. The stopping criterion should ensure that all fireflies all work around the MPP.

The convergence time and failure rate as well as the oscillations at steady-state are a function in the FFA control parameter values (β_o, γ, n, α, population size, N). The advantages of FFA is not highly affected by the initial values of duty ratio which is one of the main advantages of this MPPT technique, meanwhile, the high failure rate of this technique is counted as one of the main shortcomings [119].

Whale Optimization Algorithm (WOA)

The Whale Optimization Algorithm (WOA) is one of the modern swarm optimization technique which has been developed in 2016 [132]. This technique is inspired by the WOA. This technique used for pursuing a procedure is called bubble-net feeding strategy. Humpback whales want to chase little fishes near the surface by making a bubble net around the prey rises along a circular path. This technique has been used in MPPT of the PV systems in many studies in the literature [133–135]. The mathematical formulas that model this technique is shown in the following equations:

$$D = \left| \overrightarrow{C} \cdot \overrightarrow{X} * (t) - \overrightarrow{X}(t) \right| \tag{80}$$

$$X(t+1) = \overrightarrow{X} * (t) - \overrightarrow{A} \cdot \overrightarrow{D} \tag{81}$$

where t is the iteration number, $\vec{A}$ and $\vec{D}$ are the coefficient vectors, $\vec{X}^*$ is a vector used to represent the vector of the best solution, $\vec{X}$ represents the current position vector.

$$\vec{A} = 2\vec{a} \cdot \vec{r} - \vec{a} \tag{82}$$

$$\vec{C} = 2.\vec{r} \tag{83}$$

where $\vec{a}$ in equations is a variable linearly decrease from 2 to 0 through the progress of the iterations, $\vec{r}$ is an arbitrary vector in the range of [0, 1]. The value of $\vec{r}$is used to balance between exploitation and exploration.

The spiral path between the position of the whale (current position) and prey (best solution) can be determined as shown in (84).

$$\overrightarrow{X}(t+1) = \overrightarrow{D'} \cdot e^{bl} \cdot \cos(2\pi l) + \overrightarrow{X^*}(t) \tag{84}$$

where $\overrightarrow{D'} = \overrightarrow{X^*}(t) - \vec{X}(t)$ and demonstrates the distance between the ith whale and the prey (best solution), b is a constant for characterizing the state of the logarithmic spiral and its value is randomly chosen between −1 and 1.

Whales swim around the prey inside shrinking circle and along with a spiral form. There is a probability of half to select one of two approaches as shown in (85).

$$\vec{X}(t+1) = \begin{cases} \overrightarrow{X^*}(t) - \vec{A} \cdot \vec{D} & \text{if } p < 0.5 \\ \overrightarrow{D'} \cdot e^{bl} \cdot \cos(2\pi l) + \overrightarrow{X^*}(t) & \text{if } p \geq 0.5 \end{cases} \tag{85}$$

where p is a predefined value that can adjust the balance between exploration and exploitation and it can be selected between 0 and 1 and it can be adjusted during the progress of the iterations.

Cuttlefish Algorithm (CFA)

The cuttlefish algorithm (CFA) is one of the modern metaheuristic optimization algorithms that is inspired by a type of fish called cuttlefish that can change their skin color to mimic the surrounding environment to either seemingly disappear into its environment or to produce amazing displays. This algorithm was first introduced by Eesa et al. [134]. The algorithm undergoes the same mechanism of the cuttlefish color-changing behavior to optimize mathematical problems. There are three cells on the skin of Cuttlefish, namely, *Chromatophores, Iridophores, and Leucophores.*

The *Chromatophores* cell groups are having pigments to change the color of the Cuttlefish when it is needed. *Iridophores* cell groups are used to reflect the light which can help in concealing the Cuttlefish when it is needed. The *Leucophores* cell groups are responsible for the white spots occurring on some species of cuttlefish that are used to scatter and reflect incoming light. The behavior mechanism is mainly based on two processes which are *reflection* and *visibility*. The mathematical model of the algorithm uses different reflection and visibility factors for each group of solutions in an iteration to reach an optimum solution accurately and as fast as possible [136].

The mathematical formulation of the CFA is as follows. In general, the update of the suggested solution is presented in (86), while the reflection and visibility factors are calculated for each group using (87) and (88).

$$D_{\text{new}} = R_{ni} + V_{ni} \tag{86}$$

$$R_i = rd * (r_1 - r_2) + r_2 \tag{87}$$

$$V_i = rd * (v_1 - v_2) + v_2 \tag{88}$$

where D_{new} is the newly updated population ready to be studied in the new iteration.

R_{ni} are the new reflected population cells for the i-th group.
V_{ni} are the new visible population cells for the i-th group.
R_i is the reflection factor for the i-th group.
V_i is the visibility factor for the i-th group.
rd is a generated random value between 0 and 1.
r_1, r_2 are the upper and lower limits of the reflection factor, respectively, ($r_1 = 1$, $r_2 = -1$) [137].
v_1, v_2 are the upper and lower limits of the visibility factor, respectively, ($v_1 = 0.5$, $v_2 = -0.5$) [137].

Initially, the population values (duty ratios) for each searching agent will be initialized may be randomly ....but it is recommended to initialize it based on Eqs. (35) or (36) and divide them into the different groups equally. The population is divided into four identical groups (G_{p1}, G_{p2}, G_{p3}, and G_{p4}). Then calculate the corresponding output power for each cell (duty ratio). The duty ratio associated with the global best is selected to D_{best}. The new reflection and visibility factors can be determined from the following equations:

For G_{p1}:

$$R_{n1} = R_1 \times G_{p1} \tag{89}$$

$$V_{n1} = V_1 \times \left(D_{\text{best}} - G_{p1}\right) \tag{90}$$

$$G_{p1nw} = R_{n1} + V_{n1} \tag{91}$$

For G_{p2}:

$$R_{n2} = R_2 \times G_{p2} \tag{92}$$

$$V_{n2} = V_2 \times \left(D_{best} - G_{p2}\right) \tag{93}$$

$$G_{p2nw} = R_{n2} + V_{n2} \tag{94}$$

where

R_{n1} and R_{n2} are the arrays of the new updated reflected cells for groups 1 and 2, respectively.

R_1 and R_2 are the reflection factors set for groups 1 and 2, respectively.
V_{n1} and V_{n2} are the arrays of the new updated visible cells for groups 1 and 2, respectively.
V_1 and V_2 are the visibility factor set for groups 1 and 2, respectively.
G_{p1} and G_{p2} are the arrays of cells in groups 1 and 2 of the population, respectively.G_{p1new} and G_{p2new} are the arrays of new updated cells in groups 1 and 2 of the population, respectively.

For G_{p3} and G_{p4}:

$$R_{n3} = R_3 \times F_{\text{best}} \tag{95}$$

$$V_{n3} = V_3 \times (F_{\text{best}} - B_{av}) \tag{96}$$

$$G_{p3new} = R_{n3} + V_{n3} \tag{97}$$

$$G_{p4new} = random\ values \tag{98}$$

where

R_{n3} is the array of the new updated reflected cells for group 3.
R_3 is the reflection factor set for group 3.
V_{n3} is the array of the new updated visible cells for group 3.
V_3 is the visibility factor set for group 3.
G_{p3} is the array of cells in group 3 of the population.
G_{p3new}, G_{p4new} are the arrays of new updated cells in groups 3 and 4, respectively, of the population.

Grass Hopper Optimization (GHO)

Grass Hopper Optimization (GHO) is one of the best swarm optimization algorithms which first introduced by Saremi et al. [138]. The GHO algorithm mimics the behavior of the grasshoppers during their life cycle. The GHO algorithm consists of two sub-cycles, namely nymph and adult sub-cycles. In the nymph sub-cycle, the algorithm uses it to control the movement (Jump) to enhance the exploitation search. In the adult sub-cycle, it is characterized by fast jumps with random intervals which can help to enhance the global exploration search. The controlled GHO parameters are helping in the balance between the local and global explorations.

The movement of searching agents is given by Eq. (99) [139];

$$X_i = \omega_1 S_i + \omega_2 G_i + \omega_3 A_i \tag{99}$$

where X_i is the position of the i-th searching agent, S_i is the social interaction, G_i is the gravity factor of the i-th grasshopper, and A_i is the variable to represent the effect of wind on the movement of each particle, ω_1, ω_2, and ω_3 are the factors to represent the weighted social interaction, the gravity factor, and the advection, respectively.

The social interaction factor is shown in Eq. (100) is a very important factor that controls the behavior of convergence where it enhances the exploration of the search by sharing the information within the swarm.

$$S_i = \sum_{j=1}^{N} s\left(d_{i,j}\right).\vec{d}_{ij} \tag{100}$$

where $d_{i,j}$ is the distance between the i-th and j-th grasshoppers and s is the function used to define the social forces which can be determined from Eq. (101).

$$s(r) = f.e^{-\frac{r}{la}} - e^{-r} \tag{101}$$

where la gives the attraction length and f provides the intensity of interaction. The function s(r) shows the impact of social interaction and is fine-tuned for an optimization problem to maintain a balance between the exploitation and exploration of the search task.

The gravitational factor G_i is given by Eq. (102) and wind attraction factor is given by Eq. (103)

$$G_i = -g\,.\,\vec{e}_g \tag{102}$$

$$A_i = -u\,.\,\vec{e}_w \tag{103}$$

where $\vec{e}_g$ is a unit vector in the direction of the gravitational constant g, and $\vec{e}_w$ is a unit vector in the direction of constant drift u which is caused by wind.

Substituting the values of social interaction parameters into Eq. (99) yields

$$X_i = \omega_1 \sum_{\substack{j=1 \\ j \neq i}}^{N} s\left(\left|x_j - x_i\right|\right)\frac{x_j - x_i}{\vec{d}_{ij}} + \omega_2.g.\vec{e}_g + \omega_3.u.\vec{e}_w \tag{104}$$

The GHO converged fast using Eq. (104) and this mechanism has been improved for fast and reliable convergence by [139] as shown in Eq. (105).

$$X_i^d = c\left(\sum_{\substack{j=1 \\ j \neq i}}^{N} c\frac{ub_d - lb_d}{2}s\left(\left|x_j - x_i\right|\right)\frac{x_j - x_i}{\vec{d}_{ij}} + \vec{e}_g\right) + \vec{T}_d \tag{105}$$

$$c = c_{\max} - l\frac{c_{\max} - c_{\min}}{L} \tag{106}$$

where c is a decreasing coefficient, ub_d and lb_d are upper and lower bounds in the d-th dimension $\vec{T}_d$ is the best value of the d-th dimension of the target up to the current iteration. L is the total number of iterations and l is the current iteration.

GHO is having two main advantages compared to other swarm optimization which are the fast convergence and the lowest oscillations at the steady state. In the beginning, the population is better to be initialized as has been introduced before in (35) or (36). Here, the search space represents the duty cycle of the boost converter. The constraints are fine-tuned accordingly in the search space.

Intelligent Monkey King Evolution (IMKE)

The Intelligent Monkey King Evolution (IMKE) is a metaheuristic optimization algorithm introduced by Meng et al. [140]. This optimization algorithm is inspired by the behavior of monkeys in their superpower abilities under a challenging situation in which they divided themselves into many small groups of monkeys and start working toward the solution. After achieving the solution to the problem, another group will report the whole situation to the monkey king.

Based on these reports, the monkey king decides the most accurate solution. Based on the monkey king's decision the whole swarm will move. In the start of the operation of IMKE algorithm, the whole swarm is divided into n groups, each group contains p monkeys and get their initial position as shown in (107). The variable R_c represents the evaluation of the monkey king for the next movement.

Accurate selection for the value of R_c will improve the performance of IMKE algorithm in terms of convergence time and failure rate. Due to the importance of this factor, it has been limited by upper and lower limits $[R_{c,\min}, R_{c,\max}]$ as shown in (110). Generally this $R_{c,\min}$ and $R_{c,\max}$ lie in the range 0.1–0.9 [4, 141].

$$X^i = \begin{bmatrix} x^i_{11} \; x^i_{12} \; \ldots x^i_{1v} \\ x^i_{21} \; x^i_{22} \; \ldots x^i_{2v} \\ \ldots\ldots \\ x^i_{p1} \; x^i_{p2} \; \ldots x^i_{pv} \end{bmatrix} p * v \tag{107}$$

From the results obtained when the positions in (107) are applied to the PV system, the best solution can be obtained as shown in (108).

$$X^i_{\text{gbest}} = [X^i_{\text{gbest},1}, \; X^i_{\text{gbest},2}, \; \ldots X^i_{\text{gbest},n} \tag{108}$$

The best solution should be updated as shown in (109).

$$X^{i+1}_{\text{gbest}} = X^{i+1}_{\text{gbest}} + R_c * X_{\text{diff}} \tag{109}$$

$$R_c = R_{c,\min} + \frac{R_{c,\max} - R_{c,\min}}{i} * rand \tag{110}$$

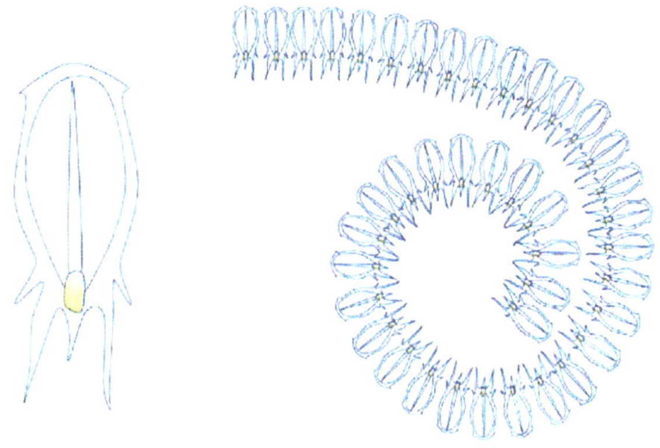

Fig. 25 **a** Salp as single and **b** a group of salps (salp chain) [146]

Salp Swarm Algorism (SSA)

The salp is a sea creature that has a transparent body like jellyfishes with a barrel-like shape. Salps move in seawater through suction and propulsion the water through their barrel body. The salps move in a group called chain as shown in Fig. 25. The shape of salps is barrel-shaped and it has a transparent body. The salps tissues are like jellyfishes. Like a jellyfish, the salps are moving and it moves forward by pushing the water through the body like propulsion. Figure 25 shows the shape of the salp chain [142]. The salp chain formation is used to improve their movement and to help them in foraging. This movement inspired the researchers to use this chain to solve nonlinear optimization problems. One of the real-world applications of SSA is PV MPPT [143].

The swarm of salps is divided into two subgroups called leaders and followers [142, 143], The leader subgroup is responsible for guiding the swarm, meanwhile, the followers obey the leaders' orders. The swarm is distributed in searching space in n dimensions. First, the leader takes their positions based on the formula shown in Eq. (111).

$$X_j^1 = \begin{cases} F_j + c_1\big((ub_j - lb_j)c_2 + lb_j\big) & c_3 \geq 0 \\ F_j - c_1\big((ub_j - lb_j)c_2 + lb_j\big) & c_3 < 0 \end{cases} \tag{111}$$

where X_j^1 are the leaders' positions, F_j are the food sources, ub_j and lb_j are the upper and lower limit, respectively, c_1, c_2, and c_3 are random numbers. As shown in Eq. (111), the position of the leader is updated about the food source. c_1 is an important parameter that can determine the exploitation and exploration performances and its value can be determined from Eq. (112).

$$c_1 = 2\,e^{-\left(\frac{4l}{L}\right)^2} \tag{112}$$

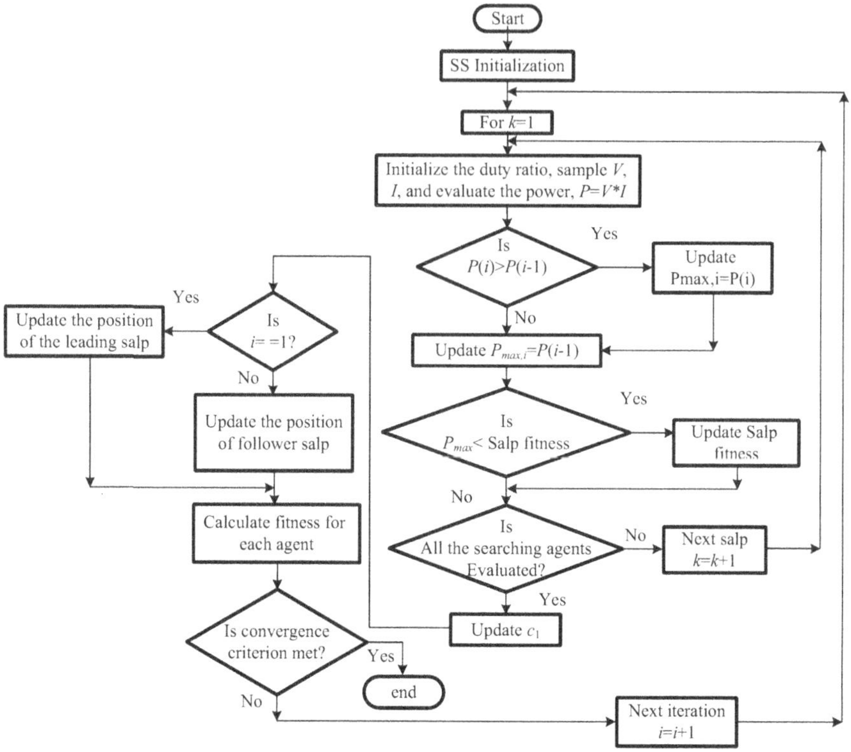

Fig. 26 The flowchart of the salp swarm MPPT technique

where L is the maximum iteration number, l is the current iteration number. c_2 and c_3 are generated randomly between [0, 1]. The follower position can be determined as shown in (113).

$$X_j^i = \frac{1}{2}at^2 + V_o t \tag{113}$$

where X_j^i are the followers' positions, V_o is the initial velocity, $a = V_{\text{final}}/V_o$, and $V = x - x_o/t$. Equation (113) is altered. The new position of salps chain is shown in Eq. (114).

$$X_j^i = \frac{1}{2}\left(X_j^i + X_j^{i-1}\right) \tag{114}$$

Application of SSA in the PV MPPT is done by equating the initial positions of salps by the duty ratio of the boost converter and use the above Eqs. (111)–(114) to keep tracking the GP of PV arrays. The flowchart of SSA when it is used as an MPPT of the PV system is shown in Fig. 26 [144].

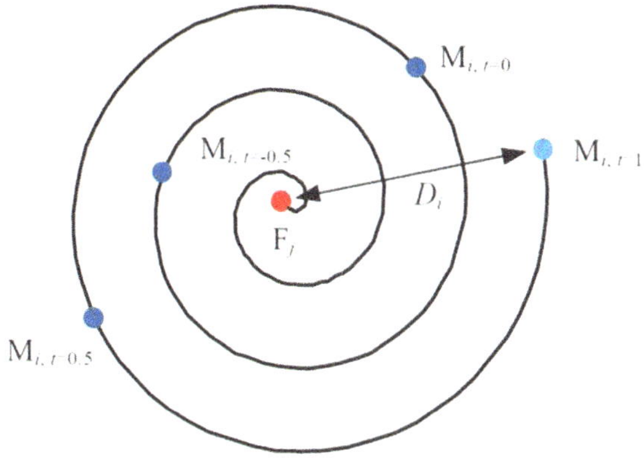

Fig. 27 Spiral flight of a moth around its corresponding flame

Moth–Flame Optimization (MFO)

Moths are types of insects similar to butterflies. There are two main milestones in their lifetime namely, larvae and adults. They have a special night navigation mechanism called transverse orientation, where they fly in a straight line having a fixed angle to the moon. In the case of the light source is switched off, they fly in a spiral shape around the previous source of light and it can capture it after a few corrections. The spiral flight of a moth around its corresponding flame is shown in Fig. 27 [145]. This flight mechanism is translated into mathematical formulas to perform the searching mechanism to capture the solution in nonlinear optimization problems. In the MFO algorithm, every moth representing a searching agent that is required to fly around a certain source of light to enhance the exploration search and a lower probability of local optima stagnation. Therefore, a set of sources of light locations can be represented in a matrix with the same dimensions to represent the moth positions. Both the moths and the lights are representing solutions. The moths and light are treated and updated in different ways During the progress of MFO. Each moth is representing a search agent which can fly the search area to get the global best solution. The lights are representing the best solutions that the moths have captured so far. In other words, flames can be considered as flags or pins that are dropped by moths when exploring a search space. Each moth searches around a flame and updates it in the case of finding a better solution. With this mechanism, a moth never loses its best solution.

The moth position can be updated using the following equation:

$$M_i = S(M_i,\ F_j) \tag{115}$$

where M_i indicates the i-th moth, F_j indicates the j-th flame, and S is the spiral function. The characteristics of the spiral function is listed in the following points:

- The initial point of a spiral is the initial moth position.
- The source of light location is the final point of a spiral and it represents the best position of the moth.
- The predefined range of the spiral should be within the searching area.

Taking the previous points into considerations, the mathematical model of the spiral function is shown in Eqs. (116) and (117).

$$S(M_i, F_j) = D_i \cdot e^{bt} \cdot \cos(2\pi t) + F_j \tag{116}$$

$$D_i = \left| F_j - M_i \right| \tag{117}$$

where D_i represents the distance between the positions of i-th moth and the j-th light, b is a constant represents the shape of the spiral, and t is a random number in $[r, 1]$, where r is a constant that linearly decreasing with iterations from -1 to -2 and is called the adaptive convergence constant.

A strategy was introduced to enhance the exploitation search of the MFO by decreasing the number of flams as shown in Eq. (118).

$$flame\ number = round\left(N - l\frac{N-1}{L}\right) \tag{118}$$

where l is the current iteration number, N is the maximum number of flames, and L indicates the maximum number of iterations.

This mechanism introduced in (118) enhances the balance performance between the exploration and exploitation in a solution area. The MFO has been used as an MPPT of the PV system by initializing the Moths (duty ratios) as has been shown in (38) or (39), in the beginning, the moths' positions are selected to be as a flam position. During the iterations, the flame positions will be equal to the best solutions of Moths. This searching mechanism provides a good balance between exploration and exploitation which makes MFO is a superior choice for MPPT of PV systems. In the case of acute change in the output power as has been discussed in Eq. (40), the Moths and flame positions should be updated.

A Comprehensive Comparison Between Swarm Intelligence Algorithms MPPT Techniques

The performance of all swarm techniques introduced above is tabulated in the following Table 4 shows the difference between them in many important factors. Based on the experience from using these techniques, BA is the fastest convergence but it has higher oscillations in steady-state conditions.

Table 4 Comprehensive comparison between Swarm Intelligence Algorithms MPPT techniques

No.	MPPT Technique	Convergence Speed	Tracking Efficiency	Oscillations at steady-state	No of sensors	Implementation complexity	PSC MPPT ability	No. of Control parameters
1	PSO	High	High	Low	2	High	Yes	3
2	BA	Very high	High	Medium	2	High	Yes	6
3	CS	High	High	Low	2	Medium	Yes	2
4	GWO	High	High	Medium	2	Medium	Yes	2
5	ABC	Medium	High	Low	2	High	Yes	2
6	CSO	High	Medium	Medium	2	High	Yes	1
7	ACO	Medium	Medium	Low	2	Medium	Yes	4
8	FFA	High	High	Medium	2	High	Yes	4
9	WOA	Medium	High	Medium	2	High	Yes	5
10	CFA	Medium	Medium	High	2	High	Yes	3
11	GHO	High	Medium	High	2	High	Yes	6
12	IMKE	Medium	Medium	High	2	Medium	Yes	2
13	SSA	Medium	High	High	2	High	Yes	4
14	MFO	High	High	Medium	2	High	Yes	3

4.2.2 Bio/Natural-Inspired Algorithms (BNIA)

Earthquake Optimization Algorithm (EOA)

The earthquake optimization algorithm (EOA) is one of the nature algorithms and sometimes it is called (geo-inspired) [146]. This technique is inspired by the behavior of *P* and *S* waves which can be generated from earthquakes. This optimization technique is introduced in 2018 and it has been used in the control of the electric machine [147]. The first time to be used as an MPPT of the PV system was in 2020 by Mendez [148].

The mathematical formulae that can model the velocities of *P* and *S* waveforms are shown in (119) and (120), respectively [148].

$$v_p = \sqrt{\frac{\lambda + 2\mu}{\rho}} \tag{119}$$

$$v_s = \sqrt{\frac{\mu}{\rho}} \tag{120}$$

where v_p and v_s are the velocities of waves, λ and μ are called the Lamé parameters, and ρ the density of the material. The optimal relation for the Lamé parameters was found to be 1.5, consequently [148]:

$$\lambda = \mu = 1.5 \text{ GPa} \tag{121}$$

The densities of the material ρ are chosen randomly between 2200 and 3300 kg/m^3 [148]. It is important to define an operating range for the S-wave or S-range, S_r to decide whether to use v_p or v_s. Searching flag (S_{flag}) is performed to be sure that the best duty cycle is within the searching positions.

The main difference between the modification shown in [148] and the original EOA is it returns to the global best duty ratio after evaluating an epicenter to have a faster reaction against irradiation changes. Figure 28 shows the flowchart of the earthquake optimization algorithm (EOA) MPPT algorithm [148].

Simulated Annealing (SA)

Simulated annealing (SA) is a metaheuristic optimization technique inspired by the annealing process that is used in nature to produce high-quality crystals. This technique uses the temperature, final temperature, and nominal cooling rate for searching the optimal solution of nonlinear optimization problems. This technique was first developed in 1970 by Pincus [149]. This technique is used in the MPPT of the PV system by many research studies [5, 135]. Many improvements were introduced to SA which improved its performance in tracking the MPP in uniform irradiance

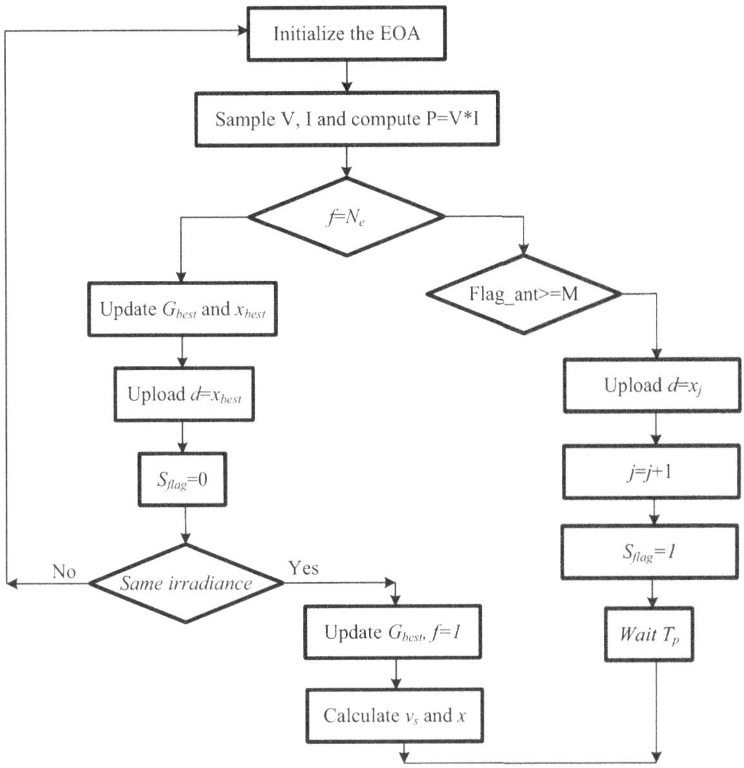

Fig. 28 The flowchart showing the logic of using the Earthquake Optimization Algorithm (EOA) as a PV MPPT algorithm [152]

and the PSC [7]. In normal operation of the SA, a random number of solutions are initialized but here in the MPPT of PV systems, it is better to initiate the duty ratio of DC/DC converter to be as stated before in Eqs. (35) or (36). The initial values of power corresponding to each duty ratio will be determined. By the neighborhood mathematical structure, the new position of particles will be determined. If the new operating point has greater power, then it will be accepted as the new operating power.

The operating point can be accepted if its new point has less power than the reference operating point based on the acceptance property shown in (122) [5]:

$$P_r = \exp\left[\frac{P_k - P_i}{T_k}\right] \tag{122}$$

where P_k is the power of the current point, P_i is the power at the previous best operating point, and T_k is the current temperature of the system.

The cooling mechanism used in the SA can be either static or adaptive as shown in (123).

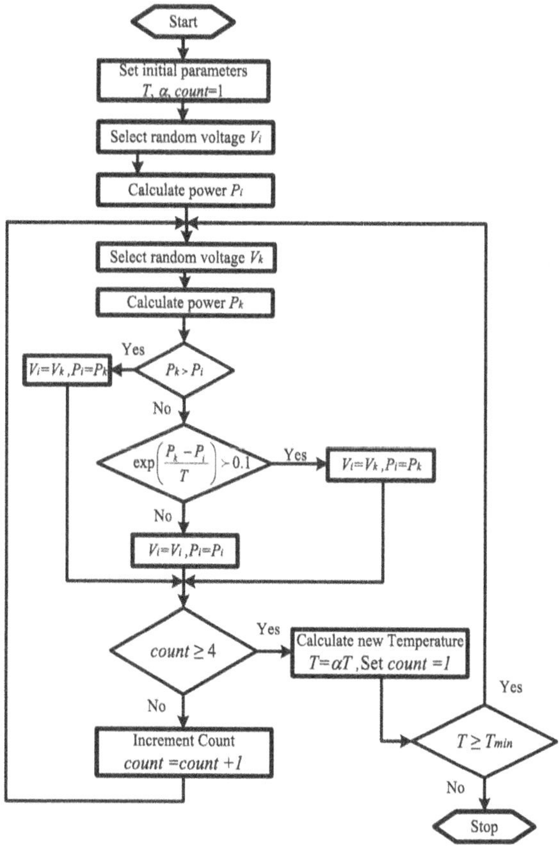

Fig. 29 Flowchart of SA based GMPPT under PSC

$$T_k = \alpha T_{k-1} \tag{123}$$

where T_k is the temperature for step k, T_{k-1} is the temperature at step $k - 1$ and α is a constant always less than 1.

The SA algorithm has been used for tracking the MPP of the PV system in PSC in [135]. The results obtained from this study showed the superiority of the SA compared to famous MPPT techniques like PSO and P&O in terms of convergence speed, failure rate, and generation efficiency. As the PSO, the SA needs reinitialization when an acute change in power is detected as has been discussed before and shown in Eq. (40). Figure 29 shows the flowchart SA-based GP under PSC.

Human Psychology Optimization (HPO)

Human psychology optimization (HPO) is one of the metaheuristic techniques inspired by the manner of thinking of humans for improving his/her situation and taking decisions in the real world. The person gets experience from his own learning and the experience of other persons which is the logic used in this technique. The first time to use HPO as an MPPT of PV systems is done in 2017 by Kumar et al. [150]. In this study, the performance evaluation of the HPO algorithm has been compared with two different PSO strategies in tracking the MPP of partially shaded PV systems using a single sensor to measure the current feeding a battery in the DC-link. The advantage of the HPO when it is used as an MPPT of the PV system is its fast convergence rate and the lower failure rate. Moreover, its performance does not depend on the initial value and the dependency on the algorithm specified parameter is very less [150]. The searching mechanism of this technique has four stages that are listed in the following points:

- *Excitement*: In this stage, the searching agents take values of duty ratios that can be determined from Eqs. (35) or (36). It gives an initial move and tries to build confidence.
- *Self-motivation*: It is a self-encouraging process that depends on the person's own experiences and achievements in which it enhances the local exploitation ability, which increases the rate of convergence.
- *Inspiration*: In this stage, the person is inspired by the experience of a successful person to follow his creative idea for success. This stage enhances global exploration ability.
- *Lesson*: The previous fail experiences learned from the person's own experience and other persons should be avoided in the future decision. This stage helps in avoiding the searching to be trapped in one of the LPs.

The results presented in the use of HPO in the MPPT of the PV system show its superiority in terms of convergence time and failure rate compared to other states of the art PSO strategies [150]. Moreover, the use of a single sensor to measure the battery current reduces the implementation cost of this system.

Flower Pollination Algorithm (FPA)

Flower pollination algorithm (FPA) is a metaheuristic optimization technique inspired by the flower pollination process in the plants. Two types of pollination, namely, self-pollination and cross-pollination. The self-pollination takes place when the same flower is pollinated internally which represents the private search of the FPA, whereas cross-pollination occurs when pollen moves from one flower to another which represents the global search of the FPA. The abiotic pollination occurs when the pollen is transferred via wind, whereas the biotic pollination occurred when the pollen transferred from the flower of male plants to the female through the insects.

The FPA has four steps which are summarized in the following points:

- The global pollination occurred in the two types of pollinations when the pollen carrying pollinators transferred in Lévy flights shape.
- The biotic and self-pollinations represent the local pollination.

- Reproduction probability is representing the flower constancy which is proportional to the similarity of two flowers involved;
- The local and global pollination has been controlled and switched via switch probability P∈ [0, 1].

The concepts shown in the above points are converted to a mathematical model as shown in the following equations. The position of the pollinators that moves using Lévy flights shape can be obtained as shown in (124).

$$\mathrm{x}_{\mathrm{i}}^{\mathrm{k+1}} = \mathrm{x}_{\mathrm{i}}^{\mathrm{k}} + \gamma \mathrm{L}(\lambda)\left(\mathrm{g}_{*} - \mathrm{x}_{\mathrm{i}}^{\mathrm{k}}\right) \tag{124}$$

where x_i^t is the solution vector x_i at iteration t and g is the best solution (duty cycle) of boost converter duty cycle. γ is a scaling factor that used for controlling the step size, $L(\lambda)$ are the Lévy flights-based step size that can be obtained from the following equation [5, 151]:

$$\mathrm{L} \approx \frac{\lambda \Gamma(\lambda) \sin\left(\frac{\Pi\lambda}{2}\right)}{\Pi} \frac{1}{\mathrm{S}^{1+\lambda}} (\mathrm{S} \geq \mathrm{S}_0 > 0) \tag{125}$$

where $\Pi(\lambda)$ represents the gamma function.

FPA has been used as an MPPT of PV system and it showed superior performance in terms of convergence time and failure rate compared to many MPPT techniques like PSO and P&O techniques [152]. The FPA required only control parameter (γ) which makes it very easy to be tuned for better performance.

Teaching Learning Algorithm (TLA)

The TLA is one of the modern metaheuristic optimization algorithms that inspired by the influence of a teacher on learners or students and it has two phases; first, the teacher phase which means leaning from the teacher; second, the learner phase which means learning by the interaction between students (cooperation). This technique uses candidate solutions (Duty ratio) as many other metaheuristic techniques to track the optimal solution (MPP in PV applications). The teacher phase uses the experience of the teacher to improve the students' level of knowledge. The learner phase is performed through the teacher and interaction between the learners to increase their knowledge. A learner interacts randomly with other learners through group discussions, presentations, formal communications, etc. A learner learns new information when the other learner has more information than him. A detailed discerption of using this technique as an MPPT of the PV system is introduced in literature [153, 154]. The results obtained from these studies showed superior performance of TLA compared to other optimization techniques in terms of convergence time and failure rate.

Water Cycle Algorithm (WCA)

The water cycle algorithm is a meta-heuristic optimization technique that imitates the flowing of streams and rivers into the sea and its idea is derived by observing the water cycle in nature. Hydrologic cycle begins when water in the river, lakes, and streams is evaporated and also plants release water during the photosynthesis process. The evaporated water is carried out into the colder layer of the atmosphere to create clouds that condense and releasing water back to the earth. The initial population of WCA is called raindrops. The best raindrop is chosen as a sea, the number of good raindrops is chosen as a river and the rest of the raindrops are considered as streams. The water cycle algorithm has been used in many fields such as water resources, civil engineering, mechanical engineering, and mathematics. In the water cycle algorithm, the variables are called Raindrops for a single solution [155]. This technique showed a fast convergence and low failure rate compared to many states of the art MPPT techniques.

4.2.3 Evolutionary Algorithms (EA)

Differential Evolution (DE)

The use of differential evaluation theory was developed by Tajuddin et al. [156], in which the optimization problem will be solved by using a different formula for the evolution of candidate solutions. The solutions that have the best fitness are allowed to remain in the population and the other solutions will be removed from the population. Four DE operations, namely, *initialization*, *mutation*, *crossover*, and *selection* are required to track the optimal solution of the optimization problem. The duty cycle represents a member of the population where its value can be determined from (126).

$$D_i(j) = D_{iL} + r\,(D_{iH} - D_{iL}) \tag{126}$$

where D_{iL}, D_{iH} are the lower and the higher limit of the duty ratios, respectively, r is a random number between 0 and 1, j is representing the iteration number, and i represents the number of the searching agent (duty ratio) inside the population, N.

In each iteration, the individuals of the current population become the target vectors. For each target vector, the *mutation* operation produces a mutant vector, by adding the weighted difference between two randomly chosen vectors as shown in (127) [157].

$$D_{i,G+1} = D_{r1.GX} + F\left(D_{r2,G} - D_{r3,G}\right) \tag{127}$$

Then the crossover operation generates a new vector, called trial vector $D_{i,G}$. This is obtained by mixing the parameters of the mutant vector $D_{i,G}$ with those of the target vector $D_{i,G}$ which is used with the PV system to get the corresponding power.

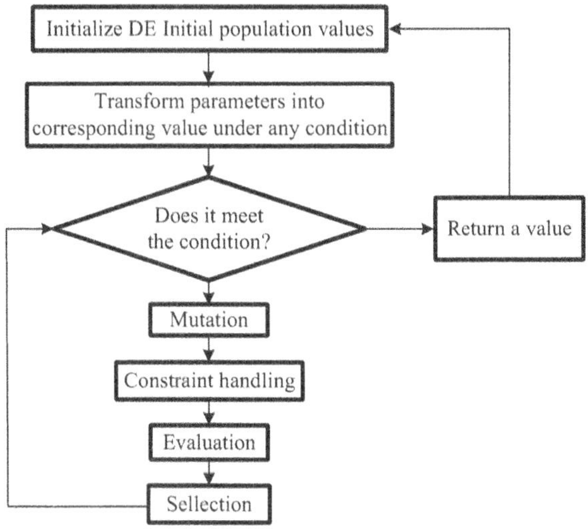

Fig. 30 The flowchart of differential Evaluation. [163]

If the trial vector obtains a better fitness value than the target vector, then the trial vector replaces the target vector in the next generation. This process will be repeated through iterations until an acute change is detected which forces the DE to reinitialize again as has been introduced before with PSO and show in Eq. (40).

Differential evaluation is reasonably it has fast convergence performance but it has high oscillations, especially in fast-changing conditions. Many modifications are introduced in the literature to further improve the performance of DE [158–160]. The flowchart of the DE is shown in Fig. 30 [159].

Genetic Algorithm (GA)

The genetic algorithm is one of the evolutionary techniques which is used to capture accurate solutions for optimization problems. This technique (GA) is inspired by biological genetics based on three operators namely, mutation, crossover, and selection. This technique is developed based on the concept of Darwin's theory of evolution in which the children having the highest fitness value means that they are powerful enough to have a higher chance for production. The GA optimization algorithm is one of the metaheuristic optimization techniques that use the generation, systematic evaluation, and enhancement of potential design solutions. The mutation operator is used to maintain the genetic diversity from one generation to the next one which can permit a stochastic variability of GA which can reduce the convergence time considerably [161].

In using the GA as an MPPT of PV systems, the duty ratio or terminal voltage is used to initially set the chromosomes to follow the GA performance for tracking the GP. These chromosomes are encoded in the form of a binary code which is used to determine the chromosome population. The *mutation* and *crossover* will be

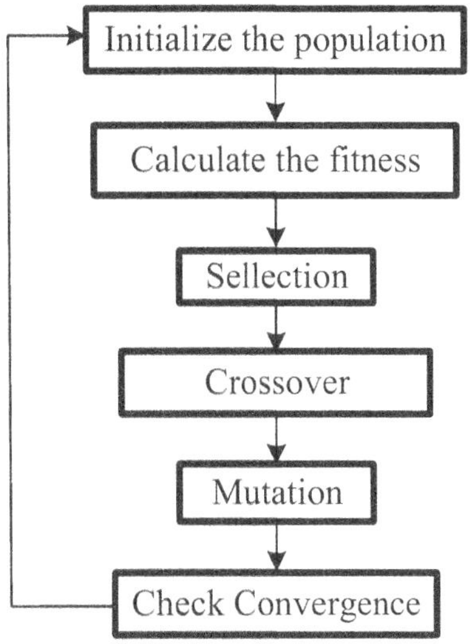

Fig. 31 The following part in GA from [166]

performed in the execution of the algorithm to produce the new generation. The new generation will be determined using the fitness function which can be obtained from the PV system. The algorithm will be repeated keeping the highest fitness value as the best solution [8]. A detailed comparison between using the GA and other traditional PV MPPT techniques is introduced in [162]. The GA has been used also to determine the optimal configuration of hybrid renewable energy systems [163]. The flowchart of the GA is shown in Fig. 31 [162].

4.2.4 Mathematical Based Algorithms (MBA)

These techniques are soft-computing-based techniques that use mathematical formulas in searching for the optimum solutions without imitating any natural, biological, or physical evolution. Some of these techniques when they are used as an MPPT of PV systems are shown in the following sections:

Fibonacci Search Algorithm (FSA)

Fibonacci search (FSA) is counted as one of the soft-computing MPPT techniques [164]. This technique is using samples derived from Fibonacci series as shown in (128):

$$F_n = F_{n-1} + F_{n-2}, \text{ where } F_0 = F_1 = 1 \text{ and } n = 2, 3, 4 \ldots \qquad (128)$$

In this case, the first 10 FS numbers are shown in the following Table:

n	0	1	2	3	4	5	6	7	8	9	10
F_n	1	1	2	3	5	8	13	21	34	55	89

The value of F_n is determining the accuracy of the solution, where the higher the value of F_n the higher the accuracy. The relation between the value of F_n and the accuracy of the results as a ratio of the exact solution is given from the following condition (129).

$$F_n \geq \frac{1}{2\varepsilon} \tag{129}$$

where ε is the acceptable tolerance, as an example if it is required the accuracy of the solution to be 0.01 from the value of the exact solution, then $F_n \geq 50$, the n should be equal to 9 as shown in the above table. Then $F_n = 55$;

Four samples will be generated as x_1, x_2, x_3, x_4, where x_3 and x_4 are selected as the lower and upper and lower limits of searching values, and x_1 and x_2 are chosen to be in between x_3 and x_4 [165]. The relation between the duty ratio and the samples x_i is used as shown in (130). These values of duty ratios will be applied to the PV system one by one and the corresponding power will be collected. The sample corresponding to the highest power is called the best sample. Based on the results obtained from the first iteration, the two conditions are shown in (131) and (132) will be determined [166].

$$D_i = \frac{x_i}{256 - 1} \tag{130}$$

If $P(x_1) > P(x_2)$, then

$$x_4^{i+1} = x_2^i,\ x_3^{i+1} = x_3^i,\ x_2^{i+1} = x_1^i,\ x_1^{i+1} = x_2^{i+1} - F_n \tag{131}$$

If $P(x_1) < P(x_2)$, then

$$x_4^{i+1} = x_4^i,\ x_3^{i+1} = x_1^i,\ x_1^{i+1} = x_2^i,\ x_2^{i+1} = x_1^{i+1} - F_n \tag{132}$$

Modified equations are introduced in [167] to reduce the convergence time as shown in (133) and (134).

If $P(x_1) > P(x_2)$, then

$$x_4^{i+1} = x_2^i,\ x_3^{i+1} = x_3^i,\ x_2^{i+1} = x_1^i,\ x_1^{i+1} = x_1^i + \frac{F_{n-1}}{F_n}\left(x_4^i - x_3^i\right) \tag{133}$$

If $P(x_1) < P(x_2)$, then

$$x_4^{i+1} = x_4^i,\ x_3^{i+1} = x_1^i,\ x_1^{i+1} = x_2^i,\ x_2^{i+1} = x_2^i + \frac{F_{n-1}}{F_n}\left(x_4^i - x_3^i\right) \tag{134}$$

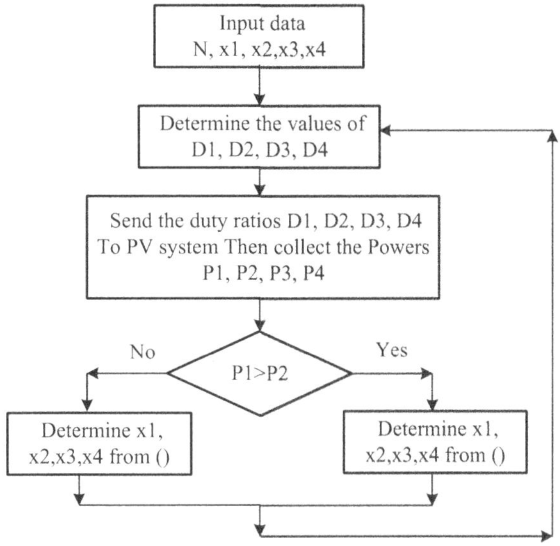

Fig. 32 Fibonacci Search Flowchart

The flowchart showing the Fibonacci Search is shown in Fig. 32. The main shortcoming of FS is its limitation to capture the GP in the case of PSC. Moreover, the complexity of the implementation of this technique which makes it not favorite to be used as an MPPT of the PV system.

Jaya Optimization Algorithm (JOA)

The generic Jaya algorithm is one of the soft-computing, mathematical-based optimization technique introduced in 2016 by Rao [168]. The operating principle of the Jaya algorithm is to iteratively update solutions for a given problem by moving them toward the best solution and away from the worst solution. Jaya algorithm does not require controlled parameters which makes Jaya is an attractive option for tracking the MPP of the PV system. Two random numbers generated from uniform distribution are used to update the candidate solutions to the optimization problem. Jaya algorithm has been used as an MPPT of partially shaded PV systems in 2017 [169]. The results obtained by this technique have been compared to two different PSO strategies. The results showed better convergence performance for the Jaya algorithm compared to the PSO in tracking the MPP of the partially shaded PV system.

The use of the Jaya algorithm in the MPPT of the PV system is done in [169] by initializing n candidate solutions and use these values to collect the corresponding power associate with each solution (Voltage). Then the best solution and worst solution can be determined and their corresponding voltage is selected as V_{best}^{0}, and V_{worst}^{0}, respectively. And use these values to determine the new generation of solutions. Use the voltage to determine the corresponding power from the PV system. If the newly generated power is greater than the one in the previous iteration, then keep the new one otherwise keep the one in the previous iteration for the same particle. The iterations should be repeated again until the system detects an acute change in power which forces the Jaya algorithm to reinitialize again.

$$V_k^{l+1} = V_k^l + rand_1\left(V_{\text{best}}^l - \left|V_{m,n}^l\right|\right) - rand_2\left(V_{\text{worst}}^l - \left|V_{m,n}^l\right|\right) \tag{135}$$

$$V_k^{l+1} = \begin{cases} V_k^{l+1} & if \ P\left(V_k^{l+1}\right) > P\left(V_k^l\right) \\ V_k^l & otherwise \end{cases} \tag{136}$$

where V_k^l is a candidate solution (PV terminal voltage) of particle k in iteration l. V_{best}^l and V_{worst}^l are the voltages of the best and worst solutions in iteration l, respectively. $rand_1$ and $rand_2$ are random numbers between 0 and 1.

To enhance the MPPT performance of PV systems in terms of faster convergence, lower oscillation, and higher efficiency, a natural cubic spline-based prediction model is incorporated into the iterative solution update of the Jaya algorithm is introduced in [170]. The utilization of the natural cubic spline model in the iterative process of the S-Jaya algorithm can avoid worse updates and thereby improves the MPPT performance. Simultaneously, the natural cubic spline model can be renewed online to maintain its prediction accuracy and produce correct decisions of updating solutions [170].

Tabu Search Algorithm (TSA)

The Tabu search algorithm (TSA) is one of the mathematical-based Algorithms (MBA) that can be used to solve nonlinear optimization problems. It is using many other optimization algorithms such as linear programming and heuristics in adaptive procedures to avoid their limitations such as their high failure rates. The TSA is one of the metaheuristics optimization techniques and it has been created in 1986 by Glover [171]. By relaxing TSA's basic rule, the TSA enhances its local search performance. First, at each step worsening moves can be accepted if no improving move is available. In addition, prohibitions are introduced to discourage the search from coming back to previously seen solutions.

The implementation of the Tabu search uses memory structures that describe the visited solutions or user-provided sets of rules. In the case of the solution that has been previously seen within a short-term period, The TSA will mark these solutions candidates to avoid the possibility to look at it again. TSA has been used in used as MPPT of the PV system in [172], where it has been divided into three sub-strategies, namely diversification search, local search, and intensification search. The diversification search is performed first by scanning the whole range of the PV array voltage for the detection of the promising voltage area. Then the hill-climbing method with a relatively large step size is employed as the local search to explore the detected promising area. These two search strategies are assembled to compose the main loop of TSA-MPPT and are repeated until the termination conditions are satisfied. Moreover, the tabu list which records the explored regions is used during the loops to avoid returning to the already visited voltage areas. Finally, after the loops are terminated, the hill-climbing method with a small step size is applied as the intensification search to refine the MPP visited so far for a more accurate result. These searches are applied in such a way that they give the TSA-MPPT method a better chance to explore among the PV array voltage range, thus avoiding the risk of trapping into the local MPP. The terms used in TSA-MPPT are illustrated first. Then the detailed procedure is described [172].

Extremum Seeking Control (ESC)

The extremum seeking control (ESC) is a nearly model-free self-optimizing control strategy that can search for the unknown and/or time-varying optimal input parameter regarding a given performance index of a nonlinear plant process. The MPPT is achieved by driving the obtained gradient by closing the search loop with an integrator. In this study, we have followed an alternative path of ESC for PV MPPT, based on the *dither–demodulation framework* described in [173]. Such an ESC scheme relies on the use of a pair of dither and demodulation signals, along with high-pass and low-pass filters, to extract gradient information. Similar to the method in [173], closing the control loop with an integrator can drive the gradient towards zero in steady state, which achieves the optimality. As the gradient information is locked to the particular dither frequency, this ESC scheme is more robust to the process noise and temporal variation of the performance map, compared to the classic ESC methods without dithering signals. This ESC method has successful applications in various systems such as axial flow compressors, jet engines, combustion, HVAC, wind turbine among others [173]. For the dither–demodulation scheme, one advantage is that the gradient information is carried by the dither harmonic, with which it is more robust against measurement noise and change in performance map. Another advantage is that particular dither action such as square wave provides transient information that can be used for fault detection [173].

4.3 Chaos Optimization Algorithm (COA)

4.3.1 Stepped-Up Chaos Optimization (SCO)

Chaos optimization Algorithm (SCO) is a very attractive technique in optimal solutions of nonlinear problems due to their ability to escape from local solutions. Many modifications are introduced to the traditional chaos optimization search techniques to improve its ability to become faster in capturing the global solution. One of these modifications is the stepped-up chaos optimization (SCO) [174] which has been used in MPPT of PV systems in order to improve its efficiency and reduce the convergence time. Also, Zhou et al. [175] proposed a dual-carrier chaotic strategy in which it selects two different chaos generations to produce chaos variables. This technique exhibits robust and accurate tracking performance when it is used in MPPT of PV system. Another strategy introduced in [174] called stepped-up SCO in which applies chaos theory and the iteration formula to produce random and ergodic chaos variables, which can be used to capture the best solution in a continuous variable searching area. Chaos mapping is shown in Eq. (137):

$$x_{n+1} = \mu \sin(\pi x_n), \quad n = 1, 2, 3, \ldots, \tag{137}$$

where μ is a control parameter. Setting $\mu = 2$, Eq. (137) is completely in chaos condition, and x_n is ergodic within $[-2, 2]$. The optimization function can be described as follows [174]:

$$f = f\left(x_i^*\right) = \max\ f(x_i),\ i = 1, 2, 3, \ldots N \quad x_i \in [c_i, d_i] \tag{138}$$

where x_i is representing the optimization variables and it can represent the duty ratio of DC/DC converter or the terminal voltage of PV array in the PV MPPT application. c_i and d_i are the lower and upper limits of x_i;$f(x_i)$ is the fitness function that represents the output power in the PV MPPT application. $f\left(x_i^*\right)$ is the maximum output power of the PV array, and x_i^* is the duty ratio or the output voltage of the PV array at the MPP. Two stepped-up SCO strategy is introduced to improve the performance of SCO in tracking the GP of PV MPPT. In this strategy, two searching strategies are implemented, namely, rough search and fine search. The rough search strategy is used to improve the performance of SCO during transient conditions meanwhile the fine search is used to reduce the search space of optimized variables and improve the convergence speed [174]. The results obtained from this work showed the superior performance of this technique in terms of convergence time and failure rate as well as the low oscillations at the steady-state condition.

4.3.2 Dual-Carrier Chaotic Search (DCCS)

The dual-carrier chaotic search (DCCS) is a modification to the traditional chaotic technique. This modification is done by using different mapping techniques than traditional chaotic techniques. In this case, if iteration continues, any variable in the optimization space can be obtained in the iterative sequence of logistic mapping. The probability distribution of the chaos point set When the logistic equation is in chaos condition, the theoretical probability distribution of the chaos point set x_n meets

Chebyshev distribution as shown in (139) [175]:

$$p(x) = \frac{1}{\pi\sqrt{x(1-x)}}, \tag{139}$$

This distribution characteristic of the chaos sequence is uniform in the middle and dense in the two ends. The probability density of x_i is given by Eq. (140)

$$p(x_i) = \frac{1}{\pi\sqrt{x_i(1 - x_i)}} \tag{140}$$

According to the Perron–Frobenious equation, defined as shown in (141).

$$p(x) = \sum_{(y_i = f^{-1}(x))} \frac{p(y_i)}{f'(y_i)} \tag{141}$$

When $\mu = 2$, and $y_i \in [0, 0.5]$, then

$$|f'(x)| = 2\pi \cos(\pi x), \; y_i = \frac{\arcsin(x/2)}{\pi} \tag{142}$$

From (141) and (142),

$$p(x) = \sum_{(y_i = f^{-1}(x))} \frac{p(y_i)}{2\pi \cos(\pi y)} = \frac{p\left(\frac{\arcsin(x/2)}{\pi}\right)}{\pi\sqrt{4 - y^2}} \tag{143}$$

$$\text{Then, } p(x) = \frac{1}{\pi\sqrt{x\left(4 - x^2\right)}} \tag{144}$$

From (144), $x_{n+1} = \mu\,(\pi yn)$, the mapping chaos point set x_n mainly centralizes in the middle and uniformly distributes in the two ends. So, this technique combines the two chaos mappings to make sure that the search is sufficient.

The results obtained from this technique shows its superiority in tracking the MPP in the case of PSC in terms of convergence time and failure rate.

5 Other Non-Model-Based Techniques

5.1 Skipping Mechanism Algorithm (SMA)

Modern techniques have been introduced recently to the field of PV MPPT called skipping mechanism technique [23, 176, 177]. This technique is working based on avoiding certain intervals from the searching space of P–V curve in which for sure the MPP does not exist. The rest of the searching area will be reduced which reduces the convergence time and increase the convergence speed of this technique. Different strategies used the same idea is introduced in literature are shown in the following points:

- Voltage Window Search (VWS) [21]
- Search–Skip–Judge (SSJ) [22]
- Maximum Power Trapezium (MPT) [23].

6 Hybrid MPPT Techniques (HMT)

The performance of MPPT is characterized using three main issues as shown in the following points:

The failure Rate: This is the factor that gives the percentage of attempts that converge to LPs with respect to the total number of attempts. This factor is very important because the convergence at the LPs reduces the generated power from the PV system considerably. So, this factor should be minimized or even become zero to have the highest generated efficiency.

The Convergence Time: The convergence time is the time required for the MPPT to reach the steady-state condition. This factor is very important especially in the case of fast change weather conditions. This factor should be minimized to improve the stability of the PV system and increase the generated efficiency.

The oscillation around steady state: This factor should be minimized to improve the stability of the PV system and to reduce the losses at a steady state.

The above three factors should be minimized to improve the performance of the PV system. These three factors have trade-off performance which means if the MPPT techniques tried to reduce one of these factors the other two may be increased. So, the idea here is to hybridize two or more MPPT techniques to work together to get the benefits of these techniques to reduce these three factors at the same time. This means that, in the transient state or fast-changing weather conditions, an accurate convergence to GP and avoidance of convergence at the LPs is required. Moreover, a fast MPPT technique (low convergence time) is required to capture the GP in a short time to improve system stability and increase the generated efficiency. Also, in the case of a steady state, it is required to have low oscillations. The idea in most of these hybrid MPPT techniques is to have a very fast and reliable convergence technique in the transient and lower oscillation technique at the steady state. The hybridizing between these techniques can be classified into four different categories as shown in the following points:

- Traditional with traditional MPPTs (T–T MPPT).
- Traditional with soft-computing MPPT (T–SC hybrid MPPT).
- Soft-computing with soft-computing (SC–SC hybrid MPPT).

These techniques are listed in the following sections:

6.1 Traditional with Traditional MPPTs (T–T Hybrid MPPT)

In this technique, the hybrid system will use two traditional techniques to work together to capture the GP very fast and reliable in the transient state and the other traditional technique is providing low oscillations at the steady-state condition. A list of these techniques are shown below:

- Beta with P&O [58]
- P&O with InCond [178]
- Fractional short-circuit current with P&O [179]
- Fractional open-circuit voltage with P & O [44].

6.2 Traditional with Soft-Computing MPPT (T–SC Hybrid MPPT)

Most of the hybrid MPPT techniques lie in this category where the soft-computing technique is used in transient to reliably capture the GP in a very short convergence time and switch the control after that to the traditional MPPT technique to reduce the oscillations at the steady-state conditions. A list of these techniques is shown in the following points:

- ACO with P&O [180]
- ANN with P&O [181–183]
- ANN with InCond [75, 184]
- BA with Beta [109]
- BA with P&O [109]
- BA with InCond [109]
- GWO with P&O, [185]
- PSO with Sliding mode controller (SMC) [186]
- PSO with P&O [187, 188]
- PSO with INC [189]
- SSA with P&O [190]
- GWO and P&O.

6.3 Soft-Computing with Soft-Computing (SC–SC Hybrid MPPT)

- ACO with FLC [191]
- CS with FLC [192]
- DE with ANN [193]
- GWO with FLC my paper [40]
- GWO-CSA [194]

- PSO with ANN [195]
- PSO with FLC [196]
- Quasi Oppositional Chaotic with GWO [197]
- Binary chaotic with CSA (BCCSO) [198]
- FLC with ANN [199, 200]
- DE with PSO [201]
- Lagrange interpolation with PSO [202]
- WOA with DE [203]
- Jaya with DE [204].

7 Conclusions and Recommendations

The MPPT techniques are very important to improve the efficiency of PV systems and increase the generated power. Three main factors are required from the MPPT which are, low failure rate, low convergence time, and low oscillations around the steady-state conditions. All these factors should be achieved with minimal cost and hardware complications. The uniform irradiance of the PV array generates only one peak in the P–V curve of the PV array. Meanwhile, in partial shading conditions, multiple peaks in the P–V curve will be generated the one with the highest generated power is called the global peak (GP), and the rest is called the local peaks (LPs). Most of the traditional MPPT techniques are able to track the peak in the uniform irradiance condition but it may be trapped in one of the LPs in the case of partial shading conditions. For this reason, soft-computing techniques are introduced to avoid this limitation of the traditional techniques. Most of the soft-computing techniques will be able to capture the GP in PSC but with a longer convergence time compared to the traditional MPPT technique. Moreover, most of the soft-computing techniques, especially swarm techniques cannot capture the GP in the case of shading pattern changes. Several improvement strategies are introduced to improve the performance of soft-computing techniques. This problem is avoided by reinitializing the searching agents if there is an acute change in generated power is detected. Another improvement strategy is introduced by initializing the searching agents at the position of anticipated peaks or at an equal distance between each searching agent. This strategy reduced the failure rate to zero and reduced the convergence time considerably. The evaluation of the proposed techniques showed that the swarm technique is reliable and fast when their control parameters are accurately tuned. From the simulation results shown in many papers reviewed in this study, the BA is recommended as one of the best MPPT technique where it can converge in less than 1 s with 0.05-s sampling time. This convergence can be reduced to less than half of the scanning criteria that have been used with the BA. The main problem of most of the soft-computing techniques is that the improvement of convergence time will increase the failure rate and oscillations at steady state. For this reason, hybrid MPPT techniques are introduced in the literature to use the fast and reliable technique in the transient state and use the techniques that have low oscillations after that like traditional techniques and fuzzy logic controllers.

References

1. Romero-Cadaval E et al (2013) Grid-connected photovoltaic generation plants: components and operation. IEEE Indus Electron Mag 7(3):6–20
2. https://www.solarnovus.com/strong-growth-trends-in-world-solar-installations_N12178.html
3. https://www.iea.org/fuels-and-technologies/solar
4. Kumar N et al A state of the art review on conventional, soft computing, and hybrid techniques for shading mitigation in photovoltaic applications. Int Trans Electric Energy Syst: e12420.
5. Eltamaly AM, Farh HM, Al-Saud MS (2019) Grade point average assessment for metaheuristic GMPP techniques of partial shaded PV systems. IET Renew Power Gener 13(8):1215–1231
6. Eltamaly AM, Farh HM, Othman MF (2018) A novel evaluation index for the photovoltaic maximum power point tracker techniques. Solar Energy 174:940–956
7. Yang B et al (2020) Comprehensive overview of maximum power point tracking algorithms of PV systems under partial shading condition. J Clean Product 121983.
8. Javed MY et al (2019) A comprehensive review on a PV based system to harvest maximum power. Electronics 8(12):1480
9. Ram JP, Babu TS, Rajasekar N (2017) A comprehensive review on solar PV maximum power point tracking techniques. Renew Sustain Energy Rev 67:826–847
10. Farh HM, Eltamaly AM (2020) Maximum power extraction from the photovoltaic system under partial shading conditions. In: Modern maximum power point tracking techniques for photovoltaic energy systems. Springer, Cham, pp 107–129
11. Bingöl, Okan, and Burçin ÖZKAYA (2019) A comprehensive overview of soft computing based MPPT techniques for partial shading conditions in PV systems. Mühendislik Bilimleri ve Tasarım Dergisi 7(4):926–939
12. Eltamaly AM (2014) Pairing between sites and wind turbines for Saudi Arabia Sites. Arab J Sci Eng 39(8):6225–6233
13. Seyedmahmoudian M et al (2016) State of the art artificial intelligence-based MPPT techniques for mitigating partial shading effects on PV systems—a review. Renew Sustain Energy Rev 64:435–455
14. Baba AO, Liu G, Chen X (2020) Classification and evaluation review of maximum power point tracking methods. Sustain Futures 2:100020.
15. Pathak PK, Yadav AK, Alvi PA (2020) Advanced solar MPPT techniques under uniform and non-uniform irradiance: a comprehensive review. J Solar Energy Eng 142(4)
16. Eltamaly AM (2013) Design and simulation of wind energy system in Saudi Arabia. In: 2013 4th International conference on intelligent systems, modelling and simulation. IEEE
17. Farh HM, Othman MF, Eltamaly AM (2017) Eltamaly. Maximum power extraction from grid-connected PV system. In: 2017 Saudi Arabia smart grid (SASG). IEEE
18. Rezk H, Eltamaly AM (2015) A comprehensive comparison of different MPPT techniques for photovoltaic systems. Solar Energy 112:1–11
19. Bawah U, Addoweesh KE, Eltamaly AM (2013) Comparative study of economic viability of rural electrification using renewable energy resources versus diesel generator option in Saudi Arabia. J Renew Sustain Energy 5(4):042701
20. El-Tamaly HH, El-Tamaly AM, El-Baset Mohammed AA (2003) Design and control strategy of utility interfaced PV/WTG hybrid system. In: The ninth international middle east power system conference, MEPCON
21. Boztepe M et al (2013) Global MPPT scheme for photovoltaic string inverters based on restricted voltage window search algorithm. IEEE Trans Indus Electron 61(7):3302–3312
22. Kermadi M et al (2018) An effective hybrid maximum power point tracker of photovoltaic arrays for complex partial shading conditions. IEEE Trans Indus Electron 66(9):6990–7000
23. Furtado AMS et al (2017) A reduced voltage range global maximum power point tracking algorithm for photovoltaic systems under partial shading conditions. IEEE Trans Indus Electron 65(4):3252–3262

24. Eltamaly AM (2018) Performance of MPPT techniques of photovoltaic systems under normal and partial shading conditions. In: Advances in renewable energies and power technologies. Elsevier, pp 115–161
25. Mahmoud YA, Xiao W, Zeineldin HH (2012) A parameterization approach for enhancing PV model accuracy. IEEE Trans Indus Electron 60(12):5708–5716
26. Eltamaly AM (2015) Performance of smart maximum power point tracker under partial shading conditions of PV systems. In: 2015 IEEE International conference on smart energy grid engineering (SEGE). IEEE
27. Ishaque K, Salam Z (2011) A comprehensive MATLAB Simulink PV system simulator with partial shading capability based on two-diode model. Solar Energy 85(9):2217–2227
28. Ma T, Yang H, Lin L (2014) Development of a model to simulate the performance characteristics of crystalline silicon photovoltaic modules/strings/arrays. Solar Energy 100:31–41
29. Eltamaly AM, Farh HM PV Characteristics, performance and modelling. In: Modern maximum power point tracking techniques for photovoltaic energy systems. Springer, Cham, pp 31–63
30. Khanna V et al (2015) A three diode model for industrial solar cells and estimation of solar cell parameters using PSO algorithm. Renew Energy 78:105–113
31. Eltamaly AM (2011) Modeling of fuzzy logic controller for photovoltaic maximum power point tracker. Trends Electric Eng 1(2)
32. Eltamaly AM, Abdelaziz AY (eds) (2019) Modern maximum power point tracking techniques for photovoltaic energy systems. Springer
33. Eltamaly AM (2015) Performance of smart maximum power point tracker under partial shading conditions of photovoltaic systems. J Renew Sustain Energy 7(4):043141
34. Zhang Q et al (2009) A novel topology for solving the partial shading problem in photovoltaic power generation system. In: 2009 IEEE 6th international power electronics and motion control conference. IEEE
35. Farh HM, Eltamaly AM, Al-Saud MS (2019) Interleaved boost converter for global maximum power extraction from the photovoltaic system under partial shading. IET Renew Power Gener 13(8):1232–1238
36. Farh MH et al (2018) Maximum power extraction from a partially shaded PV system using an interleaved boost converter. Energies 11(10):2543
37. Agamy MS et al (2012) A high efficiency DC-DC converter topology suitable for distributed large commercial and utility scale PV systems. In: 2012 15th International power electronics and motion control conference (EPE/PEMC). IEEE
38. Kouro S et al (2009) Control of a cascaded H-bridge multilevel converter for grid connection of photovoltaic systems. In: 2009 35th annual conference of IEEE industrial electronics. IEEE
39. Agarwal N, Agarwal A (2014) Mismatch losses in solar photovoltaic array and reduction techniques. MIT Int J Electric Instrum Eng 4(1):16–19
40. Eltamaly, AM, Farh. MH (2019) Dynamic global maximum power point tracking of the PV systems under variant partial shading using hybrid GWO-FLC. Solar Energy 177:306–316
41. Bhatnagar P, Nema RK (2013) Maximum power point tracking control techniques: State-of-the-art in photovoltaic applications. Renew Sustain Energy Rev 23:224–241
42. Scarpa VV, Buso S, Spiazzi G (2008) Low-complexity MPPT technique exploiting the PV module MPP locus characterization. IEEE Trans Indus Electron 56(5):1531–1538
43. Ma J et al Dem: direct estimation method for photovoltaic maximum power point tracking. Proc Comput Sci 17:537–544
44. Mroczka J, Ostrowski M (2014) A hybrid maximum power point search method using temperature measurements in partial shading conditions. Metrol Measur Syst 733–740
45. Esram T, Chapman PL (2007) Comparison of photovoltaic array maximum power point tracking techniques. IEEE Trans Energy Conver 22(2):439–449
46. Malathy S, Ramaprabha R (2013) Maximum power point tracking based on look up table approach. Adv Mater Res 768. Trans Tech Publications Ltd
47. Liu F et al (2008) Comparison of P&O and hill climbing MPPT methods for grid-connected PV converter. In: 2008 3rd IEEE conference on industrial electronics and applications. IEEE

48. Hua C-C, Fang Y-H, Chen W-T (2016) Hybrid maximum power point tracking method with variable step size for photovoltaic systems. IET Renew Power Gener 10(2):127–132
49. Strache S et al (2012) Maximum power point tracker for small number of solar cells connected in series. In: IECON 2012-38th annual conference on ieee industrial electronics society. IEEE
50. Eltamaly AM, El-Tamaly HH, Enjeti P (2002) An improved maximum power point tracker for photovoltaic energy systems. In: The 2nd Minia international conference for advanced trends in engineering
51. Saidi K, Maamoun M, Bounekhla MH (2019) A new high performance variable step size perturb-and-observe MPPT algorithm for photovoltaic system. Int J Power Electron Drive Syst 10(3):1662
52. Shobana G, Sornadeepika P, Ramaprabha R (2013) Global maximum power point tracking of photovoltaic array under partial shaded conditions. Int J Eng Res 2(3):220–224
53. Liu Y-H, Chen J-H, Huang J-W (2014) Global maximum power point tracking algorithm for PV systems operating under partially shaded conditions using the segmentation search method. Solar Energy 103:350–363
54. Ghassami AA, Sadeghzadeh SM, Soleimani A (2013) A high performance maximum power point tracker for PV systems. Int J Electric Power & Energy Systems 53:237–243
55. Tey KS, Mekhilef S (2014) Modified incremental conductance algorithm for photovoltaic system under partial shading conditions and load variation. IEEE Trans Indus Electron 61(10):5384–5392
56. Jain S, Agarwal V (2004) A new algorithm for rapid tracking of approximate maximum power point in photovoltaic systems. IEEE Power Electron Lett 2(1):16–19
57. De Brito MAG et al. (2011) Comparative analysis of MPPT techniques for PV applications. In: 2011 International conference on clean electrical power (ICCEP). IEEE
58. Li X et al. (2016) Photovoltaic modified β-parameter-based mppt method with fast tracking. J Power Electron 16(1):9–17
59. Midya P et al. (1996) Dynamic maximum power point tracker for photovoltaic applications. In: PESC Record. 27th annual IEEE power electronics specialists conference, vol 2. IEEE
60. Kitano T, Matsui M, Xu DH (2001) Power sensor-less MPPT control scheme utilizing power balance at DC link-system design to ensure stability and response. In: IECON'01. 27th annual conference of the IEEE industrial electronics society (Cat. No. 37243), vol 2. IEEE
61. Arias J et al (2004) A modular PV regulator based on microcontroller with maximum power point tracking. In: 2004 39th IAS annual meeting conference record of the 2004 IEEE industry applications conference, vol 2. IEEE
62. Shmilovitz D (2005) On the control of photovoltaic maximum power point tracker via output parameters. IEE Proc Electric Power Appl 152(2):239–248
63. Li H et al (2017) Three-point bidirectional perturbation MPPT method in PV system. In: IECON 2017, 43rd annual conference of the IEEE industrial electronics society. IEEE.
64. Wang P et al (2010) A novel approach of maximizing energy harvesting in photovoltaic systems based on bisection search theorem. In: 2010 Twenty-fifth annual IEEE applied power electronics conference and exposition (APEC). IEEE.
65. Zhang M, Wu J, Zhao H (2004) The application of slide technology in PV maximum power point tracking system. In: Fifth world congress on intelligent control and automation (IEEE Cat. No. 04EX788),vol 6. IEEE
66. Li D et al (2009) Transient maximum power point tracking for single-stage grid-tied inverter. In: 2009 IEEE energy conversion congress and exposition. IEEE.
67. Bodur M, Ermis M (1994) Maximum power point tracking for low power photovoltaic solar panels. In: Proceedings of MELECON'94. Mediterranean electrotechnical conference. IEEE
68. Eltamaly, AM (2010) Modeling of fuzzy logic controller for photovoltaic maximum power point tracker. In: Solar Future 2010 conference, Istanbul, Turkey
69. Eltamaly AM, Alolah AI, Abdulghany MY (2010) Digital implementation of general purpose fuzzy logic controller for photovoltaic maximum power point tracker. In: SPEEDAM 2010. IEEE

70. Eltamaly AM., Alolah AI, Badr BM (2010) Fuzzy controller for three phases induction motor drives. In: 2010 International conference on autonomous and intelligent systems, AIS 2010. IEEE
71. Farh HM, Eltamaly AM (2013) Fuzzy logic control of wind energy conversion system. J Renew Sustain Energy 5(2):023125
72. Eltamaly AM (2013) Design and implementation of wind energy system in Saudi Arabia. Renew Energy 60:42–52
73. Eltamaly AM, Farh HM (2013) Maximum power extraction from wind energy system based on fuzzy logic control. Electric Power Syst Res 97:144–150
74. Altas IH, Sharaf AM (2008) A novel maximum power fuzzy logic controller for photovoltaic solar energy systems. Renew Energy 33(3):388–399
75. Punitha K, Devaraj D, Sakthivel S (2013) Artificial neural network based modified incremental conductance algorithm for maximum power point tracking in photovoltaic system under partial shading conditions. Energy 62:330–340
76. Kennedy J, Eberhart R (1995) Particle swarm optimization. In: Proceedings of ICNN'95-international conference on neural networks, vol 4. IEEE
77. Eltamaly AM et al (2020) Photovoltaic maximum power point tracking under dynamic partial shading changes by novel adaptive particle swarm optimization strategy. Trans Inst Measur Control 42(1):104–115
78. Eltamaly AM, Farh HMH, Al Saud MS (2019) Impact of PSO reinitialization on the accuracy of dynamic global maximum power detection of variant partially shaded PV systems. Sustainability 11(7):2091
79. Eltamaly AM, Farh HM, Abokhalil AG (2020) A novel PSO strategy for improving dynamic change partial shading photovoltaic maximum power point tracker. Energy Sourc Part A Recovery, Utilization Environ Effects 1–15
80. Eltamaly AM, Al-Saud MS, Abo-Khalil AG (2020) Performance improvement of PV systems' maximum power point tracker based on a scanning PSO particle strategy. Sustainability 12(3):1185
81. Eltamaly AM et al (2020) Simulation and experimental validation of fast adaptive particle swarm optimization strategy for photovoltaic global peak tracker under dynamic partial shading. Renew Sustain Energy Rev 124:109719
82. Farh HMH et al (2020) A novel crow search algorithm auto-drive PSO for optimal allocation and sizing of renewable distributed generation. IEEE Access 8:27807–27820
83. Mohamed MA, Eltamaly AM, Alolah AI (2016) PSO-based smart grid application for sizing and optimization of hybrid renewable energy systems. PloS One 11(8):e0159702
84. Khaled U, Eltamaly AM, Beroual A (2017) Optimal power flow using particle swarm optimization of renewable hybrid distributed generation. Energies 10(7):1013
85. Mohamed MA, Eltamaly AM, Alolah AI (2017) Swarm intelligence-based optimization of grid-dependent hybrid renewable energy systems. Renew Sustain Energy Rev 77:515–524
86. Mohamed MA, Eltamaly AM (2018) A novel smart grid application for optimal sizing of hybrid renewable energy systems. In: Modeling and simulation of smart grid integrated with hybrid renewable energy systems. Springer, Cham, pp 39–51
87. Mohamed MA, Eltamaly AM (2018) A PSO-based smart grid application for optimum sizing of hybrid renewable energy systems. In: Modeling and simulation of smart grid integrated with hybrid renewable energy systems. Springer, Cham, pp 53–60
88. Mohamed MA, Eltamaly AM (2018) Modeling and simulation of smart grid integrated with hybrid renewable energy systems. Springer, New York
89. Mohamed MA, Eltamaly AM (2018) Modeling of hybrid renewable energy system. In: Modeling and simulation of smart grid integrated with hybrid renewable energy systems. Springer, Cham, pp 11–21
90. Eltamaly AM, Mohamed MA (2018) Optimal sizing and designing of hybrid renewable energy systems in smart grid applications. Adv Renew Energies Power Technol 231–313. Elsevier
91. Eltamaly AM, Al-Saud MS (2018) Nested multi-objective PSO for optimal allocation and sizing of renewable energy distributed generation. J Renew Sustain Energy 10(3):035302

92. Eltamaly AM, Sayed Y, Elghaffar AN (2017) Power flow control for distribution generator in Egypt using facts devices. Acta Tech Corviniensis-Bull Eng 10(2)
93. Eltamaly AM, Mohamed MA (2016) A novel software for design and optimization of hybrid power systems. J Braz Soc Mech Sci Eng 38(4):1299–1315
94. Eltamaly AM, Al-Saud MS, Abokhalil AG (2020) A novel bat algorithm strategy for maximum power point tracker of photovoltaic energy systems under dynamic partial shading. IEEE Access 8:10048–10060
95. Liu Y-H et al (2012) A particle swarm optimization-based maximum power point tracking algorithm for PV systems operating under partially shaded conditions. IEEE Trans Energy Convers 27(4):1027–1035
96. Ji B et al (2020) PV MPPT control under partial shading conditions with a particle replacement gaussian particle swarm optimization method. IEEJ J Indus Appl 9(4):418–427
97. Farh HMH et al (2019) Dynamic global power extraction from partially shaded photovoltaic using deep recurrent neural network and improved PSO techniques. Int Trans Electric Energy Syst 29(9):e12061.
98. Chowdhury SR, Saha H (2010) Maximum power point tracking of partially shaded solar photovoltaic arrays. Solar Energy Mater Solar Cells 94(9):1441–1447
99. Khare A, Rangnekar S (2013) A review of particle swarm optimization and its applications in solar photovoltaic system. Appl Soft Comput 13(5):2997–3006
100. Ishaque K, Salam Z (2012) A deterministic particle swarm optimization maximum power point tracker for photovoltaic system under partial shading condition. IEEE Trans Indus Electron 60(8):3195–3206
101. Yang X-S (2010) A new metaheuristic bat-inspired algorithm. In: Nature inspired cooperative strategies for optimization (NICSO 2010). Springer, Berlin, Heidelberg, pp 65–74
102. Oshaba AS, Ali ES, Abd Elazim SM (2015) MPPT control design of PV system supplied SRM using BAT search algorithm. Sustain Energy Grids Netw 2:51–60
103. Karagoz M, Demirel H (2017) A novel MPPT method for PV arrays based on modified bat algorithm with partial shading capability
104. Kaced K et al (2017) Bat algorithm based maximum power point tracking for photovoltaic system under partial shading conditions. Solar Energy 158:490–503
105. Oshaba AS, Ali ES, Abd Elazim SM (2017) PI controller design for MPPT of photovoltaic system supplying SRM via BAT search algorithm. Neural Comput Appl 28(4):651–667
106. Seyedmahmoudian M et al (2018) Maximum power point tracking for photovoltaic systems under partial shading conditions using bat algorithm. Sustainability 10(5):1347
107. Tey KS, Mekhilef S, Seyedmahmoudian M (2018) Implementation of BAT algorithm as maximum power point tracking technique for photovoltaic system under partial shading conditions. In: 2018 IEEE energy conversion congress and exposition (ECCE). IEEE
108. Eltamaly AM, Al-Saud MS, Abokhalil AG (2020) A novel scanning bat algorithm strategy for maximum power point tracker of partially shaded photovoltaic energy systems. Ain Shams Eng J
109. da Rocha MV, Sampaio LP, da Silva SA (2020) Comparative analysis of MPPT algorithms based on Bat algorithm for PV systems under partial shading condition. Sustain Energy Technol Assess 40:100761
110. Yang XS, Deb S (2009) Cuckoo search via Lévy flights. In: 2009 World congress on nature & biologically inspired computing, 2009 NaBIC. IEEE
111. Ahmed J, Salam Z (2014) A maximum power point tracking (MPPT) for PV system using Cuckoo Search with partial shading capability. Appl Energy 119:118–130
112. Ahmed J, Salam Z (2013) A soft computing MPPT for PV system based on Cuckoo Search algorithm. In: 4th International conference on power engineering, energy and electrical drives. IEEE
113. Anand R, Swaroop D, Kumar B (2020) Global maximum power point tracking for PV array under partial shading using Cuckoo search. In: 2020 IEEE 9th power India international conference (PIICON). IEEE

114. Ibrahim A, Obukhov S, Aboelsaud R (2019) Determination of global maximum power point tracking of PV under partial shading using cuckoo search algorithm. Appl Solar Energy 55(6):367–375
115. Ferdiansyah I et al (2019) Implementation of maximum power point tracking on solar panels using cuckoo search algorithm method. In: 2019 2nd International conference on applied information technology and innovation (ICAITI). IEEE
116. El Baset AA et al (2019) A comparative study between perturb and observe and Cuckoo search algorithm for maximum power point tracking. In: 2019 21st International middle east power systems conference (MEPCON). IEEE
117. Abo-Elyousr FK, Abdelshafy AM, Abdelaziz AY (2020) MPPT-based particle swarm and cuckoo search algorithms for PV systems. In: Modern maximum power point tracking techniques for photovoltaic energy systems. Springer, Cham, pp 379–400
118. Mohamed MA et al (2019) A novel framework-based cuckoo search algorithm for sizing and optimization of grid-independent hybrid renewable energy systems. Int J Green Energy 16(1):86–100
119. Atia DM (2020) Global maximum power point tracking-based computational intelligence techniques. In: Modern maximum power point tracking techniques for photovoltaic energy systems. Springer, Cham, pp 131–163
120. Mirjalili S, Mirjalili SM, Lewis A (2014) Grey wolf optimizer. Adv Eng Softw 69:46–61
121. Karaboga D (2005) An idea based on honey bee swarm for numerical optimization. Technical report-tr06, Erciyes university, engineering faculty, computer engineering department, vol 200
122. Motahhir S, Chouder A, El Hammoumi A, Benyoucef AS, El Ghzizal A, Kichou S, Kara K, Sanjeevikumar P, Silvestre S Optimal energy harvesting from a multistrings PV generator based on artificial bee colony algorithm. IEEE Syst J
123. Sundareswaran K et al (2014) Enhanced energy output from a PV system under partial shaded conditions through artificial bee colony. IEEE Trans Sustain Energy 6(1):198–209
124. Soufyane Benyoucef A et al (2015) Artificial bee colony based algorithm for maximum power point tracking (MPPT) for PV systems operating under partial shaded conditions. Appl Soft Comput 32:38–48
125. Chu S-C, Tsai PW, Pan JS (2006) Cat swarm optimization. In: Pacific Rim international conference on artificial intelligence. Springer, Berlin, Heidelberg
126. Guo L et al (2018) A modified cat swarm optimization based maximum power point tracking method for photovoltaic system under partially shaded condition. Energy 144:501–514
127. Dorigo M, Gambardella LM (1997) Ant colony system: a cooperative learning approach to the traveling salesman problem. IEEE Trans Evol Comput 1(1):53–66
128. Jiang LL, Maskell DL, Patra JC (2013) A novel ant colony optimization-based maximum power point tracking for photovoltaic systems under partially shaded conditions. Energy Build 58:227–236
129. Titri S et al (2017) A new MPPT controller based on the Ant colony optimization algorithm for Photovoltaic systems under partial shading conditions. Appl Soft Comput 58:465–479
130. Yang X-S (2009) Firefly algorithms for multimodal optimization. In: International symposium on stochastic algorithms. Springer, Berlin, Heidelberg
131. Sundareswaran K, Peddapati S, Palani S (2014) MPPT of PV systems under partial shaded conditions through a colony of flashing fireflies. IEEE Trans Energy Conv 29(2):463–472
132. Mirjalili S, Lewis A (2016) The whale optimization algorithm. Adv Eng Softw 95:51–67
133. Premkumar M, Sowmya R (2019) An effective maximum power point tracker for partially shaded solar photovoltaic systems. Energy Rep 5:1445–1462
134. Kumar CH, Srinivasa Rao R (2016) A novel global MPP tracking of photovoltaic system based on whale optimization algorithm. Int J Renew Energy Dev 5(3)
135. Diab AA (2020) MPPT of PV system under partial shading conditions based on hybrid whale optimization-simulated annealing algorithm (WOSA). In: Modern maximum power point tracking techniques for photovoltaic energy systems. Springer, Cham, pp 355–378

136. Eesa AS, Brifcani AM, Orman Z (2014) A new tool for global optimization problems-cuttlefish algorithm. Int J Math Comput Natural Phys Eng 8(9):1208–1211
137. Sameh MA et al (2019) Enhancing the performance of photovoltaic systems under partial shading conditions using cuttlefish algorithm. In: 2019 8th International conference on renewable energy research and applications (ICRERA). IEEE.
138. Saremi S, Mirjalili S, Lewis A (2017) Grasshopper optimisation algorithm: theory and application. Adv Eng Softw 105:30–47
139. Mansoor M et al (2020) Novel Grass Hopper optimization based MPPT of PV systems for complex partial shading conditions. Solar Energy 198:499–518
140. Meng Z, Pan J-S (2016) Monkey king evolution: a new memetic evolutionary algorithm and its application in vehicle fuel consumption optimization. Knowledge-Based Syst 97:144–157
141. Kumar N et al (2017) Maximum power peak detection of partially shaded PV panel by using intelligent monkey king evolution algorithm. IEEE Trans Indus Appl 53(6):5734–5743
142. Vahid, MZ et al (2020) Optimal, reliable and economic designing of renewable energy photovoltaic/wind system considering different storage technology using intelligent improved salp swarm optimisation algorithm, commercial application for Iran country. Int J Sustain Energy 39(5):465–485
143. Mirza AF et al (2020) A salp-swarm optimization based MPPT technique for harvesting maximum energy from PV systems under partial shading conditions. Energy Conv Manag 209:112625.
144. Premkumar M et al (2019) Analysis and simulation of bio-inspired intelligent salp swarm MPPT method for the PV systems under partial shaded conditions. Int J Comput Digital Syst 8(5):490–496
145. Shi J-Y et al (2019) Moth-flame optimization-based maximum power point tracking for photovoltaic systems under partial shading conditions. J Power Electron 19(5):1248–1258
146. Ponce-Cruz P et al (2020) A practical approach to metaheuristics using LabVIEW and MATLAB®. CRC Press
147. Mendez E et al (2018) Electric machines control optimization by a novel geo-inspired earthquake metaheuristic algorithm. In: 2018 Nanotechnology for instrumentation and measurement (NANOfIM). IEEE
148. Mendez E et al (2020) Improved MPPT algorithm for photovoltaic systems based on the earthquake optimization algorithm. Energies 13(12):3047.
149. Pincus M (1970) Letter to the editor—a Monte Carlo method for the approximate solution of certain types of constrained optimization problems. Oper Res 18(6):1225–1228
150. Kumar N et al (2017) Single sensor based MPPT for partially shaded solar photovoltaic by using human psychology optimisation algorithm. IET Gener Trans Distrib 11(10):2562–2574
151. Elbehairy NM et al (2019) Maximum power point tracking for a stand alone PV system under shading conditions using flower pollination algorithm. In: 2019 21st International middle east power systems conference (MEPCON). IEEE
152. Ram JP, Rajasekar N (2017) A new global maximum power point tracking technique for solar photovoltaic (PV) system under partial shading conditions (PSC). Energy 118:512–525
153. Chao K-H, Meng-Cheng W (2016) Global maximum power point tracking (MPPT) of a photovoltaic module array constructed through improved teaching-learning-based optimization. Energies 9(12):986
154. Nagadurga T, Narasimham PVRL, Vakula VS (2020) Global maximum power point tracking of solar PV strings using the teaching learning based optimisation technique. Int J Ambient Energy 1–12
155. Mahmoud A et al (2019) Photovoltaic array reconfiguration to reduce partial shading losses using water cycle algorithm. In: 2019 IEEE electrical power and energy conference (EPEC). IEEE.
156. Tajuddin MFN et al (2013) Evolutionary based maximum power point tracking technique using differential evolution algorithm. Energy Build 67:245–252
157. Ahmed J, Salam Z (2015) A critical evaluation on maximum power point tracking methods for partial shading in PV systems. Renew Sustain Energy Rev 47:933–953

158. Taheri H, Salam Z, Ishaque K (2010) A novel maximum power point tracking control of photovoltaic system under partial and rapidly fluctuating shadow conditions using differential evolution. In: 2010 IEEE Symposium on industrial electronics and applications (ISIEA). IEEE
159. Wei WANG, Ning LI, Shaoyuan LI (2012) A real-time modeling of photovoltaic array. Chin J Chem Eng 20(6):1154–1160
160. Ramli MAM et al (2015) A modified differential evolution based maximum power point tracker for photovoltaic system under partial shading condition. Energy Build 103:175–184
161. Ramaprabha R, Mathur BL (2011) Soft computing optimization techniques for solar photovoltaic arrays. ARPN J Eng Appl Sci 6(10):120–129
162. Shaiek Y et al (2013) Comparison between conventional methods and GA approach for maximum power point tracking of shaded solar PV generators. Solar Energy 90:107–122
163. Eltamaly AM, Al-Shamma'a AA (2016) Optimal configuration for isolated hybrid renewable energy systems. J Renew Sustain Energy 8(4):045502
164. Ramaprabha R et al (2010) Modified Fibonacci search based MPPT scheme for SPVA under partial shaded conditions. In: 2010 3rd International conference on emerging trends in engineering and technology. IEEE
165. Miyatake M et al (2004) Control characteristics of a fibonacci-search-based maximum power point tracker when a photovoltaic array is partially shaded. In: The 4th international power electronics and motion control conference, IPEMC 2004, vol. 2. IEEE
166. Ahmed NA, Miyatake M (2008) A novel maximum power point tracking for photovoltaic applications under partially shaded insolation conditions. Electric Power Syst Res 78(5):777–784
167. Ramaprabha R, Balaji M, Mathur BL (2012) Maximum power point tracking of partially shaded solar PV system using modified Fibonacci search method with fuzzy controller. Int J Electric Power Energy Syst 43(1):754–765
168. Rao R (2016) Jaya: a simple and new optimization algorithm for solving constrained and unconstrained optimization problems. Int J Indus Eng Comput 7(1):19–34
169. Huang C et al (2017) A novel global maximum power point tracking method for PV system using Jaya algorithm. In: 2017 IEEE conference on energy internet and energy system integration (EI2). IEEE.
170. Huang C et al (2017) A prediction model-guided Jaya algorithm for the PV system maximum power point tracking. IEEE Trans Sustain Energy 9(1):45–55
171. Glover F (1986) Future paths for integer programming and links to ar tifi cial intelli g en ce. Comput Oper Res 13(5):533–549
172. Zheng Y, Wei C, Lin S (2011) A maximum power point tracking method based on tabu search for PV systems under partially shaded conditions. 52–52
173. Lei P et al (2010) Extremum seeking control based integration of MPPT and degradation detection for photovoltaic arrays. In: Proceedings of the 2010 American control conference. IEEE
174. Wang L et al (2015) MPPT of PV array using stepped-up chaos optimization algorithm. Turk J Electric Eng Comput Sci 23(6):1748–1760
175. Zhou L et al (2012) Maximum power point tracking (MPPT) control of a photovoltaic system based on dual carrier chaotic search. J Control Theor Appl 10(2):244–250
176. Xu S et al (2020) A global maximum power point tracking algorithm for photovoltaic systems under partially shaded conditions using modified maximum power trapezium method. IEEE Trans Indus Electron
177. Kermadi M et al (2020) A high-performance global maximum power point tracker of PV system for rapidly changing partial shading condition. IEEE Trans Indus Electron.
178. Masood B et al (2014) Maximum power point tracking using hybrid perturb & observe and incremental conductance techniques. In: 2014 4th International conference on engineering technology and technopreneuship (ICE2T). IEEE.
179. Sher HA et al (2015) A new sensorless hybrid MPPT algorithm based on fractional short-circuit current measurement and P&O MPPT. IEEE Trans Sustain Energy 6(4):1426–1434

180. Sundareswaran K et al (2015) Development of an improved P&O algorithm assisted through a colony of foraging ants for MPPT in PV system. IEEE Trans Indus Inform 12(1):187–200
181. Jiang LL et al (2015) A hybrid maximum power point tracking for partially shaded photovoltaic systems in the tropics. Renew Energy 76:53–65
182. Mitsuya MT, de Moura Meneses AA (2019) Efficiency of hybrid MPPT techniques based on ANN and PSO for photovoltaic systems under partially shading conditions.
183. Çelik Ö, Teke A (2017) A Hybrid MPPT method for grid connected photovoltaic systems under rapidly changing atmospheric conditions. Electric Power Syst Res 152:194–210
184. Jiang LL et al (2013) A simple and efficient hybrid maximum power point tracking method for PV systems under partially shaded condition. In: IECON 2013-39th annual conference of the IEEE industrial electronics society. IEEE.
185. Mohanty S, Subudhi B, Ray PK (2016) A grey wolf-assisted perturb & observe MPPT algorithm for a PV system. IEEE Trans Energy Convers 32(1):340–347.
186. Ahmed CC, Cherkaoui M, Mokhlis M (2020) PSO-SMC controller based GMPPT technique for photovoltaic panel under partial shading effect. Int J Intell Eng Syst 13(2):307–316
187. Sundareswaran K, Palani S (2015) Application of a combined particle swarm optimization and perturb and observe method for MPPT in PV systems under partial shading conditions. Renew Energy 75:308–317
188. Kamal NA et al (2019) PSO-based adaptive perturb and observe MPPT technique for photovoltaic systems. In: International conference on advanced intelligent systems and informatics. Springer, Cham
189. Liu J et al (2017) Global MPPT algorithm with coordinated control of PSO and INC for rooftop PV array. J Eng 13:778–782
190. Balaji V, Peer Fathima A (2020) Enhancing the maximum power extraction in partially shaded PV arrays using hybrid salp swarm perturb and observe algorithm. Int J Renew Energy Res (IJRER) 10(2):898–911
191. Adly M, Besheer AH (2013) A meta-heuristics search algorithm as a solution for energy transfer maximization in stand-alone photovoltaic systems. Int J Electric Power Energy Syst 51:243–254
192. Ling C, Honghua W, Wei H (2015) Simulation study of photovoltaic power generation in maximum power point tracking based on CSA-FPI algorithm. Electric Measur Instrum 8:39–44
193. Sheraz, Muhammad, and Mohammed A. Abido. "An efficient MPPT controller using differential evolution and neural network." 2012 IEEE International Conference on Power and Energy (PECon). IEEE, 2012.
194. Davoodkhani F et al (2020) A new hybrid method based on gray wolf optimizer-crow search algorithm for maximum power point tracking of photovoltaic energy system. In: Modern maximum power point tracking techniques for photovoltaic energy systems. Springer, Cham, pp. 421–438.
195. Ngan MS, Tan CW (2011) Multiple peaks tracking algorithm using particle swarm optimization incorporated with artificial neural network. Int J Electron Commun Eng 5(10):1325–1331
196. Farh HM, Eltamaly AM, Othman MF (2018) Hybrid PSO-FLC for dynamic global peak extraction of the partially shaded photovoltaic system. PloS One 13(110):e0206171.
197. Verma P, Garg R, Mahajan P (2020) Asymmetrical interval type-2 fuzzy logic control based MPPT tuning for PV system under partial shading condition. ISA Trans
198. Satheesh Kumar S, Immanuel Selvakumar, A (2020) Maximum power point tracking and power flow management of hybrid renewable energy system with partial shading capability: a hybrid technique. Trans Inst Measur Control 0142331220909671.
199. Karatepe E, Hiyama T (2009) Polar coordinated fuzzy controller based real-time maximum-power point control of photovoltaic system. Renew Energy 34(12):2597–2606
200. Amara K et al (2018) Improved performance of a PV solar panel with adaptive neuro fuzzy inference system ANFIS based MPPT. In: 2018 7th International conference on renewable energy research and applications (ICRERA). IEEE.

201. Seyedmahmoudian M et al (2015) Simulation and hardware implementation of new maximum power point tracking technique for partially shaded PV system using hybrid DEPSO method. IEEE Trans Sustain Energy 6(3):850–862.
202. Koad RB, Zobaa AF, El-Shahat A (2016) A novel MPPT algorithm based on particle swarm optimization for photovoltaic systems. IEEE Trans Sustain Energy 8(2):468–476
203. Chaieb H, Sakly A (2018) A novel MPPT method for photovoltaic application under partial shaded conditions. Solar Energy 159:291–299
204. Kumar N et al (2017) MPPT in dynamic condition of partially shaded PV system by using WODE technique. IEEE Trans Sustain Energy 8(3):1204–1214

A Novel Hybrid Optimization Algorithm for Maximum Power Point Tracking of Partially Shaded Photovoltaic Systems

Ahmed A. Zaki Diab, Mohamed A. Mohamed, Ameena Al-Sumaiti, Hamdy Sultan, and Mahmoud Mossa

Abstract A novel hybrid optimization technique (WOA–SSA) combines whale optimization (WOA) and salp swarm (SSA) algorithms are presented. The proposed technique is designed to gather the benefits and features of SSA and WOA algorithms. The proposed technique is applied for tracking the global maximum power point (GMPP) and improve the performance of photovoltaic (PV) plants during the conditions of partial shading (PSC). The evaluation of the performance of the proposed technique is performed via MATLAB/SIMULINK. Moreover, a comparative analysis is exhibited to confirm the performance of the planned WOA–SSA technique against the conventional SSA and WOA, separately. The obtained results show the superiority of the designed WOA–SSA technique considering tracking efficiency. Moreover, the proposed WOA–SSA algorithm reaches the best solution in less time and with a better convergence speed compared to SSA and WOA. The statistical results confirm that the success rate has been enhanced from 76.6667% and 73.333%, respectively, with WOA and SSA to 95% with the proposed hybrid algorithm. Furthermore, the value of the standard deviation of 2.7877 and 2.5329 based on WOA and SSA is reduced to 0.3320 in the case of the proposed WOA–SSA.
Index Terms—Global MPPT; Partially shaded PV Systems; Hybrid optimization; Whale optimization algorithm; Salp swarm algorithm; Convergence.

A. A. Z. Diab (✉) · M. A. Mohamed · H. Sultan · M. Mossa
Electrical Engineering Department, Faculty of Engineering, Minia University, Minya, Minia 61519, Egypt
e-mail: a.diab@mu.edu.eg

M. A. Mohamed
e-mail: dr.mohamed.abdelaziz@mu.edu.eg

H. Sultan
e-mail: hamdy.soltan@mu.edu.eg

M. Mossa
e-mail: mahmoud_a_mossa@mu.edu.eg

A. Al-Sumaiti
Advanced Power and Energy Center, Electrical Engineering and Computer Science, Khalifa University, Abu Dhabi, UAE
e-mail: ameena.alsumaiti@ku.ac.ae

S. Motahhir and A. M. Eltamaly (eds.), *Advanced Technologies for Solar Photovoltaics Energy Systems*, Green Energy and Technology,
https://doi.org/10.1007/978-3-030-64565-6_7

1 Introduction

These days, research is directed toward the field of renewable energy sources (RES) and focusing on incubating energy with minimum cost [1–3]. As an example, in [1, 2, 4–10], hybrid configurations of various RES have been developed which combine solar energy, wind energy, and/or fuel cells [2, 4, 11, 12]. With the growing utilization of solar energy, many experts are intensified on evolving MPPT methods [4–7]. Because it is well known that the efficiency of generating electricity from PV systems is low as it is highly dependent on weather conditions [6, 13, 14]. Moreover, the nonlinear properties of a PV unit cause the output energy to be affected by solar radiation and temperature [11–13]. Also, the variation of the connected load has its effect [15–17]. Therefore, catching the MPP of PV plants is considered an essential solution to increase the PV system efficiency. To ensure the efficient operation of PV systems and to save the cost of generated energy, the MPP tracking of the electrical system considering the normal and irregular circumstances is a key idea of investigation [18–23].

The constructed PV systems using multi-string arrangements are a good choice considering the higher efficiency among the other configurations [24]. Moreover, parallel bypass diodes have been coupled with the PV module. These diodes are unable to affect the system considering the normal circumstances. Although with the conditions of shading effects, the bypass diodes are biased forward and flow the current instead of the PV module. For this reason, the power–voltage properties of the PV plant considering the PSC have several local peak power points while the global peak is the only one that must be cached. Therefore, this extreme global point is necessary to be tracked. Accordingly, the success of out-of-date MPPT approaches under PSC is reduced. The traditional MPPT algorithms, for instance, hill-climbing, incremental-resistance (INR), perturb and observe, and incremental conductance (INC) can only extract the local MPP and cannot separate the global and local power points. The artificial intelligence-based methods, for instance, fuzzy logic (FL) and neural network (NN) cannot separate between the global and local peaks [24–26]. References [24–26] introduced a mathematical analysis of MPPT algorithms such as P&O and INC to validate their performance during dynamic and steady-state conditions. Moreover, a neuro-fuzzy (NF) method has been implemented for the MPPT of the PV unit [7, 24]. The described problem of extracting the MPP under shading effects primes to reducing the efficiency of tracking the global MPP [27].

In recent years, several recent MPPT techniques considering optimization methods are developed to fix global MPP. The primary purpose of all such algorithms is to improve the efficiency of the PV system considering the PSC operation. Optimization techniques as a genetic algorithm (GA) [28] and particle swarm optimization (PSO) technique [29] have been used for following the MPP with the consideration of PSC. Other soft-computing techniques such as ant colony optimization [30] and cuckoo search [22] have been introduced for the same purpose to extract the global MPP. Additionally, recent optimization algorithms have also been applied with an acceptable performance such as a novel PSO algorithm that has been presented and improved the dynamic change partial shading PV MPPT [3, 31]. Moreover, the bat method is also used for tracking the MPP, considering the partial shading [32, 33]. In Ref. [34], the scanning method has been implemented to resolve the problem of PSC of the PV system. Dynamic PSO algorithm has been applied for MPPT with a battery charging system from partial shaded PV [35]. In Ref. [36], a hybrid optimization algorithm is introduced to moderate the problem of the PSC in MPPT of PV systems. The authors in [37] introduced a comprehensive analysis of many optimization techniques, which have been applied for MPPT under different partial shading patterns. A review of recent soft computing methods that are utilized to extract the MPP under different weather conditions is presented in [2, 12, 24, 36]. However, the applied algorithms, especially those which are based on one optimization algorithm, may fail to reach the MPP under some reported cases with the variation of the weather conditions [12, 36]. Therefore, more effective algorithms have to be developed and utilized for MPPT in order to improve the efficiency of PV systems considering various operation conditions, especially with PSC. Consequently, the main contribution in this chapter is to present a novel hybrid optimization algorithm for MPPT in order to improve the accuracy of power extraction as well as the efficiency of PV system.

In the present chapter, a hybrid WOA–SSA method has been developed to enhance PV system effectiveness tacking the conditions of PSC into consideration. The main purpose of the presented method is to catch and separate the global MPP from the numerous local peaks of power. Moreover, the tracking performances of trackers are evaluated and compared under different partial shading patterns using MATLAB/SIMULINK package. Furthermore, a statistical evaluation of the established algorithm is introduced.

2 PV System Under PSC

The performance of PV is tested under PSC (i.e., 4S1P configuration). The system consists of four panels, guaranteed by three modules per string. The module contains 72 series multi-crystalline silicon cells with a top capacity of 51 W. The production capacity of such a configuration is 204 Wp (i.e., 4*51 W). To determine the effect of PSC on the PV system power, five arbitrary patterns of irradiance distribution are executed besides the normal irradiance, as referred to in Table 1. G_1 refers to the radiation in the first series branch, while G_2 represents the radiation of the second series branch and so on. The Simulink model of the PV configuration tested for the design of the shadow patterns has been illustrated in Fig. 1. Figure 2 shows the PV configuration characteristics under uniform (i.e., 1 kW/m^2) and various partial shading patterns of irradiances. Moreover, the MPP details for each pattern are shown in Table 1.

Table 1 The considered PSC patterns and MPP details for each pattern

	Irradiance (W/m^2)	MPP		Irradiance (W/m^2)	MPP
Uniform irradiance	1000	$P_{max} = 203.2$ W V = 67.7 V	**PSC Pattern #4**	1000	$P_{max} = 82.01$ W V = 33.02 V
	1000			800	
	1000			400	
	1000			200	
PSC Pattern #2	1000	$P_{max} = 172$ W V = 69.49 V	**PSC Pattern #5**	1000	$P_{max} = 110.6$ W V = 71.06 V
	1000			1000	
	800			500	
	800			500	
PSC Pattern #3	1000	$P_{max} = 130.2$ W V = 51.67 V	**PSC Pattern #6**	1000	$P_{max} = 99.49$ W V = 52.62 V
	1000			800	
	800			600	
	400			400	

3 Brief Description of Whale Optimization Algorithm

The whale optimization algorithm (WOA) is a stable and robust method that can solve engineering optimization problems [38, 39]. WOA considers the whale hunting mechanism. This is pursuing an approach termed the bubble-net feeding stratagem. In this strategy, the Humpback whales need to chase small fishes swimming near to the water surface by producing a bubble net that surrounds prey and grows dramatically in a circular track. This approach has been illustrated in Fig. 3. The mathematical representation may be described as the following [38]:

$$D = \left|\vec{C} \cdot \vec{X}^*(t) - \vec{X}(t)\right| \tag{1}$$

$$X(t+1) = \vec{X}^*(t) - \vec{A}.\vec{D} \tag{2}$$

$$\vec{A} = 2\vec{a} \cdot \vec{r} - \vec{a} \tag{3}$$

$$\vec{C} = 2 \cdot \vec{r} \tag{4}$$

where t represents the current iteration, and $\vec{X}$ denotes the position vector which represents the position of the duty cycle. $\vec{A}$ and $\vec{D}$ denote coefficients vectors, $\vec{X}^*$ denotes the position vector of the finest solution and represents the optimum duty cycle solution. This position will be varied each time with iterations with the best one. $\vec{r}$ is an arbitrary random vector in [0, 1], and $\vec{a}$ is linearly limited from 2 to 0 with

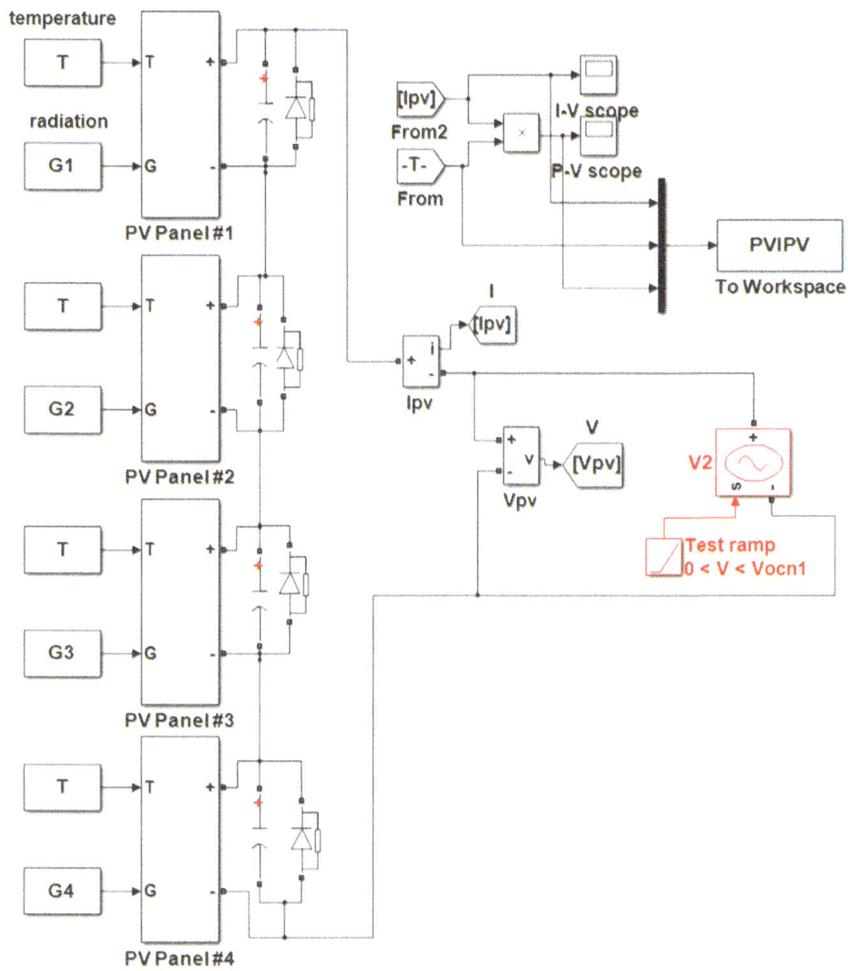

Fig. 1 The PV configuration model considering various shadow patterns

the iteration number. To describe the whole process considering the investigation stage as well as the exploitation stage; the following mathematical descriptions will be presented [39].

3.1 Mathematical Description of Exploitation Phase

This phase can be done through one of the following descriptions of air bubble net.

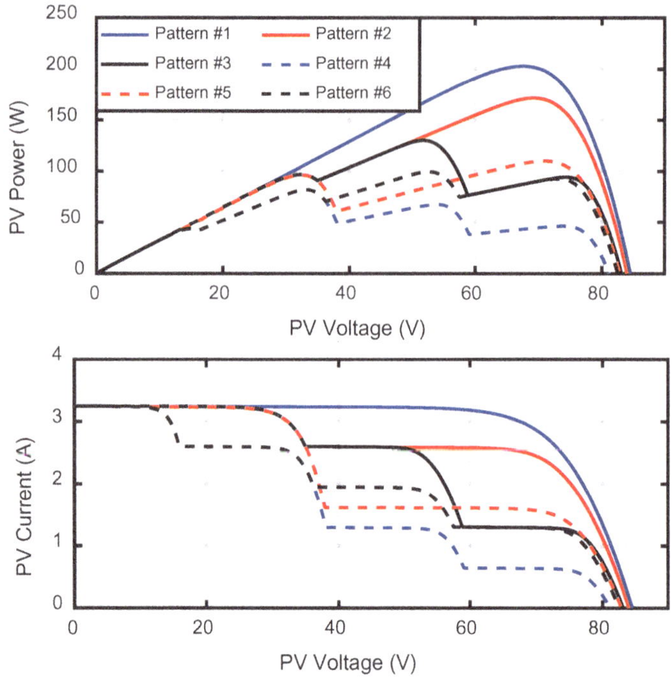

Fig. 2 The characteristics of the configurated PV

3.1.1 Shrinking Circling System

This approach is described using Eq. (3). The fluctuation scope of $\vec{A}$ maybe determined in the range of [−a, a]. $\vec{A}$ is considered in the range of [−1.0, 1.0], while the new location of $\vec{A}$ is estimated among the first agent location and the current finest agent location. Figure 3a displays the possible locations from the position of (X, Y) in the path of (X^*, Y^*), which is described by $0 \leq A \leq 1$ in a 2D space.

3.1.2 Spiral Updating Position

This method is described as exposed in Fig. 3b. The considered strategy is based on defining the distance among the whale situated at the position of (X, Y) and the prey situated at the location of (X^*, Y^*). The spiral track among the two locations of whale and prey is denoted as [40]

$$\vec{X}(t+1) = \overrightarrow{D'} \cdot e^{bl} \cdot cos(2\pi l) + \overrightarrow{X^*}(t) \tag{5}$$

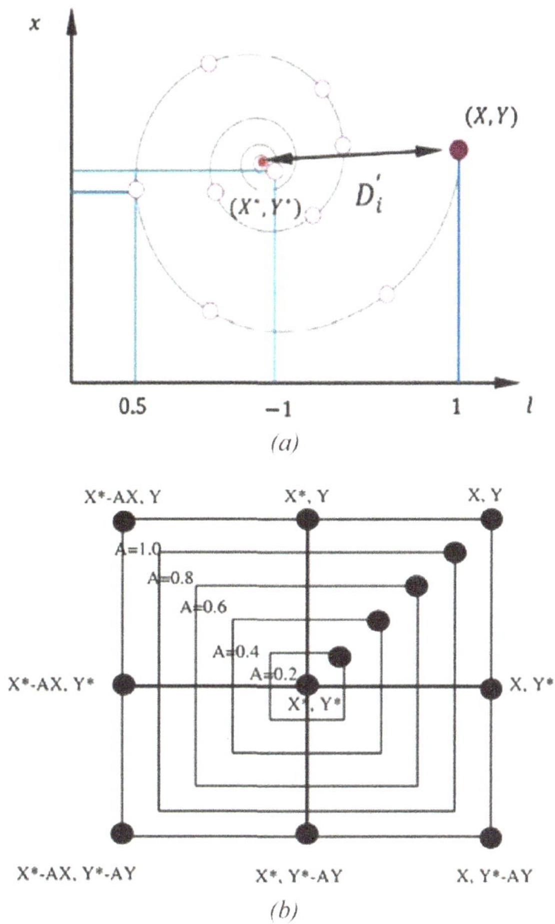

Fig. 3 Search process of the WOA technique: **a** shrinking encircling process; **b** spiral updating position process

where $\overrightarrow{D'} = \overrightarrow{X^*}(t) - \vec{X}(t)$ validates the separation of the ith whale position to the prey (finest solution). While l stands a random number in the region of $[-1, 1]$. Moreover, b denotes a constant for characterizing the state of the logarithmic spiral. Whales swim around the prey inside the shrinking circle and along with a spiral form. Additionally, a likelihood of half to indicate one of the two methods can be written as

$$\vec{X}(t+1) = \begin{cases} \vec{X}^*(t) - \vec{A} \cdot \vec{D} \ \ if p < 0.5 \\ \overrightarrow{D'} \cdot e^{bl} \cdot cos(2\pi l) + \overrightarrow{X^*}(t) \ \ if p \geq 0.5 \end{cases} \quad (6)$$

where p denotes an arbitrary number in the range of [0 and 1].

3.2 Mathematical Description of the Investigation Phase

Whales track randomly as the location of each further. Therefore, $\vec{A}$ is exploited with arbitrary numbers, which are not in the range of [−1]–[1]. This phase ensures to increase the range of searching process in order to investigate possible solutions far from the current finest position for the duty cycle. This phase, which considers the value of $\left|\vec{A}\right| > 1$, improves the search about the global pursuit. The mathematical description of this phase can be represented as in [20]:

$$D = \left|\vec{C} \cdot \vec{X}_{rand} - \vec{X}\right| \tag{7}$$

$$\vec{X}(t+1) = \vec{X}_{rand} - \vec{A} \cdot \vec{D} \tag{8}$$

Figure 4 illustrates the flowchart of the full mathematical description of the WOA procedure.

4 Salp Swarm Algorithm

The principle motivation of the salp swarm algorithm (SSA) has the same behavior as swarming conduct of salps when exploring and scavenging in oceans [41–43]. The state of a salp has been appeared in Fig. 5a. In deep oceans, salps frequently structure a swarm termed salp chain, as outlined in Fig. 5b.

To scientifically describe the salp chains, the populations will be alienated into two groups. The first group is the leader, while the other one are the followers. The leaders are at the front of the chain. Although the other salps are the followers. The location of salps has been characterized in an n-dimensional search space. Wherever n denotes the number of variables of a studied problem. Consequently, the locations of all salps remain kept in a two-dimensional matrix named x, which represents the duty cycle. Additionally, there is a food source F in the search space as the swarm's objective, which represents the power of the tested PV. The update process of the slaps can be performed as

$$x_j^1 = \begin{cases} F_j + c_1\left(\left(ub_j - lb_j\right)c_2 + lb_j\right), & c_3 \geq 0 \\ F_j - c_1\left(\left(ub_j - lb_j\right)c_2 + lb_j\right), & c_3 < 0 \end{cases} \tag{9}$$

where x_j^1 illustrates the position of the first salp (leader) in the j-th dimension. While the ub_j, lb_j indicates both the upper and lower bound of j-th dimension, respectively. Moreover, F_j denotes the location of the food source in the j-th dimension; while c_1, c_2, and c_3 stands for random numbers. Equation (9) displays the update process of the leader location, considering the food source. The coefficient c_1 denotes the

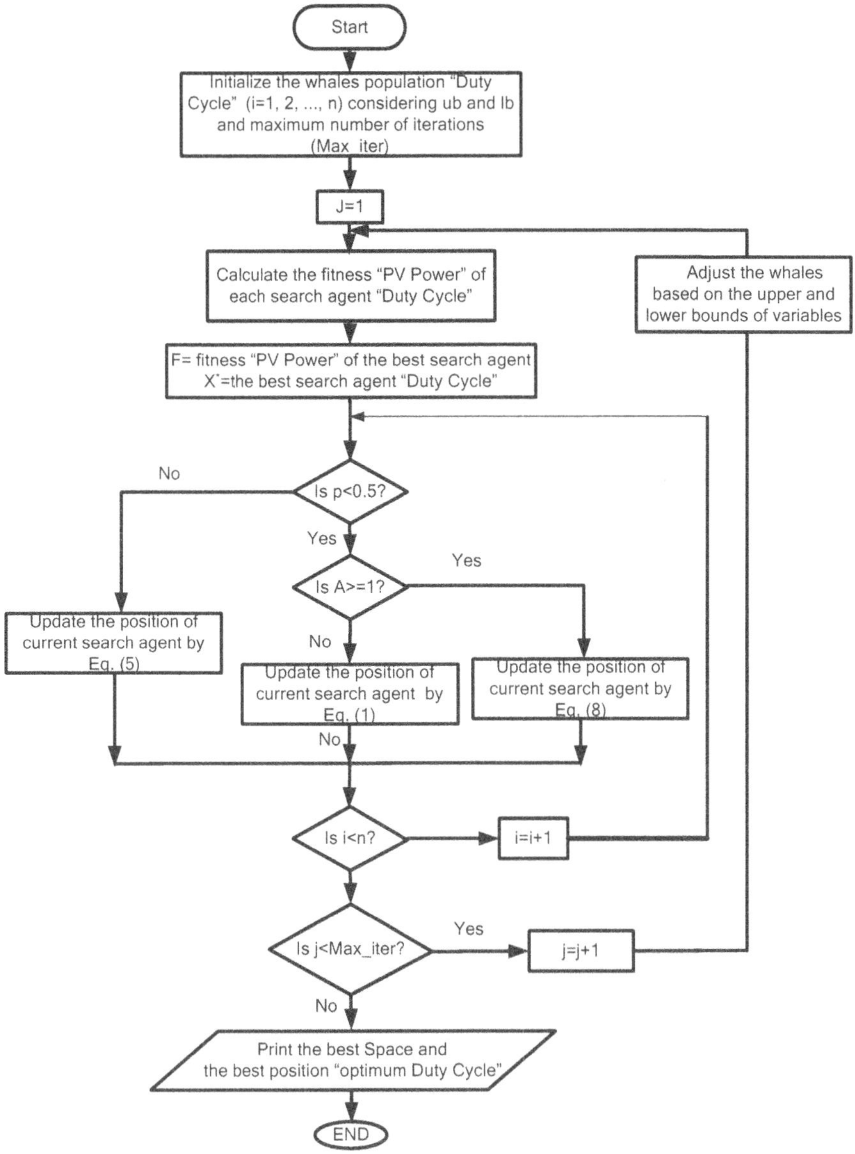

Fig. 4 The flowchart of the WOA technique

most substantial parameter in SSA. This coefficient confirms the required balance between the phases of exploration and exploitation phases. The coefficient c_1 can be calculated as

$$c_1 = 2 \cdot e^{-\left(\frac{4l}{L}\right)^2} \tag{10}$$

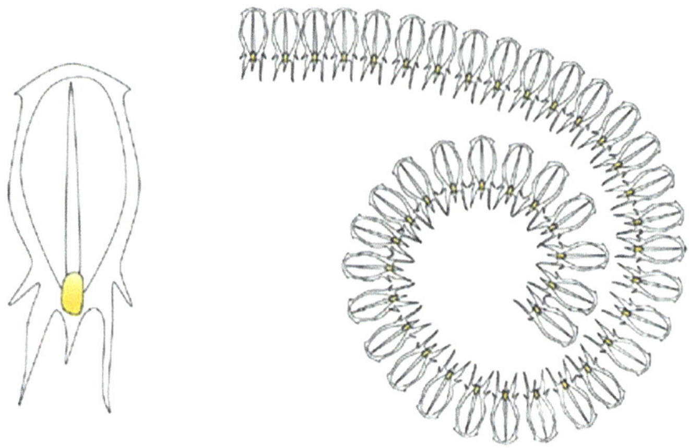

(a) (b)

Fig. 5 **a** The state of a salp **b** swarm of salps

where l denotes the present iteration, while L denotes the maximum number of iterations. The parameters c_2 and c_3 are consistently produced in the interim of [0, 1]. Actually, the parameters c_2 and c_3 are directly controlled in the following position in the j-th is to positive limitlessness or negative interminability just as the progression estimate. Furthermore, the position of the followers will be updated considering Newton's law of motion:

$$x_j^i = \frac{1}{2} \cdot a \cdot t^2 + v_0^t \tag{11}$$

where $i \geq 2$, x_j^i represents the position of i-th follower salp in j-th dimension, while t denotes time and v_0 denotes the initial speed, and $a = \frac{v_{\text{final}}}{v_0}$, where $v = \frac{x - x_0}{t}$.

Since the time in the optimization process mainly is iterative, the contradiction among iterations has been equivalent to 1, the $v_0 = 0$, and the mathematical description is as follows:

$$x_j^i = \frac{1}{2}\left(x_j^i + x_j^{i-1}\right) \tag{12}$$

The SSA algorithm has many advantages that enable it to resolve the objective optimization problems considering the obscure search spaces. Furthermore, the adaptive mechanism of this algorithm permits local solutions to be evaded and ultimately a precise assessment of the top solution obtained throughout the optimization process. The flowchart of the SSA algorithm has been shown in Fig. 6.

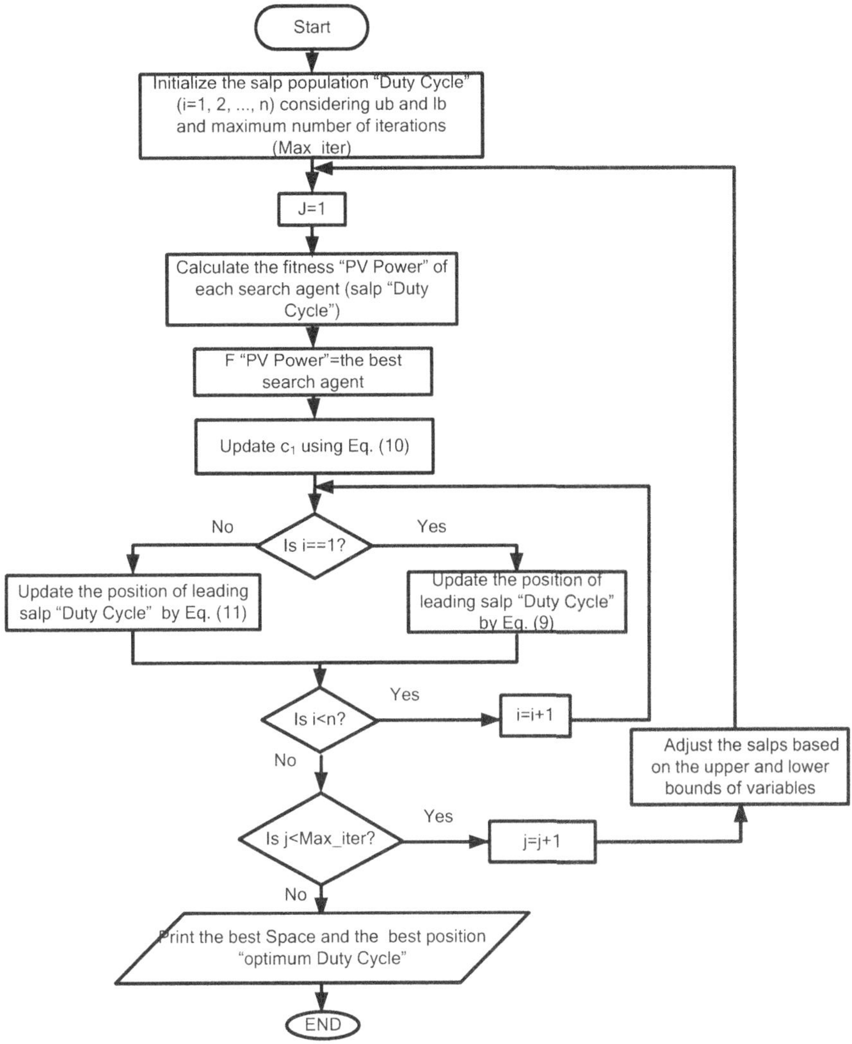

Fig. 6 The main steps of SSA

5 Hybrid WOA-SSA

The two phases of the search process for meta-heuristic optimization techniques are known as exploration and exploitation [12, 38–44]. The exploration phase is to search for global optima. Moreover, this phase is to move the search positions of the next generation as randomly as possible. On the other hand, the exploitation phase is a detailed investigating of the promising area(s) of the search space. Furthermore, the exploitation anxieties to the capability of the optimization algorithm for the local search in the region around that initiated in the exploration. The main task is to find a balance between the two phases.

The presented results in [38] showed that the WOA algorithm features the right balance of exploration and exploitation that supports the WOA technique to decide the global bests. On the other hand, the introduced results demonstrated that the WOA flops to resolve a number of the presented problems. The characteristics of the WOA technique are determined from Eq. (8) for exploration capability through the random movement of whales. Also, Eq. (6) shows that the WOA algorithm has the exploitation phase. This equation shows that the WOA permits the whales to move toward the best solution found so far. The reported results prove the power of the WOA algorithm against many techniques such as PSO and GSA.

The displayed consequences in [41] demonstrated that SSA first investigates the hunt space and afterward misuses it. Besides, SSA profoundly advances abuse utilizing the c_l parameter in the final ventures of optimization as decided from Eq. (10). Also, the outcomes demonstrated that SSA could test various areas of the hunt space in all respects adequately by coving promising locales of the inquiry space. It has been indicated that SSA can improve the nature of a lot of random solutions. At long last, the outcomes and discourse demonstrated that the exactness of the approximated global optimum is improved by SSA. This demonstrates that the SSA has an appropriate parity of investigation/local optima turning away and exploitation/convergence.

The hybrid WOA–SSA technique is proposed for the MPPT of partially shaded PV plants in this chapter in order to benefit from the advantages of both algorithms. In the hybrid WOA–SSA algorithm, to update the position of the followers which represents the duty cycle, two passes have been taken, and the average between them has been taken as the position of the next iteration. The first pass is as in the WOA algorithm as in Eqs. (1)–(8). At the same time, the second pass is as in the SSA technique considering Eqs. (9)–(11). The flowchart of the WOA–SSA technique has been illustrated in Fig. 7.

Qualitative outcomes are, for the most part, obtained from the distinctive visualization apparatuses. The furthermost well-known qualitative outcomes in the reported work are the convergence curves. Academics often record the finest result acquired so far in every iteration. The qualitative outcomes are exhibited in Appendix A. The qualitative outcomes are search chronicles of search agents in SSA, WOA, WOA–SSA algorithms through the span of iteration. The outcomes demonstrated that WOA–SSA can test various areas of the search space in all respects adequately by coving promising locales of the search space. Likewise, WOA–SSA is equipped for enhancing the quality of a lot of arbitrary results for studied issues. At long last, the outcomes demonstrated that the exactness of the approached global optimal is expanded with the WOA–SSA, which affirms the need for the proposed algorithm.

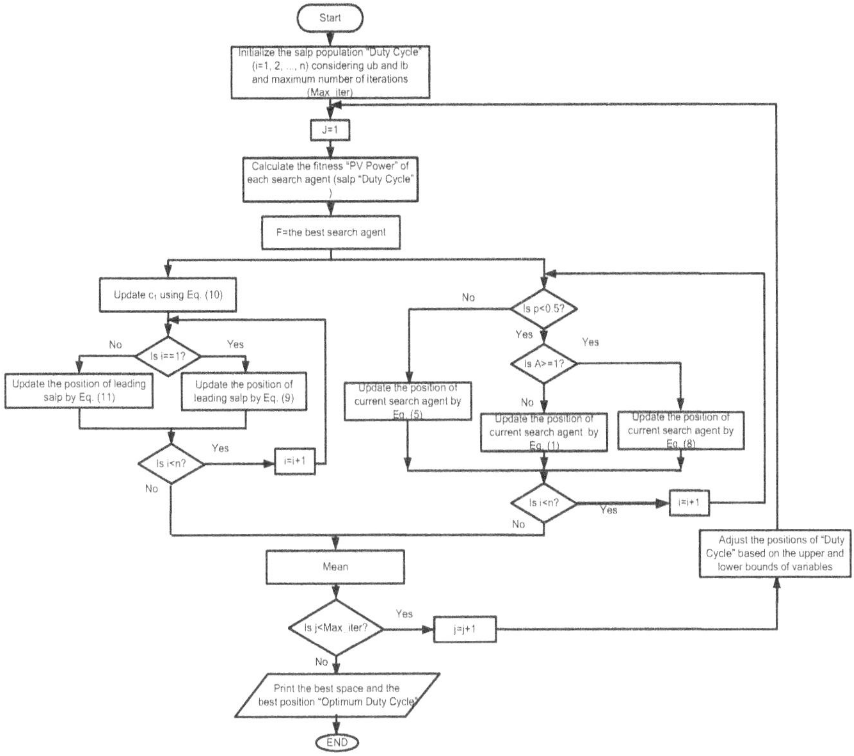

Fig. 7 The flowchart of the proposed hybrid WOA-SSA algorithm

6 Results and Discussions

The validation of the WOA–SSA technique for catching the global MPP of PV systems under PSC has been performed. The simulation of PV configuration considering various PSC patterns using the developed WOA–SSA based tracker has been performed through MATLAB/SIMULINK package. Moreover, a comprehensive comparison between the tracking performance and the efficiency of WOA–SSA-based tracker against the traditional WOA and SSA techniques have been analyzed using different PSC. This system involves LA361K51S PV panel, DC/DC boost converter with a switching frequency of 30 kHz in a continuously conducted current mode. Moreover, the input inductance is with the value of 1mH while the value of the output capacitor is taken as 47μF. Furthermore, a 60 Ω resistive load has been considered. The Simulink model of the PV system combined with WOA–SSA-based MPPT has been visualized in Fig. 8.

Case 1: Under this case study, the performance of the three algorithms have been evaluated and compared under a uniform irradiance of $G = 1000$ W/m^2 and a temperature of $T = 25$ °C. Figure 9 illustrates the performance of the developed WOA–SSA

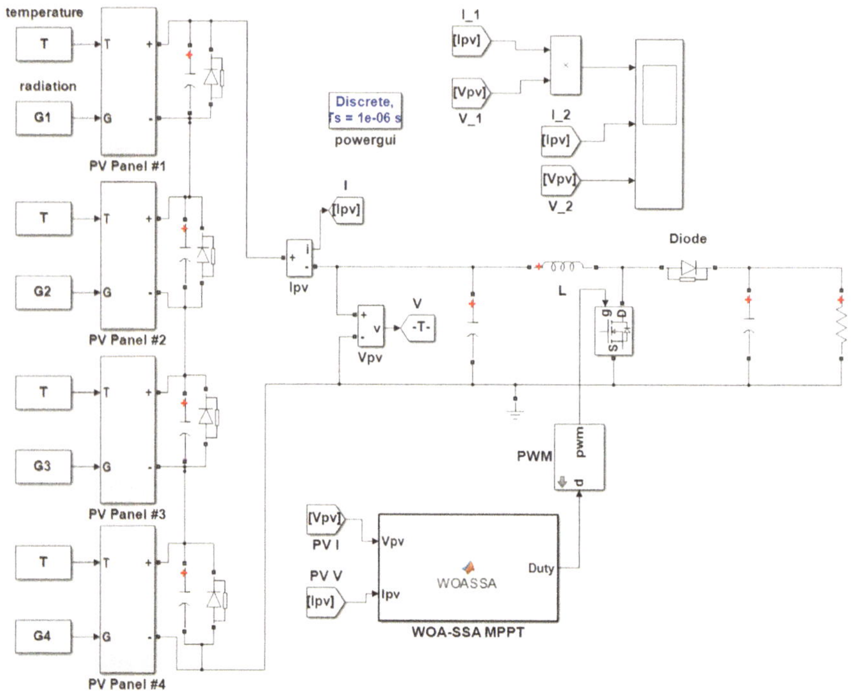

Fig. 8 The proposed configuration of PV combined with MPP tracker

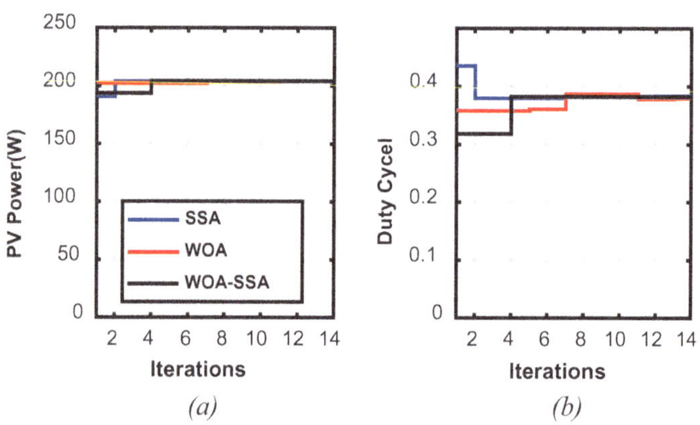

Fig. 9 **a** P–V curves **b** Duty cycle considering the uniform irradiance

algorithm against the conventional WOA and SSA algorithms. It is obvious from this figure that the three algorithms have the ability to catch the MPPT efficiently and approximately reached the same value of the maximum power. The extracted power from the PV with the three optimization algorithms is 203.2 which equals the listed one for the specified value in Table 1. The main difference among the three techniques is in the value of the convergence speed of each one to reach the MPP as exposed in Fig. 9. The speeds of convergence of the SSA, WOA, and WOA–SSA are 2, 8, and 5 iterations, respectively. The simulation results have been displayed in Fig. 10. In this figure, the duty cycle, PV power, PV voltage, and PV current of the different methods have been shown. From the figure, it is shown that the needed time of the SSA, WOA, and WOA–SSA to reach the MPP is 0.6, 0.15, and 0.1 s, respectively. Finally, it may be concluded that for the uniform irradiance case, the three algorithms can track the local MPP while the hybrid one is the best considering the required time to reach the MPP.

Case 2: For further investigation, the second case assumes the first partial shading pattern. From Fig. 2, the global MPP is at the second point on the P–V curve from two points of peaks. The ability of the optimization algorithms to determine the global

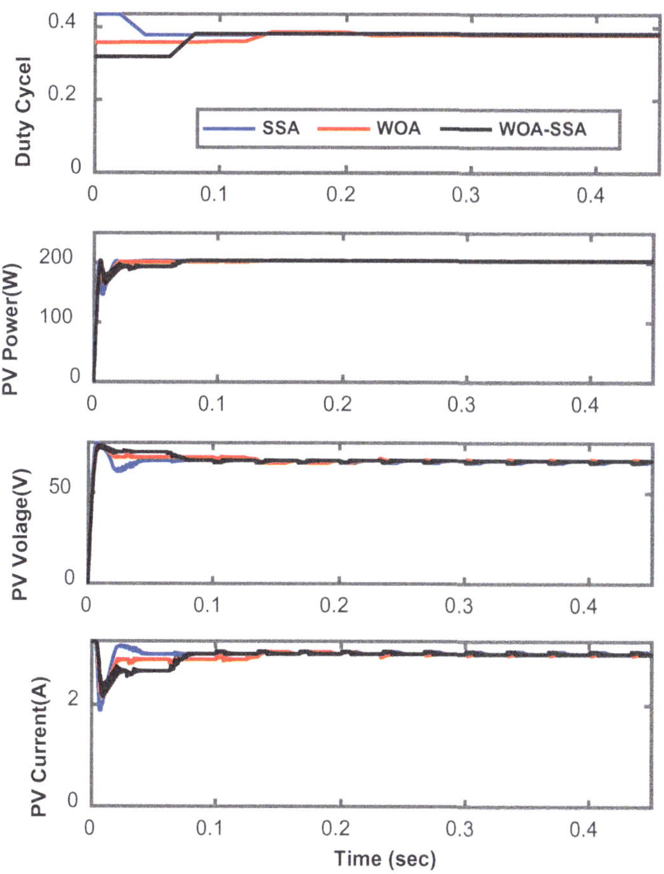

Fig. 10 Performance of PV considering the uniform irradiance

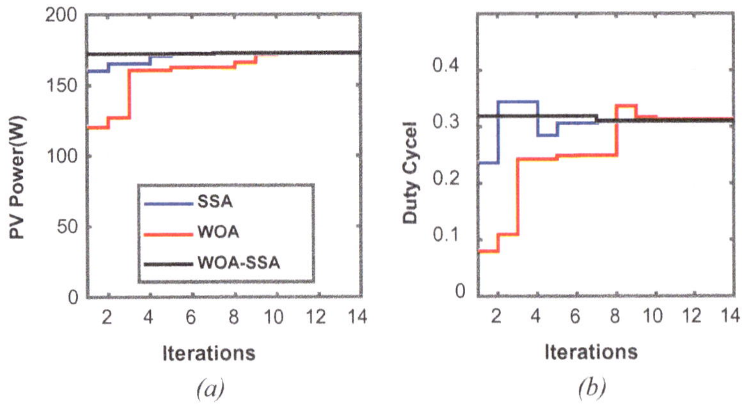

Fig. 11 **a** *P–V* curves **b** Duty cycle considering second partial shading patterns

point and to not fails in the local one is a critical task in this case of study. The performance of the three algorithms has been displayed in Fig. 11. Figure 11 demonstrates that the three algorithms can extract the MPP of 172w, which is indicated from the characteristics of Fig. 2 and Table 1. From this figure, the required iterations to reach the MPP are 10, 8, and 5 for WOA, SSA, WOA–SSA-based trackers, respectively. This confirms for the second time, the enhancement of the speed convergence of the hybrid WOA–SSA optimization algorithm over the WOA and SSA algorithms. Also, the simulation results are pictured in Fig. 12. From the figure, the WOA–SSA can locate the global MPP within the best time of 0.2 s. The introduced results confirm that the WOA–SSA has the lead of speed convergence in order to reach the global MPP.

Case 3: The third partial shading pattern is assumed in this case. Furthermore, the global MPP is at the second point on the *P-V* curve from three points of peaks, as shown in Fig. 2. The performance of three algorithms has been shown in Figs. 13 and 14. It is obvious from these figures that the three algorithms can track the MPP with reasonable accuracy. However, the other essential point is the time to reach the MPP that can be considered to judge the performance of the three algorithms. As the figures indicate, the convergence speeds are 6, 2, and 6 iterations for WOA, SSA, and WOA–SSA algorithms, respectively.

Case 4: Another case study has been considered to evaluate the performance of the three algorithms further. The fourth partial shaded pattern has been assumed. The global MPP is at the first point on the *P–V* curve from three points of peaks, as displayed in Fig. 2. The results of this case under study are shown in Figs. 15 and 16. The three algorithms have the ability to track the MPP, while the main difference is in the speed of convergence. The convergence speeds of the three methods are 13, 11, and 10 iterations for WOA, SSA, and WOA–SSA-based trackers, respectively.

Case 5: The fifth partial shaded pattern is assumed in this case. The global MPP is at the second point on the *P–V* curve from the two points of peaks, as exposed in Fig. 2. The results of this case under study are shown in Figs. 17 and 18. The

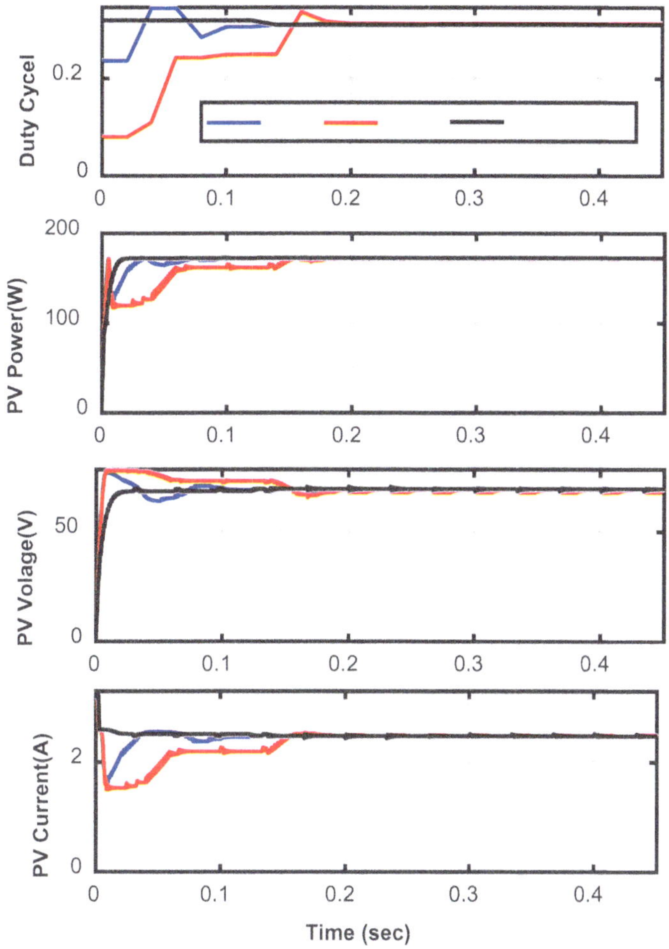

Fig. 12 Performance of PV considering the second shading pattern

convergence speeds are 14 iterations for the techniques of WOA and WOA–SSA-based trackers. From the figure, it is obvious that the SSA tracker did not catch the MPP. The WOA alone, in this case, can track the MPP, while the SSA fails to discriminate between the local and global MPPs. So, the hybridization between the two algorithms enhances tracking efficiency.

Case 6: The sixth partial shaded pattern is considered to study. The global MPP is at the third point on the *P–V* curve from three points of peaks as visualized in Fig. 2. The simulation results of this case under testing have been discovered in Figs. 19 and 20. The convergence speeds of the three trackers are 6, 9, and 5 iterations for WOA, SSA and WOA–SSA-based trackers, respectively. Also, this time, the hybrid algorithm proves its self as the best algorithm among the other investigated algorithms in this study. It is almost timeless in reaching the MPP than those other algorithms.

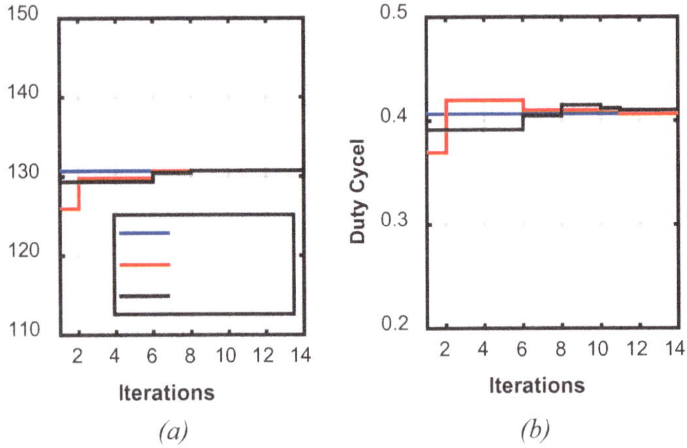

Fig. 13 **a** *P–V* curves **b** Duty cycle considering the third partial shading patterns

Fig. 14 Performance of the PV, considering the third shading pattern

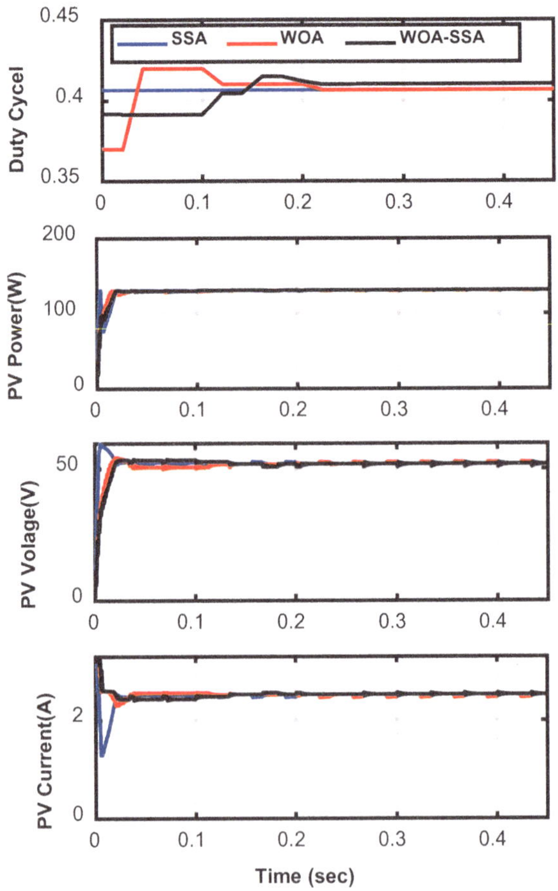

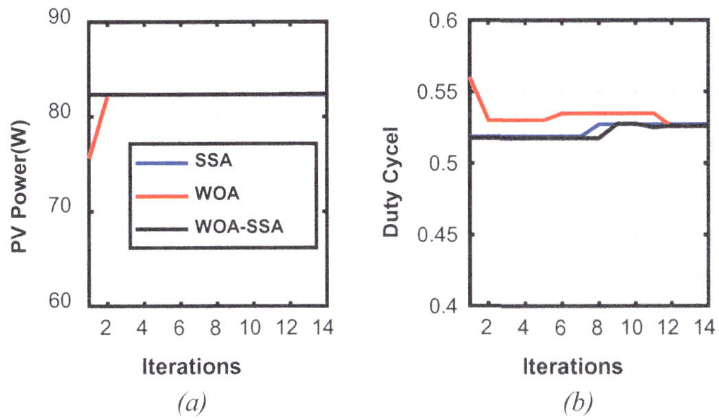

Fig. 15 **a** *P–V* curve **b** Duty Cycle considering the fourth partial shading patterns

Fig. 16 Performance of the PV system considering forth shading pattern

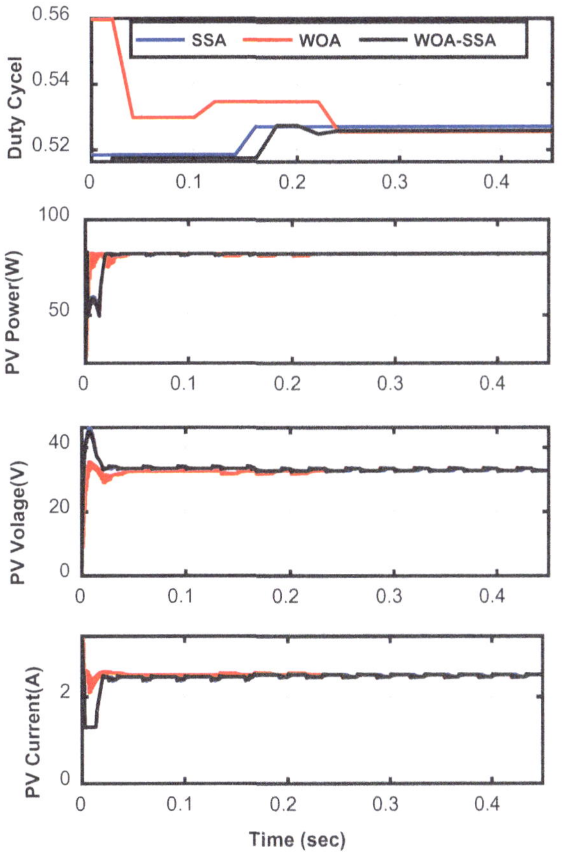

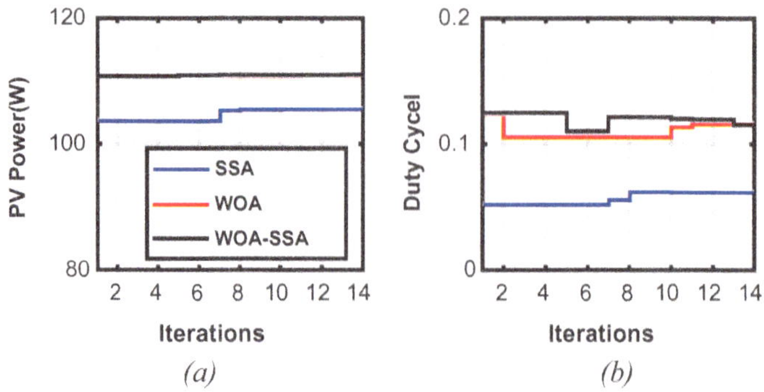

Fig. 17 **a** *P–V* curves **b** Duty cycle considering the fifth partial shading patterns

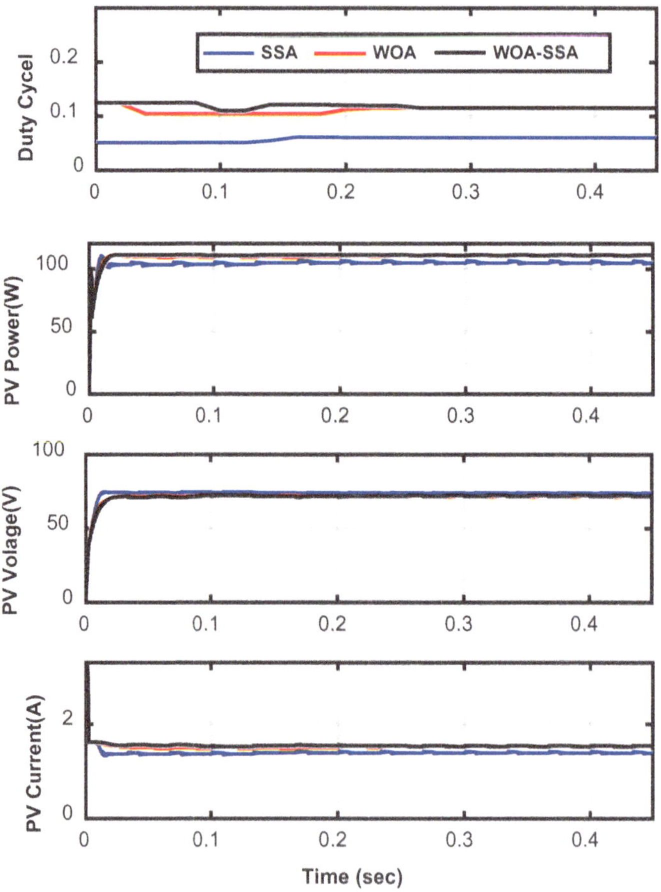

Fig. 18 Performance of the PV, considering the fifth shading pattern

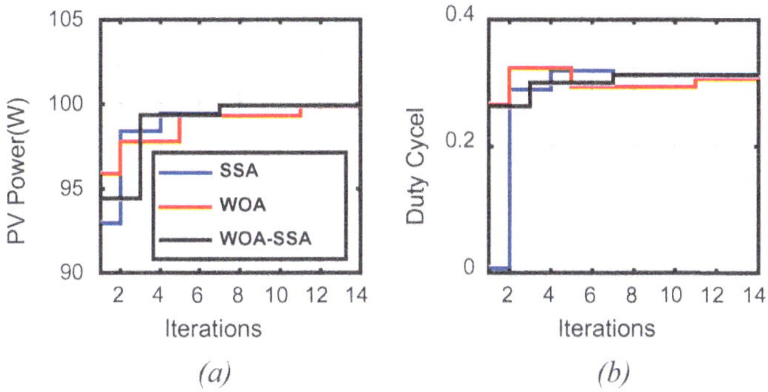

Fig. 19 **a** *P–V* curve **b** Duty cycle considering the sixth partial shading patterns

Fig. 20 Performance of the PV considering the sixth shading pattern

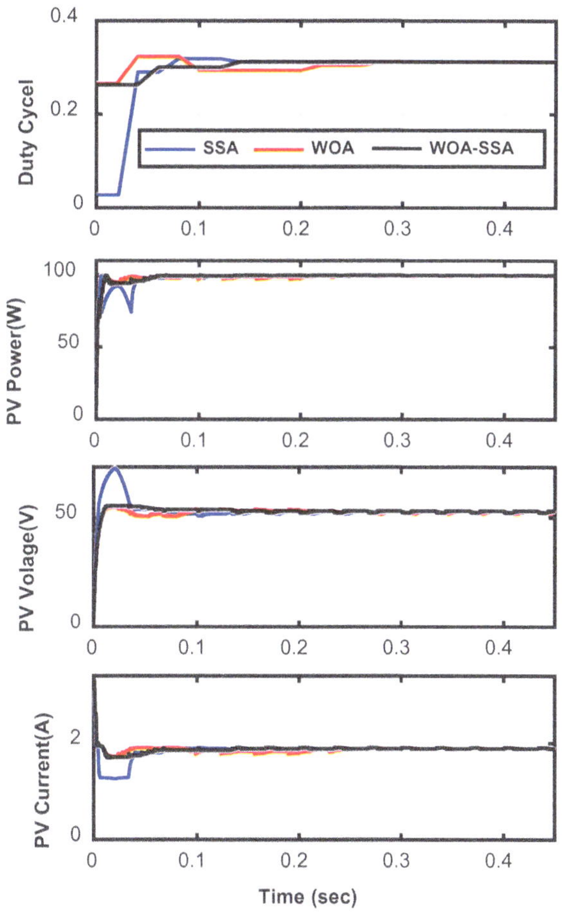

For more visualization, the PV power of WOA, SSA, and WOA–SSA algorithms under different partial shading patterns is shown in Figs. 21 and 22 besides the PV characteristics of each pattern. These figures showed that all the investigated algorithms can track the global MPP with high tracking efficiency for most of the cases under study. However, in case 5, the SSA algorithm could not catch the MPP. In most of the cases, the convergence speed of the presented WOA–SSA technique is better than that with WOA.

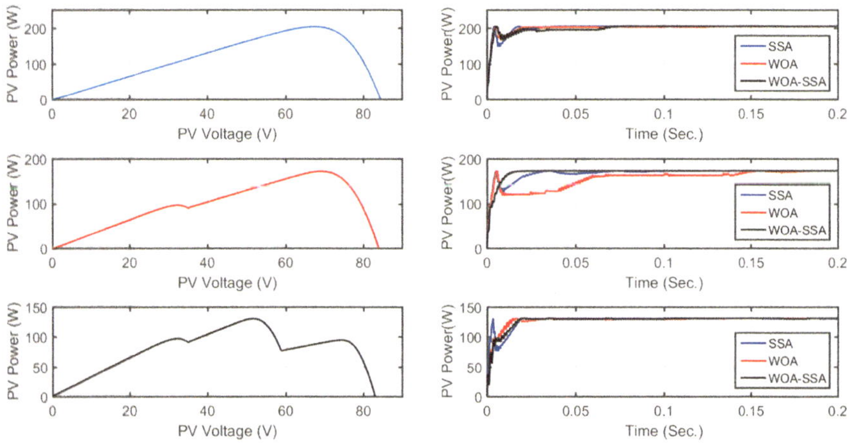

Fig. 21 **a** P–V curves **b** the PV output power for WOA, SSA, and WOA–SSA algorithms with respect to uniform patterns and Patterns #2 and 3

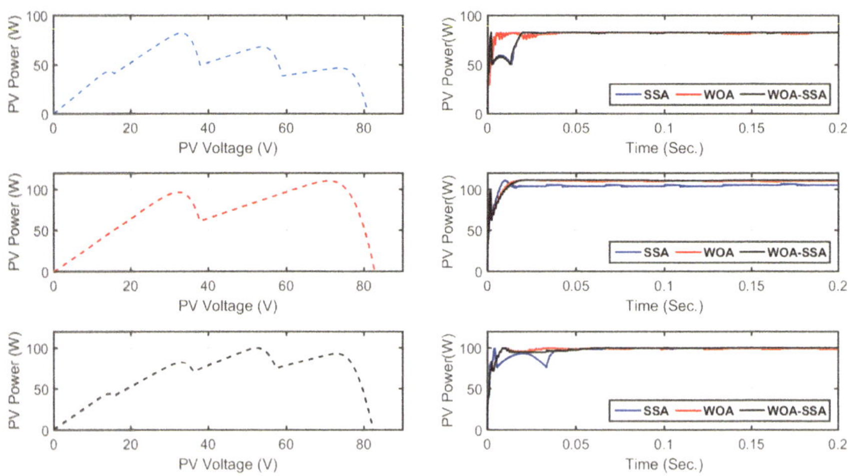

Fig. 22 **a** P–V curves **b** the PV output power for WOA, SSA, and WOA–SSA algorithms with respect to the Patterns #4, 5, and 6

For assessing the features of algorithms, the speed of convergence and efficiency of each algorithm to catch the MPP are assumed as the main criteria. The detailed results of the criteria have been displayed in Table 2. The outcomes from this table showed the priority of the WOA–SSA technique corresponding to the efficiency of tracking the maximum power. The average time and the convergence speed to track confirm the superiority of the WOA–SSA-based trackers.

7 Statistical Evaluation of the Presented Algorithms

To evaluate and validate the effectiveness of the studied algorithms, it is essential to select a set of metrics to introduce a statistical evaluation of the obtained results and therefore provide a good validation of the proposed hybridization. Table 3 shows numerous quality metrics to estimate the features of the studied algorithms. Such indices include variance, standard deviation (STD), and success rate. The latter refers to the attempts to extract the MPP. These metrics measure the true PV power ($P_{PVt)}$ gotten from the SIMULINK model shown in Fig. 1, and the PV power (P_{PVe}) found from the optimization algorithms. The parameters for the three trackers of WOA, SSA, and WOA–SSA have been established as the maximum number of iterations of (15) and population of (3). Also, each technique is performed ten times for each pattern (60 runs for each algorithm).

The performance of each algorithm for the MPP is shown in Fig. 23. From the figure, the SSA algorithm failed to discover the MPP from the PV for 16 runs for all patterns. Alternatively, the WOA technique failed to reach the MPP for the PV system for 14 runs for all patterns. Moreover, the proposed hybrid WOA–SSA algorithm failed to reach the MPP for only three runs for all patterns. The results of the figure display that the projected hybrid WOA–SSA is better than the other two techniques.

Tables 4, 5 and 6 show the performance evaluation for WOA, SSA, and WOA–SSA algorithms for different patterns. These tables showed that the hybrid WOA–SSA algorithm has an acceptable lower variance than the variance of WOA or SSA. Also, the STD confirmed that no change in the extracted maximum power at each run from the other, which designated the stability and the priority of the WOA–SSA algorithm.

The comparison among the obtained results of the three techniques has been recorded in Table 7 and Fig. 24. The indicated results of success rate, STD, and variance have been emphasized. From this table, it has been detected that the WOA–SSA technique has the lowest value of the variance. Moreover, the value of STD proved that the values of MPP did not change along the iterative process. All these results prove the stability of the WOA–SSA techniques over the algorithms of SSA and WOA. The STD is 2.7877, 2.5329, and 0.3320 for WOA, SSA, and WOA–SSA, respectively. The success rate indicates the percentage of numbers of successful tracking of the correct global MPP concerning the whole number of runs. The success rates for WOA, SSA, and WOA–SSA algorithms are 76.6667, 73.333, and 95, respectively. The superiority of the hybrid WOA–SSA has been proved from the table as well as the whole presented results.

Table 2 Comparison between the algorithms understudy

Algorithm		Pattern#1	Pattern#2	Pattern#3	Pattern#4	Pattern#5	Pattern#6	Average
WOA	Convergence Speed (Iterations)	8	10	6	13	13	10	10
	Power (Watts)	204.112	172.723	130.678	78.077	108.167	96.308	131.6775
SSA	Convergence Speed (Iterations)	3	8	2	11	## did not reach	8	##
	Power (Watts)	203.718	172.289	129.893	79.867	109.441	98.806	132.3357
WOA-SSA	Convergence Speed (Iterations)	5	5	6	10	13	6	7.5
	Power (Watts)	204.145	172.743	130.658	81.898	111.035	99.840	133.3865

Table 3 Indexes applied to assess the performance of the studied techniques

Metric	Abbreviation	Formulation
Variance	Variance	$\sigma^2 = \frac{\sum_{i=1}^{n_r}(P_{PVe,i} - \bar{P}_{pvt})^2}{n_r}$
Standard deviation	SD	$\sqrt{\frac{\sum_{i=1}^{n_r}(P_{pvt} - \bar{P}_{pvt})}{n_r}}$
Successful rate	Suc. Rate	$\frac{No.of runs which extract the MPP}{n_r} * 100\%$

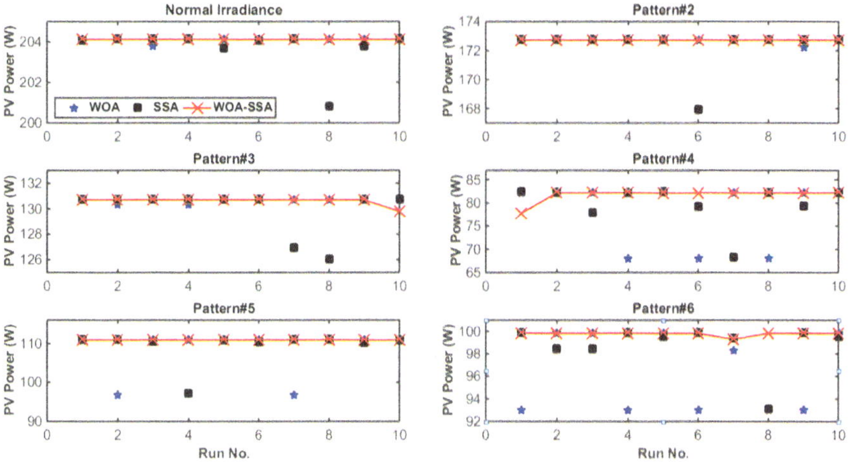

Fig. 23 The PV system output power for WOA, SSA, and WOA–SSA algorithms for 10 runs of operation under all patterns

Table 4 The performance of WOA under different shadow patterns

Cases	WOA						
	Best MPP	Worst MPP	Average MPP	Median MPP	STD	C	Suc. Rate %
Patterns 1	204.146	203.814	204.112	204.145	0.105	0.0109	90
Patterns 2	172.779	172.255	172.723	172.774	0.164	0.0269	90
Patterns 3	130.750	130.385	130.678	130.749	0.151	0.0227	80
Patterns 4	82.380	68.168	78.077	82.317	6.824	46.571	70
Patterns 5	111.049	96.776	108.167	111.019	6.003	36.0431	80
Patterns 6	99.902	93.038	96.308	95.675	3.479	12.1036	50
Average					2.7877	15.7964	76.6667

Table 5 The performance of SSA under different shadow patterns

Cases	SSA						
	Best MPP	Worst MPP	Average MPP	Median MPP	STD	Variance	Suc. Rate
Patterns 1	204.146	200.822	203.718	204.119	1.0315	1.064	70
Patterns 2	172.779	167.907	172.289	172.777	1.539	2.369	90
Patterns 3	130.7507	126.010	129.893	130.750	1.818	3.308	80
Patterns 4	82.387	68.214	79.867	82.313	4.435	19.671	60
Patterns 5	111.060	97.216	109.441	110.959	4.305	18.535	90
Patterns 6	99.908	93.133	98.806	99.532	2.069	4.2828	50
Average					2.5329	8.2050	73.333

Table 6 The performance of WOA–SSA under different shadow patterns

Cases	WOA–SSA						
	Best MPP	Worst MPP	Average MPP	Median MPP	STD	Variance	Suc. Rate
Patterns 1	204.148	204.144	204.145	204.144	0.0013	1.847e-6	100
Patterns 2	172.779	172.285	172.743	172.774	0.122	0.014	100
Patterns 3	130.750	129.835	130.658	130.749	0.288	0.083	90
Patterns 4	82.399	77.925	81.898	82.325	1.396	1.949	90
Patterns 5	111.062	111.019	111.035	111.031	0.0173	0.0003	100
Patterns 6	99.913	99.365	99.840	99.888	0.1674	0.0280	90
Average					0.3320	0.3457	95

Table 7 Comparison results between WOA, SSA, and WOA–SSA algorithms

	STD	Variance	Suc. Rate
WOA	2.7877	15.7964	76.6667
SSA	2.5329	8.2050	73.333
WOA–SSA	0.3320	0.3457	95

8 Conclusion

In this chapter, a novel hybrid WOA–SSA optimization technique has been developed and utilized as one of the meta-heuristic optimization algorithms. The core goal of the WOA–SSA is to enhance the efficiency of the PV system, considering the PSC. The proposed algorithm is utilized to determine the global MPPT from the multiple local peaks. Two other algorithms, WOA and SSA, had been implemented for the evaluation and comparison purposes with the hybrid WOA–SSA algorithm. The comprehensive evaluation of the three trackers is approved with MATLAB/SIMULINK package. The obtained results displayed the superiority of the WOA–SSA in terms of the success rate. Moreover, the results proved that the features of the WOA–SSA are faster than the other techniques considering convergence speed. Moreover, the statistical results of 10 individual run for each algorithm show that the STD is 2.7877, 2.5329, and 0.3320 for WOA, SSA, and WOA–SSA respectively, while the success rate of WOA, SSA, and WOA–SSA techniques are 76.6667, 73.333, and 95, respectively. Future work can consider the application of the presented hybrid algorithm and other recent algorithms with the grid-connected PV systems.

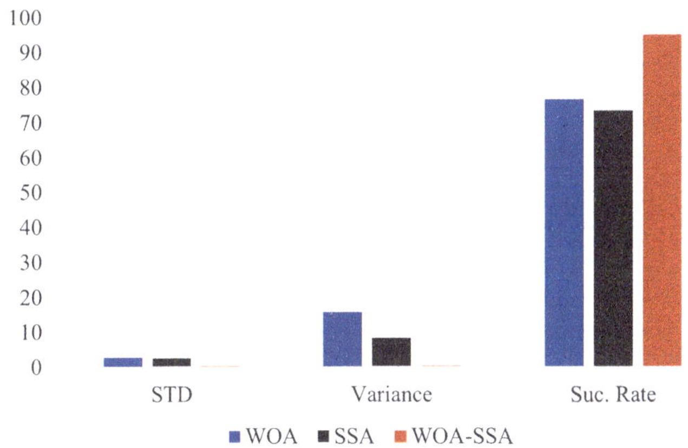

Fig. 24 Comparison results between WOA, SSA, and WOA–SSA techniques

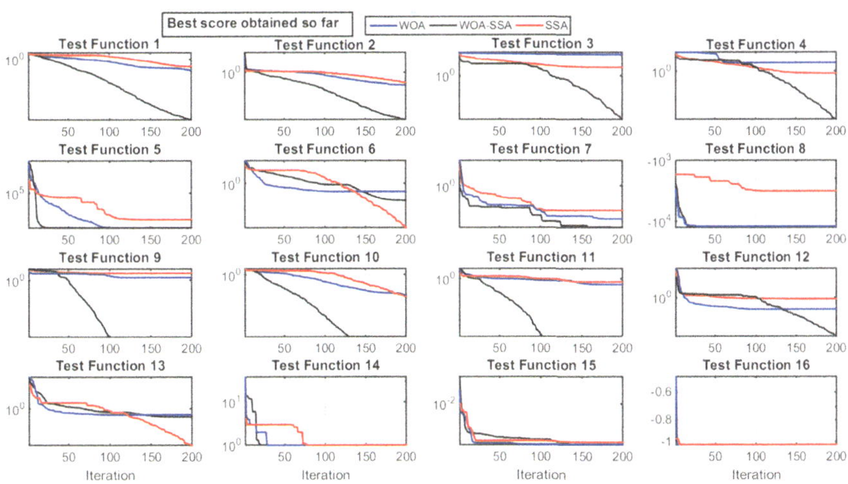

Fig. A1 WOA–SSA convergence curve for 16 test functions

Appendix A

See Fig. A1.

References

1. Diab AAZ, Sultan HM, Mohamed IS, Kuznetsov ON, Do TD (2019) Application of different optimization algorithms for optimal sizing of PV/wind/diesel/battery storage stand-alone hybrid microgrid. IEEE Access 7:119223–119245
2. Eltamaly AM, Abdelaziz AY (2019) Modern maximum power point tracking techniques for photovoltaic energy systems. Springer
3. Eltamaly AM, Farh HM, Abokhalil AG (2020) A novel PSO strategy for improving dynamic change partial shading photovoltaic maximum power point tracker. In: Energy sources, part a: recovery, utilization, and environmental effects, pp 1–15
4. Diab AAZ, El-ajmi SI, Sultan HM, Hassan YB (2019) Modified farmland fertility optimization algorithm for optimal design of a grid-connected hybrid renewable energy system with fuel cell storage: case study of Ataka, Egypt. Int J Adv Comput Sci Appl 10(8)
5. Diab AAZ, Sultan HM, Kuznetsov ON (2019) Optimal sizing of hybrid solar/wind/hydroelectric pumped storage energy system in Egypt based on different meta-heuristic techniques. Environ Sci Pollut Res 1–23
6. Motahhir et al (2020) Optimal energy harvesting from a multistrings PV generator based on Artificial Bee Colony Algorithm. IEEE Syst J
7. Narasipuram RP (2018) Optimal design and analysis of hybrid photovoltaic-fuel cell power generation system for an advanced converter technologies. Int J Math Model Numer Optim 8(3):245–276
8. Kuznetsov ON, Sultan HM, Aljendy RI, Diab AAZ (2019) Economic feasibility analysis of PV/Wind/Diesel/Battery isolated microgrid for rural electrification in south egypt. In: 2019 IEEE conference of Russian young researchers in electrical and electronic engineering (EIConRus). IEEE, pp 1001–1006
9. Sultan HM, Kuznetsov ON, Menesy AS, Kamel S (2020) Optimal configuration of a grid-connected hybrid PV/Wind/hydro-pumped storage power system based on a novel optimization algorithm. In: 2020 international youth conference on radio electronics, electrical and power engineering (REEPE). IEEE, pp 1–7
10. Saad Al-Sumaiti A, Kavousi-Fard A, Salama M, Pourbehzadi M, Reddy S, Rasheed MB (2020) Economic assessment of distributed generation technologies: a feasibility study and comparison with the literature. Energies 13(11), p 2764
11. Al-Sumaiti AS, Salama M, Konda SR, Kavousi-Fard A (2019) A guided procedure for governance institutions to regulate funding requirements of solar PV projects. IEEE Access 7:54203–54217
12. Diab AAZ, Rezk H (2017) Global MPPT based on flower pollination and differential evolution algorithms to mitigate partial shading in building integrated PV system. Solar Energy 157:171–186
13. Kırlı MS, Fahrioğlu M (2019) Sustainable development of Turkey: deployment of geothermal resources for carbon capture, utilization, and storage. Energy Sourc Part A Recover Utilizat Environ Effects 41(14):1739–1751
14. Shahid H, Kamran M, Mehmood Z, Saleem MY, Mudassar M, Haider K (2018) Implementation of the novel temperature controller and incremental conductance MPPT algorithm for indoor photovoltaic system. Solar Energy 163:235–242
15. Mohamed MA, Diab AAZ, Rezk H (2019) Partial shading mitigation of PV systems via different meta-heuristic techniques. Renew Energy 130:1159–1175

16. Mirza AF, Ling Q, Javed MY, Mansoor M (2019) Novel MPPT techniques for photovoltaic systems under uniform irradiance and partial shading. Solar Energy 184:628–648
17. Terki A, Moussi A, Betka A, Terki N (2012) An improved efficiency of fuzzy logic control of PMBLDC for PV pumping system. Appl Math Model 36(3):934–944
18. Dhimish M (2019) Assessing MPPT techniques on hot-spotted and partially shaded photovoltaic modules: comprehensive review based on experimental data. IEEE Trans Electron Dev 66(3):1132–1144
19. Reddy J, Natarajan S (2018) Control and analysis of MPPT techniques for standalone PV system with high voltage gain interleaved boost converter. Gazi Univ J Sci 31(2)
20. Dolara A, Grimaccia F, Mussetta M, Ogliari E, Leva S (2018) An evolutionary-based MPPT algorithm for photovoltaic systems under dynamic partial shading. Appl Sci 8(4):558
21. Mohamed MA, Eltamaly AM (2018) Modeling and simulation of smart grid integrated with hybrid renewable energy systems. Springer
22. Ahmed J, Salam Z (2014) A maximum power point tracking (MPPT) for PV system using Cuckoo search with partial shading capability. Appl Energy 119:118–130
23. Al-Sumaiti AS, Ahmed MH, Rivera S, El Moursi MS, Salama MMA, Alsumaiti T (2019) Stochastic PV model for power system planning applications. IET Renew Power Gener 13(16):3168–3179
24. Motahhir S, El Hammoumi A, El Ghzizal A (2020) The most used MPPT algorithms: Review and the suitable low-cost embedded board for each algorithm. J Cleaner Product 246:118983
25. Rezk H, Eltamaly AM (2015) A comprehensive comparison of different MPPT techniques for photovoltaic systems. Solar Energy 112:1–11
26. Narasipuram RP, Somu C, Yadlapalli RT, Simhadri LS (2018) Efficiency analysis of maximum power point tracking techniques for photovoltaic systems under variable conditions. Int J Innov Comput Appl 9(4):230–240
27. Tey KS, Mekhilef S, Yang H-T, Chuang M-K (2014) A differential evolution based MPPT method for photovoltaic modules under partial shading conditions. Int J Photoenergy
28. Daraban S, Petreus D, Morel C (2013) A novel global MPPT based on genetic algorithms for photovoltaic systems under the influence of partial shading. In: IECON 2013–39th annual conference of the IEEE industrial electronics society. IEEE, pp 1490–1495
29. Phimmasone V, Kondo Y, Kamejima T, Miyatake M (2010) Evaluation of extracted energy from PV with PSO-based MPPT against various types of solar irradiation changes. In: 2010 International conference on electrical machines and systems. IEEE, pp 487–492
30. Jiang LL, Maskell DL, Patra JC (2013) A novel ant colony optimization-based maximum power point tracking for photovoltaic systems under partially shaded conditions. Energy Build 58:227–236
31. Eltamaly AM, Al-Saud M, Abo-Khalil A (2020) Performance improvement of PV systems' maximum power point tracker based on a scanning PSO particle strategy. Sustainability 12(3):1185
32. Eltamaly AM, Al-Saud M, Abokhalil AG (2020) A novel scanning bat algorithm strategy for maximum power point tracker of partially shaded photovoltaic energy systems. Ain Shams Eng J
33. Eltamaly AM, Al-Saud M, Abokhalil AG (2020) A novel bat algorithm strategy for maximum power point tracker of photovoltaic energy systems under dynamic partial shading. IEEE Access 8:10048–10060
34. Chalh A, Motahhir S, Ghzizal AE, Hammoumi AE, Derouich A (2020) Global MPPT of photovoltaic system based on scanning method under partial shading condition. SN Appl Sci 2:1–5
35. Obukhov S, Ibrahim A, Diab AAZ, Al-Sumaiti AS, Aboelsaud R (2020) Optimal performance of dynamic particle swarm optimization based maximum power trackers for stand-alone PV system under partial shading conditions. IEEE Access 8:20770–20785
36. Diab AAZ (2020) MPPT of PV system under partial shading conditions based on hybrid whale optimization-simulated annealing algorithm (WOSA). In: Modern maximum power point tracking techniques for photovoltaic energy systems. Springer, pp. 355–378

37. Eltamaly AM, Farh HM, Al-Saud MS (2019) Grade point average assessment for metaheuristic GMPP techniques of partial shaded PV systems. IET Renew Power Gener 13(8):1215–1231
38. Mirjalili S, Lewis A (2016) The whale optimization algorithm. Adv Eng Softw 95:51–67
39. Kumar N, Hussain I, Singh B, Panigrahi BK (2017) MPPT in dynamic condition of partially shaded PV system by using WODE technique. IEEE Trans Sustain Energy 8(3):1204–1214
40. Hasanien HM (2017) Whale optimisation algorithm for automatic generation control of interconnected modern power systems including renewable energy sources. IET Gener Transmission Distrib 12(3):607–614
41. Mirjalili S, Gandomi AH, Mirjalili SZ, Saremi S, Faris H, Mirjalili SM (2017) Salp Swarm Algorithm: a bio-inspired optimizer for engineering design problems. Adv Eng Softw 114:163–191
42. Sutherland KR, Madin LP (2010) Comparative jet wake structure and swimming performance of salps. J Exper Biol 213(17):2967–2975
43. Hussien AG, Hassanien AE, Houssein EH (2017) Swarming behaviour of salps algorithm for predicting chemical compound activities. In: 2017 eighth international conference on intelligent computing and information systems (ICICIS). IEEE, pp 315–320
44. Belhachat F, Larbes C (2018) A review of global maximum power point tracking techniques of photovoltaic system under partial shading conditions. Renew Sustain Energy Rev 92:513–553

Distributed Maximum Power Point Tracking for Mismatched Modules of Photovoltaic Array

S. Berclin Jeyaprabha

Abstract The multiple peaks in the output *P-V* characteristics of the photovoltaic (PV) module and the complete loss of shaded module's generation due to the existing bypass diode-based scheme are eliminated through the implementation of proven distributed maximum power point tracking (DMPPT). Considering the unique behavior of each PV Module, the artificial neural network is used in the DMPPT algorithm to track the MPP at every instant by learning the unique behavior of each PV module in this chapter. This eliminates the effect of manufacturing dispersion. Though the unique MPP is identified, the inability of the DMPPT algorithm in maintaining the PV modules in its own MPP is eliminated by the compensator circuits which are introduced in the array configuration along with the DMPPT in this chapter. These compensators enabled the maintenance of each PV module in its own MPP by providing the deficient current of each module and the deficient voltage of each string. So, this configuration increases the output power by including the generation of shaded modules instead of bypassing it. The results show that the proposed configuration avoids the multiple peak condition in *P-V* characteristics and improves the efficiency of the PV array under partially shaded conditions.

Keywords Distributed maximum power point tracking · Partial shading · Compensators · PV module · Artificial neural network · Photovoltaic

1 Introduction

The growing energy demand due to the growing need of the world population, push everyone to search for an alternate energy source. Considering the environmental safety and comfort, all nations are moving toward the self-sustainable source called renewable energy system (RES). Though the governments and the industries are investing in RES to become independent, the technical problems faced by the

---

S. B. Jeyaprabha (✉)
Department of Electrical and Electronics Engineering, CHRIST (Deemed to Be University), Bangalore, India
e-mail: berclin.jeyaprabha@christuniversity.in

S. Motahhir and A. M. Eltamaly (eds.), *Advanced Technologies for Solar Photovoltaics Energy Systems*, Green Energy and Technology,
https://doi.org/10.1007/978-3-030-64565-6_8

investors are more. These problems include the higher curtailment rate of output from RES due to its fluctuating nature, losses due to the mismatch or manufacturing dispersion, partial shading from the nearby buildings, clouds, bird droppings, dust, the distribution poles, etc. Though the governments are working on the policies for the reduction of renewable power curtailment rate, the daily losses occurred due to the partial shading condition or the mismatch due to the manufacturing dispersion are huge. The researchers around the world claim that the losses under the partial shading and the mismatched condition varies from 10 to 70% [1]. The nearby structures may not be present during the initial days of renewable energy plants but built or installed after a few years. This uneven shading causes the difference in the output of PV modules which are connected in series. Similarly, the PV modules connected together may be of the same manufacturer with the same rating. But there are mismatches in the output due to the manufacturing dispersion. These differences add complications in the control of the RES which resulted in reduced yield. So, under these conditions, the investors are interested to find a solution for this sudden decrease in yield.

Generally, the RES is equipped with the higher-end power converters with inbuilt maximum power point tracker (MPPT) facilities. The MPPT is used to track the maximum power point of the *P-V* characteristics of the PV array at every instant to extract the highest power from the PV array. The number of peaks in the module's *P-V* characteristic is decided by the operating temperature, insolation, shading pattern, and the array configuration [2]. Under partial shading conditions, the PV characteristics have more peaks as the entire PV array is not able to receive the uniform solar radiation [3–6]. The conventional MPPT algorithms fail to identify the global peak or the maximum power point among the multiple peaks under the rapidly varying environments. Though the partial shading phenomena could not be predicted or avoided, the researchers around the world are working for a better MPPT algorithm which finds the highest peak at every instant among the multiple peaks to get better yield even under partially shaded conditions. The conventional MPPT algorithms like Perturb and Observe method, Incremental Conductance method, Open-Circuit Voltage method, Short Circuit Current method, Sliding Control, etc., are performing well under normal working condition of PV system but, they fail to track the maximum power point under partial shading condition due to their inability to differentiate the local and global peaks [7]. Though the soft computing approaches are able to track the maximum power point under randomly changing environments, the losses occurred in the photovoltaic (PV) module-level during the partial shading condition are more due to the unique behavior and circuit configuration. These losses are explained in this chapter. To avoid these losses, the module-level MPPT implementation and the utilization of compensating circuits are considered in this chapter. The compensating circuits are used to satisfy the requirements of the series and the parallel connections of PV modules which will enhance the operation of PV modules in their actual operating point. The proposed method will enhance the overall performance of the PV array by enabling the other healthy or unshaded modules to perform normally. The characteristics of a PV module, the MPPT algorithms, and their working at shaded, unshaded conditions, and the proposed solution for reducing the losses are explained in the following sections.

2 Effect of Partial Shading and Mismatch in PV Array

Initially, to understand the characteristics of the PV array, the structure of the same is given in detail. The fundamental PV element called solar cells or PV cells is connected together to form a PV module or a PV panel. As an example, a PV panel that can give an open-circuit voltage V_{OC} of 21 V and a short circuit current I_{SC} of 3.74 A is built by connecting 36 solar cells in series. To satisfy the required voltage and the current rating of the PV system, many such PV modules are connected in series and parallel. The series connection of many PV modules creates a PV string. Also, the parallel connection of these PV strings together forms a PV array. If the required output voltage is 210 V, then, ten such PV modules are connected in series and forming a string. So, each string is rated for an open-circuit voltage V_{OC} of 210 V and a short circuit current I_{SC} of 3.74 A. Similarly, if the required current rating is 187 A, then 50 such strings are connected in parallel to form an array. So, the array is rated for 39.27 kW or an open-circuit voltage of V_{OC} of 210 V and a short circuit current I_{SC} of 187 A. The PV array with ten modules in each string and 50 strings in parallel is given in Fig. 1.

The PV array is partially shaded in the above condition. The shading is not uniform throughout the array, because of its uneven coverage among all the strings of the array. Based on the shading pattern, the strings are divided into different groups named G1, G2, and G3. The shading pattern is given in Table 1. The G1 consists of 20 strings in parallel. All the 20 strings of G1 are having the same shading pattern. In G1, each string has four shaded modules and six unshaded modules connected in series. The G2 consists of 18 strings in parallel and each string have three shaded modules and seven unshaded modules. Similarly, in G3, there are 12 parallel strings and each string consists of one shaded module and nine unshaded modules. The characteristics of the PV array are drawn with the assumption that the shaded modules and unshaded modules are receiving the solar insolation of 400 W/m^2 and 1000 W/m^2, respectively. In Fig. 2a, b the *I*-*V* and *P*-*V* characteristics of the unshaded PV array are given. The *P*-*V* characteristics of the unshaded array have a single peak power point. Similarly in Fig. 2c, d, the *I*-*V* and *P*-*V* characteristics of the shaded PV array are given. The *P*-*V* characteristics given in Fig. 2d have multiple peaks due to shading. These multiple peaks complicate the MPPT. Even, in partial shading, there are two types. They are called as static shading and dynamic shading. If the shade on the PV array stays for a long time in the same position, then it is called as static shading. If the shade is moving fast due to the wind then it is called dynamic shading. Under this dynamic shading the maximum power point P_{MPP} varies randomly.

Generally, the PV modules are equipped with parallel-connected bypass diodes to avoid the hotspot problem under the shaded or mismatched condition as shown in Fig. 3. The role of bypass diode is to divert the current out of the shaded module

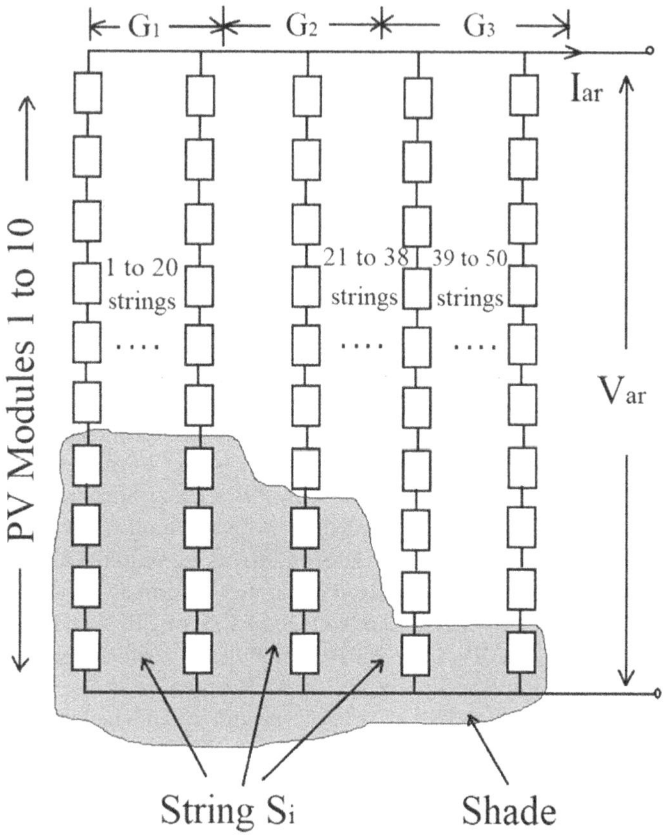

Fig. 1 Configuration of PV array with shading

Table 1 Shading pattern and the configuration of PV array

Group	No. of shaded modules in string	No. of unshaded modules in string	No. of strings in group
G1	4	6	20
G2	3	7	18
G3	1	9	12

when it is getting activated by the different shading patterns. The modules under normal solar insolation generate more current than the shaded modules. To avoid the loading effect in shaded modules, the bypass diodes which are connected across the shaded modules are activated and diverting the current through unshaded modules and bypass diodes. When the bypass diodes are activated, the shaded modules are shorted and their generation is completely lost.

Fig. 2 **a** *I-V* characteristics of array without shading **b** *P-V* characteristics of array without shading **c** *I-V* characteristics of array with shading **d** *P-V* characteristics of array with shading

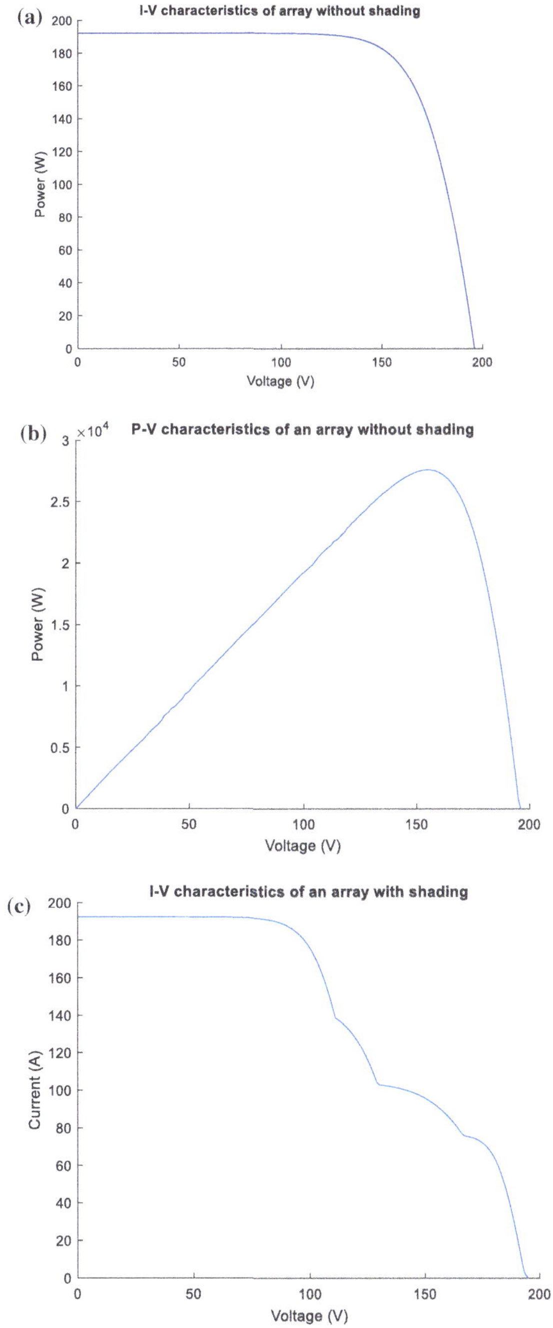

Fig. 2 (continued)

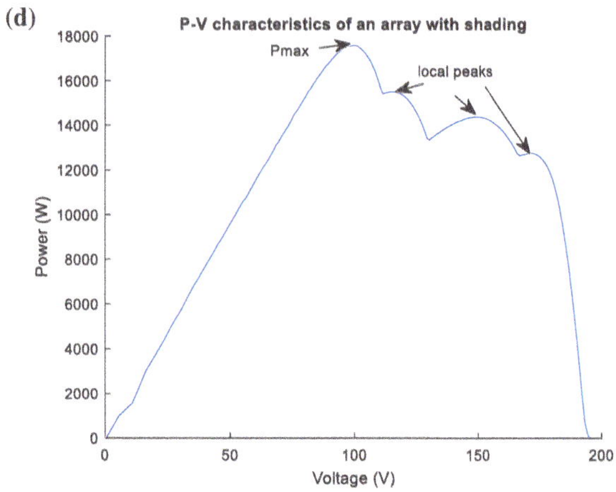

Though the hotspot problem is eliminated due to the usage of bypass diodes and reverse current flow due to the blocking diode, the overall output is drastically reduced and also there are multiple peaks in the *P-V* characteristics under the shaded condition [8]. As the conventional MPPT algorithms fail with multiple peaks, there are three different approaches considered. They are,

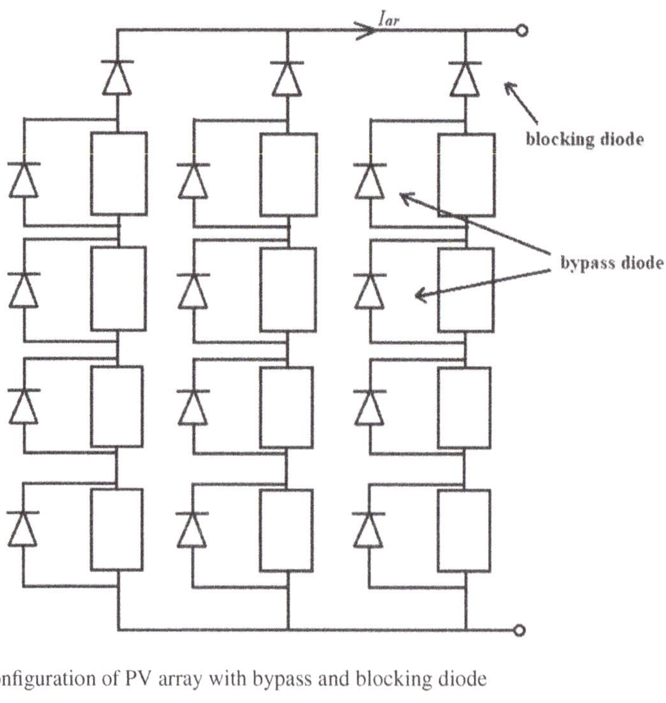

Fig. 3 Configuration of PV array with bypass and blocking diode

- Including additional power electronic circuits along with the existing converter or inverter to extract the power from partially shaded modules.
- Modified MPPT algorithm for the central inverter to work efficiently under partial shading conditions.
- Using a separate converter across each module of a PV array to implement its own MPPT.

Though the first type had shown limited improvement in the output, the additional investment on costly hardware which has to be fitted with the existing system, the related technical work, and the limitation of the technical expertise made the scheme less preferable. Though the second approach is attractive with respect to the cost and ease of implementation, the outputs of shaded modules are lost and not utilized. Due to the limitations of the first two approaches, the third approach is preferred by the investors though, the system is large and the initial cost is a little high. It is providing more yield even under partially shaded conditions by including the reduced generation from the shaded modules [9]. So, the third approach which is otherwise called a distributed maximum power point tracking (DMPPT) is considered in this chapter.

3 Distributed Maximum Power Point Tracking

As the conventional MPPT fails to track the maximum power point of the overall PV array under this multiple peak condition, the necessity of individual module-level MPPT becomes inevitable. To operate each module in its own MPP for maximum output, the module-level MPPT is used. This is called a distributed maximum power point tracking (DMPPT). The configuration of the PV array with DMPPT is given in Fig. 4. In DMPPT, each module is connected across its own DC-DC converter with the MPPT feature. Generally, the buck, boost, Cuk, and buck-boost converters are considered as suitable topologies for module integrated converters. Among the different DC-DC converter topologies, the buck-boost and Cuk converter had shown flexibility in voltage level. But, their operating efficiencies are very poor along with higher costs [10]. The boost converter is used as the best topology for DMPPT due to its promising solution [11]. In DMPPT, the module integrated DC-DC converter is controlled by the MPPT algorithms.

The conventional MPPT algorithms like perturb and observe (P&O), Incremental Conductance (IC), etc., are used for DMPPT due to the simple characteristics of each module. Considering the high performance, simplicity, and the low-cost implementation, the P&O algorithm is mostly used in DMPPT. The output from the algorithm is V_{ref} which is used to fix the operating voltage of the PV module which is connected across the DC-DC converter in the respective V_{MPP}. Once the MPPT algorithm can find the peak power, the V_{ref} is given to the control circuit which will control the duty cycle of the boost converter. The duty cycle will adjust the input side voltage of the DC-DC converter which is the output voltage of the PV module due to their parallel

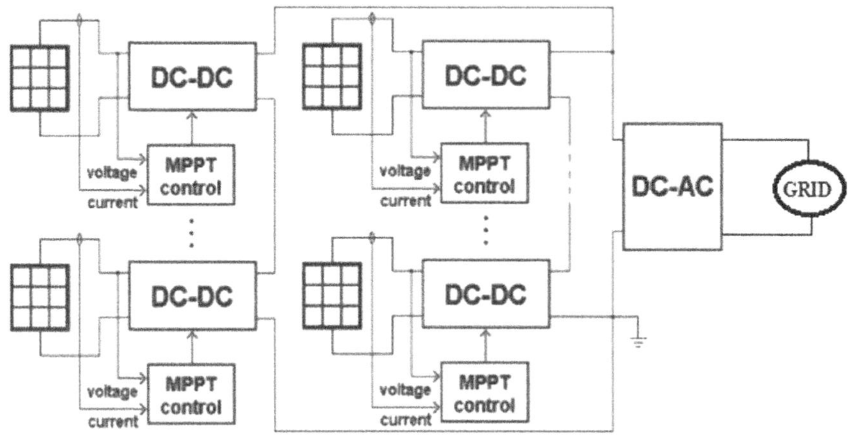

Fig. 4 Configuration of PV array with DMPPT

connection. The converter's input voltage is adjusted through the variation of the duty cycle by maintaining the output voltage of the PV panel in the same value. Due to this variation of PV Module's operating voltage to the V_{MPP}, the power produced will be P_{MPP} or P_{max}. So, the PV module will be forced to operate in its maximum power point (MPP) for that instant. In this way, the PV module is operated at its MPP and higher output is generated from each module at every instant. The control circuit which is used to maintain the DMPPT is given in Fig. 5. The instantaneous voltage and the current of the PV module are measured and given to the MPPT algorithm which decides the reference voltage V_{ref}. The control circuit decides the duty cycle of the device which is used in the converter based on the output of the MPPT.

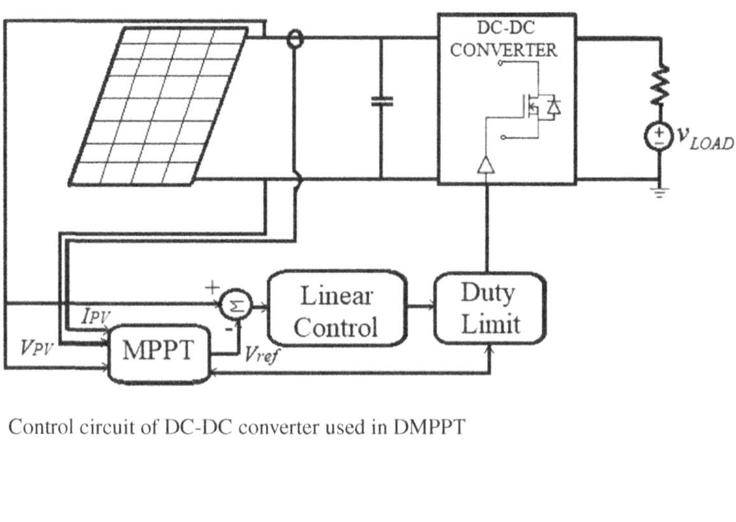

Fig. 5 Control circuit of DC-DC converter used in DMPPT

3.1 ANN Controlled DMPPT

In the rapidly changing environments, the characteristics of the PV module change due to their dependency on the environmental factors like solar insolation, the surrounding temperature, shading condition, and the array configuration [12]. Though the conventional algorithms are talented to identify the MPP in slow-changing environments, they fail at randomly varying conditions due to their iterative procedure. So, the iteration free artificial neural network (ANN) controlled tracking procedure is chosen here. The ANN is a replica of the human brain and it is designed in the way in which the brain performs the task though, the actual ability of the human brain is far greater.

> Artificial Neural Networks or ANN is an information processing paradigm that is inspired by the way the biological nervous system such as brain process information. It is composed of a large number of highly interconnected processing elements (neurons) working in unison to solve a specific problem.

The ANN accumulates knowledge through the training or learning process and saves the same knowledge in the interneuron connection weights like our human brain. These connection weights are otherwise called as synaptic. Though there are many ANN controlled MPPT algorithms [13], the new methodology is implemented here. The advantage of using this ANN controlled MPPT in this chapter is that there is no need for costly sensors for measuring the solar insolation as the proposed algorithm is independent of the solar insolation. The existing current and voltage sensors along with the rear side temperature sensor give the inputs for the ANN controlled algorithm. In DMPPT, the ANN is used to track the MPP of each PV module in every instant. So, to train the ANN of individual modules, initially, the *P-V* characteristics of each module are measured in different environments. The characteristics of the individual module are measured and used for the training process to include the unique behavior or the character of each module due to the manufacturing dispersion. The accuracy of the MPP value which is predicted through the ANN is based on the size of the training data set and also the randomness of the environmental condition during which the measurements are taken. The data set corresponding to each of the characteristics contains the current variation from zero to short circuit current I_{SC}, voltage variation from zero to open-circuit voltage V_{OC}, the temperature of the rear surface during the instant of measurement, peak power P_{MPP} or $P_{\max}$, and the corresponding V_{MPP} or $V_{\max}$. The data set consists of 50 equally spaced voltage values from zero to V_{OC} and their corresponding current values. Also, each characteristic has a single P_{MPP} and V_{MPP} to indicate the highest power received and the corresponding voltage of the PV module. In this example, 85 sets of data are collected for each PV module. The schematic diagram of ANN controlled DMPPT implementation is given in Fig. 6. To track the MPP, the ANN collects inputs like the instantaneous module voltage V, module current I, and the backside module temperature T. The voltage sensor which is used for the voltage measurement is connected across the output terminals of the PV module, the current sensor is connected in series with the positive terminal of the PV module and the temperature sensor is

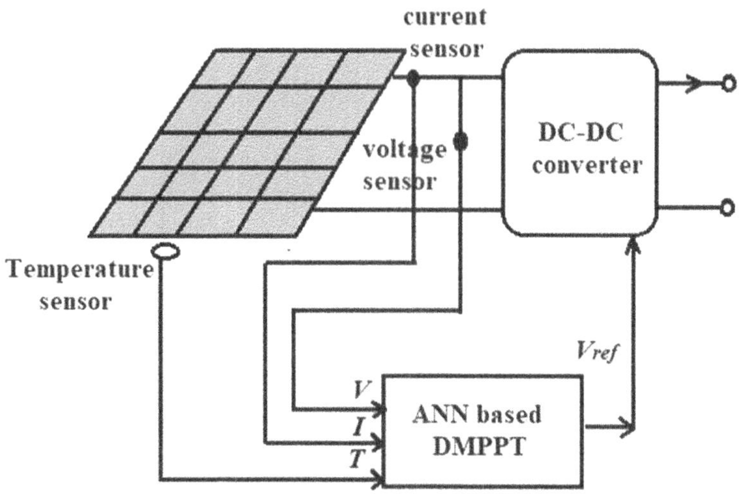

Fig. 6 Schematic diagram of ANN controlled DMPPT

fixed on the backside of the module. The temperature sensor is fixed in the backside instead of fixing it at the top of the PV module to avoid the shading effect. It is also proved that the temperature difference is negligible when the sensor is connected to the backside of the PV module [14].

The ANN is initially trained and tested for its accuracy before being used in the PV array. The accuracy of the ANN controlled tracking is based on the training. So, the ANN of each module is trained with the respective module's data sets of 70 characteristics which were measured during different conditions. The remaining 15 data sets are kept for testing purposes. The ANN adjusts its weight during the training based on the given inputs and outputs. The structure of the ANN is given in Fig. 7. The ANN consists of three layers called the input layer, hidden layer, and the output layer. The input layer consists of three neurons to accept the inputs like module voltage V, module current I, and the backside temperature of module T, respectively. The hidden layer consists of five neurons. The output consists of one neuron to provide V_{MPP} value for the converter control. The "tansig" and the "purelin" functions are used as the activation function in the input layer and the output layer, respectively, to calculate its output. So, there are three variables given as input to ANN and one variable is taken as the output.

The input variables are the instantaneous module voltage V, instantaneous module current I, and the rear side module temperature T to track the MPP at every instant. The output which is taken from the ANN is V_{ref}. Initially during the training, from each measured characteristic, the 50 sample values of voltage, current, temperature, and the V_{MPP} values are given to the ANN. Due to the single value of V_{MPP} for the entire characteristics of the 50 samples of a single characteristic, the same V_{MPP} value will be used as the output value for the training process against all the 50 sets of inputs from the same characteristics. Though there are different types of ANN,

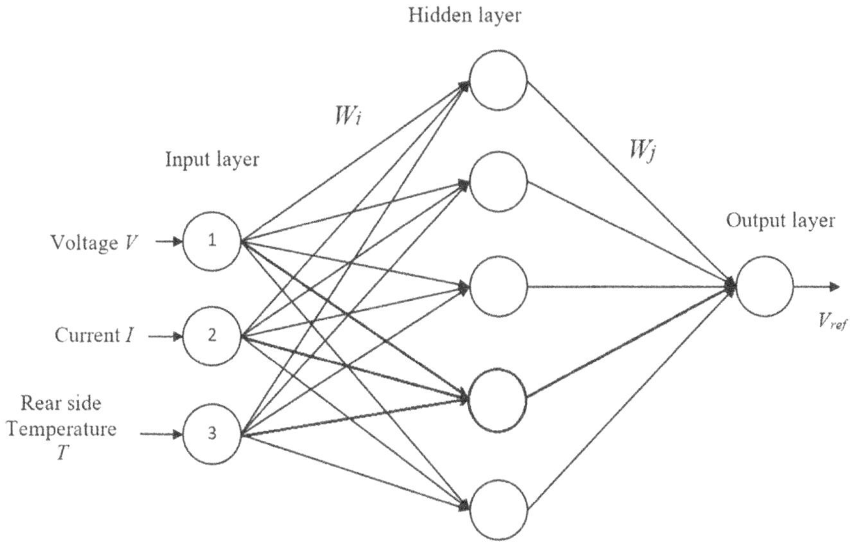

Fig. 7 Structure of ANN used for DMPPT

the backpropagation neural network (BPNN) is used due to its powerful learning process.

The steps for the ANN training is given below.

- Step 1: Build the network and initialize the network weights with randomly generated value
- Step 2: Apply the input data sets
- Step 3: Set the parameters of the ANN
- Step 4: Train the network and calculate the output from ANN
- Step 5: Calculate the error between the required output and the actual output of ANN
- Step 6: Adjust the weights of ANN and reduce the error until the required minimum error value is attained by repeating the steps from 3 to 5
- Step 7: Stop the training process once the error value becomes lesser than the acceptable minimum value
- Step 8: Save the trained ANN

The trained ANN adjusted its weight and learned the correlation between the input and output from the data set of the specific PV module using the Levenberg–Marquardt (LM) learning algorithm. To validate the ANN and its performance, the inputs from the remaining data sets corresponding to the same PV module is used. The output is predicted using ANN and the error is calculated. The performance is accessed by the root-mean-square (RMSE) error value as given in Eq. 1.

$$\mathrm{RMSE} = \frac{1}{2}\left[\Sigma_p \Sigma_i \left(t_{ip} - O_{ip}\right)^2\right]^{\frac{1}{2}} \tag{1}$$

Table 2 Prediction results of ANN controlled DMPPT

Module voltage (V)	Module current (A)	Temperature (°C)	Predicted V_{MPP} (V)	Actual V_{MPP} (V)
17.4	3.04	32	18.86	18.79
18.5	2.76	36	17.98	18.01
15.67	3.32	40	19.12	19.11

where, p is the number of input sets, i is the specific node, t is the desired output or target, and O is the actual predicted output of a network. The trained ANN is used as an MPPT algorithm for that specific PV module as shown in Fig. 6. At every minute the ANN collects the values of module voltage V, module current I, and the backside module temperature T from the respective sensors of a PV module and predicts the V_{ref} to control the operation of DC-DC converter connected across the PV module. This V_{ref} is used to vary the duty cycle of the converter to adjust the output voltage of the PV module or the input voltage of the DC-DC converter. So, the PV module is forced to operate at its V_{MPP} to produce the highest power P_{MPP} or P_{max} of that instant. The output of the ANN algorithm for a specific module at different environmental conditions are given in Table 2.

The output of the ANN controlled algorithm is very close to the actual V_{MPP} value. So, the ANN can predict the voltage value corresponding to the maximum power with a lesser number of sensors. Due to the iteration free operation of the ANN algorithm, the effectiveness of the ANN controlled DMPPT is high at randomly varying environmental conditions.

4 Current Compensation for DMPPT

In the PV array, each string consists of many PV modules in series. Though the series-connected modules are of the same rating and from the same manufacturer, due to manufacturing dispersion, some modules may generate higher current than the other modules. Similarly, when few of the PV modules in the string get shaded, the unshaded modules produce higher current than the shaded modules. Due to the series connection of all these shaded and unshaded modules in the string, the path allows only the lower current which is generated by the shaded modules. This low current forces the other unshaded modules to generate the same lower current though they can generate more current based on the received solar insolation. When the healthy modules are forced to operate in lower current, their power point is also shifted due to the change in current and voltage. This downgrades the other healthy modules in series and reduces the yield from the complete string. The bypass diodes are used to avoid this condition. But, the usage of bypass diodes makes the P-V characteristic curve multimodal.

Though, the DMPPT estimates the maximum peak and tried to maintain the module in V_{MPP} corresponding to the MPP, the law of equal voltage across the parallel-connected strings and the law of the same current in series-connected modules does not allow the module to operate in its own MPP at the mismatched condition. At, mismatched condition, the output of each module varies. The current mismatch in the shaded and unshaded modules of string1 is shown in Fig. 8. Though the manufacturing dispersion is not considered in these characteristics, due to the difference in solar isolation under shading conditions, the current from shaded modules reduced drastically. But the unshaded modules can generate high current. As these modules are serially connected, the current in the series string is equal to the current of the shaded PV module. This condition makes the unshaded module to move from its peak performance. To solve the problem of the current mismatch in a series string under mismatched condition, each module is equipped with a dedicated current compensator [15]. So, the modules can operate at its own MPP irrespective of the current mismatch. The DC-DC converter which is connected across each module to maintain the module in MPP in the DMPPT scheme is used as a current compensator. This avoids the extra compensation equipment for each module.

Also, this compensator increases the output by enabling the shaded and unshaded modules to operate in its actual MPP, by providing a compensating or deficient current for the module with low current generation in its own MPP. This maintains the same current in the series-connected string. So, the string current is equal to the operating current of the healthy module. Whereas, the conventional scheme which consists

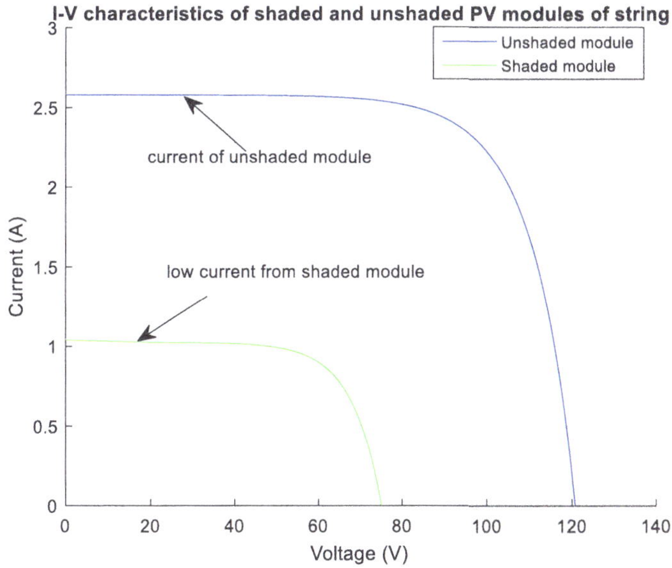

Fig. 8 Current mismatch among the series-connected PV modules due to shading

of bypass diode completely short the shaded module to maintain the string current equal to the current of healthy modules and lost the generation of shaded modules.

Though, the compensation scheme or the DMPPT consumes power for the DC-DC converters, the energy consumption is negligible compared to the yield or the profit due to the inclusion of generation from multiple shaded modules [15]. The DC-DC converters which are used as current compensators are the "*compensation power DC-DC converters instead of the full power DC-DC converters*" [15]. So the power consumption is proportional to the amount of compensation provided by the converter. The current flow in the string due to the implementation of the current compensator is shown in Fig. 9.

To maintain the current of $I_{str(1)}$ which is equal to the current I_{MPP} of the healthy module in the same string, the current compensator which is connected across the

Fig. 9 Configuration of PV array with current compensator

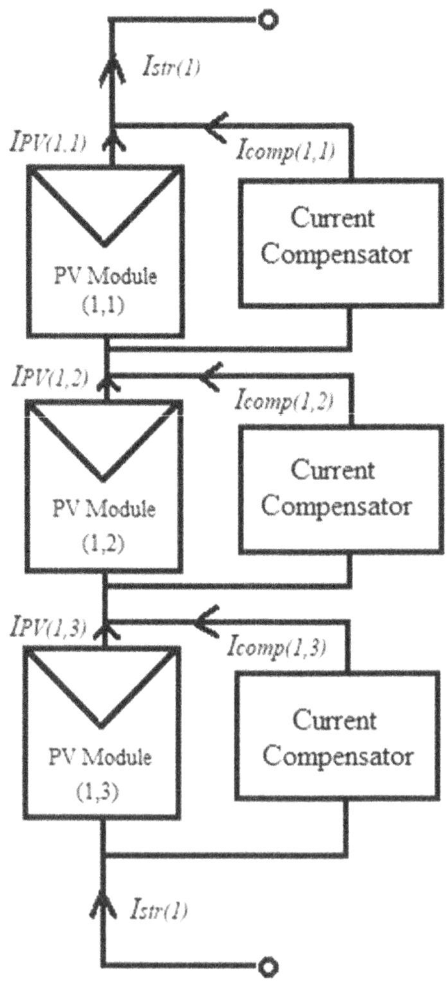

shaded PV module provides the lacking current I_{comp}. The current supplied by the compensator is provided as Eq. 2.

$$I_{comp(i,1)} = I_{str(i)} - I_{MPP(i,1)} \tag{2}$$

where, $I_{comp(i,1)}$ is the current provided by the compensator connected across the first module of ith string. $I_{str(i)}$ is the current flowing through ith string and $I_{MPP(i,1)}$ is the current of the first PV module in ith string which is working in its own MPP. The power consumed by the current compensator during compensation is given by Eq. 1.3.

$$P_{comp(i,1)} = V_{ar} \times I_{conv(i,1)} \tag{1.3}$$

where, $P_{comp(i,1)}$ is the power consumed by the current compensator connected across the first module of ith string during compensation, V_{ar} is the output voltage of PV array which is the input voltage of DC-DC converter and $I_{conv(i,1)}$ is the current consumed by the compensator or converter.

The input supply for the compensator is given by the PV array. The power loss is drastically reduced due to the lesser power consumption of the DC-DC converters during their compensation mode. The suitable design of the DC-DC converter results in reduced no-load losses. So, the parallel-connected current compensator is used to

- Maintain the PV module's voltage at V_{MPP} through MPPT algorithm
- Supply the deficient current for the shaded module to equalize the string current with the normal current of a healthy module.

5 Voltage Compensation for DMPPT

Though the implementation of current compensation aid the power output by including the generation from the shaded modules, the modules in the string will move itself away from the V_{MPP} under the condition of voltage mismatch between the parallel-connected strings due to the law of same voltage across the parallel-connected strings. Under shading conditions, the voltage corresponding to the maximum power of each module varies as shown in Fig. 10. The operating voltage V_{max} corresponding to the P_{max} of the healthy module is around 95 V. But, the operating voltage V_{max} of the shaded module is 60 V. The number of shaded modules is different in different strings.

Due to the difference in the operating voltage of each module in the string, the string voltage which is the summation of all the module voltages will differ from other string voltages. But, all the strings of PV array are connected in parallel. As per the law of parallel circuits, the voltages across all the parallel branches should be the same. To satisfy this law, the string with the highest voltage or the string with more unshaded or healthy modules tries to reduce its voltage to the lowest string

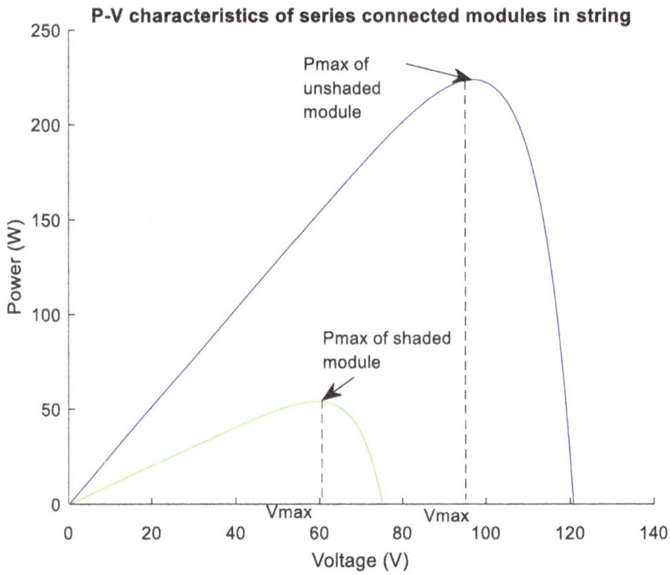

Fig. 10 Voltage mismatch among the series-connected modules of string

voltage of the array. Due to this condition, all the healthy modules in the healthy string will reduce its actual operating voltage V_{max} and generates lesser power. To avoid this loss, there is a need for additional voltage sources in series with the weak string. So, to equalize the voltages of strings under the mismatched condition and to enable the modules to operate in its own MPP the series compensator is connected in each string. The role of a single voltage compensator which is connected for each string is to provide the compensating voltage for the string to equalize the string voltage with the voltage of a healthy string of the PV array. The DC-DC converter is used as a voltage compensator. Though, the current compensation does not need any additional DC-DC converter in the DMPPT scheme, the voltage compensation needs a single DC-DC converter in addition for each string. The configuration of the string with voltage and current compensators are given in Fig. 11. The voltage and the current supplied by the voltage compensator is given in Eq. 4 and Eq. 5, respectively.

$$V_{\mathrm{comp}(i,n+1)} = V_{\mathrm{str}(j)} - V_{\mathrm{str}(i)} \tag{4}$$

$$I_{\mathrm{comp}(i,n+1)} = I_{\mathrm{str}(i)} \tag{5}$$

where, $V_{comp(i,n+1)}$ is the compensating voltage supplied by the additional (n + 1)th DC-DC converter for the ith string, $V_{str(j)}$ is the voltage of healthy string in the same PV array, $V_{str(i)}$ is the voltage of the ith string for which compensation is provided, $I_{comp(i,n+1)}$ is the current flowing through the voltage compensator and $I_{str(i)}$ is the

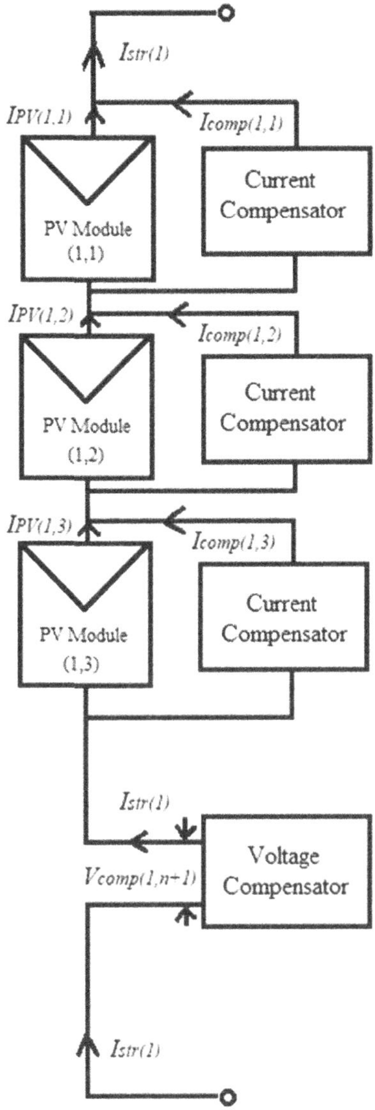

Fig. 11 Configuration of string with voltage and current compensator

current flowing through the ith string. The configuration of a string with voltage compensator given in Fig. 11 shows that the current flowing through the voltage compensator is the same as the string current. This is also equal to the current of the healthy module in the same string. But the voltage provided by the voltage compensator is equal to the difference between the voltage of healthy jth string in the array and the ith string for which the voltage compensation is provided. If jth string is the healthy string of PV array, due to the higher voltage output from jth string, all the other parallel-connected strings work to meet the voltage level of healthy jth string to satisfy the law of equal voltage by combining the voltage of voltage compensator. This allows all the modules of healthy strings and unhealthy strings to work in its actual V_{MPP} or V_{max}. The power consumption of this voltage compensator will be proportional to the voltage supplied by the compensator. The power loss due to the inclusion of voltage compensator is negligible when it is compared with the maintenance of actual MPP in the series-connected multiple modules. Also, the power consumption of individual series voltage compensators is almost zero in healthy strings. So, the serially connected voltage compensator is used to

- Supply the deficient voltage for the string to equalize its voltage with the voltage of the healthy string
- Maintain the PV module's voltage at V_{MPP} through MPPT algorithm and control circuit

The PSIM-based simulation results of the 2×2 array which have two strings and two modules in each string are given in Figs. 12 and 13. The solar insolations received in the modules of the first string are 400 W/m^2, 1000 W/m^2 and the second strings are 800 W/m^2, 800 W/m^2, respectively. The currents developed by the solar modules of string1 are 0.661 A and 0.769 A, respectively. Similarly, the currents from the PV modules of string2 are 1.13 A and 1.13 A, respectively. To maintain the current flow of 0.769 A in string1, the current compensator which is connected across the shaded module of string1 provides a compensation current of 0.11 A. But the compensator which is connected across the healthy module produces a current of 0.005 A which is negligible. The current compensators connected in string2 produces negligible current due to the healthy condition of all modules which are connected in series.

In Fig. 13, the PV modules in string1 are kept at their V_{MPP} values of 10.789 V and 19 V, respectively, due to the shading of the first module. The PV modules of the second string are maintained in their V_{MPP} values of 16.8 V and 16.8 V, respectively, due to the same insolation condition. The second string generated a total voltage of 33.6 V. But the string1 is able to produce only 29.78 V. To compensate for the voltage deficiency, the voltage compensator of string1 provides a voltage of 3.8 V. But the voltage compensator of string2 provides a negligible voltage of 64 mV which can be considered as zero. This enables all the PV modules to operate in its own MPP irrespective of the parallel connections. As the shaded module's generation is not shorted like the bypass diode scheme, the power generation from the array increases and it improves the efficiency. To prove the effectiveness of the proposed method, the results of the proposed method are compared with the other DMPPT methodology which was proposed for the partially shaded condition and known as the TEODI MPPT technique [16]. The comparison is shown in Table 3. Though, the modified TEODI MPPT method had shown improved performance under normal and shaded condition, the overall output power was lesser due to the limitation of this MPPT method in maintaining the PV modules in their actual MPP. Also, the computation time was more than the proposed ANN controlled method. So, the proposed ANN

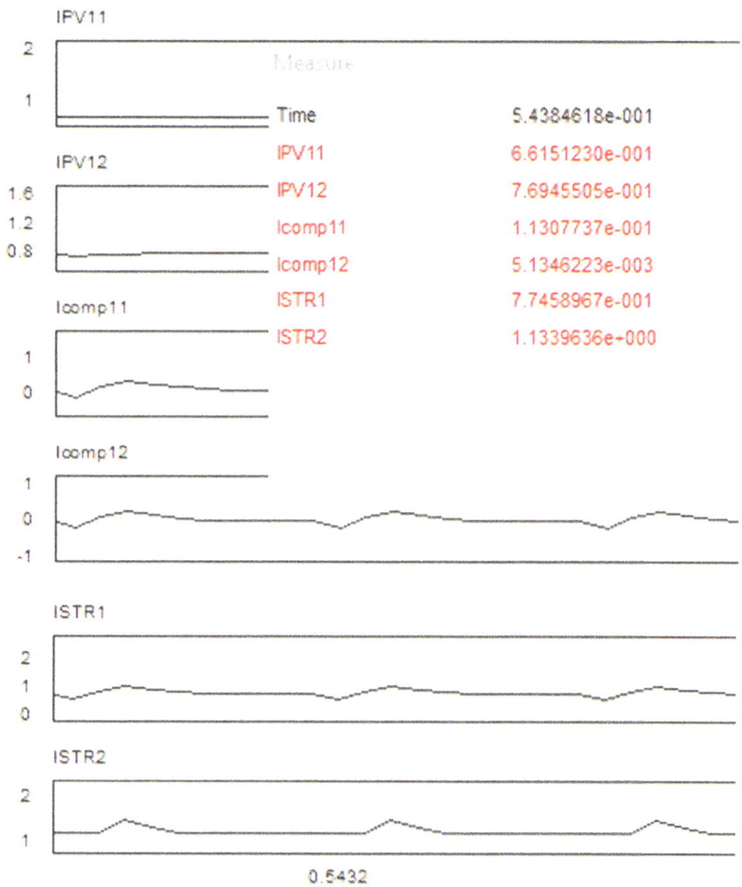

Fig. 12 Current compensation provided by the current compensators of 2 × 2 PV array

controlled method is proved as an effective method at the partially shaded and the mismatched condition of PV modules.

6 Conclusion

This chapter presented the distributed maximum power point tracking of PV modules. This reduced the problem of multiple peaks in the P-V characteristics of the PV module at partial shading and mismatched condition. The ANN controlled DMPPT was implemented and its performance was validated at rapidly changing conditions with a lesser number of sensors. But, to maintain each module in its own MPP according to the output of ANN controlled DMPPT, the current and voltage compensation schemes were implemented across each module and in series with each string,

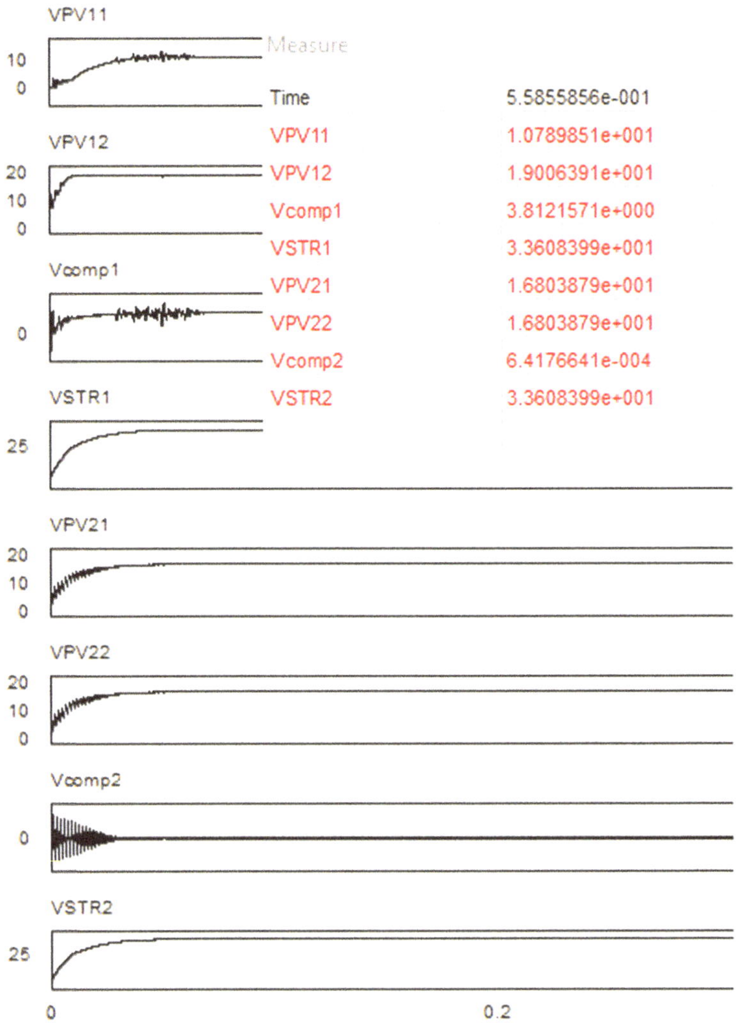

Fig. 13 Voltage compensation provided by the voltage compensators of 2 × 2 PV array

Table 3 Comparison of the proposed method with other DMPPT method

MPPT	Control circuit complexity	Speed	Efficiency (%)
Proposed method with compensators	Less	High due to the trained ANN	72.337
Modified TEODI DMPPT	More	Less compared to ANN	67.879

respectively. This utilizes the power generated by the shaded modules instead of losing it across the bypass diode. This improves the overall output of the PV array even under the partially shaded condition as compared to the conventional MPPT scheme and its implementation.

References

1. Esram T, Chapman PL (2007) Comparison of photovoltaic array maximum power point tracking techniques. IEEE Trans Energy Conv 22(2):439–449
2. Loredana C, Marco R (2014) An improved model-based maximum power point tracker for photovoltaic panels. IEEE Trans Instrum Measure 63(1):63–71
3. Motahhir S et al (2020) Optimal energy harvesting from a multistrings PV generator based on artificial bee colony algorithm. IEEE Syst J. https://doi.org/10.1109/JSYST.2020.2997744
4. Motahhir S, El Hammoumi A, El Ghzizal A (2020) The most used MPPT algorithms: review and the suitable low-cost embedded board for each algorithm. J Clean Prod. https://doi.org/10.1016/j.jclepro.2019.118983
5. Kermadi M, Salam Z, Ahmed J, Berkouk EM (2020) A high-performance global maximum power point tracker of PV system for rapidly changing partial shading condition. IEEE Trans Industr Electron. https://doi.org/10.1109/TIE.2020.2972456
6. Farh HMH, Eltamaly AM (2020) Maximum power extraction from the photovoltaic system under partial shading conditions. In modern maximum power point tracking techniques for photovoltaic energy systems. Springer, Cham, pp 107–129
7. Ahmed J, Salam Z (2015) A critical evaluation on maximum power point tracking methods for partial shading in PV systems. Renew Sustain Energy Rev 47:933–953
8. Patel H, Agarwal V (2008) Maximum power point tracking scheme for PV systems operating under partially shaded conditions. IEEE Trans on Ind Electron 55:1689–1698
9. Erickson RW, Maksimovic D (2001) Fundamentals of power electronics, 2nd edn. Springer, New York
10. Femia N, Lisi G, Petrone G, Spagnuolo G, Vitelli M (2008) Distributed maximum power point tracking of photovoltaic arrays: novel approach and system analysis. IEEE Trans Ind Electron 55(7):2610–2621
11. Sharma P, Agarwal V (2014) Exact maximum power point tracking of grid connected partially shaded PV source using current compensation concept. IEEE Trans on Power Electron 29(9):4684–4691
12. Eltamaly AM (2018) Performance of MPPT techniques of photovoltaic systems under normal and partial shading conditions. In Advances in renewable energies and power technologies. Elsevier, pp 115–161
13. Anil K Rai, Kaushika ND, Singh B, Agarwal N (2011) Simulation model of ANN based maximum power point tracking controller for solar PV system. Sol Energy Mater Sol Cells 95(2):773–778
14. Adkins DA (2010) Novel method and system for monitoring CPV cell and module temperature. In Proc. 35th IEEE PVSC 1660–1665
15. Jeyaprabha SB, Selvakumar AI (2017) Model-Based MPPT for shaded and mismatched modules of photovoltaic farm. IEEE Trans Sustain Energy 8:1763–1771
16. Balato M, Costanzo L, Marino P, Rubino G, Rubino L, Vitelli M (2017) Modified TEODI MPPT technique: theoretical analysis and experimental validation in uniform and mismatching conditions. IEEE J Photovolt 2:604–613

Design and Comprehensive Analysis of Maximum Power Point Tracking Techniques in Photovoltaic Systems

Ali M. Eltamaly, Mohamed A. Mohamed, and Ahmed G. Abo-Khalil

Abstract In this chapter the performance of various maximum power point tracking techniques for Photovoltaic (PV) systems has been presented, under uniform and non-uniform irradiance conditions. Under uniform irradiance conditions, the power-voltage curve of PV systems is nonlinear and contains one peak point whose location appertains to the irradiation and surface temperature of the PV system. Partial shading on PV modules reduces the generated power than the maximum power generated from each module separately. The traditional techniques of tracking the maximum power point are designed to track the global peak but they always failed to capture the exact point. In this chapter, different techniques of maximum power point tracking have been introduced, analyzed, and simulated. MATLAB, SIMULINK, and PSIM software have been utilized to simulate the PV systems under various shading conditions. Furthermore, the response of the different techniques of maximum power point trackers has been evaluated under different weather conditions.

Keywords Photovoltaic · Maximum power point tracking · Partial shading · PSIM · Optimal design · Solar energy

A. M. Eltamaly (✉)
Electrical Engineering Department, Faculty of Engineering, Mansoura University, Mansoura 35511, Egypt
e-mail: eltamaly@ksu.edu.sa

Sustainable Energy Technologies Center, King Saud University, Riyadh 11421, Saudi Arabia

K.A. CARE Energy Research and Innovation Center, Riyadh 11451, Saudi Arabia

M. A. Mohamed
Electrical Engineering Department, Faculty of Engineering, Minia University, Minia 61519, Egypt
e-mail: dr.mohamed.abdelaziz@mu.edu.eg

A. G. Abo-Khalil
Department of Electrical Engineering, College of Engineering, Majmaah University, Almajmaah 11952, Saudi Arabia
e-mail: a.abokhalil@mu.edu.sa

Department of Electrical Engineering, College of Engineering, Assuit University, Assuit 71515, Egypt

S. Motahhir and A. M. Eltamaly (eds.), *Advanced Technologies for Solar Photovoltaics Energy Systems*, Green Energy and Technology,
https://doi.org/10.1007/978-3-030-64565-6_9

1 Introduction

The solar PV technology is currently progressing and appearing the capability of being utilized in isolated and grid-connected applications [1, 2]. The limitations that have kept the exceptionally expansive scale use of PV so far are the initial high costs requirement and low efficiency [3]. Although, with late advances, the cost is descending by utilizing advanced technology in the manufacture of PV modules [4]. Proficiency is enhancing by utilizing multilayers of PV solar cells [5]. Current research is coordinated on progressing existing modules, for example, the thin-film and new material for crystalline cell technologies [6, 7]. Figure 1 appearing and anticipated capital expense of sun-oriented photovoltaics. Figure 2 demonstrates the world total establishment from 2004 to 2014. Before the finish of 2017, the global-introduced PV capacity hops to 303 GW [8]. Utilizing PV in Hybrid Renewable Energy Systems (HRES) is a decent choice where the sensible cost of the PV system and the great connection between the generation and the load [9]. This will diminish the extent of the storage system utilized in HRES particularly if the ideas of the smart grid are considered [10].

Solar power is converted into electricity by photovoltaic (PV) technology as shown in Fig. 3, or concentrating solar power (CSP) as shown in Fig. 4. The following sections introduce a brief review of these technologies:

A. The Technology of Solar Photovoltaic (PV)

In the late 1880s the photovoltaic technology was discovered but did not gain significance until 1954 when it was rediscovered by Bell Telephone [12]. Fundamentally, the PV technology employs silicon and some other materials device to entrap sunlight known as photons that will hit the free electrons in the silicon device to produce an electric voltage by a process referred to as the photovoltaic effect. This process produces direct current (DC) electricity power. The PV system produced power is a

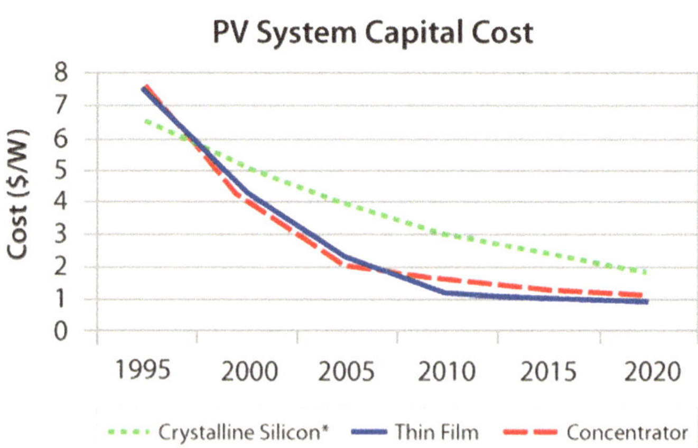

Fig. 1 The historical and projected capital cost of solar photovoltaics [9]

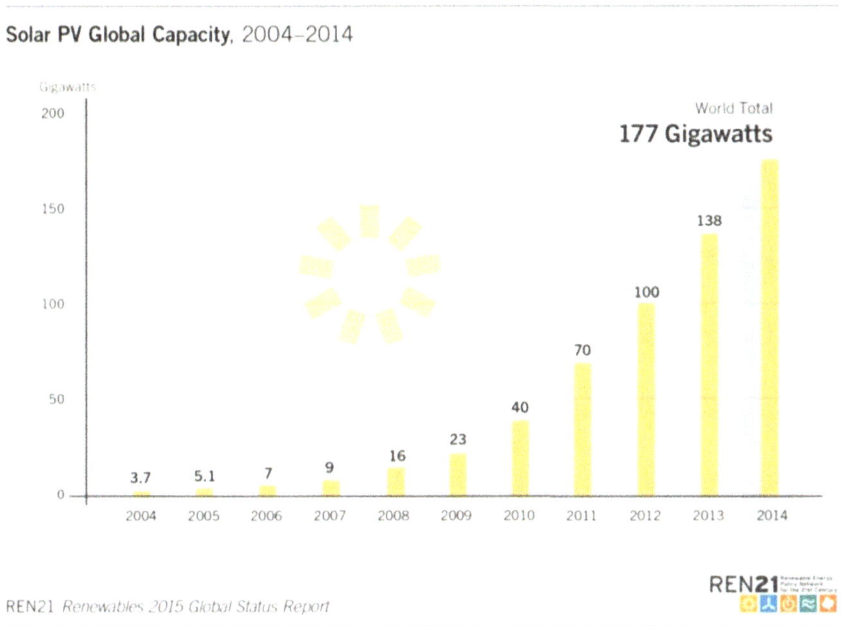

Fig. 2 World cumulative installation from 2004 to 2014 [7]

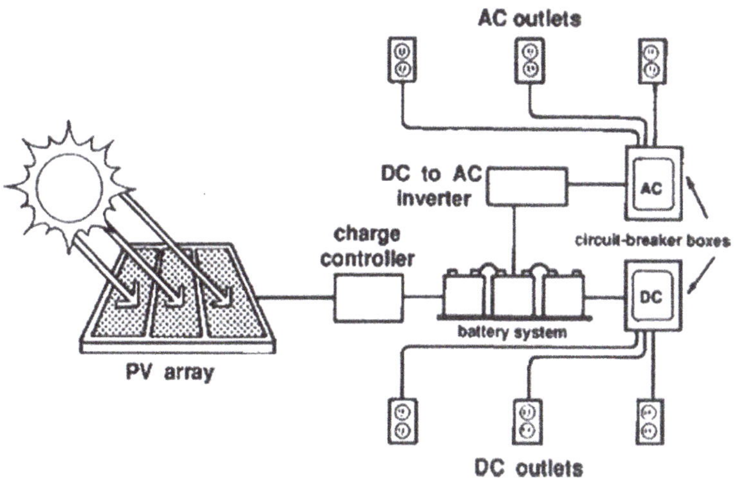

Fig. 3 Photovoltaic energy system [10]

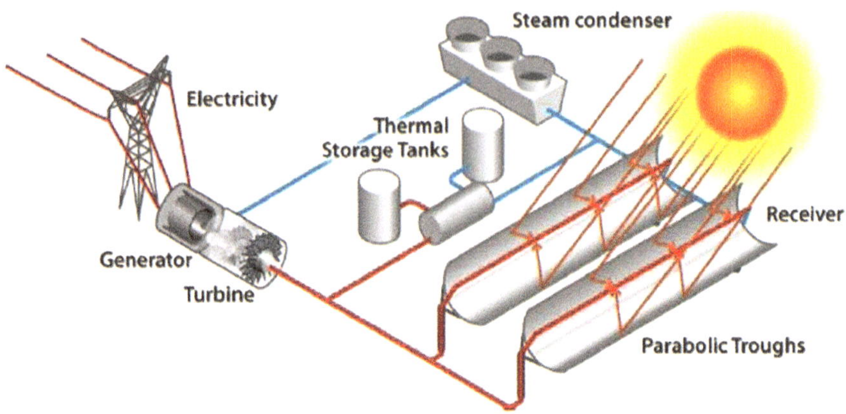

Fig. 4 Concentrated solar power tower plant [11]

function of the output voltage as shown in Fig. 5. The research literature has produced several mathematical models of the PV cell. Some aspects of the literature will be discussed in the chapters to come. Figure 5 shows the output power versus terminal voltage relation under various radiation and temperature. This figure shows that there is a maximum power point (MPP) located for each terminal voltage [13]. This is the reason why there are many variations in MPP trackers (MPPT) as introduced in many research publications [14].

Many types of materials are utilized to design the PV-cells and they are different in their characteristics. The most popularly used PV-cells are the crystalline silicon [15]. PV-cells from crystalline silicon have been in use for a long time and it is now a

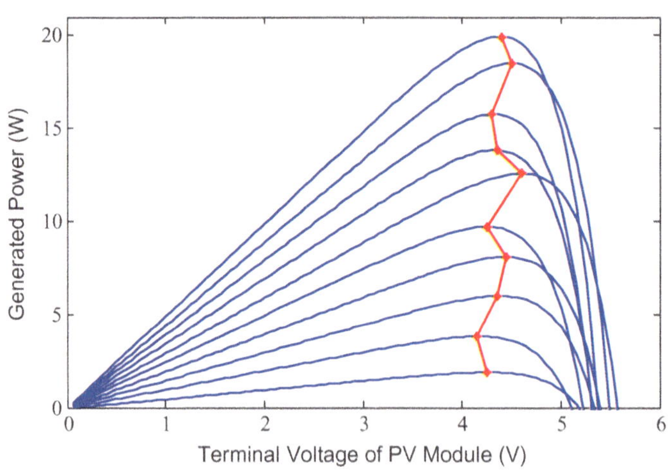

Fig. 5 The output power Versus terminal voltage relation under various radiation and temperature [14]

matured technology in the manufacturing technology with its price becoming lower and lowers especially in the presence of mass-fraction production. As a result, this chapter will concentrate more on the crystalline silicon PV-cells. PV-cells of other types have a similar analysis but with dissimilar performance characteristics [14]. The converter efficiency of PV-cell ranges from 10 to 25% subject to the materials being used in the fabrication of the PV-cells [16]. Besides, PV module's power is determined by PV modules to tilt angle and temperature [17]. The presence of dust can reduce the efficiency of the PV systems and it should be cleaned frequently using different technologies [18].

The PV system' DC-generated power is used to charge a battery or be converted into alternating current (AC) employing power electronics inverter for AC appliances. The PV system can be used as a standalone or part of a hybrid system to support remote loads far away from the electric utility or may be interconnected with the electric utility [19].

In a grid-intertie system, the PV configuration is tied directly to the electric network using DC/AC inverter. If the PV system generates power over its local load requirement, the extra generated power is transferred to the electric utility. If the PV local load has a deficiency in power from the PV system, the load can absorb extra power from the electric utility [20]. The grid-intertie with battery backup is an extended version of the former with the inclusion of battery to store power for periods when the grid is not available.

The isolated or grid-independent solar systems are utilized in decentralized applications and remote areas away from the network [21]. These systems require storage batteries facilities, DC/AC inverters, and charge controllers. The isolated HRES is usually utilized in hybrid systems where diesel generators can be used as an auxiliary when the generated power from HRES is lower than the demand requirements [22]. The following are the configurations used to mount the PV modules:

1. ***Flat-Plate Modules***

The PV systems can be used as a flat-plate configuration which requires a large number of cells and larger land areas. The flat plate should face the sun in the best way as possible and there is a compromise between this need with the cost. The best tilt angle for the PV modules is the site's latitude angle. Other systems continuously change the module's tilt angle to a monthly or seasonal best tilt angle. Another alternative is to use a solar tracking system to track the sun using one-axis or two-axis tracking systems but this comes with an increase in cost. Hence, a careful analysis of the compromise between the cost of the tracking system and the increase of energy due to using sun trackers must be analyzed in detail. The different configurations used to mount the PV modules are listed in the following points:

- Fixed tilt angle arrays
- Systems adjusted to the monthly best tilt angle
- Systems adjusted to the seasonally best tilt angle
- One-axis tracking

Fig. 6 Fixed tilt angle, FTA photovoltaic system [24]

- Two-axis tracking
- Concentrator arrays

2. ***Fixed Tilt Angle Arrays***

In this method, the modules are fixed facing north in the southern hemisphere at a tilt angle comparable to the site's latitude [14, 23]. The best tilt angle approximately equals the latitude of the site facing the equator in the north and south of the earth [24]. Figure 6 shows a Fixed Tilt Angle (FTA) photovoltaic system.

3. ***Monthly Adjusted Tilting***

An alternative to using a fixed tilt angle of the photovoltaic array is to vary the tilt monthly depending on the monthly optimum tilt angle. The literature approximate 8% [24] increase in energy captured by the photovoltaic compared to the fixed optimum tilt angle. The optimum tilt angle is either adjusted manually or by using an electromechanical system. Literature concluded that the adjustment to the tilt angles every three months increases the annual energy production by about 5% [25]. This increase may not be economically beneficial, considering the cost of implementation. In monthly adjusted tilted arrays, the monthly best tilt angle can be estimated for the site, and then the array's monthly angle can be adjusted. This method has no complexity and yields increased efficiency. Figure 7 shows Adjustable Tilt Angle, ATA photovoltaic system with monthly adjusted tilting.

4. ***One-Axis Tracking***

One-axis tracking is used to actively track the sun during the day time. The tracking is done each hour or in a lower period. This system changes the angle concerning the vertical axis and the tracking starts in the morning where the array faces the

Fig. 7 Adjustable tilt angle, ATA photovoltaic system [25]

east and at the end of the day when it faces the west. Output can be increased considerably concerning the previous techniques. One-axis tracking can increase the energy captured by 20–30% in comparison with the optimum fixed tilt angle [25] as discussed above. A single-axis tracking array, SAST photovoltaic system is shown in Fig. 8. The system installation and maintenance cost could be higher than the increase in energy captured and this is the reason why it is not recommended to use this system in commercial photovoltaic energy systems.

5. ***Two-Axis Tracking***

In this configuration, the output energy can be increased higher than the previous techniques. Nearly 30% insolation improvement relative to an optimum fixed tilt array is achievable. However, both the capital and maintenance costs are high hence only a few large systems are presently installed. Also, it needs more space to freely

Fig. 8 SAST concentrator solar cell array [26]

Fig. 9 TAST concentrator solar cell array [25]

move the two-axis systems. The two-axis tracking system, TAST photovoltaic system is shown in Fig. 9. By mounting the array on a two-axis sun tracking, from 30% [24] to 40% [25] more solar energy can be collected over the year than in fixed tilt installation. Furthermore, the gain is mostly in the early morning and late evening, when it is particularly valuable in meeting peak loads.

6. ***Concentrator PV Arrays***

The concentrator PV arrays utilize optical lenses or mirrors to focus the solar radiation on high-efficiency PV arrays. Precise tracking of the sun is required for these systems, principally when the concentration ratio is high. Tracking increases the intercepted insolation but with concentrators, the trade-off is the ability to access only the direct fraction. The overall outcome depends on the clarity of the sky at the site. The concentrators will only use the direct beam light; the diffuse light cannot be implemented. To increase total annual energy output the concentrators will require tracking devices. Although the expenses will increase, the increase in annual energy output is up to 30% in comparison to the increase by using just the tracking devices. Another additional advantage of using concentrators is that it uses a reduced number of solar cells and hence the area needed for installation. There are many types of concentrators as mentioned below [27]:

(a) ***Parabolic troughs:*** Silvered glass and polished aluminized flexible film are used in reflective materials. This type is shown in Fig. 10a.
(b) ***The ordinary lens:*** Glass lens collects the sunlight in the small solar cell area. This type is shown in Fig. 10b.
(c) ***Fresnel lens:*** Fresnel lens similarly diffracts the sunrays as would be done by a conventional lens. The advantage of a fresnel lens is that; it is much thinner and lighter than a conventional lens. The fresnel lens and parabolic trough concentrator have received the most attention for use in the photovoltaic system. This type is shown in Fig. 10c.

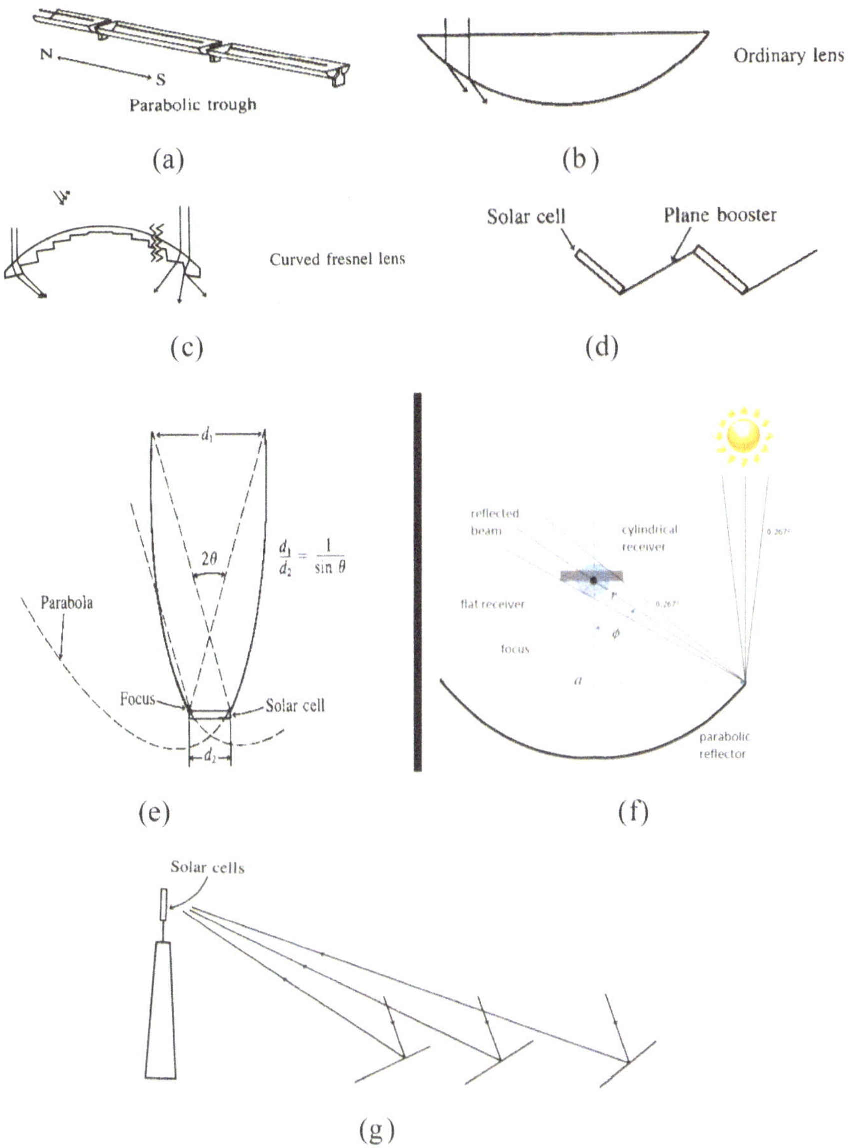

Fig. 10 Types of concentrator [27]

(d) ***Plan booster concentrator:*** This is the simplest form of flat-plate concentrator or booster. The solar cell output can be improved by about 50% [27]. This type of concentrator is proposed to be used with a space-satellite photovoltaic generation station. This type is shown in Fig. 10d.

(e) ***Compound parabolic concentrator:*** This was proposed by Winston proposes in 1974 [28]. This concentrator referred to as a Winston collector, is formed with two parabolic cups that satisfy two conditions. This type is shown in Fig. 10e.

(f) ***Parabolic Dish:*** Parabolic dish requires two axes sun tracking system and provides a very high concentration ratio. This type is shown in Fig. 10f.

(g) ***Central receiver concentrator: The total mirror surface in one module can collect*** from tens of kilowatts to tens of megawatts. The concentrator solar cell systems trade the added costs of the optical concentrator for the cost of solar cells since a concentrator system uses a much smaller quantity of solar cells than flat-plate systems of the same capacity. This type is shown in Fig. 10g.

B. The Technology of Concentrating Solar Power (CSP)

The main types of CSP are power tower systems, linear concentrators, and dish/engine CSP systems. The CSP systems are rarely utilized in isolated hybrid energy systems. Most of the applications of CSP are utility integration systems because it is economically to install it in large central power plants.

Solar energy is becoming more attractive and it is counted as future energy. Solar thermal energy or CSP is the most important option of renewable energies to provide clean electric energy shortly [26].

The operation of CSP plants depends on concentrating the sun's energy by using solar mirrors to increase the heat on a boiler which can be used to produce super steam to hit a blade of a steam turbine to convert heat energy into mechanical energy and then into electric energy using electric generators. Heat storage can store heat in the day time to be converted later to electricity at night or cloudy days. Active researchers have used certain oil; some salts materials, some types of sand as heat storage. The need for heat storage is not an important issue in the utility integration of CSP because of the correlation between most of the loads and the power generated from the sun which makes it preferable to be direct feeds the loads through the electric grid.

Many techniques have been used in the generation of electricity from CSP systems. Four main technologies can be used to generate electric power from solar thermal plants. These techniques are briefly discussed and shown in the following sections:

1. ***Trough Systems***

A trough-shaped mirror is used to reflect the solar radiation in a tube in the focus of mirrors. The pipe in the center of the trough mirror is called "absorber pipe," or "heat collection element." Figure 11 shows the trough system as a part of the solar thermal system. Most of the trough systems use a axis sun tracker system to ensure that the mirrors reflect the sun's rays on the receiver all time.

2. ***Solar Tower Systems***

Solar tower systems also called central receivers in which many flat heliostats (mirrors) are used to reflect the sun rays onto a tower called a receiver as shown

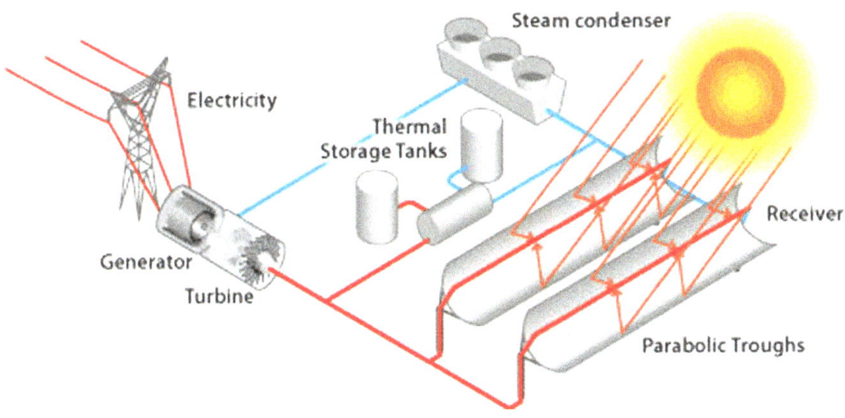

Fig. 11 Trough system [26]

in Fig. 12 [21]. The heliostats rotate with two-axis control systems that can track the sun's movement one axis to capture the sun during the day and the other one is to track it through the year. The mirrors are focusing on the receiver located at top of the tower to heat the fluid, such as molten salt, as hot as 1,050°F. The electricity generation can be achieved by direct use of the hot fluid or the heat can be stored in

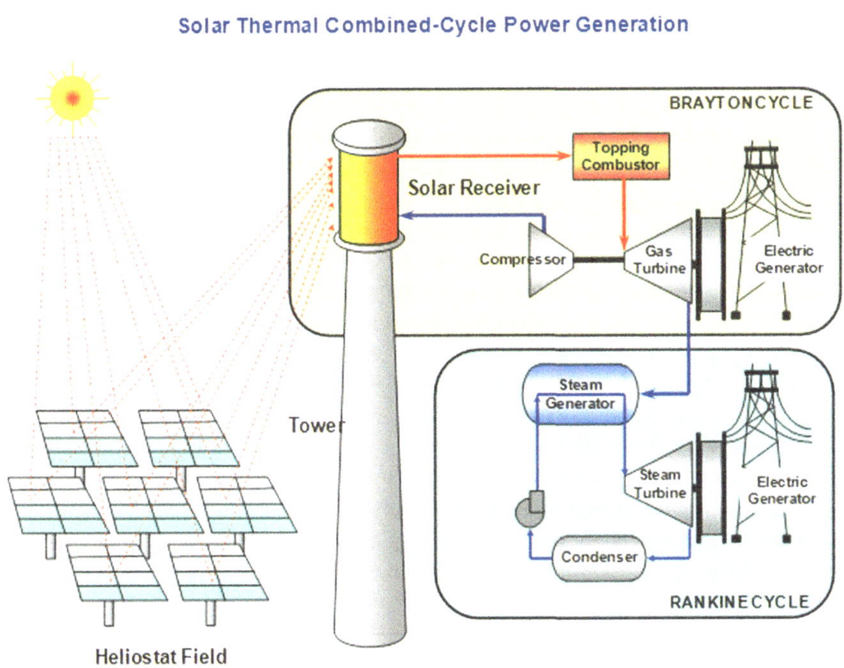

Fig. 12 Scheme of a solar tower concentrating system for electricity generation [21]

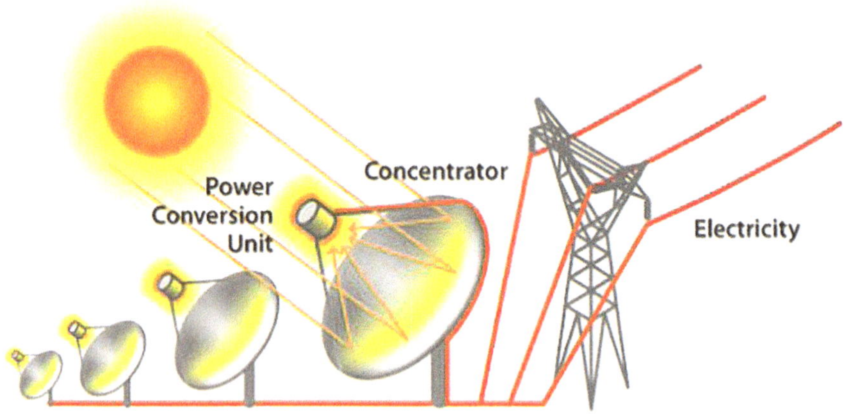

Fig. 13 Solar Dish/Engine power plant illustration [22]

heat storage for later use at night or on cloudy days. The idea behind using molten salt is its ability to store heat for a longer time than any other materials in a very efficient way. Due to the higher temperature associated with the solar tower system, it has higher efficiency, and better use of the energy storage system.

3. ***Dish Engine Systems***

Figure 13 [22] shows a schematic for the dish engine system. The dish should track the sun motion to capture the maximum possible energy. The receiver is attached to a special combustion engine through tubes containing hydrogen or helium gas that can drive a special engine to generate mechanical power. The mechanical power generated from the engine is used to generate the electric power.

4. ***Linear Fresnel Collectors (LFCs)***

LFCs utilizes a series of long flat, or slightly curved, mirrors positioned at various angles to focus sunlight on both sides of the fixed receiver [19]. A schematic for the linear Fresnel reflector power plant is shown in Fig. 14 [23].

2 MPPT of PV Systems

Much research is developed to mathematically implement the PV-cell model. The two-diode model has been utilized in numerous literature [29–31] as appeared in Fig. 15 and Eq. (1).

$$I = I_{PH} - I_{sat1} * \left[e^{\left(\frac{q}{KT}(V+R_s I)\right)} - 1\right] - I_{sat2} * \left[e^{\left(\frac{q}{2KT}(V+R_s I)\right)} - 1\right] - \frac{V + R_s I}{R_{sh}} \tag{1}$$

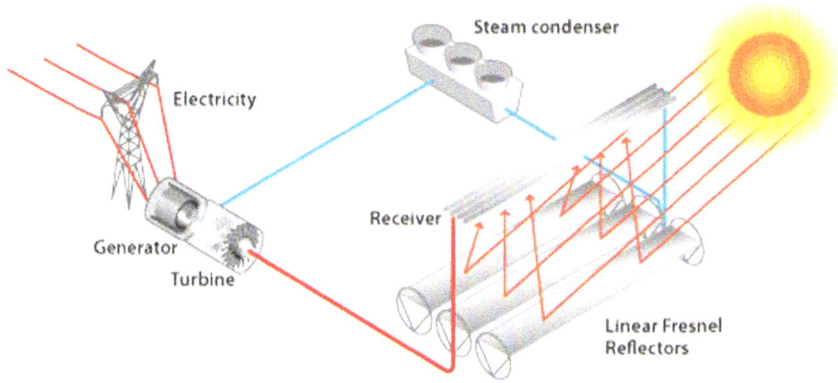

Fig. 14 A linear Fresnel reflector power plant [23]

Fig. 15 The two-diode model

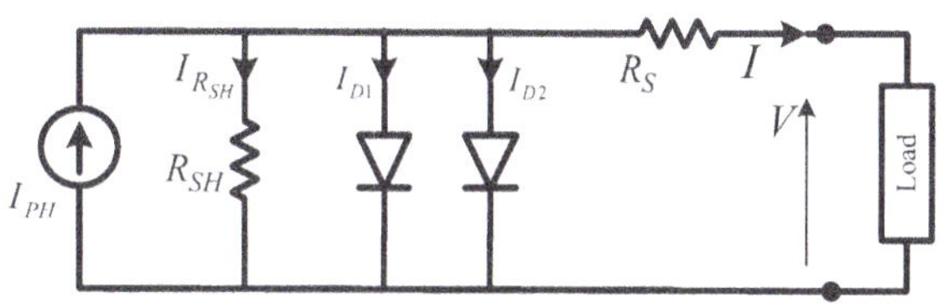

The model of one-diode is more extensively utilized than the model of two-diode for simulation due to its effortlessness with satisfactory exactness. Moreover, its parameter can be gotten tentatively with basic and exact procedures [32, 33]. Therefore this model will be utilized in this chapter. This model of PV-cell is appeared in Fig. 16 and Eq. (2).

$$I = I_{PH} - I_{sat1} * \left[e^{\left(\frac{q}{KT}(V+R_s I)\right)} - 1 \right] - \frac{V + R_s I}{R_{sh}} \tag{2}$$

MPPT needs a quick and intelligent controller to neutralize the rapid changes in weather and load. MPPT comprises the dc-dc converter and its controller as appeared in Fig. 17. Numerous approaches are acquainted with determining the MPP of PV systems as detailed in [29, 34–38].

Fig. 16 The one-diode model

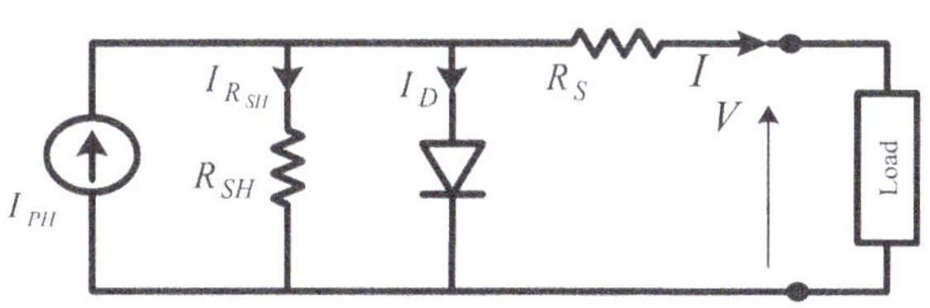

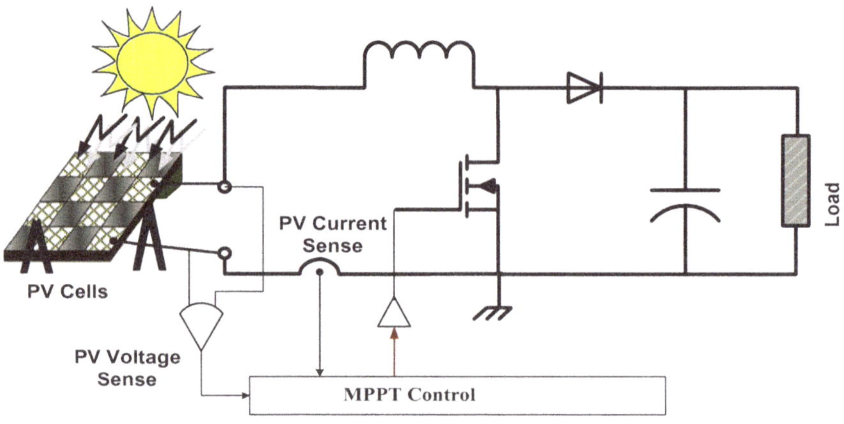

Fig. 17 The MPPT within the PV energy system [1]

1. ***Constant Ratio Technique***

From the P-V curve, it is clear that the ratio of the maximum voltage of the array energy, V_{mp}, to the open-circuit voltage, V_{oc}, is approximately constant. Therefore, the PV array can be forced to act as a ratio of open-circuit voltage. The literature indicates a 73 to 80% achievement of V_{oc} [39]. Also, the relationship between the short circuit current and the current connection to the maximum power (MP) is almost constant. Therefore, it is possible to use the constant current MPPT algorithm that approximates the current MPP as a fixed ratio of short circuit current [40, 41]. The instantaneous infiltration of continuous voltage or current can be avoided by using a pilot cell [42].

2. ***Perturb and Observe (P&O) Technique***

Perturb and observe (P&O) is one of the most popular techniques used in MPP tracking. This process is performed by periodically disrupting the system by increasing the array operation voltage and observing its effect on the output power of the array. Due to a fixed step-width, the system will experience high fluctuations, especially under unstable environmental conditions, resulting in loss of power in the PV system. This technique suffers from the wrong process, especially if multiple local maxima may occur in the case of partial shading [43, 44].

As appeared in Fig. 5 if the operating voltage of the PV module changes and the power increase the control system moves the point of operation of the PV module in that direction; otherwise the operating point is moved in the opposite direction. The flowchart appears in Fig. 18 and the simulation is included in the PSIM software package as appeared in Fig. 19. A common problem with this technique is that the final voltage of the PV unit is confused with each MPPT cycle. Therefore, when the MPP is reached, the output power oscillates around the maximum, which leads to power loss in the PV system. A modified P&O technique was introduced in [45] to solve this problem by multiplying the change in charge ratio (DR) by a dynamic constant

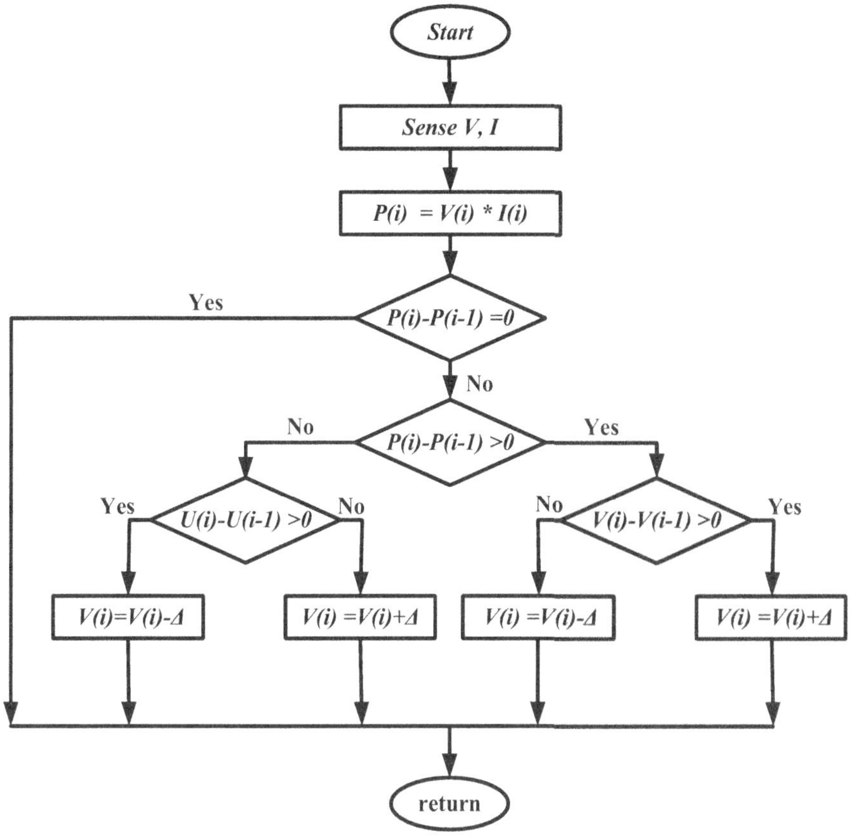

Fig. 18 State-flowchart of P&O MPPT

based on the previous change in the energy extracted as shown in Eq. (3). Another technique [46] ANN was used to predict this fixed multiplier. These technologies complicate the system and may lead to more fluctuations in stable weather conditions [29]. Several amendments were made to this technique in literature [47–49].

The adjusting factor for the change in DR of the modified P&O technique can be obtained from the following equation:

$$M = \frac{|\Delta D|}{|\Delta P|} \tag{3}$$

Where ΔP is the change of output power, ΔD is the change in DR.

3. ***Incremental Conductance Technique (INC)***

Auxiliary conductivity technology is widely used due to its high tracking accuracy instability and being able to adapt well to rapidly changing weather conditions [50].

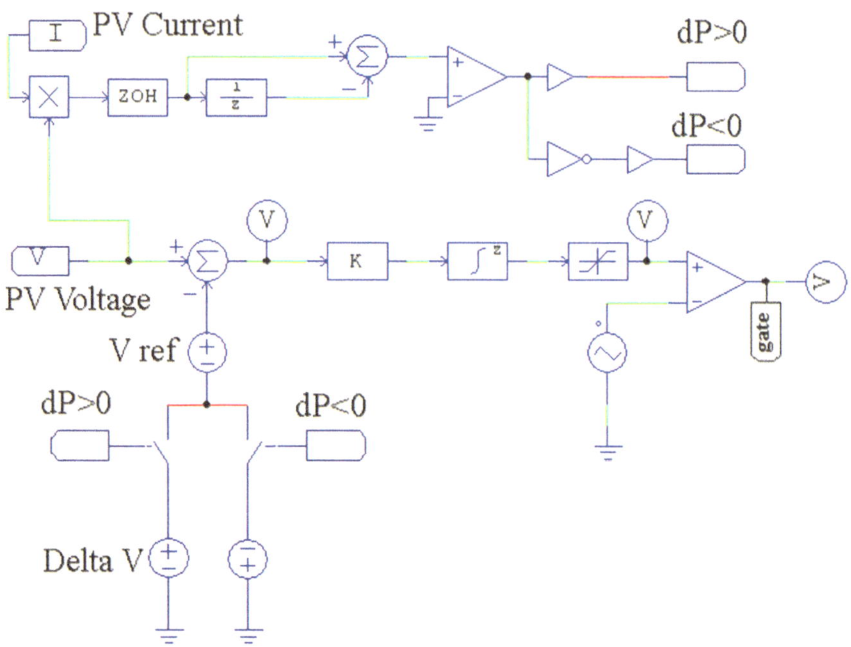

Fig. 19 P&O MPPT control with PSIM [13]

Incremental conductivity (IncCond) is based on a comparison of the instantaneous conductivity of the plate with the additional conductivity of the plate. The input impedance of the DC-DC adapter is matched with the optimum impedance of the PV board. The PV module derivative is given as in Eq. (4), the error equation (e) is the same as in Eq. (5) [51–53].

$$\frac{dP}{dV} = \frac{d(V \times I)}{dV} = I + V\frac{dI}{dV} = 0 \tag{4}$$

$$e = \frac{I(i) - I(i-1)}{V(i) - V(i-1)} + \frac{I(i)}{V(i)} \tag{5}$$

Therefore, MPP tracking requires the following procedure as shown in Fig. 20. It can be implemented by a simple integration unit with an error signal (e) which is the input and the scaling factor k. The function k is to adjust the error signal e to an appropriate range before the full compensator. Since the error signal e becomes smaller as the operating point approaches MPP. Therefore, smooth tracking and adaptation can be achieved [51].

A dynamic step-by-step change in volume was introduced for INC to effectively improve MPPT speed and accuracy simultaneously [50]. This technology improves the performance of INC technology but increases the complexity of the control

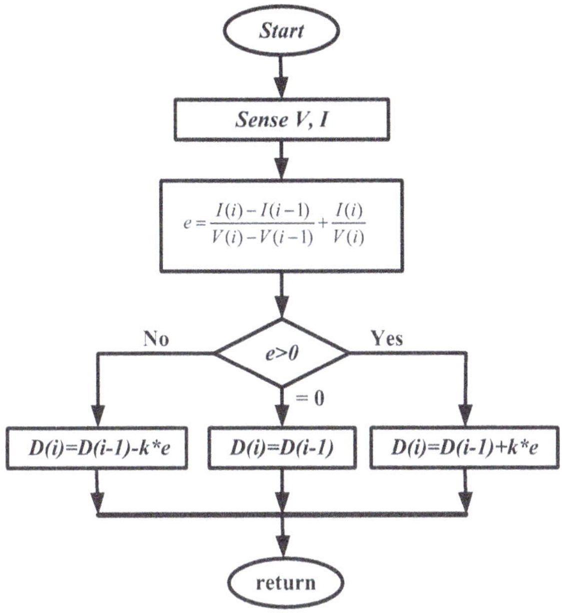

Fig. 20 State-flowchart of INC MPPT [13]

system. The INC flow chart is shown in Fig. 20 and simulation included in the PSIM software package as shown in Fig. 21.

4. ***Hill Climbing Technique (HC)***

HC technique utilizes a boost converter duty cycle as the judging parameter [54, 55]. The flow diagram of the HC technique is shown in Fig. 22. This technique was incorporated and emulated into the PSIM software as appeared in Fig. 23.

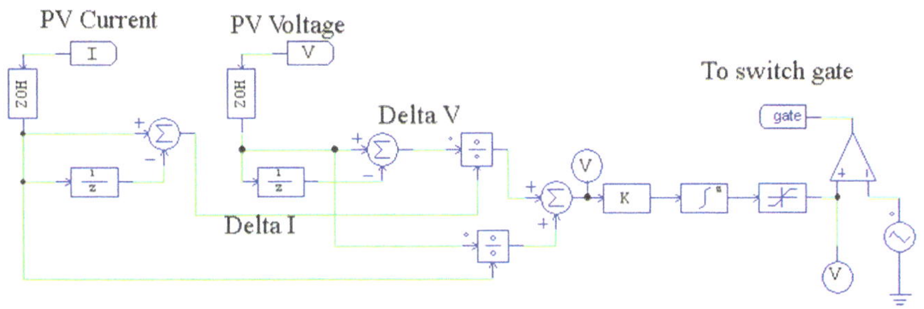

Fig. 21 INC MPPT control with PSIM [13]

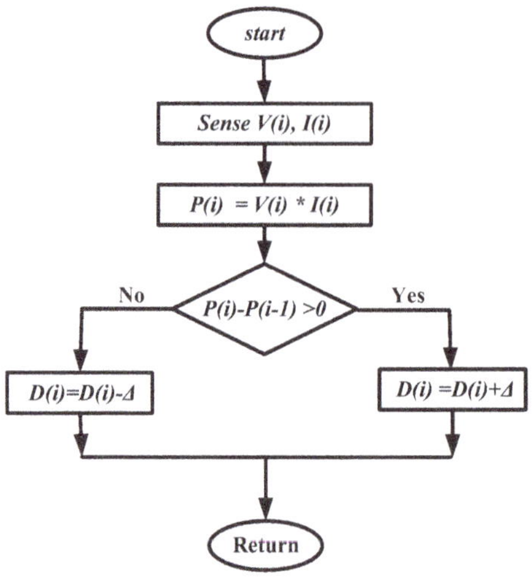

Fig. 22 State-flowchart of HC-MPPT

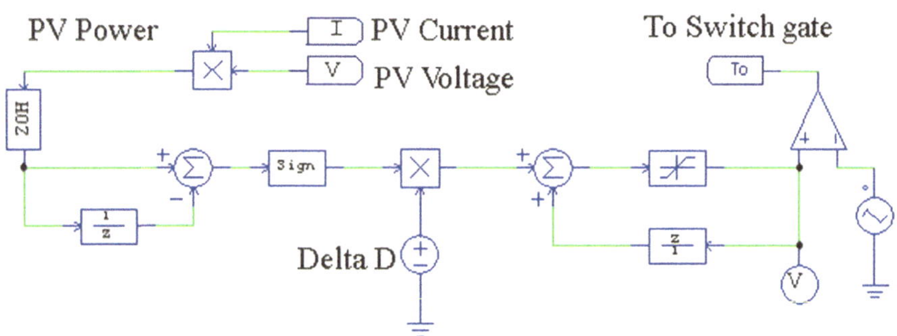

Fig. 23 HC-MPPT Control with PSIM

3 The Performance of MPPT of PV Systems Under Partial Shading Conditions

Partial shading occurs when one or more of the PV-cells in the PV array are introduced to different radiations [56–58]. When this happens, the photovoltaic cell shading will work with the current higher than the generated current and act as a load for the other photovoltaic cells. Due to the increased current flow in the shaded PV-cell higher than the generated current, the voltage ends in negative on this PV-cell as shown in Fig. 24 of PV-cells in the series with 500 and 1000 W/m^2. Figure 25 shows the relationship between the forces generated by two series of cells and the resulting PV voltage along with the shaded photocell radiation. The relationship between the shaded photoelectric cell forces of two PV series systems with a variety of their final potential with this cell radiation is shown in Fig. 26

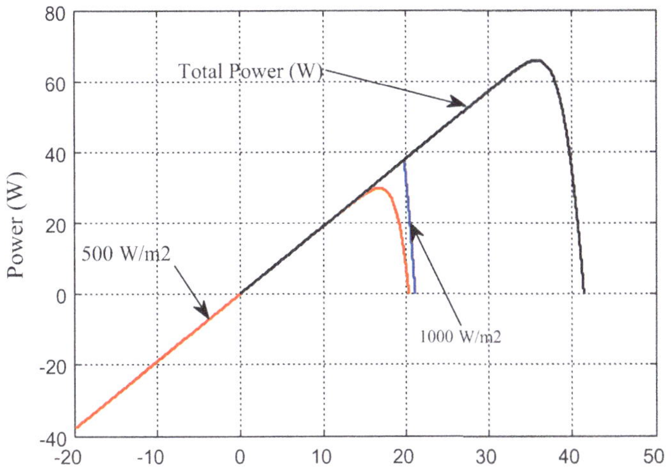

Fig. 24 PV-cell' terminal voltage of the two series PV-cells without bypass diode

Fig. 25 The two P-V cells total power without bypass

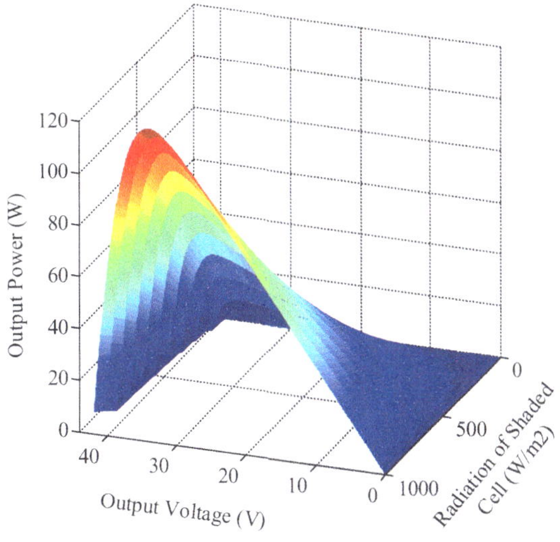

[59, 60]. Figure 27 demonstrates the variety-generated power of 20 series PV-cells with the terminal voltage and radiation on the shaded PV cell when other cells are exposed to 1000 W/m^2 radiation. Figure 28 demonstrates the power of shaded PV cell among 20 series PV-cells along with its terminal voltage and the radiation of these cells where the rest of PV-cells is 1000 W/m^2. It is clear from the above discourse that, the hotspot can be increasingly risky for a higher number of series PV cells. A few kinds of literature work in the optimum number of PV-cells on series to prevent its damage due to the hotspot [59, 60].

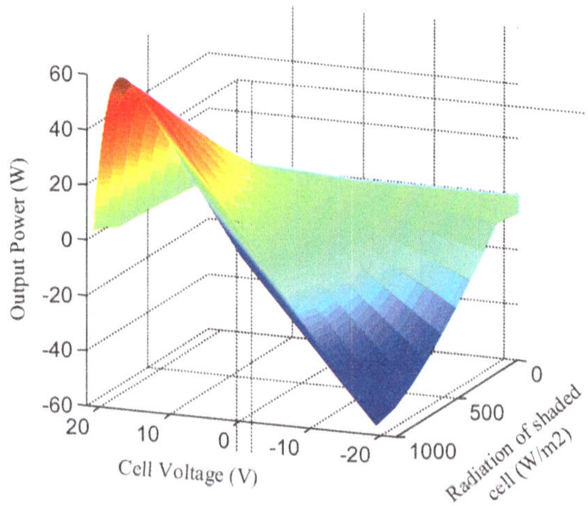

Fig. 26 The shaded cell power with its voltage and radiation

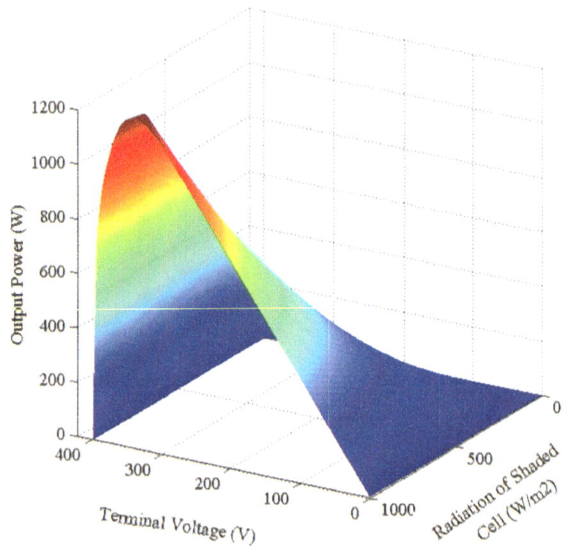

Fig. 27 Total power of 20 series with one shaded PV-cells

A wide range of literature has been produced to treat the influence of PV hotspots [61, 62]. One of the best ways is to include the side diode next to each PV-cell or it can be associated with a certain number of photovoltaics to reduce the cost of photovoltaics and reduce the losses of the PV system. In the case of two series of photovoltaic cells with bypass diodes, the relationship between the terminal voltage and the output energy of these two photovoltaic cells with radiation of 1000 and 500 W/m^2 is shown in Fig. 29. Figure 30 shows the relationship between the output

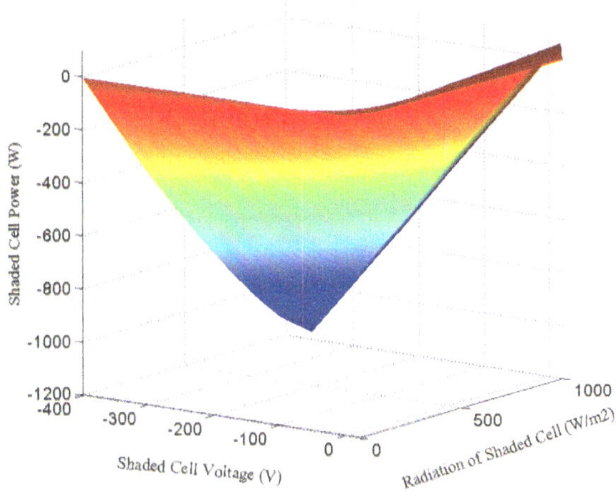

Fig. 28 The power of shaded PV-cell among 20 series PV-cells

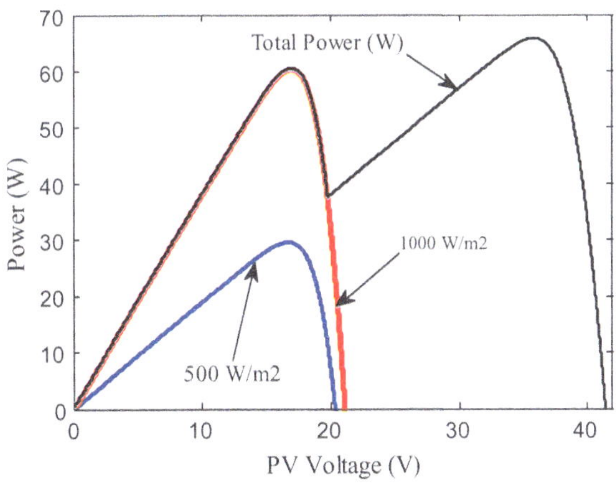

Fig. 29 The output power and terminal voltage relation for two series PV-cells with bypass diodes

energy and the terminal voltage of two series of bypass photovoltaic cells in a different shading state when the radiation of the other photovoltaic cells is 1000 W/m^2.

In the case of 20 series of diode-connected photovoltaic cells connected to each of them, the total form of the energy generated by the different radiation of the shaded photovoltaic cells is shown in Fig. 31 The rest of the radiation in other PV-cells is 1000 W/m^2. It is clear from this figure that the shaded PV-cell begins to operate when the terminal voltage forces the shaded PV-cell voltage to be positive. Figure 32 shows the relationship between the energy output of shaded photovoltaic cells and their voltage for multiple radiations for this cell.

There are three MPP showed up in Fig. 33, the global peak (GP) is the one relating to point#2. The MPPT system ought to pursue the global MPP which is #2 as appeared in Fig. 33.

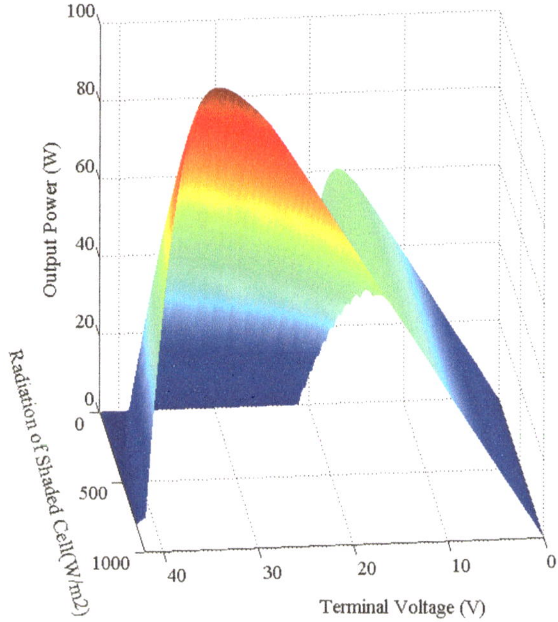

Fig. 30 The output power and terminal voltage relation for two series PV-cells with bypass diodes under various irradiances

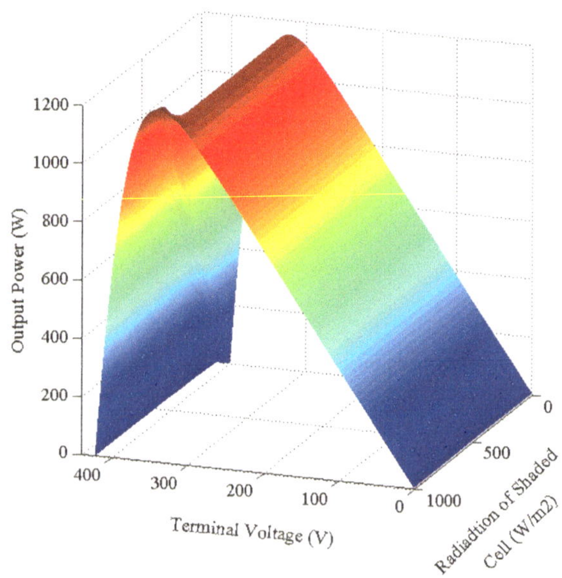

Fig. 31 The total generated power along with terminal voltage for different radiation of shaded PV-cell in case of 20 series PV-cells with bypass diodes

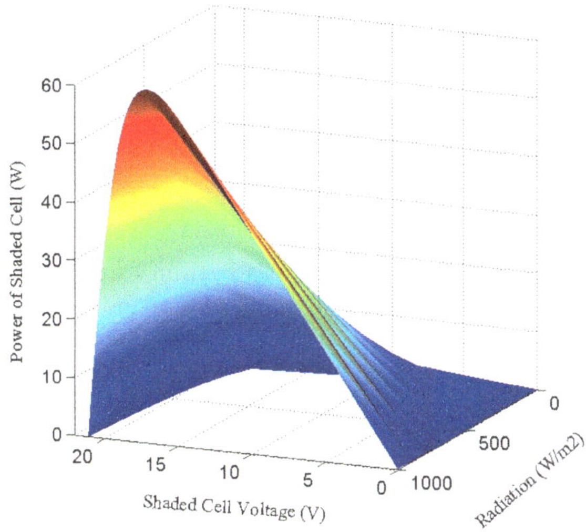

Fig. 32 The relation between the output power of shaded PV-cell and its voltage for different radiation of this cell

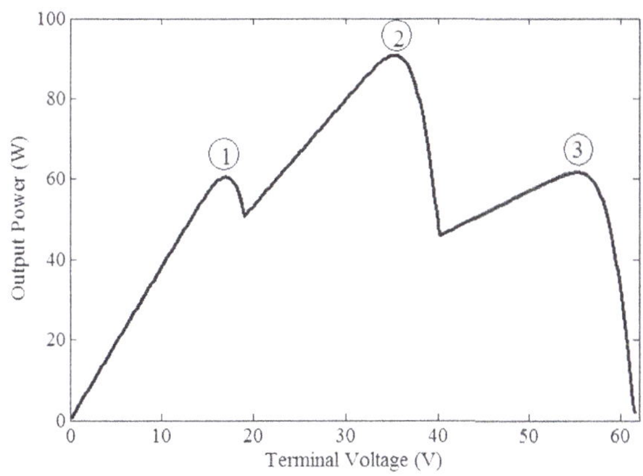

Fig. 33 The relation between the generated power and terminal voltage in case of three series PV-cells at radiation of 1000, 700, 300 W/m^2

A. Mismatch Power Loss

The relationship between the peak strength and the sum of the peaks is called a mismatch loss (MML) and can be calculated mathematically as shown in Eq. (6). Along these lines, the more MML, the more energy is generated from the photoelectric system and vice versa. If the entire photovoltaic cells had the same radiation and the entire photovoltaic system worked in the MPP and each cell operated in its MPP, then the MML factor would be 100%. The value of the MML factor can be obtained from the accompanying equation:

$$\text{MML} = \frac{\text{Maximum power of whole PV system}}{\sum_{i=1}^{N} P\max(i)} * 100 \quad (6)$$

where N is the total number of PV-cells in the PV system.

B. Fuzzy Controller MPPT Technique

Initially, the Fuzzy Logic Control (FLC) has been in focus since 1920 [63]. By 1965 another exploration [64] had introduced FLC as a console for real applications. Since that time FLC has been applied to many applications in various fields of science [65]. FLC can be effectively implemented in various digital devices, for example, microcontrollers [43, 66] digital signal processors, DSP [67] and field-programmable gate group, FPGA, [43, 68] and end up developing Innovation in industrial applications. One of the valuable uses for FLC is the MPPT for PV systems.

C. PSO MPPT Technique

PSO is one of Swarm Intelligence techniques that use randomized population-based variables to solve optimization problems. This technology was first introduced by Eberhart and Kennedy (1995) [69]. The first PSO work was used in MPPT from PV systems implemented in 2004 [70]. PSO is inspired by the behavior of social swarm from school education or bird rising. An evolutionary PSO process, potential solutions, called particles; move around the multidimensional search space by following and tracking the best current particle position in a swarm. The PSO process can be illustrated in the following [71]:

Every particle in the swarm has two variables: position vector $x_i(t)$ and velocity vector $v_i(t)$ as appeared in Eq. (7). Therefore, every particle $x_i(t)$ is described by a vector $[x_{i1}(t), x_{i2}(t),\ldots,x_{iD}(t)]$, as i is the index number of every particle, D is the dimension of the search space and t is iteration number.

$$x_i(t+1) = x_i(t) + v_i(t+1) \quad (7)$$

$x_i(t)$, $v_i(t)$, and the global best position $G_i(t)$ are utilized to set the new position of the particle by calculating the velocity as follows:

$$v_i(t+1) = \underbrace{\omega(t)v(t)}_{\text{Inertial parameters}} + \underbrace{c_1 r_1(P_i(t) - x_i(t))}_{\text{Personal best velocity components}} + \underbrace{c_2 r_2(G_i(t) - x_i(t))}_{\text{Global best velocity component}} \quad (8)$$

where, $\omega(t)$ is the inertia weight factor that controls the search space exploration. The value of $\omega(t)$ can be chosen as a constant value equal to 0.5 or as a variable value for the obtained *GP* acceleration [72]. c_1 and c_2 are acceleration constants, which provide weight to individual and social *GP* components, individually. Where c1 is self-confidence; Range: 1.5–2; and c_2 is swarm confidence. Range: 2–2.5 [73].

A modified PSO technology named deterministic PSO (DPSO) [74] was used to improve MPPT performance under a partial shading component. Also, hill climbing (HC-MPPT) has been used in the uniform distribution of insolation on the PV system but DPSO will be used in the case of PSC. The authors used three parallel branches and four PV units in the series [74]. Another research [75] has suggested a great idea to use some particles to locate the LP and some other particles to find a global peak (GP).

A modified PSO (MPSO) technique [76] was used to adjust the weight of inertia, using the principles of a genetic algorithm (GA) to improve its strength while searching for a GP. The adjusted inertial weight ω can be obtained from the Eq. (9) which uses the main GA principles by increasing the inertial weight value ω at the beginning (in the global research phase) and reducing it to a precise increase at the end of the improvement when the particles approach the optimal solution.

$$\omega(k) = \omega_s - (\omega_s - \omega_e)(T_m - k)/T_m \tag{9}$$

where$_s$ is the initial inertia weight, ω_e is the inertia weight when reaching maximum inertia times, and T_m is the maximum inertia times.

D. Simulation of Proposed Systems

A simulation model of three PV units and a boost transformer used in FLC and MPSO is shown in Fig. 34. The simulation model for the system proposed in SIMULINK is shown in Fig. 35. The simulation parameters for the PV unit are shown in Table 1.

The results of MPSO and FLC simulation are shown in Fig. 36. The response of different MPPT techniques is evaluated in rapidly changing weather conditions. The simulation time shown in this figure is divided into six periods of about two seconds each. Every two seconds, the radiation will be changed in two PV units out of three and the first will be stationary (1000 W/m2) throughout the simulation time (Fig. 37).

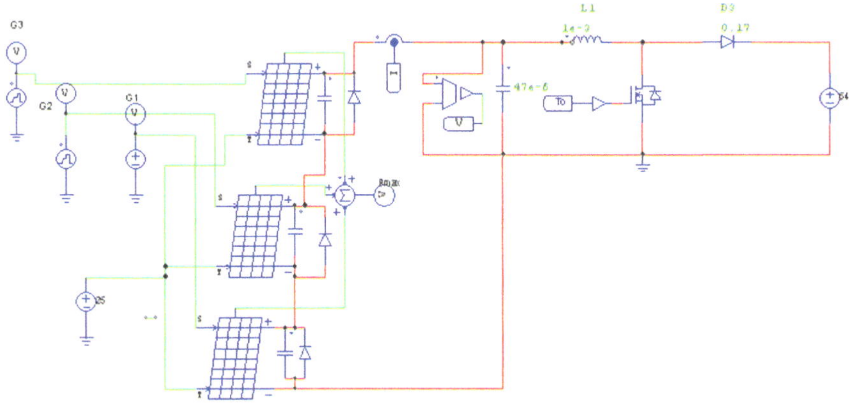

Fig. 34 The simulation model for PV modules and boost converter

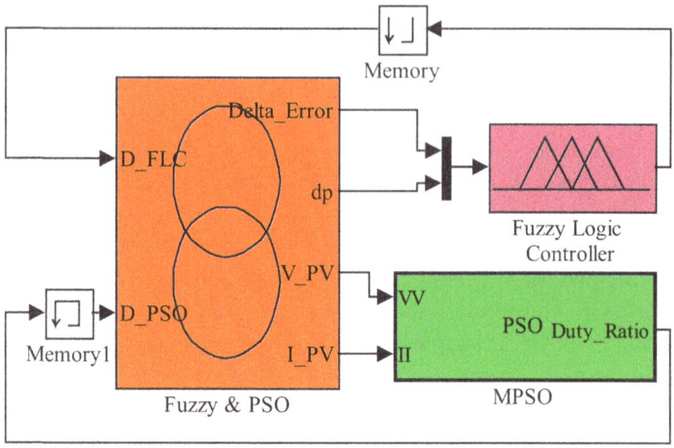

Fig. 35 The SIMULINK model showing the co-simulation between SIMULINK and PSIM

Table 1 Simulation parameters of each PV Module in PSIM [54]

Solar Module (physical model)

Parameters | Other Info | Color

Solar module (physical model) — Help

Parameter	Value	Display
Name	SCP2	
Number of Cells Ns	36	
Standard Light Intensity S0	1000	
Ref. Temperature Tref	25	
Series Resistance Rs	0.008	
Shunt Resistance Rsh	1000	
Short Circuit Current Isc0	3.8	
Saturation Current Is0	2.16e-8	
Band Energy Eg	1.12	
Ideality Factor A	1.2	
Temperature Coefficient Ct	0.0024	
Coefficient Ks	0	

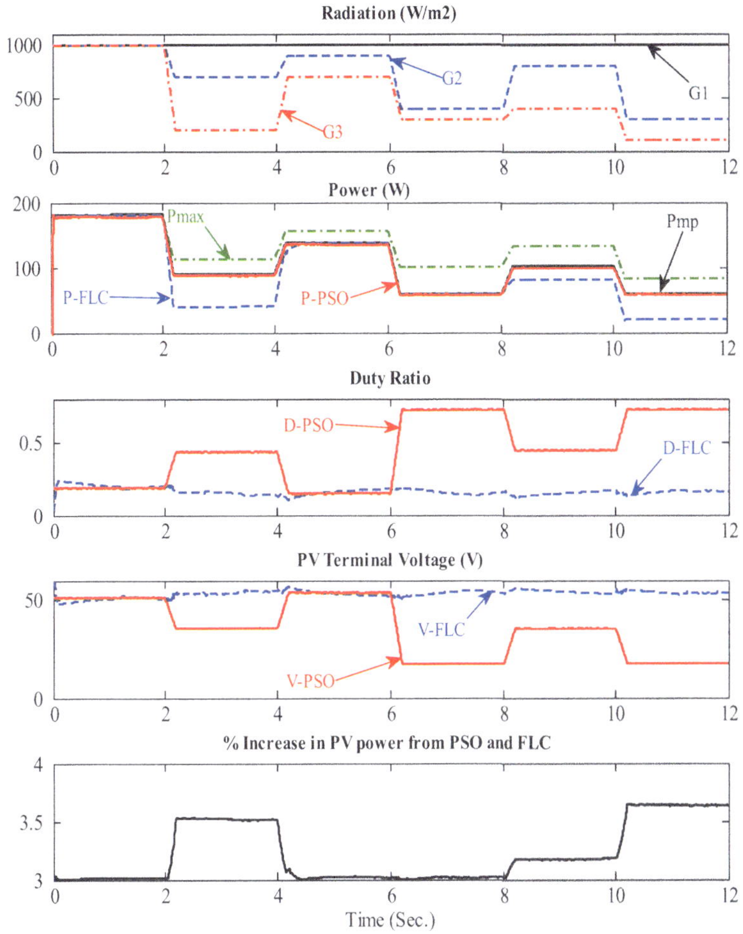

Fig. 36 The simulation results of MPSO and FLC

4 Conclusion

In this chapter, various techniques for tracking, analyzing, simulating, and comparing maximum power points are demonstrated. MATLAB/SIMULINK/PSIM is used to model the photoelectric system under partial shading conditions. Moreover, the response of various power point tracking techniques in rapidly changing weather conditions is evaluated.

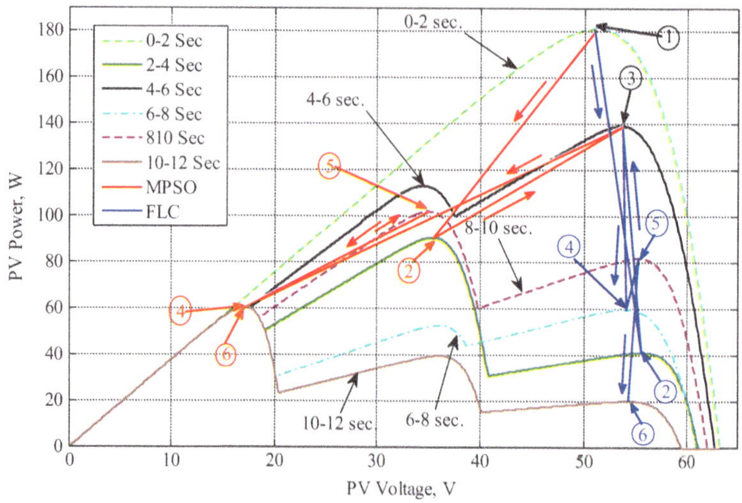

Fig. 37 The relation between the generated power along with terminal voltage and the MPPT response for MPSO and FLC

References

1. Eltamaly AM, Al-Saud MS, Abokhalil AG, Farh HMH (2020) "Simulation and experimental validation of fast adaptive particle swarm optimization strategy for photovoltaic global peak tracker under dynamic partial shading. Renew Sustain Energy Rev 124:109719
2. Eltamaly AM, Al-Saud MS, Abokhalil AG (2020) A novel bat algorithm strategy for maximum power point tracker of photovoltaic energy systems under dynamic partial shading. IEEE Access 8:10048–10060
3. Eltamaly AM, Farh HMH (2020) PV Characteristics, Performance and Modelling. In: Modern maximum power point tracking techniques for photovoltaic energy systems. Springer, Cham, pp 31–63
4. Gong X, Dong F, Mohamed MA, Awwad EM, Abdullah HM, Ali ZM (2020) Towards distributed based energy transaction in a clean smart island. J Clea Prod 122768
5. Eltamaly AM, Farh HMH, Al Saud MS (2019) Impact of PSO reinitialization on the accuracy of dynamic global maximum power detection of variant partially shaded PV systems. Sustainability 11(7):2091
6. Lee TD, Ebong AU (2017) A review of thin film solar cell technologies and challenges. Renew Sustain Energy Rev 70:1286–1297
7. Adib R, Murdock HE, Appavou F, Brown A, Epp B, Leidreiter A, Lins C et al (2015) Renewables 2015 global status report. REN21 Secretariat, Paris, France
8. Eltamaly AM, Mohamed MA (2018) Optimal sizing and designing of hybrid renewable energy systems in smart grid applications. In: Advances in renewable energies and power technologies. Elsevier, pp 231–313
9. Mohamed MA, Eltamaly AM (2018) A novel smart grid application for optimal sizing of hybrid renewable energy systems. In: Modeling and simulation of smart grid integrated with hybrid renewable energy systems. Springer, Cham, pp 39–51
10. Eltamaly AM, Mohamed MA, Alolah AI (2016) A novel smart grid theory for optimal sizing of hybrid renewable energy systems. Sol Energy 124:26–38

11. A Review of Concentrated Solar Power in 2014. https://www.engineering.com/ElectronicsDesign/ElectronicsDesignArticles/ArticleID/9286/A-Review-of-Concentrated-Solar-Power-in-2014.aspx
12. Motahhir S, Chouder A, El Hammoumi A, Benyoucef AS, El Ghzizal A, Kichou S, Kamel K, Sanjeevikumar P, Silvestre S (2020) Optimal energy harvesting from a multistrings PV generator based on artificial bee colony algorithm. IEEE Syst J
13. Rezk H, Eltamaly AM (2015) A comprehensive comparison of different MPPT techniques for photovoltaic systems. Sol Energy 112:1–11
14. Eltamaly AM (2015) Performance of smart maximum power point tracker under partial shading conditions of photovoltaic systems. In: J Renew Sustain Energy 7(4): 043141
15. Kroposki B, Margolis R, Ton D (2009) Harnessing the sun. IEEE Power Energ Mag 7(3):22–33
16. Eltamaly AM (2018) Performance of MPPT techniques of photovoltaic systems under normal and partial shading conditions. In: Advances in renewable energies and power technologies. Elsevier
17. Mohamed MA, Eltamaly AM, Alolah AI (2017) Swarm intelligence-based optimization of grid-dependent hybrid renewable energy systems. Renew Sustain Energy Rev 77:515–524
18. Mohamed MA, Eltamaly AM, Alolah AI (2016) PSO-based smart grid application for sizing and optimization of hybrid renewable energy systems. PLoS ONE 11(8):e0159702
19. Eltamaly AM, Al-Shamma'a AA (2016) Optimal configuration for isolated hybrid renewable energy systems. J Renew Sustain Energy 8(4):045502
20. Mohamed MA, Eltamaly AM (2018) Modeling and simulation of smart grid integrated with hybrid renewable energy systems. Springer, New York
21. Eltamaly AM, Mohamed MA, Al-Saud MS, Alolah AI (2017) Load management as a smart grid concept for sizing and designing of hybrid renewable energy systems. Eng Optimi 49(10):1813–1828
22. Mohamed MA, Eltamaly AM, Alolah AI (2015) Sizing and techno-economic analysis of stand-alone hybrid photovoltaic/wind/diesel/battery power generation systems. J Renew Sustain Energy 7(6):063128
23. Eltamaly AM (2015) Performance of smart maximum power point tracker under partial shading conditions of PV systems. In: 2015 IEEE International Conference on Smart Energy Grid Engineering (SEGE). IEEE, pp 1–8
24. Al-Saud M, Eltamaly AM, Mohamed MA, Kavousi Fard A (2019) An intelligent data-driven model to secure intra-vehicle communications based on machine learning. IEEE Trans Ind Electr
25. Eltamaly AM, Farh HMH (2019) Dynamic global maximum power point tracking of the PV systems under variant partial shading using hybrid GWO-FLC. Solar Energy 177: 306–316
26. Mohamed MA, Eltamaly AM, Farh HM, Alolah AI (2015) Energy management and renewable energy integration in smart grid system. In: 2015 IEEE international conference on smart energy grid engineering (SEGE). IEEE, pp 1–6
27. Eltamaly AM, Addoweesh KE, Bawa U, Mohamed MA (2014) Economic modeling of hybrid renewable energy system: a case study in Saudi Arabia. Arab J Sci Eng 39(5):3827–3839
28. Farh HMH, Othman MF, Eltamaly AM, Al-Saud MS (2018) Maximum power extraction from a partially shaded PV system using an interleaved boost converter. Energies 11(10):2543
29. Eltamaly AM, Farh HMH, Othman MF (2018) A novel evaluation index for the photovoltaic maximum power point tracker techniques. Solar Energy 174:940–956
30. Farh HMH, Eltamaly AM, Othman MF (2018) Hybrid PSO-FLC for dynamic global peak extraction of the partially shaded photovoltaic system. PloS one 13(11):e0206171
31. Eltamaly AM, Farh HMH, Al-Saud MS (2019) Grade point average assessment for meta-heuristic GMPP techniques of partial shaded PV systems. IET Renew Power Gener 13(8):1215–1231
32. Farh HMH, Othman MF, Eltamaly AM (2017) Maximum power extraction from grid-connected PV system. In 2017 Saudi Arabia Smart Grid (SASG). IEEE, pp 1–6
33. Farh HMH, Eltamaly AM, Al-Saud MS (2019) Interleaved boost converter for global maximum power extraction from the photovoltaic system under partial shading. IET Renew Power Gener 13(8):1232–1238

34. Motahhir S, Hammoumi AEL, Ghzizal AEL, Derouich A (2019) Open hardware/software test bench for solar tracker with virtual instrumentation. Sustain Energy Technol Assess 31: 9–16
35. Etamaly AM, Mohamed MA, Alolah AI (2015) A smart technique for optimization and simulation of hybrid photovoltaic/wind/diesel/battery energy systems. In: 2015 IEEE International conference on smart energy grid engineering (SEGE). IEEE, pp 1–6
36. Eltamaly AM, Al-Saud MS, Abo-Khalil AG (2020) Performance improvement of pv systems' maximum power point tracker based on a scanning PSO particle strategy. Sustainability 12(3):1185
37. Balouktsis A, Karapantsios TD, Anastasiou K, Antoniadis A, Balouktsis I (2006) Load matching in a direct-coupled photovoltaic system-application to Thevenin's equivalent loads. Int J Photoenergy
38. Eltamaly AM, El-Tamaly HH, Enjeti P (2002) An improved maximum power point tracker for photovoltaic energy systems. In: The 2nd Minia international conference for advanced trends in engineering
39. Eltamaly AM, Al-Saud MS, Abokhalil AG (2020) A novel scanning bat algorithm strategy for maximum power point tracker of partially shaded photovoltaic energy systems. Ain Shams Eng J
40. Eltamaly AM, Al-Saud MS, Abokhalil AG, Farh HMH (2020) Photovoltaic maximum power point tracking under dynamic partial shading changes by novel adaptive particle swarm optimization strategy. Trans Inst Measurement Control 42(1):104–115
41. Yuvarajan S, Xu S (2003) Photo-voltaic power converter with a simple maximum-power-point-tracker. In: Proceedings of the 2003 international symposium on circuits and systems, vol 3. ISCAS'03. IEEE, pp III–III
42. Veerachary M, Senjyu T, Uezato K (2002) Voltage-based maximum power point tracking control of PV system. IEEE Trans Aerosp Electron Syst 38(1):262–270
43. Eltamaly AM, Alolah AI, Abdulghany MY (2010) Digital implementation of general purpose fuzzy logic controller for photovoltaic maximum power point tracker. In: SPEEDAM 2010. IEEE, pp 622–627
44. Farh HMH, Eltamaly AM (2020) Maximum power extraction from the photovoltaic system under partial shading conditions. In: Modern maximum power point tracking techniques for photovoltaic energy systems. Springer, Cham, pp 107–129
45. Al-Diab A, Sourkounis C (2010) Variable step size P&O MPPT algorithm for PV systems. In: 2010 12th International conference on optimization of electrical and electronic equipment. IEEE, pp 1097–1102
46. Mohamed MA, Eltamaly AM (2018) Sizing and techno-economic analysis of stand-alone hybrid photovoltaic/wind/diesel/battery energy systems. In: Modeling and simulation of smart grid integrated with hybrid renewable energy systems. Springer, Cham, pp 23–38
47. Femia N, Granozio D, Petrone G, Spagnuolo G, Vitelli M (2007) Predictive & adaptive MPPT perturb and observe method. IEEE Trans Aerosp Electron Syst 43(3):934–950
48. Eltamaly AM, Mohamed MA (2016) A novel software for design and optimization of hybrid power systems. J Braz Soc Mech Sci Eng 38(4):1299–1315
49. El-Tamaly HH, El-Tamaly AM, El-Baset Mohammed AA (2003) Design and control strategy of utility interfaced PV/WTG hybrid system. In: The ninth international middle east power system conference. MEPCON, pp 16–18
50. Liu F, Duan S, Liu F, Liu B, Kang Y (2008) A variable step size INC MPPT method for PV systems. IEEE Trans Industr Electron 55(7):2622–2628
51. Abouobaida H, Cherkaoui M (2012) Comparative study of maximum power point trackers for fast changing environmental conditions. In: 2012 International conference on multimedia computing and systems. IEEE, pp 1131–1136
52. Eltamaly AM, Mohamed MA (2014) A novel design and optimization software for autonomous PV/wind/battery hybrid power systems. Math Probl Eng

53. Siri K, Conner KA (2001) Fault-tolerant scaleable solar power bus architectures with maximum power tracking. In: APEC 2001. Sixteenth annual IEEE applied power electronics conference and exposition (Cat. No. 01CH37181), vol 2. IEEE, pp 1009–1014
54. Farh HMH, Eltamaly AM, Ibrahim AB, Othman MF, Al-Saud MS (2019) Dynamic global power extraction from partially shaded photovoltaic using deep recurrent neural network and improved PSO techniques. Int Trans Electr Energy Syst 29(9): e12061
55. Eltamaly AM, Addoweesh KE, Bawah U, Mohamed MA (2013) New software for hybrid renewable energy assessment for ten locations in Saudi Arabia. J Renew Sustain Energy 5(3):033126
56. Rajesh R, Carolin Mabel M (2014) Efficiency analysis of a multi-fuzzy logic controller for the determination of operating points in a PV system. Solar Energy 99:77–87
57. Kimball JW, Krein PT (2007) Digital ripple correlation control for photovoltaic applications. In: 2007 IEEE power electronics specialists conference. IEEE, pp 1690–1694
58. Mohamed MA, Eltamaly Am (2018) Introduction and literature review. In: Modeling and simulation of smart grid integrated with hybrid renewable energy systems. Springer, Cham, pp 1–10
59. Rossi D, Omaña M, Giaffreda D, Metra C (2014) Modeling and detection of hotspot in shaded photovoltaic cells. IEEE Trans Very Large Scale Integr (VLSI) Syst 23(6):1031–1039
60. Solórzano J, Egido MA (2014) Hot-spot mitigation in PV arrays with distributed MPPT (DMPPT). Sol Energy 101:131–137
61. Eltamaly AM, Abdelaziz AY eds (2019) Modern maximum power point tracking techniques for photovoltaic energy systems. Springer
62. Kim K (2014) Hot spot detection and protection methods for photovoltaic systems. PhD diss., University of Illinois at Urbana-Champaign
63. Libkin L (2016) SQL's three-valued logic and certain answers. ACM Trans Database Syst (TODS) 41(1):1–28
64. Mendel JM (2007) Type-2 fuzzy sets and systems: an overview. IEEE Comput Intell Mag 2(1):20–29
65. Eltamaly Am (2010) Modeling of fuzzy logic controller for photovoltaic maximum power point tracker. In: Solar Future 2010 Conference, Istanbul, Turkey
66. Boukens M, Boukabou A (2013) PD with fuzzy compensator control of robot manipulators: Experimental study. In: 3rd International conference on systems and control. IEEE, pp 973–978
67. Eltamaly AM (2011) Modeling of fuzzy logic controller for photovoltaic maximum power point tracker. Trends Electr Eng 1(2)
68. Tipsuwanpom R, Runghimmawan T, Intajag S, Krongratana V (2004) Fuzzy logic PID controller based on FPGA for process control. In: 2004 IEEE international symposium on industrial electronics, vol 2. IEEE, pp 1495–1500
69. Farh HM, Eltamaly AM, Mohamed MA (2012). Wind energy assessment for five locations in Saudi Arabia. In: International conference on future environment and energy (ICFEE 2012,) IPCBEE (2012) © (2012) IACSIT Press, February 26–28. Singapoore
70. Mohamed MA, Eltamaly AM, Alolah AI, Hatata AY (2019). A novel framework-based cuckoo search algorithm for sizing and optimization of grid-independent hybrid renewable energy systems. Int J Green Energy 16(1):86–100
71. Eltamaly AM, Farh HMH, Abokhalil AG (2020). A novel PSO strategy for improving dynamic change partial shading photovoltaic maximum power point tracker. Energy Sour Part A Recovery, Utilization, Environ Effects, 1–15
72. Mohamed MA, Eltamaly AM (2018) A PSO-based smart grid application for optimum sizing of hybrid renewable energy systems. In: Modeling and simulation of smart grid integrated with hybrid renewable energy systems. Springer, Cham, pp 53–60
73. Ishaque K, Salam Z, Shamsudin A, Amjad M (2012) A direct control based maximum power point tracking method for photovoltaic system under partial shading conditions using particle swarm optimization algorithm. Appl Energy 99:414–422

74. Fu Q, Tong N (2010) A new PSO algorithm based on adaptive grouping for photovoltaic MPP prediction. In: 2010 2nd International workshop on intelligent systems and applications. IEEE, pp 1–5
75. Mohamed MA, Eltamaly AM (2018) Modeling of hybrid renewable energy system. In: Modeling and simulation of smart grid integrated with hybrid renewable energy systems. Springer, Cham, pp 11–21
76. PSIM (2020) Psim User Manual of PSIM computer Simulation. Accessed June 2020. http://powersimtech.com/wp-content/uploads/2015/05/PSIM-User-Manual.pdf

Enhancement Techniques to Design a Standalone PV System for Residential Application

R. Ramaprabha and S. Malathy

Abstract Standard design procedures are available to design a standalone solar photovoltaic (SSPV) system. However, it does not include the impact of partial shading on the PV array. The harvested power may reduce appreciably if the PV array is partly or completely shaded. The reduction is to be compensated by increasing the size of the PV array and this increases the overall cost of the system. It is therefore necessary to devise techniques to mitigate the impact of partial shading and enhance the power generation under such conditions. This chapter introduces and explains various strategies (shade resilient arrangement, global peak detecting algorithm, and reduced device count inverters) to incorporate along with the standard design procedures to increase the power generated by the SSPV system. Enhanced performance reduces the return on investment (ROI) period, thereby making the standalone PV system more attractive.

Keywords Photovoltaic · Partial shading · Shade tolerant configuration · GMPPT · Multilevel inverter

1 Introduction

The unrelenting focus among the countries on diversifying their energy mix away from imported oil and fossil fuels has widened the market for power generation using renewable energy sources. Power generation by photovoltaic (PV) systems has gained much attention than other renewable energy sources due to its abundant availability and nonpolluting nature. Further, the subsidies/incentives offered by the government and the technological advancements which have reduced the cost per watt have significantly increased the reach of PV technology among the general population.

R. Ramaprabha (✉) · S. Malathy
Sri Sivasubramaniya Nadar College of Engineering, Kalavakkam 603110, Tamil Nadu, India
e-mail: ramaprabhar@ssn.edu.in

S. Malathy
e-mail: smalathy@gmail.com

S. Motahhir and A. M. Eltamaly (eds.), *Advanced Technologies for Solar Photovoltaics Energy Systems*, Green Energy and Technology,
https://doi.org/10.1007/978-3-030-64565-6_10

The large-scale commercial installations are installed in shade-free areas. But, the small-scale residential SSPV systems are installed on the rooftops or facades of the buildings that are susceptible to shades due to space limitations. The shade is often caused by nearby buildings or structures in the same building.

Partial shading is a condition where all the panels in a PV array do not receive the same amount of irradiation. Clouds, bird litters, nearby buildings, or structures in the same building as chimneys and overhead tanks may cast their shade on the PV panels. When a PV array is shaded partially, the shaded panels generate lesser photon current and impose a current limitation on the other serially connected non-shaded panels. If this limitation is violated, the shaded panels get reverse biased, and hot spots may develop due to increased thermal stress. Bypass diodes prevent hot spot development by offering an alternate path for the current flow [1]. However, the shaded panel is completely bypassed and the residual power generated by the panel remains uncollected. The net power generated is significantly reduced under such circumstances and to meet the desired specifications, the size of the PV array is to be increased. This in turn increases the ROI that makes power generation by PV systems less attractive.

It is evident that partial shading is the major issue that needs to be addressed in an SSPV system. The simple solution is to provide bypass diodes to prevent the panels from getting damaged. However, the inclusion of bypass diodes causes multiple power peaks in the voltage-power characteristic curve, and to address this issue global maximum peak tracking (GMPPT) algorithms are to be included in the system. These algorithms track the global peak and make sure that maximum power is transferred from the array to the load under all environmental conditions. GMPPT algorithms based on soft computing techniques, artificial intelligence, and other search methodologies have been reported in the literature [2–5]. Some algorithms require advanced processors to do complex computations, while others require huge data set to train. The choice of the GMPPT algorithm relies on the user's requirements and availability of resources.

Further, the reduction in output power of partially shaded PV array is not directly proportional to the shade intensity but relies highly on the number and configuration of bypass diodes embedded in the panels, shade pattern and the interconnection scheme adopted in the array. The reduction in power is mainly due to mismatch in currents and it can be reduced to a greater extent by shade dispersion or irradiation equivalence [6]. Many schemes based on algorithms (online and offline) and puzzle patterns like Sudoku, magic square, and Latin square [7–9] have been proposed in the literature to equalize the irradiation. The underlying principle in all these strategies is to disperse the shade evenly all over the PV array to reduce the mismatch in currents among the serially connected panels. The online algorithms are dynamic and involve several sensors to assess the prevailing shading conditions [10–12]. The switches provided with each of the panels are then triggered appropriately to change the interconnectivity between the panels in the array. The offline algorithms are static and adopt puzzle patterns or thumb rules to fix the position of the individual panels in the array.

These two strategies (irradiation equivalence and global peak detecting) are discussed and analyzed separately by most of the researchers to address partial shading issues.

This chapter tends to unify these strategies into the standard design procedure to enhance the power generation under partially shaded conditions and to utilize the installed PV system efficiently.

The standard design procedure calculates the size of the array and the rating of other components to be included in the system. It does not take into account the impact of partial shading. The efficiency of the SSPV system can be enhanced appreciably if the irradiance equivalence algorithm is incorporated in the design stage to fix the location of panels within the array. One such strategy is discussed in Sect. 4. The irradiation cannot be equalized perfectly for all the shading conditions and in such cases; the electrical characteristics show multiple peaks. Several peak detecting algorithms are reported in the literature and two such algorithms are discussed in Sect. 5. The PV inverter failure rate is higher compared to the other components in the PV system and Sect. 6 discusses the strategy that can be adopted to improve the reliability. The design of an SSPV system for a residential unit is considered in this chapter to demonstrate the significance of improvisation strategies.

2 Standalone PV System

A standalone PV system has a PV array that is sized to meet the load requirements. The schematic of the typical SSPV system is depicted in Fig. 1.

The major components of the system include PV array, maximum power point tracking (MPPT) controller, battery bank, and inverter. The output power of the PV array is not constant and it depends greatly on the prevailing environmental

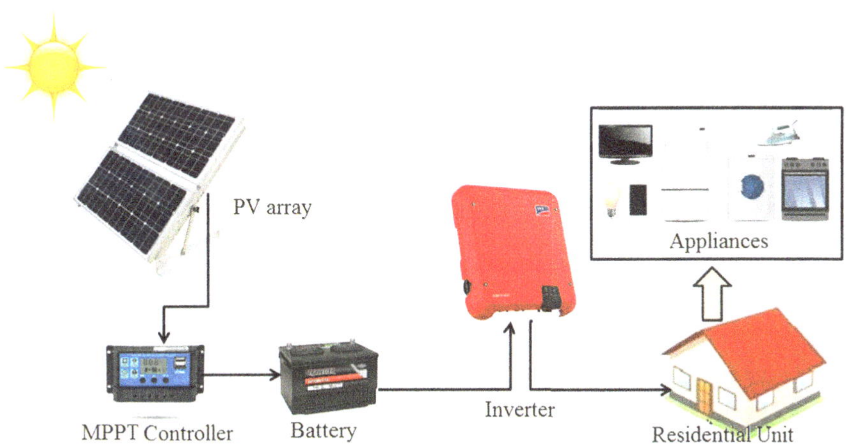

Fig. 1 Schematic of SSPV system

conditions. The MPPT controller operates the PV array at its optimal power point and ensures the transfer of maximum power from the source to the load under all environmental conditions. Most of the residential load operates on ac and hence an inverter is integrated into the standalone PV system. Commercial inverters have an inbuilt conventional MPPT controller. However, the conventional controller can't track the peak effectively under partially shaded conditions as it can't differentiate between global and local peaks, and consequently, the power extracted from the PV source is significantly less. The battery bank is essential in the SSPV system to enable the operation of the loads during night times. The battery bank size is determined considering the load requirements and days of autonomy.

Nowadays, batteryless systems are gaining popularity given the cost, size, and limited lifetime of the battery besides the impacts on the environment. Hence, in this work, the SSPV system that is directly connected to the residential load during the day (with a minimal battery bank to support short time mandatory loads) and connected to the grid during the night is considered.

As a design example, a residential unit located in Chennai, India is considered. The residential building is a single-family home with two floors and a built-up area of 1900 square feet. The photovoltaic power potential for the location is 4 kWh/kW$_p$ (daily) and 1461 kWh/kW$_p$ (yearly sum) [13] inclusive of the power conversion losses and losses due to dirt. Also, the power plant availability is assumed to be 100%. The design is given below.

- After efficiency improvements, the annual energy demand for the building is determined to be 12780 kWh/year. That is, E = 12780 kWh/year.

(Note: Energy demand can be calculated by considering either all possible loads of the residential building or the average annual energy consumed by the residential building from its electricity bill)

- Considering the solar energy resource for the location as 5.5 kWh/m^2/day, the average annual sunlight hours is calculated to be 2008 ($5.5 \times 365 = 2007.5$) full sun hours.
- Required PV panel wattage rating = (12780 kWh/2008 h) = 6.365 kW.
- Considering 25% loss, PV Panel wattage = 6.365 kW/0.75 $\approx$ 8488 W.
- Considering 250 W$_p$ panel with 31.5 V and 7.94 A at peak point, number of panels required = 8488/250 = 33.95 $\approx$ 35 panels.
- Number of panels in series = 7 (to meet the 220/230 V requirement).
- Number of strings in parallel = 5 ($7 \times 5 = 35$).
- Voltage and current of 7×5 array are 220.5 V and 39.7 A (at max. power).
- Maximum power produced by 7×5 array $\approx$ 8750 W.

Appropriate power converter (power conditioning unit—PCU) can be introduced between the PV array and load to regulate voltage and current.

These calculations are done assuming standard test conditions (STC). But, the energy yield of the PV array may significantly fall if the array is shaded partially or completely. The PV array is usually installed on the rooftop (flat roof) and it is likely to be shaded partially by the parapet wall, overhead water tanks, antenna, cables, and

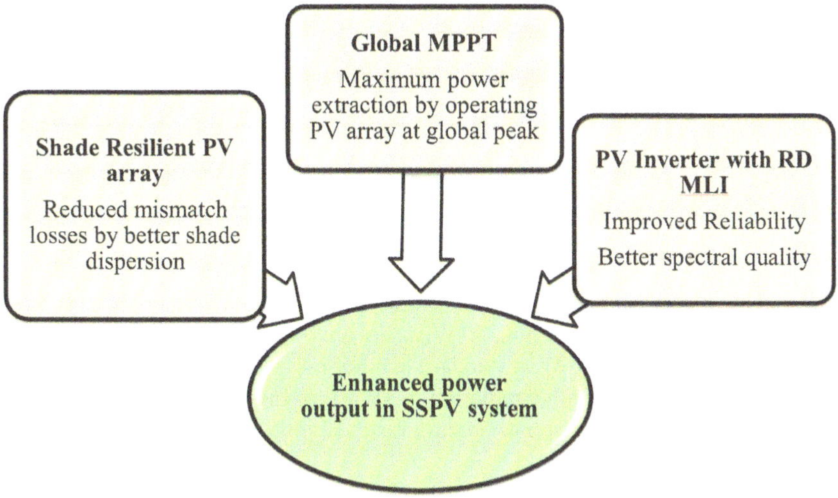

Fig. 2 Enhancement strategies for SSPV system

trees and nearby buildings beside the shading by bird litters and dirt accumulation. These shading conditions are caused by fixed objects/structures and hence can be categorized as fixed shadings.

Shading/partial shading is unavoidable and will prevail for a considerable duration in most of the urban residential installations. This results in reduced power generation and, to meet the load, the PV array is to be upsized and that is not a favorable option. This issue can be addressed by arranging the PV array in a shade resilient (SR) configuration where the shade tolerance of the PV array is enhanced by dispersing the shade uniformly all over the array. If the shaded geometry is short and narrow as in most of the cases, the voltage-power (V-P) characteristics of the SR PV array may exhibit multiple peaks, and incorporation of the global peak detecting algorithm will track and operate the PV array at its optimal power point. Further, the reliability of the SSPV system can be enhanced by employing a multilevel inverter with reduced device count (RD MLI) instead of the conventional inverter. The performance of the residential SSPV system can be enhanced with these three enhancement strategies (SR arrangement, GMPPT, and RD MLI). The enhancement strategies proposed in this chapter for the SSPV system are depicted in Fig. 2.

3 PV Array and Interconnection Schemes

The study considers a 7×5 array of 250 W_p panels. The mathematical model (single diode PV model) of the panel is developed based on the standard equations [14] to assess the power generation under partially shaded conditions. The specifications of

Table 1 Specifications of the 250 W_p PV Panel

Specifications	Values
Open circuit voltage (V_{oc})	37.8 V
Short circuit current (I_{sc})	8.7 A
Max. power (P_m)	250 W
Voltage at P_m (V_m)	31.5 V
Current at P_m (I_m)	7.94 A

the panel are presented in Table 1. The simulated electrical characteristics (voltage-power and voltage-current) of the 250 W_p panel are presented in Fig. 3. The peak or the maximum power is 250 W at standard conditions.

The peak power falls with irradiation as shown in Fig. 4. However, the voltage (V_m) at which the maximum power occurs varies slightly with irradiation. It is often neglected to ease calculations.

The panels in the PV array are connected in series and parallel to meet the voltage and current specifications. Series and parallel are the basic interconnection schemes and it has been proved in the literature that the parallel scheme results in optimal output under all environmental conditions [15]. However, higher array current and lower array voltage resulting from the parallel configuration are not desirable. Hence,

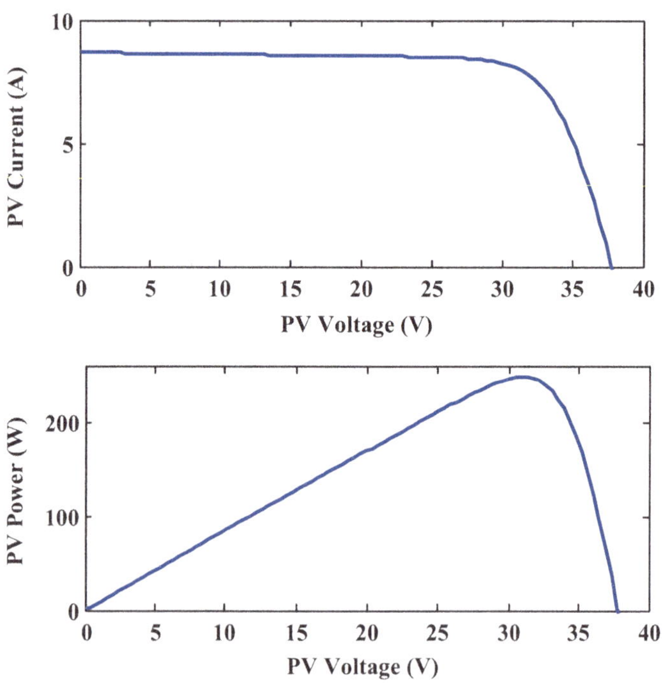

Fig. 3 Characteristics of 250 W_p panel

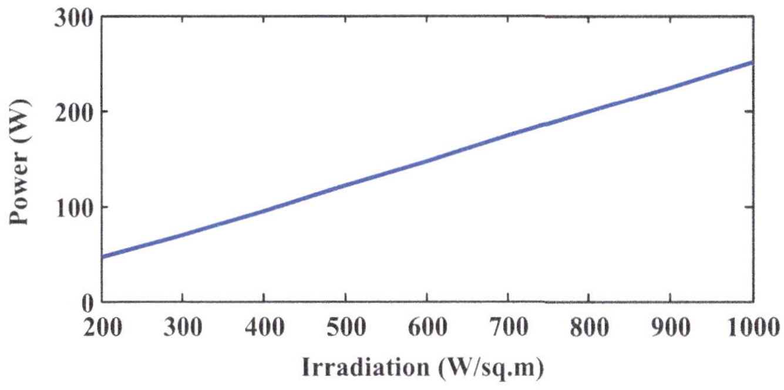

Fig. 4 Change in peak power (P_m) with irradiation

new interconnection schemes like the honey comb (HC), bridge link (BL), and total cross-tied (TCT) are derived from the basic schemes with the inclusion of cross-ties. The basic and derived interconnection schemes for the 7 × 5 PV array are presented in Fig. 5.

The simulated electrical characteristics of the 7 × 5 array for various interconnection schemes are presented in Fig. 6. The characteristics are simulated under standard test conditions. The series configuration results in a lesser array current at higher voltage levels. It is due to the fact that the array current is equal to the panel current in series connection. In the parallel configuration, the array current is equal to the sum of all the PV panel currents and this results in higher current at low voltage levels. In the case of the derived configurations, the array voltage depends on the number of panels connected in series, and the array current depends on the number of panels connected in parallel. The voltage and current levels of the array remain the same for the derived interconnection schemes under full irradiation conditions as shown in the current-voltage characteristics of Fig. 6. The peak power of the 7 × 5 array is the same for both the basic and derived interconnection schemes as shown in the power-voltage characteristics of Fig. 6.

It can be seen from the figure that all the derived schemes resulted in a similar characteristic curve at STC. However, the characteristics differ much under shaded or partial shaded conditions due to the difference in the count and location of the cross-ties [16]. The impact of partial shading on various interconnection schemes of the 7 × 5 array is assessed by the utilization factor. The ratio between the maximum power generated by the PV array and the sum of the maximum power generated by the individual PV modules is defined as the utilization factor (UF).

$$\text{UF} = \frac{\text{maximum power generated by PV array}}{\text{sum of maximum power generated by individual panels}} \tag{1}$$

The basic and derived configurations are analyzed under various test shading conditions (short, narrow, long, wide, and diagonal shade patterns) that mimic the real world shading scenario and the corresponding utilization factor is depicted in Fig. 7.

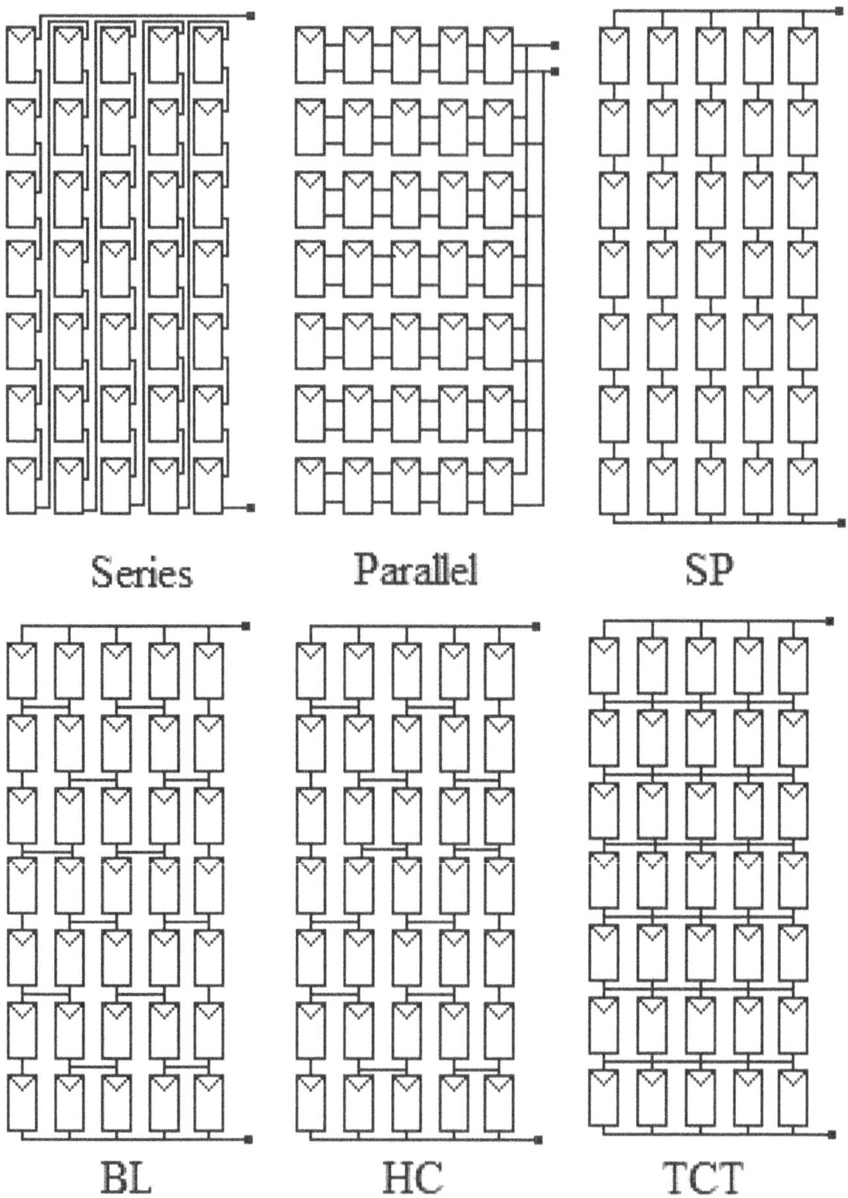

Fig. 5 Interconnection schemes of PV panels

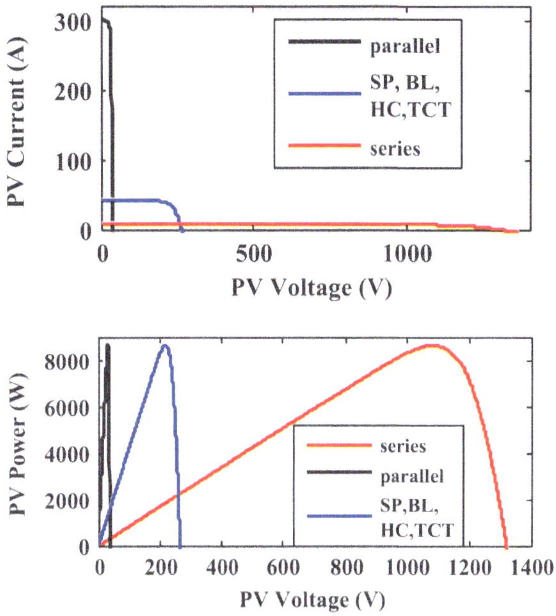

Fig. 6 Characteristics of 7 × 5 PV array under various interconnection schemes

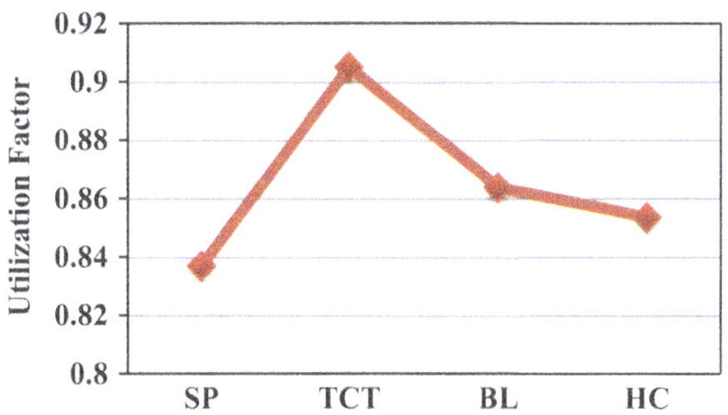

Fig. 7 Performance of interconnection schemes under various shading conditions

The presence of more cross-ties in the TCT scheme reduces the shading losses and hence it is the preferred scheme. The utilization factor can further be improved by making the PV array shade tolerant and this strategy is explained in the following section.

4 Shade Resilient PV Array (Strategy 1)

The 7 × 5 array can generate 8750 W at STC. The array is subjected to the shading pattern shown in Fig. 8 where the shade is concentrated at the corner of the array. The shade dispersion is not uniform and causes four different row currents. The first four

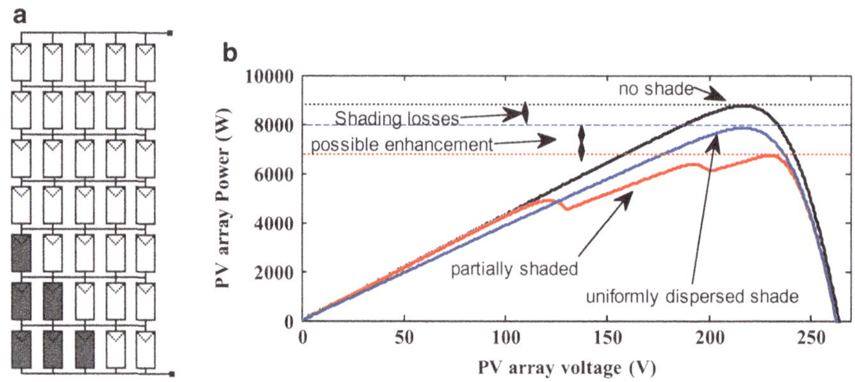

Fig. 8 **a**. Test shade pattern **b**. characteristics with and without shade dispersion

rows could generate a current of $5I_m$ (considering I_m to be the current at maximum power point). The currents generated by the fourth, fifth, and sixth rows are $4.3I_m$, $3.6I_m$, and $2.9I_m$, respectively.

The array generates a maximum power P_m of 6758 W. If the shade is uniformly dispersed all over the array, the losses due to mismatching row currents can be reduced to a greater extent. All the rows will have one shaded panel and generate a current of $4.3I_m$. The maximum generated power, in that case, will be 7853 W. Thus it is possible to enhance the power generation by dispersing the shade uniformly all over the array. This can be achieved by planning the placement of the panels within the array by shade resilient algorithm.

The arrangement of panels in a conventional TCT array is shown in Fig. 9. Each of the panels is identified by a two-digit number. The index represents the row and column numbers. For example, the index 53 points to the panel that is located in the fifth row and third column. In the conventional TCT arrangement, the panels are arranged sequentially. For example, the panels 11, 12, 13, 14, and 15 are placed next to each other in a row and are connected in parallel. The seven such paralleled strings are connected in series to form the TCT array.

The shade resilient PV array (SR PVA) places each of its panels in predetermined locations to disperse the shade uniformly all over the array. The location of the panels in the 7×5 array is determined according to the SR algorithm [17]. The SR algorithm is an offline algorithm and it uses the size of the array to calculate the displacement between the panels. The arrangement is presented in Fig. 10.

In the SR TCT arrangement, the panels that are to be physically placed in the first row are determined based on the displacement factor (DPF). The way the DPF is calculated leads to various SR arrangements [18]. One way to calculate the DPF is to divide the number of rows in the array by 2 and round it off to the floor (3 in this case).

The panel with index 11 is placed in the first row, the first column. The second panel to be physically placed next to it in the same row is determined by adding the

Fig. 9 The conventional arrangement of PV panels in 7×5 array

Fig. 10 Arrangement of panels in SR TCT

Table 2 Determination of panel location in SR TCT

Column number	Row number	Panel index
1	1	11
2	$1 + 3 = 4$	42
3	$4 + 3 = 7$	73
4	$7 + 3 = 10$; $10 > 7$; Hence, $10-7 = 3$;	34
5	$3 + 4 = 6$	65

DPF with the row number of the previous panel ($1 + 3 = 4$). This panel is to be placed in the first row, the second column, and hence the index is 42. The next panel to be placed in the first row, the third column is 73 ($4 + 3 = 7$). The next panel to be placed in the fourth column of the first row is 10 ($7 + 3 = 10$) and as 10 is greater than the array row number, 7 is subtracted from 10 ($10-7 = 3$). The resulting panel index is 34. The panel in the last column of the first row is 65 ($3 + 3 = 6$). The calculation is tabulated in Table 2.

The other rows are arranged sequentially as shown in Fig. 10. The panels 11, 12, 13, 14, and 15 are placed in different rows as highlighted and they are electrically connected in parallel. Thus, in SR arrangement, the five panels that electrically belong to a row (parallel string) are physically placed in different positions in five rows. The shade dispersion in the conventional arrangement and the SR TCT arrangement are analyzed under test shade patterns to assess the possible enhancement in power extraction.

The fixed shading is classified based on its geometry as long, short, narrow, and wide. The PV array is installed on the rooftop of residential units after careful inspection and, usually, they are shaded by parapet walls or structures on the rooftop. The most common shading pattern is short and/or narrow. The TCT and SR TCT arrays are assessed under three such fixed shading conditions.

The first shading pattern has seven shaded panels (400 W/m^2) and the shade is concentrated in four of the seven rows. The shade dispersion in conventional and SR arrangement is shown in Fig. 11a.

In the conventional arrangement, three rows are not shaded, one row has one shaded panel and three of the rows have two shaded panels each. This results in three different row currents ($5I_m$, $4.4I_m$, and $3.8I_m$) and hence the V-P curve exhibits three peaks. The maximum power that can be extracted is 7020 W as shown in Fig. 11b. The shade is dispersed in SR TCT arrangements and each of the rows has one shaded panel. The shade dispersion is uniform and each row generates a similar current ($4.4\ I_m$). The shade dispersion and the resulting V-P curve with a single peak are shown in Fig. 11. The maximum power generated by the SR TCT array is 7671 W. The enhancement is 40 W and would be significant under heavy shading conditions.

The second shading case is shown in Fig. 12 is a short one with four shaded panels. This type of shading is commonly caused by overhead structures.

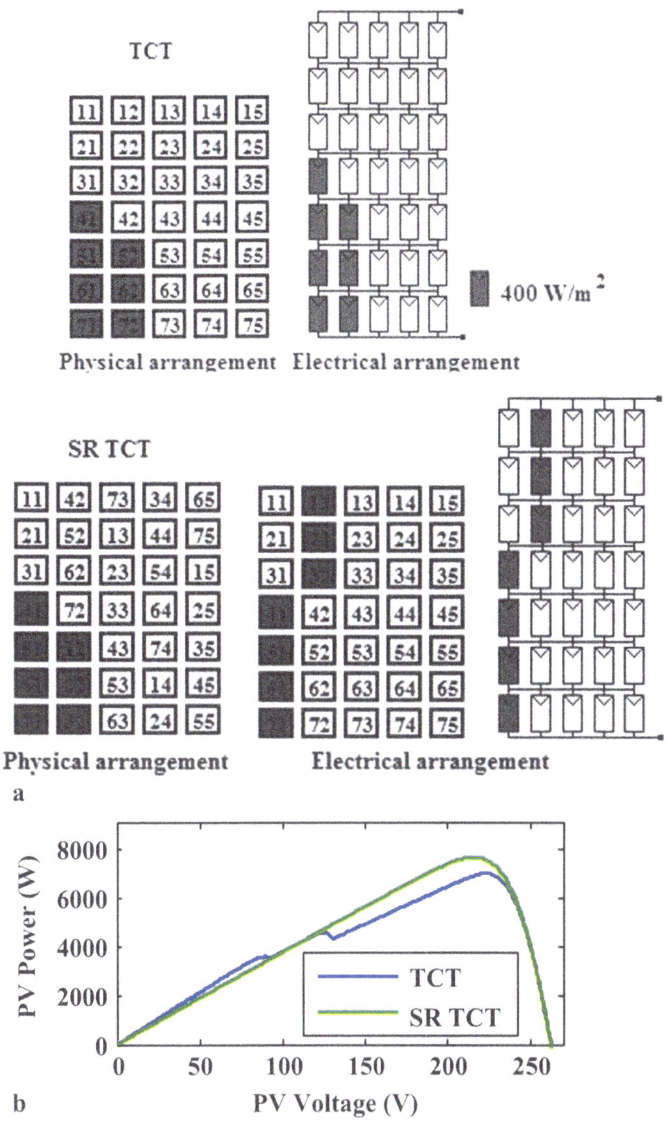

Fig. 11 Shade pattern 1 **a**. shade dispersion in TCT and SR TCT **b**. characteristics

In conventional TCT, five rows are not shaded and two rows have shaded panels. The first five rows generate a current of $5I_m$, the sixth row generates a current of $4.2I_m$ and the seventh row generates a current of $4I_m$. The three different row currents result in three peaks in the V-P curve. The maximum power generated by the conventional TCT array is 7620 W. The shade is dispersed in SR TCT and dispersion is not uniform as the shade is too short. The rows 1, 4, and 5 generate a current of $5I_m$, rows 2, 3, and

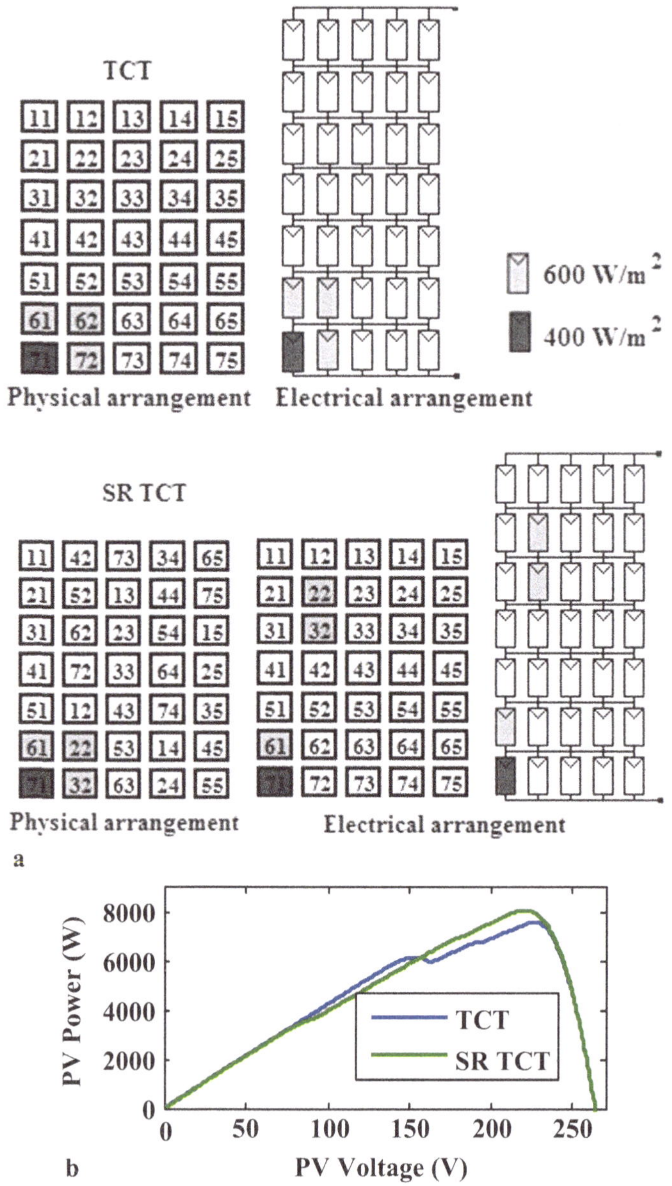

Fig. 12 Shade pattern 2 **a**. shade dispersion in TCT and SR TCT. **b**. characteristics

6 generate a current of $4.6I_m$ and the seventh row generates a current of $4.4I_m$. The V-P curve exhibits three peaks due to three different row currents. The peak power generated by the SR TCT array is 8112 W. The enhancement is significant in this case (492 W).

In the third case, all the five last row panels are shaded at 500 W/m^2 as shown in Fig. 13. This type of shade pattern is commonly caused by the parapet walls and the shading intensity remains uniform.

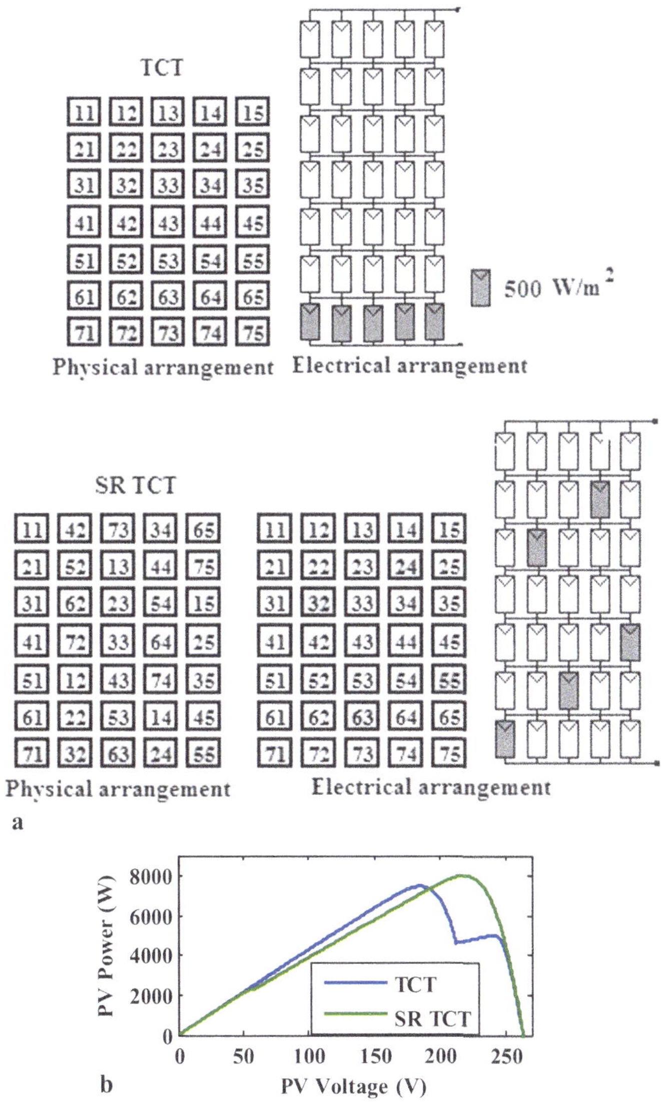

Fig. 13 Shade pattern 3 **a**. shade dispersion in TCT and SR TCT **b**. characteristics

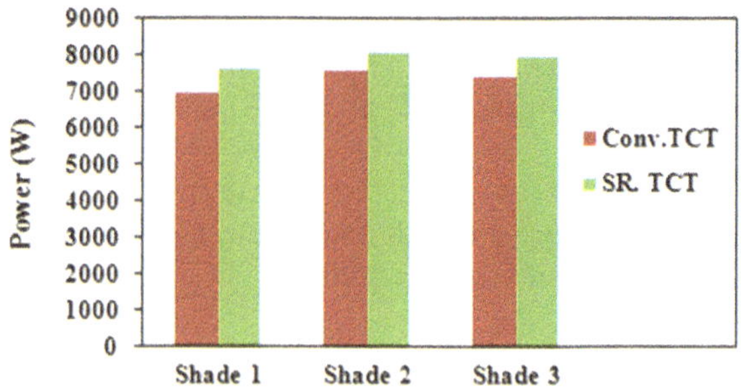

Fig. 14 Power generated by TCT and SR TCT

The conventional TCT array generates two different row currents. The first six rows generate a current of $5I_m$ and the last shaded row generates $2.5I_m$. The shaded row limits the PV array current and results in two power peaks in its V-P curve. The peak power generated by the TCT arrangement is 7468 W.

The shade is dispersed in the SR TCT arrangement as shown in Fig. 13a. Uniform shade dispersion is not possible as the shade is narrow. Five of the seven rows have one shaded panel and two rows are shade-free. Rows 1 and 4 generate a current of 5 I_m and the other rows (2, 3, 5, 6 and 7) generate a current of 4.5 I_m. The V-P curve exhibits two peaks. The peak power generated by the SR TCT is 8000 W which is significantly higher than the conventional TCT arrangement (532 W).

The peak power generated by the conventional TCT and the SR TCT arrays for the three shading conditions are compared in Fig. 14. The shade patterns considered for assessment and comparison belong to the short and narrow category as this type of shading is predominant in the residential PV system. The shade that is concentrated in a conventional arrangement is dispersed all over the array when arranged in SR TCT.

It can be inferred from Fig. 14 that the shade tolerance is better in SR TCT arrangement due to better shade dispersion. The power generation is thus enhanced under partially shaded conditions by this irradiation equivalence strategy. This is an offline strategy that does not involve switches and sensors and hence it can be adopted in residential installations without any additional cost.

This is the first strategy suggested to enhance the performance of the residential SSPV system under all environmental conditions. The SR TCT arrangement tends to disperse the shade all over the array. However, uniform dispersion is not possible if the shade is too short and/or narrow. Yet, the SR TCT arrangement enhances the output under such conditions, but multiple peaks in the V-P curve are often unavoidable. This demands the inclusion of a fast global peak detecting algorithm (GMPPT) that can discard local peaks and identify the global peak in the system.

5 Global Peak Detecting Algorithm (Strategy 2)

The commercial PV inverter has an inbuilt MPPT algorithm that can operate the PV array at its optimal power point. These algorithms are mostly conventional and cannot differentiate between the local and global peaks. The presence of multiple peaks in the V-P curve due to partial shading requires more sophisticated algorithms that can identify the global maximum.

In this study, two-line-search (golden section search and dichotomous search)-based GMPPT algorithms are analyzed and compared [19, 20]. These algorithms employ two stages. The first stage locates the region of the global peak by shrinking the search interval and the later stage identifies the global peak accurately. The first stage of both algorithms is based on the conventional fractional voltage algorithm (FVA). The fractional voltage algorithm searches for a peak at the voltage kV_{oc}. The value of k depends on environmental conditions and usually varies between 0.7 and 0.85 [21].

Partial shaded conditions are characterized by multiple peaks and hence the GMPPT algorithm has to search the entire range (0 to mV_{oc}) for the global peak. Searching each point in the search interval for the possible location of the global peak is time-consuming. To speed up the process, the first stage of the GMPPT algorithm which is based on the FVA, searches the entire search interval only at strategic voltage points (kV_{oc}, $2kV_{oc}$, $3kV_{oc}$… $7kV_{oc}$) for possible global peaks as shown in Fig. 15.

The powers at the strategic voltage points are compared and the region of the global peak is identified. The second stage of the algorithm has to locate the global peak accurately with lesser iteration by adopting 'split and remove strategy' around the region identified by the first stage. Two-line search algorithms (golden section search and dichotomous search) are assessed in this section for the 7 × 5 SR TCT array. The pseudo-codes of the two GMPPT algorithms are presented in Table 3.

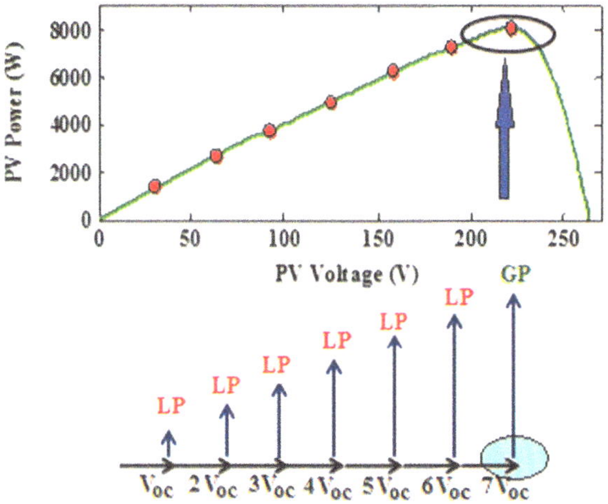

Fig. 15 First stage of GMPPT algorithm

Table 3 Pseudo-code of GMPPT algorithms

Algorithm 1 GS GMPPT	Algorithm 2 DS GMPPT
Measure PV array power and compute the change in power ΔP If $\Delta P > P_c$, initiate new tracking; P_c - critical power Stage 1 Initialize search interval $(0, 7 \times 31.5)$ Set $V_i = 0.83 \times 31.5$ and measure power (p_i) Apply voltage perturbation $V_{i+1} = 2 \times 0.83 \times 31.5$ and measure Power (P_{i+1}) $V_{ref-} = V_i$; if $P_i > P_{i+1}$ else $V_{ref-} = V_{i+1}$ Continue measuring power till $i = 7$;	
Stage 2 Fix search interval around V_{ref-} Fix search points $x_{1j} = b\text{-}L_j{}^*$ and $x_{2j} = a + L_j{}^*$ $L_j{}^* = L_o/\gamma^j$; Lo - initial search range; γ - golden ratio Measure corresponding powers P_1 and P_2 If $(P_1 - P_2) > \Delta$, If $P_1 > P_2$, discard right side section; Else discard left Fix new searching interval and probe points Measure Powers End Else GMPP located End of stage 2	Fix search interval around V_{ref-} Fix midpoint c and search points $x_1 = c\text{-}\ \varepsilon$, $x_2 = c + \varepsilon$; ε - tolerance value, Measure corresponding powers P_1 and P_2 If $(P_1 - P_2) > \Delta$, If $P_1 > P_2$, discard right side section; Else discard left Fix new searching interval and probe points Measure Powers End Else GMPP located End of stage 2

The GS GMPPT algorithm is based on the golden ratio and hence the name golden section search. This algorithm can detect the peak of unimodal function accurately and hence it is employed in the second stage. The initial search interval (a, b) is fixed around the region identified by the first stage. The search points x_1 and x_2 that are determined based on the golden ratio divides the search interval into three regions (a, x_1), (x_1, x_2), and (x_2, b). The corresponding powers P_1 and P_2 are measured and compared. If the power P_1 is lesser than P_2, the region (a, x_1) is rejected. The search interval is reduced from (a, b) to (x_1, b). The search continues with the new search interval as depicted in Fig. 16a and at the end of the iteration, the search interval shrinks further. The search continues until the global peak is identified.

In case of the DS GMPPT algorithm also, the search interval is fixed based on the first stage. The midpoint 'c' of the interval is fixed and the search points x_1 and x_2 are introduced on either side of 'c'. The corresponding powers P_1 and P_2 are measured and compared. If the power P_2 is greater than P_1, the region (a, x_1) is rejected. The search interval is reduced from (a, b) to (x_1, b). The search continues with the new search interval as depicted in Fig. 16b and at the end of the iteration, the search interval shrinks further. The search continues until the global peak is identified.

The second shading pattern is shown in Fig. 12 of the 7×5 array is considered to assess the performance of the two GMPPT algorithms. Both the algorithms generate the reference voltages as shown in Fig. 17 and the controller ensures accurate tracking

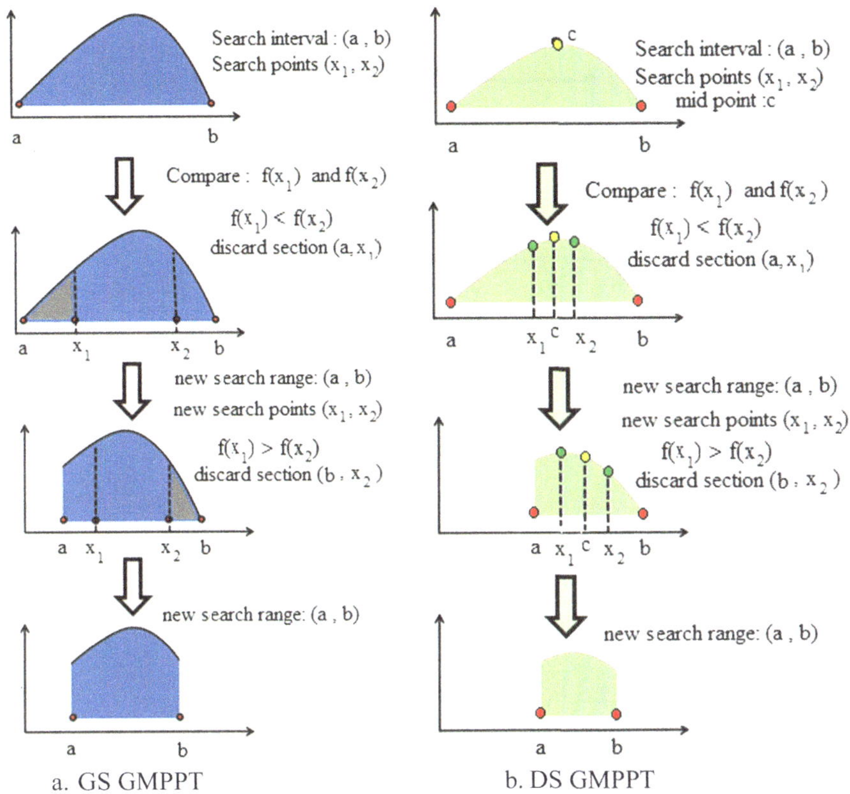

Fig. 16 **a**. The second stage of the GMPPT algorithm a. GS GMPPT **b**. DS GMPPT

of the global peak. The reference voltage is compared with the actual PV array voltage. The resulting error signal is fed to the controller, which in turn adjusts the duty cycle of the power converter so as to match the source and the load impedances. When the impedances match, maximum power is extracted from the PV array.

Both the algorithms effectively tracked the global peak of 8112 W and the array is operated at 220.4 V as depicted in Fig. 17. The GMPPT algorithm waits for a significant change in power (power changes with a change in environmental or shading conditions). If the change is greater than the critical value, the algorithm initiates a new search. The first and second stages of the GMPPT algorithms then track down the global peak.

The DS GMPPT converges faster as nearly half of the search interval is discarded at the end of each of the iterations. It is quantified by a factor called the rejection ratio. The reduction ratio is defined as

$$RR = \frac{L_n}{L_o} \tag{2}$$

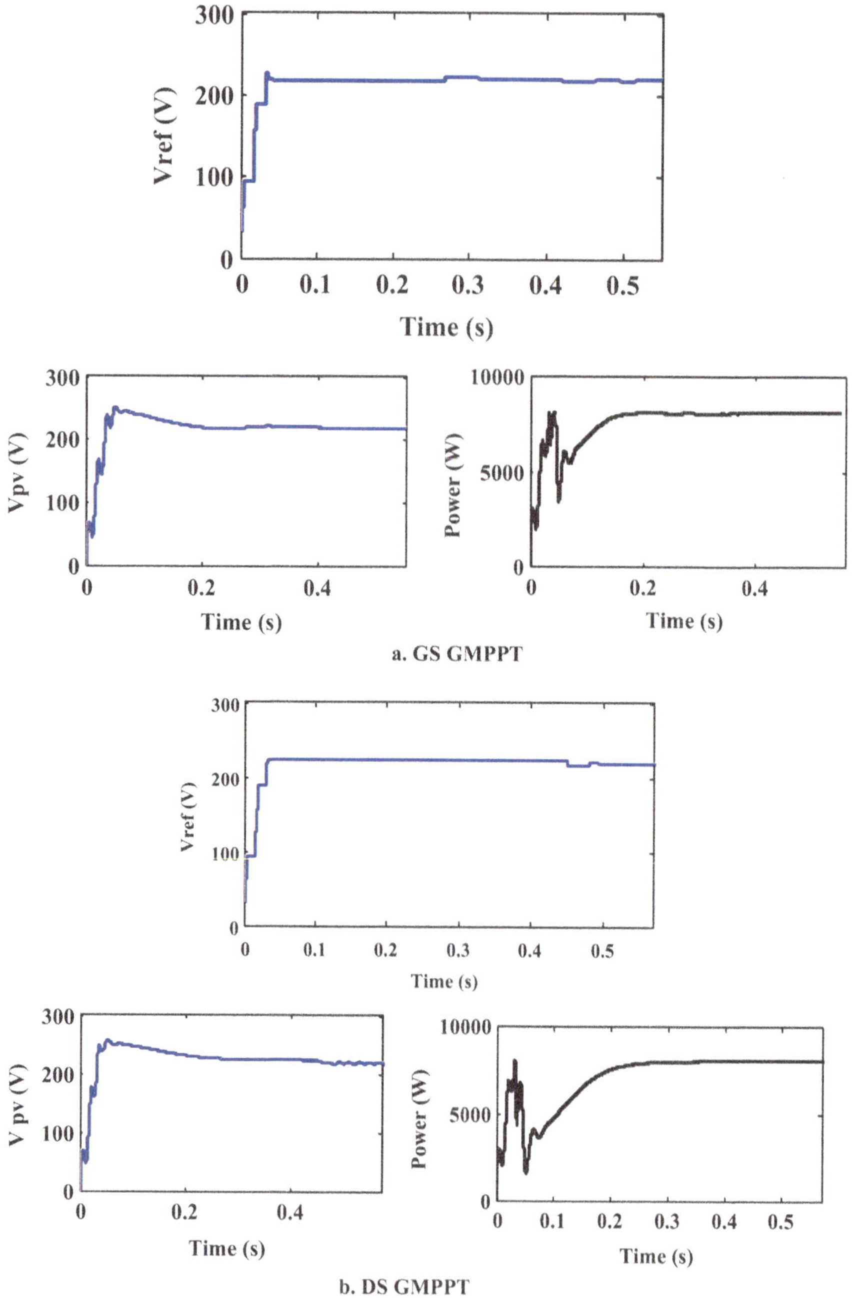

Fig. 17 Simulated results of the GMPPT algorithms **a**. GS GMPPT **b**. DS GMPPT

where L_n is the length of uncertainty after n experiments and L_o is the length of the initial interval of uncertainty. The reduction ratio of GS GMPPT is 0.618^{n-1} and that of DS GMPPT is $(2^{0.5n})^{-1}$ where n is the number of iterations. Stage one of both the algorithms locate the region of the global peak in 7 iterations. The second stage requires 14 iterations (GS GMPPT) and 12 iterations (DS GMPPT) to converge within $\pm 1\%$ of error tolerance. Hence, the DS GMPPT algorithm tracks down the global peak with 19 iterations while the GS GMPPT requires 21 iterations to converge.

Thus, with the appropriate GMPPT algorithm, the global peak of the PV array can be detected under all shading conditions and maximum power can be extracted from the 7 × 5 SR TCT PV array under shaded and partially shaded conditions. The extracted power is stored in a battery bank and fed to inverters to run the load in case of battery-based systems. In the case of the batteryless system, the extracted power is directly fed to the inverter.

6 Reduced Device Multilevel Inverter (Strategy 3)

The dc power extracted from the PV array is to be converted to ac to operate the utilities in the residential unit. The battery bank can be connected to the RDC MLI to convert the stored dc power to ac in the case of a battery-based system. In the batteryless system, the PV array directly feeds the MLI. The way the PV inverters are fed has led to the classification as central, string, multi-string, and micro-inverters. The central inverter has single MPPT and the output of the PV array is fed to it. If the inverter fails, the entire system shuts down. This drawback is addressed in the later developed architectures like string and multi-string structures that tend to decentralize. These structures include a dedicated MPPT for each of its strings. The provision of dedicated MPPT per string will not only improve the reliability of the system considerably but also the SSPV system's efficiency under shaded/partly shaded conditions. The arrangement is flexible and scalable as it is easy to integrate additional modules. The latest development is micro-inverters or AC Modules where the inverter is an integrated part of the PV-module. The mismatch losses and hot spot risks are totally eliminated resulting in better efficiency. The modular structure also enables easy enlargement of the system. But, the cost involved is higher than the other architectures. Generally, string/multi-string inverters are preferred over central inverters as they offer better reliability, efficiency, and scalability.

The PV inverters thus play a vital role in the PV system and often inverter failure is the major reason for the system failure. Replacing the failed inverters adds to the capital cost of the overall system and in turn, increases the return on investment period which is undesirable. It is therefore necessary to enhance the reliability of the PV inverter and in turn the reliability of the SSPV system. Reliability is assessed by counting the number of parts involved or by computing the stress on the parts. Hence, the reliability can be improved either by reducing the number of parts involved in a circuit or by reducing the stress across it.

Most of the commercial inverters adopt conventional two-level inverter or five-level inverter. The multilevel inverters have been reported as a possible solution for PV systems, with several inherent benefits, such as high efficiency, low distortion ac waveforms, and low leakage currents [22]. Additionally, in high power applications, the ac power can be synthesized from several low-level cells, i.e., low voltage semiconductors could be employed. The modular, scalable, and simpler structure of cascaded MLI (CMLI) topology makes it a preferred structure over the other conventional MLI structures like neutral point MLI and flying capacitor MLI for interfacing with the renewable sources. The major disadvantage of CMLI structure is the use of more number of semiconductor switches which reduces the reliability of the PV inverter.

Many MLI topologies that use a lesser number of switches have been proposed in the literature [23] and employing one such topology may not only improve the reliability of the inverter but also improves the spectral quality.

In this section, one such RDC MLI that has cascaded basic units and a full-bridge inverter as shown in Fig. 18 is considered. The basic unit has a dc source, a unidirectional switch, and a diode. This unit can generate one positive level $+V_1$ when the switch S_1 is closed. When the stitch is opened, the source is disconnected from the rest of the circuit. Several basic units are cascaded to generate the desired number of voltage steps at the output.

The cascaded structure is connected to a full-bridge (FB) inverter circuit to generate both positive and negative levels. The FB circuit generates a positive level if the switches T_1 and T_3 are turned on. It generates a negative level if switches T_2 and T_4 are turned on. The number of voltage levels or steps at the output is more if the inverter operates in asymmetric mode. The metrics of the RD MLI are tabulated in Table 4.

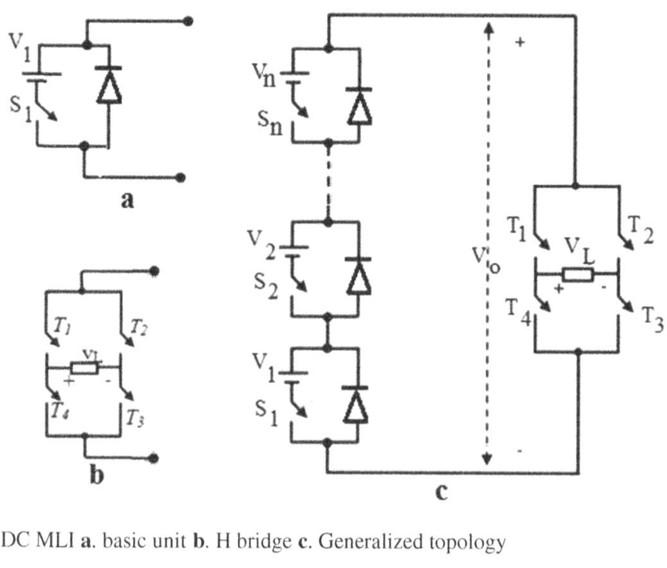

Fig. 18 RDC MLI **a**. basic unit **b**. H bridge **c**. Generalized topology

Table 4 Metrics of the RD MLI presented in Fig. 18

	RD MLI
Basic units	N
Levels generated	2n + 1; Symmetric mode $2^{n+1}-1$; Asymmetric mode
Switches required	n + 4
Voltage sources	N
Magnitudes of dc sources	V_{dc}; Symmetric mode $2^{n-1}V_{dc}$; Asymmetric mode
Maximum V_{out}	n V_{dc}; Symmetric mode 2^n-1V_{dc}; Asymmetric mode
Driver circuits required	n + 4

The conventional CMLI can be operated in two asymmetric modes (binary and trinary) and the trinary mode generates more voltage steps at the output. Hence, the level to switch ratio is better for that mode. The RD MLI (asymmetric mode) suggested in this section is compared with that of the CMLI in the trinary mode as in Table 5.

The RD MLI can generate 31 voltage levels with 8 switches whereas; the CMLI (trinary mode) can generate only 9 levels. The level to switch ratio of the RD MLI is relatively higher than that of the CMLI in trinary mode.

A 15-level inverter with a reduced number of devices is presented in Fig. 19. It has three cascaded basic units and a FB inverter unit. The circuit uses seven switches, three sources, and three diodes. The source V_1 is connected to the FB when switch S_1 is turned on. The sources V_1 and V_2 add up and appear across the FB by switching

Table 5 Comparison of RD MLI presented in Fig. 18 and CMLI

	RD MLI (Asymmetric)	CMLI (Trinary)
Basic units	n	N
Levels generated	2^{n+1-1}	3^n
Switches required	n + 4	4n
Voltage sources required	n	n
Magnitudes of dc sources	$2^{n-1}V_{dc}$; Asymmetric mode	3^{n-1} V_{dc}
Maximum V_{out}	2^n-1V_{dc}; Asymmetric mode	$(3^n-1)/2$ V_{dc}
Driver circuits required	n + 4	4n
Level to switch ratio	$2^{n+1-1}/n+4$	$3^n/4n$

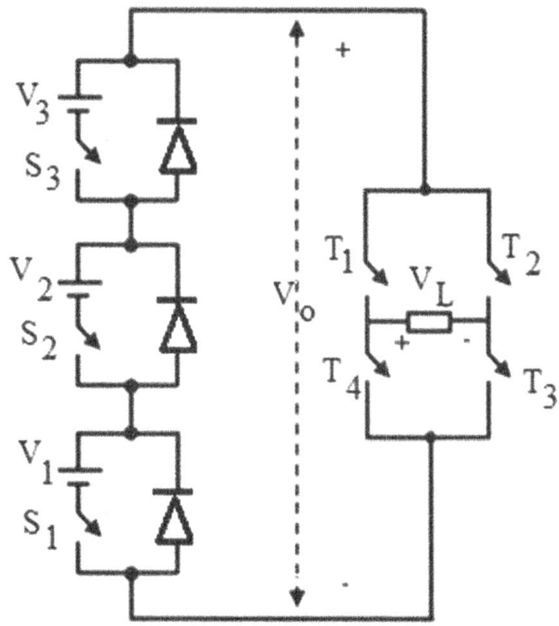

Fig. 19 15-level RD MLI

on S_1 and S_2. The voltage across the load is positive if switches T_1 and T_3 are turned on and it is negative if the other two switches are turned on.

The magnitudes of the dc sources are chosen to be V_{dc}, $2V_{dc}$, and $4V_{dc}$. Capacitors can also be used to split the voltage from the source. The required gating pulses are generated by a fundamental frequency modulation strategy called half-height PWM scheme [24] which determines the instant at which the switches are to be triggered. The switches are triggered sequentially as presented in Table 6.

The simulated output voltage [25] and the harmonic profile of the 15-level RD MLI are presented in Fig. 20. The output voltage waveform shows 15 distinct voltage steps and the % THD is found to be 1.88% without using filters. The Spectral quality improves with the number of voltage steps. The more the number of levels, the more the output waveform gets closer to the sine wave. But, the number of switches and the allied circuit increases with the levels and it, in turn, will affect the reliability of the inverter due to increased part count. The 15-level RD MLI presented in Fig. 19 has produced a stepped voltage waveform with a lesser harmonic distortion (1.88%) that is within the desirable limits. Hence, it is not required to extend the number of levels further compromising reliability.

The conventional cascaded MLI can generate 15 voltage levels when operated in binary mode. The CMLI cannot generate 15 levels in trinary mode. Hence, the metrics of the RD MLI presented in Fig. 19 are compared with CMLI (binary mode) in Table 7.

The 15-level RD MLI has used only 7 switches as compared to 12 in conventional CMLI (Binary) to generate 15 unique voltage steps at the output. The %THD (1.88%)

Table 6 Status of switches and the output voltage

States	Status of switches							Output Voltage
	S_1	S_2	S_3	T_1	T_2	T_3	T_4	
1				✓	✓			0
2	✓			✓		✓		V_{dc}
3		✓		✓		✓		$2V_{dc}$
4	✓	✓		✓		✓		$3V_{dc}$
5			✓	✓		✓		$4V_{dc}$
6	✓		✓	✓		✓		$5V_{dc}$
7		✓✓	✓	✓		✓		$6V_{dc}$
8	✓	✓	✓	✓		✓		$7V_{dc}$
9	✓✓				✓		✓	$-V_{dc}$
10		✓			✓		✓	$-2V_{dc}$
11	✓	✓			✓		✓	$-3V_{dc}$
12			✓		✓		✓	$-4V_{dc}$
13	✓		✓		✓		✓	$-5V_{dc}$
14		✓	✓		✓		✓	$-6V_{dc}$
15	✓	✓	✓		✓		✓	$-7V_{dc}$

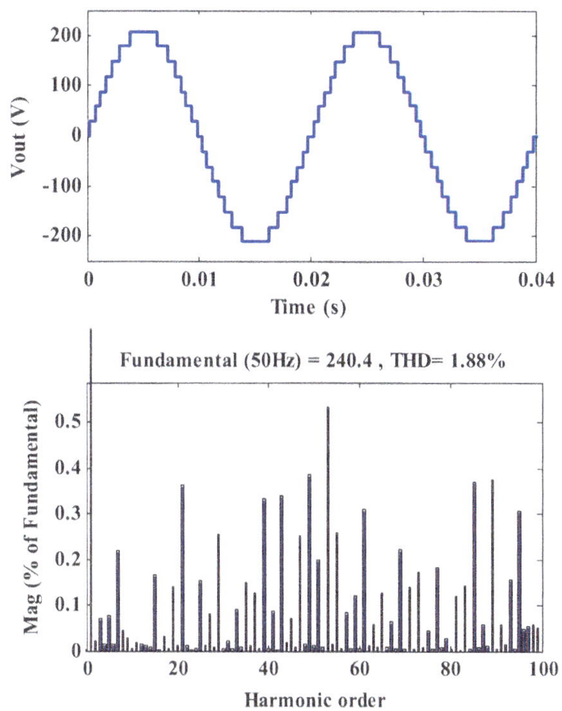

Fig. 20 Simulated output and harmonic profile of the 15-level RD MLI

Table 7 Comparison between CMLI (B) and RD MLI in Fig. 19

	CMLI (Binary)	RD MLI
Number of levels	15	15
Number of switches	12	7
Number of dc sources	3	3
Magnitudes of dc sources	V_{dc}, $2V_{dc}$, $4\ V_{dc}$	V_{dc}, $2V_{dc}$, $4\ V_{dc}$
Maximum V_{out}	$7\ V_{dc}$	$7\ V_{dc}$
Number of gate driver circuits	12	7
Level to switch ratio	1.25	2.14

is also well within the desirable limits. The reduction in the switches involved reduces the driver circuit requirements and it in turn reduces the overall part count. Reliability can be assessed based on the number of components involved. As this RD MLI adopts fewer components than the conventional topology, the reliability is much better. Further, the number of switches that are turned on at any given instant is less compared to the conventional topologies and hence the losses involved are also reduced.

Thus, the performance of the SSPV system can be improved by incorporating all three enhancement strategies, SR TCT, GMPPT, and RD MLI along with the standard design procedure. The enhanced SSPV system is depicted in Fig. 21.

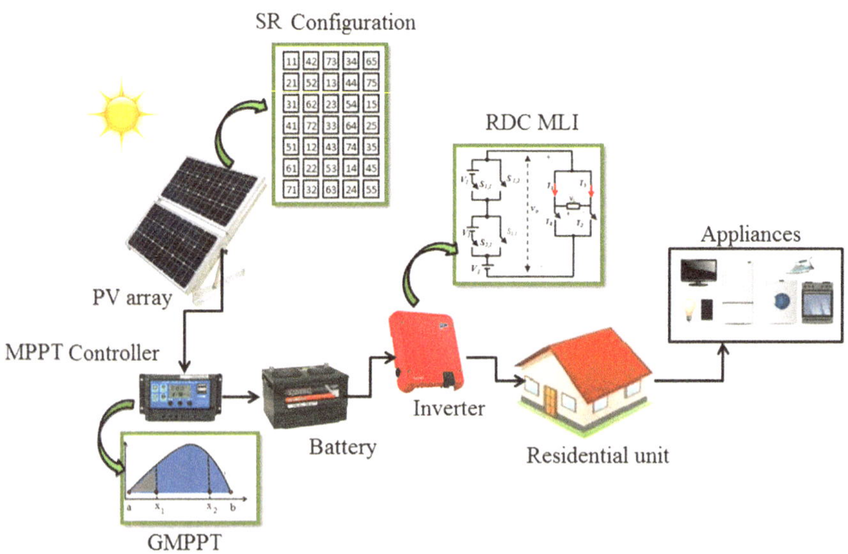

Fig. 21 Enhanced SSPV system

The power extracted from the PV array can be enhanced by improving equalizing the irradiation received by the panels. This is achieved by arranging the panels according to the SR algorithm and adopting TCT interconnection. This arrangement enhances the output under all environmental conditions by minimizing the mismatch losses due to shading/partial shading. Incorporating a fast and accurate GMPPT algorithm operates the PV array at its optimal point and ensures the transfer of maximum power from the source to the load. The reliability of the inverter and in turn the SSPV system can be improved by employing RD MLI.

7 Conclusion

This chapter has elaborately explained the impact of partial shading on the performance of the PV array and has suggested three strategies to improve the power generation under such conditions. The SSPV system is designed for a residential unit using standard procedures and the suggested strategies are demonstrated in the design. The role of interconnection schemes of a partially shaded PV array is assessed in terms of UF and it is suggested that the TCT scheme is a better option. Further, shade resilience is included in the TCT PV array by adopting the irradiation equivalence strategy. The panels are arranged according to an offline algorithm in SR TCT array and this arrangement tends to disperse the shade evenly among the rows. The mismatch in row currents is thus reduced and the power generation is enhanced without additional cost. This strategy is best suited for small residential installations. Shade dispersion may not be uniform if the shade is short and narrow as in most of the practical cases and results in multiple peaks in the V-P curve. Many algorithms are reported in the literature and the selection depends on the requirements and the availability of resources. This chapter has suggested two-line-search-based GMPPT algorithms (GSS and DS) that can detect the global peak accurately and quickly. The DS GMPPT algorithm is faster compared to the GSS algorithm as the search interval is almost reduced to half at the end of each of the iterations. RD MLI is suggested as the third strategy to improve the reliability of the system. Reduced device count improves reliability, as the part count is reduced. A 15-level inverter is developed based on the suggested topology and its working is explained in detail. Many new topologies are reported in the literature and appropriate structure can be selected based on the requirements and size of the PV array and/or battery bank. All three suggested strategies can be considered while designing an SSPV system to enhance its performance under all environmental conditions.

References

1. Ibrahim H, Anani N (2020) A Study of the effect of different configurations of bypass diodes on the performance of a PV string. Sustainability in energy and buildings, pp 593–600
2. Motamarri R, Nagu B (2020) GMPPT by using PSO based on Lévy flight for photovoltaic system under partial shading conditions. IET Renew Power Gener 14(7):1143–1155
3. Nguyen BN, Nguyen VT, Duong MQ, Le KH, Nguyen HH, Doan AT (2020) Propose a MPPT algorithm based on Thevenin equivalent circuit for improving photovoltaic system operation. Front Energy Res 8:14
4. Zou Y, Yan F, Wang X, Zhang J (2020) An efficient fuzzy logic control algorithm for photovoltaic maximum power point tracking under partial shading condition. J Franklin Inst 357(6):3135–3149
5. Eltamaly AM (2015) Performance of smart maximum power point tracker under partial shading conditions of PV systems. In: 2015 IEEE International conference on smart energy grid engineering (SEGE). IEEE
6. Ajmal AM, Babu TS, Ramachandaramurthy VK, Yousri D, Ekanayake JB (2020) Static and dynamic reconfiguration approaches for mitigation of partial shading influence in photovoltaic arrays. Sustain Energy Technol Assess 40:100738
7. Sagar G, Pathak D, Gaur P, Jain V (2020) A Su Do Ku puzzle based shade dispersion for maximum power enhancement of partially shaded hybrid bridge-link-total-cross-tied PV array. Solar Energy 204:161–180
8. El Iysaouy L, Lahbabi M, Baskys A, Oumnad A (2020) Performance analysis of partially shaded photovoltaic array using magic square view configuration for shade dispersion. J Solar Energy Eng 142(6)
9. Yadav AS, Mukherjee V (2018) Line losses reduction techniques in puzzled PV array configuration under different shading conditions. Solar Energy 171:77–783
10. Farh HMH, Eltamaly AM, Othman MF (2018) Hybrid PSO-FLC for dynamic global peak extraction of the partially shaded photovoltaic system. PloS one 13(11): e0206171
11. Akrami M, Pourhossein K (2018) A novel reconfiguration procedure to extract maximum power from partially-shaded photovoltaic arrays. Solar Energy 173:110–119
12. Schettino G, Pellitteri F, Ala G, Miceli R, Romano P, Viola F (2020) Dynamic reconfiguration systems for PV plant: Technical and economic analysis. Energies 13(8)
13. https://solargis.com/maps-and-gis-data/download/india
14. Walker L, Hofer J, Schlueter A (2019) High-resolution, parametric BIPV and electrical systems modeling and design. Appl Energy 238:164–179
15. Sahoo SK, Shah M, Dawlatzai NA, Amalorpavaraj RAJ (2020) Assessment of mismatching in series and parallel connection of the PV modules of different technologies and electrical parameters. In 2020 International conference on computer communication and informatics, pp 1–5
16. Malathy S, Ramaprabha R (2015) Comprehensive analysis on the role of array size and configuration on energy yield of photovoltaic systems under shaded conditions. Renew Sustain Energy Rev 49:672–679
17. Malathy S, Ramaprabha R (2020) Shade resilient total cross tied configurations to enhance energy yield of photovoltaic array under partial shaded conditions. Emerg Trends Comput Expert Technol 35:122–133. Springer
18. Malathy S, Ramaprabha R (2018) Reconfiguration strategies to extract maximum power from photovoltaic array under partially shaded conditions. Renew Sustain Energy Rev 81:2922–2934
19. Mostafa HH, Ibrahim AM, Anis WR (2019) A performance analysis of a hybrid golden section search methodology and a nature-inspired algorithm for MPPT in a solar PV system. Arch Electr Eng 611–627
20. Malathy S, Ramaprabha R (2020) Tracking the maximum power point of PV array using dichotomous search. U.P.B. Sci Bull Ser C82(1):179–188
21. Baimel D, Tapuchi S, Levron Y, Belikov J (2019) Improved fractional open circuit voltage MPPT methods for PV systems. Electronics 8(3):321

22. Kumar N, Saha TK, Dey J (2020) Multilevel inverter (MLI)-based stand-alone photovoltaic system: modeling, analysis, and control. IEEE Syst J 14(1):909–915
23. Salem A, Robbersmyr KG, Norambuena M, Rodriguez J (2020) Voltage source multilevel inverters with reduced device count: topological review and novel comparative factor. IEEE Trans Power Electron
24. Malathy S, Ramaprabha R (2019) Reliability and performance assessment of reduced component count multilevel inverter for PV systems. AIP Conf Proc 2161:020017
25. www.mathworks.com

Controlling the Hybrid PV/T System Self-heating Using Extrinsic Cell Resistance

A. A. Aminou Moussavou, A. K. Raji, and M. Adonis

Abstract Water and space heating represent major energy consumption in a domestic residence. The heating system is supplied from different energy sources such as electrical heating systems, biomass, kerosene, propane, fuel oil, and natural gas. These energy sources are expensive, diminishing, and often pollute the environment. Hybrid PV/T systems have proven to be economical and utilize renewable energy for space heating and domestic hot water applications. However, one of its limitations is the low thermal efficiency. This study presents an innovative means of merging the photovoltaic module and solar collector (PV/T) system. Emphasis is placed on the solar energy conversion strategy to modulate the ratio of thermal to electrical produced from the PV panel, which balances the energy (electrical and thermal) based on the user requirement. Analysis of electrical power and thermal performance of photovoltaic is carried out to highlight the critical parameters influencing the PV system behavior using MATLAB/Simulink. The simulation results show an effective balance of thermal and electrical power. When varying the extrinsic cell resistance, the PV cell temperature ranges from 45 to 62 °C in the PV/T system while proportionally generating 2800 to 110 W to the load. The PV cell can produce electricity and may also be useful for domestic hot water preheating and space heating applications.

Keywords Efficiency · Photovoltaic systems · Solar thermal · Modeling and simulation · Power production · Thermal energy · Electrical energy

1 Introduction

The use of hot water and space heating in residential sectors represents the primary energy consumed in most countries [1, 2]. This energy comes from different sources, such as electrical heating systems, propane, fuel oil, and natural gas. For economic reasons, the hybrid PV/T system can be utilized to cover the demand of the useful heat

A. A. Aminou Moussavou (✉) · A. K. Raji · M. Adonis
Department of Electrical Electronic and Computer Engineering, Cape Peninsula University of Technology, Cape Town, South Africa
e-mail: akdech80@yahoo.fr

S. Motahhir and A. M. Eltamaly (eds.), *Advanced Technologies for Solar Photovoltaics Energy Systems*, Green Energy and Technology,
https://doi.org/10.1007/978-3-030-64565-6_11

(for hot water and space heating) and electricity [3–7]. However, this system presents some limitations [8]. For instance, flat-plate PV/T collectors warm working liquids to a minimum temperature above surrounding temperatures, relying upon the structure and the number of the glass cover. However, more than three glass covers may not be used, because their electrical efficiency is extremely low due to triple glazing cover and low transmittance [8, 9]. Kostic et al. studied the reflection efficiency of a flat plate composed of an aluminum sheet and aluminum foil collectors and found that both materials have the same total reflectance, but the specular reflector was dominated by the aluminum foil concentrator, increasing the solar radiation intensity. The energy produced by a c-PV/T collector using aluminum foil is higher than the energy produced using an aluminum sheet [10]. Budihardjo and Morrison studied the properties of glass cooling tubes for solar water heaters such as optical and thermal losses, comparing the effects of flat solar collectors to solar panels. It was reported that the evacuated tube collector array generally has lower thermal performance than the two flat-plate PV/T in the domestic water heating. Solar water heaters with vacuum tubes in the glass were the largest and the most widely used form of vacuum tube collector due to the high thermal efficiency [11].

PV/T systems are integrated into buildings and offer a multi-power generation [12]. Some researchers developed an air-based PV/T system on a building facade to determine the thermal performance and radiation gain factors. However, it was complicated to estimate the convective heat transfer coefficient due to parameters such as natural flow convection, laminar and turbulent flow, and internal and external wind load on the plates [13]. Vokas et al. analyzed the theoretical performance of air-based PV/T collector for domestic heating and cooling. However, the results show that the solar PV/T hybrid air collector had an efficiency of 9% lower than the standard solar thermal collector. The solar collector with selective absorber had the highest thermal efficiency (about 75%) [14].

Theoretical and experimental evaluation studies of natural and mechanical ventilation of solar air-based PV/T collector at steady-state have been proposed [3]. The evaluation shows that the new recovery of thermal energy allows about an 18% improvement in the overall performance of air-based PV/T collector. Chow et al. developed a water-based PV/T simulation model for building integrated PV (BIPV/T) and heating systems, and then analyzed the annual energy performance of the water-based PV/T system in natural and forced circulation modes. Both operation modes presented a thermal transmission efficiency reduction of 72% and 71%, respectively [15].

A study had analyzed PV/T systems associated with two water channels in which water flows through the upper channel and returns through the lower channel. This system has high thermal efficiency; however, the geometric complexity makes it difficult to manufacture [16, 17]. Another part of the design transfers heat from the water layer to the glass plate on top of the water channel that has been analyzed by Xu et al. and Musallam et al. [18, 19]. However, each glass cover creates additional reflections, decreasing substantial electricity efficiency in the PV/T collector [19, 20].

Some authors proposed the heat pipe-based PV/T, consisting of three parts (an evaporator, an adiabatic, and a condenser) as a viable solution for removing and transferring heat [21–23]. However, one of the difficulties of high-temperature heat pipes is the corrosion from the incompatibility between the working fluid and the shell material. This incompatibility results in a chemical reaction, producing non-condensable gases, which considerably affect the thermal performance of the heat pipe [24].

Numerous studies have been conducted on PV/T systems with limited information and without parameters such as weather conditions and geographic location. Excepted for the annual performances of the PV/T system that are available [25, 26]. A statistical study conducted by Brottier and Brennacer addressed the shortage of reliable PV/T system thermal performance data. This study was achieved by evaluating the efficiency of 28 PV/T system installations in-ground, fitted with new non-overglazed PV/T systems in Western Europe. The robustness of the non-overglazed PV/T was highlighted, and the irregularity on PV/T system thermal performance was considerable [27]. Simulation models have also been developed in TRNSYS software to study the losses occurring in the PV/T for hot water. It has been indicated that up to 52 °C of thermal losses are from the total energy, and the circulation system is affected by the return temperature [28].

Valuable studies on the optimization techniques using maximum power point tracker (MPPT) have been published in PV systems like [29, 30]; but this is not discussed here.

In view of these PV/T system limitations, it appears that the improvement of energy production is needed. Therefore, it is an essential topic for new research perspectives and technological development. The novelty of this research study is in the solar energy conversion strategy to modulate the ratio of thermal to electrical produced from the PV panel to preheat domestic hot water systems. Although, the PV/T systems mainly prioritize electrical efficiency. This study proposes a controllable self-heating (useful heat) of the PV cell using an external parameter that balances the energy (electrical and thermal) produced according to the need. This focuses on the thermal behavior of a photovoltaic cell, and the heat transfer via radiation, convection, and particularly conduction (useful heat). The modeling is based on the calculation of the heat balance, heat transfer occurring in the PV cell, the temperature within the PV cell, thermal and electrical power, and other efficiencies that will be accessed using the following parameters. A comprehensive review of the performance characteristics of PV modules with an emphasis on temperature effects is conducted in this study. Also, the development of a method to optimize combined photovoltaic and thermal efficiency by controlling the heat flow is undertaken.

2 Photovoltaic Modules

Photovoltaic modules are devices that are made to transform the sun's rays into electricity and are characterized by their peak power under certain conditions in the laboratory. The PV module performance is associated with its operating temperature.

Generally, the generated electricity decreases when the temperature increases above the threshold value [31].

The solar module converts photons of various wavelengths into electrical energy following its spectral response (SR). SR is defined based on its bandgap, cell width, and transport properties of the carriers in the material. The amount of current generated by a module depends on the wavelength of the spectral response to the provided input energy in the form of illumination. The spectral response defines a short-circuit current that is obtained at the wavelength of the photovoltaic modules, and it is calculated in watts per square meter per nanometre (W·m^{-2}·nm^{-1}). The short-circuit current depends on the bandgap of the PV module, which is determined by the PV cell technology used. The short-circuit current (I_{sc}) of the photovoltaic module is calculated through the integral product of the external SR and the required spectrum [32, 33].

2.1 Photovoltaic Module Performance Parameters

A photovoltaic module is a semiconductor device that can be approached by a simple dark diode equation, obtained from the current–voltage (I–V) characteristic curve. It allows the current to flow in the direction of forward-bias and blocks any reverse current flowing in the circuit, or so-called reverse bias, as shown in Eq. 1.

$$I_D = I_0\left(e^{\frac{q(V_j)}{nkT}} - 1\right) \quad (1)$$

where: I_0 is known as the saturation current of the diode, (A); n represents the ideal the diode's factor;q is equal to 1,602 × 10^{-19}C, the absolute value of electron charge;k represents the Boltzmann constant, equivalent to 1.38 × 10^{-23} J/K; T stands for the temperature of the device in Kelvin (K); and V_j stands for the voltage across the diode junction, (V).

A diode equation has a significant influence on the circuit, as it affects the operation of the diode. The junction voltage is represented by the following equation:

$$V_j = V + IR_s \quad (2)$$

The diode is a good correspondent of the photovoltaic cell. Figure 1 shows the photovoltaic cell model: the current source in the circuit shows the photon-current, I_L. However, when considering a practical device, it has some limitations. This modification entails expressions by adding parallel and series resistance:

$$I = I_L - I_0\left(e^{\frac{q(V_j)}{nkT}} - 1\right) - \frac{V_j}{R_{sh}} \quad (3)$$

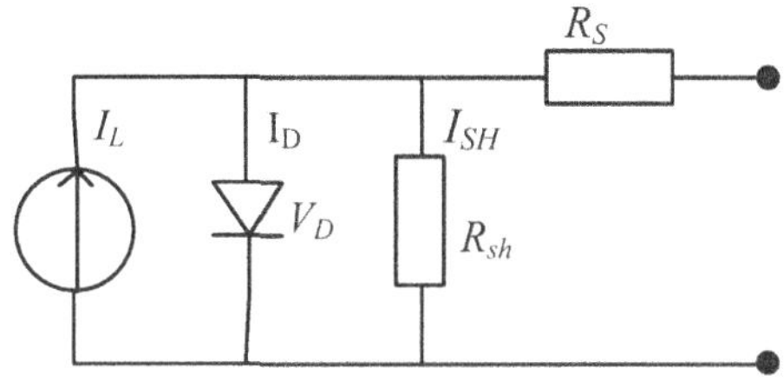

Fig. 1 PV cell model

The fill factor (FF) represents the ratio between the maximum power of the PV module (P_{mp}) and the ideal power ($I_{sc} \cdot V_{oc}$). FF can provide information about efficiency and can be used as an indicator of the module quality. Also, it shows the effect of parasitic resistance on the PV module. The P_{mp} is a state of voltage and current specific to the power of the solar cell and reaches a maximum when the derivative of the power is equivalent to zero. The equation is given as follows:

$$FF = \frac{V_m I_m}{V_{oc} I_{sc}} \tag{4}$$

where: V_m is the voltage at the maximum power point, (V); I_m is the current at the maximum power point, (A); I_{sc} is the short-circuit current, (A); V_{oc} is the open voltage (V).

The conversion efficiency of the module is the ratio of maximum power transfer of the PV module and the incident of the solar radiation on the surface of the photovoltaic module given by Eq. 5. It is determined under standard test conditions (STC): 1000 W/m^2 irradiance; air mass 1.5; the module temperature 25 °C; and an incident angle 0 °C. Under these conditions, the power of the module is referred to as peak power (W_p). The equation is given as follows:

$$n = \frac{V_{oc} I_{sc} FF}{AG} \tag{5}$$

where: A is the surface of the module, (m^2); G is the irradiance falling on the surface of the module (W/m^2).

2.2 Influence of the Operating Temperature of the PV Module

The PV module performance is mainly determined by ambient temperature and solar radiation. The temperature range depends heavily on the local wind speed, the PV

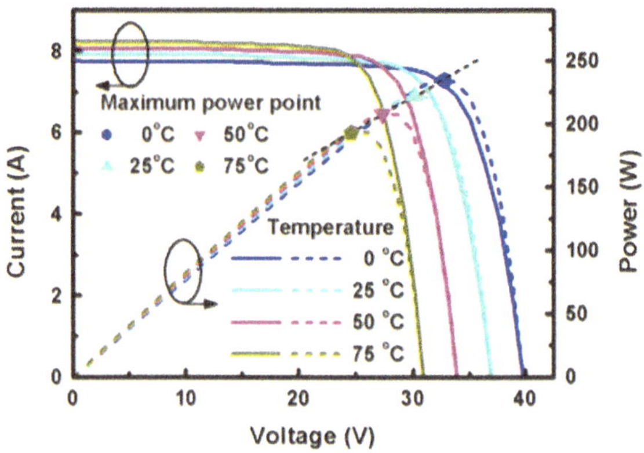

Fig. 2 The PV array characteristic under various temperatures and an irradiation intensity of 1000 W/m^2 [40]

material, and the structure of the photovoltaic module, such as glazing-cover transmittance and absorbance [34, 35]. A photovoltaic module is directly affected by temperature as it reduces the open-circuit voltage (V_{oc}) and consequently, the maximum power (P_{mp}). High temperatures reduce the effectiveness of the module bandgap [36, 37]. When the temperature increases due to the formation of electron-holes which require less energy, the open-circuit voltage, and bandgap decreases, and the dark current saturation increases [38, 39]. Photovoltaic performance is characterized by the I–V characteristic curve shown in Fig. 2.

The curve is used to obtain PV parameters, such as I_{sc}, V_{oc}, FF, and efficiency. Figure 2 shows that when the temperature increases, the V_{oc} of the photovoltaic module decreases, with a small rise in I_{sc} [41–43]. As temperature varies, the effect on the I_{sc} is marginal, while the impact on V_{oc} is substantial [38, 39]. The equation is given as follows:

$$V_{oc} = \frac{nkT}{q} \ln\left(\frac{I_L}{I_o}\right) \tag{6}$$

2.3 Losses Due to Extrinsic and Intrinsic in a Solar Cell

The power losses in the solar cell have been reported and classified according to extrinsic and intrinsic losses, optical and electrical losses [44, 45]. These are shown in Fig. 3.

Extrinsic losses are due to external factors and attributed to reflection, shading, series resistance, incomplete collection of generated photocarriers, absorption in the window layer, and non-radiative recombination.

Intrinsic losses are due to two factors:

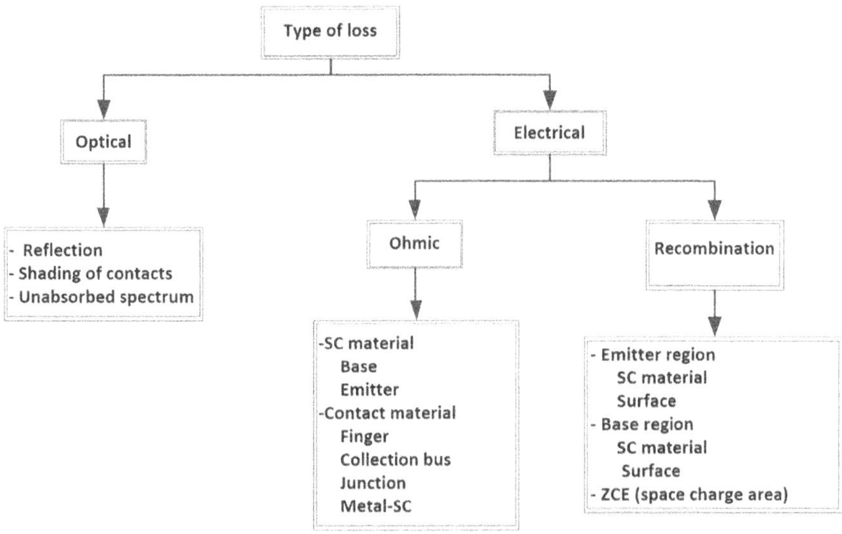

Fig. 3 Different sources of losses [44]

- The inability of the single-junction solar cell to respond efficiently to all wavelengths of the solar spectrum. The solar cell becomes transparent for photons whose energy is less than the energy of the forbidden band of the semiconductor when $E_{ph} < E_g$. On the other hand, provided that the photons have an energy greater than the forbidden band when $E_{ph} > E_g$, the extra energy is dissipated in the form of heat.
- The second type is a result of radiative recombination in the solar cell.

2.3.1 The Recombination Loss

Recombination is defined as the reverse mechanism of generation. It engages a loss of energy; excess carriers disappear by restoring their thermodynamic equilibrium. A disturbed semiconductor restores its thermodynamic equilibrium through several possible mechanisms:

- *Radiative recombination*: direct bandgap materials are often restricted by radiative recombination;
- *Indirect or assisted recombination*: SHR constrains solar cell with low material quality;
- *Auger recombination*: solar cells with high V_{oc} are regularly restricted by Auger recombination; and
- *Surface recombination.*

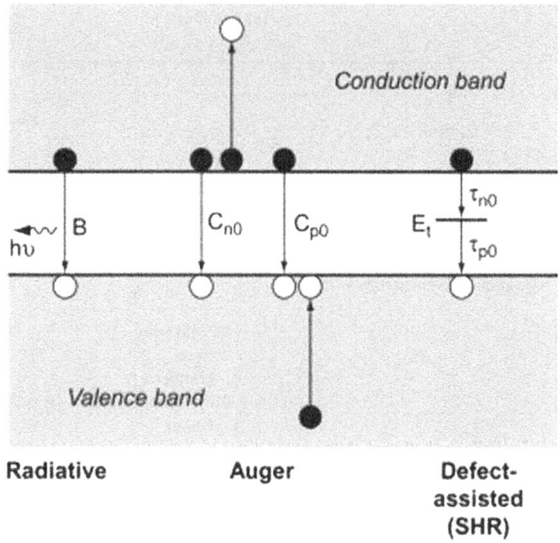

Fig. 4 Diagram of the main recombination processes in a semiconductor [44]

The radiative and the Auger recombination are intrinsic processes. In contrast, SRH and surface recombination are extrinsic processes assisted by a defect level in the bandgap and therefore are at the origin of heating during their realization. Figure 4 presents the primary recombination process in a semiconductor.

2.3.2 Radiative Recombination

This mechanism is dominant in pure semiconductors, with a direct gap structure such as GaAs. Radiative recombination is described as the reverse mechanism of optical absorption; an electron from the conduction band combines with a hole in the valence band. The energy of the emitted photon will correspond to the gap energy of the forbidden band of the material. This phenomenon is characterized by the short life of the minority carriers. The spontaneous radiative recombination is shown in Fig. 5.

The rate of total recombination is relative to the concentration in the material. In the absence of a generation phenomenon, the overall recombination rate is calculated by the following formula [46]:

$$U_{\text{rad}} = B(np - n_i^2) \tag{7}$$

where: B is a constant that depends on the semiconductor used; n_i the intrinsic concentration; n and p represent the concentrations of electrons and holes at equilibrium.

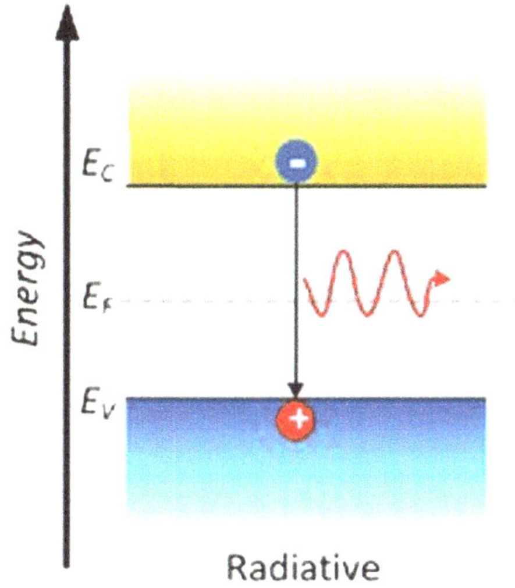

Fig. 5 Radiation recombination

2.3.3 Indirect or Assisted Recombination (SHR)

The Shockley–Read–Hall (SRH) recombination process relies on the material quality and imperfections of the material. Assisted recombination occurs in indirect gap semiconductors. This mechanism introduces a step in the transition between the conduction and valence (the depletion zone). This is located in a trap zone in the forbidden band, as seen in Fig. 6.

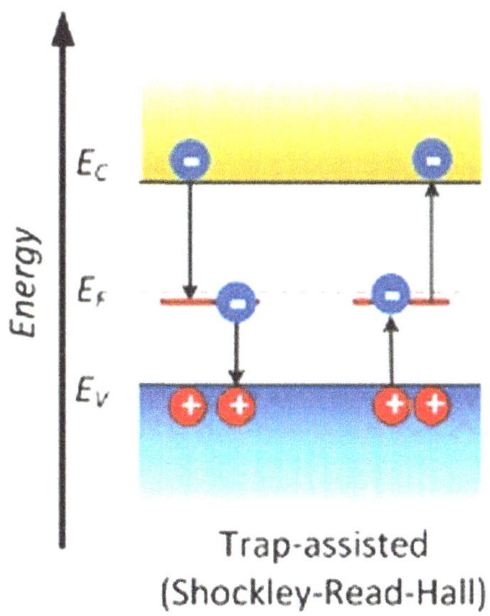

Fig. 6 SRH recombination process

The defect comes from impurities or the structure of the network.

The indirect recombination rate for N_t defect concentration located at an energy level and in the forbidden band is represented by the Shockley–Read–Hall formula [46]:

$$U_{\text{SHR}} = \frac{\sigma_p \cdot \sigma_n \cdot V_{th} \cdot N_t \cdot (np - n_i^2)}{\sigma_n \cdot (n + n_1) + \sigma_p \cdot (p + p_1)} \tag{8}$$

It has been diminished using watchful creation strategies that reduce surrenders in the semiconductor precious stone. The equation is given as follows:

$$n_1 = n_i \cdot \exp\left(\frac{E_t - E_i}{kT}\right) \tag{9}$$

$$p_1 = p_i \cdot \exp\left(\frac{E_i - E_i}{kT}\right) \tag{10}$$

where: N_t: is the concentration of defects; σ_p, σ_n sections of effective capture by the hole and the electron; V_{th} is the thermal speed; E_i is the intrinsic Fermi level; and U_{SHR} >> is given in free carriers, cm^{-3}. s^{-1}.

2.3.4 Surface Recombination

The surface of the semiconductor has many defections after being treated during polishing and etching. The local oxide increases in a matter of minutes over several tens of angstroms in the environment, interfering with the crystalline structure near the surface. When the material is uniformly illuminated, the recombination of the surface results in an excess carrier concentration on the surface rather than in the volume. Figure 7 shows the surface recombination through one gap state within defect density.

The equation is given as the following [47]:

For the material of type P:

$$J_{\text{surface}} = q \cdot S_p \cdot (_n - p_{n0}) \tag{11}$$

For the material of type N:

$$J_{\text{surface}} = q \cdot S_p \cdot (n_p - n_{n0}) \tag{12}$$

where: S is the surface recombination rate (cm/s).

The recombination of the surface is documented here as a significant parameter in solar cell research, influencing the saturation of the current and quantum efficiency of the cells. The velocity depends on the properties of the semiconductor material used. The velocity is high for the direct gap of semiconductors. Some semiconductor

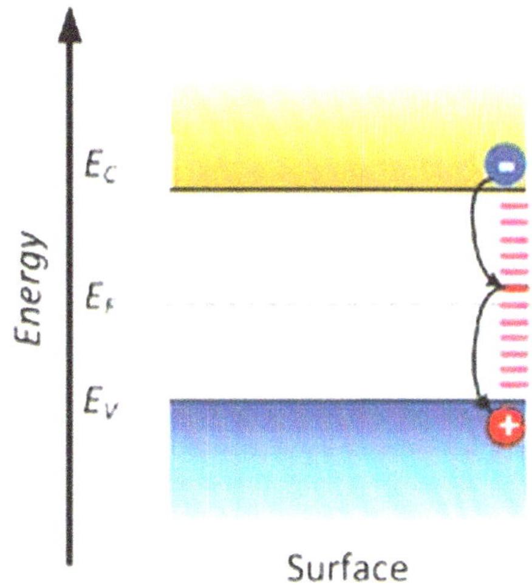

Fig. 7 Surface recombination

materials are stimulated to reduce their recombination rate. For instance, in the silicon cells, an oxide layer is added, and the GaAs cell increases the window layer using the AlGaAs layer, thereby reducing the surface recombination rate, which represents the surface layer recombination rate.

2.3.5 Auger Recombination

The Auger recombination is the mechanism that occurs during a high concentration of free carriers, as in the indirect gap semiconductors [48]. The energy released by recombination transfers an electron to a greater level of the conduction band or a hole at a deeper level of the valence band. The Auger recombination is illustrated in Fig. 8.

The total recombination rate is calculated by the following formula [44]:

$$\mathrm{U}_{\mathrm{Auger}} = (C_{p0} + C_{n0})(np - n_i^2) \tag{13}$$

where: C_{p0},C_{n0} are Auger capture coefficients of the hole and the electron.

2.3.6 Joule Effect

Concerning the influence of Joule, it is reasonable to consider the heat dissipation due to the passage of the charge carriers in a resistivity material. It is used to determine the thermoelectric phenomenon found when charge carriers pass through an electric field. Joule heating losses associated with the internal series resistance of the module depends on the electrical properties of the metals used. Figure 9 illustrates the power

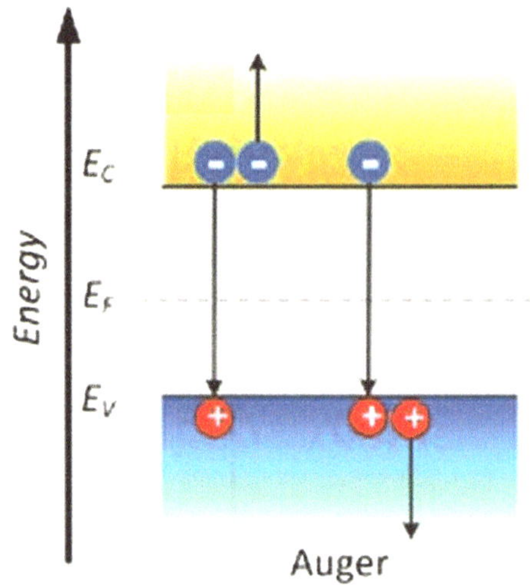

Fig. 8 Auger recombination

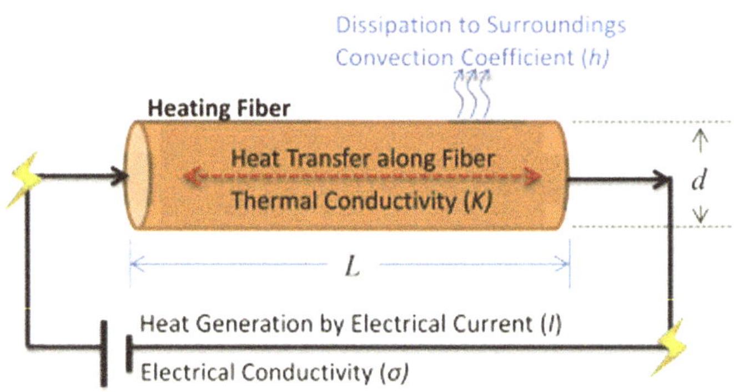

Fig. 9 Combined loss due to Joule heating phenomenon

lost through the cable due to Joule heating. The combined loss in cable resistance, the contact resistance in the terminals, fuses, connectors, shunt resistance, and the diode voltage drop in the array.

The equation of the Joule effect leads to Eq. 14 [49]:

$$Q(x) = \vec{I}_L(x) \cdot \vec{E}(x) \tag{14}$$

In n-type material, the charge distribution function of Fermi–Dirac statistic reveals that there are electrons in the conduction band in larger quantities than in the case of p-type material and that in thermal models, and, all these electrons do not have the same energy.

In the case where the diode is in direct polarization, the electrons of the n-doped zone move toward the p-doped region under the effect of the electron concentration gradient between these two zones. This natural movement is due to the gradient as opposed to the potential barrier born from the *PN* junction. Among the electrons, only those with enough energy will be able to cross the potential barrier and conserve the electronic distribution of the doped material. These will repopulate the high energy levels and therefore maintain the electronic balance of the doped material. In the case of a photovoltaic cell in direct polarization, the photo-generated minority carriers are injected into the zone where they become the majority. During this transfer, they give up some of the energy they carry in thermal form by emitting photons. This process induces a release of energy, heating the cell.

2.3.7 Peltier Effect

As the current flows through the metal-semiconductor interface, the difference in potential between the working output level of the metal and the valence or conduction band induces heating. It is essential to observe that despite the simplicity of the metal-semiconductor interface, the amount of energy loss in the model under the Peltier effect is the loss in real contact in thermal form. Even with a more extensive interface, the semiconductor Fermi level in the surface is aligned with the output work of the metal. Therefore, the charge carriers move between the semiconductor and the metal. This charge will always lose energy between the valence or conduction band and which results in a thermal form, in the Peltier effect, or by Joule effect, provided that the surface of the semiconductor is highly doped, or by tunnel resistance. If the interface is a tunnel diode and no longer a metal-semiconductor interface without defect [50].

2.3.8 Parasite Resistances

Practically, photovoltaic modules are not perfect; they have extra parasitic resistance (series resistance, R_s and parallel resistance, R_p). Both high series resistance value and low parallel resistance value influence the fill factor of the solar module. The solar cell efficiency is susceptible to parasitic resistance parameter, by dissipating the solar cell power, so their effect varies depending on the geometry of the solar cell and the manufacturing technology. Both resistors have a substantial impact on solar cell characteristics and performance [51].

(1) **Series resistance**

The series resistance (R_s) in a PV module represents the resistance in front, rear, and terminal contacts and cell resistivity of the PV module interconnection; caused by the impedance of the cathodes (gate metallization and the back face) and the deficient doping in the relatively neutral regions (emitter-base). The low estimation of the series resistance does not influence the short-circuit current and the open-circuit voltage, but slightly diminishes the value of the shape factor. On an excellent cell quality, R_s must be less than 1 Ω. The influence of the series resistance is noticeable in the I–V characteristic curve, with R_s diminishing the short-circuit current without influencing the open-circuit voltage. The transmitter and the upper grid of the solar cell characterize the main factors of loss due to the series resistance. The optimization is achieved by reducing the thickness of the emitter and the metal-semiconductor contact area, which will reduce the resistivity of the material [51, 52].

(2) **Parallel resistance, R_p**

Manufacturing defects commonly cause losses due to parallel resistance (R_p), occurring when the diffusion of high-temperature metals that puncture the transmitter, insufficient materials for photovoltaic modules, or the resistance between the terminals and the path of the PV unit not passing through the junction. This relates to the intersection resistance, which becomes significant when the PN junction is placed close to the surface. Likewise, it restricts the separation distance between grid lines in thin-film solar [51, 52].

3 Mathematical Model and Thermal Performance of PV Array

The numerical analysis of thermal and electrical energy from the PV array is presented in this section. The PV array is assumed to lay perfectly flat on the Earth's surface and receive full solar radiation. The solar radiation is converted to electrical/thermal energy by the PV array.

To simplify the physical structure of the proposed PV array, the following assumptions should be ensured when configuring the model:

- The overall efficiency of the system is in a quasi-steady-state condition.
- All surfaces of layers incorporate uniform temperature.
- The temperature of the gradient around the tube is considered insignificant.
- Heat dissipation is one dimensional.
- Edge loss is considered insignificant because of insulation.
- Due to the high resistance to heat, the heat loss through the insulation is low.
- The PV module is in a horizontal orientation upward, with solar radiation perpendicular to the module.

The design and numerical equations are given in this study may not necessarily be the most well-defined due to the assumptions. However, they are generally used in the literature and are straightforward to apply and acceptable for most design calculations.

3.1 Thermal Modeling

Duffie and Beckman [54] described the equation of the energy balance in solar energy thermal processes.

The useful energy Q_u is extracted under conditions of stability, proportional to the useful energy absorbed by the collector, minus the amount lost by the collector, and it is expressed as follows equation [54]:

$$Q_u = Q_{\text{in}} - Q_{\text{loss}} \tag{15}$$

where: Q_u, Q_{in}, and Q_{loss} are respectively the useful energy, the absorbed energy, and the energy losses.

The solar cell energy increases when the solar radiation reaches the absorber plate surface. Energy gained by the absorber is the active optical fraction of the solar radiation; the following equation expresses the amount of solar radiation received by the solar cell [54]:

$$Q_{in} = (\alpha\tau) * G * A_c \tag{16}$$

where: G is the intensity of the solar radiation, (W/m^2). A_c is the surface area of the solar cell, (m^2). τ is the transmission coefficient of the glazing; and α is the energy absorbed in a fraction. The amount of incoming radiation absorbed is given by $\alpha\tau$.

The mode of loss—convection, radiation, and conduction—is dependent on the temperature difference between the solar cell, the environment, and the geometry of the solar cell. The formula is given in the following equation:

$$Q_{\text{loss}} = Q_{\text{conv}} + Q_{\text{rad}} \tag{17}$$

Q_{conv} is the heat convection to the external environment. The convective exchange losses depend on the linear wind speed function and the environment temperature. These losses are proportional to the ambient and the solar cell temperature, and is provided by the following equation [54]:

$$Q_{\text{conv}} = A \times h \times (T_{\text{ext}} - T_s) \tag{18}$$

where: T_s is the temperature surface, T_{ext} is the external temperature, and h is the heat transfer coefficient.

Most of these coefficients a, b, and n are in the form of the following equation:

$$h = a + b \times v^n \tag{19}$$

Thus, this is proposed as the most used equation to study the relationship between flat solar collectors:

$$h = 5.7 + 3.8 \times v^n \tag{20}$$

This transfer mode occurs through the emission, and electromagnetic absorption between objects at different temperatures is called thermal radiation. The perfect emitter of blackbody emission power or thermal radiation, given by Stefan–Boltzmann, is proportional to the fourth power of its temperature. This is described by Eq. 4.7 [54, 55]:

$$Q_{\text{rad}} = \sigma \cdot A \cdot T^4 \tag{21}$$

The completion of the radiation between the surfaces involves losses from surface radiation and lost into the environment. This is affected by the physical and geometric properties of the surface. And it is quantified by the parameters known as the shape factor. To exchange radiation between the two surfaces 1 and 2, the net heat energy of surface 1 can be approximately transferred to surface 2, in the following equation [54, 55]:

$$q_r^{1\leftrightarrow 2} = -q_r^{1\leftrightarrow 2} = \frac{\sigma \cdot \left(T_1^4 - T_2^4\right)}{\frac{1-\varepsilon_1}{\varepsilon_1 \cdot A_1} + \frac{1}{A_1 \cdot F_{12}} + \frac{(1-\varepsilon_2)}{\varepsilon_2 \cdot A_1}} \tag{22}$$

The general form of the radiation heat transfer coefficient between surfaces is:

$$h^r = \frac{\sigma \cdot \left(T_1^4 - T_2^4\right)(T_1 + T_2)}{\frac{1-\varepsilon_1}{\varepsilon_1} + \frac{1}{F_{12}} + \frac{(1-\varepsilon_2)A_1}{\varepsilon_2 A_2}} \tag{23}$$

The radiation exchange between two surfaces (convex objects and large shells) is applicable when a sizeable concave surface surrounds a curved object. Under these circumstances, $A_l/A_2 \rightarrow 0$, and mainly no reflection of the radiation is emitted by the object, $F_{12} \rightarrow 1$. The rate of radiation heat loss is proportionally related to the emittance of the surface and the temperature difference in the power of four. The equation is given as follows [54, 55]:

$$Q_{\text{rad}} = \sigma \cdot \varepsilon_1 \cdot A_1 \cdot \left(T_1^4 - T_2^4\right) \tag{24}$$

$$h_r = \sigma \cdot \varepsilon_1 \cdot \left(T_1^2 + T_2^2\right)(T_1 + T_2) \tag{25}$$

The expression is used in the case of coverage of a solar cell surface at a temperature (T). Therefore, the equation concerning the ambient temperature (T_a) is rewritten in the following:

$$h^r_{c-s} = \sigma \cdot \varepsilon_c \cdot \frac{\left(T_c^4 + T_{sky}^4\right)}{(T_c - T_a)} \tag{26}$$

The accumulated heat in the solar collector is transmitted to the atmosphere through convection and radiation. The collector heat loss coefficient (U_L) explains how much of the energy is lost to the environment by the top with excellent backing insulation. The rate of heat loss depends on the overall heat transfer coefficient of the collector, as in the following equation:

$$Q_{\text{out}} = h_{\text{cv}} \cdot (T_c - T_a) + \varepsilon \cdot \sigma \cdot (T_c^4 - T_s^4) \tag{27}$$

where: Q_{out} is the heat loss (W); U_L is the heat loss coefficient, W/(m^2·K); T_c is the average temperature of the collector (°C), and T_a is the ambient temperature (°C).

3.2 Modeling of Temperature Effects on PV Array

The PV array only converts a part of the solar radiation into electrical energy, but at the same time, the rest is transformed into heat. The PV module operating temperature has a significant impact on its performance. However, this depends on many factors such as the semiconductor material used, the manufacturer of the PV module, and the installation on site. The effect of temperature on the PV module in this section, including the factors that affect the PV cell operating temperature and their results, is discussed.

3.2.1 Effect of the Operating Temperature of the Solar Cell

The solar cell temperature increases while functioning adversely affecting its performance. The equations governing the PV cell performance depend on parameters such as temperature, bandgap, diffusion coefficient, integral voltage, intrinsic carrier concentration, a width of the depletion zone, and the limitations of models used. The temperature affects V_{oc} as described by Eq. 28 [39, 56].

$$V_{\text{oc}} = \frac{n \cdot k \cdot T}{q} \ln\left(\frac{I_{\text{sc}}}{I_0} + 1\right) = V_t \cdot \ln\left(\frac{I_{\text{sc}}}{I_0} + 1\right) \tag{28}$$

where: V_{oc} is logarithmically proportional to the I_{SC} (V); I_0 the inverse of the saturation current (A); q is electron charge (C), and k is the Boltzmann constant.

Regarding the temperature effect, the V_{oc} changes linearly and is inversely proportional to the temperature. Under PV cell operating conditions, the short-circuit current is relatively independent of the temperature. When electrons are excited by the thermal effect rather than the electrical properties of the semiconductor, V_{oc} and FF are reduced. The temperature tends to decrease the performance of a PV cell by decreasing V_{oc}.

However, the saturation current varies with temperature. Also, the bandgap energy, E_g, of a semiconductor cell decreases as the temperature rises, and is modeled as follows [39, 56]:

$$E_g(T) = E_g(0) - \frac{\alpha T^2}{T + \beta} \tag{29}$$

where: $E_g(0)$ refers to the bandgap value at zero Kelvin, $E_g(T)$ is the bandgap of a semiconductor at temperature (T), and α and β are specific constants of the semiconductors.

Due to the improvement of absorption and photocurrent, the decrease in the bandgap causes a slight increase in the I_{sc}. However, the heating has an undesirable effect on the V_{oc}. From PV materials, it is observed that the linear decrease in V_{oc} and a linear increase in I_{sc}, slightly offset FF.

3.2.2 Calculation of Temperature of PV Cells

Most of the formulas consider the linear character of the PV cell performance to reproduce its temperature dependence and based on its efficiency or its maximum power. The widely used expression which relates the efficiency of a cell to its operating temperature is given in the following formula [39]:

$$\eta_c = \eta_{\text{Tref}} \cdot [1 - \beta_{\text{ref}} \cdot (T_c - T_{\text{ref}})] \tag{30}$$

where: η_{Tref} is the efficiency of the cell (or module) at the reference temperature; T_{ref} is the reference temperature of 1000 W/m^2; T_c represents the operating temperature of the module; and β_{ref} represents the temperature coefficient.

The coefficients β_{ref} and η_{Tref} are dependent on the properties of the material, representing the linear variation of the yield as a function of the PV cell temperature.

The linear equations equivalent to those proposed equations above can be applied when calculating the power loss of the PV module as a function of its operating temperature. Again, very few models consider elements that aid in cooling the module (power gain) by convection such as wind or fluid that passes to the back of the PV module.

3.2.3 Consideration of Temperature in Commercial PV Models

STCs are used to compare the performance of different cells or modules but are not representative of the real-world operation of a PV module. A high operating temperature is a factor that detracts the PV module performance.

A more complex and energy rating model, which considers the effect of the spectral disagreement and the angle of incidence of the light rays, has been developed by the National Renewable Energy Laboratory (NREL). I_{sc} and V_{oc} are defined in Eqs. 31 and 32, respectively.

$$I_{sc} = \frac{G}{G_{ref}} \times I_{SC_{ref}} \times [1 + \alpha(T_c - T_{ref})] \tag{31}$$

$$V_{oc} = V_{oc,ref} \times [1 + \beta.G_{ref}(T_c - T_{ref})] \times \left[1 + \delta(T_c).\ln\left(\frac{G}{G_{ref}}\right)\right] \tag{32}$$

where α, β, and δ are the temperature coefficients for the current, the voltage, and the luminous intensity, respectively. Moreover, G is the light intensity (W/m^2) [53, 54].

The index refers to STC, Eq. 32 takes into account the temperature dependence of the module and solar radiation. The models rely on the empirical coefficients such as α, β, and δ and, are generally more reliable than models that determine power or efficiency from solar radiation, environmental temperature, and in some cases, wind speed. However, the influence of the solar spectrum distribution and thermal effects depends on the location and might be responsible for a difference of 10% losses.

The thermal losses are determined by an energy balance between the ambient temperature and the solar cell temperature in the following equation:

$$U(T_c - T_a) = \alpha G(1 - \eta_c) \tag{33}$$

where: α represents the absorption coefficient of the solar radiation of the PV module; this coefficient is usually set at 0.9, and U represents the thermal exchange by natural convection of the module with its external environment W/(m^2·K).

The energy balance per unit area of the PV module, which is cooled by a loss due to the external environment, can be expressed as follows:

$$\tau \cdot \alpha \cdot G_T = \eta_c \cdot G_T + U_L \cdot (T_c - T_a) \tag{34}$$

where: τ represents the transmittance of any cover that may be over the cells, α refers to the fraction of the incident radiation absorbed on the surface of the cell, and η_c is the efficiency of the module for converting the incident.

Nominal operating cell temperature (NOCT) is determined as a module temperature under the conditions as listed below [57]: solar radiation of 800 W/m^2; wind speed 1 m/s; ambient temperature of 20 °C; and no-load operation ($\eta_c = 0$).

The solar cell is exploited using the linear radiation or a high circular concentrator to define the loss coefficient. The loss coefficient includes convection and radiation from the top and bottom of the module, mounting frame losses at any ambient temperature, T_a.

To determine the $\frac{\tau\alpha}{U_L}$ ratio, the PV cell temperature, ambient temperature, and solar radiation can be used in Eq. 35, which is expressed as follows [54, 57]:

$$\frac{\tau \cdot \alpha}{U_L} = \frac{(T_{c,NOCT} - T_a)}{G_{T,NOCT}} \tag{35}$$

Assuming that the temperature under any other condition is constant, T_c can be obtained from the following Eq. 36:

$$T_c = T_a + \left(\frac{G_T \cdot \tau \cdot \alpha}{U_L}\right)\left(\frac{1 - \eta_c}{\tau \cdot \alpha}\right) \tag{36}$$

The factor $\tau\alpha$ in Eq. 36 is generally unknown but is estimated to be 0.9, and there are no serious errors in using this value because the term $\frac{\tau\alpha}{U_L}$ is smaller than unity.

3.3 *Influence of Extrinsic Cell Resistance on PV Performance*

This section explores the impact of extrinsic cell resistance on PV cell performance. The internal series resistance of the PV cells is an important parameter influencing the maximum power and the *FF*. It is also a parameter demonstrating the quality of the device.

There are several resistivity losses in a PV cell (Fig. 10). The thick black line represents the metallic strip, and the slender black stripes represent the grid conductors. The pink color represents the PV cell emitter, and the ground is a pale blue colour. The symbols L, H, W, and t_E denote the length, the height, the space between grid conductors, and the emitter width, respectively. Figure 10a presents the view from the top of the solar cell, while Fig. 10b presents the current flow. The grey lines and the arrows show the appropriate path and direction of the current flow.

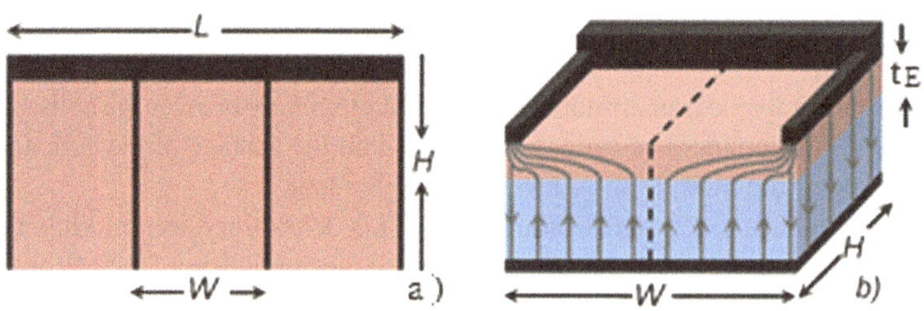

Fig. 10 Resistive losses in solar cells [58]

There are several ways to evaluate the internal series resistance of a PV cell from the I–V characteristic curve. As the temperature of the PV cell increases, the resistance of most materials also increases. Therefore, when optimizing the design of a PV cell, it is essential to consider the loss due to the resistance at the expected operating temperature.

The R_{si} of a PV cell is composed of a variety of internal resistances such as contact resistance (R_c), the resistance of the bus bars (R_b), grid resistance (R_g), and (R_e) lateral conduction in the emitter, (Remitter). This includes back conductor loss, internal and emitter loss, front and back contact loss, front grid conductor, busbar loss, and bias point loss. The total extrinsic series resistance, R_t of the photovoltaic cell [58] is as follows:

$$R_{si} = R_e + R_b + R_g + R_c + R_a \tag{37}$$

where: R_e represents the sum of the emitter series resistance, including accounts for the resistive losses inside of the transparent conducting layers (TCL) and emitter layers; R_b presents the bus bar resistance, including any resistive loss from the terminal contact pad to the grid electrodes; R_g presents grid electrode resistance, including the resistive losses caused by the grid electrodes; R_c represents a contact resistance, including the resistive losses between the grid electrodes and the front lateral conducting layers (LCL); and $R_{a\text{ is}}$ semiconductor resistance, including any resistive losses between the emitter and the back contact.

4 Simulation Results and Discussion

The simulation will study several parameter models such as the ambient temperature, solar irradiance, convective heat coefficient, extrinsic cell resistance, and also to predict the thermal behavior, the power generated, and the efficiency of PV cell models. The total power dissipated by the PV cell model composed of a diode (representing the semiconductor material of the PV cell), an internal series resistance (formed of contact resistance, the busbars resistance, grid resistance, and lateral conduction in the emitter) as given by R_{si}, and an internal series resistance (representing the emitter at the edge of the cells, cracks, recombination sites at the cell edge, Schottky-type shunts below grid lines, scratches, and aluminum particles at the surface), as given by $R_{pi.}$

Energy analysis is essential to evaluate the electro-thermal performance of a PV system and to assess the energies generated. It is viable to determine the magnitude of thermal losses and the energy principle in terms of the second law of thermodynamics (Akyuz et al. 2012). The internal heat generated in the photovoltaic cell is ascertained depending on a mathematical equation, as well as the sum of the $I^2 \cdot R$ losses for each of the resistance and the losses in each of the diodes.

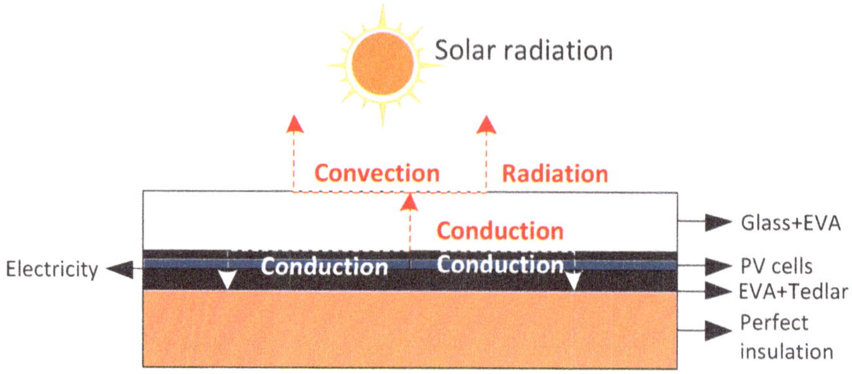

Fig. 11 Heat transfer characteristic of the PV system

(A) Simulation setup

It should be noted that during the simulation time, non-cloudy periods occurred, preventing the PV cell from reaching a high air temperature at the exit of the absorber duct. For the simulation of a physical phenomenon like the problem of heat transfer in Simulink/Simscape, there is a need to develop all the equations of the heat transfer phenomenon in this study. Figure 11 depicts the heat transfer characterization of the PV system.

The main objective of this section is to assess the distribution of power dissipation caused by R_{se}. The study further highlights the thermal behavior of photovoltaic cells. For this, the electro-thermo-radiative behavior of PV for different values of R_{se} has been modeled, keeping the rest of the parameters fixed (for example, solar radiation at 1000 W/m^2, ambient temperature at 20 °C, and convective heat transfer of 20 W/(m^2·K)). Electrical power, heat loss, efficiency, and thermal were measured for different extrinsic cell resistance values from 0 to 100 Ω.

In this section, a set of values defining extrinsic cell resistance (R_{se}) values is added to the existed internal series resistance of the PV cell, as shown in Fig. 12. The performance of the PV cell was evaluated through calibration of the R_{se} value; this, however, dissipates power that is in the form of heat, thus creating useful heat energy.

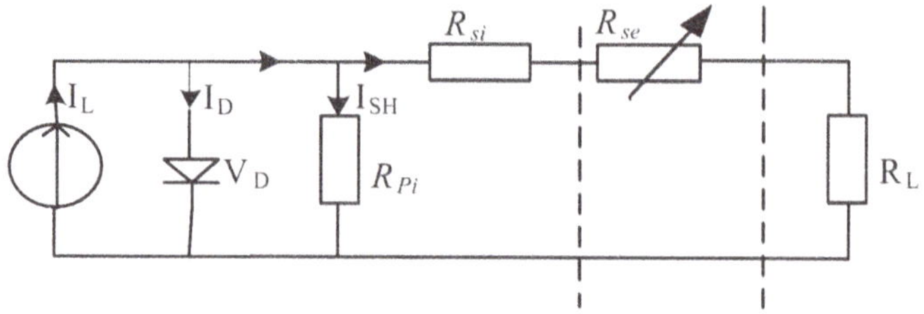

Fig. 12 Evaluation of PV performance under extrinsic cell resistance

Table 1 PV module parameters

Component	Parameter	Value
PV modules	Cell type	Mono-crystalline
	Packing factor	0.91
	Conversion efficiency	16%
	Module peak power	3.25 kW
	Maximum voltage, V_m	255 V
	Maximum current, I_m	12.4 A
	Open-circuit voltage, V_{oc}	310 V
	Short-circuit current, I_{sc}	14.64 A
	Internal Series resistance R_{si}/cell	0.0042 Ω
	Parallel resistance R_{pi}/cell	10.1 Ω

Table 2 Optical parameters of PV cells

	Absorbance α	Emissivity ε	Thermal conductivity	Thickness δ	Temperature coefficient	Energy gap EG
PV module type	0.8	0.75	840	0.003	0.000905	1.11

(B) **Design parameters of the PV array**

The photovoltaic is selected from Simscape, and it offers a solar cell model with a thermal option, which is made up of thermal mass. This allows for heat flow and displays the internally generated heat from the PV models. A PV module is generally made of solar cells connected in series (for example, 36 or 72 cells connected in series). The PV modules parameters appear in Tables 1 and 2.

The entire system consists of two PV arrays assumed to perform identically and in a parallel configuration; the system has a capacity of 3.24 kW_p at 1000 W/m^2 (Fig. 13).

4.1 PV Cell Power Dissipation as a Function of R_{se}

The performance evaluation of the PV cell composed of the diode, $R_{si,}$ and R_{pi} based on the influence of varying R_{se} was conducted. Figure 14 shows a logarithmic growth curve of the total power dissipated (P_d) of the PV cell, which increases from 990 to 3490 W, as R_{se} increases. R_{si} marginally increases, while R_{pi} remains almost the same. Much of the total power dissipated is attributed to the diode (due to recombination current of the semiconductor material property used to produce the PV cell). The model in the PV cell ranges from 750 to 3480 W, R_{pi} and R_{si} resistivity losses reduce

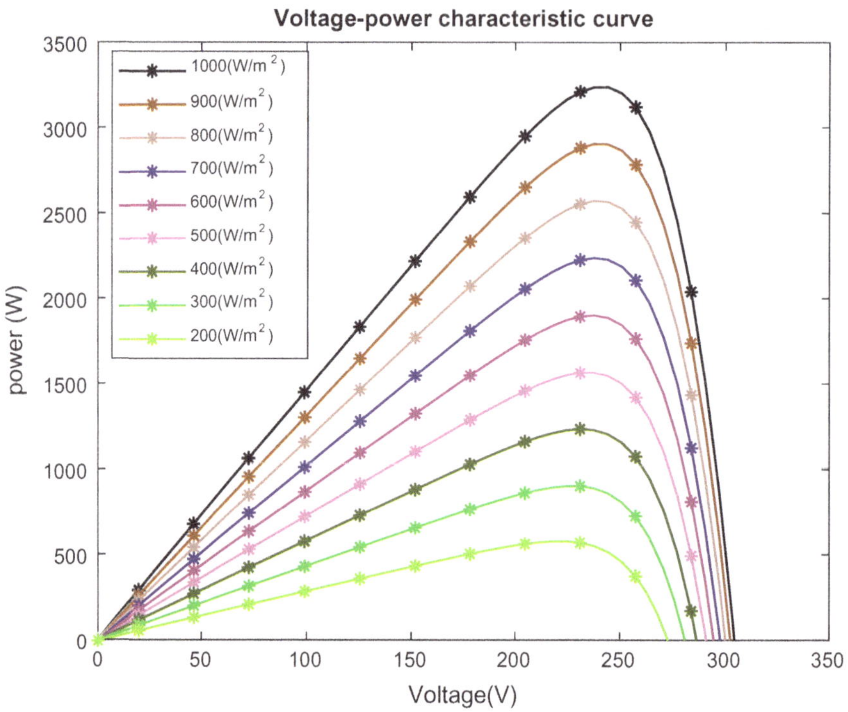

Fig. 13 PV module voltage-power characteristic curve

as less current flow through them. This power dissipated in the PV cells is due to a large reverse current across the PV cell, there is low current flow in the cell, the overall current becomes limited because of the power dissipated in the PV cell in the form of heat. The power dissipation occurring in a PV cell leads to local overheating, which in turn results in heat conduction. The thermal resistance of the PV cell depends on the thickness of the material and its thermal resistivity.

At the point when a photon is absorbed by a PV cell, it produces an electron and a hole pair, with an energy below the bandgap which hypothetically contributes to the current. However, the reverse process occurs when the carriers recombine with no net contribution to cell current and emit the energy as heat or as a photon. Any energy lost in a PV cell is turned into heat, so any inefficiency in the cell increases the PV cell temperature when it is placed in solar radiation. Relatively a photon contribution to several of the mechanisms that contribute to the total heat dissipation in a solar cell. The rate of recombination loss changes with the variation of R_{se}.

The following phenomenon can be explained as: an R_{se} assistant controls the PV self-generated useful heat by partially converting the output of the PV cell into useful thermal energy (dissipating power in the form of useful heat), thereby reducing the output power generated.

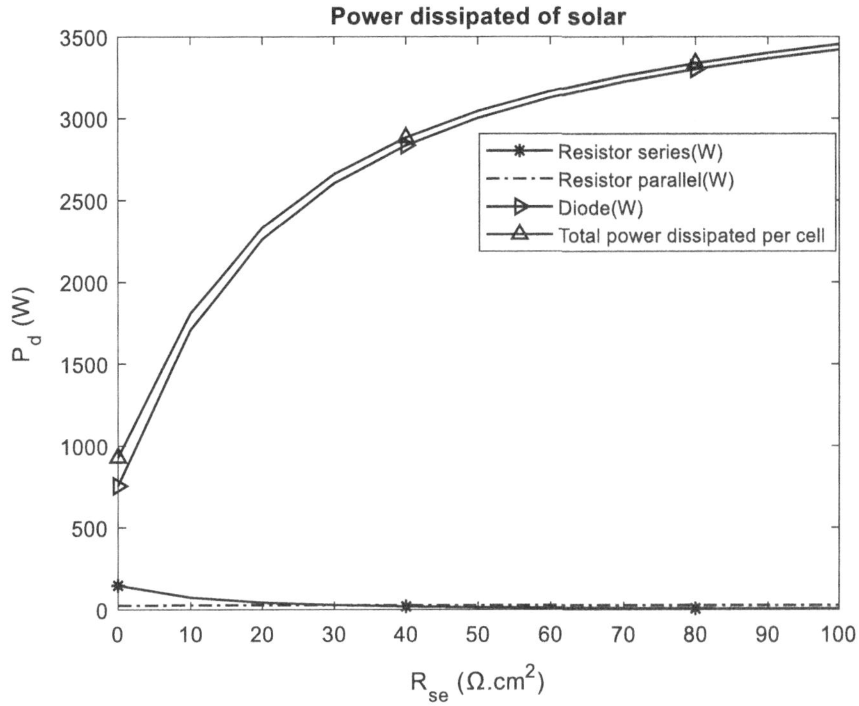

Fig. 14 Dissipated power by PV cell versus R_{se}

This study concentrates on the internal heat generation and electrical power generated from the PV cell based on the R_{se}. The method relies on the linear regression equation curve to model the behavior of different types of power as a function of R_{se} in the PV cell being studied. The subsection below presents the linear regression equation calculation of figures as a function of extrinsic cell resistance. By means of the linear regression equation, a graphical model can be derived; including the goodness of the fit and the confidence interval of the coefficients for the derived equation.

4.1.1 Estimation of Heat Generated by the PV Cell

Figure 15 presents the heat generated by Q_{cond} within the PV cell. Q_{cond} increases from 425 to 1715 W (in magnitude) as R_{se} moves from 0 to 100 Ω. This heat was attributed to the electrical power dissipated in the PV cell in Fig. 16, and part of the power dissipated turns to useful energy within the PV cell. However, the temperature differential is the main impetus behind the conductive flow of heat in a material with a given thermal resistance, and the Fourier law governs the transfer. The estimation graph related to useful heat by conduction is described in the following equation:

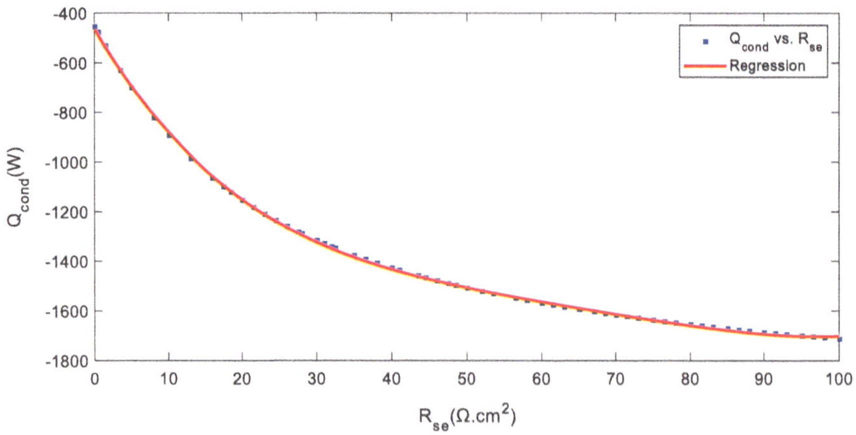

Fig. 15 Conduction heat transfer versus R_{se}

$$Q_{cond}(R_{se}) = p1 \cdot R_{se}^4 + p2 \cdot R_{se}^3 + p3 \cdot R_{se}^2 + p4 \cdot R_{se} + p5 \quad (38)$$

where: p1, p2, p3, p4, and p5 are coefficients; Q_{cond} is the thermal transfer coefficient by conduction (W); R_{se} is the extrinsic cell resistance (Ω).

Table 3 introduces a polynomial interpretation of heat generated by conduction simulation results in Fig. 15; the coefficient of determination (R^2) is 0.9996. R^2 is a marker to pass judgment on the nature of linear regression. The sum of squared errors (SSE) is 2313 and presents the deviations predicted from the actual simulation data. Root Mean Square Error (RMSE) is a good indicator of how accurately the model predicts the response, and it is determined to be 6.37.

Figure 16 presents a continuous increase of Q_{conv} from 3100 to 5300 W when the R_{se} goes from 0 to 100 Ω, as the heat is taken away to the environment. The thermal effect on the photovoltaic cell is due to the high generated electrical power dissipation of R_{se}, and by the heat lost by conduction occurring in the PV cell. The

Table 3 Linear model Poly4 of Fig. 15

Description of Eq. 38	Goodness of fit
$Q_{cond}(R_{se}) = p1 \cdot R_{se}^4 + p2 \cdot R_{se}^3 + p3 \cdot R_{se}^2 + p4 \cdot R_{se} + p5$	SSE: 2313
Coefficients (with 95% confidence bounds):	R-square: 0.9996
	Adjusted R-square: 0.9996
p1 = 3.547e-05 (3.232e-05, 3.863e-05)	RMSE: 6.37
p2 = −0.009531(− 0.01016, − 0.008902)	
p3 = 0.9784(0.9375, 1.019)	
p4 = −50.36(− 51.32, − 49.41)	
p5 = −461.2(− 467.8, − 454.6)	

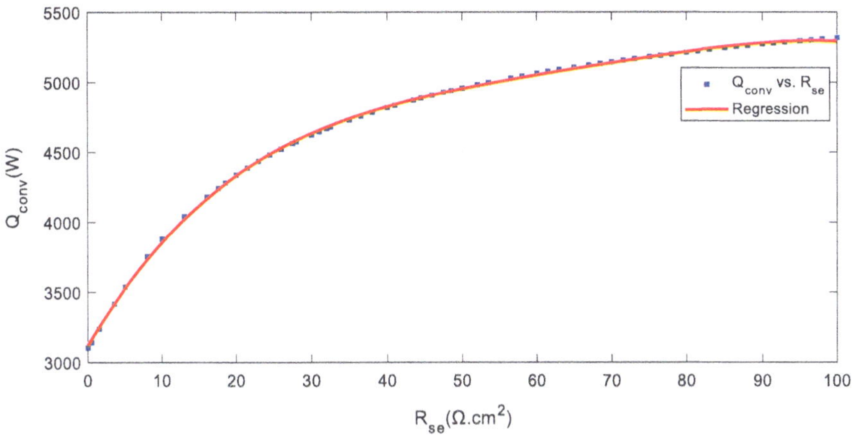

Fig. 16 Convection heat transfer versus R_{se}

thermal loss by Q_{conv} increases faster when R_{se} varies between 0 and 50 Ω; however, it slows down and closes to saturation when R_{se} was above 50 Ω.

Figure 17 presents the incremental change of Q_{rad} from 350 to 660 W when R_{se} goes from 0 to 100 Ω. This shows a radiative thermal transfer to the high emissivity of the PV cell, and the free carriers increase with the R_{se}. The PV cell emits radiation based on its temperature. Also, the losses depend on the absorptivity of the covering glass.

The outline of these results is presented in Figs. 15, 16, and 17, exhibiting that the development of heat loss by Q_{conv} is more noticeable than that in Q_{rad}. Both were measured as a positive value indicating that they are taken away into the surrounding environment. In contrast, the Q_{cond} is measured negative and transferred to the back

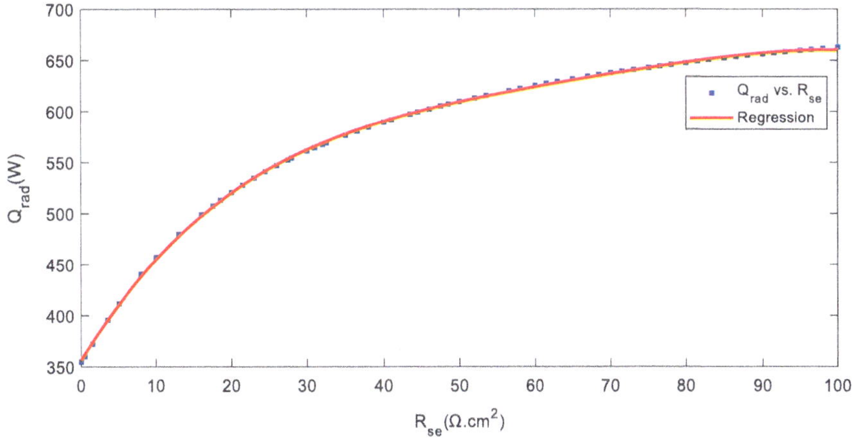

Fig. 17 Radiation heat transfer versus R_{se}

of the PV cell. Comparing the heat transfer occurring at the PV cell, Q_{conv}, Q_{cond}, and Q_{rad}, were 2200 W, 1290 W, and 310 W, respectively, as a function of R_{se}.

4.1.2 Estimation of Generated Power as a Function of R_{se}

It can be seen in Fig. 18 that the generated PV power decreases as R_{se} increases. The power rapidly falls exponentially from 2800 to 260 W when R_{se} increases from 0 to 50 Ω, and beyond 50 Ω, the power decreases more slowly from 255 to 110 W. The power degradation of PV cell is due to recombination accordingly to R_{se} variation, leading to electrical power dissipation in the form of heat by conduction. Therefore, the generated electrical power decreases proportionally to the increase in power dissipated in the PV cell as R_{se} varies in Fig. 14. The estimation graph is expressed by Eq. 42.

$$\mathrm{P_{PV}(R_{se}) = p21 \cdot R_{se}^4 + p22 \cdot R_{se}^3 + p23 \cdot R_{se}^2 + p24 \cdot R_{se} + p25} \tag{42}$$

where: p21, p22, p23, p24, and p25 are coefficients; P_{PV} is the generated PV power; R_{se} is the extrinsic cell resistance (Ω).

Table 4 presents a polynomial interpretation of generated electrical power as varying R_{se} in Fig. 18, the computation of the coefficient of determination (R^2) is determined to be 9564; it was a marker to assess the accuracy of the linear regression. SSE is 7510 and represents the deviations anticipated from the simulation exploratory data. RMSE is described as 40.96 and is a good measure of how correctly the model predicts the response.

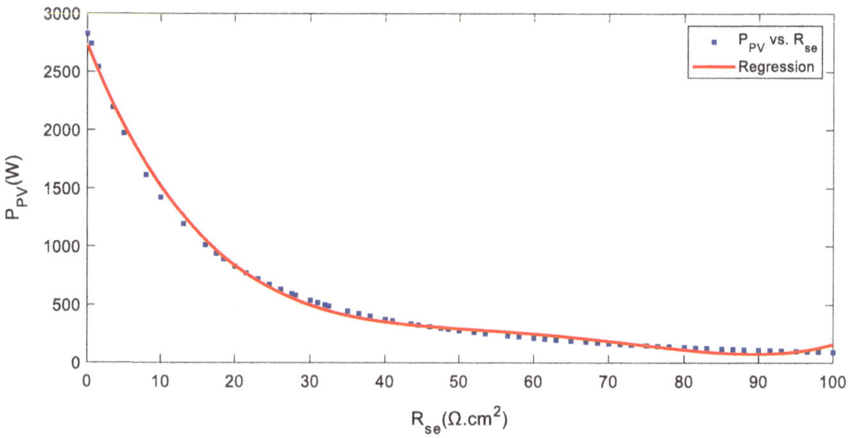

Fig. 18 PV power versus R_{se}

Table 4 Linear model Poly4 of Fig. 18

Description of Eq. 42	Goodness of fit
$P_{PV}(R_{se}) = p21 \cdot R_{se}^4 + p22 \cdot R_{se}^3 + p23 \cdot R_{se}^2 + p24 \cdot R_{se} + p25$	SSE: 9.564e+04
Coefficients (with 95% confidence bounds):	R-square: 0.9964
	Adjusted R-square: 0.9962
$p21 = 0.0001633(0.000143, 0.0001835)$	RMSE: 40.96
$p22 = -0.04122(-0.04527, -0.03718)$	
$p23 = 3.788(3.525, 4.05)$	
$p24 = -155.5(-161.7, -149.4)$	
$p25 = 2738(2695, 2780)$	

4.1.3 The PV Cell Temperature as a Function of R_{se}

Figure 19 shows a logarithmic growth of cell temperature as a function of R_{se}. An increase in the R_{se} from 0 to 50 Ω, brings about an increase in T_c for temperature from 45 to 59 °C. While increasing the R_{se} value from 50 to 100 Ω, brings about an increase in T_c for temperature from 59 to 62 °C. The increase in the temperature is a result of built-up electrical power dissipated in the form of heat in Fig. 14. The output of the PV cell can be turned partially to a heat source and provide additional heat to the PV/T system. The PV cell reached up to 62 °C may potentially perform better than the flat PV/T system, once the solar thermal plant is attached to the rear of the PV cell.

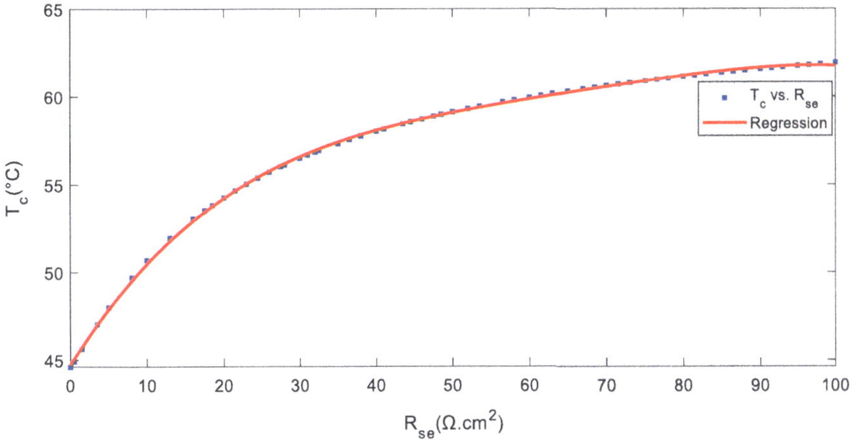

Fig. 19 PV cell temperature versus R_{se}

5 Conclusion

This study developed a model of a hybrid PV/T system that employed innovative energy balancing to produce thermal and electrical energy to suit end-user requirements. This technique employed the PV cell's self-heating properties using a variation of the extrinsic cell resistance to obtain the PV/T hybrid model.

Design and modeling of a PV cell system were developed in MATLAB/Simulink to validate the heat transfer occurring in the PV cell model, which converts the radiation solar into electricity as well as the heat. This partially converts the output of the internal heat generation of the PV cell into useful thermal energy (for domestic hot water, for space heating). A variation of R_{se} could be an effective way of controlling the amount of thermal and electrical energy. The method depends on the linear regression equation curve to model the behavior of different types of power as a function of $R_{se,}$ including the goodness of the fit and the confidence interval of the coefficients for the derived equation.

The results obtained in this study are useful for domestic water heating applications. R_{se} can be adjusted to generate auxiliary heat while the working fluid transports the generated heat to the load. The work demonstrated in this study that the usefulness of a PV cell can be extended such that it can be used to generate not only electricity but can be additionally applied for domestic hot water preheating and space heating.

References

1. Ferrantelli A, Ahmed K, Pylsy P, Kurnitski J (2017) Analytical modelling and prediction formulas for domestic hot water consumption in residential Finnish apartments. Energy Build 143:53–60. https://doi.org/10.1016/j.enbuild.2017.03.021
2. Rahmatmand A, Vratonjic M, Sullivan PE (2020) Energy and thermal comfort performance evaluation of thermostatic and electronic mixing valves used to provide domestic hot water of buildings. Energy Build 212:109830. https://doi.org/10.1016/j.enbuild.2020.109830
3. Tiwari A, Sodha MS, Chandra A, Joshi JC (2006) Performance evaluation of photovoltaic thermal solar air collector for composite climate of India. Sol Energy Mater Sol Cells 90(2):175–189. https://doi.org/10.1016/j.solmat.2005.03.002
4. Brittman S, Adhyaksa GWP, Garnett EC (2015) The expanding world of hybrid perovskites: materials properties and emerging applications. Mrs Commun 5(1):7–26. https://doi.org/10.1557/mrc.2015.6
5. Li K, Liu C, Chen Y, Liu GL (2016) Upgrading both Geothermal and Solar Energy. In: 1st workshop on geothermal reservoir engineering Stanford University, California, pp 1–17
6. Ramos A, Chatzopoulou MA, Guarracino I, Freeman J, Markides CN (2017) Hybrid photovoltaic-thermal solar systems for combined heating, cooling and power provision in the urban environment. Energy Convers Manag 150:838–850. https://doi.org/10.1016/j.enconman.2017.03.024
7. Moussavou AA, Raji A, Adonis M (2019) Controllable and flexible energy production in a water-based photovoltaic/thermal system. Int J Eng Technol 8(4):473–477. https://doi.org/10.14419/ijet.v8i4.29485
8. Wu J et al (2017) A review of thermal absorbers and their integration methods for the combined solar photovoltaic/thermal (PV/T) modules. Renew Sustain Energy Rev 75:839–854. https://doi.org/10.1016/j.rser.2016.11.063

9. Zondag HA, de Vries DW, van Helden WGJ, van Zolingen RJC, van Steenhoven AA (2003) The yield of different combined PV-thermal collector designs. Sol Energy 74(3):253–269. https://doi.org/10.1016/S0038-092X(03)00121-X
10. Kostic LT, Pavlovic TM, Pavlovic ZT (2010) Influence of reflectance from flat aluminum concentrators on energy efficiency of PV/Thermal collector. Appl Energy 87(2):410–416. https://doi.org/10.1016/j.apenergy.2009.05.038
11. Budihardjo I, Morrison GL (2009) Performance of water-in-glass evacuated tube solar water heaters. Sol Energy 83(1):49–56. https://doi.org/10.1016/j.solener.2008.06.010
12. Chel A, Kaushik G (2018) Renewable energy technologies for sustainable development of energy efficient building. Alex Eng J 57(2):655–669. https://doi.org/10.1016/j.aej.2017.02.027
13. Infield D, Mei L, Eicker U (2004) Thermal performance estimation for ventilated PV facades. Sol Energy 76(1–3):93–98. https://doi.org/10.1016/j.solener.2003.08.010
14. Vokas G, Christandonis N, Skittides F (2006) Hybrid photovoltaic–thermal systems for domestic heating and cooling—a theoretical approach. Sol Energy 80(5):607–615. https://doi.org/10.1016/j.solener.2005.03.011
15. Chow TT, Pei G, Fong KF, Lin Z, Chan ALS, Ji J (2009) Energy and exergy analysis of photovoltaic–thermal collector with and without glass cover. Appl Energy 86(3):310–316. https://doi.org/10.1016/j.apenergy.2008.04.016
16. Kim J-H, Kim J-T, Kim J-H, Kim J-T (2012) The experimental performance of an Unglazed PVT collector with two different absorber types, the experimental performance of an Unglazed PVT collector with two different absorber types. Int J Photoenergy 2012(2012):e312168. https://doi.org/10.1155/2012/312168
17. Zondag HA (2008) Flat-plate PV-Thermal collectors and systems: a review. Renew Sustain Energy Rev 12(4):891–959. https://doi.org/10.1016/j.rser.2005.12.012
18. Xu X, Meyers MM, Sammakia BG, Murray BT (2012) Thermal modeling of hybrid concentrating PV/T collectors with tree-shaped channel networks cooling system. In: 2012 13th IEEE intersociety conference on thermal and thermomechanical phenomena in electronic systems (ITherm), May 2012, pp 1131–1138. https://doi.org/10.1109/itherm.2012.6231550
19. Musallam AT, Mohd YO, MohdHafidz R (2014) A review paper on PV/T combination flat plate collector system and design. World Essays J 1(1):1–5
20. Khelifa A, Touafek K, Benmoussa H, Tabet I, Adouane M (2014) Hot water system based on the hybrid solar collector photovoltaic/thermal PVT. In: 2014 15th international conference on sciences and techniques of automatic control and computer engineering (STA), December 2014, pp 537–540. https://doi.org/10.1109/sta.2014.7086668
21. Tang X, Quan Z, Zhao Y (2010) Experimental investigation of solar panel cooling by a novel micro heat pipe array. Energy Power Eng 02(03):171. https://doi.org/10.4236/epe.2010.23025
22. Shan F, Tang F, Cao L, Fang G (2014) Performance evaluations and applications of photovoltaic–thermal collectors and systems. Renew Sustain Energy Rev 33(Supplement C):467–483. https://doi.org/10.1016/j.rser.2014.02.018
23. Bellos E, Tzivanidis C (2018) A review of concentrating solar thermal collectors with and without nanofluids. J Therm Anal Calorim. https://doi.org/10.1007/s10973-018-7183-1
24. Jouhara H, Chauhan A, Nannou T, Almahmoud S, Delpech B, Wrobel LC (2017) Heat pipe based systems—advances and applications. Energy 128:729–754. https://doi.org/10.1016/j.energy.2017.04.028
25. Chow TT, He W, Ji J (2007) An experimental study of façade-integrated photovoltaic/water-heating system. Appl Therm Eng 27(1):37–45. https://doi.org/10.1016/j.applthermaleng.2006.05.015
26. Aste N, Del Pero C, Leonforte F, Manfren M (2016) Performance monitoring and modeling of an uncovered photovoltaic-thermal (PVT) water collector. Sol Energy 135:551–568. https://doi.org/10.1016/j.solener.2016.06.029
27. Brottier L, Bennacer R (2020) Thermal performance analysis of 28 PVT solar domestic hot water installations in Western Europe. Renew Energy 160:196–210. https://doi.org/10.1016/j.renene.2020.06.072

28. Braas H, Jordan U, Best I, Orozaliev J, Vajen K (2020) District heating load profiles for domestic hot water preparation with realistic simultaneity using DHWcalc and TRNSYS. Energy 201:117552. https://doi.org/10.1016/j.energy.2020.117552
29. Motahhir S et al Optimal energy harvesting from a multistrings PV generator based on Artificial Bee Colony algorithm. IEEE Syst. J 1–8. https://doi.org/10.1109/jsyst.2020.2997744
30. Eltamaly AM (2015) Performance of smart maximum power point tracker under partial shading conditions of photovoltaic systems. J Renew Sustain Energy 7(4):043141
31. Maghami MR, Hizam H, Gomes C, Radzi MA, Rezadad MI, Hajighorbani S (2016) Power loss due to soiling on solar panel: a review. Renew Sustain Energy Rev 59:1307–1316. https://doi.org/10.1016/j.rser.2016.01.044
32. Tanabe K (2009) A review of ultrahigh efficiency III-V semiconductor compound solar cells: multijunction tandem, lower dimensional, photonic up/down conversion and plasmonic nanometallic structures. Energies 2(3):504–530. https://doi.org/10.3390/en20300504
33. Hsu H-J, Hsu C-H, Tsai C-C (2013) The effect of bandgap graded absorber on the performance of a-Si1−xGex:H single-junction cells with μc-SiOx:H n-type layer. Int J Photoenergy. https://www.hindawi.com/journals/ijp/2013/364638/. Accessed 08 Feb 2019
34. Dupré O, Niesen B, De Wolf S, Ballif C (2018) Field performance versus standard test condition efficiency of tandem solar cells and the singular case of Perovskites/Silicon devices. J Phys Chem Lett 9(2):446–458. https://doi.org/10.1021/acs.jpclett.7b02277
35. Moussavou AAA, Raji A, Adonis M (2018) Impact study of partial shading phenomenon on solar PV module performance. In: 2018 international conference on the Industrial and Commercial Use of Energy (ICUE), pp 1–7
36. Zaraket J, Khalil T, Aillerie M, Vokas GA, Salame C (2017) The effect of electrical stress under temperature in the characteristics of PV solar modules. Energy Proc 119:579–601. https://doi.org/10.1016/j.egypro.2017.07.083
37. Warta W, Schubert MC, Rein S, Fertig F (2011) Impact of junction breakdown in multi-crystalline silicon solar cells on hot spot formation and module performance. In: 26th European Photovoltaic Solar Energy Conference and Exhibition, pp 1168–1178, October 2011. https://doi.org/10.4229/26theupvsec2011-2do.3.1
38. Löper P et al (2012) Analysis of the temperature dependence of the open-circuit voltage. Energy Proc 27:135–142. https://doi.org/10.1016/j.egypro.2012.07.041
39. Singh P, Ravindra NM (2012) Temperature dependence of solar cell performance—an analysis. Sol Energy Mater Sol Cells 101:36–45. https://doi.org/10.1016/j.solmat.2012.02.019
40. Wang J-C, Su Y-L, Shieh J-C, Jiang J-A (2011) High-accuracy maximum power point estimation for photovoltaic arrays. Sol Energy Mater Sol Cells 95(3):843–851. https://doi.org/10.1016/j.solmat.2010.10.032
41. Teo J, Tan R, Mok V, Ramachandaramurthy V, Tan C (2018) Impact of partial shading on the P-V characteristics and the maximum power of a photovoltaic string. Energies 11(7):1860. https://doi.org/10.3390/en11071860
42. Swart AJ, Hertzog PE (2016) Varying percentages of full uniform shading of a PV module in a controlled environment yields linear power reduction. J Energy South Afr 27(3):28–38
43. Arjyadhara P (2013) Analysis of solar PV cell performance with changing irradiance and temperature. Int J Eng Comput Sci 2(1):214–220
44. Jha AR (2009) Solar cell technology and applications. Auerbach Publications
45. Henry CH (1980) Limiting efficiencies of ideal single and multiple energy gap terrestrial solar cells. J Appl Phys 51(8):4494–4500. https://doi.org/10.1063/1.328272
46. Vázquez M, Rey-Stolle I (2008) Photovoltaic module reliability model based on field degradation studies. Prog Photovolt Res Appl 16(5):419–433. https://doi.org/10.1002/pip.825
47. Sze SM, Lee M-K (2012) Semiconductor devices: physics and technology, 3rd edn. Wiley, Hoboken, N.J
48. Haug A (1983) Auger recombination in direct-gap semiconductors: band-structure effects. J Phys C Solid State Phys 16(21):4159. https://doi.org/10.1088/0022-3719/16/21/017
49. Lindefelt U (1994) Heat generation in semiconductor devices. J Appl Phys 75(2):942–957. https://doi.org/10.1063/1.356450

50. Straube H, Wagner J-M, Breitenstein O (2009) Measurement of the Peltier coefficient of semiconductors by lock-in thermography. Appl Phys Lett 95(5):052107. https://doi.org/10.1063/1.3194156
51. Dhass AD, Lakshmi P, Natarajan E (2016) Investigation of performance parameters of different photovoltaic cell materials using the Lambert-W function. Energy Proc 90:566–573. https://doi.org/10.1016/j.egypro.2016.11.225
52. Bensalem S, Chegaar M (2013) Thermal behavior of parasitic resistances. Rev Energ Renouvelables 16(1):171–176
53. Ramaprabha R, Mathur BL (2017) A comprehensive review and analysis of solar photovoltaic array configurations under partial shaded conditions. Int J Photoenergy 2012. Accessed 02 August 2017. https://www.hindawi.com/journals/ijp/2012/120214/abs/
54. Duffie JA, Beckman WA (1991) Solar engineering of thermal processes, 2nd edn. Wiley-Interscience, New York
55. Kreith F, Manglik RM, Bohn MS (2010) Principles of heat transfer, 7 edn. Cengage Learning
56. Saadah M, Hernandez E, Balandin AA (2017) Thermal management of concentrated multi-junction solar cells with graphene-enhanced thermal interface materials. Appl Sci 7(6):589. https://doi.org/10.3390/app7060589
57. Tian H, Mancilla-David F, Ellis K, Muljadi E, Jenkins P (2012) A cell-to-module-to-array detailed model for photovoltaic panels. Sol Energy 86(9):2695–2706. https://doi.org/10.1016/j.solener.2012.06.004
58. Wilcox JR, Gray JL (2012) A distributed emitter model for solar cells: extracting a temperature dependent lumped series resistance. In: 2012 38th IEEE photovoltaic specialists conference, June 2012, pp 002122–002127. https://doi.org/10.1109/pvsc.2012.6318016

A Review on Vehicle-Integrated Photovoltaic Panels

Marwa Ben Said-Romdhane and Sondes Skander-Mustapha

Abstract Vehicular transport is considered as the most important origin of air pollution in cities. However, despite the commercial growing achievement of electric vehicles, there had been no detectable reduction in energy consumption and CO_2 emission, at least in a short-term scenario. Solar vehicles can be considered as an alternative to this problem. Indeed, they are considered as a restricted, but promising technology and they are slowly gaining the interest of several automotive companies and researchers. This is due to several factors such as the rapid rise of photovoltaic technologies due to decreasing cost and improvement of their efficiency in addition to the increasing development of electric vehicles considering their environmental friendliness and their reduced dependence on fossil fuels. This manuscript highlights various aspects, challenges, and problems for solar vehicle development. In fact, this chapter widely reviews vehicle-integrated photovoltaic panels where different power train architectures are highlighted. In addition, a review of different power structures of vehicle-integrated PV is exposed. Also, energy storage system solutions are detailed with possible recommendations. Furthermore, energy management systems for vehicle-integrated photovoltaic panels are discussed and evaluated.

Keywords Vehicle-integrated PV · Photovoltaic technologies · Energy storage system · Power converters · Energy management system

M. Ben Said-Romdhane (✉) · S. Skander-Mustapha
Université de Tunis El Manar, Ecole Nationale d'Ingénieurs de Tunis, LR11ES15, Laboratoire des Systèmes Electriques, 1002 Tunis, Tunisie
e-mail: marwa.bensaidromdhane@enit.utm.tn

S. Skander-Mustapha
e-mail: sondes.skander@enit.utm.tn

M. Ben Said-Romdhane
Université de Gabès, Institut Supérieur des Sciences Appliquées et de Technologie de Gabès, 6029 Gabès, Tunisie

S. Skander-Mustapha
Université de Carthage, Ecole Nationale d'Architecture et d'Urbanisme, 2026 Sidi Bousaid, Tunisie

S. Motahhir and A. M. Eltamaly (eds.), *Advanced Technologies for Solar Photovoltaics Energy Systems*, Green Energy and Technology,
https://doi.org/10.1007/978-3-030-64565-6_12

1 Introduction

The automotive sector has been developed and prospered thanks to abundant, cheap, and energy-efficient oil. For more than a century, its domination was unchallenged. Likewise, for decades, the rise in the power of vehicles was made without taking into account the environmental dimension [1, 2].

Then the tide turned. The time of scarcity and the high price of raw materials has arrived, due to an uninterrupted rise in world demand and the realization that reserves may be depleted. The time to take environmental protection into account has also come, putting oil and its discharges on the spot [3, 4, 5].

It was, therefore, necessary to innovate and find solutions. Subject to increasingly stringent regulations, manufacturers have improved the performance of thermal vehicles, reducing fuel consumption and polluting emissions. At the same time, the automotive industry has invested in developing new engines like hybrid cars and 100% electric cars [2, 6, 7].

Since the invention of photovoltaic cells, engineers around the world have started to explore various prototypes of solar cars. These electric cars use batteries that can be recharged by natural light. When there is insufficient natural light, the car uses the energy stored in the batteries [8]. They are based on the concept that an integrated PV system supplies an electric power train. The electrical energy extracted from solar energy is transformed on motion, so there is no need for the combustion process [7, 9, 10, 11].

One of the biggest hurdles that need to be addressed is the current power of cars, which is limited by the efficiency of photovoltaic cells. The ability of batteries to store a large amount of energy causes also problems. For the vast majority of solar cars, manufacturers prefer to circumvent these problems by reducing the energy requirements of the vehicles, using lightweight materials to reduce weight, as well as an aerodynamic design to achieve less air resistance when in motion [12, 13, 14, 15].

This chapter highlights various aspects, challenges, and problems for solar vehicle development. It is organized as follows, first, in Sect. 2, the electrical vehicle's classification and terminologies are presented. Then, in Sect. 3, the challenges of solar vehicles including emission reduction, as well as the problems of the electric vehicle charging station are detailed. Section 4 present the issues that affect the solar vehicle's performance. These issues are mainly the fast irradiance variability and partial shading of the PV cells, the limited surface area for PV panels, the operating requirements, and the driving constraints in urban traffic. Section 5 describes the vehicle-integrated PV powertrain architecture which is mainly divided into two groups: all-electric architecture and hybrid electric architecture. The most used hybrid electric vehicles are parallel hybrid, series hybrid, series-parallel hybrid, and complex hybrid. Section 6 presents the global power structure of the vehicle's integrated photovoltaic panels. It includes the electric vehicle drives, the power converters in addition to the energy storage system. Finally, Sect. 7 reviews the control strategies and the energy management systems for electric vehicle applications.

2 Electrical Vehicles Classification and Terminologies

The vehicle-integrated PV (VIPV) are vehicles that incorporate PV cells on the roof and body of the vehicle with additional power converters to charge batteries. The PV system is considered as the main source and batteries as an auxiliary source. Based on the classification of electric vehicles (EV) presented in [7], a classification of Vehicle-integrated PV is presented in Fig. 1. Indeed, VIPV can be classified into two groups: hybrid electric vehicles (HEV) and all-electric vehicles (AEV).

- The group of HEV incorporates several propulsion motors (internal combustion and electrical motors [16]) and includes plug-in hybrid electric vehicles (PHEV) which incorporate batteries and plugs for external charging and Hydrogen fuel cell plug-in hybrid vehicles.
- The group of AEV comprises fuel cell electric vehicles (FCEV) and battery electric vehicles (BEV).

PV integrated with EV can be used in varying degrees depending on installation characteristics; it can be just useful for supplying vehicle auxiliary devices such as fan, audio players, etc., or feeding air conditioning systems. But the final object is to charge batteries, this can be done while parking or driving as exposed in Fig. 2 [7, 12, 16]. There are several challenges of VIPV, such as CO_2 emission reduction, no longer need batteries charging stations, etc. But VIPV also faces several problems such as the fast irradiance variability and partial shading of the PV cells, the limited surface area for PV panels, as will be detailed in the next section.

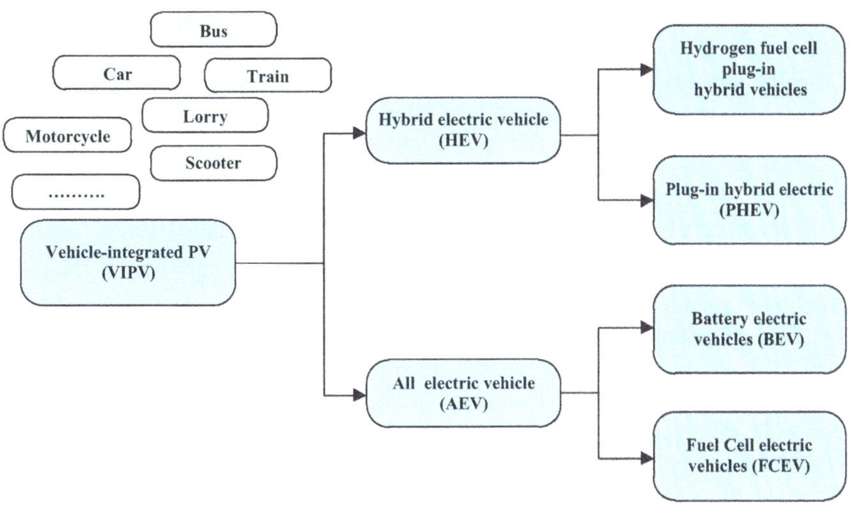

Fig. 1 Classification of vehicle-integrated PV

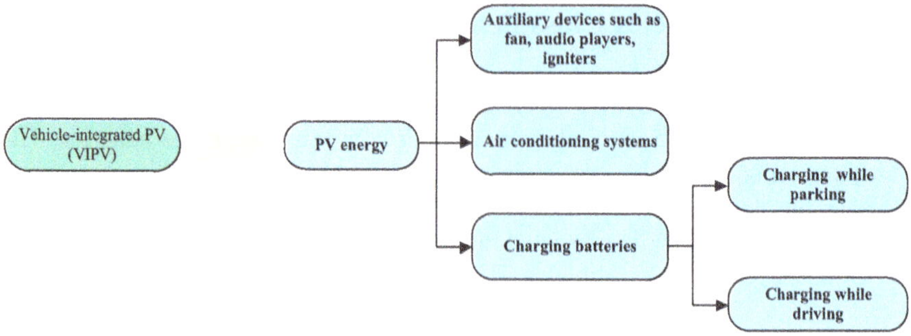

Fig. 2 Use of PV energy in VIPV

3 Challenges of Solar Vehicles

3.1 Emissions Reduction

This challenge concerns environmental pollution taking into account the vehicle's CO_2 emission. In 2015, the transport sector has delivered about 22.9% of total world CO_2 emission [17], in 2018, about 24% [18], and with current circumstances, it is expected to increase to 60% by 2050 [18]. EVs are developed to take the place of the conventional ones gradually, due to their energy-savings and emissions reduction [3]. However, despite their growing commercial achievement, there had been no detectable reduction in CO_2 emission, at least in a short-term scenario [19]. The VIPV presents a long term solution for this issue [20]. The results presented in [21] confirm that the combination generator battery diminishes the environmental effect of the medium-sized ship, and in [4] the authors confirm the necessity to introduce renewable energy to achieve the actual reduction of CO_2.

3.2 Bypassing the Problems of the EV Charging Station

EV charging stations present several problems. Indeed, they amplify the electric load. Consequently, potentially intensify the peak load or produce other peaks. Moreover, EV charging stations can boost load side uncertainties, overload distribution grid devices which reduce their lifetime, augment power loss, and induce important voltage deviations compared to their nominal value [5, 22, 23, 24]. Integrating PV panels in the vehicle will allow the charging battery autonomously. This means no longer needing charging stations or at least reducing their use as much as possible.

4 Issues Affecting Solar Vehicles Performance

4.1 The Fast Irradiance Variability and Partial Shading of the PV Cells

The fact that vehicles are in continuous motion generates variable irradiance, mainly caused by the partial shading of the photovoltaic panels [6] due to the structures close to the road such as poles, chimneys, raised buildings, etc. Consequently, a large changeability in the DC voltage of the solar panel is recorded and PV array efficiency is decreased [8, 16].

4.2 Limited Surface Area for PV Panels

The variable solar irradiance added to the vehicle's curved shape has a big influence on the resulting energy. To overcome this problem, it is imperative to measure and model solar irradiance for the vehicle. Some studies suggest applying the correction of the curved surface of the PV modules in order to take into account the random distribution of shading things and vehicle direction. In [13], the authors propose to install five pyranometers in different axes during one year to deduce the solar irradiance model.

4.3 Operating Requirements

The weight, cooling process, and power conversion are fundamental points to be considered when integrating PV structure in vehicles. Alternative carbon-fiber-reinforced plastic structures were investigated in [25] by finite elements using static and modal analyses, to evaluate numerous proposed approaches considering these criteria: natural frequencies, deformations, flexural stiffness, torsional stiffness, and heat exchange plane. A roof section was tested to verify the theoretical model. An important enhancement compared to the pre-existing solar roof was noticed. Light composites are a good option for solar cars, because the lighter the vehicle, the less energy is used to overcome inertia [26]. In [27, 28, 29], authors investigate a composite monocoque chassis in order to ensure lightweight solar car, for example, in [27] the authors propose an iterative finite element analysis process.

4.4 Driving Constraints in Urban Traffic

Urban driving is characterized by transient traffic situations. Which induces frequent starts and stops [30, 31]. Consequently, the electric vehicle presents a significantly random and fluctuant current. The element most concerned with this problem is the storage item. Indeed, it must be able to follow these fluctuations and adapt to this behavior [32, 33]. This problem is, therefore, not specific to VIPV but it concerns EV in general. But it must be taken into account for VIPV especially if the battery charge source is limited to PV modules.

5 Vehicle-Integrated PV Powertrain Architectures

5.1 All-Electric Vehicle

The all-electric VIPV powertrain architecture is presented in Fig. 3. As it is shown in this figure, this kind of vehicle employs electric power as the only source to move the vehicle.

5.2 Hybrid Electric Vehicle

Nowadays, the most common hybrid VIPV architecture includes the electric motor and the internal combustion engines (ICE). The combination of these energy converters allows to have several possible configurations of the powertrain. The most used ones are parallel hybrid, series hybrid, series-parallel hybrid, and complex hybrid as shown in Fig. 4 [34–37].

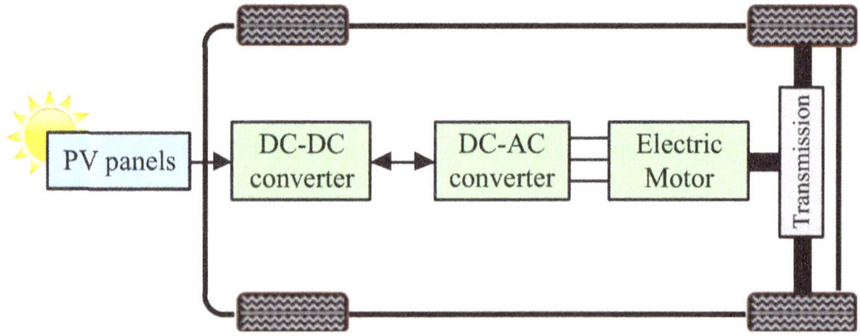

Fig. 3 All-electric vehicle powertrain architecture

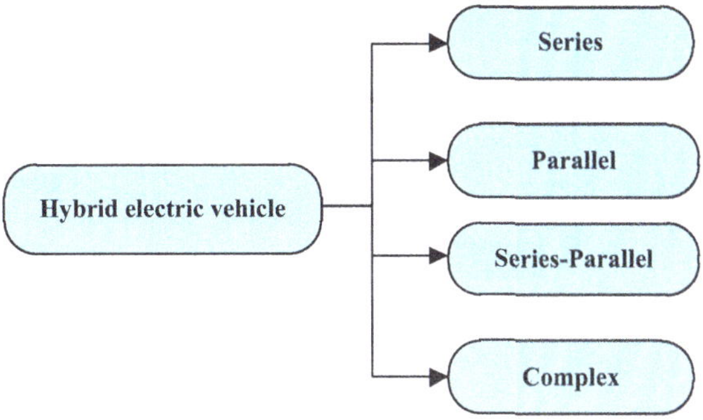

Fig. 4 Hybrid VIPV powertrain architecture classification

5.2.1 Series Hybrid VIPV

The series hybrid VIPV (SHVIPV) powertrain configuration (Fig. 5) is composed of an ICE, a generator, and an electric motor [38]. In this configuration, the SHVIPV is powered solely by the electric motor which can be supplied either from the battery or from the ICE generator unit, or even both. In this case, the ICE can't directly power the vehicle since it has no mechanical link with the traction load. The electric current produced by the ICE can be either provided to the electric motor or stored in the batteries. In case more power is needed, the electric motor acquires energy from both the ICE and the batteries. The advantages of the series configuration are: (i) Increased flexibility due to no common interaction between ICE and electrical motor [16]; (ii) efficiency in driving cycles that require frequent stops and start [39]. The shortcomings of the SHVIPV powertrain configurations are: (i) high losses [16]; (ii) expensive configuration due to the need for a generator [39, 40].

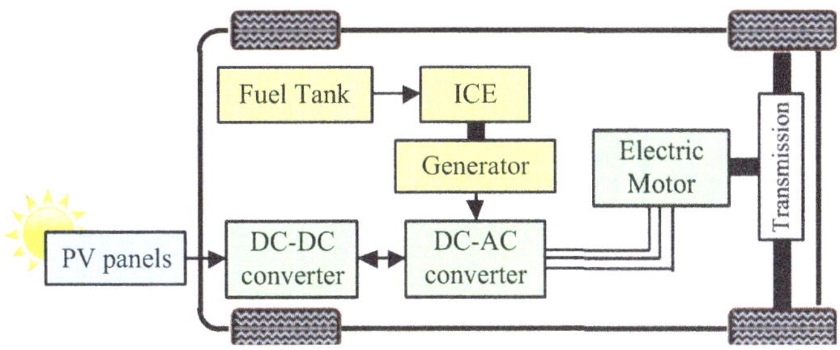

Fig. 5 Series hybrid VIPV powertrain architecture

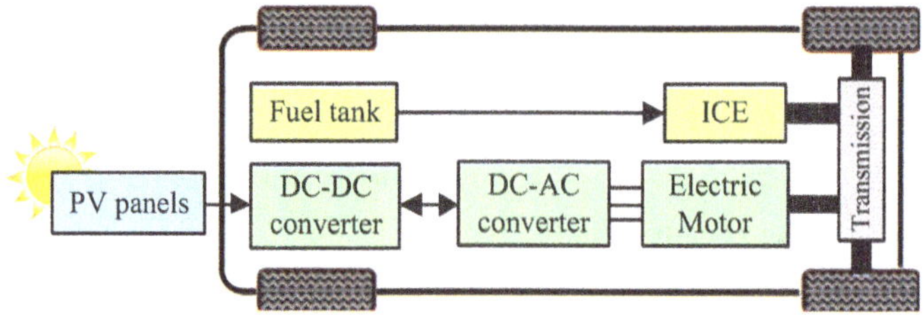

Fig. 6 Parallel hybrid VIPV powertrain architecture

5.2.2 Parallel Hybrid VIPV

The parallel hybrid powertrain configuration (Fig. 6) is composed of an ICE and an electric motor [7, 40]. The electric motor and the ICE are coupled together by a mechanical device. Consequently, during low traction power demand, they can individually propel the vehicle and during high power demand, they jointly propel it. Moreover, this configuration has an electric traction motor that rolls the wheels and can recover part of the braking energy, in order to recharge the batteries (regenerative braking) or to help the ICE in the conditions of acceleration. This configuration makes it possible to reduce the size of the ICE and the electric motor. This helps reduce consumption while maintaining good performance [36].

5.2.3 Series-Parallel Hybrid VIPV

The series-parallel hybrid powertrain architecture (Fig. 7) joins the advantages of both series and parallel architectures. In fact, the ICE can supply the electrical motor and thanks to a generator it can also charge the battery. Although this configuration combines the advantages of series and parallel configurations, it is relatively expensive and more complicated and requires a very complex control system [7, 39, 41].

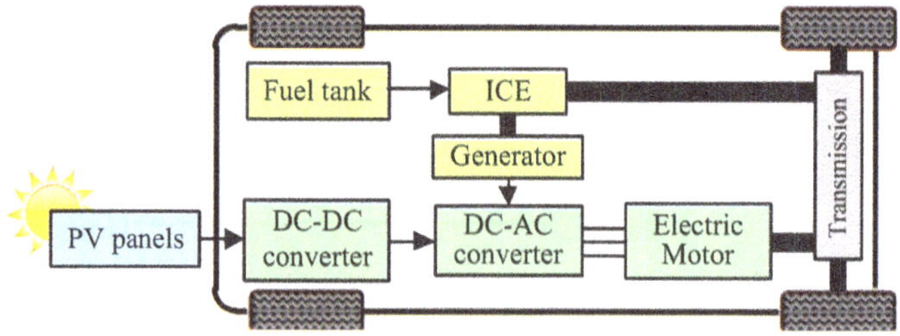

Fig. 7 Series-parallel hybrid VIPV powertrain architecture

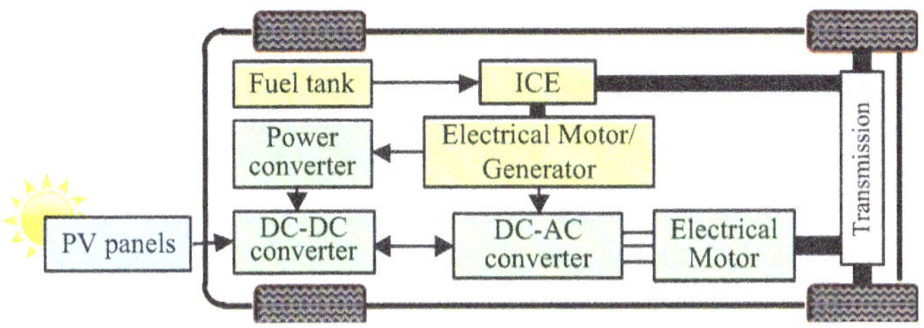

Fig. 8 Complex hybrid VIPV powertrain architecture

5.2.4 Complex Hybrid VIPV

The complex hybrid powertrain configuration (Fig. 8) is similar to a series-parallel powertrain configuration. The difference is that it uses more complicated architectures of many motors and generators. This makes this configuration more controllable and reliable than the other configurations [7, 39].

6 Global Power Structure

6.1 Electric Vehicle Drives

The most essential component in the Electric vehicle is, of course, the motor. Its fundamental characteristics are as follows [42]:

- High torque and power density to pull the load;
- Large speed range;
- High efficiency for a wide variation of torque and speed;
- Extensive ability to work in constant-power;
- Wide ability of the torque for electric startup and raised area climbing;
- Elevated alternating overload ability for overtaking;
- Small acoustic noise

Many classifications of EV drives are given in literature [2, 37, 42, 43]. The main two groups are AC and DC motors including a large variety as Brushless DC Motor (BLDCM) [6, 20, 44, 45], regenerative brushless DC motor [9], permanent magnet brushless DC motor drives (PMBDCM) [20, 44], Induction motor (IM) [2], switched reluctance motors (SRM) [45], permanent magnet synchronous motors (PMSM) [45], and permanent magnet hybrid motor drives [2]. Among these, PMSM is appropriate in terms of power density, reliability, and efficiency [20]. BLDCM is frequently included in EV due to their elevated efficiency and power density in addition to their great starting torque, and their better performance regarding noise [45]. As to SRM, they are considered as an attractive option due to their reduced material costs, high efficiency, and simple control algorithm [46]. The structures integrating two types of motors are also present on the market, such as HEV driven by an Internal Combustion Engine and a PMSM for this structure PV system and energy storage devices supply the PMSM [47]. Furthermore, research regarding EV drives is very diverse. In [48], the authors focus on the EV driving range parameters. In [49], the authors investigate the use of just one electric machine which switches between the two modes: motor and generator. In [50], the issue is to resolve the straight-line driving stability question.

6.2 *Power Converter Structures*

Highly developed technology of power converters has an important impact on VIPV advancement in terms of energy-saving and control efficiency. The general configuration of VIPV incorporates two major power converters units which are DC-DC and DC-AC converters. In fact, AC motors used in VIPV are fed by DC-AC converters which are fed by DC-DC converters. Figure 9 presents a classification of the DC-AC and DC-DC power converters integrated with the VIPV.

6.2.1 DC-AC Converter Topologies

The bidirectional DC-AC converter is an essential element for VIPV. It is used to convert the DC power from the supercapacitors, the fuel cell, the battery, or their combination into AC power that will be provided to the electric motor. The most used DC-AC converters topologies in VIPV are impedance source converter (ZSI), current source inverter (CSI), and voltage source inverter (VSI). The ZSI is considered

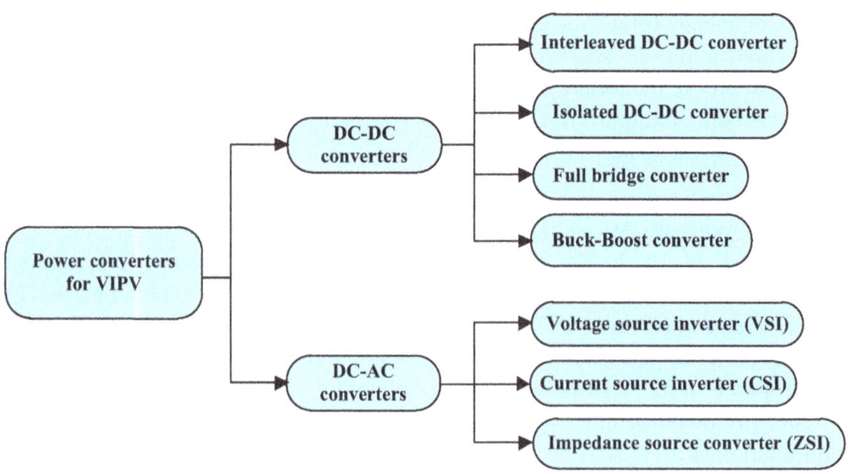

Fig. 9 Vehicle-integrated PV power converters classification

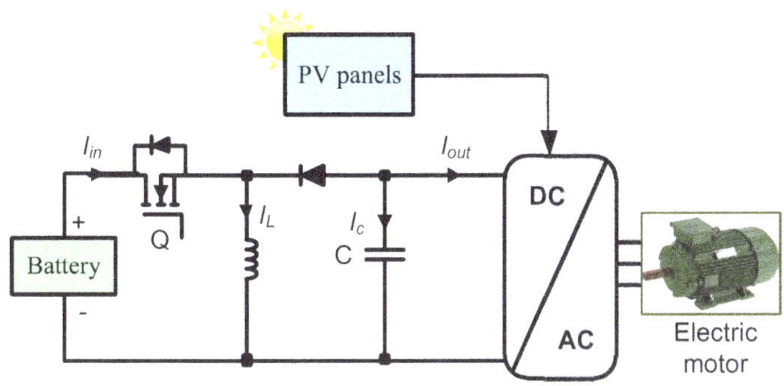

Fig. 10 DC-DC buck-boost converter

to be one of the most promising power electronics converter topologies suitable for motor drive applications. For VIPV, the CSI is employed for the speed control of AC motors, and the VSI is characterized by its multiple motor controls, as well as its good speed range [2, 10].

6.2.2 DC-DC Converter Topologies

The different VIPV powertrain architectures include at least one DC-DC converter. This converter is employed in order to interface between the supercapacitors, the fuel cell, the battery, or their combination to the DC-link. This converter is an electric circuit used to convert a source of direct current (DC) from one voltage level to another. It can be unidirectional or bidirectional. The bidirectional DC-DC converter is very useful for vehicles mainly in regenerative braking since it can move power in either direction. For VIPV, several types of DC-DC converters have been proposed in the literature. Among which, we can cite: boost, buck, full-bridge, isolated DC-DC converter, interleaved DC-DC converter, etc. [10, 51].

Buck-Boost Converter

A buck-Boost converter (Fig. 10) is a power converter which produces a DC voltage that can be either higher or lower than the input voltage [52, 53].

Full Bridge Converter

The full-bridge converter (Fig. 11) is composed of three stages: a DC-AC converter, a high-frequency transformer, and an AC-DC converter [52, 54]. This converter is employed to overcome the drawbacks of the boost converter which are mainly the

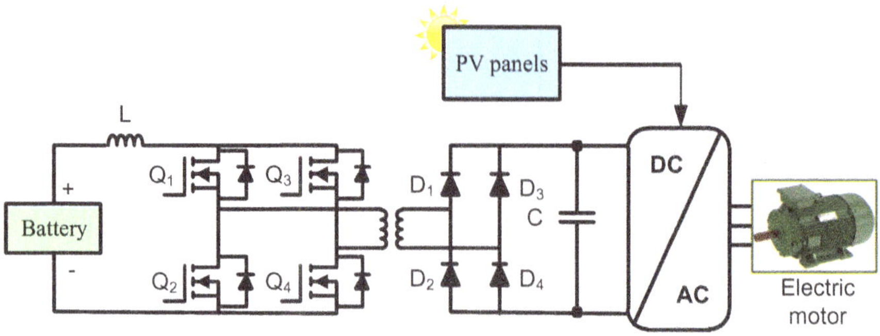

Fig. 11 Full bridge converter

high voltage and current ripples, no electrical isolation, as well as the great weight and volume [55–57].

Isolated DC-DC Converter

In this converter, a high-frequency transformer is employed in order to ensure that the output is completely isolated from the input [52, 58, 59]. Many studies propose this converter for electrical vehicle applications. In [60], a high-frequency isolated bidirectional DC-DC converter is suggested. The proposed converter is based on the grouping of an H-bridge, a three-level half-bridge, and a three-phase full-bridge topology, the voltage rise from 24 V DC to 144 V DC. In [61], the authors suggest using GaN in an isolated step-down DC-DC converter, and the voltage rises from 13.6 to 200 V.

Interleaved DC-DC Converter

The interleaved DC-DC converter (Fig. 12) is a good option to interface the low voltage of the ESS with the DC high voltage of the VIPV. It allows reducing voltage stress, as well as the size of the input filter. However, its elevated number of magnetic cores presents an obstacle for VIPV application given the size constraints. Some researchers suggest a modified configuration of this converter to overcome this problem, for example, in [62], an integrated interleaved ultra-high step-up DC-DC converter incorporating dual coupled-inductors is proposed. In literature, many configurations of the interleaved DC-DC converter are proposed. In [63], the authors propose a system based on three-phase interleaved topology, and in [64], a two legs topology is proposed.

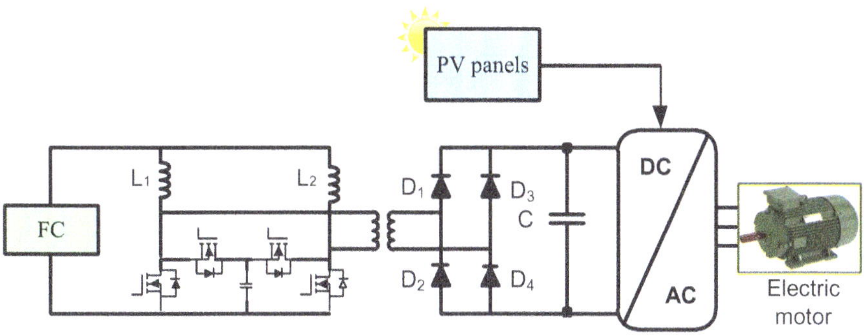

Fig. 12 Interleaved boost DC-DC converter

6.3 *Energy Storage System Solution*

The energy storage system (ESS) presents a key component for VIPV. To select an adequate solution, many items should be checked, such as the safety, the size, the cost, as well as the management system, etc. Nowadays EVs are mostly supplied by lithium-ion batteries which have the greatest grouping of best properties concerning energy density and cycle life [65–67]. Nevertheless, the solution to mixing several storage devices like batteries, supercapacitors, and fuel cells presents a promising solution for VIPV. Figure 13 exposes the most used ESS for EV, indeed batteries, supercapacitors, and fuel cells, as well as hybrid solutions are frequently adopted for EV [68].

Based on the spider diagram of different ESS characteristics given by [69] a comparative graphic is exposed in Fig. 14. Such a plot will permit the identification of the strengths and weaknesses of some ESS.

6.3.1 Batteries

The use of batteries for EVs has evolved from lead-acid to nickel and presently to lithium, seeking in all this to reach high specific energy, less chemical leakage, and better temperature performance [68, 70–72]. The preference of lithium batteries for EV is due to their elevated energy efficiency and power density, in addition to their fast charging ability and small self-discharge rate [20]. On the other hand, they have a wide range of working temperatures, as well as a compact and lighter weight [65–67]. It is to note that, lithium battery encompasses an extensive variety of chemical substances (LiFePO, Li–titanate, Li–S…) [68, 70, 71]. Safety worry is a major difficulty that encounters manufacturers and users. In fact, the largest detriment of lithium-ion batteries is thermal runaway [66]. The Li-Ion are also characterized by major battery aging factors [69, 73]. Authors in [74] propose some recommendations to reach the maturity of lithium batteries. They suggest optimizing the lithium electrode material to enhance the management expertise while taking into account the

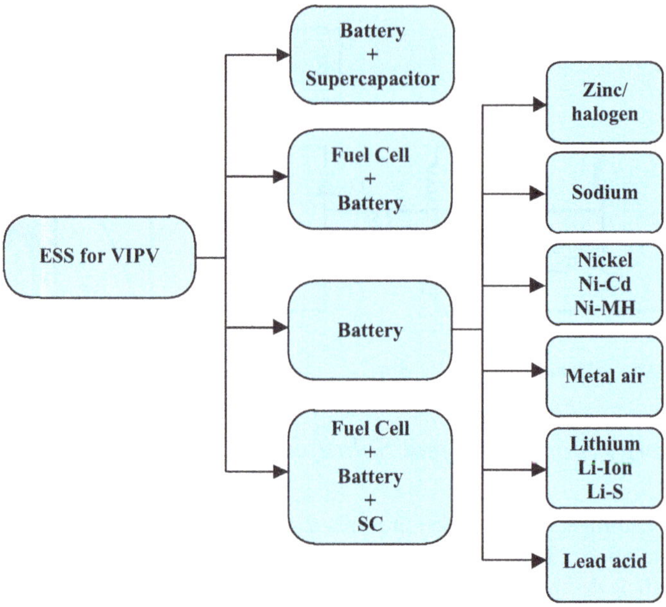

Fig. 13 Energy storage systems the most used for electrical vehicles

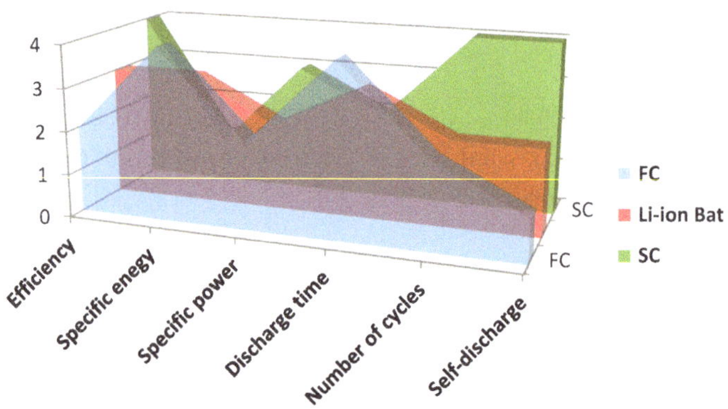

Fig. 14 Comparative study of different ESS

cost, the maintenance in addition to the life cycle. It should also reduce the memory effect and promote second-hand battery employment.

6.3.2 Hybrid Energy Storage System

For VIPV, to get the best from each storage device, hybrid ESS can be considered as an excellent alternative. This solution requires among others an adequate energy

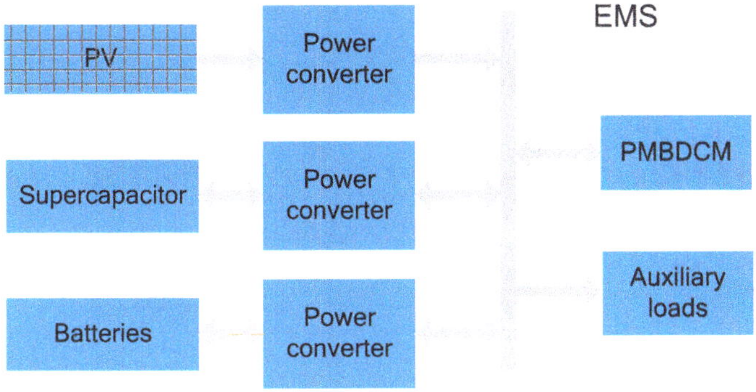

Fig. 15 Example of VIPV equipments including hybrid ESS based on battery and SC

management system (EMS) to maximize PV power and simultaneously respect the dynamic of each storage element. Frequency separation technique is the most popular method to ensure this purpose [46, 75, 76].

Hybrid ESS Based on Batteries and Supercapacitors

Concerning the load mission, the vehicle presents a specific profile. Indeed the peak current can reach 6 to 10 times the nominal current during each startup [77]. Supercapacitors (SC) are specified by their higher power density and lower energy density. These characteristics are complementary to those of batteries. Which supports the solution of the energy storage system based on the association of batteries and SCs [44, 71, 78]. For hybrid ESS, the function of the SC is to smooth out the energy supplied by batteries. The battery supplies steady-state energy and SC supplies the peak power [77, 78] which boost the efficiency of regeneration and support the EV acceleration [71, 78]. Consequently, it reduces battery current fluctuation and increases its lifetime, and leads to meet the limit space and weight restrictions, in addition to better vehicle dynamic [69]. Figure 15 presents an example of the integration of hybrid ESS for VIPV [44, 71]. As demonstrated in [44], the hybrid ESS based on batteries and supercapacitors monitored by an appropriate energy management algorithm is able to reduce the losses of the EV DC motor starter, further, a total disconnection of the batteries is possible for the duration of the regenerative braking.

Hybrid ESS Based on Batteries and FC and Eventually SCs

FCs present several variants, such as proton exchange membrane FC (PEMFC), alkaline FC, and solid oxide FC. The more suitable variety for automotive propulsion applications is the PEMFC [2]. This does not alter the fact that the PEMFC presents a major disadvantage, indeed they have slow responses to ensure the power demand

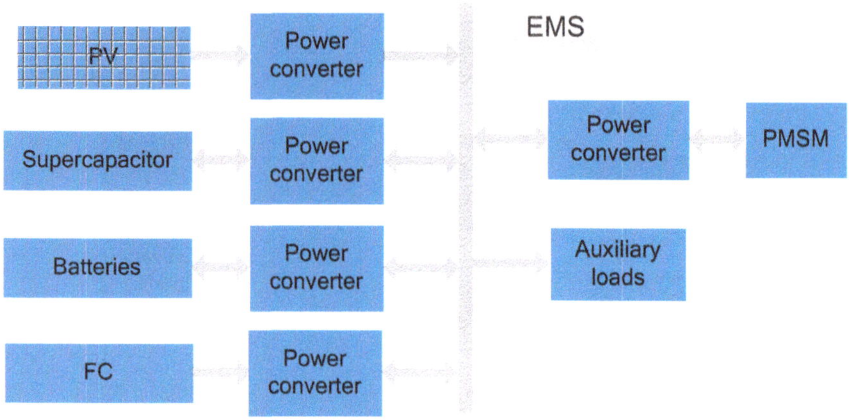

Fig. 16 Example of VIPV equipment incorporating hybrid ESS based on batteries, FC, and SC

of vehicles [46, 79, 80] To overcome this, in general, FC is associated with other types of storage devices. The use of hybrid ESS based on batteries, FC, and SCs has been investigated with accomplishment in several areas of automobile sector research and it is expected to be valuable as an onboard power generation for EV in the near prospect [69, 81–83]. For VIPV, PV system and FC are considered as primary power sources, and batteries with SCs are considered as secondary sources. Figure 16 presents an example of hybrid ESS that power VIPV, all storage devices are connected to the DC bus in parallel via their DC-DC converter, The motor is connected to the same bus via a DC-AC converter, This design presents additional flexibility in the control of the DC bus voltage that must be preserved stable during function [47, 69]. According to [84], the general objective for vehicles including PEMFC is to reach the cost, the durability, and the performance of conventional automobiles. Authors confirm that both performance and durability of the main FC stack components are considerably enhanced in the previous decade and it is currently possible that the cost and the sustainability purposes will be achieved during the upcoming decade.

7 VIPV Control and Energy Management System

For the VIPV application, the energy management system (EMS) is introduced to meet all power requirements while making the most of the PV panels. The EMS takes into account many other constraints such as reducing the overall weight of the vehicle and extending the life of energy storage devices. Fixed targets and constraints depend mainly on the adopted storage technologies. The EMS is based on several control levels with a wide variety of algorithms [85]. In literature, diverse control strategies are exposed, but in general nonlinear controllers have better performances, because these controllers take into account the model parameters variation which enlarges the operation range and guarantees the system overall stability [86]. For example, fuzzy techniques added to artificial intelligence are commonly proposed for EV [87]. Results presented in [88] demonstrate a reduction of fuel consumption from 0.46% to 3.39% when applying EMS based on the fuzzy technique. Lyapunov and sliding mode controllers are also proposed in the literature [86]. Moreover, many studies propose predictive algorithms for VIPV. In [89], deadbeat predictive controllers are investigated to control a bidirectional three-level cascaded converter connected to the used hybrid energy storage system. The control ensures the power management between PV and energy storage devices in addition to control the DC bus voltage. In [90], the MPC predictive controllers are used, authors propose a hierarchical control process through the virtual droop control. On another side, the Maximum Power Point Tracking (MPPT) algorithm including its different varieties is typically used for this application [91]. Indeed, the VIPV presents a continuous moving partial shading situation which includes a rapid variation of the irradiation applied to the vehicle solar panels. In [11], the authors investigate a modified incremental conductance MPPT process in order to better follow quick-shifting irradiation parameters. In [14], a fast MPPT algorithm is proposed and then compared to conventional P&O MPPT one [75]. In [15], a fuzzy logic based MPPT approach is considered. In [92], the authors propose the MPPT technique to control the proton exchange membrane fuel cell integrated into the EV. The proposed control is based on a radial basis function network algorithm. Furthermore, the Bandwidth allocation technique (Fig. 17) is commonly used to synchronize the hybrid power supply of the electrical vehicle [47, 93, 79]. This technique is based on the principle of respecting the specific dynamics of each component. In general, supercapacitors are dedicated to supply pulse demand, fuel cell and batteries provide the rest according to the adopted strategy.

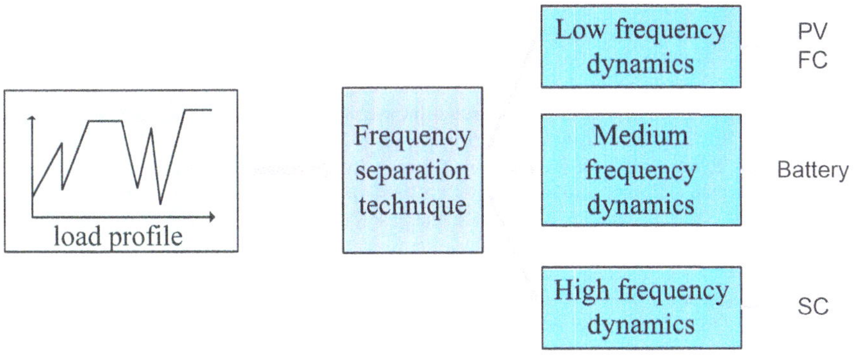

Fig. 17 Example of Frequency separation technique applied to VIPV

8 Conclusion

With the rapid expansion of the world's population, production, and consumption, the demand for transport has registered significant growth. Then, driving in city traffic induces repeated starts and stops which causes additional consumption of fuel and consequently less efficiency. In addition to gas emission, the VIPV presents a solution to all these problems. Actually, the attention of car manufacturers for vehicles incorporating photovoltaic panels remains ambiguous and still does not respond to large production. They are facing several obstacles as global size reduction, batteries location, luggage compartment, the variability of PV power, etc. But with the scientific and industrial developments in PV panels, control algorithms, and storage system devices, the VIPVs are expected to grow rapidly in the coming years.

Acknowledgements This work was supported by the Tunisian Ministry of Higher Education and Scientific Research under Grant LSE-ENIT-LR 11ES15.

References

1. Minak G, Lukovic M, Maglio S, Kojic S (2019) Toward a sustainable mobility a solar vehicle for a new quality of life. In: IOP conference series: materials science and engineering, p 659
2. Krithika V, Subramani C (2017) A comprehensive review on choice of hybrid vehicles and power converters, control strategies for hybrid electric vehicles. Int J Energy Res
3. Pang W, Yu H, Zhang Y, Yan H (2019) Solar photovoltaic based air cooling system for vehicles. Renew Energy 130:25–31
4. Abas AEP, Yong J, Mahlia TMI, Hannan MA (2019) Techno-economic analysis and environmental impact of electric vehicle. IEEE Access 7:98565–98578
5. Domínguez-Navarro JA, Dufo-López R, Yusta-Loyo JM, Artal-Sevil JS, Bernal-Agustín JL (2019) Design of an electric vehicle fast-charging station with integration of renewable energy and storage systems. Int J Electr Power Energy Syst 105:46–58
6. Al-Saud Mamdooh, Eltamaly Ali M, Mohamed Mohamed A, Kavousi-Fard Abdollah (2019) An intelligent data-driven model to secure intravehicle communications based on machine learning. IEEE Trans Industr Electron 67(6):5112–5119
7. Siang Fui T, Wei Tan C (2013) A review of energy sources and energy management system in electric vehicles. Renew Sustain Energy Rev 20:82–102
8. Maddukuri SVPK, Visvakumar A, Renduchintala UK (2016) An intelligent closed loop single-switch DC/DC converter with high voltage step-up ratio for roof-mounted solar cells electric vehicle. IEEE international conference on power electronics, drives and energy systems, pp 1–6. Trivandrum
9. Koyuncu T (2017) Practical efficiency of photovoltaic panel used for solar vehicles. In: IOP conference series: earth and environmental science, p 83
10. ElMenshawy M, Massoud A, Gastli A (2016) Solar car efficient power converters' design. In: 2016 IEEE symposium on computer applications & industrial electronics (ISCAIE)
11. Bhattacharya S, Samanta S (2019) Modified incremental conductance MPPT algorithm for very fast changing atmospheric condition for solar electric vehicle application. In: 2019 IEEE 16th India Council International Conference (INDICON), Rajkot, India, pp 1–4
12. Bhatti AR, Salam Z, Aziz MJBA, Yee KP, Ashique RH (2016) Electric vehicles charging using photovoltaic: Status and technological review. Renew Sustain Energy Rev 54:34–47

13. Araki K, Ota Y, Yamaguchi M (2020) Measurement and modeling of 3D solar irradiance for vehicle-integrated photovoltaic. Appl Sci 10(3):872
14. Raizada S, Verma V (2018) Step up gain converter with fast MPPT control under moving partial shading for train rooftop PV-DC-μG. In: 2018 IEEE international conference on electrical systems for aircraft, railway, ship propulsion and road vehicles & international transportation electrification conference (ESARS-ITEC), Nottingham, pp 1–6
15. Bendjedia B, Rizoug N, Boukhnifer M, Bouchafaa F (2017) Intelligent energy management of a multisource power supply for electric vehicle application. In: 2017 IEEE vehicle power and propulsion conference (VPPC), Belfort, pp 1–6
16. Sameeullah M, Chandel S (2016) Design and analysis of solar electric rickshaw: a green transport model. In: 2016 international conference on energy efficient technologies for sustainability (ICEETS), Nagercoil, pp 206–211
17. Shrivastava P, Alam MS, Asghar MSJ (2019) Design and techno-economic analysis of plug-in electric vehicle-integrated solar PV charging system for India. In: IET Smart Grid
18. https://www.planete-energies.com/en/medias/close/global-transportation-sector-co2-emissions-ise
19. Rizzo G, Naddeo M, Pisanti C (2018) Upgrading conventional cars to solar hybrid vehicles. Int J Powertrains 7(1/2/3):249
20. Mohan K, Sankaranarayanan S, Devi Prasad SS, Sivasubramaniam V, Sairam V Solar powered Hybrid vehicle. In: IOP conference series: materials science and engineering, vol 390, p 012102
21. Kim K, Park K, Lee J, Chun K, Lee S (2018) Analysis of battery/generator hybrid container ship for CO_2 reduction. IEEE Access 6:14537–14543
22. Gan L, Topcu U, Low SH (2013) Optimal decentralized protocol for electric vehicle charging. IEEE Trans Power Syst 28(2):940–951
23. Wang X, Yuen C, Hassan NU, An N, Wu W (2017) Electric vehicle charging station placement for urban public bus systems. IEEE Trans Intell Transp Syst 18(1):128–139
24. Shaaban MF, Mohamed S, Ismail M, Qaraqe KA, Serpedin E (2019) Joint planning of smart EV charging stations and DGs in eco-friendly remote hybrid microgrids. IEEE Trans Smart Grid 10(5):5819–5830
25. Pavlovic A, Sintoni D, Fragassa C, Minak G (2020) Multi-objective design optimization of the reinforced composite roof in a solar vehicle. Appl Sci 10:2665
26. Camargo FV, Giacometti M, Pavlovic A (2017) Increasing the energy efficiency in solar vehicles by using composite materials in the front suspension. Sustain Design Manufact 68
27. Denny J, Veale K, Adali S, Leverone F (2018) Conceptual design and numerical validation of a composite monocoque solar passenger vehicle chassis. Int J Eng Sci Technol 21:1067–1077
28. Minak G, Brugo TM, Fragassa C, Pavlovic A, Camargo FV, Zavatta N (2019) Structural design and manufacturing of a cruiser class solar vehicle. J Vis Exp 143
29. Howell E, Neal D, Kieffer D (2018) Changing the paradigm of transportation: lightweight composites used in solar car in intercollegiate competition. Reinf Plast 62(4):190–193
30. Manzie C, Watson H, Halgamuge S (2007) Fuel economy improvements for urban driving: Hybrid versus intelligent vehicles. Transp Res Part C Emerg Technol 15(1):1–16
31. Fernández RA, Caraballo SC, López FC (2019) A probabilistic approach for determining the influence of urban traffic management policies on energy consumption and greenhouse gas emissions from a battery electric vehicle. J Clean Product
32. Slouma S, Skander Mustapha S, Slama Belkhodja I, Orabi M (2015) An improved simple fuel cell model for energy management in residential buildings. J Electric Syst 11(2):154–159
33. Jaafar A, Akli CR, Sareni B, Roboam X, Jeunesse A (2009) Sizing and energy management of a hybrid locomotive based on flywheel and accumulators. IEEE Trans Veh Technol 58(8):3947–3958
34. Rizzo G, Arsie I, Sorrentino M (2010) Hybrid solar vehicles, solar collectors and panels, theory and applications. IntechOpen
35. Indira D, Venmathi M (2020) A comprehensive survey on hybrid electric vehicle technology with multiport converters. In: Emerging trends in computing and expert technology. COMET 2019. Lecture notes on data engineering and communications technologies, vol 35

36. Devaiah MV, Subramaniyam RS, Rakesh S (2018) Solar hybrid vehicle, solar collectors and panels. In: International conference on sustainable engineering and technology
37. Singh KV, Om Bansal H, Singh D (2019) A comprehensive review on hybrid electric vehicles: architectures and components. J Mod Transp 27(2):77–107
38. Shabbir W, Evangelou SA (2014) Efficiency analysis of a continuously variable transmission with linear control for a series hybrid electric vehicle. IFAC Proc Vol 47(3):6264–6269
39. Vidyanandan KV (2018) Overview of electric and hybrid vehicles. Energy Scan 3:7–14
40. Prajapati KC, Patel R, Sagar R (2014) Hybrid vehicle: a study on technology. Int J Eng Res Technol 3:1076–1082
41. Chen L, Xi G, Sun J (2012) Torque coordination control during mode transition for a series–parallel hybrid electric vehicle. IEEE Trans Veh Technol 61(7):2936–2949
42. Chau KT, Chan CC, Chunhua L (2008) Overview of permanent-magnet brushless drives for electric and hybrid electric vehicles. IEEE Trans Industr Electron 55(6):2246–2257
43. Khan F, Husin ZA, Soomro HA, Mazlan MA, Sulaiman E (2016) Deterministic optimization of single phase 8S-4P field excitation flux switching motor for hybrid electric vehicle. ARPN J Eng Appl Sci 11(8)
44. Bhargav P, Kaushik S (2019) Real-time energy management scheme for dual converter-based hybrid solar/battery/ultra-capacitor vehicular system. In: Proceedings of the third international conference on microelectronics, computing and communication systems, pp 369–385
45. Joseph Godfrey A, Sankaranarayanan V (2018) A new electric braking system with energy regeneration for a BLDC motor driven electric vehicle. Int J Eng Sci Technol 21(4):704–713
46. Lee CHT, Kirtley JL, Angle M (2017) Switched reluctance motor drives for hybrid electric vehicles. Switched reluctance motor—concept, control and applications
47. Snoussi J, Elghali SB, Benbouzid M, Mimouni MF (2018) Optimal sizing of energy storage systems using frequency-separation-based energy management for fuel cell hybrid electric vehicles. IEEE Trans Veh Technol 67(10):9337–9346
48. Tian L, Wu L, Huang X, Fang Y (2019) Driving range parametric analysis of electric vehicles driven by interior permanent magnet motors considering driving cycles. CES Trans Electric Mach Syst 3(4):377–381
49. Xu W, Zhu J, Guo Y, Li Y, Wang Y, Wang S (2009) Performance analysis of electric machine drives for plug-in hybrid electric vehicles. In: 2009 international conference on applied superconductivity and electromagnetic devices, Chengdu, pp 60–63
50. Lingfei W, Lifang W, Yong L, Junfeng L (2014) Torque coordination control of distributed drive electric vehicle for straight line driving. In: 2014 IEEE conference and expo transportation electrification Asia-Pacific (ITEC Asia-Pacific), Beijing, pp 1–6
51. Sakka M, Mierlo JV, Gualous H (2011) DC/DC converters for electric vehicles, electric vehicles—modelling and simulations. IntechOpen
52. Qiao H, Zhang Y, Yao Y, Wei L (2006) Analysis of buck-boost converters for fuel cell electric vehicles. In: Proceedings of the 2006 IEEE international conference on vehicular electronics and safety, Shanghai, China, 13–15, pp 109–113
53. Northcott DR, Filizadeh S, Chevrefils AR (2009) Design of a bidirectional buck-boost DC/DC converter for a series hybrid electric vehicle using PSCAD/EMTDC. In: Vehicle power and propulsion conference (VPPC), pp 1561–1566
54. Gu B, Lai J-S, Kees N, Zheng C (2013) Hybrid-switching full-bridge DC–DC converter with minimal voltage stress of bridge rectifier, reduced circulating losses, and filter requirement for electric vehicle battery chargers. IEEE Trans Power Electron 28(3):1132–1144
55. Lim C, Jeong Y, Lee M, Yi K, Moon G (2020) Half-bridge integrated phase-shifted full-bridge converter with high efficiency using center-tapped clamp circuit for battery charging systems in electric vehicles. IEEE Trans Power Electron 35(5):4934–4945
56. Al Ameen M, Prasanna Moorthy V (2019) Half-bridge integrated phase-shifted full-bridge converter with high efficiency using center-tapped clamp circuit for battery charging systems in electric vehicles. Int Res J Eng Technol (IRJET) 6(5)
57. Lee M, Lim C, Kim K, Park M, Moon G (2019) A phase-shift full-bridge converter with novel voltage oscillation clamping circuit for electric vehicle on-board charger. In: 2019 10th

international conference on power electronics and ECCE Asia (ICPE 2019—ECCE Asia), pp 2040–2045
58. Parida A, Barai M, Mothukuri KR (2019) Study of a soft switched isolated DC-DC bidirectional converter for electric vehicles. In: TENCON 2019—2019 IEEE region 10 conference (TENCON), India, pp 1136–1141
59. Martinez WH, Cortes CA (2013) High power density interleaved DC-DC converter for a high performance electric vehicle. In: Workshop on power electronics and power quality applications (PEPQA)
60. Chao Z, Zhigang G, Te C, Jie Y (2019) Isolated DC/DC converter with three-level high-frequency link and bidirectional power flow ability for electric vehicles. IET Power Electron 12(7):1742–1751
61. Matsumori H, Kosaka T, Sekido K, Kim K, Egawa T, Matsui N (2019) Isolated DC-DC converter utilizing GaN power device for automotive application. In: 2019 IEEE applied power electronics conference and exposition (APEC), Anaheim, CA, USA, pp 1704–1709
62. Moradisizkoohi H, Elsayad N, Mohammed OA (2020) An integrated interleaved ultrahigh step-up DC–DC converter using dual cross-coupled inductors with built-in input current balancing for electric vehicles. IEEE J Emerge Select Topics Power Electron 8(1):644–657
63. Yuan Z, Wang J, Yuan X, Zhang Q (2019) High efficiency and high power density interleaved DC-DC converter for electric vehicles. In: 22nd international conference on electrical machines and systems (ICEMS), Harbin, China, pp 1–5
64. Farh, HMH, Eltamaly AM, Al-Saud MS (2019) Interleaved boost converter for global maximum power extraction from the photovoltaic system under partial shading. IET Renew Power Generat 13(8):1232–1238
65. Cano ZP, Banham D, Ye S, Hintennach A, Lu J, Fowler M, Chen Z (2018) Batteries and fuel cells for emerging electric vehicle markets. Nat Energy 3(4):279–289
66. Feng X, Ouyang M, Liu X, Lu L, Xia Y, He X (2018) Thermal runaway mechanism of lithium ion battery for electric vehicles: a review. Energy Storage Mater 10:246–267
67. Grosso M, Lena D, Bocca A, Macii A, Rinaudo S (2016) Energy-efficient battery charging in electric vehicles with solar panels. In: IEEE 2nd international forum on research and technologies for society and industry leveraging a better tomorrow (RTSI), pp 1–5
68. Hannan MA, Hoque MM, Mohamed A, Ayob A (2017) Review of energy storage systems for electric vehicle applications: issues and challenges. Renew Sustain Energy Rev 69:771–789
69. Snoussi J, Ben Elghali S, Benbouzid M, Mimouni M (2018) Auto-adaptive filtering-based energy management strategy for fuel cell hybrid electric vehicles. Energies 11(8)
70. Fotouhi A, Auger DJ, Propp K, Longo S, Wild M (2016) A review on electric vehicle battery modelling: from lithium-ion toward lithium-sulphur. Renew Sustain Energy Rev 56:1008–1021
71. Kouchachvili L, Yaïci W, Entchev E (2018) Hybrid battery/supercapacitor energy storage system for the electric vehicles. J Power Sources 374:237–248
72. Stübler T, Lahyani A, Allah Zayoud A (2020) Lithium-ion battery modeling using CC-CV and impedance spectroscopy characterizations. Appl Sci 2(5):817 (Springer)
73. Yong JY, Ramachandaramurthy VK, Tan KM, Mithulananthan N (2015) A review on the state-of-the-art technologies of electric vehicle, its impacts and prospects. Renew Sustain Energy Rev 49:365–385
74. Hannan MA, Hoque MM, Hussain A, Yusof Y, Ker PJ (2018) State-of-the-art and energy management system of lithium-ion batteries in electric vehicle applications: issues and recommendations. IEEE Access 6:19362–19378
75. Slouma S, Skander-Mustapha S, Slama-Belkhodja I, Machmoum M (2019) Frequency separation model based on infinite-impulse response filter applied to hybrid power generation intended for residential sector. Int J Renew Energy Res IJRER 9(1):118–128
76. Marzougui H, Kadri A, Amari M, Bacha F (2018) Frequency separation based energy management strategy for fuel cell electrical vehicle with super-capacitor storage system. In: 2018 9th international renewable energy congress (IREC), Hammamet, pp 1–6

77. Shadman M, Sundar V, Dave R, Pal A (2018) Hybrid energy storage system containing bidirectional DC convertor, battery, super capacitor, solar panel for increasing the performance of electric vehicles. In: 4th international conference on advances in electrical, electronics, information, communication and bio-informatics (AEEICB-18)
78. Mahmoudzadeh Andwari A, Pesiridis A, Rajoo S, Martinez-Botas R, Esfahanian V (2017) A review of battery electric vehicle technology and readiness levels. Renew Sustain Energy Rev 78:414–430
79. Marzougui H, Kadri A, Amari M, Bacha F (2019) Energy management of fuel cell vehicle with hybrid storage system: a frequency based distribution. In: 2019 6th international conference on control, decision and information technologies (CoDIT), Paris, France, pp 1853–1858
80. Alloui H, Khoucha F, Rizoug N, Benbouzid M, Kheloui A (2017) Comparative study between rule-based and frequency separation energy management strategies within fuel-cell/battery electric vehicle. In: 2017 IEEE international conference on environment and electrical engineering and 2017 IEEE industrial and commercial power systems Europe (EEEIC/I&CPS Europe), Milan, pp 1–5
81. Song Z, Li J, Hou J, Hofmann H, Ouyang M, Du J (2018) The battery-supercapacitor hybrid energy storage system in electric vehicle applications: a case study. Energy 154:433–441
82. Marzougui H, Amari M, Kadri A, Bacha F, Ghouili J (2017) Integration of batteries with ultracapacitors for a fuel cell hybrid transit bus. Int J Hydrog Energy 42:8857–8869
83. Pablo G, Juan P, Torreglosa LMFFJ (2013) Control strategies for high-power electric vehicles powered by hydrogen fuel cell, battery and super capacitor. Expert Syst Appl 40:4791–4804
84. Pollet BG, Kocha SS, Staffell I (2019) Current status of automotive fuel cells for sustainable transport. Current Opinion Electrochem 16:90–95
85. Duan C et al (2018) A solar power-assisted battery balancing system for electric vehicles. IEEE Trans Transp Electrific 4(2):432–443
86. Siffat SA, Ahmad I, Ur Rahman A, Islam Y (2020) Robust integral backstepping control for unified model of hybrid electric vehicles. IEEE Access 8:49038–49052
87. Fitri Desanti A, Uta Nugraha Y, Nur Yuniarto M, Wikarta A (2019) Review of the topology and energy management hybrid energy storage on electric vehicle. In: IOP conference series: materials science and engineering, p 694
88. Wang S, Huang X, López JM, Xu X, Dong P (2019) Fuzzy adaptive-equivalent consumption minimization strategy for a parallel hybrid electric vehicle. IEEE Access 7:133290–133303
89. Wang B, Zhang X, Manandhar U, Gooi HB, Liu Y, Tan X (2019) Bidirectional three-level cascaded converter with deadbeat control for hess in solar-assisted electric vehicles. IEEE Trans Transp Electrific 5(4):1190–1201
90. Banaei MR, Alizadeh R (2016) Simulation-based modeling and power management of all-electric ships based on renewable energy generation using model predictive control strategy. IEEE Intell Transp Syst Mag 8(2):90–103
91. Khoucha F, Benrabah A, Herizi O, Kheloui A, Benbouzid MEH (2013) An improved MPPT interleaved boost converter for solar electric vehicle application. In: 4th international conference on power engineering, energy and electrical drives, Istanbul, pp 1076–1081
92. Jyotheeswara Reddy K, Sudhakar N (2018) High voltage gain interleaved boost converter with neural network based MPPT controller for fuel cell based electric vehicle applications. IEEE Access 6:3899–3908
93. Traoré B, Doumiati M, Morel C, Olivier J, Soumaoro O (2019) Energy management strategy design based on frequency separation, fuzzy logic and Lyapunov control for multi-sources electric vehicles. In: IECON 2019—45th annual conference of the IEEE industrial electronics society, Lisbon, Portugal, pp 2676–2681

Improvement of the Power Quality in Single Phase Grid Connected Photovoltaic System Supplying Nonlinear Load

Chiraz Khomsi, Monia Bouzid, Gérard Champenois, and Khaled Jelassi

Abstract This chapter is focused on the improvement of the grid current quality at the Point of Common Coupling (PCC) of a Single-Phase Grid-Connected Photovoltaic System (GCPVS) supplying the nonlinear load. Thus, with the implementation of a specific control strategy applied to the PV inverter which is used essentially to inject the solar power into the grid, the compensation of the disturbing current introduced by the nonlinear load can be performed. Furthermore, the efficiency of this control strategy is related to an algorithm aimed to extract the correct disturbing currents and to the performance of the used current controllers. Consequently, two methods of control strategy of the PV inverter are investigated and compared in this chapter. The first method comes from those found in the bibliography and the second is original due to use simple PI controllers. The effectiveness of each control strategy is verified by simulation using Matlab/Simulink and validated experimentally through an experimental platform. Therefore, basing on simulation and experimental results, the comparative study shows better performances of the second proposed methods.

C. Khomsi (✉) · M. Bouzid · K. Jelassi
LSE-ENIT, University of Tunis EL Manar ENIT-L.S.E, LR 11 ES 15, BP 37, 1002 Tunis, Tunisia
e-mail: chiraz.khomsi@enit.utm.tn

M. Bouzid
e-mail: monia.bouzid@yahoo.fr

K. Jelassi
e-mail: khaled.jelassi@enit.utm.tn

G. Champenois
LIAS, University of Poitiers - ENSIP, 2 Rue Pierre Brousse – Bat. B25, TSA 41105, cedex 9, 86073 Poitiers, France
e-mail: gerard.champenois@univ-poitiers.fr

S. Motahhir and A. M. Eltamaly (eds.), *Advanced Technologies for Solar Photovoltaics Energy Systems*, Green Energy and Technology,
https://doi.org/10.1007/978-3-030-64565-6_13

1 Introduction

Continuously rising demand for electric power in the world and environmental pollution problems of fossil energy has been conducted to increasing the penetration of renewable energy sources (RES) into power distribution systems [1]. The Photovoltaic (PV) source is one of the RES that provides a reliable, sustainable, and clean energy supply [2]. Nowadays, due to their cost-effective application, the PV systems are largely operated as connected to the grid. Single-phase Grid-Connected Photovoltaic systems (GCPVS) are widely used since they can be installed on the building roofs to supply residential loads and inject the surplus of the PV generated power to the grid. However, with the intensive use of nonlinear loads, different disturbances caused by the injection of harmonic and reactive current affect the grid current quality at the Point of Common Coupling (PCC). This introduces negative effects on the efficiency of the power distribution system. Therefore, Power Quality (PQ) problems caused by these disturbing currents appear as important as environmental problems of fossil sources. Consequently, it is mandatory to limit the injection of this disturbing current as well as their negative effects.

Therefore, to cancel the disturbing grid current in the PCC, and in the aim to ensure a better optimization, several proposed works are focused on the use of the PV inverter as a shunt active filter in addition to its main role of power injection. This is can be achieved due to a specific control strategy applied to the PV inverter [3, 4].

The efficiency of this control strategy to mitigate disturbing grid current introduced by a nonlinear load is related to the algorithm used to extract this correct disturbing current. Several algorithms are proposed in the literature. Due to their efficiency and simplicity, Instantaneous Reactive Power (IRP) theory [5, 6] and the Synchronous Reference Frame (SRF) theory [7, 8] are the most widely used since they were adapted to be applied in single-phase systems by including some proposed techniques able to produce an imaginary axis in order to obtain a virtual orthogonal frame. This solution has a problem with a significant delay time which can affect the dynamic response of the system.

Many adaptive techniques are also proposed for disturbing current compensation such as the LMS (Least Mean Square) method [9], the LMMN (Least Mean Mixed-Norm) method [10], the DNLMS method (Decorrelation Normalized Least Mean Square) [11]. Although these techniques are efficient for estimating the grid current harmonics, their concepts remain complicated and require significant computing time. In [12], an improved method for harmonic identification based on the adaptive noise cancellation principle is proposed. This technique using a variable step is able to overcome the problem of traditional adaptive techniques which is the conflict between the steady-state accuracy and the convergence speed.

Some of Phase-Locked Loop (PLL) techniques are also extended to detect harmonic current components. Authors of [13] are used a cascaded association of a Second Order Generalized Integrators (SOGI) and a Synchronous Reference Frame PLL (SRF-PLL) structure to identify current harmonics to subtract them from the total grid current. This detection technique can be used for both single and three-phase systems to compensate selected highest harmonic current components.

Consequently, in this chapter, two new algorithms aimed to extract the correct disturbing currents are exposed and their performances are compared. Therefore, the first algorithm is based on the use of the FFT technique to detect the fundamental grid current component which is subtracted subsequently from the total grid current [12]. The Proportional-Resonant (PR) controller is used to control the detected current [4, 10, 13]. On the other hand, the principle of the mitigation of the disturbing current in the second algorithm is based on the extraction of the most predominant disturbing grid current by identifying, each active and reactive grid current amplitude to be regulated then to a null signal using a PI controller. In this case, the use of simple PI controllers is sufficient since reference signals are dc components. The effectiveness of these two control strategies is verified through the good simulation results obtained using Matlab/Simulink and validated experimentally through an experimental platform. Then, a comparative study between these two methods will be presented in this chapter to highlight the originality of the algorithm of the second method comparing to the first method which is based on the use of multiple PR controllers. This control method is frequently used in recent various algorithms and techniques of literature such as in [12, 13], which employs multiple PR controllers to control the extracted harmonic current components [4, 10, 13].

2 Description of the Grid Connected Photovoltaic System (GCPVS)

The global structure of the GCPVS considered in this work is illustrated in Fig. 1. It consists of a PV generator connected to a single-phase grid-tied inverter via a dc-dc boost converter. The connection of the PV inverter to the grid is performed by the intermediate of an inductive filter L_F to eliminate the high-frequency grid current components.

To guarantee the operation of the PV generator at its maximum power point, the dc-dc boost inverter is controlled by the Maximum Power Point Tracking (MPPT) algorithm based on the Perturb and Observe (P&O) method due to its simplicity and its ability to reach the exact point of maximum power in a short time [14–16].

The dc-link between the boost converter and the PV inverter is performed through the capacitor C_{dc} which is used to create a constant voltage source useful for supplying the photovoltaic inverter and controlling the power flow between the grid and the photovoltaic system.

The considered GCPVS is simulated according to the parameters showed in Table 1.

As it is presented in Fig. 1, the nonlinear load L_{NL} is modeled as a single-phase full wave rectifier L_1 connected in parallel to an inductive load L_2 to conceive a load with spectral content rich in harmonics. The GCPVS is simulated with two cases of

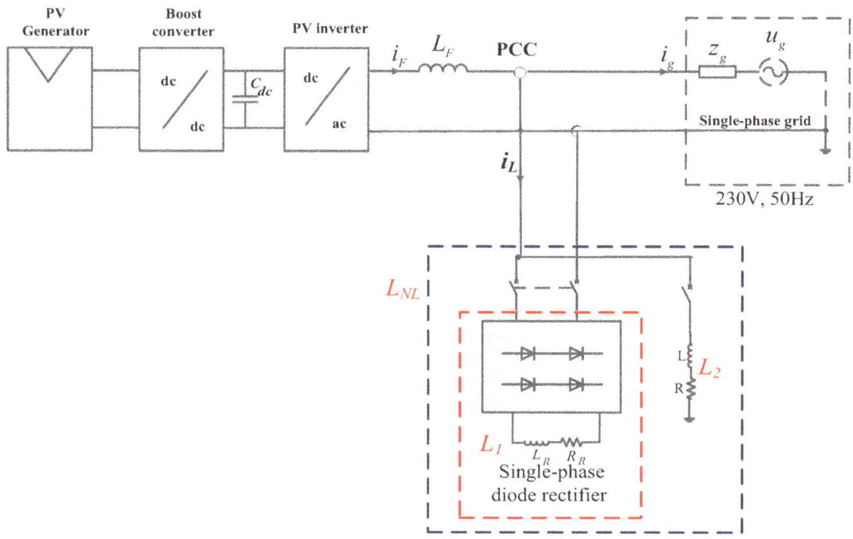

Fig. 1 General structure of the considered Grid Connected PV System (GCPVS)

Table 1 Parameters of the simulated grid-connected photovoltaic system (GCPVS)

Parameters	Values
Grid phase voltage (rms)	$U_g = 180$ V
Grid frequency	$f = 50$ Hz
Boost input capacitor	$C_b = 4700$ μF
Boost inductance	$L_b = 0.625$ mH
dc-bus capacitor	$C_{dc} = 4450$ μF
dc-bus voltage (average)	$V_{dc} = 350$ V
Switching frequency of the PV inverter	$f_{sw} = 10$ kHz
Inductive filter	$L_F = 20$ mH
Inductive load (L_2)	L = 0.5 H, R = 2Ω

load (L_{NL1} and L_{NL2}) composed by (L_{11}, L_2) and (L_{12}, L_2) respectively, to investigate the impact of the load variation. Each rectifier (L_{11} and L_{12}) supplies an inductive load (L_{R1}, R_{R1}) and (L_{R2}, R_{R2}) respectively. The active and reactive powers of the simulated loads (L_{NL1} and L_{NL2}) are presented in Table 2.

The frequency spectra of the conceived nonlinear loads are also observed in order to investigate their effect on the spectral content of the grid current. As it is depicted in

Table 2 Active and reactive powers of the simulated nonlinear loads

Loads	L_{NL1}	L_{NL2}
Active power (W)	650	480
Reactive power (VAR)	360	310

Fig. 2, for the two cases of the nonlinear loads, the absorbed currents are significantly distorted. They present an important level of THD equal to 32.33% with the load L_{NL1} and 29.05% with the load L_{NL2}. In addition, the frequency representations of these two load currents are composed of odd harmonic components of which the most dominant components are limited to the 13th order as it is shown in Fig. 2.

The study of the simulated system consists to evaluate the quality of the grid current under load condition variation (two considered nonlinear loads) and under different cases of the generated PV power which depends on the climatic conditions. Therefore, three levels of the PV inverter power (P_{inv}) have been fixed according to a chosen solar irradiance (G) profile presented in Fig. 3. Thus three modes are considered. **Mode 1** corresponds to solar irradiance (G) equal to 100 W/m^2 for a period of time "t" between 0 s and 2 s. In mode 2, G increases to 230 W/m^2 for "t" between 2 s and 4 s and in mode 3, G decreases to 170 W/m^2 when "t" is between 4 s and 6 s.

In this chapter, a detailed comparative study of the grid current quality without and with the implementation of the two proposed harmonic compensating algorithms is performed. This comparative study is aimed to evaluate the effectiveness of the

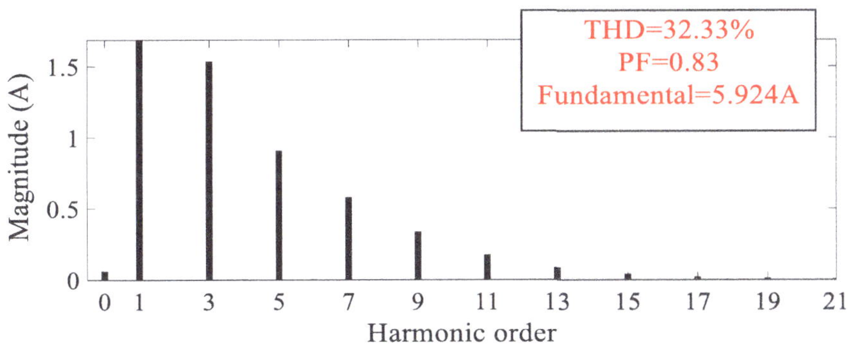

(a)

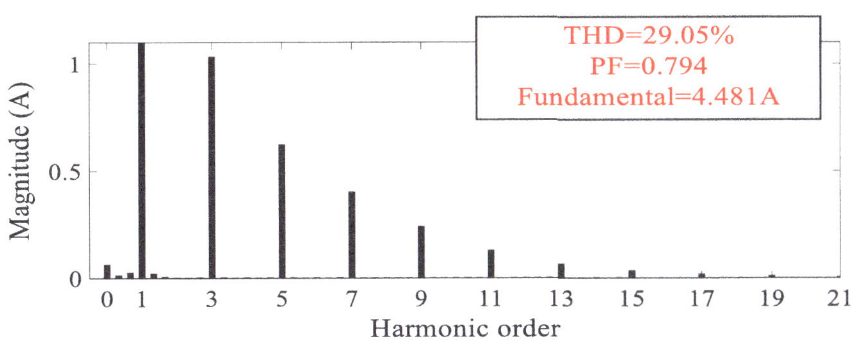

(b)

Fig. 2 Spectrum of the load current i_L (**a**) in the case of L_{NL1}

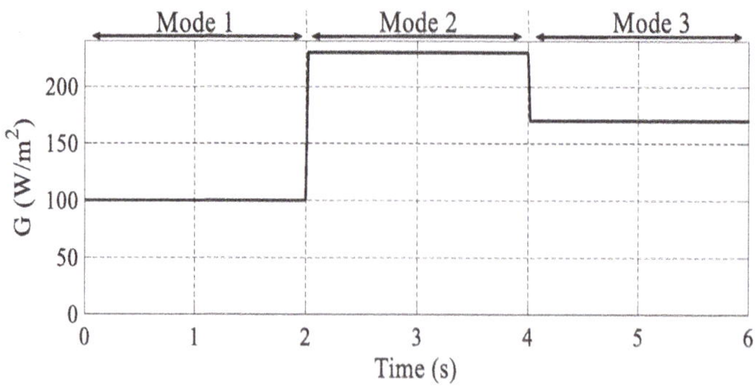

Fig. 3 Simulated profile of the solar irradiance (G)

two proposed algorithms to improve the grid current quality which is affected by nonlinear load. Thereafter, the following part of this paragraph will be focused on the investigation of the quality of the grid current simulated without the proposed algorithms. Consequently, the time and the frequency representations as well as the THD index of the grid current simulated without the two proposed algorithms for the two loads under the three levels of the fixed solar irradiance (G) are presented in Fig. 4. It is worth noting that without a harmonic compensating algorithm, the grid current simulated with each considered nonlinear load is highly distorted for the three levels of the solar irradiance (G). This explains the significant THD obtained in any case of operating mode (mode 1, 2, or 3) of the photovoltaic system. Comparing Fig. 2, 3 and 4, for a specific nonlinear load, the obtained spectra of the grid current under the three

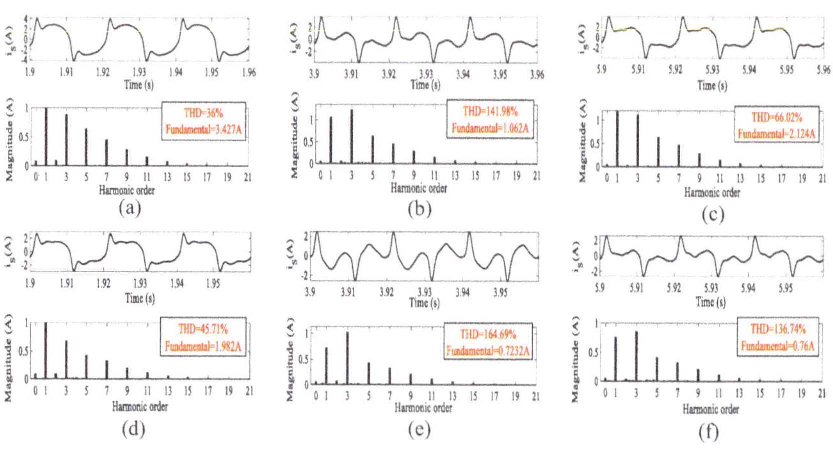

Fig. 4 Time and frequency representations of the simulated grid current (i_g) without the proposed algorithm (**a**) in mode 1 with the load L_{NL1}, (**b**) in mode 2 with the load L_{NL1} (**c**) in mode 3 with the load L_{NL1}, (**d**) in mode 1 with the load L_{NL2}, (**e**) in mode 2 with the load L_{NL2}, (**f**) in mode 3 with the load L_{NL2}

chosen solar irradiance (G) are constituted of the same harmonic content. They have the same orders of harmonic. The most dominants of these harmonic components are limited to the 13th order. Thus, it can be concluded that the harmonic components of the grid current are provided from the used load current.

Furthermore, it can be noted from Fig. 4a–f that for each case of the simulated load, the magnitudes of the harmonic components remain unchanged under the three chosen levels of the solar irradiance (G), but it is only the fundamental components which are affected by the variation of G. Consequently, for a specific nonlinear load, the variation of the THD of the grid current is basically due to the variation of the fundamental component magnitude. If the magnitude of the fundamental component decreases, the THD value will increases, and vice versa.

Thereafter, a detailed theoretical study on the principle of the PV inverter control strategy based on disturbing grid current extraction methods will be presented.

3 Proposed Methods Used for the Improvement of the Power Quality at the PCC of the GCPVS

The grid-tied inverter which is a single-phase voltage source inverter is used in the PV system mainly to control the power flow between the PV system, the utility grid, and the nonlinear load connected to the PCC. In addition to this main function, the PV inverter is used as a shunt active filter in order to attenuate disturbing grid currents introduced by nonlinear loads and then guarantee a grid current with a sinusoidal form and low THD value.

To perform these two functions, the PV inverter is properly controlled. The principle of the control scheme of the PV inverter is shown in Fig. 5. It is based on two parallel control loops aimed to generate two signals c_l and c_2. The sum of these signals represents the reference signal useful to generate the PWM signal to control the switched devices of the PV inverter. The first loop is aimed to create the first signal c_l representing the dc voltage loop. This loop has the task to maintain the dc voltage to the desired value equal to 350 V in this work. To do that, a PI regulator is used since the input signal is a continuous one. Furthermore, the output signal of the PI regulator will be multiplied by a unitary sinusoidal signal which is synchronized with the frequency of the grid voltage. To obtain this frequency, the PLL technique was then applied.

On the other hand, as it is showed in Fig. 5, the second loop aims to control the disturbing grid current which is obtained from the disturbing current extraction block. This is performed by comparing the extracted disturbing grid currents introduced by the nonlinear load to a zero signal to cancel them from the total grid current to obtain a sinusoidal form with good quality. According to Fig. 5, the current control loop is composed of two main blocks. The first one is the disturbing grid current extracting block which consists of the proposed algorithm useful to extract accurately

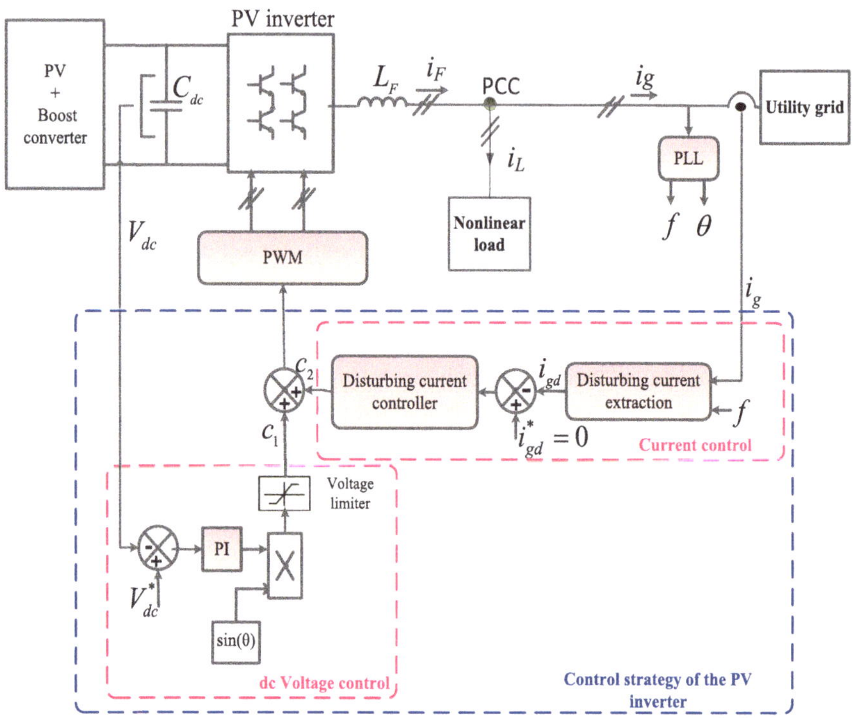

Fig. 5 The considered control scheme of the PV inverter

the disturbing current and the second block represents the disturbing grid current controller.

In this chapter, we propose two methods for controlling the disturbing grid current. Each method is specified by its own disturbing-current extraction algorithm and disturbing-current controller.

3.1 Investigation of the First Proposed Method of the PV Inverter Control

This paragraph is focused on the investigation of the first algorithm proposed to improve the grid current quality affected by the nonlinear load. The PV inverter control scheme based on the first method is presented in Fig. 6. The principle of each block of the current controller will be explained in the following subsections.

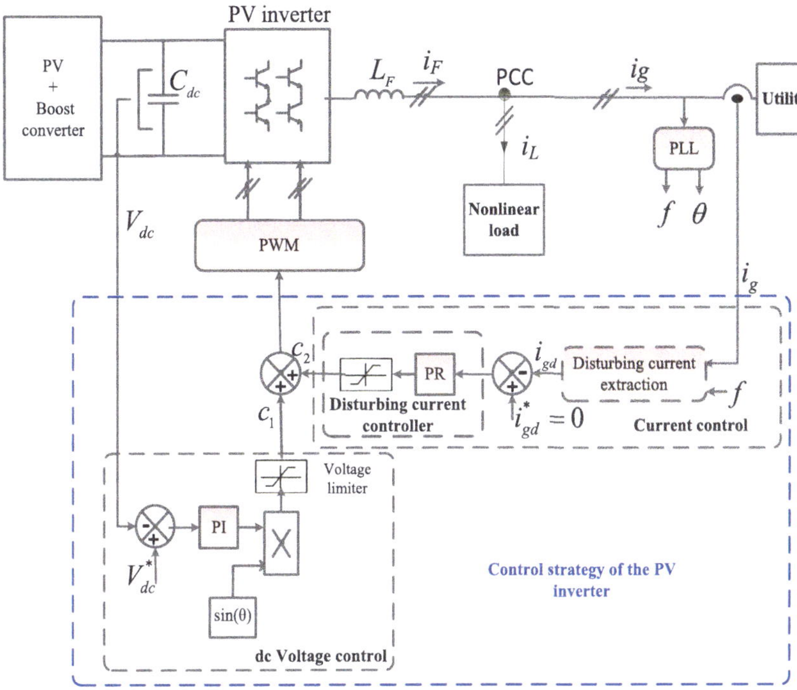

Fig. 6 The first proposed control strategy of the PV inverter

3.1.1 The Proposed Algorithm for the Extraction of the Disturbing Grid Current with the First Method

With the first method, the principle of the proposed algorithm is demonstrated as follows.

The distorted grid current expressed by (1) is composed of the sum of the fundamental component i_{gl} and harmonic components i_{gh} according to (2) and (3).

$$i_g(t) = \sum_{h=1}^{\infty} I_{gh} \sin(h\omega t + \theta_h) \tag{1}$$

$$i_g(t) = I_{g1} \sin(\omega t + \theta_1) + \sum_{h=2}^{\infty} I_{gh} \sin(h\omega t + \theta_h) \tag{2}$$

$$i_g(t) = i_{g1}(t) + i_{gh}(t) \tag{3}$$

On the other hand, according to (4), the fundamental component of the grid current can be expressed as the sum of the active and reactive components.

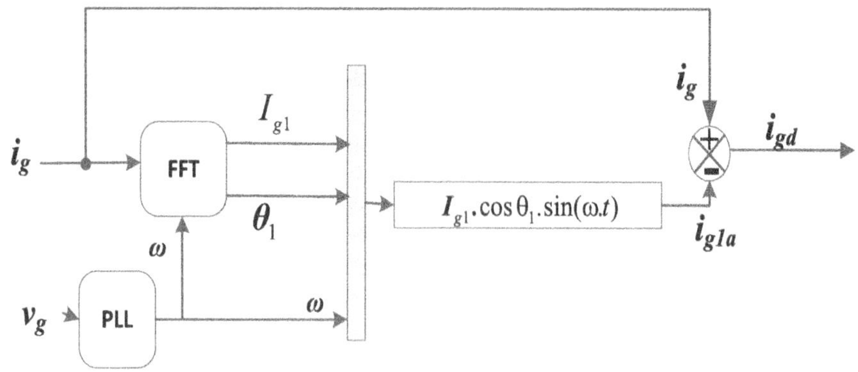

Fig. 7 Principe of harmonic grid current (i_{gd}) extraction block

$$i_{g1}(t) = i_{g1a}(t) + i_{g1r}(t) \tag{4}$$

Therefore, (3) can be expressed as follows.

$$i_g(t) = i_{g1a}(t) + i_{gd}(t) \tag{5}$$

where $i_{gd}(t)$ is the disturbing grid current which represents the sum of the reactive fundamental component and harmonic components of the grid current as indicated by (6).

$$i_{gd}(t) = i_{g1r}(t) + i_{gh}(t) \tag{6}$$

Consequently, referring to (5), the disturbing grid current $i_{gd}(t)$ can be obtained by subtracting the total grid current $i_g(t)$ from the active grid current $i_{1ga}(t)$. Based on this principle, the algorithm for the extraction of the disturbing grid current illustrated in Fig. 7 was implemented. This algorithm is then based on the calculation of the active grid current to subtract it thereafter from the total grid current.

From Eq. (7) which illustrates the expression of the fundamental grid current $i_{g1}(t)$, the active grid current $i_{g1a}(t)$ and reactive grid current $i_{g1r}(t)$ can be identified according to (8) and (9).

$$\begin{aligned} i_{g1}(t) &= I_{g1}\sin(\omega t + \theta_1) \\ &= I_{g1}\cos(\theta_1)\sin(\omega t) + I_{g1}\sin(\theta_1)\cos(\omega t) \end{aligned} \tag{7}$$

$$i_{g1a}(t) = I_{g1}\cos(\theta_1)\sin(\omega t) \tag{8}$$

$$i_{g1r}(t) = I_{g1}\sin(\theta_1)\cos(\omega t) \tag{9}$$

Basing on (8), the active grid current $i_{g1a}(t)$ is then calculated. It requires the amplitude I_{g1} and the phase angle θ_1 of the fundamental grid current. To extract them, the FFT method was applied on the grid current at the fundamental frequency of the grid using the PLL technique as it is showed in Fig. 7.

3.1.2 Description of the Current Regulation Loop

As it has been mentioned above, the disturbing grid current is regulated by comparing it to a zero signal. In this work, the Proportional Resonant (PR) controller is then used to set the error of this comparison to zero. Consequently, the PV inverter is able to force the disturbing grid current to zero.

The choice of the PR controller is explained by its effectiveness to track a reference signal having a sinusoidal form. Therefore, by implementing several blocks of cascading PR controllers adjusted to the low frequencies of the harmonic components of the current, selective harmonic compensation can be obtained since the disturbing current is considered as the sum of sinusoidal currents with different frequencies. In this work, the transfer function implemented by the PR controller which is expressed by (10), is designed to attenuate the harmonic components of the grid current of order 2–13. The choice of 13th order is explained by the fact that the order harmonic components of the nonlinear load current considered in this work are limited to 13 as it is explained in the previous paragraph.

$$G_S(s) = \sum_{13}^{h=1} \frac{2K_{ih}\omega_{cut}s}{s^2 + 2\omega_{cut}s + (h\omega_0)^2} \tag{10}$$

where, ω_{cut} is the cutoff frequency, ω_0 is the grid angular frequency, h is the harmonic order of the grid current to be controlled, K_{ih} is a constant gain.

Figure 8 shows the implemented structure of the multiple PR controllers in this work.

3.1.3 Simulation Results Obtained Using the First Method

The considered GCPVS was simulated using the proposed first method under the two considered nonlinear loads (L_{NL1} and L_{NL2}) and with the chosen solar irradiance (G) profile presented in Fig. 3. The transfer of the power flow between the three main elements of the considered PV system: the PV inverter, the grid, and the user load are investigated in order to study the behavior of the considered GCPVS.

Figure 9a, b show the active power of these three elements simulated with the loads L_{NL1} and L_{NL2} respectively, under the three operating modes given by the solar irradiance (G) profile. In each mode, Table 3 represents the values of the three active powers. It can be noted from the corresponding power flow that with the two used loads, the sum of the PV inverter and the grid active powers represent the load active

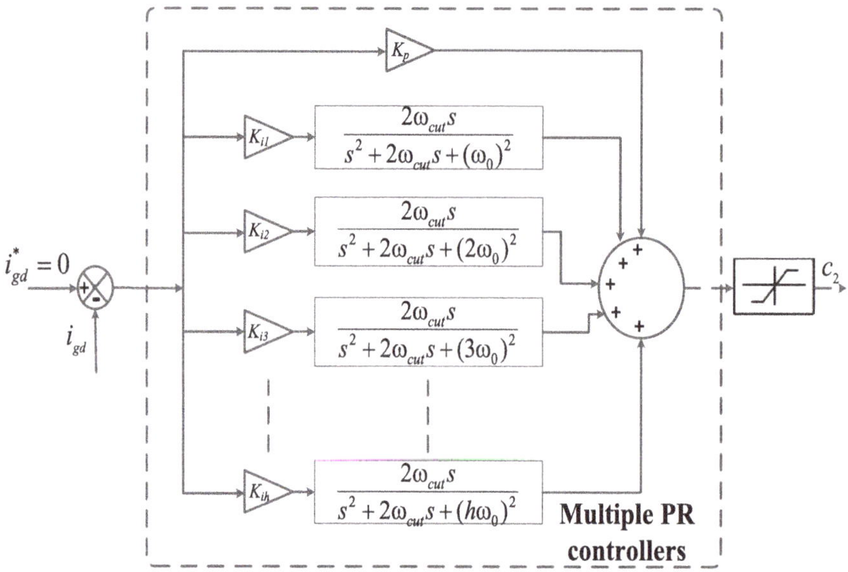

Fig. 8 Structure of the multiple PR controllers implemented in the first method

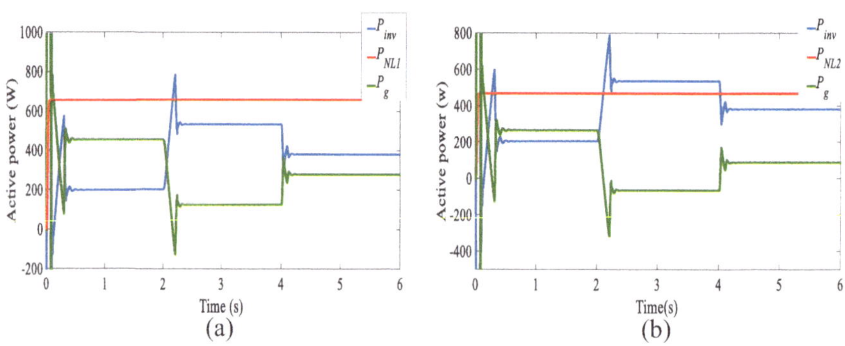

Fig. 9 Active grid power (P_g), active load power (P_{NL1} and P_{NL2}), and active PV inverter power (P_{inv}) simulated in the three operating modes of the PV (**a**) with L_{NL1} and (**b**) with L_{NL2}

Table 3 Simulated active powers of the grid, the PV inverter and the two used loads L_{NL1} and L_{NL2} with the first method

	Solar Irradiance (W/m^2)	100	230	170
	PV inverter power (W)	227	540	397
L_{NL1}	Grid power (W)	427	118	260
	Load power (W)	650	650	650
L_{NL2}	Grid power (W)	253	−60	83
	Load power (W)	480	480	480

power for any case of the used load and the operating mode of the PV system. This means that the supply of each load is correctly ensured by both the utility grid and the PV inverter.

After verifying the operation of the designed GCPVS with the proposed first method, the quality of the grid current was then evaluated for the two cases of load conditions considering the solar irradiance (G) profile shown in Fig. 3. Therefore, the time and the frequency representations of the grid current as well as the THD simulated in the three operation modes of the PV inverter with two cases of load (L_{NL1} and L_{NL2}) are presented in Fig. 10. The values of the obtained THD are indicated in Table 4.

Comparing Fig. 10 that shows the simulated grid current with the proposed algorithm to Fig. 4, the quality of the waveform of the grid current was significantly improved and the THD values were attenuated in any case of operating mode of the GCPVS and under the two cases of nonlinear loads. It is worth noting that for the three cases of (Fig. 10b, e, f), although the simulated grid current has undergone

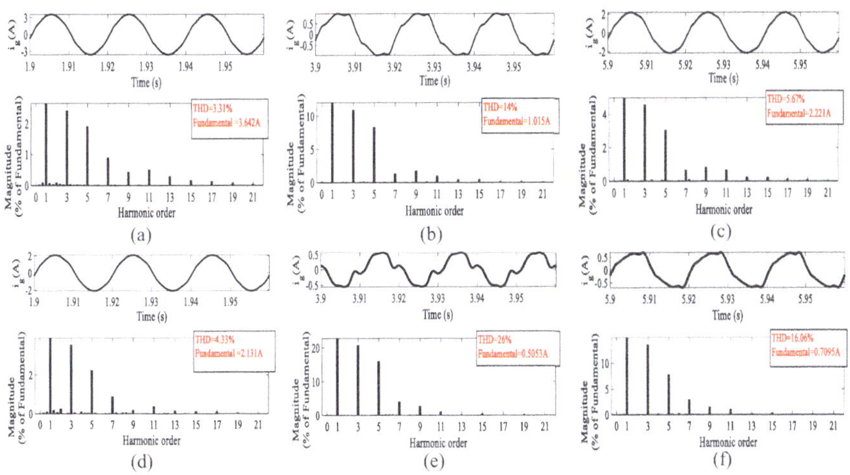

Fig. 10 Time and frequency spectrum representations of the simulated grid current (i_g) with the first proposed algorithm (**a**) in mode 1 with the load L_{NL1}, (**b**) in mode 2 with the load L_{NL1}, (**c**) in mode 3 with the load L_{NL1}, (**d**) in mode 1 with the load L_{NL2}, (**e**) in mode 2 with the load L_{NL2}, (**f**) in mode 3 with the load L_{NL2}

Table 4 THD values obtained with the first method and without improvement

Solar Irradiance (W/m^2)		100	230	170
THD (%)	L_{NL1} without improvement	36	141.98	66.02
	L_{NL1} with first method	3.31	14	5.67
	L_{NL2} without improvement	45.71	164.69	136.74
	L_{NL2} with first method	4.33	26	16.06

an important attenuation of the THD, the waveform is yet distorted. However, the quality of the grid current remains acceptable since the amplitude of the fundamental component is low.

Comparing the simulation results to those obtained without the proposed algorithm, we can conclude that the proposed technique is efficient to compensate the harmonic current introduced by the nonlinear load and to obtain a grid current with a low THD even under a variation of the PV power and the load conditions (see Table 4).

3.2 *Investigation of the Second Proposed Method of the PV Inverter Control*

The control strategy of the PV inverter based on the second method of disturbing current control is showed in Fig. 11.

3.2.1 Proposed Algorithm in the Second Method for the Extraction of the Disturbing Grid Current

The principle of the second algorithm aimed to extract the disturbing grid current is essentially based on the extraction of each active and reactive disturbing current

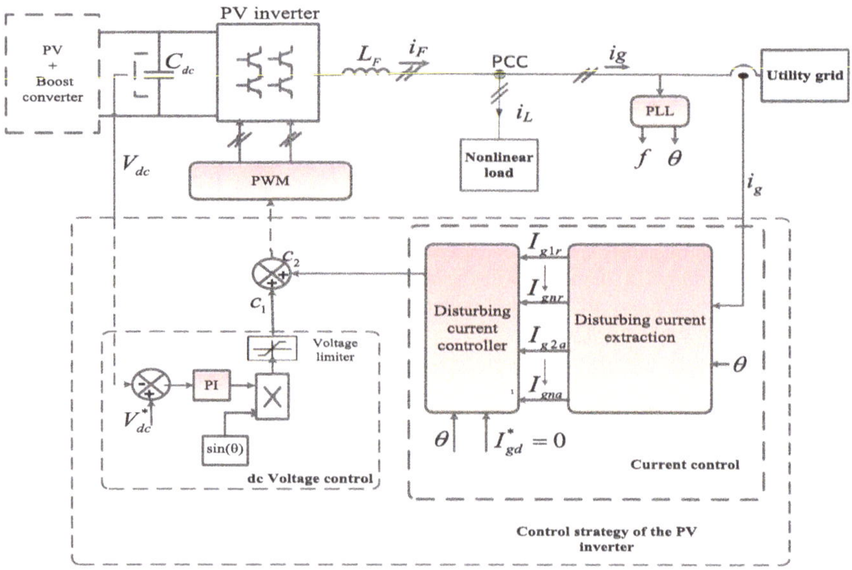

Fig. 11 The second proposed control strategy of the PV inverter

vector from the total grid current, except the vector of the active component of the fundamental current. Referring to (8) and (9), the amplitudes of active I_{g1a} and reactive I_{g1r} grid currents of the fundamental component can be expressed respectively by (11) and (12).

$$I_{g1a} = I_{g1}\cos(\theta_1) \tag{11}$$

$$I_{g1r} = I_{g1}\sin(\theta_1) \tag{12}$$

Similarly, each harmonic component of the grid current can be decomposed on active and reactive components as showed by (13) and (14). The expressions of the active I_{gna} and reactive I_{gnr} harmonic components of the grid current of order n are defined respectively by (15) and (16).

$$i_{gh}(t) = \sum_{n=2}^{\infty}\left[I_{gn}.\cos(\theta_n).\sin(n\omega t) + I_{gn}.\sin(\theta_n).\cos(n\omega t)\right] \tag{13}$$

$$i_{gh}(t) = \sum_{n=2}^{\infty}\left[i_{gna}(t) + i_{gnr}(t)\right] \tag{14}$$

With

$$I_{gna} = I_{gn}\cos(\theta_n) \tag{15}$$

$$I_{gnr} = I_{gn}\sin(\theta_n) \tag{16}$$

Therefore, the disturbing grid current $i_{gd}(t)$ expressed by (6) can be reformulated as follows

$$\begin{aligned} i_{gd}(t) &= i_{g1r}(t) + i_{gh}(t) \\ &= i_{g1r}(t) + \sum_{n=2}^{\infty}\left[i_{gna}(t) + i_{gnr}(t)\right] \\ &= \sum_{n=2}^{\infty} I_{gna}\sin(n\theta) + \sum_{n=1}^{\infty} I_{gnr}\cos(n\theta) \\ &= i_{ga-d}(t) + i_{gr-d}(t) \end{aligned} \tag{17}$$

Thus, according to (5), to obtain a grid current with a sinusoidal form composed only by the active current $i_{g1a}(t)$, the disturbing grid current $i_{gd}(t)$ must be set to zero. This means that the amplitudes of all disturbing components of the grid current (I_{g1r}, I_{gna}, I_{gnr}) expressed by (12), (14) and (16) must be set to zero using a current regulation loop. In this aim, the amplitudes (I_{g2a}, ..., I_{gna}) and (I_{g1r},..., I_{gnr}) of different disturbing grid current components must be extracted.

In this work, an efficient algorithm having the task to extract the amplitudes of active and reactive disturbing components of the grid current is proposed. The principle of this method is illustrated in Fig. 12. As showed in this figure, to extract an amplitude of a component with order "n" among the different reactive components (I_{g1r},..., I_{gnr}) or the different active components (I_{g2a},..., I_{gna}) of the total grid current, the grid current expressed by (2) is multiplied respectively by $y_n(t) = \cos(n\theta)$(for I_{g1r},..., I_{gnr} extraction) or $x_n(t) = \sin(n\theta)$ (for I_{g2a},..., I_{gna} extraction) with a phase angle which has the same order "n" of the extracted considered component. As a result, a corresponding dc component equal to $I_{g1r}/2$,..., $I_{gnr}/2$ (for I_{g1r},..., I_{gnr} extraction) and $I_{g2a}/2$,..., $I_{gna}/2$ (for I_{g2a},..., I_{gna} extraction) are obtained as well as a variable term. This result is demonstrated by (18) and (19) to

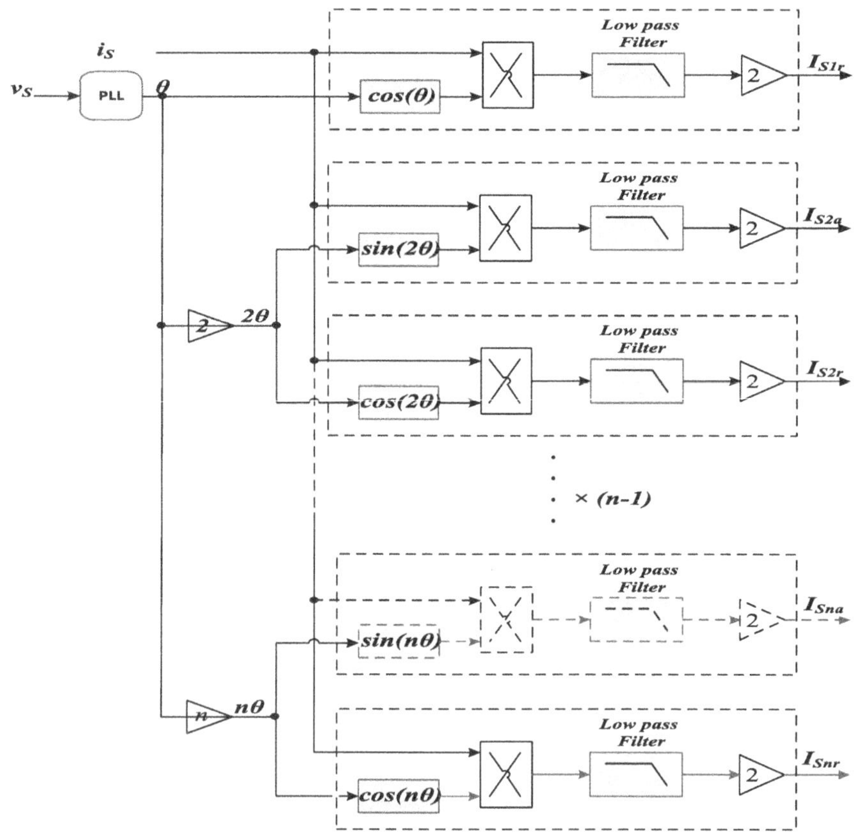

Fig. 12 Second algorithm scheme used to extract the amplitudes of different active and reactive components of the harmonic grid currents

extract respectively the amplitude of the reactive current component and the active current component of order 2.

$$\begin{aligned}
i_g(t).\sin(2\omega t) =& I_{g1}\sin(\omega t+\theta_1).\sin(2\omega t)+I_{g2}\sin(2\omega t+\theta_2).\sin(2\omega t))\\
&+\sum_{n=3}^{\infty}(I_{gn}\sin(n\omega t+\theta_n).\sin(2\omega t))\\
&=\frac{I_{g2}}{2}.\cos(\theta_2)-\frac{I_{g2}}{2}.\cos(4\omega t+\theta_2)+I_{g1}\sin(\omega t+\theta_1)\times\sin(2\omega t)\\
&+\sum_{n=3}^{\infty}(I_{gn}.\sin(n\omega t+\theta_n).\sin(2\omega t))\\
&=\frac{I_{g2a}}{2}-\frac{I_{g2}}{2}.\cos(4\omega t+\theta_2)+I_{g1}\sin(\omega t+\theta_1)\times\sin(2\omega t)\\
&+\sum_{n=3}^{\infty}(I_{gn}.\sin(n\omega t+\theta_n).\sin(2\omega t))
\end{aligned}\tag{18}$$

$$\begin{aligned}
i_g(t).\cos(\omega t) =& I_{g1}.\sin(\omega t+\theta_1).\cos(\omega t)+\sum_{n=2}^{\infty}(I_{gn}\sin(n\omega t+\theta_n).\cos(\omega t))\\
&=\frac{I_{g1}}{2}.\sin(\theta_1)+\frac{I_{g1}}{2}.(\sin(2\omega t+\theta_1)\\
&+\sum_{n=2}^{\infty}(I_{gn}\sin(n\omega t+\theta_n)\times\cos(\omega t))\\
&=\frac{I_{g1r}}{2}+\left[\frac{I_{g1}}{2}.\sin(2\omega t+\theta_1)+\sum_{n=2}^{\infty}(I_{gn}\sin(n\omega t+\theta_n)\times\cos(\omega t))\right]
\end{aligned}\tag{19}$$

Consequently, to extract each dc component $I_{g1r}/2,\ldots,I_{gnr}/2$ and $I_{g2a}/2,\ldots,I_{gna}/2$, a low pass filter is used. This leads to obtaining each amplitude of the reactive components $I_{g1r},\ldots,I_{gnr}$, and the active components $I_{g1a},\ldots,I_{gna}$ after multiplying the output component of the low pass filter by again equal to 2.

3.2.2 Description of the Current Regulation Loop Used in the Second Method

As it is mentioned, to improve the grid current quality at the PCC of the considered GCPVS, the active current must be isolated from the total grid current. This is performed by canceling all the dominant components of the disturbing grid current. In this work, once the amplitudes of these components $I_{g1r},\ldots,I_{gnr}$ and $I_{g1a},\ldots,I_{gna}$ are extracted using the appropriate algorithm described above, they are compared to a reference signal equal to zero. To regulate each amplitude to zero, the result of each comparison is then presented to a PI regulator, as shown in Fig. 13. The choice

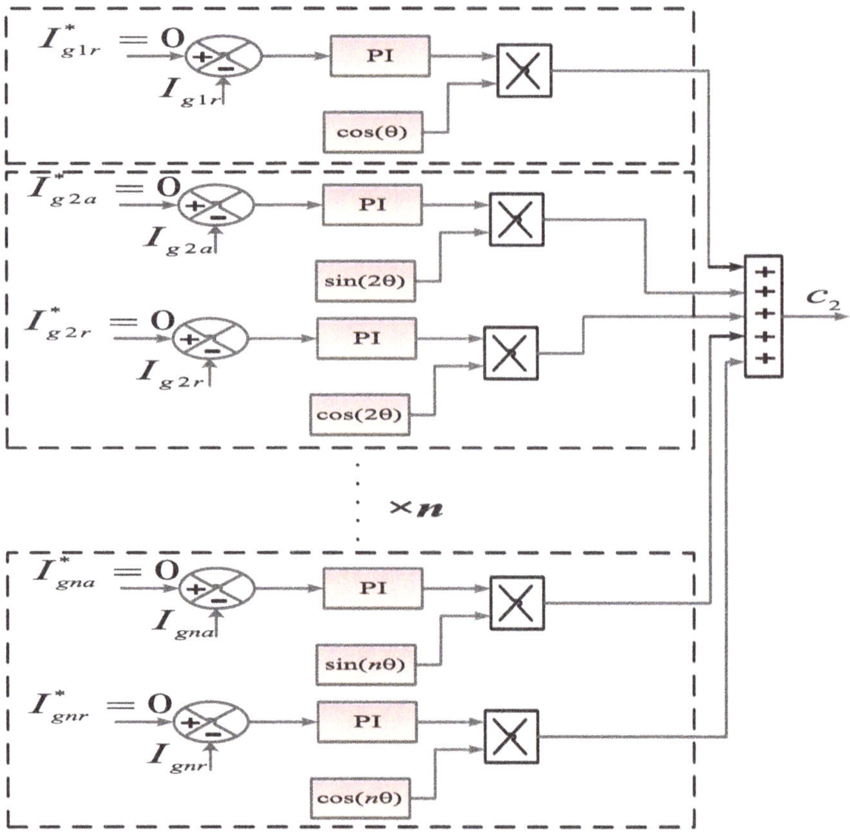

Fig. 13 The current control block scheme used for the cancellation of selected harmonic components

of the PI regulator is justified by the fact that it is more efficient and gives a zero steady-state error in case of dc component regulation. Thereafter, each output signal of the PI regulator is multiplied by $y_n(t) = \cos(n\theta)$ in case of the control of ($I_{g1r},\ldots, I_{gnr}$) and $x_n(t) = \sin(n\theta)$ case of the control of ($I_{g2a},\ldots, I_{gna}$), having a phase angle with the same order "n" of the controlled considered component. The sum of the obtained signals represents the reference signal c_2 generated by the current control loops as illustrated in Fig. 13.

3.2.3 Simulation Results Obtained Using the Second Method

The second proposed method was implemented with the GCPVS simulated according to the parameters presented in Table 1, in order to investigate its effectiveness for disturbing grid current compensation using the second algorithm. Furthermore, the two nonlinear loads (L_{NL1} and L_{NL2}) as well as the solar irradiance (G) profile presented in Fig. 3 are considered to compare the performances of the proposed first and second methods.

As mentioned, to examine the behavior of the simulated GCPVS system, the power flow between the PV inverter, the utility grid, and each used load must be investigated. Thus, the active power of each element is simulated with the second proposed method and showed in Fig. 14a, b. Comparing this simulation curves to that obtained with the first proposed method, it can be noted that the dynamic response of these elements is slower since the active power of each element takes an important time to reach its steady-state. On the other hand, referring to Table 5 indicating the values of their steady-state active power, the GCPVS has the same behavior as in the case of the first method. This proves the good operation of the GCPVs system with the second proposed method but with rather significant response time.

Now, to evaluate the performance of the second proposed method for the grid current quality improvement, the time and the frequency representations, as well as the THD of the grid current obtained for the three levels of solar irradiation (G) and under the two nonlinear load conditions, are shown in Fig. 15. Comparing this figure to Fig. 10, it can be noted that the quality of the grid current was improved with the second proposed method since the waveform of the grid current has a sinusoidal shape for any case of operating mode of the GCPVS. In addition, from the THD

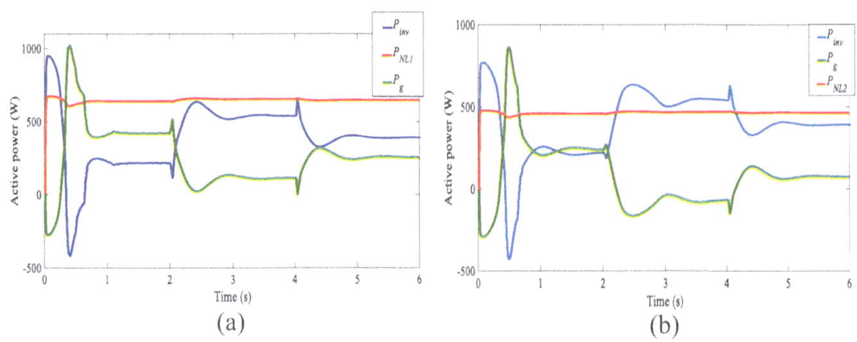

Fig. 14 Active grid power (P_g), active load power (P_{NL1} and P_{NL2}) and active PV inverter power (P_{inv}) simulated in the three operating modes of the PV (a) with (**a**) and (**b**) in the case of L_{NL2}

Table 5 Active powers of the grid, the PV inverter and the two used loads L_{NL1} and L_{NL2} simulated with the second method

	Solar Irradiance (W/m^2)	100	230	170
	PV inverter power (W)	227	540	397
L_{NL1}	Grid power (W)	427	118	260
	Load power (W)	650	650	650
L_{NL2}	Grid power (W)	253	−60	83
	Load power (W)	480	480	480

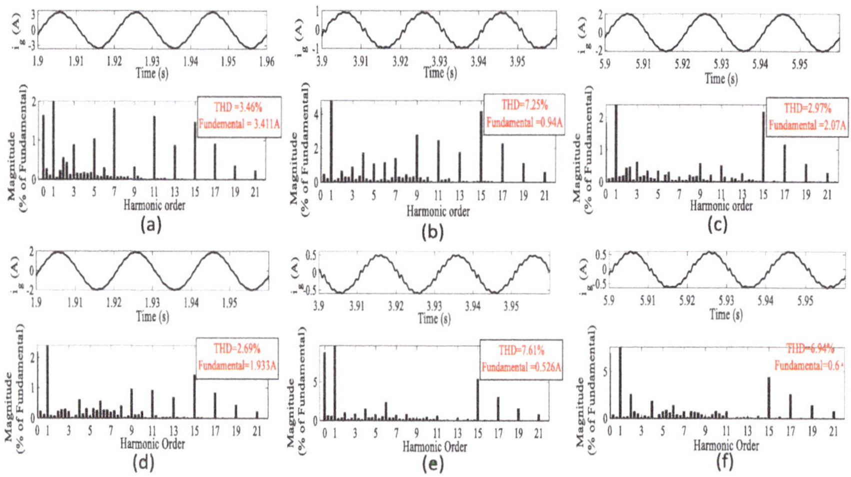

Fig. 15 Time representation and frequency spectrum of the simulated grid current (i_g) with the second proposed algorithm (**a**) in mode 1 with the load L_{NL1}, (**b**) in mode 2 with the load L_{NL1}, (**c**) in mode 3 with the load L_{NL1}, (**d**) in mode 1 with the load L_{NL2}, (**e**) in mode 2 with the load L_{NL2}, (**f**) in mode 3 with the load L_{NL2}

values obtained with the first and the second methods in the three operating modes of the GCPVS using the two cases of loads (L_{NL1} and L_{NL2}), we can conclude that the second control method is more efficient for disturbing grid current mitigation than the first method. The THD values are lower than those obtained with the first method. Therefore, for any case of GCPVS operating mode, with the second method, the THD value did not exceed 8% contrary to the first method in which the THD value remains important and the waveform of the grid current has not a sinusoidal form in some cases (Fig. 10b, e, f).

Furthermore, referring to Fig. 16 that shows the THD curves as a function of fundamental grid current for the two proposed methods and with the two used loads, note that the THD value decreases when the fundamental grid current increases. This is explained by the fact that the harmonic components have the same amplitudes for any case of the operating mode of the PV system with the same nonlinear load. Then, it can be concluded that the THD value is related only to the fundamental grid current amplitude for a considered load.

We can conclude then, that the proposed second algorithm is more efficient than the first algorithm since it has successfully compensated the disturbing grid current at any operating point of the PV system even under the variation of the load conditions. But it has a slower dynamic response and requires more computing time.

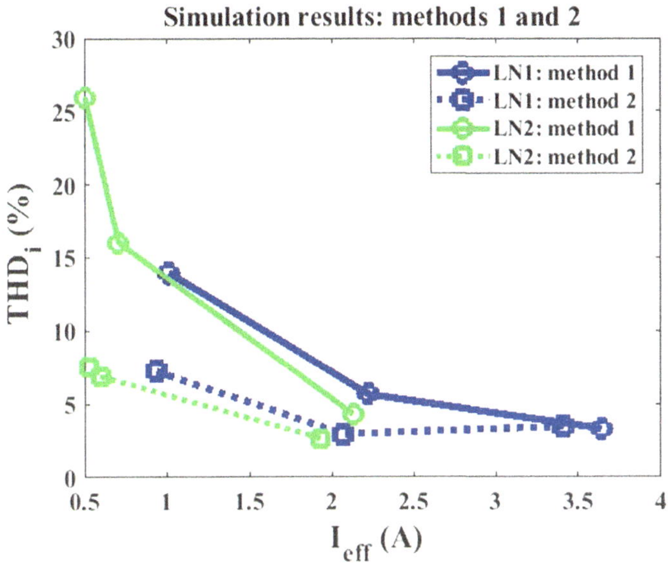

Fig. 16 Simulated THD as a function of the fundamental grid current obtained with the two proposed methods

3.3 Experimental Results of the Two Used Methods

To validate the obtained simulation results of the two proposed algorithms, experimental tests have been achieved on a testbed presented in Fig. 17. This testbed is composed of a Chroma 62020H-150 s programmable power supply used to model the PV generator and thus to provide a variable power to a dc-dc boost converter which is connected then to a voltage inverter. This voltage inverter is connected to an autotransformer via an inductive filter L_F equal to 20 mH to deliver a single-phase voltage with an amplitude equal to 260 V. On the other hand, two parallel loads having the task to cause a problem of power quality are installed at the PCC. The first load is a single phase full wave rectifier supplying a variable inductive load with a maximum power P equal to 656 W. Therefore, two values of this loaded power were considered during the experimental tests. The first one is equal to $P_{Ll} = 656$ W while the second value is equal to $P_{L2} = 492$ W. On the other hand, the second load is considered as an inductive load with R = 2 Ω and L = 0.5 H.

Furthermore, the P&O algorithm which is used to control the dc-dc boost converter and the two proposed algorithms to compensate disturbing current introduced by nonlinear load were implemented on the dSPACE card under the Simulink/Matlab environment.

During all the experimental tests of the control algorithms, samples of voltage and current are acquired with a sample step of 1.1e-4 s.

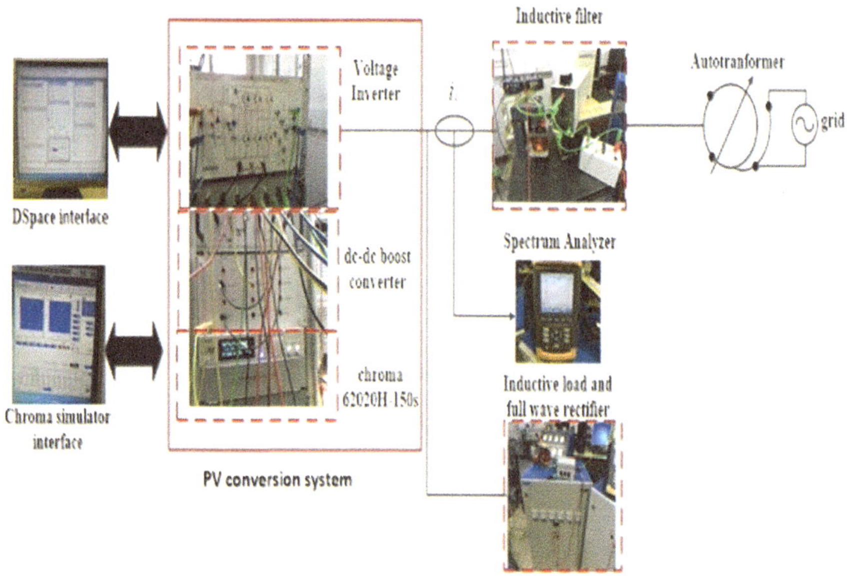

Fig. 17 Experimental testbed

3.3.1 Experimental Results During the P&O Algorithm Test

As first step, the P&O algorithm which is aimed to control the dc-dc boost converter was tested and validated. Table 6 presents the electrical power parameters of a PV module at the Standard Test Conditions (STC), which are fixed using the Chroma 62020H-150 s graphic interface. Consequently, the two characteristics of $I_{PV}-V_{PV}$ and $P_{PV}-V_{PV}$ are obtained by the Chroma 62020H-150 s programmable power supply as showed in Fig. 18. This P_{PV}-V_{PV} curve represents a maximum power point equal to P_{mpp} = 335,3 W corresponding to an optimum voltage equal to V_{mpp} = 144,31 V. While the corresponding optimum current is equal to I_{mpp} = 2.324A, referring to the $I_{PV}-V_{PV}$ curve shown in Fig. 18.

To validate the two proposed control algorithms of the voltage inverter with variable power, the Chroma 62020H-150 s programmable power supply was programmed to generate a sequential three levels of optimal maximum power (P_{mpp}). Figure 19

Table 6 Electrical parameters of a PV module set in the Chroma 62020H-150 s programmable power supply

Parameters	Values
Short circuit current: Isc (A)	2.494
Open circuit voltage: V_{SC} (V)	*169.1*
Voltage at MPP: V_{mpp} (V)	143.4
Current at MPP: I_{mpp} (A)	2.34
Power at MPP: P_{mpp} (W)	335.7

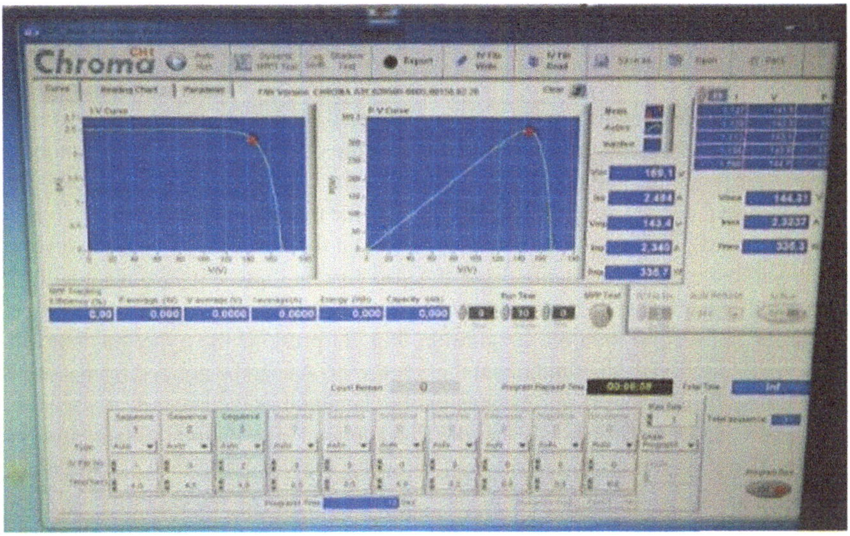

Fig. 18 I_{PV}-V_{PV} and P_{PV}-V_{PV} caracteristics programed in the Chroma 62020H-150 s programmable power supply

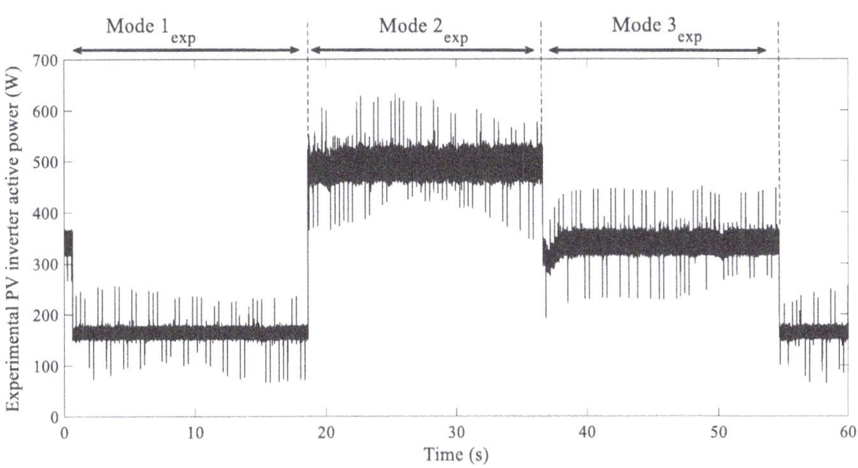

Fig. 19 Experimental active power of the PV inverter

illustrates the three cases of the active power injected by the voltage inverter called as the three experimental modes. In experimental mode 1 ($Mode1_{exp}$), the active power of the voltage inverter P_1 is equal to 165 W. Then, in the second experimental model 2 ($Mode2_{exp}$), P_2 is raised to 500 W, while in the third experimental mode 3 ($Mode3_{exp}$), the power is decreased to P_3 equal to 335 W.

3.3.2 Experimental Results During the Two Proposed Algorithms Test

The performances of the two proposed control algorithms are verified experimentally using the same testbed presented in Fig. 17. The time and the frequency representations obtained with the two proposed algorithms according to the experimental PV inverter power showed in Fig. 19 and under the two considered load conditions (L_{NL1_exp} and L_{NL2_exp}) are presented in Fig. 20 (for the first method) and 21 (for the second method). Consequently, it can be noted that for each case of both the inverter power and the used experimental load, the grid current represents a sinusoidal form with a low THD value between 3.5 and 11.6% for the first method and 1.88 and 6.94% for the second method. In addition, Fig. 22 presents the experimental THD curves as a function of the fundamental grid current. As can be noticed from simulation results, the THD level raises when the fundamental grid current decreases for a used nonlinear load. This proves that the THD depend only on the fundamental grid current. Moreover, compared to simulation results, the second control method is also more efficient for disturbing grid current compensation since the experimental THD values are lower, for all fundamental grid current than the first method. Therefore, the experimental results validate the obtained simulated results and prove that the proposed second method is more efficient to ensure good grid current quality for different PV inverter power even with the presence of a nonlinear load.

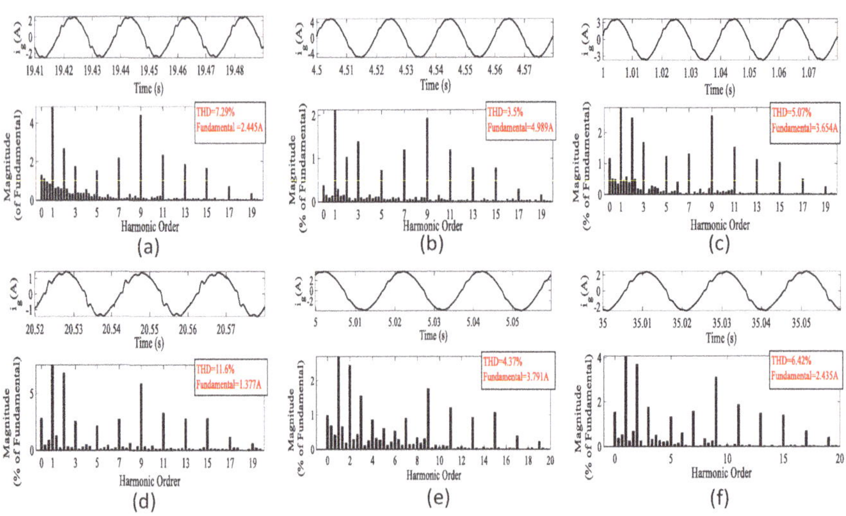

Fig. 20 Time and frequency representations of the grid current with the first control method in (**a**) $\text{Mode1}_{\text{exp}}$ with the load L_{NL1_exp}, (**b**) $\text{Mode2}_{\text{exp}}$ with the load L_{LN1_exp}, (**c**) $\text{Mode3}_{\text{exp}}$ with the load L_{NL1_exp}, (**d**) $\text{Mode1}_{\text{exp}}$ with the load L_{NL2_exp}, (**e**) $\text{Mode2}_{\text{exp}}$ with the load L_{NL2_exp}, (**c**) $\text{Mode3}_{\text{exp}}$ with the load L_{NL2_exp}

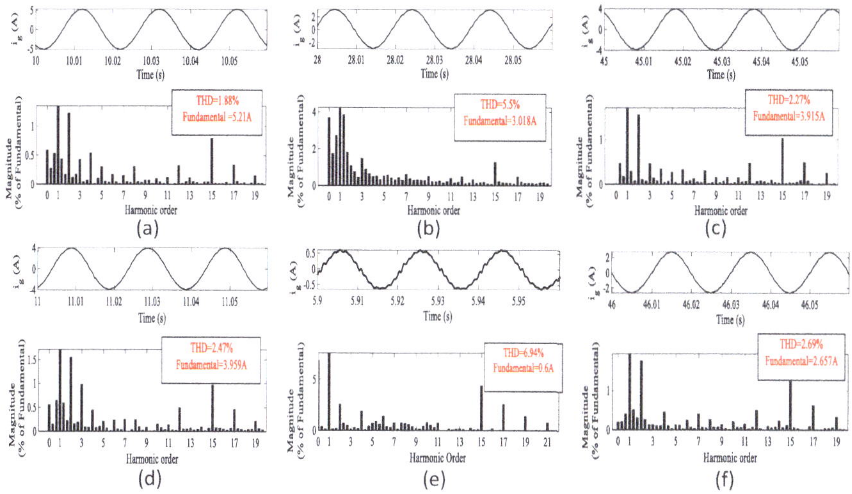

Fig. 21 Time and frequency representations of the grid current with the second control method in (**a**) Mode1$_{exp}$ with the load L_{NL1_exp}, (**b**) Mode2$_{exp}$ with the load L_{LN1_exp}, (**c**) Mode3$_{exp}$ with the load L_{NL1_exp}, (**d**) Mode1$_{exp}$ with the load L_{NL2_exp}, (**e**) Mode2$_{exp}$ with the load L_{NL2_exp}, (**c**) Mode3$_{exp}$ with the load L_{NL2_exp}

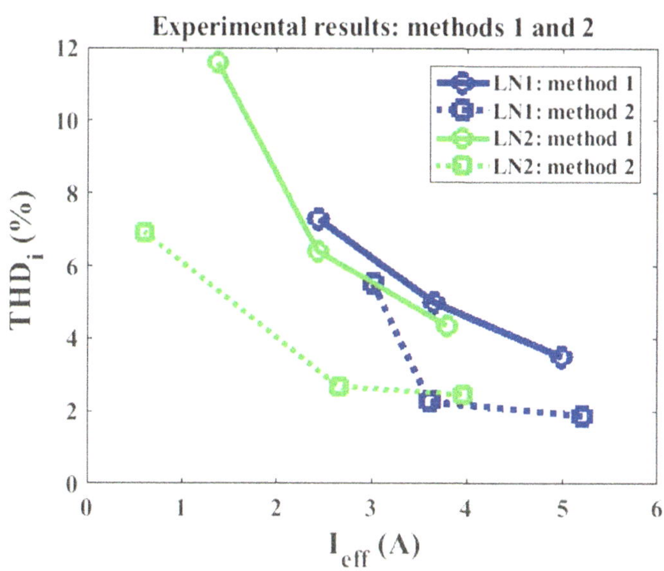

Fig. 22 Experimental THD as a function of fundamental grid current obtained with the two proposed methods

4 Conclusion

In this chapter, two new efficient current control techniques are proposed for the control strategy of the voltage inverter connected in a single-phase grid-connected PV system supplying nonlinear load. Each technique is based on a novel algorithm aimed to extract the disturbing grid current introduced by a nonlinear load at the point of common coupling. To cancel the disturbing current, the resonant controller is used in the first method while the second method is based on the use of several PI controllers. The performances of the two proposed techniques were studied by simulation for two different load cases and with a variable PV inverter power. Thus, it has been proved that with the two proposed techniques, the PV inverter is able to inject the solar power into the grid and to improve the grid current quality simultaneously, even under the variation of the load and climatic conditions. The effectiveness of the two proposed methods have also been verified and validated experimentally on an experimental platform. It has been shown from the simulation and the experimental results that the second algorithm is more efficient in improving the quality of the grid current.

References

1. Srinivas VL, Singh B, Mishra S (2019) Fault ride-through strategy for two-stage GPV system enabling load compensation capabilities using EKF algorithm. IEEE Trans Industr Electron 66(11):8913–8924
2. Hassaine L, OLias E, Quintero J, Salas V (2014) Overview of power inverter topologies and control structures for grid connected photovoltaic systems. Renew Sustain Energy Rev 30:796–807
3. Singh Y, Hussain I, Singh B, Mishra S (2017) Single-phase solar grid interfaced system with active filtering using adaptive linear combiner filter-based control scheme. IET Gener Transm Distrib 11(8):1976–1984
4. Eltamaly AM (2009) Harmonics reduction techniques in renewable energy interfacing converters. In: Renew Energy. Intechweb
5. Herrera RS, Salmeron P, Kim H (2008) Instantaneous reactive power theory applied to active power filter compensation: different approaches, assessment, and experimental results. IEEE Trans Ind Electron 55(1):184–196
6. Akagi H, Watanabe E, Aredes M (2007) Instantaneous power theory and applications to power conditioning. Wiley-IEEE Press, NewYork, NY
7. Wu L, Zhao Z, Liu J (2007) A single-stage three-phase grid-connected photovoltaic system with modified MPPT method and reactive power compensation. IEEE Trans Energy Convers 22(4):881–886
8. Kesler M, Ozdemir E (2011) Synchronous-reference-frame-based control method for UPQC under unbalanced and distorted load conditions. IEEE Trans Industr Electron 58(9):3967–3975
9. Abhijit K, Vinod J (2013) Mitigation of lower order harmonics in a grid connected single phase PV inverter. IEEE Trans Power Electron 28(11):5024–5037
10. Chilipi R, Al Sayari N, Alsawalhi J (2020) Control of single-phase solar power generation system with universal active power filter capabilities using least mean mixed-norm (LMMN)-based adaptive filtering method. IEEE Trans Sustain Energy 11(2):879–893
11. Pradhan S, Hussain I, Singh B, Panigrahi BK (2019) Performance improvement of grid-integrated solar PV system using DNLMS control algorithm. IEEE Trans Ind Appl 55(1):78–91

12. Li Z, Wang L, Wang Y, Li G (2020) Harmonic detection method based on adaptive noise cancellation and its application in photovoltaic—active power filter system. Electric Power Syst Res 184
13. Pereira HA, da Mata GLE, Xavier LS, Cupertino AF (2019) Flexible harmonic current compensation strategy applied in single and three-phase photovoltaic inverters. Electrical Power Energy Syst 104:358–369
14. Agarwal S, Jamil M (2015) A comparison of photovoltaic maximum power point techniques. In: 2015 annual IEEE India conference (INDICON), pp 1–6, 17–20 December 2015
15. Tampubolon M, Purnama I, Chi PC, Lin JY, Hsieh YC, Chiu HJ (2015) A DSP-Based differential boost inverter with maximum power point tracking. In: 9th international conference on power electronics-ECCE Asia, pp 309–314, 1–5 June 2015
16. Eltamaly AM (2018) Performance of MPPT techniques of photovoltaic systems under normal and partial shading conditions. In: Advances in renewable energies and power technologies, pp 115–161. Elsevier

Toward a Sustainable Agriculture in Morocco Based on Standalone PV Pumping Systems: A Comprehensive Approach

AA. Mana, A. Allouhi, K. Ouazzani, and A. Jamil

Abstract This paper involves a multi-level perspective to capture potential drivers and barriers of the solar water pumping in Moroccan agriculture. The effective design of such systems should consider many parameters, including the local climatic conditions, crops water requirements, and the suitable system configuration. Orderly, to gain an in-depth understanding of the conditions influencing the sector's transition toward the best energy management, three investigations are carried out: (i) a case study analysis, (ii) a comparative cost analysis, and (iii) a Strength-weakness-opportunities-treats analysis (SWOT). A general procedure for designing solar photovoltaic water pumping systems (SPVWP) is presented. To operationalize the followed theoretical approach, technical requirements, pump flow, produced energy losses, and monthly performances are estimated. Then, the simulation and non-linearity between water flow and radiation were tested to find the well-adapted system on a small scale. Illustratively, this work is strengthened through a case study to properly establish the relevance of photovoltaic systems and their performance under different utilization scenarios. The results show that an increase of 30% in annual performance and a decrease of 10% of system losses are observed when using MPPT DC converter for medium-sized crops. In terms of water use efficiency, the use of drip irrigation coupled with the direct coupling; the selected configuration is the best method to

AA. Mana (✉) · K. Ouazzani
Department of Energetic Engineering, Laboratory of innovative technologies (LTI), U.S.M.B.A, FES, Route D'Imouzzer, Morocco
e-mail: abdelali.mana@usmba.ac.ma

K. Ouazzani
e-mail: kamar.ouazzani@usmba.ac.ma

AA. Mana · A. Allouhi · A. Jamil
Ecole Supérieure de Technologie de Fès, U.S.M.B.A, FES, Route D'Imouzzer BP 242, Morocco
e-mail: amine.allouhi@usmba.ac.ma

A. Jamil
e-mail: abdelmajid.jamil@usmba.ac.ma

S. Motahhir and A. M. Eltamaly (eds.), *Advanced Technologies for Solar Photovoltaics Energy Systems*, Green Energy and Technology,
https://doi.org/10.1007/978-3-030-64565-6_14

save energy and manage water, especially for small crops like tomatoes. Besides, cost comparative analysis revealed that levelized cost of water (LCOW) of SPVWP is significant, approximately in the range of 0.08US/ m^3, which is very competitive comparing to other sources.

Keywords Solar pumps · Sustainability · Water · Agriculture · Energy · SWOT analysis · LCOW

1 Introduction

Over the last decade, climate change has been the topic of interest. Currently, it is considered an evident reality and impacts greatly agriculture, water supply, and our ecological sphere [1, 2]. In fact, the impacts are strongly related and agriculture is situated at the heart of this challenge [3]. Agriculture is also the world's largest driver of global warming [4] and, at the same time, the most affected by these changes [5]. Thus, a global agricultural transition is urgently needed to overcome the linearity of conventional evaluation methodologies that perceive farms as factories and counts plants and animals as industrial units [6]. This energy transition is the key step to rely on energy efficiency and restores social metabolism [7]. The challenge is further complicated by the need not only to produce more but also to sustain the entire food supply chain much more efficiently and reduce waste which has reached unacceptable proportions (estimated at 30%) [8].

Generally, recent studies reveal that intensive agriculture transforms landscapes, degrades biodiversity, and boosts genetic erosion. It pollutes the air, hydraulic sources, and puts human and animal health at peril [9, 10]. In this sense, sustainable agriculture is the result of an equilibrated solution to many productive, technological, environmental, and economic issues [11]. Among these, improving energy efficiency and reducing greenhouse gas emissions is vital. As agriculture becomes an energy-intensive consumer, an emergency call must be launched to attain a circular economy and sustainable goals [12]. Interestingly, sustainable agriculture has the unique potential to mitigate global warming and fortify the resilience of renewable solutions to face the impacts of climate change. Furthermore, several international organizations and initiatives are monitoring food security by creating the infrastructure for separate groups of stakeholders to come together to address economic, social, and environmental synergies [13, 14]. Agriculture in developing countries depends strongly on rain and is negatively affected if the water is insufficient. Thus, to improve water management, groundwater is pumped to reduce this dependency. Irrigation and rural water supply mainly use classic pumping systems; however, the unavailability or erroneous supply and high cost of diesel pumping remain the main problems that require special attention [15]. Closely, water use efficiency in agriculture will require an increase in crop water productivity and a reduction in water losses and pollution [16]. Many favorable policies for obtaining water use efficiency (WUE) are available.

These include suitable integrated water management practices such as efficient recycling of agricultural wastewater, bio-fertilizers, and solar irrigation [17]. Farmers can choose many habits to progress their sustainability by local resources to guarantee long-term farm effectiveness, environmental management, and improved quality of life [18]. Water conservation has become an important part of agricultural stewardship [19]. Solar pumping systems can be the most profitable answer when they are designed and sized accurately [20]. Moreover, they can be easily installed without needing long pipelines, and they are highly environmentally friendly [21]. Despite this fact, SPVWP systems currently have a shorter life cycle cost compared to diesel systems [22].

Solar water pumps are often viewed as an expensive technology, unable to pump enough water and sustainably throughout the year. In Morocco, Photovoltaic systems are the symbol of renewable energies and play a driving role in the energy transition. In 2020, renewable energies are expected to account for 42% of the energy consumed in Morocco and 52% in 2030. However, agriculture remains the biggest forgotten part of this transition since COP22. It depends largely on the rains and is affected by successive droughts. Poverty, famines, and power shortages remain the major problem in remote areas of the country. Subsidized butane water pumping systems are used mostly in Morocco for irrigation and pumping groundwater.

In the same context, the use of fossil-based irrigation methods infects and over-exploits the groundwater in an abusive manner, which explains the unavailability of water in this country.

The accelerated socio-economic development has resulted in increasing pressure on resources in water. Following the unprecedented increase in water needs in the agricultural sector, creating regional disparities and the emergence of acute pollution problems of water.

Although some researchers in the MENA countries have focused on solar irrigation and underlined SPVWP as an appropriate solution for a sustainable agriculture sector, few examinations have been carried out under the Moroccan context for the development of SPVWP. To fill this research gap, the study seeks to examine the potential of a local area under regional meteorological conditions to integrate SPVWP. However, pumping of photovoltaic water wins importance in recent years compared to conventional pumping systems. In addition, the pumping of water usually depends on conventional electricity or electricity produced with diesel and especially in African countries [23]. Solar water pumping minimizes reliance on costly and polluting conventional sources. Solar pumping systems are environmentally friendly and reduce maintenance costs and increasing fuel costs [24].

The use of SPVWP can contribute to socio-economic development [25]. It is the nominated solution for the present energy crisis for Moroccan farmers. This system conserves electricity by reducing the usage of conventional methods and conserves water by reducing water losses [26]. The proposed model ensures to work for small and medium irrigation needs to take advantage of the power produced by the system. Optimizing this type of PV irrigation involves considering both PV subsystems and irrigation requirements [27].

To sum up, a conceptual framework is proposed to figure out the most adapted SPVWP system for poor farmers and small and medium scales applications. Therefore, the purpose of this article is to introduce a detailed approach for the design, technical and economic evaluation of SPVWP systems and to give a guideline to local installers and decision-makers for integrating renewable sources in Moroccan agriculture. A first case study dealing with the irrigation of tomatoes and their water supply, to consolidate family agriculture is investigated. In addition, a second case study focuses on the irrigation of medium-sized crops (6 ha in olive trees) using the drip method to improve the performance of SPVWP systems. A complete performance analysis based on dynamic simulations is developed to highlight the most viable coupling configuration for local irrigation. In this regard, we adopted a SWOT approach to investigate the internal strengths and weaknesses, as well as the external opportunities and threats for PV solar pump development in Morocco.

This new advanced regional study will make a decisive contribution to the economic and social development of the country, through the enhancement of water potential and resources, the mobilization of various local actors, and participation in the development and implementation of structuring projects, and the strengthening of agricultural attractiveness."

Finally, based on the obtained results and according to available energy resources, this work revealed a significant contribution in the selected region to hold more than 17,000 solar pumps with capacities between 0.6 up to 40 kW, and the possibility to convert more than 80% of existing butane-powered pumps to PVWP. This proves that solar pumping must be widely considered as the most suitable solution to manage water and reduce almost agricultural CO_2 emissions.

2 Solar Photovoltaic Water Pumping System

Several works affirm that the PV pumping system has some leads to saving operating costs and reducing CO_2 emissions in comparison to conventional methods. Also, the investment payback period is found to be 4–6 years [28], when the energy is from the sunlight. Apparently, photovoltaic is one of the main applications of the autonomous pumping systems [29]. It consists of a water pump driven by an electric DC motor and powered directly by solar panels through an MPPT system. PV pumping systems are widely used in areas for agricultural purposes. However, there are many disadvantages of PV systems [30]. They have high costs, unreliable operation, and maintenance which are considerably higher if the batteries are used [31–33]. Recent works have focused on system modeling for optimum management of energy to store the monthly irrigation needs [34]. SPVWP systems can include a water tank [35], or electrical storage of energy [36, 37]. Most of the studies dealt with remote-controlled water storage [38]. In these studies, valid areas for the implementation of SPVWP were evaluated through the processing of local spatial datasets. Many kinds of research on the sizing of the photovoltaic pumping system have been published [39, 40]. These studies are centered on the simulation of the functioning of

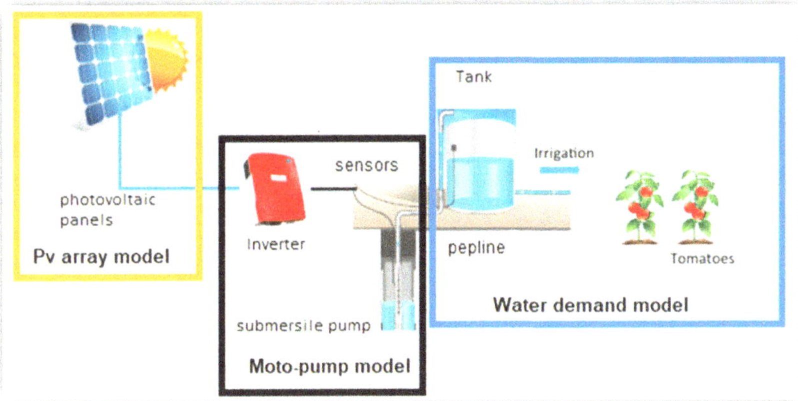

Fig. 1 Schematic diagram of PV water pumping [41]

each component of these SPVWP systems. However, the design of any cost-effective solar energy system requires experimental data that are not willingly available. The widely proposed system consists of the use of the directly coupled system to produce electricity. For more flexible irrigation planning, the system can supply water to a reservoir or a collection basin, then water runs from the basin to the crops through a drip irrigation facility. The direct-coupled solar water pumping system to supply water for drip irrigation is shown in Fig. 1.

The PV modules are linked in series and parallel for the working of a pump motor subsystem that will demand power to generate a certain total of pressure and water flow. The pump speed depends strongly on the solar radiation received by the PV generator.

One should state that, if an MPPT is used, the electrical energy delivered to the motor is nearly stable. DC or AC motors generally use the centrifugal pumps in PV pumping systems, which have generally a long-term reliability and hydraulic efficiency differing from 25 to 35% [42].

Direct-coupled DC solar pumps are simple and consistent [43], but cannot run at the maximum power point of the PV generator as the solar radiation varies during the day [44]. It is increasingly evident that SPVWP systems are a key criterion for water efficiency in Agriculture. Many researchers have investigated those systems in the MENA and sub-Saharan region, but little literature analyzed the case of Morocco. To review more subsequent literature, Table 1 gives a summary of regional investigations.

It has been concluded that SPVWP systems can be used efficiently for irrigation in agriculture and rural areas. The price of the water pumped by those systems is much less classic methods. Conventionally, almost all studies related to solar pumping systems reported positive impacts on their widespread integration.

Table 1 Summary of SPVWP performance investigation in MENA and sub-Saharan region

Ref.	Location	System specifications	Outcome
[45]	Nigeria	Pumping rate: 2.6 m^3/h Pv array: 1.6 kW_p Borehole: 30 m	– Solar pumps can replace conventional AC pumps – Solar pumps can meet the water demand of 20 m3/day at the average solar radiation level of 5 kWh/m2/day
[46]	Africa	39 cases studied Medium Heads: 50 m Power range: 0.7–4 kW_p	– Due to their technical ability and their cost-effectiveness, solar PVP system can replace the diesel engines for drinking water supply system and irrigation in Africa and rural areas – Average price 30cents/m^3
[47]	Algeria	Pv array: 2.4 kW_p Pumping head: 0–120 m Flow rate: 0–30 m3/h	Multistage centrifugal solar PV pumps coupled to DC motors show significant efficiency for high head applications (40~70 m)
[48]	Algeria	Pumping head: 12.5~13.5 m Tomato needs: 1012 mm/day	PVWP used efficiently for water pumping in agricultural sectors. Ensuring Suitability for small-scale irrigation (2 ha)
[49]	Tunisia	Pv array: 85 m^2 (100Wc, 24 V) Pump system efficiency: 45(%) Water pumped: 40 m^3/ha/day	– Studied the economic viability of SPVWPS to satisfy water requirement in Tunisia desert – Cost of water: 0.18 \$/m^3
[50]	Egypt	water head = 40, 80, 120 m water flow: 15 m^3/feddan/day (for drip irrigation)	– The cost of the water unit pumped by PV systems (0.11~0.33\$/m^3) is much less than that pumped using diesel systems (0.51~0.94\$/m^3) – The water cost is more sensitive to the PV cell's prices than the life time periods
[51]	S.Arabia	Batteries: two batteries of 12 V Pv: 2 panels (35 V generated)	This study presents the usage of photovoltaic electricity in an automated irrigation system, which optimize water requirement
[52]	Algeria	Pv array: 1.5 kW_p Pump capacity: 80 l/min Head: 33 m Moto-pump efficiency: 30%	– The system runs without battery and complex electronic control, as a result the initial cost is low and Maintenance, costs are saved – Directly coupled photovoltaic water pumping systems are suitable for low head applications
[53]	Algeria	–	– It is proposed to integrate an intelligent control method for MPPT of PV systems, to ensure high flexibility.

3 National Background and Target Area

Morocco is situated in the Maghreb region of North Africa, separates by the Mediterranean Sea from Spain to the north. Recent global warming studies confirmed that Morocco is among the countries more menaced by climatic change [54]; in 2008 the Moroccan government announced an agricultural strategy called "Plan Maroc Vert" [55]. The objective of this strategy is to promote the productivity of agriculture by addressing climate change as well as over-exploitation of groundwater, and poverty [56]. In Morocco, irrigation is very water-intensive; it is estimated that about 83% of available resources are used by agriculture with an efficiency of less than 50%. Morocco is a dependent nation on agriculture as it is responsible for 20% of the Gross Domestic Product. This sector plays an important role in food security and sustainable development. According to the Ministry of Agriculture and Maritime Fisheries data, the total area of Morocco is 71.085 million hectares. Agricultural, pastoral, and forest lands represent 38.7 million hectares or 54.4% of the total area. The UAA lands cover almost 9 million hectares, forest occupies 5.8 million, and rangelands represent 21 million hectares. Irrigated agriculture in Morocco occupies only 19% of the UAA, as illustrated in Fig. 2.

The agricultural activity is exemplified in the Sebou River, which is responsible for half of Morocco's sugar production, as well as most of the country's olive production. The most fertile part of the Sebou is the Saiss Basin (Fig. 3). It holds about a quarter of Morocco's arable land but only uses a little of the national water reserve. Moreover, this water is often overexploited and poorly managed. The potential region under investigation Saiss (Fez-Meknes) has a strategic geographical position. Given the beneficial climatic conditions for most crops, the region of Fez-Meknes knows a great diversity of plant and animal production, in addition to the existence of significant activity of the food industry. Fortunately, the Fes-Meknes Region covers an area of 40,075 km^2, representing 5.7% of the Kingdom's surface area (i.e., 14% of soluble

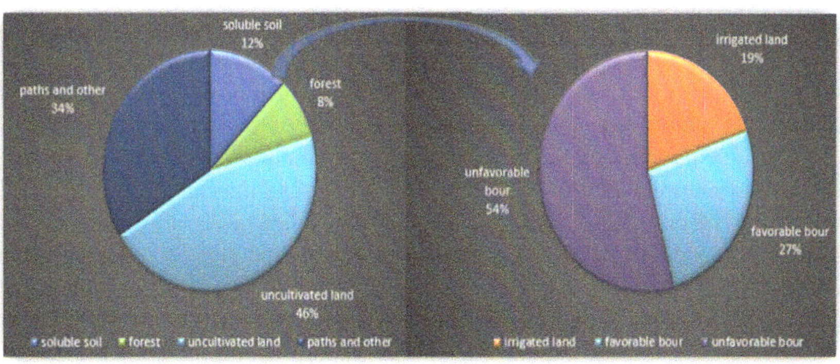

Fig. 2 Part of irrigated agriculture in morocco[a]. [a]according to annual reports of the Ministry for agriculture (http://www.agriculture.gov.ma/)

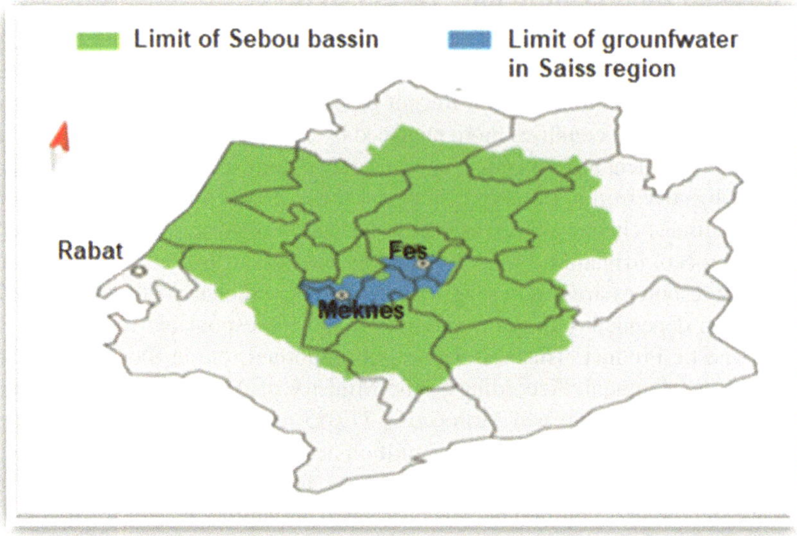

Fig. 3 Boundaries of the Sebou basin, and Sais's region [57]

soils). The land use is relatively varied with a dominance of cereals (60%), the rest is occupied by fruit plants (14.4%), legumes (6.6%), industrial crops, beet and sugar (4.2%), oilseed crops (3.6%), vegetable crops (3.1%), and fodder crops (1.7%).

4 Objective and Research Inputs

The objective of this work is to present a low-cost solution and assess the performance of photovoltaic pumping to power small-scale activities to consolidate familial agriculture in Morocco. The solar radiation data were collected from local databases and then imported to PVsys software database using its meteorological tool [58]. The Irrigation water requirements were evaluated using the software Cropwat [59].

It is designed as a useful tool for carrying out estimations of reference evapotranspiration, crop water, and more explicitly the design and management of irrigation. Calculations of crop water requirements and irrigation needs are generated with inputs of climate data, crops, soils, and local geographic data. It provides recommendations for improving irrigation practices, planning irrigation schedules under changing water supply conditions, and valuing production under a scarcity of rainfall [60].

This paper represents the theoretical design and computer simulations analysis, as a method for optimizing direct coupling systems and evaluating their performance. The life cycle cost analysis is also carried out. Generally, optimal sizing methods use "the worst month" or average values. Consequently, systems can be oversized or undersized. In analytical methods, mathematical equations are used for the sizing of the system, but those methods are limited and less exact. In contrast, numerical methods are most applied due to their accurate dynamic simulations.

A new contribution of this work includes an assessment of configurations according to the system of the regulator of the installation of pumping:

- Direct coupling,
- MPPT DC-DC regulation.

In this study, two systems in different locations were considered to deliver a daily need of 36 and 134 m^3/day with good depth in the range of 20–50 m. For external environmental data, it must be used as input during water needs calculation as well as SPVWP design.

Thus, the work follows an observation of regional meteorological data before the use of METEONORM [61]. The Saiss region is characterized by the diversity of its soils where three main types of soils can be identified. It should be noted that Saiss is characterized by rich soils and is full of important agricultural potentialities.

In terms of precipitations, the region is characterized by a continental climate in the north, cold and humid in the mountains, and a semi-arid climate in the Missourian highlands. This variability of climate leads to a diversity of cultures that adapt better to the specificities of this region. The average annual precipitation is 520 mm, which places the zone in a favorable market. It should be noted, however, those inter-annual variations can be very important. The rainy period extends from November to April. The biologically dry period lasts 5 months (May to September) as shown in Fig. 4.

While Fig. 5 shows the temperature profile observed in the period 2000–2018 with the uncertainty of yearly values T = 0.3 °C. Locally, the average annual temperature is 18 °C.

The annual averages of minimum and maximum temperatures are 11.3 °C and 26.5 °C, respectively. The coldest month is January, followed by December and February. The average temperature of the coldest month is 11 °C and the minimum

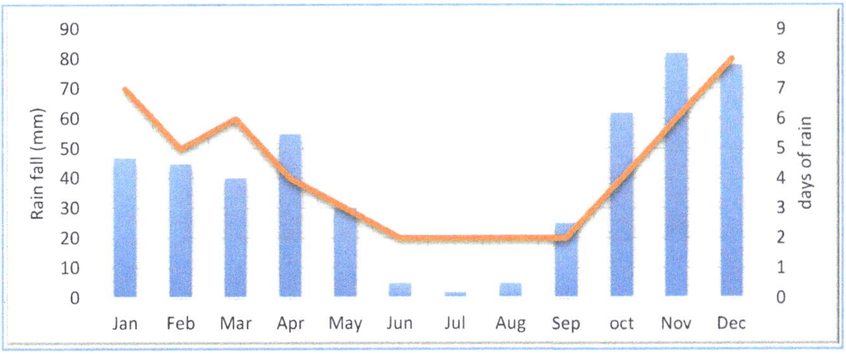

Fig. 4 Rainfall in Saiss region. 2000–2018, 4% uncertainty of yearly values

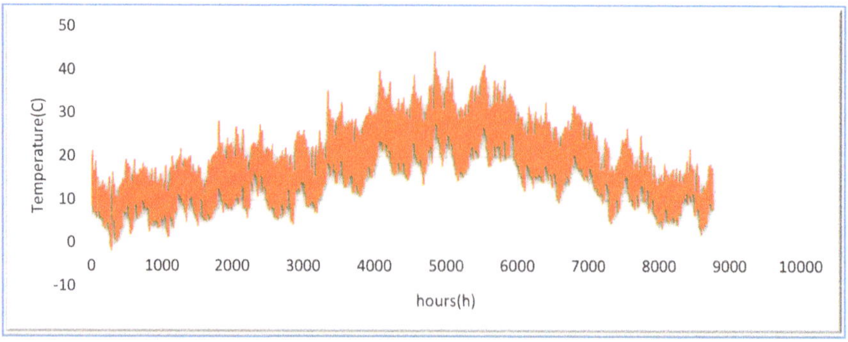

Fig. 5 Annual temperature profile

temperature is 5.5 °C. The hottest month is August, followed by July. The mean exceptionally high temperature of the warmest month is 38.5 °C. It coincides with the high radiation registered in the region. Figure 6 shows the monthly distribution of global irradiance, the global solar radiation is maximum during the dry season (May to August) and distributed from 9 to 16 am of the day, which particularly fits the irrigation period.

To summarize observations during the year 2016, Table 2 gives an overlook of climatic parameters that can influence directly or indirectly solar pumping systems. It provides illustrations of detailed weather data for the area of Fes-Meknes temperature, the predictable number of sunny days, evapotranspiration, and rainfall status to

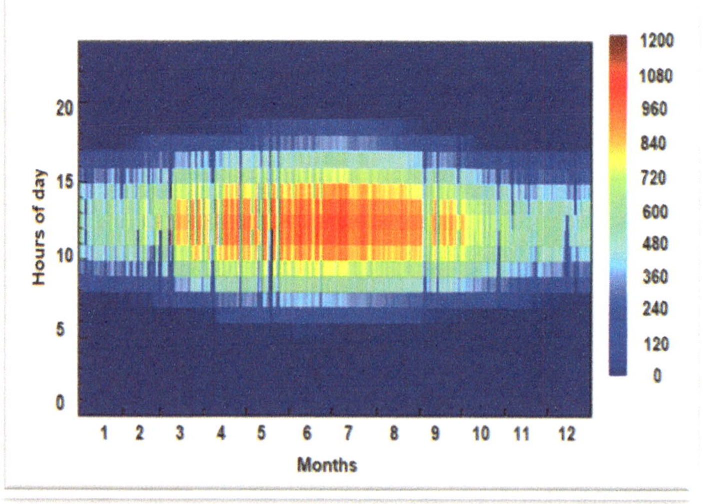

Fig. 6 Irradiance GHI calculated (W/m^2) for the Saiss region

Table 2 Monthly climate parameters in Saiss region[a]

Month	T_{max} (°C)	Humidity (%)	Wind (km/day)	Sun (Hours)	ETo (mm/day)
January	20.0	74	233	6.7	2.06
February	18.0	67	311	7.3	2.74
March	19.0	64	294	8.0	3.35
April	23.0	67	302	8.9	4.23
May	27.0	64	294	9.6	5.27
June	33.0	52	328	12.1	7.52
July	38.0	49	294	12.1	8.37
August	38.0	50	302	11.9	8.23
September	33.0	58	277	10.7	6.09
October	29.0	62	294	8.9	4.58
November	20.0	66	302	7.7	2.80
December	18.0	72	233	6.8	1.87

[a]Own estimations by comparing METEONORM and local data from different sources

lead more researches in this region. The literal data for water pumping requirements were examined, measured, then compared and presented in the results section.

5 Conceptual Framework

In this section, a brief description of the sizing process and conventional mathematical model are reported. The performance and economic indices are also carried out.

5.1 Sizing Procedure

- **Water requirements**

The assessment of crop water requirements is a critical step to sizing every irrigation system since it depends on climatic and crop parameters [62]. In theory, the water requirements are precisely defined as the difference between the amount of rainfall and the cultural evapotranspiration (ETP) [63].

Theoretical methods can be used (e.g., the Blaney–Criddle method) to determine the reference crop evapotranspiration (ET_0) [64], especially under "extreme" conditions, the ET_0 is undervalued (up to 60%), while in calm, humid, clouded areas, the ET_0 is overrated (up to 40%) [65]. There are many other equations more or less simple, in our case, evapotranspiration can be estimated from several years weather data from the nearest meteorological station or through the maximum temperature (T_{max}) using the regression following equation, [66]:

$$ET_o = 0.16T_{\max} + 0.14 \quad (1)$$

The evapotranspiration is calculated from ET_o taking into consideration a specific cultural coefficient K_c dependent on the type of crop and growing phase [67].

$$ET_c = ET_o.K_c \quad (2)$$

The net requirement CWR and the gross requirement IWR are calculated by the following equations, respectively, [68]:

$$CWR = (ET_c - P_u) \quad (3)$$

$$IWR = \frac{CWR}{\eta_r} \quad (4)$$

with P_u represents 60% of the total monthly rainfall, and η_r is the overall irrigation efficiency (for Drip irrigation, $\eta r \cong 95\%$).

Referring to local authorities, carrying out on-site interviews with farmers and monitoring chronological archives of water demand are usually the main means of validating the calculated demand for water [69].

- **Selection of the pump motor**

For the application of PV water pumping, there are two main types of pump technologies: Pumps positive displacement and centrifugal. Positive displacement pumps are used in low volumes and cost-effective systems. Centrifugal pumps are generally exploited for applications with photovoltaic energy because of their feeble initial couple drive, and it runs with very low sunshine, for medium and high flows [70].

The daily hydraulic energy of the pump and its efficiency can be expressed by the following equations:

$$E_{hyd} = \rho.g.H.IWR \quad (5)$$

$$\eta_p = \frac{P_{hyd}}{P_{mec}} \quad (6)$$

where E_{hyd} the hydraulic and H is the total head.

The hourly rate is calculated by the following formula:

$$\phi_h = \frac{IWR}{\Delta t} \quad (7)$$

The global system efficiency η_g is equal to the product of the PV array efficiency and the subsystem (motor and pump) efficiency η_{sys} and given by the following expression:

$$\eta_g = \eta_{pv}\eta_{sys} \tag{8}$$

where η_{pv} is the efficiency of the PV array under running conditions, given as following:

$$\eta_{pv} = f_m[1 - \alpha(T_c - T_r)].\eta_{pr} \tag{9}$$

where f_m is the matching factor ($f_m = 0.90$), η_{pr} is the efficiency of the PV array at a reference temperature T_r. Factor α is the temperature coefficient for cell efficiency.

- **Photovoltaic field**

Two procedures are handled for sizing pumping systems: Analytical method and the graphical method [71]. The sizing of the photovoltaic field proper is the series-parallel arrangement of the panels which is determined in conjunction with the choice of the pump to ensure compatibility between supply and energy requirements [72]. The effective area of the modules can be found by following:

$$A_{pv} = \frac{(\rho g H \phi)}{G_t \eta_g} \tag{10}$$

where G_t is the solar radiation received, and global system efficiency η_g.

The peak power P_c regarding hydraulic power efficiency can be expressed by the following equation:

$$P_c = \eta_r A_{pv} \tag{11}$$

with η_r overall efficiency of the PV system, assuming 20% losses due to temperature. And A_{pv} is the total PV array.

The total number of modules N_{pv} constituting the Generator PV is calculated by the following formula [73]

$$\boldsymbol{N}_{pv} = \frac{P_c}{\boldsymbol{P}_{pv}} \tag{12}$$

with P_{pv} is the power of one PV module.

5.2 Examined Configurations and Simulation Process

The purpose of this section is to examine the optimal configuration of the PV system able of providing a submersible solar pump to meet the needs of remote crops.

Consequently, two approaches are supported in search of an optimal design according to the size of the plots, the energy and hydraulic needs, and the depth of the wells.

Firstly, the direct coupling configuration, which is most common in small-scale practices because of its technological simplicity. Secondly, an improved configuration with a DC-DC converter equipped with an MPPT regulation assisting as a linear current booster.

Indeed, to perform the PV system, PVSYST was used. It is designed to consider all the parameters and the avoidance of the losses of the system and the calculation of its performance.

5.2.1 Performances Parameters

The performance of PVWP systems is evaluated on a monthly basis using many significant parameters including energy output at the pump, system array and system energy losses, system efficiencies, and especially performance ratio which represents the ratio of the energy produced compared to the energy that would be produced by a system under standard conditions [74]. To calculate the performance ratio, two critical parameters are calculated, the system's final yield Y_f and reference yield Y_r as expressed, respectively, in the following equations:

$$Y_f = \frac{E_{net}}{P_o} \tag{13}$$

with E_{net} is net energy output in kWh and P_o the Installed PV array in kW. For PVWP, the system yield is the effective energy at Pump.

$$Y_r = \frac{H_i}{G_r} \tag{14}$$

with H_i is Total plane irradiance in kWh/m^2 and G_r is PV reference irradiance in kW/m^2.

The PR is the important factor to evaluate the performance of PVWPs and their continual guarantee and it is stated as a percentage [75].

$$PR\% = \frac{Y_f}{Y_r}100 \tag{15}$$

Fig. 7 Location of farms, the village of Kasbah Hartal (33.930342, −0.60077°)

5.2.2 Field Survey and Cost Analysis

In Saiss region, the investigation shows that groundwater depth is between 9 and 19 m, it is s decreasing with an acceptable average of 0.5 m/year; which represents an opportunity for solar pumps and smart irrigation to replace classic and abusive methods. The investment and operating costs can be very different for the same type of energy used. This study focused on small and medium-sized plots of 1–6 ha to encourage the use of the PVWP systems for familial agriculture. This investigation helps to underline the most adapted system for the region. Crops selected are near to farmers and their homes (Fig. 7). For a long-term performance for pumping systems so they can pump water for trees and animals, dry ailments, and generate electricity for locals.

In order to acquire the necessary information, the approach used does not seek to compare farms with each other, but comparison for one potential farm. Information is analyzed and evaluated using triangulation. It is a method of analysis where potentially opposing results from multiple sources are compared to find valid data.

Investment and operating costs would be compared according to different energy systems for estimating the cost of 1 m^3 of water. The cost of installing a solar power system was evaluated from PVSYST and compared with market prices. It is used to calculate installation costs, from several operating parameters such as flow and well depth or borehole. The software determines the number and cost of solar panels and the cost of the electric pump and the drive to install [76]. Maintenance costs were estimated from the farmers surveyed. The cost of installing an electrical energy system includes the cost of installing and the cost of the connection. The latter was therefore not considered in the investment calculations because it depends on the distance from the electricity network. The operating cost is the cost of using energy, by the farmer, for pumping irrigation water over an average rainfall year. It is estimated by the following formula:

$$OC = \beta.Q.t \tag{16}$$

where β is the hourly pumping cost per cubic meter, with $t = 1700$ h/year estimated for the region; And Q is the hourly flow rate.

The economic feasibility focuses on the levelized cost of water as an economic parameter to evaluate the cost-effectiveness of the designed SPVWP system.

- Levelized cost of water (LCOW) in $/m^3 can be calculated as [77]:

$$LCOW = \frac{\text{Annual operating costs}}{365 \times Q} \tag{17}$$

6 Cases Studies Analysis

The purpose of this paper is to develop a decision-making approach for optimum implementation of solar pumps in the country within various regions. Thus, the reflective methodology is applied for two scenarios using two possible configurations. The main aim is to prove the opportunity for reliable integration of PV systems. In this regard, the first case corresponds to 1 ha of tomatoes adapted for small farmers. The field survey in the region shows that most farmers involved in market gardening have small plots. 70% have farms that do not exceed 6 ha. The second case concerns the irrigation of 6 ha of olive trees in Meknes-Fes, which is the emblematic region of olives in Morocco.

6.1 Comparison of Two Configurations for Seasonal Irrigations and Small Crops

- **Determination of the water requirements**

It was experimentally investigated that tomatoes need an average of 145 days to produce. It can be grown 2 times even 3 times in the year. However, it is necessary to study the worst case when the tomato is planted in May, for the development phase coincides with the dry season.

The assessment of irrigation needs remains a critical step before modeling pumping systems. For this reason, ET_0 estimation and water requirements for vegetable crops with accurate meteorological information were imported and calculated using the CROPWAT tool, then compared to the analytical calculation for the region studied.

According to the results, the reference evapotranspiration (ET_0) is higher from May to September and highest in August with 9 mm/day (Fig. 8a). This coincides with a decrease in precipitation (Fig. 8b). Using the information provided in (Fig. 4) and the calculation of evapotranspiration, the effective rainfall is estimated at 80% of the total monthly rainfall.

Notably, (ET_0) is maximal with an average of 7 mm/day at the time of the year when the higher values of irradiance were recorded and the scarcity of rainfalls.

It is well known that irrigation needs depend on many parameters: climatic conditions, type of plantation, and irrigation technology. As necessary and critical to

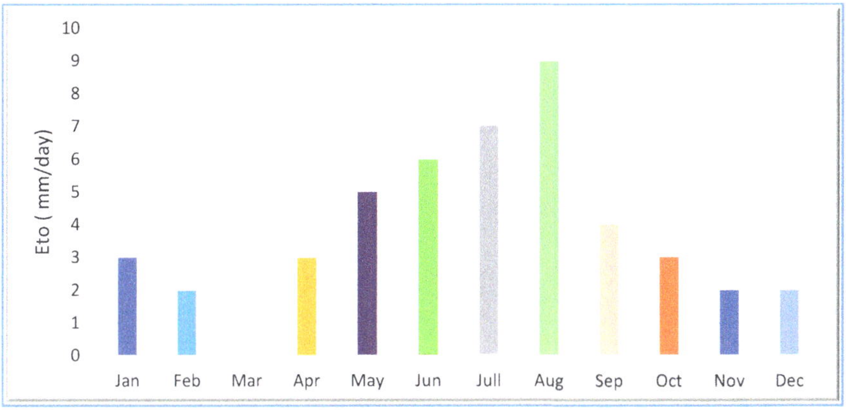

a

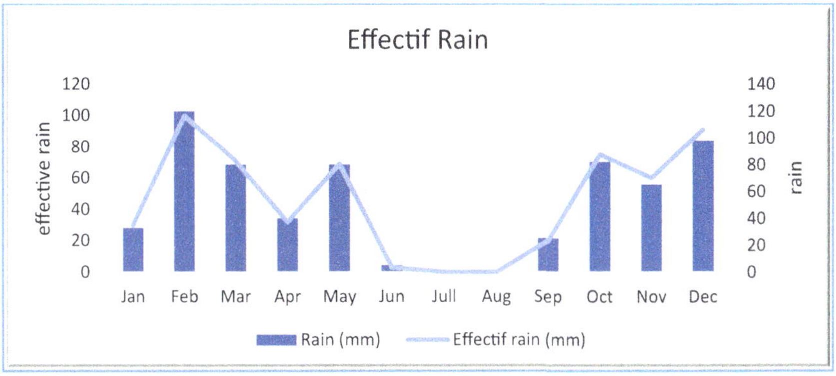

b

Fig. 8 **a** Reference evapotranspiration calculation for Sais Location. **b** Rain and effective rain calculation determining specific cultural coefficient (Kc) during the entire cultivation phase. As shown in Fig. 9, for tomatoes planted in May, Kc varied between 0.6 and 1.57.

Additionally, the monthly average water requirements for growing tomatoes and the trend of specific evapotranspiration are presented in Fig. 10. The water demand registered the peak value in August when the evapotranspiration is maximal too. Thus, to reach efficient irrigation, it is necessary to determine the water requirement per decade.

Drip irrigation is suggested to ensure the needs of tomatoes. As shown, the applied model and calculations estimate the daily average water demand for tomatoes at 36.8 m^3/ha/day. As an example of the first decade of September, in which irrigation requirement is in the order of 100 mm/dec, an average of 10 mm/day, also 100 m^3/day.

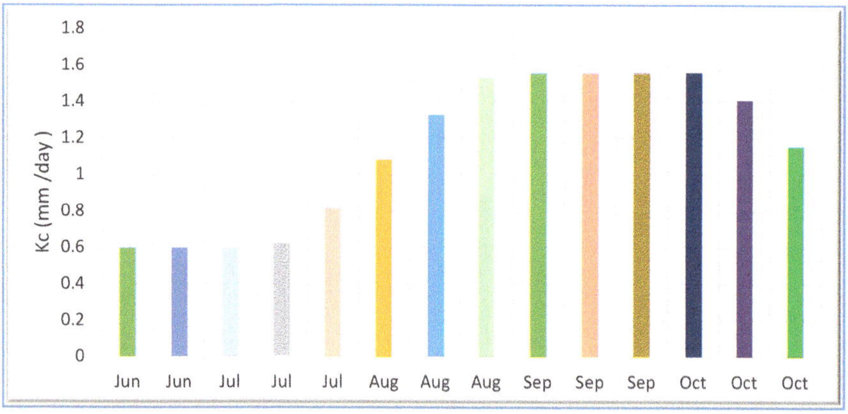

Fig. 9 Tomato K_c coefficient progression between Jun and November (Planting date in May)

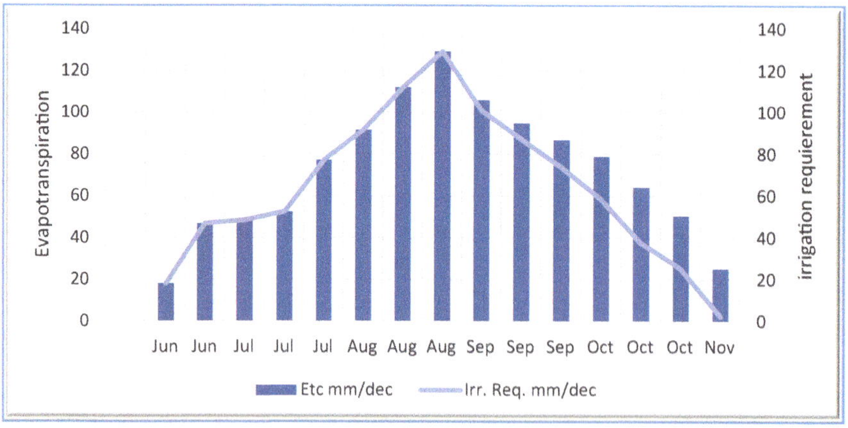

Fig. 10 Irrigation water requirement per decade for tomatoes. Plant expressed in millimeters (1 mm = 10 m3/ha)

Table 3 Water requirements in Sais region

Crops	Practical (m^3/ha/day)	Calculated (m^3/ha/day)
Olive	17–20	22
Vegetable	30–70	60
Forage plant	20–60	50
Cereals	20–40	15–30

Table 3 presents the calculated water requirements for the most frequent crops compared to practical water needs. Furthermore, there is not much difference are observed which confirms the feasibility of current calculations.

- **PV Sizing results**

Results reveal that to cover such needs, it is proposed to use a solar pump of about 2.9 kW, with a peak power of 3.7 kW_p of the surface PV and 20% of accepted missing. The characteristics of the pump and PV module to be installed for this case are presented in Table 4.

Considering the daily need in peak months. The pump selected was tested without regulation for many depths. The results show the performance of the direct coupling for low heads (34 and 43 m). As shown in Fig. 11, for example at 43 m deep well the system can attend 320 m^3/decade, with a daily flow of 34 m^3/day, at the average power of 1.5 kW and with hours of daily irrigation variably are between 3 and 6 h.

At reference conditions of Total Irradiance = 1000 W/m^2, Cell temp = 25 C, the nominal efficiency of the module is about 13.23% with a maximal power of 169.970 W, where the max power voltage is 23 V and the max power current is 7.4 amps and the short circuit current is 8.1 amps as shown in Fig. 12. The system proposed can provide an average of 8 m^3/h when using an MPPT converter. Under usual conditions in the region, it is possible to attain 26 m^3/ day. The system was examined through the whole year with dynamic simulation in order to estimate hourly outputs. It can be obviously realized from Fig. 13 that for the system with an MPPT/DC converter, the pump starts to operate with a minimum of 180 W, and the flow rate increase until the input power reaches 1.4 kW.

Table 4 The main characteristics of the system to be installed

Component	Characteristics
Centrifugal pump	Model PS4000 C-SJ8-15, Lorentz Pump Technology Centrifugal Multistage Deep well pump Motor DC motor, brushless Associated or Integrated MPPT converter with voltage range 220–340 V
PV Array	PV module Si-poly Model YL175P-23b Manufacturer Yingli Solar
PV modules	Coupling 15 PV modules, with a total area of 19.5 m^2
Pipe and well	The length of pipe was estimated at 40 m with 3 elbows, it is a pipe of 6/8 mm in diameter. The well depth is 37 m
Well	The well depth is about 37 m

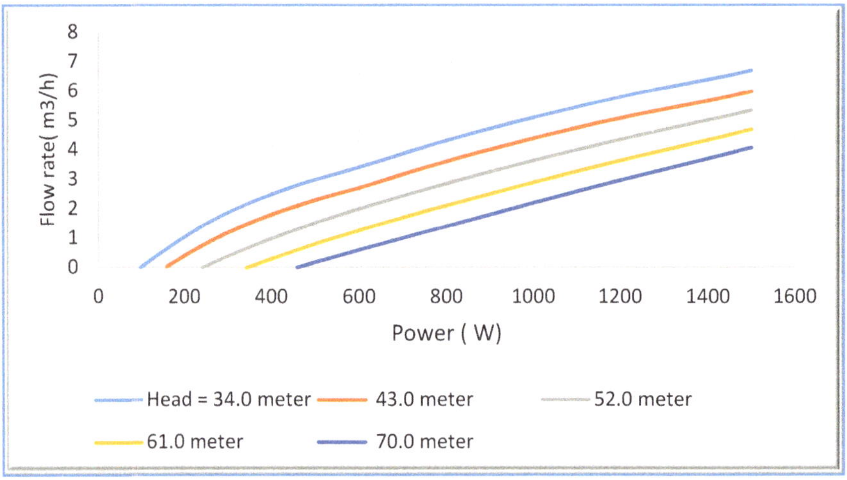

Fig. 11 Daily water flow for many heads depth

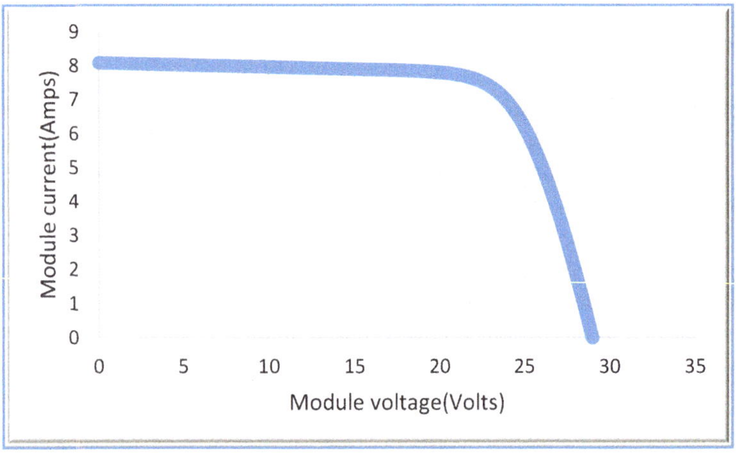

Fig. 12 PV electric characteristics

For both configuration direct coupling and MPPT/DC, the hourly flow in the function of available energy at the pump are illustrated in Fig. 14. The system equipped with the MPPT DC converter is capable to provide water for low pump effective energy as faced to the direct coupling configuration. This could actually justify that configuration with MPPT DC converter starts to pump water at lower radiation and can also reach a high flow during the day.

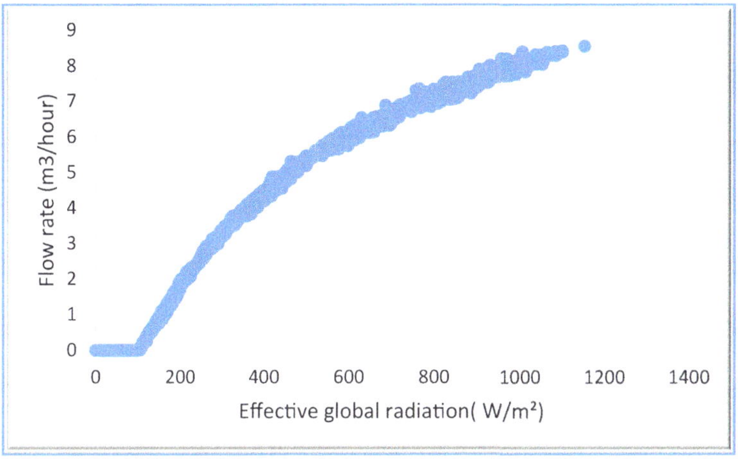

Fig. 13 Flow volume rate by radiation

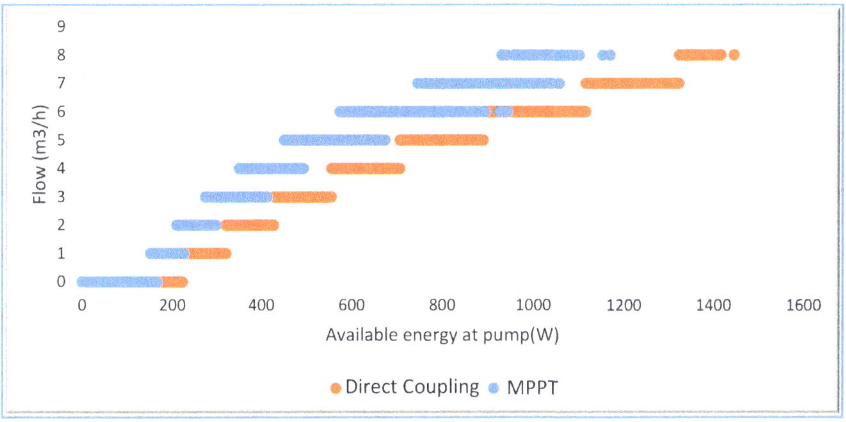

Fig. 14 Flow as a function of available energy

- **Overall system performance**

The comparison of configurations analyzed predicts the performance of systems with MPPT/DC. Figure 15 illustrates the monthly energy available at the pump and the losses of the system for both configurations. The numerical results indicate that the monthly values were expressively better, and the losses are reduced for the installation using a DC MPPT converter. In the dry season, they were generally less than 4.2 kWh/kWp/day for the direct-coupled installation and greater than 5 kWh/kWp/day for the regulated.

Based on the conducted analysis, the PR gives a monthly review of whole system efficiency, it includes PV array losses, module quality, mismatch, cabling, etc.), and

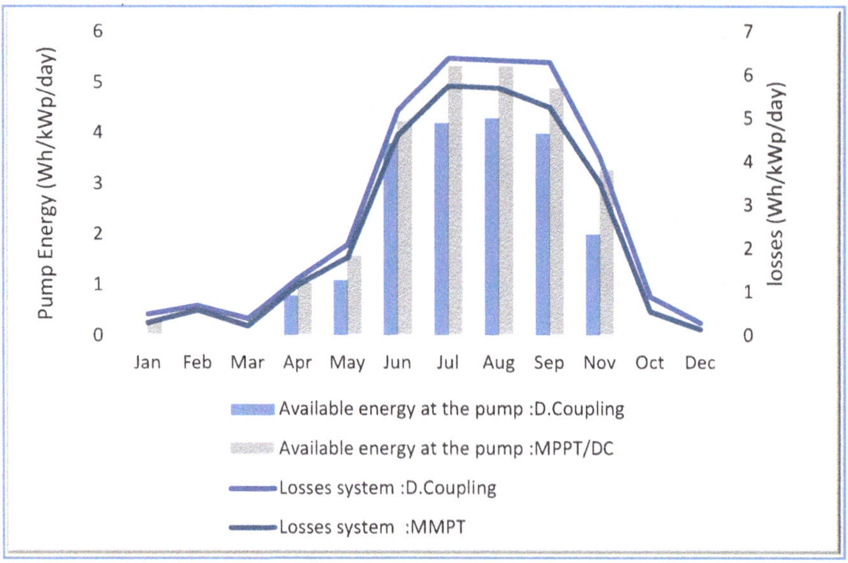

Fig. 15 Pumped energy and losses for the two configurations

system losses (Solar pump, controller efficiency or storage/ battery/standby losses, etc.). Clearly, in Fig. 16, it can be seen a significant variability of the PR for both configurations. The PR diverged between 58 and 73% in the dry season for the MPPT structure. While it varies between 40 and 55% for the direct coupling design.

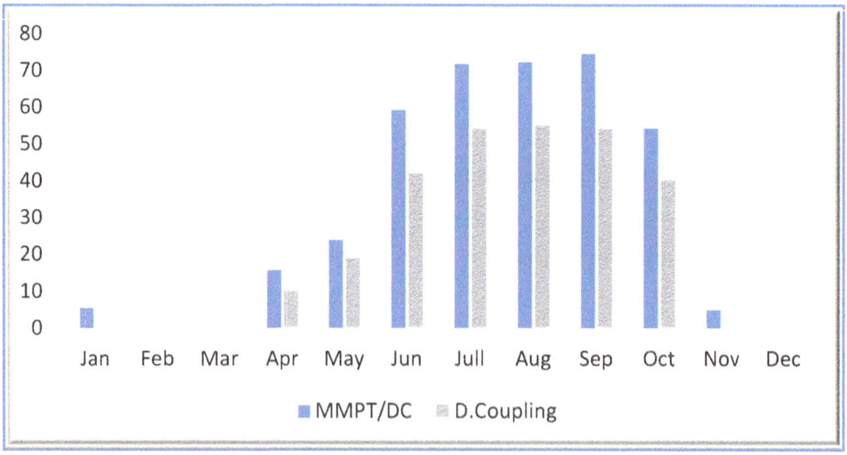

Fig. 16 Performance ratio PR of both configurations

7 Configuration with MPPT Coupled to Drip Irrigation for Olives

In this section, the results are substantially better when regarding requirements, assumptions and olives characteristics. The choice of olive was taken by consideration, to give an opportunity for the symbolic tree in all MENA countries. Through a real case study, the operation of coupling drip irrigation was underlined as the most favorable with SPVWP with MPPT/DC. Considering this, it minimizes evapotranspiration, waste of water, and assure reasoned irrigation (Fig. 17).

- **Water needs**

However, even better results are achieved when using photovoltaic and drip irrigation, coupled as a smart and reliable solution for smalls and Medium crops. The second case selected in the area investigated is about 6 ha of olive trees. For reasons of optimization, it is proposed to divide the crop into 4 sectors of irrigation. The trees and the lines are separated by 5 m, which explains a density of 400 trees/ha. Each shaft is irrigated by 4 drippers of 2(l/h) at a pressure of 1 bar; for 7–9 h a day, and 4 days a week. Therefore, the hourly flow of the area of the exploitation will be the product of the number of the tree by the flow rate of the drippers/trees, which equals 19.200 L/ h equivalent to 134400 L/ day. Tus, the daily flow is 134.4 m^3/ day. A pump with a flow rate of 20 m^3, immersed at 20 meters depth. Optionally, the water storage tank allows containing a certain amount of water, in case of bad weather. For a plot of 6 Ha, consuming 134 m^3 per day, for a period of 2 days, this gives 268 m^3.

- **Solar Pump and water pumped**

The analytical method shows that a solar pump of about 2.7 kW is required to cover such requirements; with a peak power of 3.75 kW of the PV surface. In another hand, numerical analysis underlines a solar pump of about 3.2 kW, with a peak power of 4.1 kW of the surface PV with 20% accepted missing. The use of a submersible pump can be coupled to an MPPT-DC regulation. For the simulation, it was decided

Fig. 17 Drip irrigation for Olives in Morocco

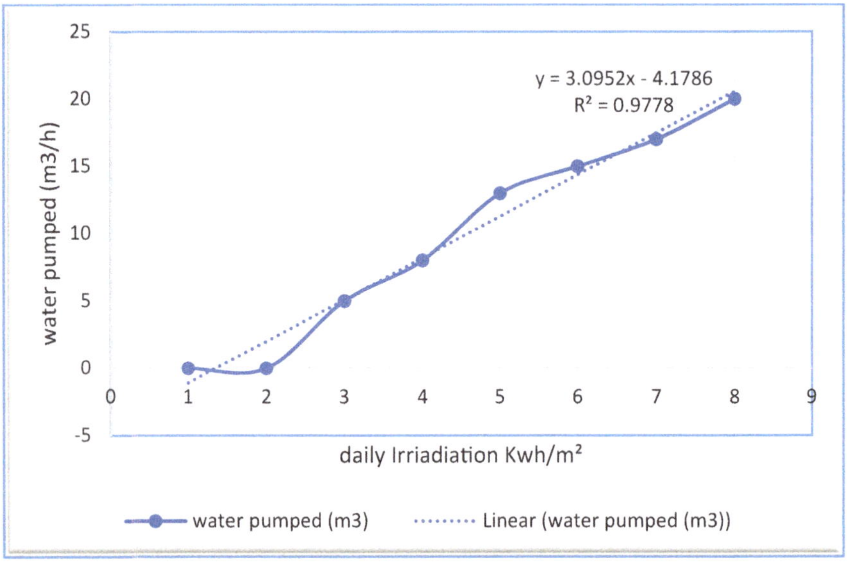

Fig. 18 The non-linearity of water pumped with daily effective global irradiation

that the pump was placed at a depth of 20 m in a well with water at a depth of 17 m, with no variation during the year. The pumped water is injected at the top of a tank, at 1 m of soil. The length of the pipe was estimated at 26 m with 3 elbows, it is a pipe of 12 mm in diameter. The main constraint is the ability to pump enough quantity of water. The flow depends strongly on solar radiation. Therefore, in the local condition, the average daily water pumped is 127 m^3, in terms of daily effective global irradiation (5.5 kWh/m^2). The output power of the PV system has a linear relationship with solar radiation as shown in Fig. 18. Practically, the system starts to pump water from 0.5 to 1 kWh/m^2. In optimal conditions, the flow rate can attain 18 m^3/h.

As shown in Fig. 19, with drip irrigation, unused energy is more considerable. Solar pumps assure the autonomy of water, approximately 50% of the energy is stored as water. During the dry period it is concentrated the highest values of produced electrical energy and during the rainy period was registered the lowest values with a little decline in the performance ratio. Generally, PR illustration shows how close a system approaches ideal performance. The system shows its performance in the dry season with an average of 61% (Fig. 20). The performance of the present system varies from 54 to 62%, which is very significant for medium crops, comparing to accepted performance in literature.

Considering the obtained results and field exigencies, the next table resumes many simulation studies and verification for many scenarios in the main region. To underline the well-adapted system, MMPT regulation is always recommended with direct coupling, considerably to optimize the system while operation and reduce the PV array, as shown in Table 5.

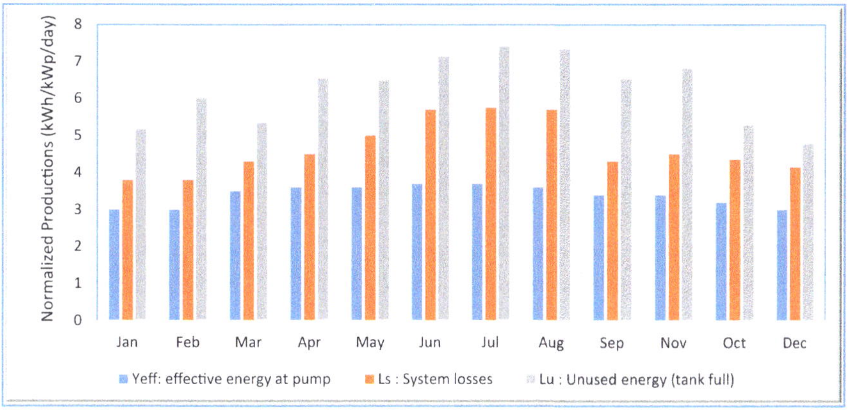

Fig. 19 Normalized production

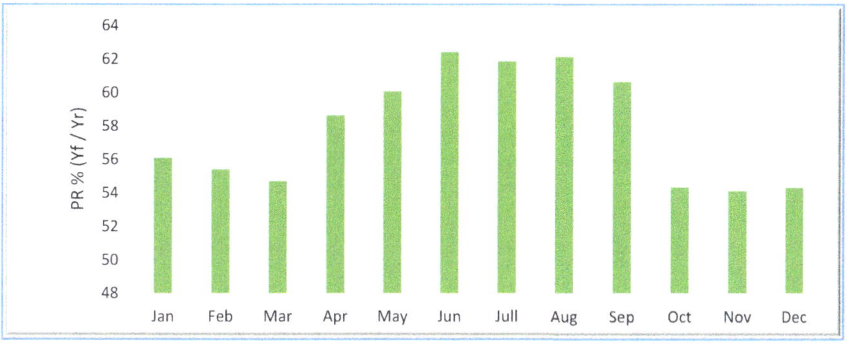

Fig. 20 Monthly performance ratio for the system selected

Table 5 Capacity and land distribution by farm size

Land (ha)	Total head	P_{pump}(kW)	% area	Proposed coupling
1–5	17–50	0.15–4.5	25	D. Coupling/ MPPT-DC
5–10	25–50	4.5–12.5	45	MPPT-DC
Up to 10	25–50	12.5≤	30	MPPT-DC/AC

From technical viewpoint, the MMPT configuration is the most suitable solution to pump water for agriculture. For economic analysis, the next section gathers to compare solar systems and classic methods.

8 Summary of Cost Analysis

According to calculations and the Benchmarking, photovoltaic systems prove its priority, due to the absence of operating and low maintenance costs (Fig. 21). Practically, the comparative analysis of solar pumps with diesel pumps and electrified pumps revealed that LCOW of photovoltaic water is significant, it is approximately in the range of 0.08US/ m^3 (Fig. 22).

Considering the first scenario to irrigate tomatoes in the main region selected; Fig. 23 shows the result of the after-tax cash-flow of the SPVWP system for 25 years. It is clear the system is profitable from the third year. Thus, considering long-term benefits and the high performance in the dry season, photovoltaic systems must be considered as the first choice for farmers in the Sais region to replace unsustainable pumps.

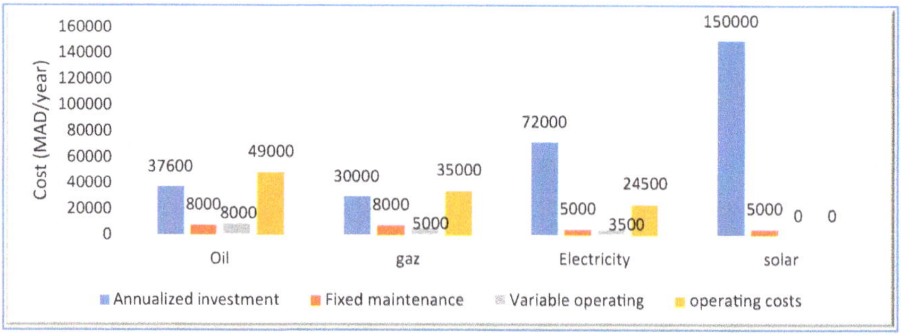

Fig. 21 Different costs (MAD/year) by using energy for Moroccan case

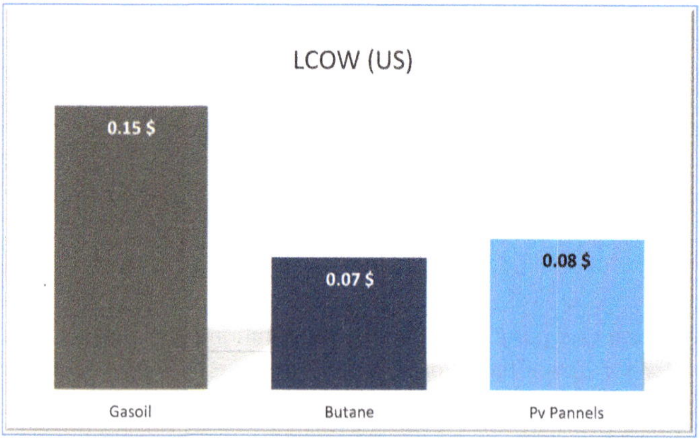

Fig. 22 Levelized cost of one cubic meter of water

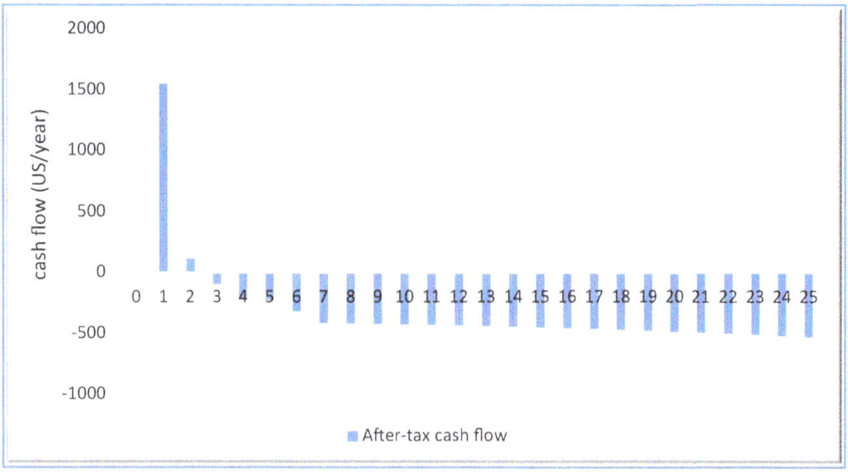

Fig. 23 MPPT coupling lifetime

9 Discussions on the SWOT Analysis of PVSWP

Based on the analysis of the case studies, the comparative cost analysis, and the local regularities, a summary of the attractiveness of water pumping projects in agriculture, can be presented in a SWOT analysis. As is well known, the objective of any SWOT analysis is to classify the main internal and external issues that are critical for attaining the given goals. It a large strategic challenge for governments and policymakers to promote solar water pumping implementation. The outline of the SWOT analysis is given in the form of a quadrant map regarding the water solar pumping development in morocco. The section progresses in comprehensive discussions on those aspects. It should be noticed that such a SWOT analysis may comprehend more elements that are not considered in this work. However, the elements considered are all relevant for constituting preferential decision-making.

9.1 Strength Analysis

a. **Rich natural resources in the region studies**: The agricultural activity is exemplified in the Sebou River. The most fertile part of the Sebou is the Saiss Basin. It holds about a quarter of Morocco's arable land but only uses a little of the national water reserve. This region is characterized by the diversity of soils and their fertility. Additionally, it holds an important solar potential with average sunshine of 3000 h/year and a yearly average solar radiation of about 5 kWh/m^2/day. The disposal water resources are stocked in the region with a total head of fewer than 30 m.

Generally, in Saiss region the total head varies between 9 and 20 m depth, fortunately, due to the disponibility of precipitations with an annual average of 520 mm which places the zone in the favorable market.

b. **Environment-friendly**: Among all other forms of renewable technologies, solar energy has the lowest GHG emissions ranging. Likewise, SPVWP that substitutes a fossil-based system will avoid about 1 kg of CO_2 per kilowatt-hour of energy output.
Practically, SPVWP can offer benefits with respect to air pollution and human health compared to fossil fuels.

c. **Minimum O&M and LCOE/W costs**: The operating and maintenance costs increase because of the substitution of materials. The inverters are warranted for 10 years, the periodic maintenance and the cleaning are done twice a year. Notably, For the SPVWP, LCOE/W is very competitive. Moreover, the current comparative analysis of solar pumps with diesel pumps and electrified pumps revealed that LCOW of photovoltaic water is significant, it is approximately in the range of 0.08 US/ m^3.

d. **Simplified system design**: As proven before, direct coupling remains as the possible and most cost-effective solution for small and medium-scale applications. Otherwise, to provide more flexibility to the SPVWP system and to make it more autonomous, a simple controller and tank full of sensors could be introduced, which is still cheaper and less complicated than an inverter for AC motors.

9.2 Weaknesses

a. **High initial investment**: The high investment cost is the major obstacle in the development of SPVWP in rural areas. It is the first limitation of the use of SPVWP applications. While the reduction of LCOE can be controlled if the balance of system costs were individually reduced. But generally, SPVWP systems are still a cost-effective application for the remote population in morocco.

b. **Lack of cooperation and materials**: For the defensible development of the solar energy DC systems need to be imported and developed comparing to AC devices. As well as the performance of DC Systems is developed, public recognition will grow toward a large-scale use of Nevertheless, the lack of coordination and structuration is still a real weakness in rural areas.

c. **The poverty of rural population**: The fundamental human needs are food, water, health care, hygiene, sanitation, and education. All these conditions are linked to pure water. The solar pumping systems could guarantee an unceasing water supply and outstanding agricultural production. But, in almost all Moroccan rural communities, drinking water is still the major problem for most groups, thus the solar systems have not attained a noticeable success yet.

9.3 *Opportunities*

a. **Grid parity achievement**: All over the world, PV module costs are falling swiftly to such an extent that some of them had already achieved solar grid parity i.e., the price of electricity from solar PV is equal to or lower than the conventional fossil fuel-based electricity. The opportunity of achieving grid parity also lies in the mitigation of the cost of solar supportive components like solar inverters and installation costs.
b. **Increasing efficiency of solar pumping configuration**: With the rising efficiency of the PV materials, prices of DC pumps and PV panels are rapidly decreasing. In morocco, direct coupling with MPPT regulation showed their efficiency especially by using more efficient materials and drip irrigation. Those configurations offer a cost-effective opportunity to solar energy development in the region.
c. **Financing for the solar system**: The "Credit Agricole Maroc" developed the program financing around 3000 solar pumping systems off-grid. Moreover, new financing instruments are presented by the Moroccan Sustainable Energy Financing Facility for rural communities (BERD).

9.4 *Threats*

a. **Conventional technologies**: Among the most ferocious threats to any renewable project, still remains fossil-based technologies as the real obstacle. For the reason that they produce energy at low prices, as they have enormous experienced experts, operators, and technical staff, looked to renewable energy which is still slightly integrated at a large-scale.
b. **Lack of balance of systems**: A second major threat to the development of solar energy is the absence of support products in the Moroccan market. Products such as regulators, inverters, and DC terminal devices.

Following our SWOT analysis, it is possible to transform weaknesses into strengths and convert threats into opportunities. On the other hand, strengths and opportunities must be combined to optimize water resources and properly integrate solar energy into agricultural practices. The Summary of the SWOT analysis of solar energy is given in Fig. 24.

As a source of renewable energy and opportunities for rural and remote populations, solar irrigation offers a potential solution to agriculture and small businesses to resize rural metabolism. This strategic intent puts citizens, farmers, factories, and laboratories at the heart of this multifaceted transition. It gives a new perspective to our society and a key model for MENA countries and African agriculture.

Under the Moroccan sun, irrigation is needed for most of the year, but irrigation is expensive for many farmers. Solar energy pumps offer an environmentally friendly, economical, and robust solution; there is a growing interest in photovoltaic water. An effective way to reduce costs in agriculture is to use renewable energy.

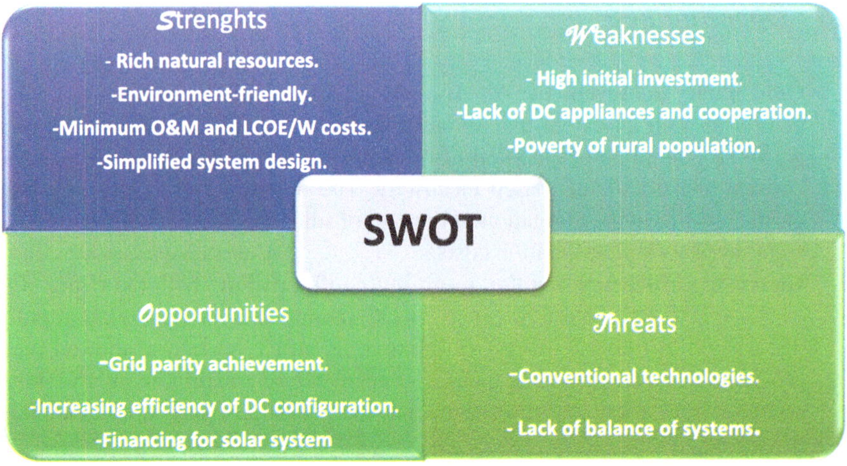

Fig. 24 SWOT analysis summary

Based on our empirical results, we developed recommendations that would allow the positive effects of solar pumping and validate direct coupling systems as being suitable for small and medium-sized crops. In this study, drip irrigation costs little maintenance. However, direct-coupled solar systems need to be integrated with solar-based MPPT and electronic control systems and other climate variability data to improve system performance.

In general, for small-scale irrigation, direct linkage with MMPT regulation seems most appropriate for family farming and the investment capacity of farmers.

Solar pumps have many benefits and can have a significant impact on rural society. This is the first application to familiarize farmers with renewable energies and to reorient agriculture toward sustainability for more renewable energy integrations.

Then a solar pump can provide lighting, irrigation, livestock; or water to agricultural biomass. The hybrid system should be seen as the next step in transforming farms into self-sufficiency and surplus production.

As the results show, the unused energy accounts for almost 50% of the total energy produced at the pump. That can be stored electrically and used to illuminate and raise animals, or used to provide water to a biomass plant. The results of the simulation include a large amount of significant data and quantify losses at all levels of the system, which helps identify weaknesses in system design. The recorded CO_2 is also produced and calculated using the PVSYST digital tool.

In both cases study, PVSPWs show their annual performance, the analysis of losses, It's necessary to indicate that head losses such as head loss due to pipe elbows, tank entrance and exit losses, and the head loss across banks of collectors in series shall also be included in the system head loss. Consequently, researchers should be aware of that fact.

Some limitations might be related to collecting our data, like the absence and the omission of some important variables.

National interest was accorded with a colossal nation investment through Green Plan Morocco to cover Initial costs. But, it seems familial agriculture suffer the poor infrastructure and weak management of water and matter.

Moreover, socio-economic development varies from one country to another, even in the same country. Although an appropriate technical solution in one region may not be appropriate in another. In developed countries, SPVWPs are generally intended for livestock and holiday homes, but in developing countries, applications are severely limited to drinking water and micro-irrigation applications. Thus, photovoltaic solar water pumping is a cost-effective application in remote areas. The reliability of solar pumps makes them the best choice for pumping agricultural water in developing countries.

In future works, big interest will be accorded to biomass and hybrid systems to consolidate the part of Green technologies in Moroccan agriculture. This makes clear that sustainable agriculture is the first key stage to assess "Industrial ecology and Social metabolism."

The links between agriculture and industrial ecology is the energy future of Morocco, while reasoned irrigation can raise questions and establish the objectives of biomass and energy vegetation (colza, corn…).

Furthermore, Green technologies have the ability to familiarize and integrate rural populations and educate their children. Moreover, agriculture can teach urban sphere ways on how to conserve water and energy.

10 Conclusion

As a source of renewable energy and opportunity for rural and remote populations, solar irrigation offers a potential solution to agriculture and small businesses to resize rural metabolism. Under the Moroccan sun, irrigation is needed for most of the year, but irrigation is expensive for many farmers. Solar energy pumps offer an environmentally friendly, economical, and robust solution. The results derived from a series of simulations reveal that an increase of 30% in annual performance and a decrease of 10% of system losses are observed when using an MPPT DC converter for small and medium-sized crops. In addition, cost comparative analysis revealed that LCOW of SPVWP is significant, approximately in the range of 0.08US/m^3, which is very competitive compared to other sources.

Based on empirical results, the author recommendations may allow the positive effects of solar pumping and validate direct coupling systems as being suitable for small and medium-sized crops. Moreover, direct-coupled solar systems need to be cohesive with solar-based MPPT and electronic control systems to improve system performance. In general, for small-scale irrigation, DC with MPPT regulation seems most appropriate for family farming and the investment capacity of farmers.

Globally, this work revealed a significant potential of the agro-pole Meknes-Fez to hold more than 17,000 solar pumps with capacities between 0.6 up to 40 kW. This proves that solar pumping must be widely considered as the most suitable solution to manage water and reduce almost agricultural CO_2 emissions.

Solar pumps have many benefits and can have a significant impact on rural society. It is the first step to familiarize Moroccan farmers with renewable energies and to reorient agriculture toward sustainability. From an economic and environmental point of view, those systems prove to be a key answer to develop one of the most ambitious renewable energy suppliers for water pumping in Morocco and especially to promote the main economic sector which is agriculture. The projected approach can fulfil as a reference for evaluation studies of water management in Morocco, in particular for ambitious social and agro-economic development programs in remote areas.

References

1. Steffen W, Broadgate W, Deutsch L, Gaffney O, Ludwig C (2015) The trajectory of the Anthropocene: the great acceleration. Anthr Rev 2:81–98
2. Rockström J, Williams J, Daily G, Noble A, Matthews N, Gordon L, Wetterstrand H, DeClerck F, Shah M, Steduto P (2017) Sustainable intensification of agriculture for human prosperity and global sustainability. Ambio 46:4–17
3. Bazilian M, Rogner H, Howells M, Hermann S, Arent D, Gielen D, Steduto P, Mueller A, Komor P, Tol RSJ (2011) Considering the energy, water and food nexus: towards an integrated modelling approach. Energy Policy 39:7896–7906
4. Godfray HCJ, Garnett T (2014) Food security and sustainable intensification. Philos Trans R Soc B Biol Sci 369:20120273
5. Juhola S, Klein N, Käyhkö J, Neset T-SS (2017) Climate change transformations in Nordic agriculture? J Rural Stud 51:28–36
6. Pardey PG, Beddow JM, Hurley TM, Beatty TKM, Eidman VR (2014) A bounds analysis of world food futures: global agriculture through to 2050. Aust J Agric Resour Econ 58:571–589
7. Tilman D, Fargione J, Wolff B, D'Antonio C, Dobson A, Howarth R, Schindler D, Schlesinger WH, Simberloff D, Swackhamer D (2001) Forecasting agriculturally driven global environmental change. Science 292(80-):281–284
8. Gustafsson J, Cederberg C, Sonesson U, Emanuelsson A (2013) The methodology of the FAO study: global food losses and food waste-extent, causes and prevention. FAO
9. Betts MG, Wolf C, Ripple WJ, Phalan B, Millers KA, Duarte A, Butchart SHM, Levi T (2017) Global forest loss disproportionately erodes biodiversity in intact landscapes. Nature 547:441–444
10. Qi X, Fu Y, Wang RY, Ng CN, Dang H, He Y (2018) Improving the sustainability of agricultural land use: an integrated framework for the conflict between food security and environmental deterioration. Appl Geogr 90:214–223
11. Lu Y, Nakicenovic N, Visbeck M, Stevance A-S (2015) Policy: five priorities for the UN sustainable development goals. Nature 520:432–433
12. Helander CA, Delin K (2004) Evaluation of farming systems according to valuation indices developed within a European network on integrated and ecological arable farming systems. Eur J Agron 21:53–67
13. Umesha S, Manukumar HMG, Chandrasekhar B (2018) Sustainable agriculture and food security. Biotechnology for sustainable agriculture. Elsevier, pp 67–92
14. Fresco LO (2009) Challenges for food system adaptation today and tomorrow. Environ Sci Policy 12:378–385

15. Mackay RJ (1992) Colonization by lotic macroinvertebrates: a review of processes and patterns. Can J Fish Aquat Sci 49:617–628
16. Heydari N (2014) Water productivity in agriculture: challenges in concepts, terms and values. Irrig Drain 63:22–28
17. Molden D, Oweis T, Steduto P, Bindraban P, Hanjra MA, Kijne J (2010) Improving agricultural water productivity: between optimism and caution. Agric Water Manag 97:528–535
18. Chel A, Kaushik G (2011) Renewable energy for sustainable agriculture. Agron Sustain Dev 31:91–118
19. Meah K, Ula S, Barrett S (2008) Solar photovoltaic water pumping—opportunities and challenges. Renew Sustain Energy Rev 12:1162–1175
20. Eker B (2005) Solar powered water pumping systems. Trakia J Sci 3:7–11
21. Narale PD, Rathore NS, Kothari S (2013) Study of solar PV water pumping system for irrigation of horticulture crops. Int J Eng Sci Invent 2:54–60
22. Odeh I, Yohanis YG, Norton B (2006) Economic viability of photovoltaic water pumping systems. Sol Energy 80:850–860
23. Chilundo RJ, Maure GA, Mahanjane US (2019) Dynamic mathematical model design of photovoltaic water pumping systems for horticultural crops irrigation: a guide to electrical energy potential assessment for increase access to electrical energy. J Clean Prod 238:117878
24. Eltamaly AM, Mohamed MA (2016) A novel software for design and optimization of hybrid power systems. J Braz Soc Mech Sci Eng 38(4):1299–1315
25. McGahey AM, Mansuri N, Carlstrom OM (2020) Evaluating solar photovoltaic water pumping systems in Khon Kaen, Thailand
26. Guan X, Mascaro G, Sampson D, Maciejewski R (2020) A metropolitan scale water management analysis of the food-energy-water nexus. Sci Total Environ 701:134478
27. Zavala V, López-Luque R, Reca J, Martínez J, Lao MT (2020) Optimal management of a multisector standalone direct pumping photovoltaic irrigation system. Appl Energy 260:114261
28. Abu-Aligah M (2011) Design of photovoltaic water pumping system and compare it with diesel powered pump. Jordan J Mech Ind Eng 5
29. Kelley LC, Gilbertson E, Sheikh A, Eppinger SD, Dubowsky S (2010) On the feasibility of solar-powered irrigation. Renew Sustain Energy Rev 14:2669–2682
30. Kappali M, Kumar RYU (2010) An approach to reduce the size and cost of PV panel in solar water pumping. In: 2010 5th international conference *on* industrial and information systems. IEEE, pp 608–613
31. Jin C, Jiang (2011) Design of a digital controlled solar water pump drive system for a nanofiltration system. In: 2011 international conference *on* power electronics and drive systems. IEEE, pp 982–986
32. Kolhe M, Joshi JC, Kothari DP (2004) Performance analysis of a directly coupled photovoltaic water-pumping system. IEEE Trans Energy Convers 19:613–618
33. Elgendy MA, Zahawi B, Atkinson DJ (2010) Comparison of directly connected and constant voltage controlled photovoltaic pumping systems. IEEE Trans Sustain Energy 1:184–192
34. Yahyaoui I, Tadeo F, Segatto MV (2017) Energy and water management for drip-irrigation of tomatoes in a semi-arid district. Agric Water Manag 183:4–15
35. Ghavidel S, Aghaei J, Muttaqi KM, Heidari A (2016) Renewable energy management in a remote area using modified gravitational search algorithm. Energy 97:391–399
36. Reges JP, Braga EJ, dos LC, Mazza S, De Alexandria AR (2016) Inserting photovoltaic solar energy to an automated irrigation system. Int J Comput Appl 975:8887
37. Hamidat A, Benyoucef B (2009) Systematic procedures for sizing photovoltaic pumping system, using water tank storage. Energy Policy 37:1489–1501
38. Kabalci Y, Kabalci E, Canbaz R, Calpbinici A (2016) Design and implementation of a solar plant and irrigation system with remote monitoring and remote control infrastructures. Sol Energy 139:506–517
39. Arab AH, Chenlo F, Mukadam K, Balenzategui JL (1999) Performance of PV water pumping systems. Renew Energy 18:191–204

40. Arab AH, Chenlo F, Benghanem M (2004) Loss-of-load probability of photovoltaic water pumping systems. Sol Energy 76:713–723
41. Yahyaoui I, Atieh A, Serna A, Tadeo F (2017) Sensitivity analysis for photovoltaic water pumping systems: energetic and economic studies. Energy Convers Manag 135:402–415
42. Chandel SS, Naik MN, Chandel R (2017) Review of performance studies of direct coupled photovoltaic water pumping systems and case study. Renew Sustain Energy Rev 76:163–175
43. Kou Q, Klein SA, Beckman WA (1998) A method for estimating the long-term performance of direct-coupled PV pumping systems. Sol Energy 64:33–40
44. Khatib T (2010) Deign of photovoltaic water pumping systems at minimum. J Appl Sci 10:2773–2784
45. Yahya HN, Sambo AS (1995) Design and installation of solar photovoltaic powered water pumping system at Usmanu Danfodiyo University, Sokoto. Renew Energy 6:311–312
46. Posorski R (1996) Photovoltaic water pumps, an attractive tool for rural drinking water supply. Sol Energy 58:155–163
47. Gherbi AD, Arab AH, Salhi H (2017) Improvement and validation of PV motor-pump model for PV pumping system performance analysis. Sol Energy 144:310–320
48. Hamidat A, Benyoucef B, Hartani T (2003) Small-scale irrigation with photovoltaic water pumping system in Sahara regions. Renew Energy 28:1081–1096
49. Mahjoubi A, Mechlouch RF, Ben Brahim A (2010) Economic viability of photovoltaic water pumping systems in the desert of Tunisia. In: International Renewable Energy Congress IREC2019, Soussa, Tunis, pp 5–7
50. Mahmoud E, El Nather H (2003) Renewable energy and sustainable developments in Egypt: photovoltaic water pumping in remote areas. Appl Energy 74:141–147
51. Al-Ali AR, Rehman S, Al-Agili S, Al-Omari MH, Al-Fayezi M (2001) Usage of photovoltaics in an automated irrigation system. Renew Energy 23:17–26
52. Mokeddem A, Midoun A, Kadri D, Hiadsi S, Raja IA (2011) Performance of a directly-coupled PV water pumping system. Energy Convers Manage 52:3089–3095
53. Moulay-Idriss C, Mohamed B (2013) Application of the DTC control in the photovoltaic pumping system. Energy Convers Manage 65:655–662
54. Bennani A, Buret J, Senhaji F (2001) Communication Nationale Initiale à la Convention Cadre des Nations Unies sur les changements climatiques [National Initial Communication to the UN Framework Convention on Climate Change], Rabat, Morocco. Ministère l'Aménagement Du Territ. l'Urbanisme l'Habitat l'Environnement, p 101
55. Rigourd C, Kemmoun H, Claus J-C, Errahj M, Dugué P, Bekkar Y (2013) Etude de faisabilité pour la mise en oeuvre d'actions pilotes de conseil de gestion agricole dans le cadre du Plan Maroc Vert
56. Kousksou T, Allouhi A, Belattar M, Jamil A, El Rhafiki T, Arid A, Zeraouli Y (2015) Renewable energy potential and national policy directions for sustainable development in Morocco. Renew Sustain Energy Rev 47:46–57
57. monographie Meknes Fes (n.d.) http://www.equipement.gov.ma/Carte-Region/RegionFes/Presentation-de-la-region/Monographie/Pages/Monographie-de-la-region.aspx. Accessed 26 Apr 2020
58. Pescod MB (1992) Wastewater treatment and use in agriculture
59. Allen RG, Pereira LS, Raes D, Smith M (1998) Crop evapotranspiration-Guidelines for computing crop water requirements-FAO Irrigation and drainage paper 56, Fao, Rome. 300, D05109
60. Shen Y, Li S, Chen Y, Qi Y, Zhang S (2013) Estimation of regional irrigation water requirement and water supply risk in the arid region of Northwestern China 1989–2010. Agric Water Manage 128:55–64
61. Nordin ND, Rahman HA (2016) A novel optimization method for designing stand alone photovoltaic system. Renew Energy 89:706–715
62. Bouzidi B, Haddadi M, Belmokhtar O (2009) Assessment of a photovoltaic pumping system in the areas of the Algerian Sahara. Renew Sustain Energy Rev 13:879–886

63. Hosseinzadehtalaei P, Tabari H, Willems P (2017) Quantification of uncertainty in reference evapotranspiration climate change signals in Belgium. Hydrol Res 48:1391–1401
64. Allen RG, Walter IA, Elliott RL, Howell TA, Itenfisu D, Jensen ME, Snyder RL (2005) The ASCE standardized reference evapotranspiration equation. In: American Society of Civil Engineers
65. Espadafor M, Lorite IJ, Gavilán P, Berengena J (2011) An analysis of the tendency of reference evapotranspiration estimates and other climate variables during the last 45 years in Southern Spain. Agric Water Manage 98:1045–1061
66. Inass Z, Karima B, Mohammed A, Laïla BA (2019) Agroforestry: smart practice for sustainable agricultural development. In: International conference on information, intelligence, systems and applications. Springer, pp 36–47
67. Tanasijevic L, Todorovic M, Pereira LS, Pizzigalli C, Lionello P (2014) Impacts of climate change on olive crop evapotranspiration and irrigation requirements in the Mediterranean region. Agric Water Manage 144:54–68
68. Campana PE, Li H, Yan J (2015) Techno-economic feasibility of the irrigation system for the grassland and farmland conservation in China: photovoltaic vs. wind power water pumping. Energy Convers Manage 103:311–320
69. Allouhi A, Buker MS, El-houari H, Boharb A, Amine MB, Kousksou T, Jamil A (2019) PV water pumping systems for domestic uses in remote areas: Sizing process, simulation and economic evaluation. Renew Energy 132:798–812
70. Senol R (2012) An analysis of solar energy and irrigation systems in Turkey. Energy Policy 47:478–486
71. Corkish R, Green MA, Watt ME, Wenham SR (2013) Applied photovoltaics. Routledge
72. Daud A-K, Mahmoud MM (2005) Solar powered induction motor-driven water pump operating on a desert well, simulation and field tests. Renew Energy 30:701–714
73. Belgacem BG (2012) Performance of submersible PV water pumping systems in Tunisia. Energy Sustain Dev 16:415–420
74. Buker MS, Mempouo B, Riffat SB (2014) Performance evaluation and techno-economic analysis of a novel building integrated PV/T roof collector: An experimental validation. Energy Build 76:164–175
75. Humada AM, Hojabri M, Hamada HM, Samsuri FB, Ahmed MN (2016) Performance evaluation of two PV technologies (c-Si and CIS) for building integrated photovoltaic based on tropical climate condition: A case study in Malaysia. Energy Build 119:233–241
76. Axaopoulos PJ, Fylladitakis ED, Gkarakis K (2014) Accuracy analysis of software for the estimation and planning of photovoltaic installations. Int J Energy Environ Eng 5:1
77. Mohamed MA, Eltamaly AM (2018) Modeling and simulation of smart grid integrated with hybrid renewable energy systems. Springer, New York

Embedded Implementation of Improved IFOC for Solar Photovoltaic Water Pumping System Using dSpace

Mustapha Errouha, Babak Nahid-Mobarakeh, Saad Motahhir, Quentin Combe, and Aziz Derouich

Abstract In this chapter, a proposed indirect field oriented control (IFOC) based induction motor drive for solar PV water pumping system (SPVWPS) powered by the two-level inverter is presented. The proposed scheme aimed at enhancing the performance and the dynamic response of the PV system. It consists of operating the induction motor at optimal reference flux which leads to improve the overall efficiency of the SPVWPS. The proposed IFOC scheme will be investigated experimental implementation using the dSpace DS1005 board. The proposed SPVWPS is evaluated and compared with the conventional IFOC under reel climatic conditions. Experimental results illustrate the effectiveness of the suggested control strategy in terms of pumped water, minimization of absorbed current, and overall efficiency.

Keywords Solar photovoltaic water pumping system (SPVWPS) · Indirect field oriented control (IFOC) · dSpace DS1005 board · Optimal flux

M. Errouha (✉) · A. Derouich
Laboratory of Industrial Technologies and Services, Higher School of Technology, SMBA University, Fes, Morocco
e-mail: mustapha.errouha@usmba.ac.ma

A. Derouich
e-mail: aziz.derouich@usmba.ac.ma

B. Nahid-Mobarakeh
GREEN, University of Lorraine, Vandœuvre-lès-Nancy, France
e-mail: babak.nahid@univ-lorraine.fr

S. Motahhir
Engineering, Systems and Applications Laboratory, ENSA, SMBA University, Fes 30000, Morocco
e-mail: saad.motahhir@usmba.ac.ma

Q. Combe
LEMTA, University of Lorraine, Vandœuvre-lès-Nancy, France
e-mail: quentin.combe@univ-lorraine.fr

S. Motahhir and A. M. Eltamaly (eds.), *Advanced Technologies for Solar Photovoltaics Energy Systems*, Green Energy and Technology,
https://doi.org/10.1007/978-3-030-64565-6_15

1 Introduction

Renewable energy sources based water pumping system are being used increasingly because they reduce the CO2 emissions and environmental pollution. Among these renewable energy sources, solar photovoltaic, solar thermal, wind, biomass and hybrid forms of energy [1, 2]. The use of solar photovoltaic can be considered as the best way to feed the water pumping system because it is boundless, clean, and freely available [3, 4].

Generally, two categories of motors are used for the water pumping system: DC and AC motors. Each one has its pros and cons [5, 6]. AC motor, especially the induction motor (IM) is often utilized due to its robustness and low cost [7, 8]. Several control strategies have been suggested in the literature to drive the IM [9–11]. Indirect field oriented control (IFOC) is the most used technique to control the IM because of its robustness and faster response [12]. Therefore, several studies have been developed and investigated by many researchers to improve the IFOC [13, 14].

Tir et al. [15] introduced a fuzzy logic controller to enhance the performance of the IFOC. Son t Nguyen et al. [16] proposed IFOC incorporating the artificial neural network. Boujoudi et al. [17] proposed indirect field oriented control with the sliding mode control. El Bourhichi et al. [18] integrated multilevel inverter to enhance the performance of the IM controlled by IFOC. In this chapter, an optimal operation design of a solar PV water pumping system using an induction motor is proposed. Most of the previous studies use a constant value of reference flux [19, 20]; however, this constant value doesn't provide the optimal operating of the system. Therefore, a proposed technique which consists of selecting the optimal reference flux based on the copper losses to improve the efficiency of the SVPWPS.

Then, a hardware implementation is effectuated to prove the real-time performance of the proposed PV system. Different controllers can be used to validate the efficacy of the control system [21, 22]. Many authors used DSP TMS320F2812 and FPGA to implement their developed control strategies [23–25]. However, software programming must be developed by the users, which requires time-consuming process. In this work, an embedded implementation based on dSpace 1005 board without any developed complex coding is effectuated. The dSpace DS1005 board and MATLAB/Simulink tools allow us to generate the control algorithm. Moreover, the programming task becomes simpler using the library blocksets. Then, the generated code can be loaded directly to the dSpace 1005 board for the real-time hardware process.

2 Design of the Proposed Solar Photovoltaic Water Pumping System

The configuration of the proposed solar PV water pumping system is illustrated in Fig. 1. An induction motor which its parameters are provided in Table 1 is selected for experimental validation. The design of the PV system values is grouped in Table 2.

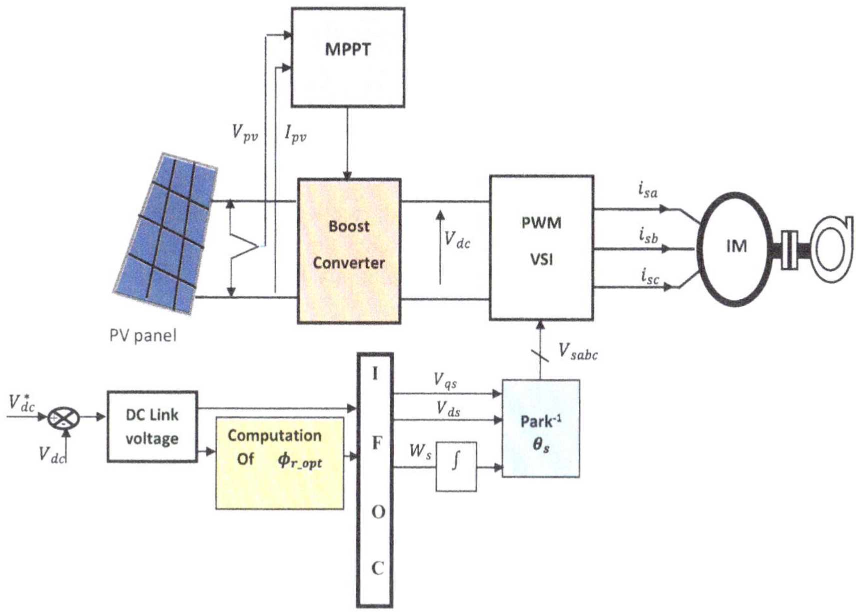

Fig. 1 Block scheme of the proposed SPVWPS

Table 1 Motor parameters

Stator resistance R_s and inductance l_s	1.8 [Ω] and 0.104 [H]
Rotor resistance R_r and inductance l_r	2.227 [Ω] and 0.104 [H]
Inertia J	0.0588 [Kg.m^2]
Mutual inductance **M**	0.0959 [H]
Number of pole pairs **p**	4
Viscous Friction f_v	0.1352 [N.m.s/rad]

Table 2 Parameters design of PV system

Component	Symbols	Expression
DC link voltage	V^*_{dc}	$V^*_{dc} = \frac{2\sqrt{2}}{\sqrt{3}} V_{LL}$
DC link capacitor	C_{dc}	$C_{dc} = \frac{6aV_{LL}I_l}{[V^{*2}_{dc} - V^2_{dc}]}$
Duty ratio	α	$\alpha = \frac{V_{dc} - V_{mp}}{V_{dc}}$
Inductance of the boost converter	L_{pv}	$L_{pv} = \frac{\alpha V_{mp}}{\Delta I f_s}$

3 Control Strategies for Proposed PV System

3.1 Indirect Field Oriented Control

When the indirect field oriented control is used, the flux is not regulated. This is given by the setpoint value and oriented from the angle θs which is obtained from the stator pulsation ωs. Hence, this method eliminates the use of the sensor or observer.

In the synchronous reference frame, the equations of the stator and rotor voltages of the induction machine can be expressed by [15]:

$$\begin{cases} V_{ds} = R_s i_{ds} + \frac{d}{dt}\phi_{ds} - w_s\phi_{qs} \\ V_{qs} = R_s i_{qs} + \frac{d}{dt}\phi_{qs} + w_s\phi_{ds} \end{cases} \tag{1}$$

$$\begin{cases} 0 = R_r i_{dr} + \frac{d}{dt}\phi_{dr} + w_{sr}\phi_{qr} \\ 0 = R_r i_{qr} + \frac{d}{dt}\phi_{qr} + w_{sr}\phi_{dr} \end{cases} \tag{2}$$

The implementation of rotor flux-oriented vector control is based on the orientation of the rotating coordinate system of axes d-q, such that the axis d is coincident with the direction of ϕ_r (Fig. 2).

This means that the flux component is oriented on the d-axis, that implies [26]:

$$\begin{cases} \phi_{dr} = \phi_r \\ \phi_{qr} = 0 \end{cases} \tag{3}$$

Therefore, the rotor voltages are expressed by:

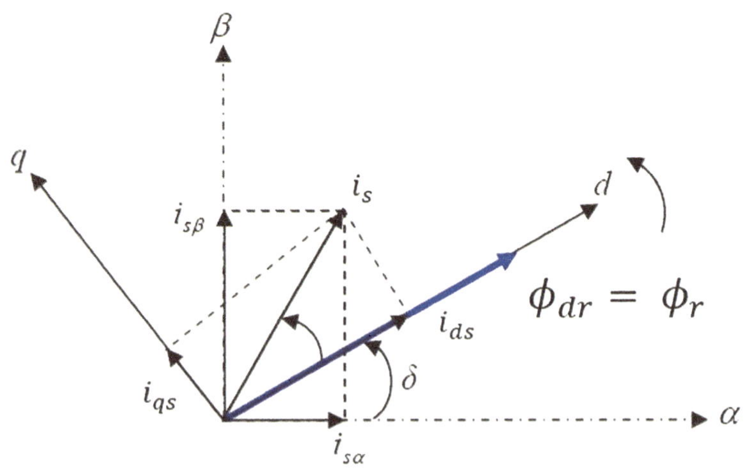

Fig. 2 Vector diagram in stationary and rotating reference frame

$$\begin{cases} 0 = R_r i_{dr} + \frac{d}{dt}\phi_{dr} \\ 0 = R_r i_{qr} + w_{sr}\phi_{dr} \end{cases} \tag{4}$$

And the components of flux are given by:

$$\begin{cases} \phi_{ds} = l_s \sigma i_{ds} + \frac{M}{l_r}\phi_{dr} \\ \phi_{qs} = l_s \sigma i_{qs} \end{cases} \tag{5}$$

Hence, the stator voltages are expressed by:

$$\begin{cases} V_{ds} = R_s i_{ds} + \sigma l_s \frac{d}{dt} i_{ds} + \frac{M}{l_r}\frac{d}{dt}\phi_{dr} - w_s \sigma l_s i_{qs} \\ V_{qs} = R_s i_{qs} + \sigma l_s \frac{d}{dt} i_{qs} + w_s \frac{M}{l_r}\phi_{dr} + w_s \sigma l_s i_{ds} \end{cases} \tag{6}$$

3.1.1 Estimation of w_s and θ_s

In this control strategy, the stator pulsation is determined indirectly from the measurement of the mechanical speed as follows:

$$w_{sr} = \frac{M}{T_r}\frac{i_{qs}}{\phi_{dr}} \tag{7}$$

Therefore, we obtain:

$$w_s = p\Omega + \frac{M}{T_r}\frac{i_{qs}}{\phi_{dr}} \tag{8}$$

We can note that the appearance of the rotor time constant (T_r) influences on the performance of this control strategy.

The position of stator flux is given by:

$$\theta_s = \int w_s dt \tag{9}$$

3.1.2 Electromagnetic Torque

The expression of the torque becomes [27]:

$$T_{em} = p\frac{M}{L_r}\phi_r i_{qs} \tag{10}$$

Equation (10) shows that by applying the indirect field oriented control, the electromagnetic torque can be controlled with q-axis stator current.

The equations of the IM can be expressed as:

$$\begin{cases} V_{ds} = \left(R_s + R_r \frac{M^2}{l_r^2}\right) i_{ds} + \sigma l_s \frac{d}{dt} i_{ds} - \frac{M}{l_r^2} R_r \phi_{dr} - w_s \sigma l_s i_{qs} \\ V_{qs} = \left(R_s + R_r \frac{M^2}{l_r^2}\right) i_{qs} + \sigma l_s \frac{d}{dt} i_{qs} + \frac{M}{l_r} w \phi_{dr} + w_s \sigma l_s i_{ds} \\ w_s = p\Omega + \frac{M}{T_r} \frac{i_{qs}}{\phi_{dr}} \\ T_r \frac{d}{dt} \phi_{dr} + \phi_{dr} = M i_{ds} \\ T_{em} = p \frac{L_m}{L_r} \phi_r i_{qs} \\ J \frac{d\Omega}{dt} = T_{em} - T_l - f_v \Omega \end{cases} \tag{11}$$

The equations of the IM can be also written by considering $\phi_{qr} = 0$ as follows:

$$\begin{cases} \frac{di_{ds}}{dt} = -\frac{1}{\sigma l_s}\left(R_s + R_r \frac{M^2}{l_r^2}\right) i_{ds} + w_s i_{qs} + \frac{1}{\sigma l_s}\left(\frac{M}{l_r^2} R_r\right) \phi_{dr} + \frac{1}{\sigma l_s} V_{ds} \\ \frac{di_{qs}}{dt} = -w_s i_{ds} - \frac{1}{\sigma l_s}\left(R_s + R_r \frac{M^2}{l_r^2}\right) i_{qs} - \frac{1}{\sigma l_s}\left(\frac{M}{l_r}\right) w \phi_{dr} + \frac{1}{\sigma l_s} V_{qs} \\ \frac{d\phi_{dr}}{dt} = \frac{M}{l_r} R_r i_{ds} - \frac{R_r}{l_r} \phi_{dr} \\ J \frac{d\Omega}{dt} = T_{em} - T_l - f_v \Omega \end{cases} \tag{12}$$

From Eq (12), it can be seen that the stator voltages V_{ds} and V_{qs} depend on both current components i_{ds} and i_{qs} which leads to an influence on the flux and the torque. Therefore, a decoupling process must be introduced

3.1.3 Decoupling Controller

Different decoupling techniques are proposed in the literature. In this paper, the decoupling compensator technique is used. Therefore, Eq (6) can be written as follows [28]:

$$\begin{cases} V_{ds} = (R_s + s\sigma l_s) i_{ds} - w_s \sigma l_s i_{qs} \\ V_{qs} = (R_s + s\sigma l_s) i_{qs} + w_s \frac{M}{l_r} \phi_r + w_s \sigma l_s i_{ds} \end{cases} \tag{13}$$

where:

s is the derivative operator

The decoupling is achieved by choosing voltage commands such that:

$$\begin{cases} V_{ds}^* = (R_s + s\sigma l_s) i_{ds} = V_{ds} + w_s \sigma l_s i_{qs} = V_{ds} + e_{ds} \\ V_{qs}^* = (R_s + s\sigma l_s) i_{qs} = V_{qs} - \left(w_s \frac{M}{l_r} \phi_r + w_s \sigma l_s i_{ds}\right) = V_{qs} - e_{qs} \end{cases} \tag{14}$$

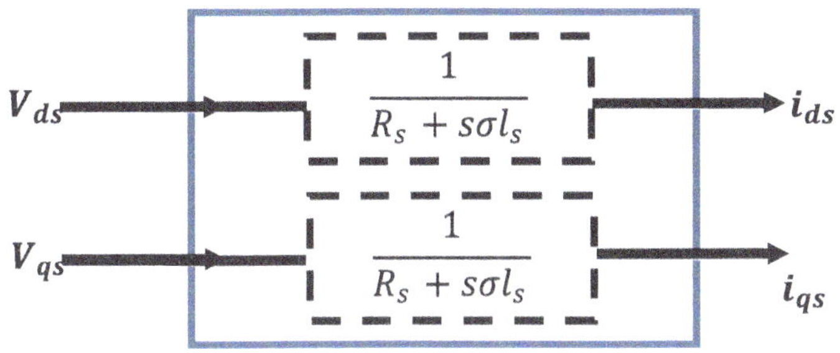

Fig. 3 The obtained controls

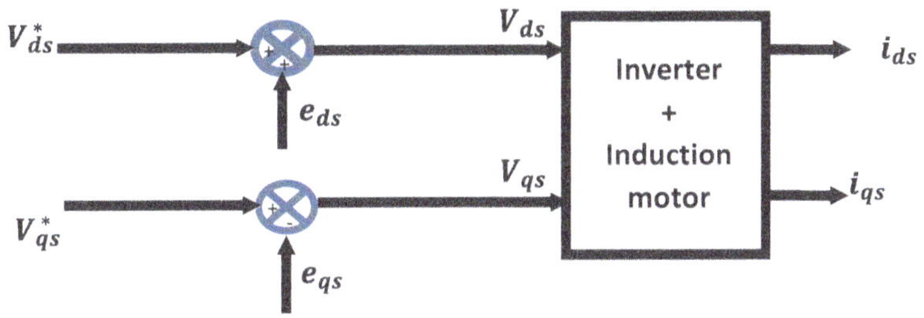

Fig. 4 Reconstruction of V_{sd} and V_{sq}

Hence, the actions on the (d, q) axis are decoupled and illustrated in the figure (Fig. 3).

The voltages V_{sd} and V_{sq} are obtained from the voltages V_{sd} * and V_{sq}* (Fig. 4).

3.2 Global Scheme of IFOC

The global control scheme of IFOC is shown in Fig. 5. It's composed of:

- Coordinate transformations
- PWM algorithm
- Current controllers
- Speed regulation loop using PI controller.
- An association of VSI-IM.

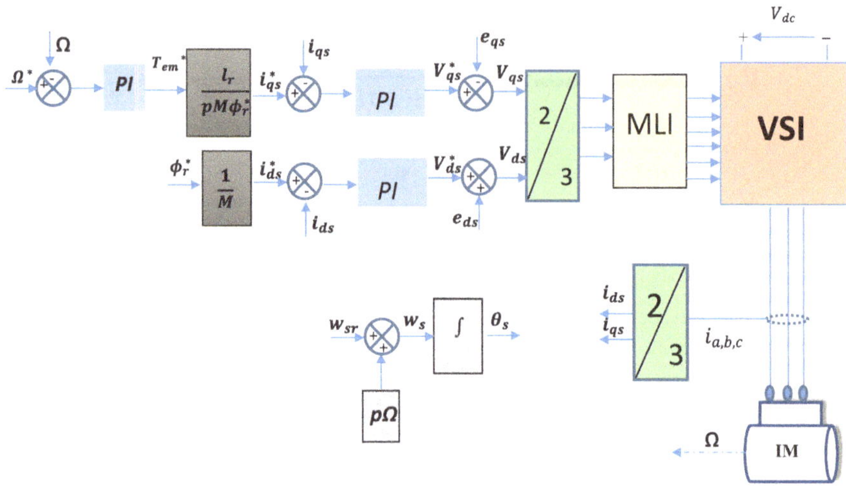

Fig. 5 The global control scheme of indirect field oriented control

3.3 Pulse Width Modulation (PWM)

3.3.1 The 3-Phase Inverter

The scheme of a two-level voltage source inverter is shown in Fig. 6, where V_a, V_b, V_c are the voltages applied to the star-connected motor windings, and V_{dc} presents the continuous inverter input voltage.

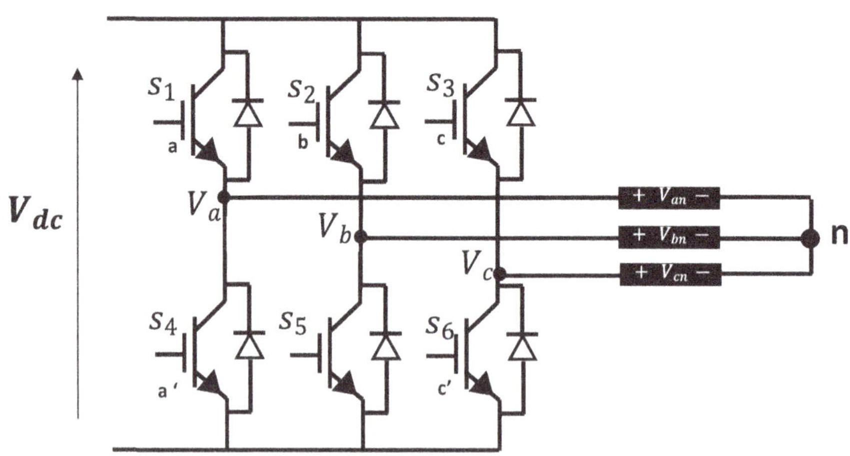

Fig. 6 Two-level VSI

The two-level VSI is composed of six switches that can be BJT, GTO, IGBT… depending on the used system and the operating of these devices must respect the following conditions:

- Three switches must all-time be ON and three all-time OFF.
- In order to avoid the vertical conduction and to guarantee that there is no overlap in the power switch transitions, the upper and the lower switches located in the same leg is controlled by two complementary pulsed signals.

Different PWM methods have been elaborated to control the VSI. The choice of the technique depends on several parameters such as the type of machine that will be controlled, the semiconductors used in the inverter, the value of the power.

Figure 7 illustrates the inverter voltage vectors (V0–V7) [29].

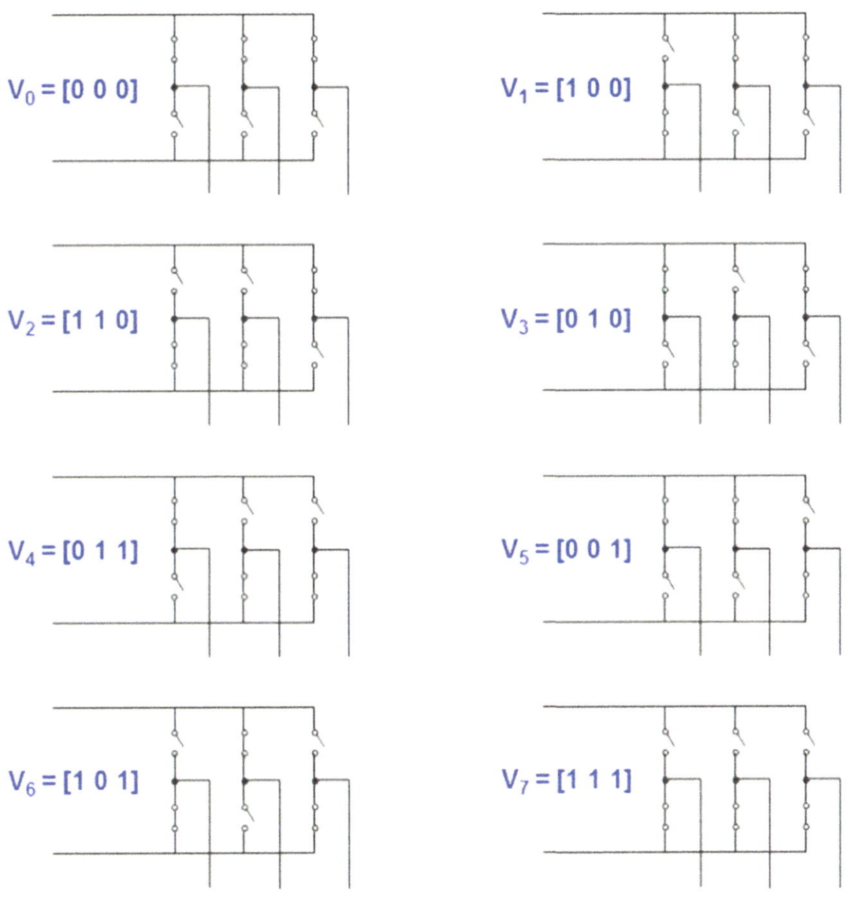

Fig. 7 The eight inverter voltage vectors (V0–V7)

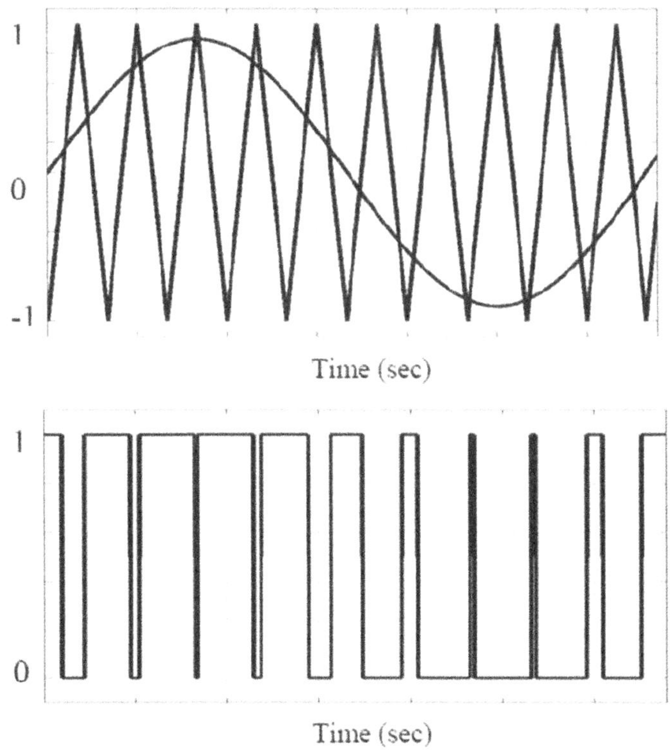

Fig. 8 PWM method

3.3.2 Sinusoidal PWM (SPWM)

To obtain the required control signals, we use the technique based on a comparison between a sine wave at low frequency (modulation signal) and a high-frequency carrier triangle signal (Fig. 9). The output control signal is equal to 1 when the modulation signal is greater than the carrier signal and zero otherwise (Fig. 8) [30]:

This sinusoidal PWM process is shown in Fig. 8.

The use of the SPWM technique is characterized by its simplicity and allows it to eliminate specific harmonics (Fig. 9).

3.4 Proposed Control Strategy

The proposed method consists of operating the PV water pumping system at the optimal reference flux. This is achieved by minimizing the losses of the induction motor. In this paper, based on the copper losses of the IM, the efficiency optimization is obtained. The equivalent circuit of the IM in (d, q) frame is illustrated in Fig. 10.

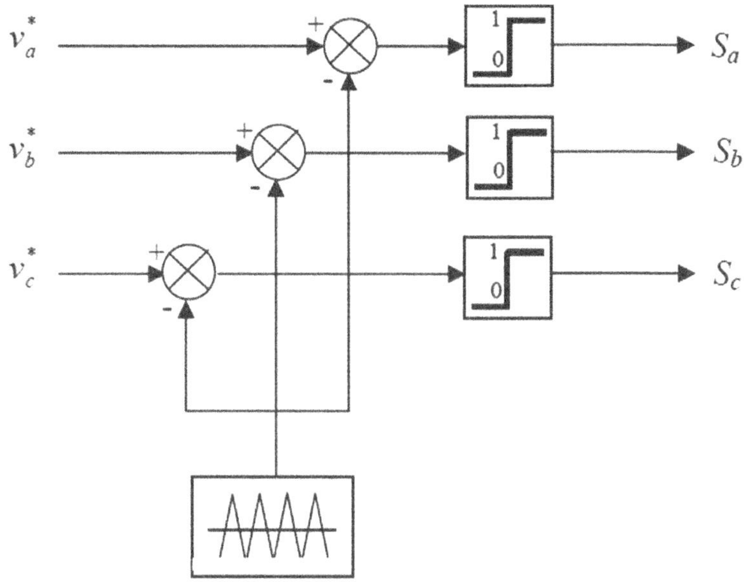

Fig. 9 SPWM technique

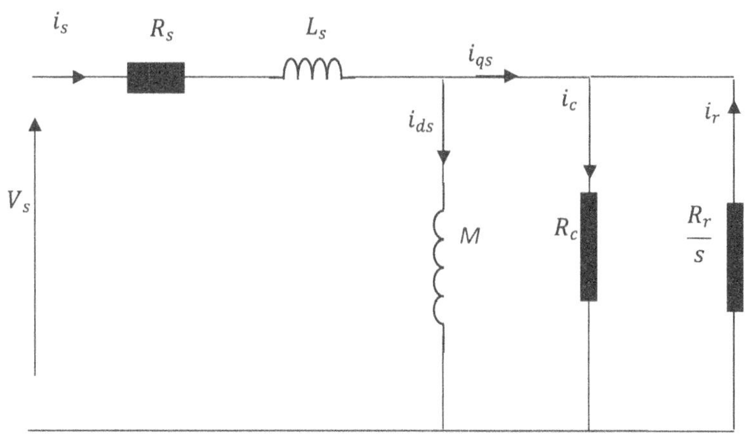

Fig. 10 Equivalent circuit in of IM

Based on the stator and rotor currents, the copper losses can be expressed as:

$$P_c = P_{sc} + P_{rc} = R_s i_s^2 + R_r i_r^2 \tag{15}$$

The expression of the stator copper losses can be written as follows:

$$P_{sc} = R_s(i_{ds}^2 + i_{qs}^2) \tag{16}$$

The expression of the rotor copper losses can be written as follows:

$$P_{rc} = R_r(i_{dr}^2 + i_{qr}^2) \tag{17}$$

The rotor currents in terms of rotor flux and components of stator current are given by:

$$i_{dr} = \frac{1}{L_r}\phi_r - \frac{M}{L_r}i_{ds} \tag{18}$$

$$i_{qr} = \frac{M}{L_r}i_{qs} \tag{19}$$

Therefore, the expression of the rotor copper losses is given by

$$P_{cur} = R_r\left(\frac{M}{L_r}\right)^2 i_{qs}^2 \tag{20}$$

On the other hand, the stator current components can be expressed by:

$$i_{ds} = \frac{\phi_r}{M} \tag{21}$$

$$i_{qs} = T_{em}\frac{L_r}{Mp\phi_r} \tag{22}$$

Hence, the total copper losses are given by:

$$P_c = R_s i_{ds}^2 + \left((R_r\left(\frac{M}{L_r}\right)^2 + R_s\right)\left(T_{em}\frac{L_r}{pM^2 i_{ds}}\right)^2 \tag{23}$$

$$\frac{dP_c}{di_{ds}} = 0 \tag{24}$$

By solving the Eq. (24), the direct current ensuring the optimal flux is expressed by eq (25)

$$I_{dsopt} = K_{opt} i_{qs} \tag{25}$$

where

$$K_{opt} = \sqrt{1 + \left(\frac{M}{L_r}\right)^2 \frac{R_r}{R_s}} \tag{26}$$

4 dSpace 1005 Controller

4.1 Overview

An experimental implementation is conducted to validate the proposed technique based on real-time interface using dSpace 1005 board. The bloc diagram of dSpace card DS1005 is illustrated in Fig. 11. It contains a PowerPC 750GX processor operating at 480 MHz and has 128 MByte SDRAM global memory (64 MB for the PowerPC separately arbitrated), 16 MByte Flash memory (1 MB reserved for booting) and 1 MByte, level 2 external cache memory. The dSpace controller provides also an interrupt handling. Moreover, three timers are provided to the PowerPC 750GX. The data transmission between the dSpace 1005 and host-PC can be effectuated by using the ISA port or PCI via an adapter.

The dSpace system incorporating the used test bench includes other boards to ensure the control of the proposed PV system (Fig. 12).

The DS2102 board ensures the D/A conversion to control the actuators. It consists of:

- 6 parallel D/A converters
- 16-bit resolution
- ±5 V, ±10 V or 0 ... +10 V output voltage range (programmable)

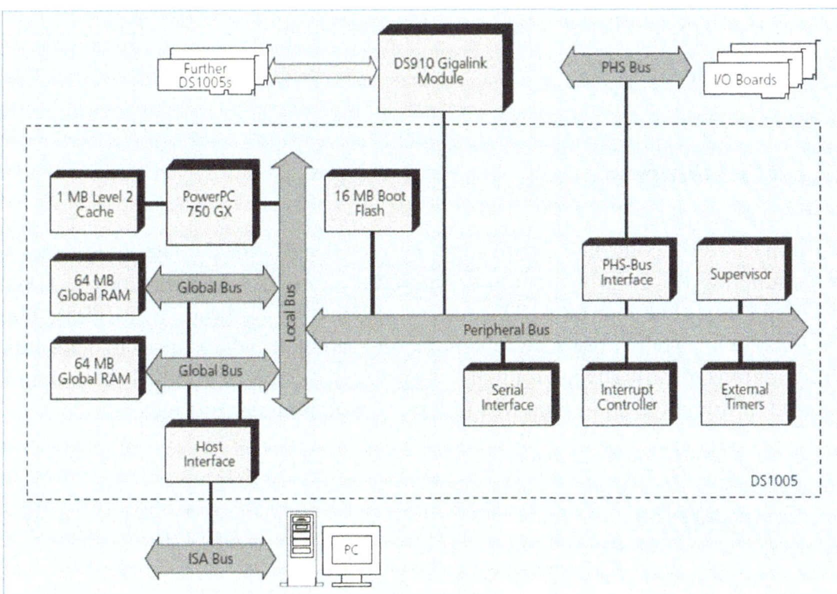

Fig. 11 Bloc diagram of dSpace 1005 board

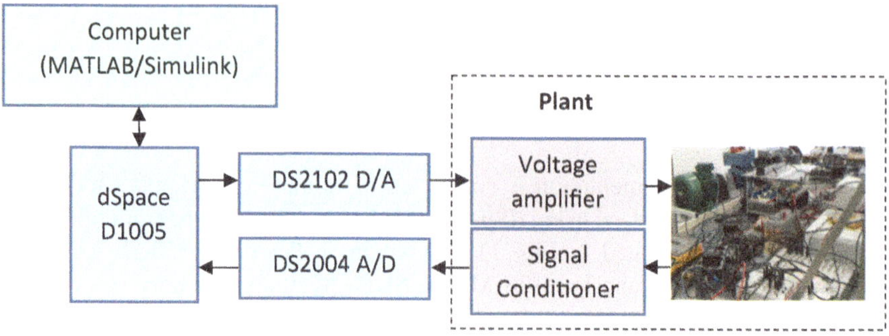

Fig. 12 Configuration diagram of the dSpace controller with SPVWPS

The DS2004 board is utilized in dSpace modular to ensure the A/D conversion at high sample rates. It consists of:

- 16 A/D input channels (differential)
- 16-bit resolution
- ±5 V or ±10 V input voltage range (programmable for each channel)

The DS5101 board is used to produce PWM signals. It consists of:

- Time resolution of 25 ns.
- 16 PWM outputs

The DS3002 speed board is introduced for incremental encoders. It consists of:

- 6 input channels

4.2 RTI Library

The dSpace 1005 controller board must be connected with the simulation model for completing the control processes. Therefore, a control system block is simulated using real-time interface (RTI) library of DS 1005 card (i.e., rtilib1005). This library contains the RTI blocks to obtain the I/O functions which allow to determine configurations of the hardware for real-time processes. Moreover, rtilib1005 provides required information, additional RTI blocks, and demo models (Fig. 13).

4.3 ControlDesk

ControlDesk software is used as a real-time interface and data storing the tool. This software provides the user with a working environment during all experimentation process and executes all required tasks. To validate the simulation program realized in

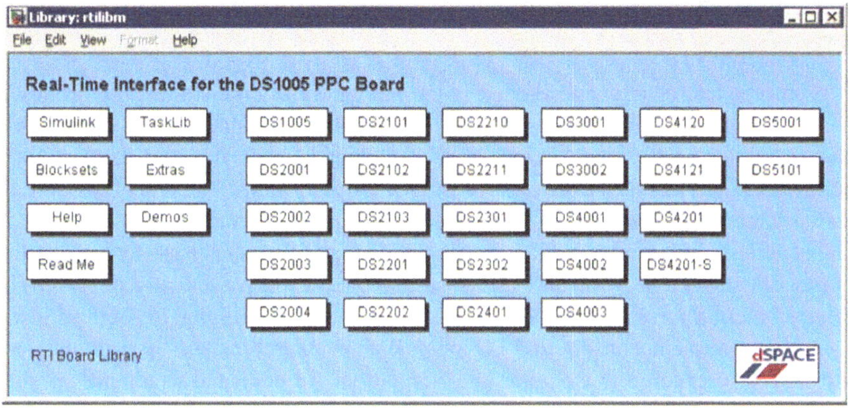

Fig. 13 RTI board library

Matlab/Simulink, the control part of the simulation is incorporated into ControlDesk to carry out the experimental part using the real-time dSpace 1005 applications (Fig. 14). To create the ControlDesk graphical user interface, online help can be used.

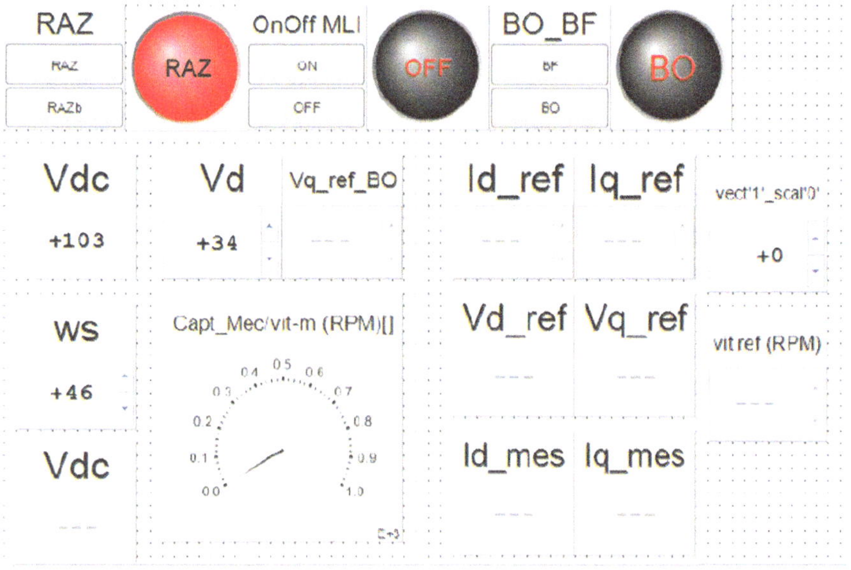

Fig. 14 ControlDesk interface

5 Experimental Validation

A prototype test bench is developed in the laboratory in order to test the proposed control strategy. It is composed of a solar simulator to produce the programmable characteristics of PV panels, a DC-DC converter, a VSI, IM, and dSpace 1005 controller board to execute the control and MPPT algorithms. The hall effect current and voltage sensors are also utilized to sense the required currents and voltages. A DC generator and resistance load are employed to emulate the characteristics of the pump. The dSpace board also gives the gating pulses to the power converters. IGBTs drivers function according to CMOS logic (0–15 V) while the digital I/O of the dSpace controller operate according to TTL logic (0–5 V), therefore, the optocouplers are introduced to ensure the isolation and adaptation. An emulated irradiance profile (Fig. 15) using PV emulator is utilized to evaluate the behavior of the proposed SPVWPS. Perturb & Observe technique is employed to provide the duty cycle which ensures the maximum power. During this experience, the proposed method and IFOC are compared.

Figures 16 and 17 illustrate the extracted power and the rotor speed of the IM, respectively. Figures 18a and 19a show the mechanical power of the induction motor for both techniques. According to Fig. 19.a the power mechanical is higher using the proposed method which proves the minimization of the motor losses. Figures 18b and 19b show the direct and quadratic stator currents ($i_{sd,}$ i_{sq}). The direct stator current is reduced using the proposed technique which leads to reduce the flux, therefore, the copper losses is minimized and energy saving. Contrary to IFOC, the direct current remains constant in all operating process. Figures 18c and 19c show the

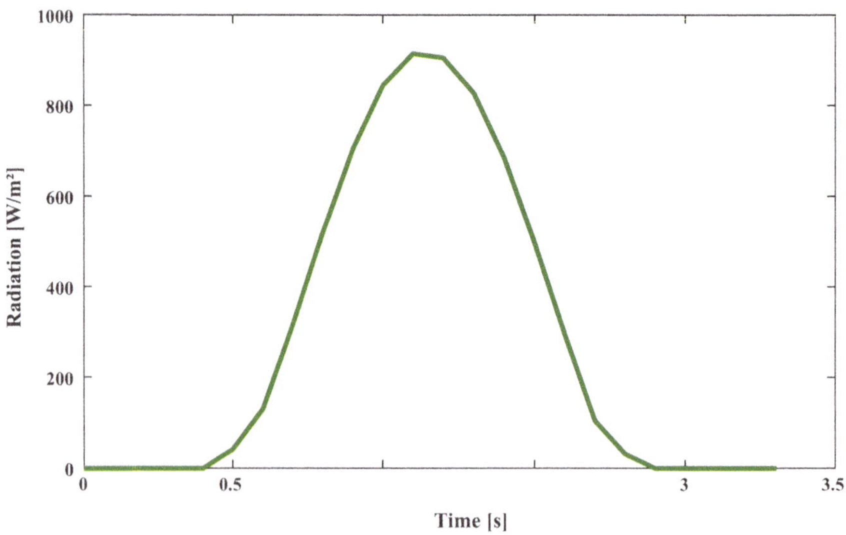

Fig. 15 Irradiation profile

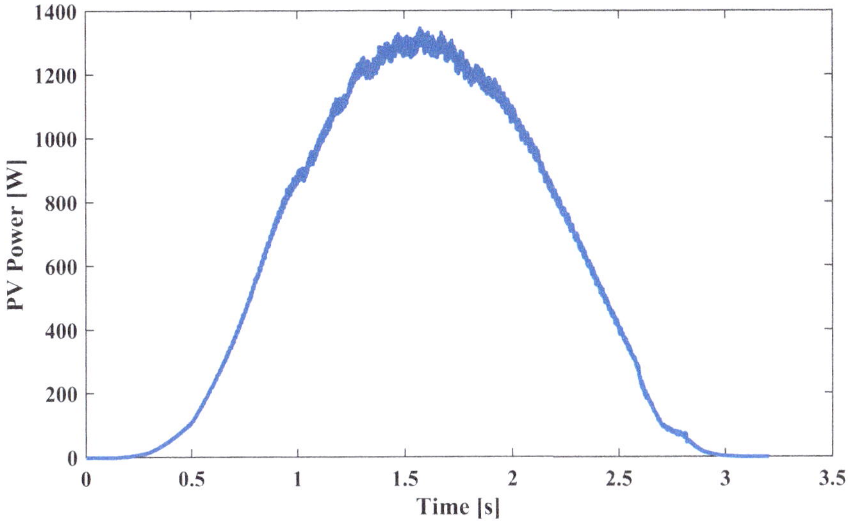

Fig. 16 Photovoltaic power

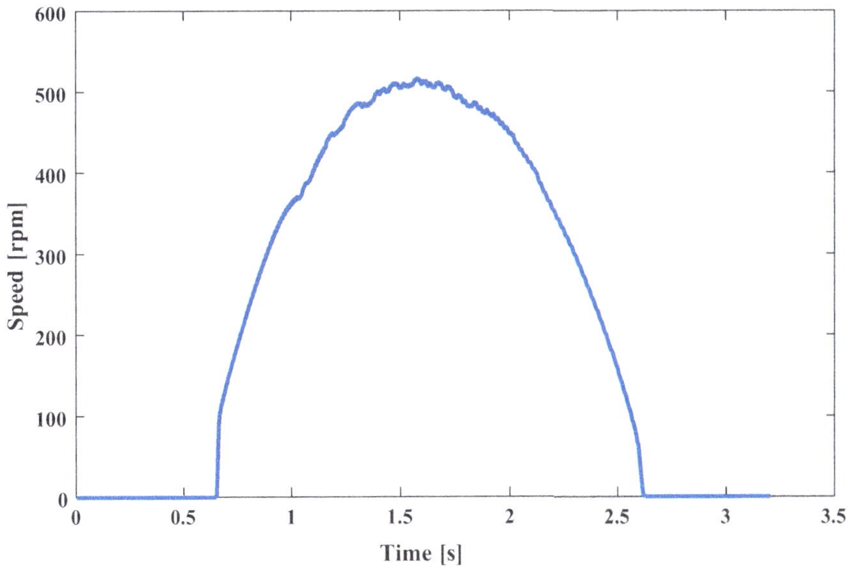

Fig. 17 Rotation speed

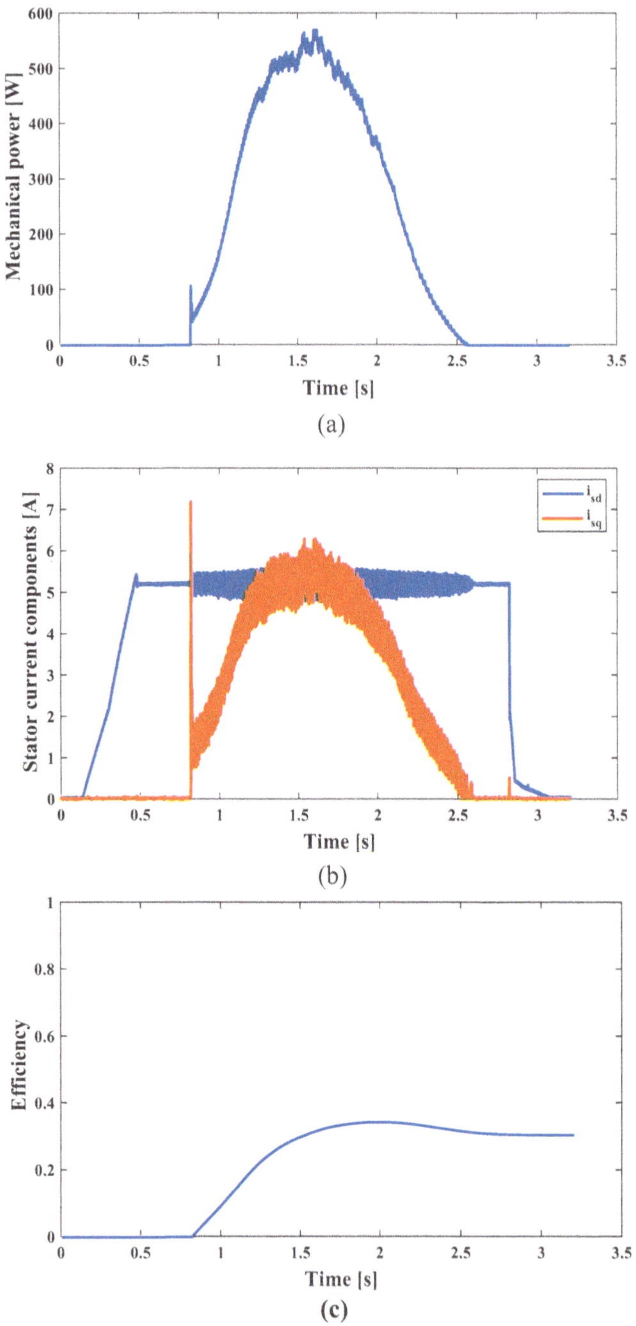

Fig. 18 Experimental results of IFOC

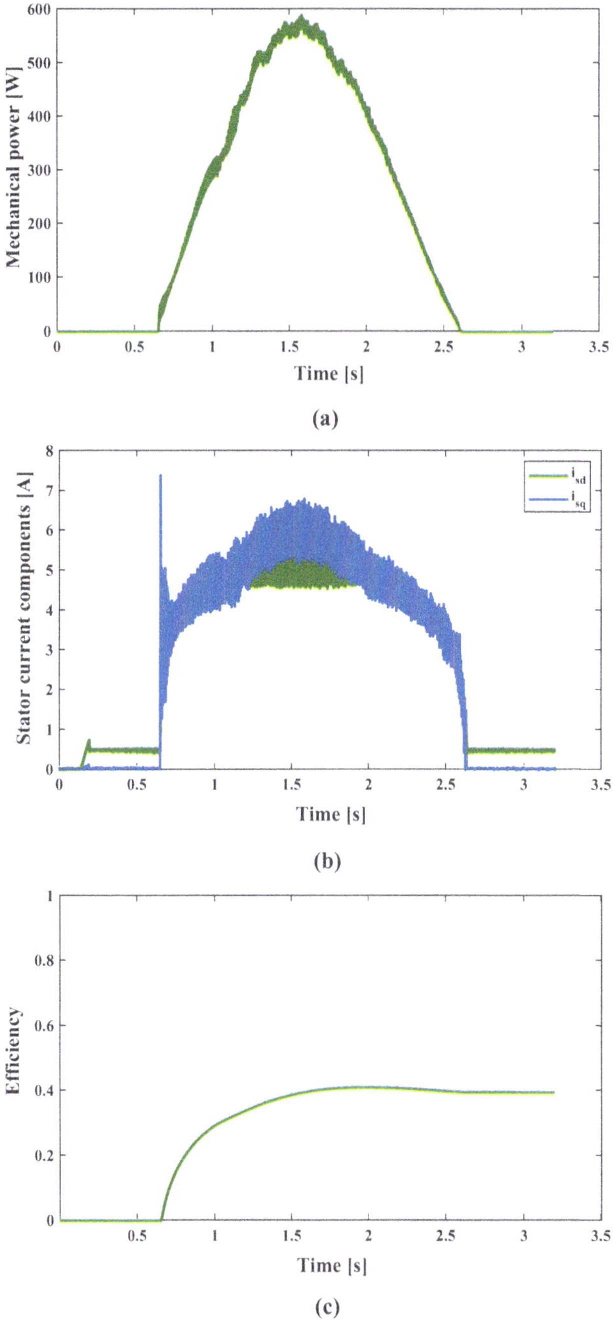

Fig. 19 Experimental results of the proposed method

Table 3 Performance comparison

Publication	Control strategy	Direct Stator current (reference flux)	Loss minimization
[15]	Fuzzy logic controller	Fixed value	No
[16]	Artificial neural network	Fixed value	No
[17]	Sliding mode	Fixed value	No
[18]	Multilevel inverter	Fixed value	No
Proposed method	Improved IFOC	Variable value (optimal reference)	Yes

overall efficiency of the solar PV water pumping system calculated using Eq (27). Using IFOC, the efficiency of the SPVWPS attains 30%. Contrary to the proposed technique, the efficiency of the SPVWPS reaches 40% which proves the robustness of the proposed control strategy. Table 3 illustrates a comparison between the proposed method and other techniques used to improve the IFOC.

$$\mathbf{Efficiency} = \frac{\mathbf{Mechanical\ Power}}{\mathbf{PV\ Power}} \tag{27}$$

6 Conclusion

In this chapter, the optimization of an indirect field oriented based IM drive in terms of absorbed current reduction and energy consumption is performed. Thus, the proposed method is designed for SPVWPS applications. In addition, an embedded implementation using dSpace 1005 board is effectuated. Experimental results illustrate that the proposed IFOC gives a good dynamic response, minimizes the motor losses, and improves efficiency. Moreover, during all the operating of the SPVWPS, the proposed method keeps searching the optimal flux which ensures the efficiency improvement. Therefore, the proposed optimal IFOC can be utilized successfully in SPVWPS applications in remote areas.

References

1. Gopal C, Mohanraj M, Chandramohan P, Chandrasekar P (2013) Renewable energy source water pumping systems-a literature review. Renew Sustain Energy Rev 25:351–370. https://doi.org/10.1016/j.rser.2013.04.012
2. Parameters extraction of single diode PV model and application in solar pumping
3. Eltamaly AM, Al-saud M, Abo-khalil AG (2020) Performance improvement of PV systems' maximum power point tracker based on a scanning PSO particle strategy. Sustainability, no Feb 2020. https://doi.org/10.3390/su12031185

4. Sontake VC, Kalamkar VR (2016) Solar photovoltaic water pumping system-a comprehensive review. Renew Sustain Energy Rev 59:1038–1067. https://doi.org/10.1016/j.rser.2016.01.021
5. Errouha M, Derouich A, Motahhir S, Zamzoum O, El Ouanjli N, El Ghzizal A (2019) Optimization and control of water pumping PV systems using fuzzy logic controller. Energy Reports 5:853–865. https://doi.org/10.1016/j.egyr.2019.07.001
6. Parveen H, Sharma U, Singh B (2020) Pole reduction concept for control of SyRM based solar PV water pumping system for improved performance. IEEE Trans Ind Electron 0046(c):1–1. https://doi.org/10.1109/tie.2020.3000089
7. Errouha M, Derouich A, Motahhir S, Zamzoum O (2020) Optimal control of induction motor for photovoltaic water pumping system. Technol Econ Smart Grids Sustain Energy 5(1). https://doi.org/10.1007/s40866-020-0078-9
8. Errouha M, Derouich A, El Ouanjli N, Motahhir S (2020) High-performance standalone photovoltaic water pumping system using induction motor. Int J Photoenergy 2020
9. Jannati M et al (2017) A review on variable speed control techniques for efficient control of single-phase induction motors: evolution, classification, comparison. Renew Sustain Energy Rev 75(Nov):1306–1319. https://doi.org/10.1016/j.rser.2016.11.115
10. Reza CMFS, Islam MD, Mekhilef S (2014) A review of reliable and energy efficient direct torque controlled induction motor drives. Renew Sustain Energy Rev 37:919–932. https://doi.org/10.1016/j.rser.2014.05.067
11. Direct Torque Control of Doubly Fed Induction Motor (DFIM)
12. Study and comparison results of the field oriented control for photovoltaic water pumping system applied on two cities in Morocco
13. Chihi A, Ben Azza H, Jemli M, Sellami A (2017) Nonlinear integral sliding mode control design of photovoltaic pumping system: real time implementation. ISA Trans 70:475–485. https://doi.org/10.1016/j.isatra.2017.06.023
14. Active and reactive power control of wind turbine based on doubly fed induction generator using adaptive sliding mode approach
15. Tir Z, Malik OP, Eltamaly AM (2016) Fuzzy logic based speed control of indirect field oriented controlled Double Star Induction Motors connected in parallel to a single six-phase inverter supply. Electr Power Syst Res 134:126–133. https://doi.org/10.1016/j.epsr.2016.01.013
16. Nguyen A, et al (2017) A sensorless three-phase induction motor drive using indirect field oriented control and artificial neural network, vol 1, no 1, pp 1–10. https://doi.org/10.1037/0022-3514.51.6.1173
17. Boujoudi B, Kheddioui E, Machkour N, Bezza M (2019) A comparative study between PI and sliding mode control for the DFIG of a wind turbine. Green Energy Technol, 205–226. https://doi.org/10.1007/978-981-13-1945-7_10
18. El Bourhichi S, Oukassi A, El Adnani M (2019) Indirect vector control of induction motor using a five-level cascaded H-bridge inverter. In: 2018 International Symposium on Advanced Electrical and Communication Technologies (ISAECT)-Proceeding, pp. 1–6. https://doi.org/10.1109/isaect.2018.8618841
19. El Ouanjli N, Motahhir S, Derouich A, El Ghzizal A, Chebabhi A, Taoussi M (2019) Improved DTC strategy of doubly fed induction motor using fuzzy logic controller. Energy Reports 5(February):271–279. https://doi.org/10.1016/j.egyr.2019.02.001
20. Errouha M, Motahhir S, Combe Q, Derouich A, El Ghzizal A (2019) Fuzzy-PI controller for photovoltaic water pumping systems, pp 0–5
21. Hannan MA, Abd Ghani Z, Mohamed A (2010) An enhanced inverter controller for PV applications using the dSPACE platform. Int J Photoenergy 2010. https://doi.org/10.1155/2010/457562
22. Eltamaly AM, Al-Saud MS, Abokhalil AG, Farh HM (2020) Simulation and experimental validation of fast adaptive particle swarm optimization strategy for photovoltaic global peak tracker under dynamic partial shading. Renew Sustain Energy Rev 124:109719
23. Venkadesan A, Himavathi S, Sedhuraman K, Muthuramalingam A (2016) Design and FPGA implementation of cascade neural network based flux estimator for speed estimation in IM drives. IET Electr Power Appl, 1–32

24. Diao L, et al (2017) An efficient DSP-FPGA-based implementation of hybrid PWM for electric rail traction induction motor control. IEEE Trans Power Electron 8993(c). https://doi.org/10.1109/tpel.2017.2707639
25. Eltamaly AM, Alolah AI, Abdulghany MY (2010) Digital implementation of general purpose fuzzy logic controller for photovoltaic maximum power point tracker. In: SPEEDAM 2010. IEEE, pp 622–627
26. Guo Z, Zhang J, Sun Z, Zheng C (2017) Indirect field oriented control of three-phase induction motor based on current-source inverter. Procedia Eng 174:588–594. https://doi.org/10.1016/j.proeng.2017.01.192
27. Abu-rub H, Iqbal A (2020) Field oriented control, no Im, pp 16–44
28. Mon-Nzongo DL, Jin T, Ekemb G, Bitjoka L (2017) Decoupling network of field-oriented control in variable-frequency drives. IEEE Trans Ind Electron 64(7):5746–5750. https://doi.org/10.1109/TIE.2017.2674614
29. Jin-Woo J (2005) Space vector PWM inverter
30. It SF (2020) Model of three-phase inverter, pp 58–80. https://www.tntech.edu/files/cesr/StudThesis/asuri/Chapter4.pdf

Single-Phase Grid-Connected Photovoltaic H-Bridge N-Level Inverter Control Strategy

Abdelaziz Fri, Rachid El Bachtiri, and Salah-Eddine Lhafdaoui

Abstract In this chapter, we present a novel control strategy for a single-phase cascaded H-bridge multilevel inverter in a grid-connected solar PV system. Unlike the known grid-connected inverters controls using a DC/DC converter for the MPPT pursuit, our control technique offers an MPPT algorithm for each PV module by handling its corresponding inverter only. The inverter connection of the grid is made through an LCL filter due to its ability to reduce harmonics instead of a simple inductance filter. To relieve the resonance phenomenon through the capacitors and inductors, a passive damping resistor is connected in series with the capacitor. We have used both controllers PI and PR, respectively, on the synchronous reference frame dq and the stationary reference frame $\alpha\beta$. The main purpose is to eliminate significant harmonics (3rd, 5th, and 7th) using our novel controller in order to improve the quality of the injected energy into the grid. By controlling the active and reactive power, we have proved that the PI controllers cannot track the sinusoidal references without a steady error in the synchronous reference frame dq. Furthermore, the PR controller has better performances and is more interesting for controlling the DC/AC converters. Indeed, in the stationary reference frame $\alpha\beta$, the PR controller gives a static error near zero for sinusoidal currents. Moreover, injected current quality improvement due to the most dominant odd harmonics elimination (3rd, 5th, and 7th) is insured by the HC. The adequate drive pulses are generated by a proportional modulator algorithm.

Keywords PV inverter · Multilevel · MPPT control · LCL filter · Harmonics compensator

---

A. Fri (✉) · R. El Bachtiri
Industrial Technologies and Services Laboratory, High School of Technology, USMBA University, Fez, Morocco
e-mail: Abdelaziz.fri@usmba.ac.ma

R. El Bachtiri
e-mail: Rachid.elbachtiri@usmba.ac.ma

S.-E. Lhafdaoui
Study Center: CDE STI, FST, USMBA University, Fez, Morocco
e-mail: salaheddine.lhafdaoui@usmba.ac.ma

S. Motahhir and A. M. Eltamaly (eds.), *Advanced Technologies for Solar Photovoltaics Energy Systems*, Green Energy and Technology,
https://doi.org/10.1007/978-3-030-64565-6_16

1 Introduction

Electrical production from photovoltaic panels (PV) gives DC voltage. So, the use of inverters is a compelling solution to convert the output voltage to the alternative form. The increase of the electric power, in stand-alone or grid-connected PV systems, leads to increase in the switched current. That's why, the increase in voltage is often privileged, to improve the performance of the installation. In all cases, it's difficult to handle semiconductors who undergo deterioration of their dynamic and static performances [1–3].

The use of conventional inverters with two levels and high switching frequencies is limited due to the large switching lasses in the devices. In addition, the inverter voltage waveform has a significant distortion, which requires harmonic filtering. To overcome this problem, multilevel inverters in medium and high powers applications is suggested by several studies [4–6].

Multilevel inverters reduce stress on the power switches of the structure on one hand and improve waveforms (harmonic spectrum) of the output voltage. On the other hand, concerning the quality of the output multilevel inverters voltage, some works, comparing different topologies, have shown that H-bridge inverter is the most suitable for photovoltaic systems [4–7].

Active and reactive power of a grid-connected multilevel inverter could be controlled by acting on the voltage at the point of common coupling PCC (Voltage-mode control) or on the inverter output current (Current-mode control). In the last case, i_{in} current is influenced by v_{in} voltage (Fig. 1). Actually, power is controlled by the phase angle and the current magnitude in regard to the voltage v_g at the PCC.

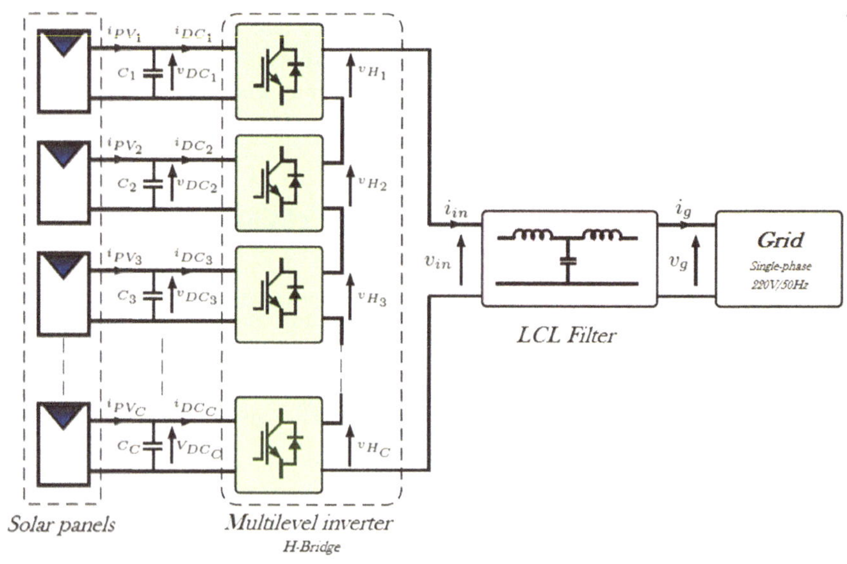

Fig. 1 PV system with a grid-connected multilevel H-bridge inverter

Thus, the inverter is protected against overloads finks regulation of the current. In addition, this control mode has more advantages such as stoutness toward the PV system and the grid parameters, advanced dynamic performances, and high control precision [8, 9].

In this chapter, we present a novel control strategy for a cascaded H-bridge multilevel inverter for grid-connected PV systems. It is the multicarrier pulse width modulation strategies (MCSPWM), a proportional method (Fig. 5). Unlike the known grid-connected inverters control based on the DC/DC converter between the inverter and the PV module for the MPPT pursuit, our command mode offers for an MPPT algorithm using the H-bridge inverter only.

Otherwise, the main motivations of choosing an LCL filter instead of a simple inductance are mentioned in Sect. 2. The resonance (of the capacity) of the electromagnetic interference filter is passively amortized due to a resistor [10, 11].

We have used PI and PR controllers to drive a multilevel inverter, respectively, on the synchronous reference frame dq and the stationary reference frame αβ. The main purpose is to eliminate significant harmonics (3rd, 5th and 7th) using the phase-shifted carrier PWM (PSCPWM) technic to improve the energy quality [12–14]. Index modulation for each inverter cell is generated as the ratio between the produced power by the corresponding PV panel P_{PVi} (P_{PVi}; i = 1..C) and the PV total power PT.

The remaining of the presentation proceeds as the second section presents the modeling and frequency analysis of the LCL filter. The third section shows the global structure of the suggested system in this work. The fourth section details the control strategy proposed for injecting current into the grid: Firstly, we introduce the expression of the dynamic modeling for de current control. Secondly, we present the current control in the dq reference frame and in the stationary reference frame αβ, respectively. Based on the quality of the resonant proportional PR controller, we win detail in our design position of an adaptive PR controller. Finally, processes the DC bus voltage control at the input of each H-bridge celli. The last section is a representation of our design PSCPWM proportional modulator. Finally, we will give our conclusions and remarks.

2 The LCL Filter

We have used a passive filter to connect the inverter to the grid. This LCL filter was designed in order to:

- Allow the current dynamic regulation, defined as:

$$\frac{\partial i_g(t)}{\partial t} = \frac{\partial i_{ref}(t)}{\partial t} \tag{1}$$

- Stop the harmonics generated by the switching, in the grid.

LCL filter can easily decrease the switching harmonics caused by the MCSPWM controller. However, the filter frequency response presents a resonance peak (Fig. 4). The passive filter design depends on how a reduction rate is needed in the line current harmonics.

According the international standard IEC-1000-3-4 related to current harmonics, we have used our filter to reduce harmonics above 33rd rank under 0.6% of rated current [10, 11].

2.1 Modeling and Transfer Function

To design the LCL filter we have taken into consideration the current ripple, the reduction of the size, the attenuation of the switching effect, and the reactive power diminution [11]. A resonance frequency peak could lead to an unstable operation due to capacitor C_f (Fig. 4). Therefore, a resistor R_d is connected in series with the capacitor for passive damping.

Figures 2 and 3 show, respectively, the electrical circuit and the block diagram of the LCL filter.

Based on the Kirchhoff laws, the filter model can be expressed as follows:

$$i_{in} - i_c - i_g = 0 \quad (2)$$

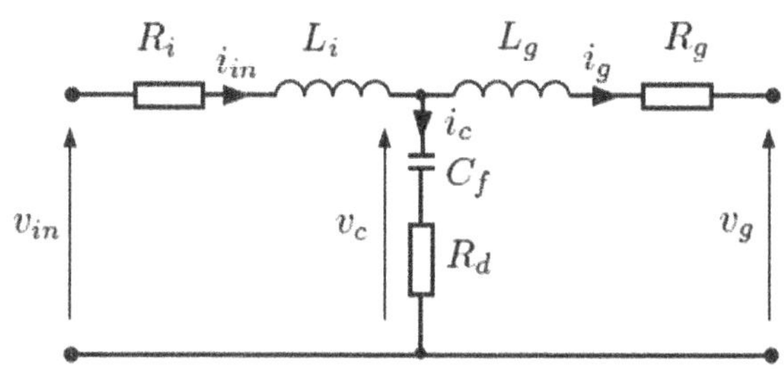

Fig. 2 Electrical circuit of the LCL filter

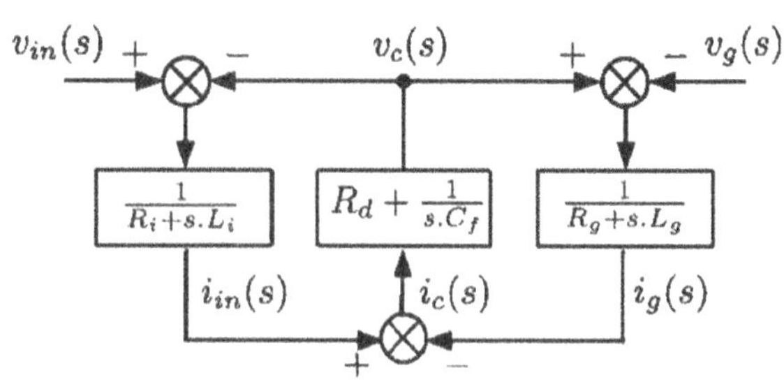

Fig. 3 Block diagram of the LCL filter

$$v_{in} - v_c = i_{in}(sL_i + R_i) \tag{3}$$

$$v_c - v_g = i_g(sL_g + R_g) \tag{4}$$

$$v_c = i_c\left(\frac{1}{sC_f} + R_d\right) \tag{5}$$

The transfer function of the LCL filter $H_{LCL}(s)$ is:

$$H_{LCL}(s) = \frac{i_g(s)}{v_{in}(s)} \tag{6}$$

To calculate $H_{LCL}(s)$, some mathematical calculations are required. Vg voltage is assumed to sinusoidal, and it can be short-circuited for harmonics. In this case since $Vg = 0$, we have from Eqs. (4) and (5):

$$i_g(sL_g + R_g) = i_c\left(\frac{1}{sC_f} + R_d\right) \Rightarrow i_c = i_g\frac{sC_f R_g + s^2 C_f L_g}{1 + sC_f R_d} \tag{7}$$

Inserting Eqs. (4), (5) and (7) in Eq. (3), the inverter output voltage can be written as follows:

$$v_{in} = i_g(sL_g + R_g) + (i_g + i_c)(sL_i + R_i) = i_g(sL_g + R_g) + \left(i_g + i_g\frac{sC_f R_g + s^2 C_f L_g}{1 + sC_f R_d}\right)(R_i + sL_i) \tag{8}$$

$$\Rightarrow v_{in} = i_g(R_i + R_g + s(L_i + L_g) + \left(\frac{(sL_i + R_i)(sC_f R_g + s^2 C_f L_g)}{1 + sC_f R_d}\right) \tag{9}$$

Considering Eq. (7), the transfer function can be given as:

$$H_{LCL}(s) = \frac{sR_d C_f + 1}{s^3 L_i L_g C_f + s^2 C_f (L_g(R_d + R_i) + L_i(R_d + R_g)) + s(L_i + L_g + C_f(R_d R_i + R_d R_g + R_i R_g)) + R_i + R_g} \tag{10}$$

If we neglected the inductor's resistances, $H_{LCL}(s)$ becomes:

$$H'_{LCL}(s) = \frac{sR_d C_f + 1}{s^3 L_i L_g C_f + s^2 C_f R_d(L_i + L_g) + s(L_i + L_g)} \tag{11}$$

If we neglect also the damping resistor R_d then $H'_{LCL}(s)$ becomes:

$$H''_{LCL}(s) = \frac{\frac{1}{L_g C_f}}{sL_i\left(s^2 + \frac{L_i + L_g}{L_i L_g C_f}\right)} = \frac{\frac{1}{L_g C_f}}{sL_i\left(s^2 + \omega_{res}^2\right)} \quad (12)$$

ω_{res} is the resonance pulse.

$$\omega_{res}^2 = \frac{L_i + L_g}{L_i L_g C_f} \Rightarrow f_{res} = \frac{1}{2\pi}\sqrt{\frac{L_i + L_g}{L_i L_g C_f}} \quad (13)$$

2.2 Frequency Analysis

The frequency transfer function of the filter is gathered by replacing s with $jn\omega$ in Eq. (14), n is the harmonic rank. The modulus of the transfer function is written as follows:

$$\left|H''_{LCL}(jn\omega)\right| = \left|\frac{i_g(jn\omega)}{v_{in}(jn\omega)}\right| = \frac{\frac{1}{L_i L_g C_f}}{n\omega\left(-(n\omega)^2 + \frac{L_i + L_g}{L_i L_g C_f}\right)} \quad (14)$$

According to the Bode diagram (Fig. 4), if the damping resistor R_d increases, the admittance magnitude decreases, so, the filter impedance increases. A highly damped filter is able to reduce high-rank harmonics and stop them to spread in the grid. The filter output current can be used as a control variable, implying that the LCL filter is properly damped.

3 Grid-Connected H-Bridge Multilevel Inverter Control

Figure 5 shows the general structure of the suggested system in this work. It represents the blocks and loops for the control of a grid-connected multilevel inverter:

- MPPT blocks;
- Grid synchronization and dq values calculator block (Grid observer);
- DC_i voltage regulation loop;
- Current control loop;
- PSCPWM proportional modulator.

Based on each PV voltage and current measurements, the P_{PVi} power is calculated, and consequently the total power P_T is determined. The ratio $\alpha_i = \frac{P_{PV_i}}{P_T}$ depends on the modulation index M_i of the matched H-bridge $cell_i$. The MPPT blocks generate references voltages V_{refi} that matches the MPP_i in the power control external loop. The current regulation block provides the control law of the grid-injected current.

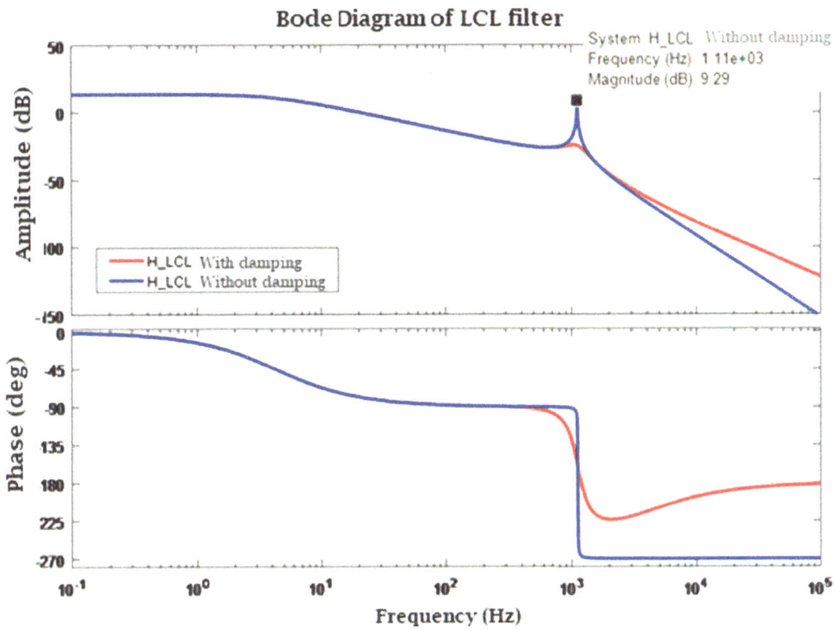

Fig. 4 Bode diagram with and without the damping resistor

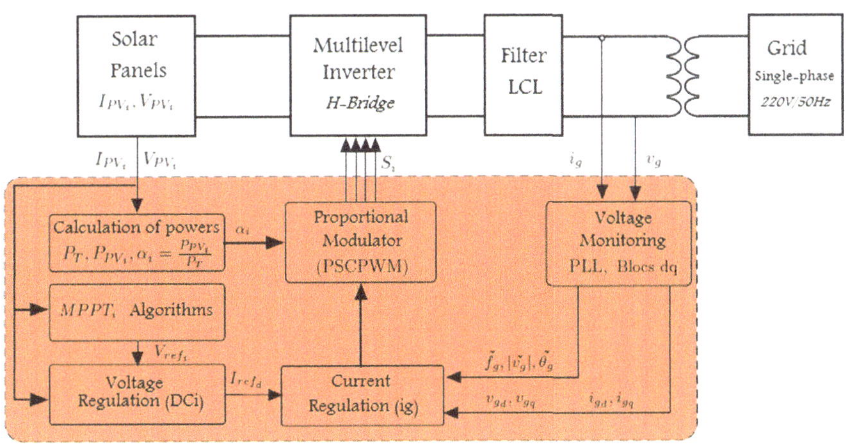

Fig. 5 Global scheme of the system control

Figure 6 presents our novel input power control strategy of the system without using the DC/DC converter. The power downstream correction is one of the new features in this proposition [15].

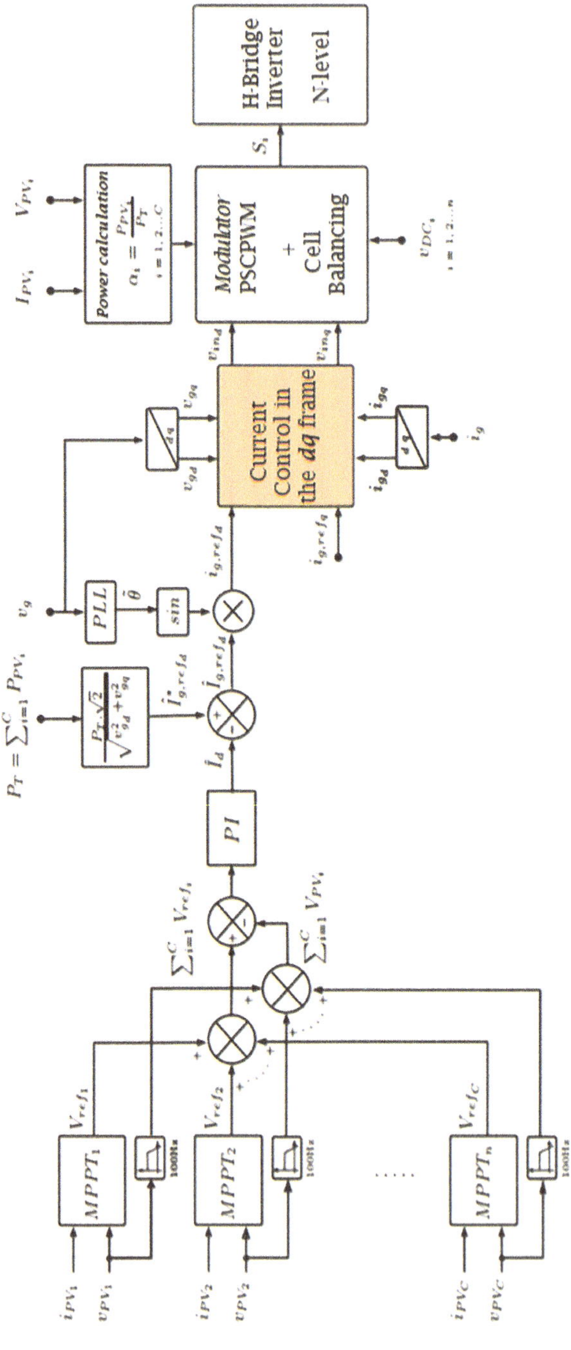

Fig. 6 Suggested strategy Control of a multilevel H-bridge inverter

4 Control of the Injected Current into the Grid

Two types of controllers will be detailed in this section. A PI controller in synchronous reference frame dq rotating at the speed ωg whose $\theta g = \omega g.t$ is the rotation angle, and a proportional resonant controller in the stationary reference frame αβ. Our contribution is based on the last one, eliminating the 3rd, 5th, and the 7th current harmonics [16].

4.1 Dynamic Modeling for the Current Control

Some simplifications must be taken into account to establish the current controller. For example, the LCL filter can be considered as an inductor $L = Li + L_g$ in series with $R = R_i + R_g$, while neglecting the damping resistor and capacitor. The inverter output voltage is:

$$v_{in} = L\frac{\partial i_g}{\partial t} + Ri_g + v_g \tag{15}$$

$$\Rightarrow \frac{\partial i_g}{\partial t} = \frac{1}{L}v_{in} - \frac{R}{L}i_g - \frac{1}{L}v_g \tag{16}$$

An approach often adopted for the analysis of the electrical system uses the synchronous reference frame dq or the stationary reference frame $\alpha\beta$ [6].

Considering that specific feature of the dq reference frame is that the space vectors have constant magnitude parameters rotate at the same speed as the reference frame. Then $\omega_g i_{gd}$ and $\omega_g i_{gq}$ must be added to the dynamic currents i_{gd} and i_{gq}.

The mathematical model in the dq reference frame for the pulse ω_g is given as follows:

$$\begin{cases} \frac{\partial i_{gd}}{\partial t} = \frac{1}{L}(v_{in_d} - Ri_{g_d} - v_{g_d}) + \omega_g i_{g_q} \\ \frac{\partial i_{gq}}{\partial t} = \frac{1}{L}(v_{in_q} - Ri_{g_q} - v_{g_q}) - \omega_g i_{g_d} \end{cases} \tag{17}$$

The matrix form of the multilevel inverter's output voltage is:

$$\begin{pmatrix} v_{in_d} \\ v_{in_q} \end{pmatrix} = R\begin{pmatrix} i_{g_d} \\ i_{g_q} \end{pmatrix} + L\frac{\partial}{\partial t}\begin{pmatrix} i_{g_d} \\ i_{g_q} \end{pmatrix} + L\omega_g\begin{pmatrix} -i_{g_q} \\ i_{g_d} \end{pmatrix} + \begin{pmatrix} v_{g_d} \\ v_{g_q} \end{pmatrix} \tag{18}$$

The PI controller block diagram of a grid-connected inverter in the dq reference frame is shown in Fig. 7.

The mathematical model of the $\alpha\beta$ reference frame is:

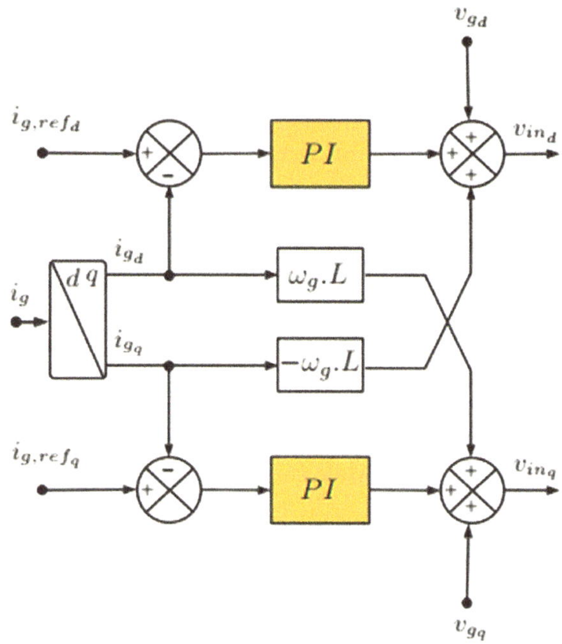

Fig. 7 Current control at the d_q reference frame

$$\begin{cases} \frac{\partial i_{g_\alpha}}{\partial t} = \frac{1}{L}(v_{in_\alpha} - Ri_{g_\alpha} - v_{g_\alpha}) \\ \frac{\partial i_{g_\beta}}{\partial t} = \frac{1}{L}(v_{in_\beta} - Ri_{g_\beta} - v_{g_\beta}) \end{cases} \tag{19}$$

The PR+HC controller block diagram of a grid-connected inverter in the αβ reference frame is shown in Fig. 8.

The transfer matrix from the dq reference frame to the $\alpha\beta$ reference frame is:

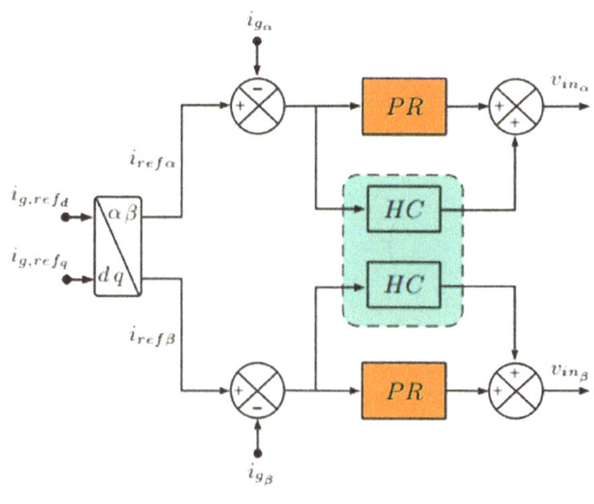

Fig. 8 Current control in the αβ reference frame

$$\begin{pmatrix} v_{g_\alpha} \\ v_g \end{pmatrix} = \begin{pmatrix} cos(\theta_g) & -sin(\theta_g) \\ \sin(\theta_g) & \cos(\theta_g) \end{pmatrix} \cdot \begin{pmatrix} v_{g_d} \\ v_{g_q} \end{pmatrix} \quad (20)$$

4.2 *Reference Current Generation*

The input power control enables the generation of the reference maximal direct current $\hat{I}_{g,refd}$. While neglecting the inverter internal losses and the LCL filter inductors stored energy, the total instant power provided by the PV panels is the sum of all the instant powers gathered by the capacitors and sent to the grid.

$$P_T = \sum_{i=1}^{C} V_{PV_i} \times I_{PV_i} = \sum_{i=1}^{C} p_{ci} + P_g \quad (21)$$

$$P_g = P_T - \sum_{i=1}^{C} p_{c_i} = \sqrt{2} \times V_g \times \hat{I}_{g,ref_d} \quad (22)$$

We assume a unity power factor ($cos(\phi) = 1$), the reference current is calculated by:

$$\Rightarrow \hat{I}_{g,ref_d} = \frac{P_T}{\sqrt{2} \times V_g} - \hat{I}_d \quad (23)$$

The second term of Eq. (23) presents the stored power inside the capacitors C_i. Those variations are due to the MPPT blocks expressed by a direct current $\hat{I}_d$, Eq. (23) can be written as follows:

$$\Rightarrow \hat{I}_{g,ref_d} = \hat{I}^*_{g,ref_d} - \hat{I}_d \quad (24)$$

We use the synchronization angle $\theta^{\sim}$ delivered by the PLL block, the inverter reference current $i_{g,refd}$ (Eq. 23)

$$i_{g,ref} = \hat{I}_{g,ref_d} \times \sin\left(\tilde{\theta}\right) \quad (25)$$

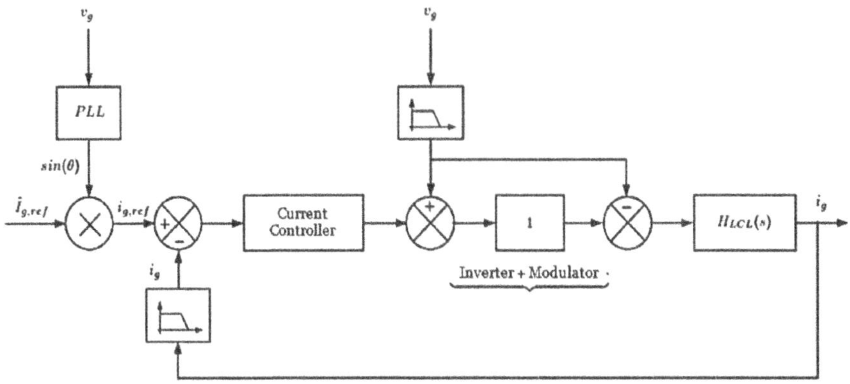

Fig. 9 Current control loop

4.3 Direct Control of the Grid-Injected Current

The direct current control (Fig. 9), needs both the magnitude and the phase of the injected current. This type of controllers does not allow the reactive power control because the reference current is considered indecomposable [7].

4.4 Current Control in the Dq Reference Frame

We consider the command strategy given in Fig. 7, where the reference current $i_{g,refd}$ and $i_{g,refq}$ gathered from the external control loop, the current $i_{g,refq}$ is chosen to be zero to eliminate the reactive power injected into the grid.

In the ideal case, where the PV system generator provides active power only, the control loops d and q have the same dynamic. The proportional and the integral gains (Kp and Ki) settings of the current adjustment are done for the d axis considering the q axis ones are the same. Thus, the current control loop in the d_q reference frame is given as follows (Fig. 10).

The transfer functions of the block diagram above are:

- PI controller transfer function:

$$C_{PI}(s) = K_p + \frac{K_i}{s} \tag{26}$$

- Transfer function of calculation time:

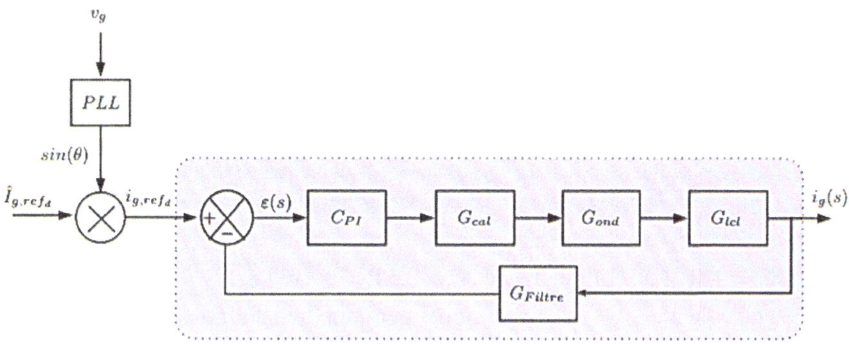

Fig. 10 The direct current control loop by the PI controller

$$G_{cal}(s) = \frac{1}{1 + sT_s} \tag{27}$$

$T_s = \frac{1}{f_s}$ and $f_s = 2.5$ kHz is the breakdown frequency.

- Inverter's transfer function:

$$G_{ond}(s) = \frac{1}{1 + 0.5sT_{dec}} \tag{28}$$

$T_{dec} = \frac{1}{f_{dec}}$ and $f_{dec} = 2.5$ kHz is the breakdown frequency.

- Filter block transfer function (only inductors and parasitic resistors are considered):

$$G_{lcl}(s) = \frac{1}{R + sL} \tag{29}$$

$$G_{lcl}(s) \approx H_{lcl}(s), L = L_i + L_g et R = R_i + R_g$$

- Low-pass filter transfer function:

$$G_{Filter}(s) = \frac{1}{1 + 0.5sT_{dec}} \tag{30}$$

Thus, the open-loop transfer function is given as follows:

$$G_{OL_c} = C_{PI} C_{cal} G_{ond} G_{lcl} G_{Filter} \tag{31}$$

4.4.1 PI Factors Determination

From Eqs. (26), (27), (28), (29), and (30), the overall transfer function of the current loop G_{OL_c} can be written as follows:

$$G_{OL_c}(s) = \frac{K_i + sK_p}{s} \times \frac{1}{1 + sT_{\sum}} \times \frac{K_e}{1 + sT_e} \tag{32}$$

where $K_e = 1/R$, $T_c = L/R$ et $T_{\sum} = T_s + 0.5T_{dec} + 0.5T_{dec}$

If we take factor $K_i = \frac{K_p}{T_c}$, the current open-loop transfer function becomes:

$$G_{OL_c}(s) = \frac{K_p}{sT_e} \times \frac{K_e}{1 + sT_{\sum}} \tag{33}$$

And the current closed-loop transfer function is:

$$G_{CL_c} = \frac{K_p K_e}{K_p K_e + sT_e + s^2 T_c T_{\sum}} \tag{34}$$

Using the modulus optimum criteria described in [8], the closed-loop gain $G_{BF}(s) = 1$, implies that $|G_{BF}(j\omega)| = 1$.

The square of this gain gives:

$$\begin{aligned} |G_{CL}(j\omega)|^2 &= \frac{(K_p K_e)^2}{(K_p K_e - \omega^2 T_e T_{\sum})^2 + (\omega K_e)^2} \\ &= \frac{(K_p K_e)^2}{(K_p K_e)^2 + \omega^2 (T_e^2 - 2K_p K_e T_e T_{\sum}) - \omega^4 (T_e T_{\sum})^2} \end{aligned} \tag{35}$$

$|G_{CL}(j\omega)|^2 = 1$for a low values of ω, $\omega^4(T_e T_{\sum})^2 \approx 0$

$$\Rightarrow |G_{CL}(j\omega)|^2 = 1 \text{ for } T_C^2 - 2K_P K_C T_\Sigma = 0 \tag{36}$$

The PI controller parameters are:

$$\begin{cases} K_p = \frac{T_e}{2K_e T_\Sigma} = \frac{L}{2T_\Sigma} = 4.625 \\ K_i = \frac{K_p}{T_e} = 125 \end{cases} \tag{37}$$

These values are used to start the analysis using the Matlab/Simulink Tool, PID Tuner. We consider some requirements for the controller design:

- Gain margin upper than 6 dB and a phase margin upper than 45°, to ensure stability.
- minimum bandwidth of 500 rad/s

Figures 11 and 12 give, respectively, the step response and the Bode diagram of the function $G_{BF}(j\omega)$.

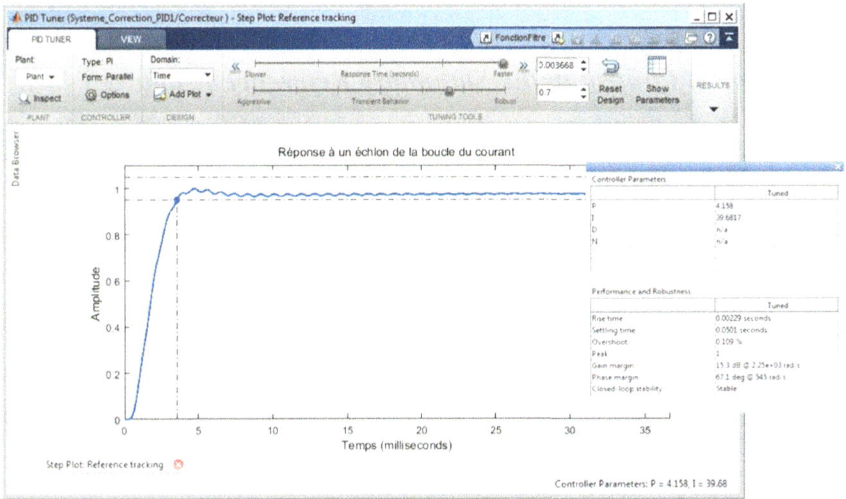

Fig. 11 Step response of the current loop and PI controller factors

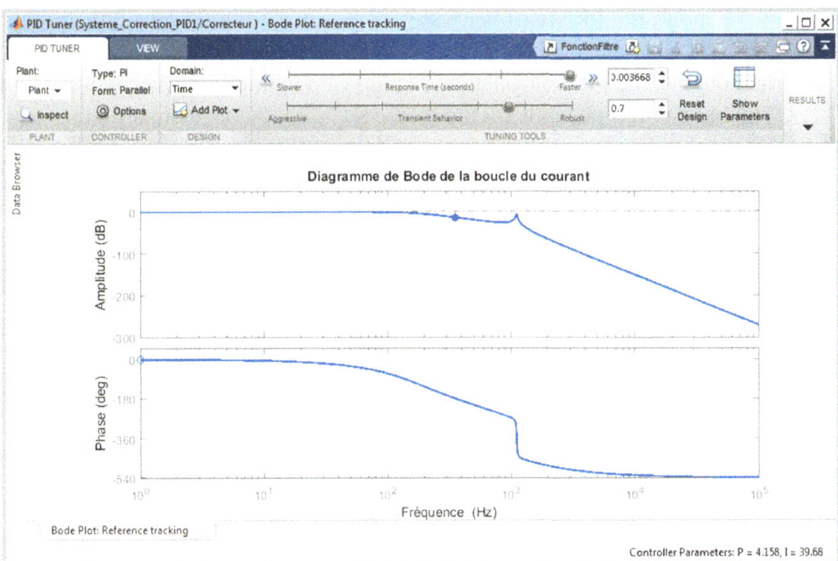

Fig. 12 Bode diagram of the current loop

The PI controller correction factors ($K_p = 4.158$ and $K_i = 39.681$) are selected for: a damping factor of $\xi = 0.7$; a bandwidth of 545.3 rad/s; a phase margin of 67.1°; and a gain margin of 15.3 dB.

4.5 Current Control in the Stationary Reference Frame $\alpha\beta$

The proportional resonant controller (PR) is suggested for the current control in the stationary reference frame αβ. Figure 13 gives the current control block diagram based on the PR controller with harmonics compensation.

4.5.1 Proportional-Resonant Controller: PR

The main advantage of the PR controllers is its infinite gain at the grid frequency, i.e., a unity gain in the closed-loop. This unity gain guarantees a good sinusoidal reference tracking.

The transfer function of the PR controller is:

$$C_{PR}(s) = K_p + \frac{K_r.s}{s^2 + \omega^2} \tag{38}$$

Figure 14 shows the PR controller bode diagram for different integral gains and a resonant frequency of 50 Hz.

We can note that the PR controller gets a high gain in thin frequency band around 50 Hz. The band width depends on the factor K_r, a low K_r leads to a thinner band while a high value of K_r leads to a larger band.

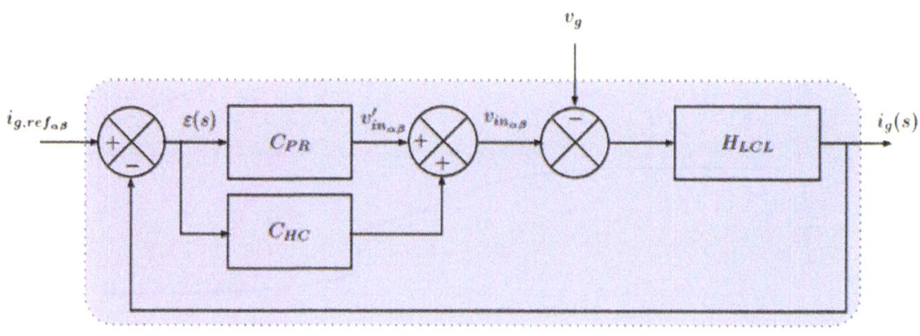

Fig. 13 Current loop based on the PR controller with harmonics compensation

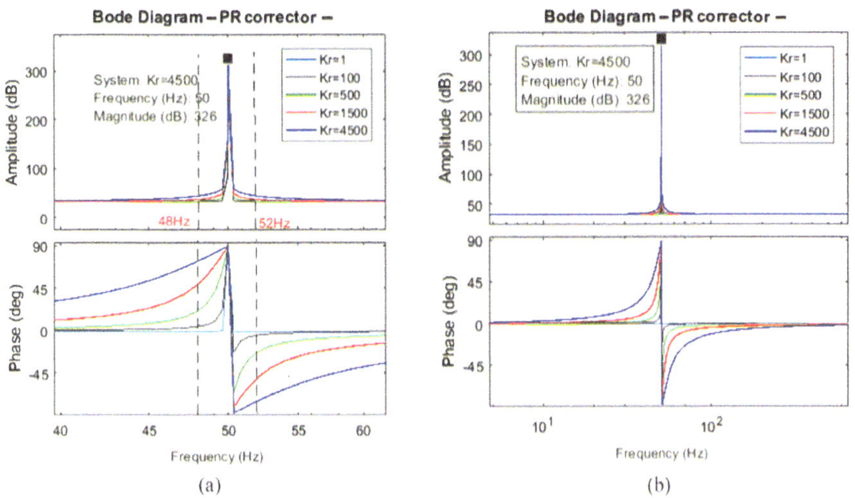

Fig. 14 PR controller Bode diagram. **a** For different Kr values **b** More details closer to 48 and 52 Hz

4.5.2 Adaptive Proportional-Resonant Controller

In the case of grid frequency variation, the PR controller performances are significantly reduced. Indeed, the PR controller uses a static resonance frequency and cannot remove the steady-state error in such case of operation. In addition, the harmonics compensation (HC) related to the PR controller based on the fundamental frequency is more sensitive to the grid frequency variation.

Normally, HC uses the multiples of the nominal grid pulse ω_g, thus, in case of grid frequency variation, the PR and HC performances are reduced. To rectify this issue, we suggest a controller improvement using a pulse provided by a PLL block to maintain the PR and HC performances (Fig. 15).

4.6 Harmonics Compensator

The harmonics compensation is defined by:

$$C_{HC}(s) = \sum_{h=3,5,7,..} \frac{K_{ih}.s}{s^2 + (\omega.h)^2} \tag{39}$$

The most important harmonics in the current spectrum are the 3rd, 5th, and the 7th ones. Thus, the harmonics compensator is designed to compensate those harmonics. Figure 16 shows the compensator gains at the frequency of the harmonics we want to eliminate.

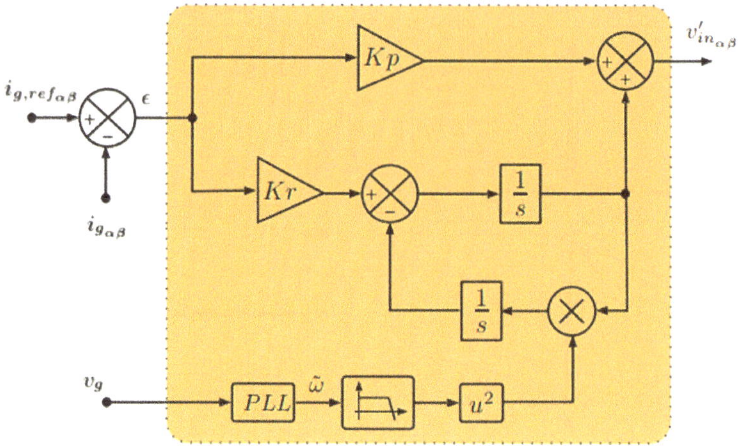

Fig. 15 PR controller with frequency adaptation

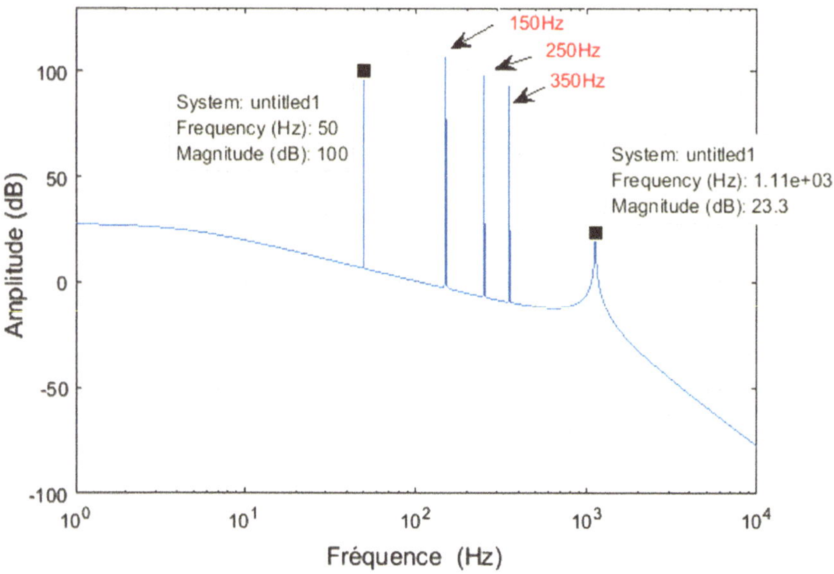

Fig. 16 PR + HC controllers with harmonics compensator

4.7 PR Controller Performances

Using the root locus method, proportional gains and resonant factors are selected to have a dominant poles damping factor of 0.7.

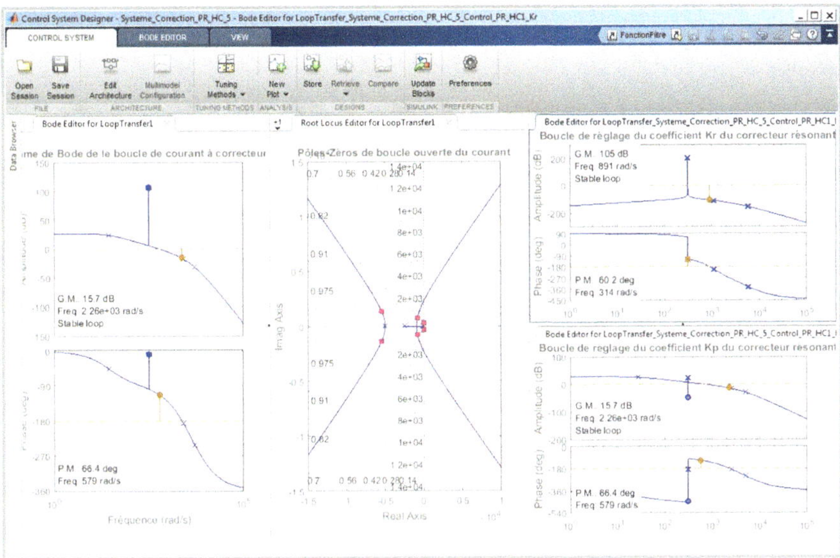

Fig. 17 Open loop system performances settings with Kp and K_r factors choice

As Fig. 17 shows, with a K_p value of 3555, the resonant poles damping reaches the value of 0.719. An integral factor K_r of 884.5 is chosen for its good dynamic and noise rejection. The zero-pole location and the bode diagram of the current control open loop are shown in Fig. 17.

The step response in the resonant regulator based closed-loop system is given in Fig. 18. The selected performances are: The response time 29 ms; the bandwidth; 579 rad/s; phase margin of 66.4° which involves the system stability; and gain margin 15.7 dB.

4.8 *Voltage VDC Control Loop*

The DC bus voltage V_{DC} of each H-bridge cell is influenced by the current variations i_{PVi} corresponding the solar panel (panel$_i$), and depends on the MPPT and on external conditions (T°, illumination intensity…). This voltage can also be increased in the case of injection of current saturation (voltage dip). In this study, we are interested in normal operation state when the current stays between acceptable limits. When a short circuit happens in the grid side, the voltage v_g falls, so the injected power P_g falls likewise. However the PV panels keep providing the same constant power P_T, the difference between the input and the output power is stored in the capacitors C_i causing an overvoltage.

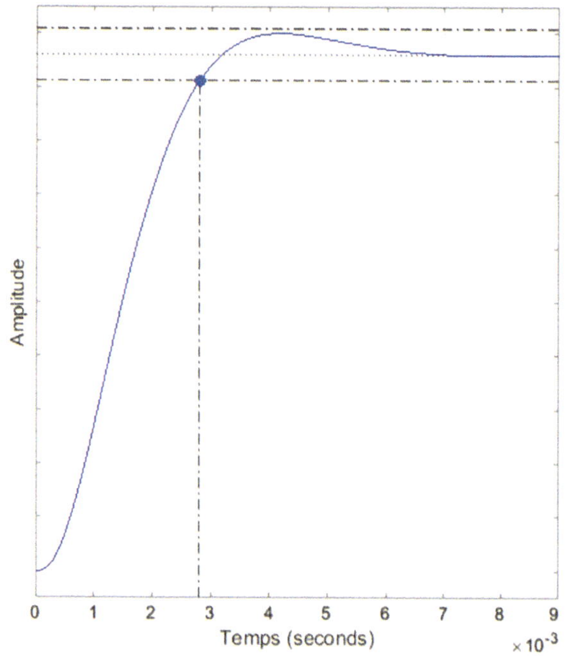

Fig. 18 Current loop step response based on a resonant controller

For this reason, DC voltages increase sometimes until exceeding the admissible limits. After the fault elimination, the powers balance and the voltages stabilize at high values. Therefore, a control loop must be added to adjust these voltages. The control concept of the voltages v_{DCi} is to discharge the residual power to reduce v_{DCi} as shown in Fig. 19.

This voltage control is carried out by supplying or absorbing active power from the grid. This voltage adjustment must be done by adding the fundamental active current to the reference current.

The transfer functions of the block diagram are shown in the Fig. 20.

- PI controller:

$$C(s) = K_{pt_i} + \frac{K_{it_i}}{s} \tag{40}$$

- G_{BF} internal current loop is assumed to be equal to 1;
- The loops G_{cal} et G_{Filtre} are the same as the current;
- The function $\frac{1}{C_i.s}$ represents the bus capacitor impedance;
- The voltage transfer function of the opened-loop is:

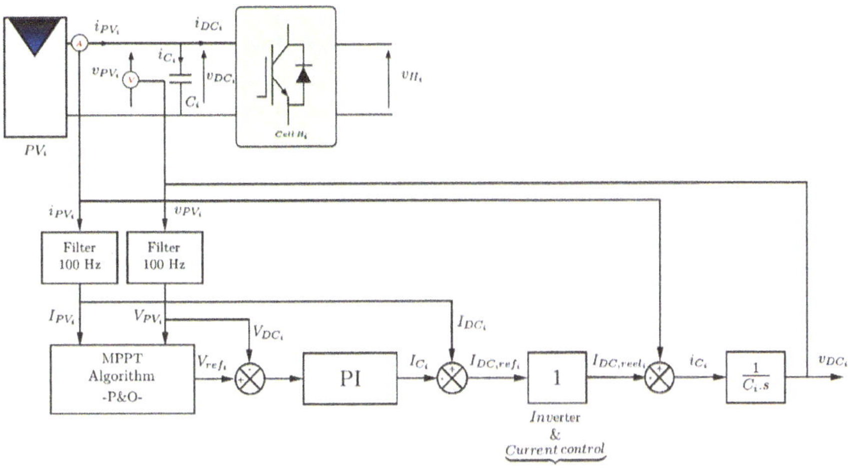

Fig. 19 Voltage control loop (for each DC bus)

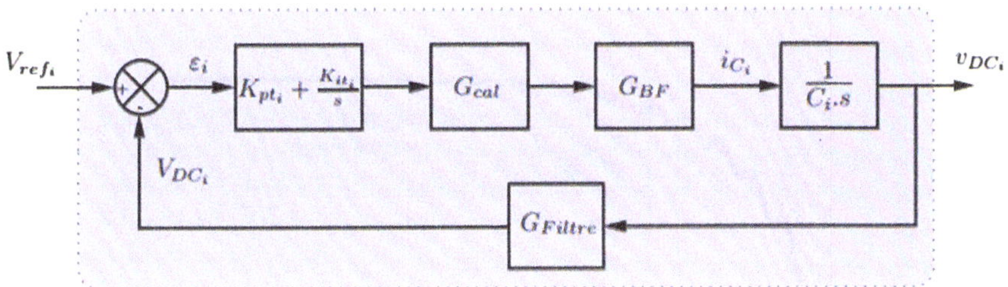

Fig. 20 Loop blocks diagram for each bus_i voltage control

$$G_{BO_T}(s) = C(s).G_{cal}.G_{BF}.\frac{1}{C_i.s}.G_{Filter} \tag{41}$$

We use the control system designer tool in Matlab/Simulink (Fig. 21), pole-zero location in opened-loop allows the PI controller factors adjustment ($K_{pti} = 78.58e^{-03}$ *and* $K_{iti} = 4.64e^{-03}$) for the following performances: a time response of 0.744 s; a bandwidth of 11.46 Hz. to ensure a good stability a gain margin of 6.23; a phase margin equal to 88.4°.

The step response of the voltage closed-loop is given in Fig. 22.

5 PSCPWM Proportional Modulator

The main modulator task is the inverter input voltage synthesis. Due to cells series connection, it is verified that:

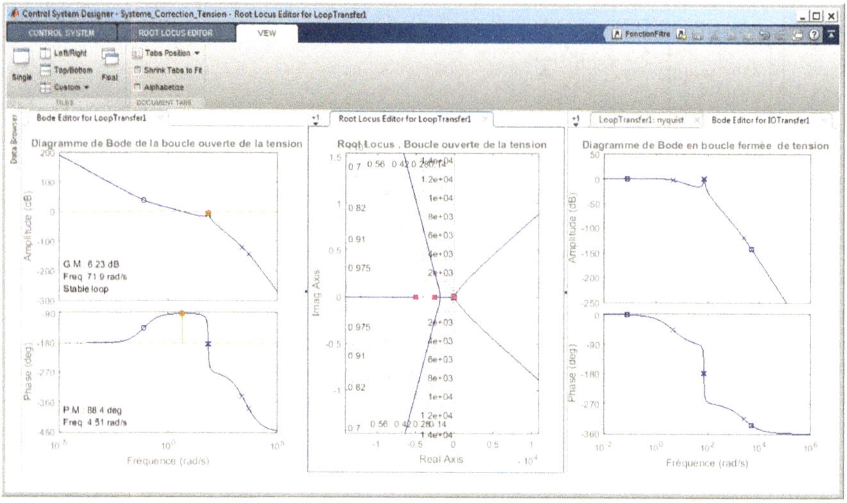

Fig. 21 PI controller factors adjustment

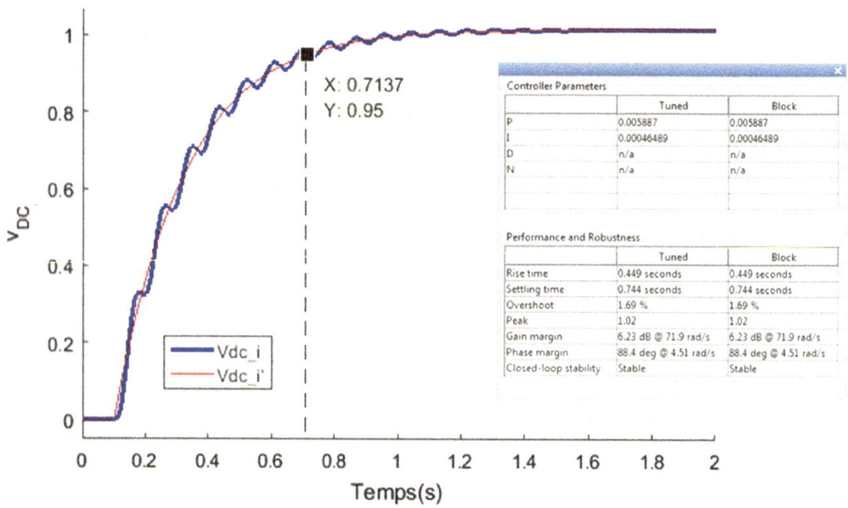

Fig. 22 Step response of the DC_i voltage loop

$$v_{in} = \sum_{i=1}^{C} v_{H_i} \tag{42}$$

Defining the proportional factors as the ratio between the individual voltage v_{Hi} and the total voltage v_{in}:

$$\alpha_i = \frac{v_{H_i}}{v_{in}} \qquad (i = 1, 2, \ldots, C) \tag{43}$$

For this inverter, each celli has the same current i_{in}. Therefore, the output voltage is proportional to the power, verifying then the following formula:

$$i_{in} = \frac{P_{PV_i}}{v_{H_i}} \Rightarrow \frac{P_{PV_i}}{v_{H_i}} = \frac{P_{PV_1}}{v_{H_1}} = \frac{P_{PV_2}}{v_{H_2}} = \cdots = \frac{P_{PV_n}}{v_{H_n}} \tag{44}$$

$$\Rightarrow \frac{P_{PV_i}}{P_T} = \frac{v_{H_i}}{v_{in}} = \alpha_i \qquad (i = 1, 2, \ldots, C) \tag{45}$$

The PSCPWM modulator, of (N-1) one signal carriers A_m and a common modulation index M for all the cells, and adapted to the control pulse generation. Our suggestion is to implement a PSCPWM proportional modulator, according to the power of the PV_i module, the duty cycle d_i and then the modulation index M_i of the corresponding $cell_i$ and proportional to the ratios $\frac{P_{PV_i}}{P_T}$ and $\frac{v_{H_i}}{v_{in}}$ (Fig. 23).

Considering that PSCPWM modulator is linear for a modulation index less than the unit [9], the duty cycle of each cell can be calculated as follows:

$$d_i = \frac{v_{H_i}}{v_{DC_i}} \Rightarrow d_i = \frac{v_{in}}{v_{DC_i}} \times \alpha_i \quad (i = 1, 2, \ldots, C) \tag{46}$$

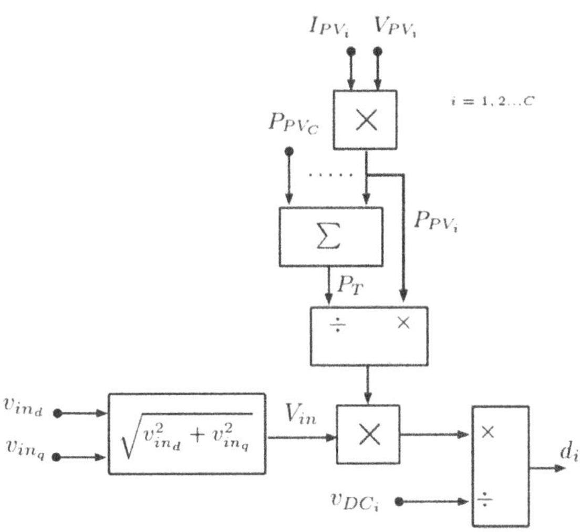

Fig. 23 Suggested PSCPWM proportional modulator

6 Conclusion

In this chapter, we have suggested a novel control strategy of a multilevel H-bridge inverter. The inverter connected to the grid is made through an LCL filter due to its ability to reduce harmonics (−60 dB/dec). For this reason, we have chosen it instead of a simple inductance filter. To rectify the resonance phenomenon through the capacitors and inductors, we have connected a passive damping resistor in series with the capacitor.

By controlling active and reactive powers in current mode, we have proved that the PI controller cannot track sinusoidal reference with the elimination of the static error in the synchronous reference frame *dq*. Therefore, a PR controller is obviously necessary every time we want to control sinusoidal current. In the stationary reference frame $\alpha\beta$, the PR controller shows its effectiveness to control sinusoidal current (static error remains almost null). Moreover, we have added a harmonics compensator HC that improves the current quality by eliminating dominant odd harmonics (3rd, 5th, and 7th).

To have a simple DC voltage for each H-bridge $cell_i$, we have used a local PI controller. Furthermore, we have suggested a proportional modulator to generate the drive pulses. This generation is made according to the rate of each module and the total power.

References

1. Kang F-S, Park S-J, Cho SE, Kim C-U, Ise T (2005) Multilevel PWM inverters suitable for the use of stand-alone photovoltaic power systems. IEEE Trans Energy Convers 20(4):906–915
2. Tolbert LM, Peng FZ (2000) Multilevel converters as a utility interface for renewable energy systems. In: Power engineering society summer meeting, 2000, vol 1272. IEEE, pp 1271–1274
3. Alonso O et al (2003) Cascaded H-bridge multilevel converter for grid connected photovoltaic generators with independent maximum power point tracking of each solar array. In: Proceedings of IEEE power electronics specialists conference, 2003, vol 2, pp 731–735
4. Rodriguez J, Bernet S, Wu B, Pontt J, Kouro S (2007) Multi-level voltage-source-converter topologies for industrial medium-voltage drives. IEEE Trans Ind Electron 54(6):2930–2945
5. Noriega-Pineda D, Espinosa-Perez G (2007) Passivity-based control of multilevel cascade inverters: high performance with reduced switching frequency. In: Proceedings of IEEE ISIE, pp 3403–3408 (Jun 2007)
6. Krishna RA, Padma Suresh L (2016) A brief review on multi level inverter topologies. In: IEEE 2016 international conference on circuit, power and computing technologies (ICCPCT), pp 18–19 (March 2016)
7. Satheesh Reddy J, Choudhury RA et al (2019) Hardware design of cascaded multi-level inverter using multi-sine PWM technique. In: IEEE, international conference on electrical, computer and communication technologies (ICECCT) 20–22 Feb 2019
8. Anil Kumar V, Arounassalame M (2018) Comparison of CHB Multi level inverters using Level shifted Modulation techniques with closed loop PI control. In: IEEE, 2018 4th international conference on electrical energy systems (ICEES), 7–9 Feb 2018
9. Agoro S, Balogun A et al (2018) Control of a three-phase multi-string five-level inverter for grid integration of PV SystEms with unbalanced DC-link voltages. In: IEEE, 2018 9th IEEE

international symposium on power electronics for distributed generation systems (PEDG), 25–28 June 2018
10. Feloups CES, Ali AIM, Mohamed EEM (2018) Single-phase seven-level PWM inverter for PV systems employing multi-level boost converter. In: IEEE, 2018 international conference on innovative trends in computer engineering (ITCE), 19–21 Feb 2018
11. El-Tamaly HH, El-Tamaly AM, El-Baset Mohammed AA (2003) Design and control strategy of utility interfaced PV/WTG hybrid system. In: The ninth international middle east power system conference, MEPCON, pp 16–18
12. Satheesh Reddy J, Choudhury RA et al (2019) Hardware design of cascaded multi-level inverter using multi-sine PWM technique. In: IEEE, 2019 IEEE international conference on electrical, computer and communication technologies (ICECCT), 20–22 Feb 2019
13. Yazdani A, Iravani R (2010) Voltage-sourced converters in power systems: modeling, control, and applications. Wiley-IEEE Press, March 2010
14. Kjaer SB (2005) Design and control of an inverter for photovoltaic applications. PhD dissertation, Institute Energy Technology, Aalborg University, Aalborg East, Denmark, pp 1237 (vol These)
15. Abdelaziz F, El Bachtiri R, El Ghzizal A (2014) Modélisation Et Contrôle d'un Onduleur H-Pont Multiniveaux Alimente Par Des Panneaux PV Et Raccorde Au Réseau Monophasé. In Maroc Tetouan, editeur, Journée Doctorale, JDP'14, 10 Mai 2014
16. Abdelaziz FRI, El Bachtiri R, El Ghzizal A (2015) Contrôle MPPT d'un Onduleur PV De Topologie H-Pont Avec Régulation De Courant Injecte Dans Le Réseau Electrique. In Maroc ENSA Oujda, editeur, Workshop International sur les Systèmes Avancés en Génie Electrique et Energie Renouvelable, SAGEER'15, 21–23 Mars 2015
17. Teodorescu R, Rodriguez P, Liserre M (2011) Grid converters for photovoltaic and wind power systems, volume Book. Wiley-IEEE Press, pp 1–416, Jan 2011
18. Abdelaziz FRI, El Bachtiri R, El Ghzizal A (2015). Amélioration De Commande D'Un Onduleur Pv A Pont Complet Raccorde Au Réseau Electrique Par Le Contrôle De La Puissance Active. In Maroc Tetouan, editeur, Journée Doctorale, JDP'15, 16 Mai 2015
19. Bajracharya C, Suul JA, Molinas M, Undeland TM (2008) Understanding of tuning techniques of converter controllers for VSC-HVDC. In: Nordic workshop on power and industrial electronics, NORPIE, June/2008
20. Fri A, El Bachtiri R, El Ghzizal A, Naamane A (2014) Triphase symmetrical cascaded multilevel inverter (5L) for PV systems controlled by various multicarrier PWM strategies. In: Energy Procedia, vol 62, ScienceDirect
21. Rajanand Patnaik N, Tagore YR, Chaitanya S (2017) Advanced seven level transformerless multilevel inverter topology for PV application. In: 2017 third international conference on advances in electrical, electronics, information, communication and bio-informatics (AEEICB), Chennai, 2017, pp 111–116. https://doi.org/10.1109/aeeicb.2017.7972393
22. Wu W, Li Y, He Y et al (2017) Damping methods for resonances caused by LCL-filter-based current-controlled grid-tied power inverters: an overview. IEEE Trans Ind Electron 46(c):1–1
23. Pena-Alzola R, Roldan Perez J, Bueno E, Huerta F, CamposGaona D, Liserre M, Burt GM (2018) Robust active damping in LCL-filter based medium-voltage parallel grid-inverters for wind turbines. IEEE Trans Power Electron 8993(c):1–1
24. Narasipuram RP, Somu C, Yadlapalli RT, Simhadri LS (2018) Efficiency analysis of maximum power point tracking techniques for photovoltaic systems under variable conditions. Int J Innov Comput Appl 9(4):230–240
25. Narasipuram RP (2018) Optimal design and analysis of hybrid photovoltaic-fuel cell power generation system for an advanced converter technologies. Int J Math Model Numer Opti 8(3):245–276

Off-Grid PV-Based Hybrid Renewable Energy Systems for Electricity Generation in Remote Areas

H. El-houari, A. Allouhi, M. S. Buker, T. Kousksou, A. Jamil, and B. El Amrani

Abstract Wind, solar, biomass, and geothermal energy are renewable energy sources known as promising alternatives for electricity generation, especially with the depletion of fossil fuels. Renewable energy sources are present, with huge quantities, free access, and without a negative impact on the environment. The electricity produced by renewable energies is gradually becoming economically and technically advantageous. In most cases, integrating a single source of renewable energy can lead to over-sizing and therefore a very expensive implementation. To remedy this, systems consisting of one or numerous renewable energy sources are adopted in order to guarantee maximum electricity production as well as practical reliability at encouraging costs. In addition, the electricity generated by renewable energy sources such as wind turbines, solar, biomass, geothermal energy … has a particular interest in isolated spaces. This chapter provides an updated literature review about Off-grid PV-Based Hybrid Renewable Energy System for electricity generation in remote areas. First, after the introduction, a presentation of the world energy situation was discussed in order to see the progress of the implementation of renewable energies on a global scale. Secondly, a section was reserved for renewable energy alternatives in order to discuss each source separately, before starting the part devoted to off-grid hybrid renewable energy structures. These structures have been examined in a large number of researches works with the aim of electrifying remote areas in several countries of the world. Finally, a detailed presentation of the main reliability, economic, and environmental performance indices is given for the evaluation of these systems.

H. El-houari (✉) · A. Allouhi · A. Jamil
Ecole Supérieure de Technologie de Fès, U.S.M.B.A, Route d'Imouzzer, BP 242, Fes, Morocco
e-mail: haytham.elhouari@usmba.ac.ma

H. El-houari · B. El Amrani
Laboratoire de Mathématiques, Modélisation en Physique Appliquée, Ecole Normale Supérieure de Fès, U.S.M.B.A, Route Bensouda, BP. 5206, Fes, Morocco

M. S. Buker
Energy and Semiconductors Research Group, Necmettin Erbakan University, Konya, Turkey

T. Kousksou
Laboratoire des Sciences de l'Ingénieur Appliquées à la Mécanique et au Génie Electrique (SIAME), Université de Pau et des Pays de l'Adour – IFR – A. Jules Ferry, 64000 Pau, France

S. Motahhir and A. M. Eltamaly (eds.), *Advanced Technologies for Solar Photovoltaics Energy Systems*, Green Energy and Technology,
https://doi.org/10.1007/978-3-030-64565-6_17

Keywords Hybrid renewable energy systems (HRESs) · PV · Isolated areas · Performance indexes

1 Introduction

A growing number of the world population, development of civilization, progress in the technological fields, and other factors contribute to growing energy needs in order to increase welfare on all scales. Electricity is a fundamental factor for economic development, urbanization, and industry in countries [1–3]. Production of electricity originates mostly from burning large quantities of fossil fuels. This excessive use of fossil fuels causes its exhaustion and consequently destroy the environment through greenhouse gas emissions, which has led to negative climate change experienced today [4, 5].

Under the current status of high energy demand, it is very necessary to seek other types of energy sources to be able to meet the energy demand in big cities also in isolated and remote regions. These sources should be environmentally friendly, with good performances at competitive costs [6]. The worldwide population without access to electricity is around 1.2 billion and 48% of them are located in the developing countries of Sub-Saharan Africa [7]. The adoption of multiple renewable energy is a good way to produce clean electricity and can be tailored for each geographic zone with respect to its renewable energy potential [8–11]. Furthermore, the overall efficiency and economic performance presented by these off-grid Hybrid Renewable Energy Systems (HRESs) were proven to be encouraging as highlighted by several studies [12–14]. Solar power generations including photovoltaics (PV) and Concentrating Solar Power (CSP), hydroelectric (small and large), wind turbines (onshore and offshore), biogas, and biomass have experienced a particular interest in recent years [15–17].

The main shortcoming of standalone renewable energy sources is that such systems produce electricity intermittently. Hybridization solves this problem since two or more technologies complement each other to avoid intermittency of energy supply [18, 19]. In this context, the HRESs can be exploited in two different ways either grid-connected; or operating autonomously with the storage systems. The latter option is more suitable for villages, regions, areas, and isolated islands. In the case of energy surplus, the excess energy can be stored for later use when the hybrid system cannot satisfy the energy demand.

In the case of HRES connected to the network in cities, towns, universities, factories, etc., the excess energy can be directly injected into the electrical network. In this method, the cost of electricity produced by the HRES is lower than the autonomous system [20].

Therefore, several studies suggested that HRES with storage implement in isolated zones requires a specific regulation as well as an adequate policy to accompany such standalone systems [21, 22]. These regulations must be diverged from already-adopted ones for HRES connected to the network and must support standalone HRES to face technical, regulatory, and financial challenges and constraints [23, 24]. These regulations and policies should in addition, respect and agree with the local and national conditions of each country [25, 26]. Having acknowledged these challenges and ensuring that standalone Hybrid Renewable Energy Systems can offer significant advantages socially (sustainable development and reduction of poverty), environmentally (reduction of greenhouse gases), and economically (predictable energy price). All these can satisfy the energy demand without a network with easy integration and simple installation of the system [20, 27].

Since the optimal design of HRES is a primary factor with likely combinations, therefore, several studies have already been carried out by using various software. The evaluation of HRESs is performed based on numerous criteria for looking into the sizing problem comprehensively [13, 28, 29].

This chapter focuses on standalone PV-based HRES for power generation in rural areas, villages, and remote islands by reviewing various HRESs architectures, formulating basic mathematical background for modeling multiple energy source systems, and proposing key performance indicators for the techno-economic assessment of such systems. Although there are numerous studies about the hybrid renewable technologies in the existing literature, the sizing procedures regarding various renewable systems with energy storage are not investigated extensively. Therefore, the novelty of this article is to present a comprehensive discussion about various HRES configurations with energy storage and demonstrate detailed reliability, economic, and environmental performance indices of the evaluated systems. In line with the objectives, this manuscript is structured as follows: Sect. 2 provides an overview of the global renewable energy sector; Sect. 3 presents the modeling of the subsystems. Section 4 details various combinations of the standalone hybrid renewable energy systems, as well as multi-criteria performance indexes, are discussed in Sect. 5.

2 Overview of the Worldwide Renewable Energy Sector

The sector of renewable energy (RE) as well as their widespread use is at the top of the worldwide energy policy, especially for the many environmental and energy outcomes they are providing [30–32]. The whole world needs to increase the share of renewable energies for electricity production, especially with the increase in population and industrialization, the exhaustion of fossil fuels, and the environmental damage they have caused.

Keeping in mind these considerations, several countries adopted national energy strategies in order to accompany the world energy transition. In recent years, renewable energies have experienced rapid development and growth, taking profit from reduced production costs of technologies (especially wind turbines and photovoltaic panels) as well as the adopted policies and strategies.

Table 1 Production growth achieved by each technology in 2018 and average annual growth 2018–2030

Renewable technologies	Production growth achieved in 2018 (%)	Average annual growth 2018–2030 (%)
Solar PV	31	16
Onshore Wind	12	7
Offshore Wind	20	–
Hydropower	3	2.5
Bioenergy	5	6
Concentrating solar power (CSP)	17	26
Geothermal	5	10
Ocean power	16	24

According to the **International Energy Agency**, electricity which comes from renewable energies represents 1/5 of the world's energy needs.

Moreover, on a global scale, 2018 saw 26% of global electricity production via renewable energies, which indicates an increment of 7% compared to the year 2017 [33]. It is interesting to highlight that the production of WT, solar, and hydroelectric represents 90% of the electricity originating from the total renewable energy in 2018 [34].

To reach the objective for the Sustainable Development Scenario 2000–2030, it is necessary to ensure that there is a global growth of 7% of production via renewable energies. The following Table 1 shows the production growth achieved by each technology in 2018 [34], as well as the average annual growth that each of these technologies will reach between 2018 and 2030 so that at the end, the annual growth of 7% can be achieved.

However, this growth in renewable energies is not uniform across the world. Today, it is in Europe that renewable energies represent the most important part of energy consumption, with 17% on average according to the International Energy Agency. South America is also doing well with its large hydroelectric infrastructures: in Brazil, for example, 42% of energy consumption is from renewable sources. But this is not the case in all the countries of the region, some of which still have an energy mix with a very poor share of renewable energy.

For the United States, only 10% of the energy consumption is obtained from renewable energies. China and India are capped at 10 and 11%, respectively.

Morocco is one of the countries that have acceded to international conventions to fight global climate change. It was the organizer of the Conference Of the Parties (COP22) in 2016 in Marrakech. Since 2009, Morocco has adopted an energy strategy to improve the effectiveness of its power sector. The ambitious strategy adopted had a challenge of increasing the production of electricity by exploiting renewable energies to reach 42% at the end of 2020 and 52% by 2030 as shown in Fig. 1 [35].

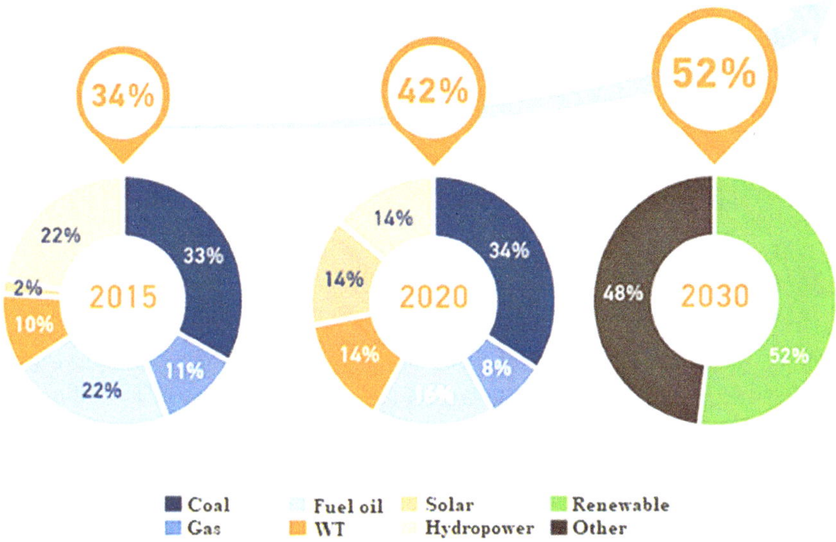

Fig. 1 Increase in the Moroccan energy mix between 2009 and 2030

According to the Ministry of Energy of Morocco, in 2018 the country has already installed power of 3700 MW of renewable energy which represents a rate of 34% of the national need for electricity.

3 Renewable Energy Alternatives

For modeling the various subsystems of an HRESs, there are several methods used by researchers. The most widely and simple modeling approaches are treated in the next section.

3.1 Photovoltaic

The power generated P_{PV} (kW) by a PV panel is illustrated in the **equation** [36–39]. The PV panel contains cells placed in parallel and in series with each other to produce the required power. The energy produced is dependent on solar radiation and the cell temperature:

$$P_{PV}(t) = P_{rated} \times Y_{PV} \times \left(\frac{G}{G_{ref}}\right) \times \left[1 + K_T\left(T_c - T_{ref}\right)\right] \tag{1}$$

P_{rated} is the estimated power that the PV panel can generate under standard conditions, Y_{PV}(-) means the derating factor of the Photovoltaic panels, $G(\frac{w}{m^2})$ and $G_{ref}\left(\frac{1000}{m^2}\right)$ means the incident solar radiations and the conditions of the test standards respectively, K_T is taken around 0.4 and 0.6% according to [40] is the Photovoltaic temperature factor; $T_c(C)$ is the temperature of the cells that make up the PV panel and T_{ref} (25 °C) means their temperature under the conditions of the test.

3.2 Wind Turbine

Multiple mathematical models of the wind energy production process have been employed by the research community. The power produced by a WT is influenced by three parameters which are the wind speed, the hub height, and the power curve of the wind turbine. The estimate of the power generated by the wind turbine $P_{WT}(kW)$ is explained as follows [41]:

$$P_{WT} = \begin{cases} 0 & V < V_{cut-in} \text{ and } V \geq V_{cut-out} \\ \frac{P_r(V - V_{cut-in})}{(V - V_{cut-in})} & V_{cut-in} \leq V < V_r \\ P_r & V \leq V < V_{cut-out} \end{cases} . \tag{2}$$

Or

$$P_{WT} = \left(\frac{\varphi}{\varphi_0}\right) \times P_{WT,STP} \tag{3}$$

where V_r and P_r represent the nominal wind speed and the nominal power, respectively. V_{Cut-in} and $V_{Cut-out}$ means the speed in and out, respectively. $P_{WT,STP}$ means the estimated power of the wind turbine under standard conditions (kW), φ and φ_0 represents the real density of the air and the density of the air in the test conditions (1.225 kg/m^3).

3.3 CSP

CSP referring to concentrated solar power, is a power plant that concentrates the rays of the incident sun using mirrors in order to heat a heat transfer fluid which generally makes it possible to produce electricity via a thermodynamic cycle. This type of power plant allows, by storing this fluid in a tank, to prolong the operation of the power plant for several hours beyond sunset.

3.4 Hydropower

Hydroelectricity is the conversion of hydraulic energy into electricity. The kinetic energy of the water current, natural or generated by the level difference, is transformed into mechanical energy by a hydraulic turbine, then into electrical energy by a synchronous electric generator. The hydroelectricity power $P_{hydro}(kW)$ can be calculated by the following equation.

$$P_{hydro} = \frac{\eta_{hydro} \times \varphi_{water} \times Q_{turbine} \times g \times H_{net}}{1000W/kW} \tag{4}$$

where $\eta_{hydro}(\%)$ is the turbine efficiency, $Q_{turbine}$ represents the nominal water flow of the turbine (l/s), φ_{water} means the water density (1000 kg/m^3), g means the acceleration due to universal gravitation (9.81 m/s^2), and $H_{net}(m)$ represents the net height.

3.5 Biomass

The generation of electrical energy from biomass can be carried out according to two mechanisms either by the thermochemical process (pyrolysis or direct combustion, gasification) or using the biochemical mechanism (anaerobic digestion or gasification). At the end of the two mechanisms, there is a recovery of a fuel which can be either in the gaseous state or in the liquid state [42]. The gasifier transforms biomass (solid) into fuel (gas), the output power of the biomass generator $P_{ge,bio}$ is illustrated in the following equation [43–46].

$$P_{ge,bio} = \eta_{ge,bio} \times Q_{ge,bio}(t) \times LHV_{bio} \tag{5}$$

where $\eta_{ge,bio}$ (%) is the electrical conversion efficiency, $Q_{ge,bio}$ (m^3/h) represents the flow rate of the fuel consumed LHV_{bio} (kWh/m^3) means the low calorific value of the biogas.

4 Off-Grid Hybrid Renewable Energy Structures

HRESs are generally divided into two broad categories: on-grid and standalone. The second category is the main focus of this study in order to design an optimal system for the electrical needs of remote areas. As hybrid renewable energy systems are the combination of two or more energy sources, at least two essential elements must be taken into account to structure a hybrid renewable energy system. The first criterion is the existence or absence of a storage system. Storage systems will store energy

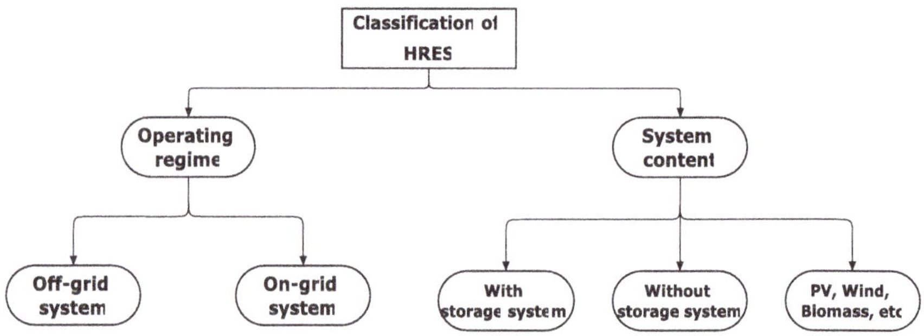

Fig. 2 General classification of hybrid renewable energy systems

during excess production and reuse when needed. The second criterion is the structure of the HRES, comprising various renewable technologies (PV, Wind, Hydro, etc.). Figure 2 shows the general classification of HRESs.

To successfully satisfy the energy demand, the configuration of the proposed system is key. It should be carefully tailored considering available energy sources and the capacity of the system as well as the specifications of the storage system. Thus, the cost of energy will be reduced, CO_2 emissions will be minimized and energy requirements will be met. Given the number of configurations for HRESs (see Fig. 3) and their complexity, the palpable approach is the optimization of these systems [28]. Favorably, several algorithms and software have been developed [47] to help to optimize the energy systems.

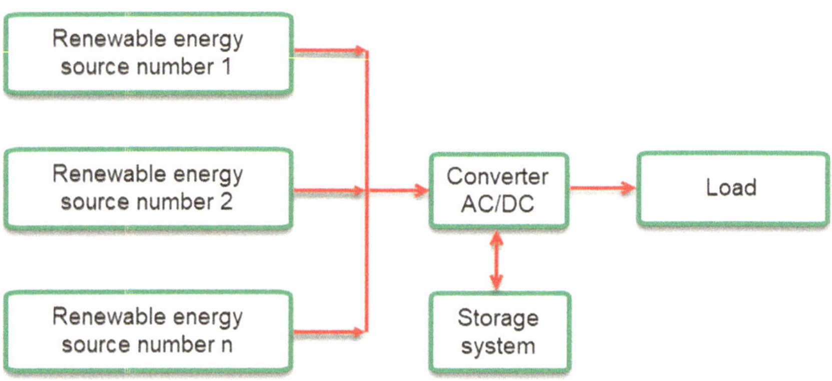

Fig. 3 Renewable hybrid energy system with n sources

4.1 Hybrid System with Two Renewable Energy Source and Storage System

4.1.1 Off-Grid PV/Wind

The PV–Wind off-grid system is a mixture of a wind turbine, solar panels, converter, and storage system, as shown in Fig. 4. Photovoltaic solar is considered to be a random and variable power [48], the solar radiation is variable during the day and all seasons. On the other hand, the wind turbine is considered a reliable source of energy taking into account the presence of constant wind blows during day and night despite the variation in wind speed [49]. In detail, electricity production is obtained mainly from PV panels during the day and from windmills during the night [50]. Hence, this PV–WT combination is massively adopted given the complementary energy production through this configuration, its reliability, and adaptability in any weather condition [51–53]. Moreover, the energy storage system will store excess energy production from hybrid PV–WT combination and meet the energy demand when electricity supply through the system is insufficient.

A significant number of studies have been carried out in the literature concerning these configurations including PV–Wind, PV, and Wind with a storage system. In [54, 55], the system that combines the two technologies (PV–Wind with a storage system) is the most profitable for isolated areas. Another study is conducted aiming to optimize the number of wind power units, PV panels, and batteries in order to reduce the capital cost of the system while securing the reliability of the hybrid renewable system [56]. Ghorbani et al. [57] presented a specific optimization and analysis process for a wind–solar system in isolated areas in the south of Brazil, with the aim of reducing the initial costs of the system. Store et al. [58] studied two energy management methods for a standalone Wind–PV system for a residential site in Denmark. Custom models have been set up as well as the actual renewable sources have been used. Papadopoulos et al. [59] have shown that the use factor of the electrolyzer is influenced by the configuration containing renewable energy sources,

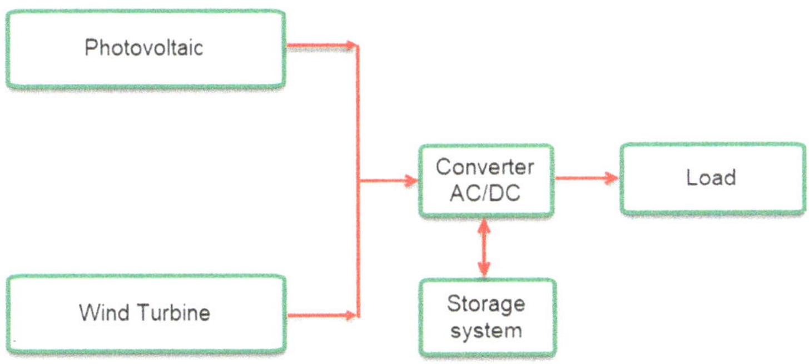

Fig. 4 Standalone PV–WT hybrid system

for this purpose several combinations of PV–Wind and storage technologies have been developed. Belouda et al. [60] presented a study focused on optimal design of the PV–Wind–Battery multi-objective system to satisfy the electrical load of a remote area. The two optimization points discussed are the loss of power supply probability (LPSP) and the general cost of the system using a multi-objective NSGA-II algorithm. Zhang et al. [61] have carried out a study on a PV–Wind–Battery system by adopting NSGA-II genetic algorithm which has as the objective of determining the number of PV, wind, and battery as well as the LPSP and general cost of the system for an isolated island. The system was tested under four scenarios of weather conditions. A study on a system composed of PV, wind turbines, and batteries for an island of Jeju was carried out by [62]. The study showed that LCC and LCOE equal to 84.3 BUSD and 0.42 $/kWh, respectively. Samy et al. [63] have set out a study of three combinations of a PV–FC, PV–Wind–FC, and Wind–FC hybrid system in Remote district, Beni-Suef, Egypt. The study showed the PV–Wind–FC system is the best combination with an energy cost worth 0.47 $/kWh. Another study which was carried out by [64] to satisfy the energy demand of Makadi Bay, Red Sea, Hurgada, Egypt in the presence of PV, Wind, and battery as well as a comparison was performed between PV–Wind–Battery and PV only and Wind only scenarios. The results showed that the PV–Wind–Battery system is the best configuration in terms of installation cost of the three systems. Benavente et al. [65] presented a modeling and simulation of a PV–Battery system for a rural area containing a school, household, and a health center in the highlands of the Bolivian Altiplano. The simulation was carried out on the basis of real consumption data as well as meteorological data, the results found that to avoid the SD effect and ensure that system is profitable, it is necessary to have a greater power of the PV field rather than increasing the battery capacity.

4.1.2 Off-Grid PV/Biomass

The second off-grid HRESs configuration is the PV, Biomass, converter, and a storage system as shown in Fig. 5. The proposed configuration is a promising alternative for

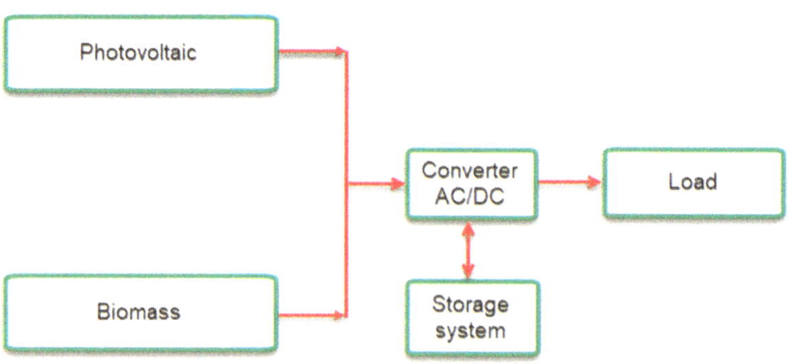

Fig. 5 Standalone PV–Biomass hybrid system

electricity generation, especially in remote areas. In most rural areas there is the abundant presence of an animal, agricultural, municipal waste, etc. [66] can be well utilized as a fuel for biomass generator. The advantage is that biomass is controllable (present when needed) which is considered to be the 4th biggest source of energy total in 8.5% of global energy consumption. The drawback of biomass is that when the fuel is not available, it must be somewhat purchased [67]. In operation, anaerobic digestion is considered to be one of the best-known techniques of energy extraction. Moreover, biogas is considered to be one of the end products that is used to generate electricity [68–70] using a biogas generator [43, 71].

To name a few recent studies about PV, Biomass, converter, and a storage system configurations, Shahzad et al. [72] represented a technical-economic study of a sytem composed of PV–Biomass–Battery for a small village (farm and residential) in Pakistan. The optimal configuration obtained in the results is 10 kW of PV, 8 kW of biogas generator, 12 kW for the converter, and 32 unit of batteries the recovery period was estimated to be 9.5 years and the cost of electricity was found to be equal to 5.51 PKR/kWh. Ganthia et al. [73] carried out a study on energy demand of the Khalardda village located in Odisha and the system chosen for this objective is PV–Biomass. Kohsri et al. [74] analyzed a PV–Biomass–Battery system in a rural area of an educational institute in Thailand. The results showed that optimal configuration comprises three converters each with a capacity of 12 kW, a biomass generator with a capacity of 33.7 kW, battery with a capacity of 60.9 kWh, 12,285 kW of PV, and 13.8 kW of three BDIs. Heydari and Askarzadeh [45] have designed an off-grid PV–Biomass system to meet the electrical needs of agricultural wells located in Bardsir, Iran. As several scenarios were analyzed, the PV–Biomass system was found to be better in comparison to PV alone or biomass alone system. The cost of electricity was estimated at 0.1855 $/kWh and the capacity of the Biomass and PV system was found to be 180 kW and 75.2 kW respectively. Singh and Baredar [75] presented an off-grid system composing Biomass–PV–FC–Battery at Maulana Azad National Institute of Technology, Bhopal in India. The simulations were carried out via Homer, the cost of electricity and the NPC were found equal to 15.064 Rs/kWh and 51,89003 Rs, respectively. Eteiba et al. [76] used three algorithms to find an optimal solution for the standalone biomass, PV, and battery system for a Monshaet Taher village in Egypt. The results pointed out that the system must include 200 kW of biomass, 131.04 kW of PV, and 298 batteries. Ghenai and Janajreh [77] presented a system composed of Biomass–PV to satisfy the need for electricity in the city of Sharjah in the United Emirates. The COE found equal to 0.328 $/kWh. Pradhan et al. [78] analyzed the performance of a standalone Biomass–PV–Battery system to satisfy the electrical charge of around 20 kWh/day for a house located in a remote area. Chowdhury et al. [79] developed an autonomous mini-network made of Biomass–PV for a remote area of Ashuganj, in Bangladesh to be able to meet the annual energy demand of 14,161 MWh. Analysis showed that the return on investment period was found to be 6.9 years as well as the amount of CO_2 avoided is approximately 3.81 tons/year.

4.1.3 PV/Hydro

In this section off-grid, PV–Hydro hybrid renewable system will be analyzed to meet the energy needs for isolated areas. The combination comprises photovoltaic solar and hydropower (see Fig. 6). Since PV panels are highly affected by ambient conditions, hydroelectricity is, on the other hand, a more flexible, stable, and adjustable sort of energy source [80, 81]. This is another complementary system as there is less solar irradiation but a high level of water during the winter and there is a risk of dry dams but abundant solar irradiation during the summer. Recently, the price of the PV system has dropped drastically, therefore the PV–hydro system seems economically feasible [82].

Yibo and Honghua [82] fitted out a study on a standalone PV–hydro system in Yushu in China, the structure of the PV panels belongs to the MW class as well as the battery of the DC bus. Das and Akella [83] presented a PV–hydro–battery hybrid system with a management control strategy and simulated using MATLAB/Simulink. The results found show that the battery life will be extended once the SOC is born between its limits. Singh [84] developed an autonomous PV–Hydro system to meet the energy needs of remote areas. In this study, a VSC control model was proposed. Kougias et al. [85] presented an optimization algorithm to ensure complementarity between the two renewable technologies of PV–Hydro. Also, the proposed strategy was tested in a case study. Shan et al. [86] performed a study regarding the complementary between PV and hydroelectric in California Independent System Operator (CAISO). The results showed that when the share of electricity production of PV increases by 1%, on the other side hydroelectricity increases between 0.01% and 0.06% this correlation is important to meet peak energy needs. Silvério et al. [87] studied a technical-economic design approach for PV–Hydro hybrid systems, as well as a case study of the São Francisco hydroelectric station in Brazil which suffers from intense drought. The results showed that PV panels should have been tilted at 3° and the proposed system can be generated from electricity at low cost.

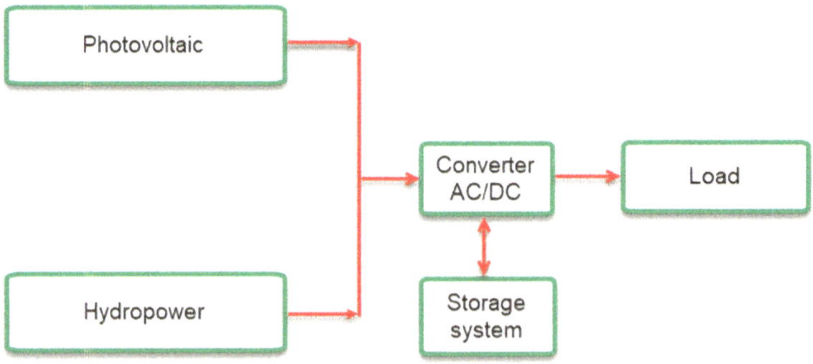

Fig. 6 Off-grid PV–Hydropower hybrid system

4.1.4 CSP/PV

The fourth combination to study is the CSP/PV hybrid standalone renewable energy system. This is the combination of two different technologies fed by the same energy source as shown in Fig. 7. The CSP technology is still immature compared to PV technologies, worldwide the capacity of the installed CSP does not exceed 5.5 GW until the end of 2018 [88]. The common preference of the two technologies is the abundance of solar radiation. The electrical conversion efficiency by these two technologies is greatly influenced by dirt, clogging, and the deposit of dust [89, 90].

CSP is suggested as a continuous and stable power generation technology in parallel with a thermal storage system [88]. But so far CSP was among the least deployed technologies compared to other solar technologies as the installed capacity of CSP by the world in 2018 was around 5.5 GW [91]. The cost of electricity generated via CSP is higher than that of conventional sources, on the other hand, PV is in the same range [92]. The PV–CSP combination has several advantages as a whole, especially in terms of the cost of electricity [93] and also the large capacities provided through these two technologies. This combination has been tested in the Atacama desert in Chile, the results showed that the capacity generated by this system can exceed 85% and have a reduced LCOE compared to CSP-only power plant [94–96].

In another study about off-grid PV–CSP–battery system [97] an optimal configuration was proposed for two sites in Italy and Morocco based on the analysis of several strategies. Also, a parametric study has been done in [98] on a large scale of a hybrid CSP–PV system with two types of BESS storage implemented in working operation. Also, Zhai et al. [99] optimized a PV–CSP system in order to achieve the lowest possible cost of electricity by integrating a small capacity battery and improving its use. Zurita et al. [88] conducted a study to assess the effect of time resolution on the modeling of the PV–CSP system with thermal storage and batteries for two sites in Carrera Pinto and Santiago in Chile. Hernández-Moro and Martinez-Duart [100] developed a mathematical method for estimating the cost of energy produced by the PV–CSP system on the basis of other inputs. Aguilar-Jiménez et al. [101] presented a technical-economic study of a PV–CSP system for a remote area of

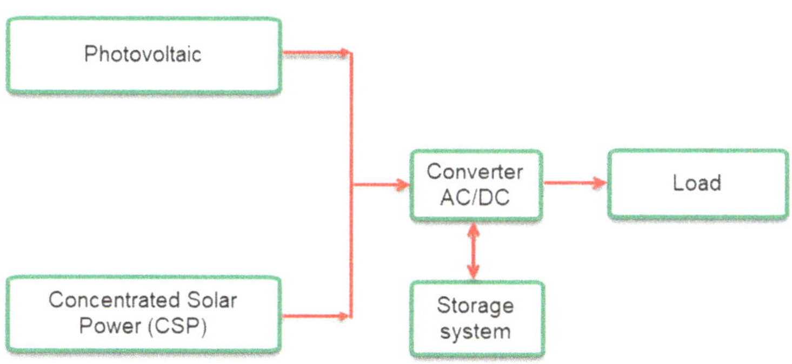

Fig. 7 Off-grid PV–CSP hybrid system

Puertecitos in Mexico. The cost of electricity generated by this system is estimated as 0.524 USD/kWh. Technical and economic analysis of a PV–CSP system by [102] was carried out in three different sites Tabuk, Majmaah, and Najran in Saudi Arabia. The system was designed at the same power output of 100 MW. The results showed that the cost of electricity by the hybrid system is estimated to be 0.105 $/kWh, 0.110 $/kWh, and 0.1 $/kWh for Najran, Majmaah, and Tabuk, respectively. Table 2 summarizes some of the recent research works.

4.2 Hybrid System with Three and More Renewable Energy Sources and a Storage System

4.2.1 PV/WT/Hydro

Another scenario is an off-grid system, constituted of PV–Wind–Hydro energy with a storage system. Solar technology and wind power are naturally intermittent due to depending on the weather conditions. However, as hydroelectricity is controllable, this increases the level of reliability and stability of this configuration. This type of HRESs has been a special focus in the existing research literature as it offers certain advantages including higher electric power supply, high profitability, and efficiency comparing the mono source systems [106].

Bekele and Tadesse [107] treated a feasibility study of a system composed of PV–Wind–Hydro in the district of Dejen, Ethiopia as the study concerned 23 different villages. In the Taba region the price of electricity was found equal to 0.16 $/kWh. Zhang et al. [108] proposed an estimation-simulation approach in order to mark the uncertainties of the systems tester on three optimal short-term operating model as well as the estimation of the probability density function of the operating mechanism. Wei et al. [109] presented a study of a PV–Wind–Hydro system in southwest China with the integration of deep neural networks, the findings showed that the performance of the hybrid plant must be improved. Ye et al. [110] proposed a feasibility study and a simulation of the PV–Hydro–Wind system in the EMTP/ATP platform.

4.2.2 PV/WT/Biomass

This section presents an off-grid system composed of Biomass–Wind–PV sources. Biomass is an energy source that has become very popular especially in remote areas [111]. The combination of these three technologies with storage system is robust in terms of high power output.

In literature, Balamurugan et al. [112] presented a hybrid system composed of Biomass–PV–Wind for remote areas in India, a feasibility and economy analysis was provided by the authors. Dhass and Santhanam [113] quantified a Biomass–Wind–PV system for electrified remote areas on the basis of the life cycle price. Rahman

Table 2 Summary of research work

PV	WT	Bio	Hyd	CSP	Place/Contry	Outcome	References
✓		✓			Bardsir, Iran	The minimum Total Net Present Cost obtained is $ 1672296.6881 with a PV and Biomass portion of 15 and 85% respectively. The COE and perste of electrical supply are 0.1885 $/kWh and 1.9997%, respectively	[45]
✓	✓				House, Tehran, Iran	In this study, a comparison was made between the results found by Homer and hybrid genetic algorithm with particle swarm optimization (GA-PSO) with the aim of minimizing the current total cost. The results found of 0.502 of the discounted cost of energy for the system being studied is considered to be the best result of the two methods	[57]
✓	✓				Zelzate, East Flanders, Belgium	Studied the effect on the employment factor of the electrolyzer when using a renewable energy source instead of the electric grid. Four scenarios were studied with an unchangeable 1 MW electrolyzer: (A) 15 MW PV, (B) 15 MW PV, 2 MW Wind, (C) 15 MW PV, 2 MW Wind, Battery, (D) 15 MW PV, Wind of 15 MW plus the factor is found equal to (A) 41.5%, (B) 65.5%, (C) 66.0e86.0%, (D) 82.0% respectively	[59]
✓	✓				Remote area Borj Cedria, Tunisia	The primary question in this research is the optimal dimensioning of the components taking into account the irradiations, charge cycles, and wind speed based on NSGA-II multi-objective algorithm, the cost of the system, and the LPSP to be optimized	[60]

(continued)

Table 2 (continued)

PV	WT	Bio	Hyd	CSP	Place/Contry	Outcome	References
✓	✓				Island, southeast coast of China	Analysis of the variability of the components of the hybrid system in this study is done via the genetic algorithm (NSGA-II) *with two objective functions* the loss of electrical power and the total cost of the hybrid system. The results revealed in the case of power demand of 24 V DC, 3500 W, 220 V AC, 2200 W, 2 sets of 6000 W wind turbines, five banks of 24 V 1000 Ah batteries, and 78 PV modules with a peak power of 100 W have been selected	[61]
✓	✓				Jeju Island, South Korea	The COE and life cycle costs were found at 0.42 $/kWh 84.3 BUSD for the optimal capacities of onshore and offshore wind turbines, photovoltaic solar panels, and Li-ion batteries are rated at 16 MW, 1,532 MW, 6,076 MW, and 14,651 MWh, respectively	[62]
✓	✓				Remote district, Beni-Suef, Egypt	The optimization of the total cost of the three combinations of HRESs and the performances were optimized via the Firefly algorithm, the results obtained were compared to those found from the Shuffled Flog Leaping algorithm and the optimization of the particle swarms (PSO). The PV/WT/fuel cell combination incorporated into an electrolyzer. The COE found is 0.47 $/kWh for this hybrid system	[63]

(continued)

Table 2 (continued)

PV	WT	Bio	Hyd	CSP	Place/Contry	Outcome	References
✓	✓				Makadi Bay, Egypte	The optimal dimensioning of the HRESs in this study is made by the TORCHE technique. The results of the analysis showed that the PV/WT/Battery system is economically cheaper than installing each technology individually	[64]
✓		✓			Agricultural farm and the small community, Pakistan	A technical-economic study is carried out by Homer of the PV–Biomass–Battery system. The results found of the optimal solution composed of 10 kW of PV, 8 kW of the bioagas generator and 32 batteries are the initial capital investment cost of 2.64 million PKR, NPC of 4.48 million PKR, the COE 5.51 PKR/kWh the investment return time is estimated at 9.5 years	[72]
✓		✓			Bhopal, Indian state of Madhya Pradesh	The Net Present Cost (NPC) found to be Rs.51, 89003 and the COE raised equal to 15.064 Rs/kWh. The loss of power found is 0%	[75]
✓		✓			Monshaet Taher, Egypt	The optimal configuration is that which contains the Nickel Iron (Ni-Fe) battery with PV-Biomass, with 298 Ni-Fe batteries, 24 PV Panels, and four biomass power systems. The COE and the NPC are 0.084 $/kWh and 2.408.895 $ respectively	[76]
✓		✓			Sharjah, United Arab Emirates	The portion of renewable energy in the HRESs is 74% and 26 of PV and biogas generator, respectively. The COE of the optimal system is estimated at 0.328 $/kWh	[77]

(continued)

Table 2 (continued)

PV	WT	Bio	Hyd	CSP	Place/Contry	Outcome	References
✓		✓			Remote area	In this article the simulation of the HRESs PV-Biomass is carried out by Homer. The portion of electrical production is 62% and 38% of PV and Biomass, respectively, the cost of HRESs is 900,000 Rs /—and the period of return on investment is found equal to six years	[78]
✓			✓		California Independent System Operator (CAISO)	If the production that comes from PV increases by 1% the small hydraulic power plant increases production from 0.01–0.06%	[86]
✓				⊖	Puertecitos Baja California, Mexico	The leveled COE is 0.524 $/kWh for the CSP-PV hybrid system	[101]
✓				⊖	Tabuk, Majmaah and Najran in Saudi Arabia	The LCOE of the three Majmaah photovoltaic power plants, Tabuk, and Najran are 0.038 $/kWh, 0.036 $/kWh, and 0.037 $/kWh, respectively. Conversely, the LCOE of the three CSP factories is 0.110 $/kWh, 0.100 $/kWh, and 0.105 $/kWh in Majmaah, Tabuk, and Najran, respectively	[102]
✓	✓				Remote island, Jiuduansha, Shanghai	In this study a mathematical model is proposed to study the effect of the saturation of a resource by decreasing the other resources, on the scale of the battery park, state of charge, excess energy, NPC, COE, and loss of electrical power. The results showed that a 2 kW wind turbine was chosen with 90% wind saturation penetration on the island studied. The result also reveals that wind power, charge, and battery cost can impact the COE compared to other factors	[103]

(continued)

Table 2 (continued)

PV	WT	Bio	Hyd	CSP	Place/Contry	Outcome	References
✓	✓				Çeşme, Izmir province, Turkey	In this study 24 simulations were made via Homer in order to find the lowest COE. In the results found, the COE of the hybrid system outside the networks and above the electric grid, also battery storage and economically places than hydrogen storage	[104]
✓	✓				Lafarge factory, Al-Tafilah, Jordan	The results found are an optimal system of 26 MW wind power, 20.75 MW PV, and 16.8 MWH Lithium-Ion batteries. The COE is 0.203 $/kWh, the return on investment is estimated at 3.44 years and the net present value is 206.63 M $. The proposed system is capable of reducing the energy bill by 21.58 M $ as well as 71,373 tons of CO_2 emissions avoided	[105]

et al. [114] proposed a design using Homer software of a Biomass–Wind–PV system for a site in Bangladesh. Singh et al. [111] proposed a hybrid Biomass–Wind–PV sizing system with a system to satisfy the electrical charge of a narrow area. The authors used an artificial bee colony (ABC) algorithm to size the components of the central hybrid. The results obtained were compared with those found by Homer and with the particle swarm optimization algorithm.

5 Performance Indexes

Intensive capacity can increase the initial cost of an HRESs. Nevertheless, it is very complex to supply reliability. Therefore, it is very essential to optimize and adjust the capacity of the system to ensure the load [30, 115].

First is to have the design of the HRESs, choosing the right type of renewable energy sources to integrate, is crucial. The second step is to model the subsystem taking into account all the subsystems. Objective functions must be traced. Then, a design of dimensioning is chosen correlating the constraints and specificity of the problems. Finally, the right capacity scheme taking into account the objectives is set. The objective functions are managed by multiple indicators, which have a very great impact on the system. In general, the reliability of the system, the economic and environmental indices are considered. These indicators will be dealt with in the subsections.

5.1 Reliability Index

The electricity production of HRESs systems is very much influenced by weather conditions and the climate, which does not leave the electrical supply considered to be reliable and safe. Reliability indices were taken into account in order to examine the capacity of the integrated system to meet the desired electrical charge.

The reliability indices essentially include the Possibility of Failling Power Supply (PFSP), Possibility of Failing Load (PFL), Hoped Energy not Provided (HEP), Possibility Lack of Power Supply (PLPS), Loss of Load Desired (LLD).

5.1.1 Possibility of Falling Power Supply (PFSP)

PFSP is determined to be an electricity supply margin which is not taken into account to meet the electrical demand [116]. This indicator is considered one of the most important factors for measuring the reliability of the HRESs. There are two ways to calculate this index either based on the chronological simulation, which is difficult on the computer scale because it requires temporary data or by a commitment to apply technical probabilities based on the instability of the energy and charge [117,

118]. The following equation [118, 119] has been used to optimize the capacity of a system not connected to the network in Iran [120].

$$PFSP = \frac{\sum_{t=1}^{T} DE(t)}{\sum_{t=1}^{T} P_{load}(t)\Delta t} \tag{6}$$

$$PFSP = \frac{\sum_{t}^{T} Power\ Failure\ Time(P_{supplied}(t) < P_{load}(t))}{T} \tag{7}$$

where $DE(t)$ means the energy-efficient at date t (kWh), $P_{supplied}(t)$ is the energy produced by the hybrid power plant at date t (kW), $P_{load}(t)$ is the load requested at date t (kW), Δt is the time margin (h), T is the total number of periods in specific calculation time.

5.1.2 Possibility of Falling Load (PFL)

PFL is the division of global energy insufficiency and the need for global electrical charge during a well-defined period [116], it can be measured by the following equation [117, 121]:

$$PFL = \frac{\sum_{t=1}^{8760} ES(t)}{\sum_{t=1}^{8760} LD(t)} \tag{8}$$

where $ES(t)$ means the electricity failure at time t (kWh), $LD(t)$ represents the electrical requirement at t hour (kWh).

5.1.3 Hoped Energy not Provided (HEP)

HEP is the desired energy which is not delivered to the electrical demand under state when demand exceeds available power generation capacity [55, 122, 123], as mentioned in [117, 122]:

$$HEP = \sum_{t=1}^{8760} E_{not-delivered}(t) \tag{9}$$

$$EIR = 1 - \frac{HEP}{E_0} \tag{10}$$

where $E_{not-delivered}(t)$ is the energy that will not be used at t hour over the full year (kWh); E_0 means the overall energy load of the system (kWh); EIR represents the energy index agreement.

5.1.4 Possibility Lack of Power Supply (PLPS)

The PLPS is determined as the division between the totalities of values LPS (lack of power supply) on the summation of load requirements for a specific period, and this describes the contingency of a missing power supply situation when the HRESs cannot satisfy the electrical demand [122] as mentioned in [117, 121].

$$PLPS = \frac{\sum_{t=1}^{T} LPS(t)}{\sum_{t=1}^{T} LD(t)} \tag{11}$$

$$LPS = \sum_{t=1}^{T} LD(t) - E_{sist}(t) \tag{12}$$

with *LPS* (lack of electrical supply) represents a frequent element when the central hybrid cannot satisfy the electrical load, $E_{sist}(t)$ means supplied with energy by the central hybrid at t hour (kWh/year).

5.1.5 Loss of Load Desired (LLD)

LLD indicates the sum of hours in a full year where the load exceeds the electricity production available in the year (h/year) [117, 121]:

$$LLD = \sum_{h=1}^{H} \sum_{i \varepsilon S} T_i \times P_i \tag{13}$$

where S means the total loss of the states of charge of the system, H is the number of hours in the year 8760 h, T_i represents the time of a charge exceeds the production (hour), P_i means the probability of the system meeting state i.

5.2 *Economic Indexes*

The economic order is embodied in the concept of HRESs, taking into account the installation of the system, its maintenance, and operating phases of the HRESs.

The improvement of the HRESs requires finding the relationship between the cost and the optimal yield of the system. Therefore, economic indices considering the initial cost, maintenance costs, operating cost, and other system costs have become paramount indicators to study the feasibility of HRESs. The most widely used economic indices are the annual global cost (AGC), the annualized cost of the HRESs (ACHRES), (COE), life cycle cost (LCC), LCOE, NPV, etc., are discussed below.

5.2.1 Annual Global Cost (AGC)

AGC is the addition of capital cost, maintenance, and annual cost of the HRESs [124, 125] reported in [124] as below:

$$TGC = C_{acap} + C_{amain} \tag{14}$$

with C_{acap} means the annual fundamental cost of HRESs; C_{amain} is the annual HRESs maintenance cost.

5.2.2 Annualized Cost of the HRESs

Annualized cost of the HRESs (ACHRESs) is the summation of the annual replacement cost, the annual capital cost, and the annual maintenance cost of the HRESs.

$$ACHRES = C_{acap} + C_{amain} + C_{arep} \tag{15}$$

with C_{arep} means the annual replacement cost of the HRESs.

5.2.3 Cost of Energy

The cost of energy (COE) is the division of the annualized cost of the hybrid renewable energy system (ACHRESs) on the annual overall energy production (AOEP). It sets out the cost per unit of energy produced by HRESs [121]:

$$COE = \frac{\sum_{i=1}^{n} ACHRES}{\sum_{i=1}^{n} AOEP_i} \tag{16}$$

5.2.4 Net Present Value (NPV)

The net present value can be estimated by adding the values of all the costs of the discounted revenues including capital cost, maintenance, and operating expenses by subtracting the current revenues during the lifetime of the HRESs that it can earn [116, 126, 127].

$$NPV = \sum NPV_{end} - C_{investment} - \sum NPV_{OM} - \sum NPV_r \tag{17}$$

with NPV_{end} is the currently discounted reward of income from the residual value of the subsystems at the end of the life of the HRESs; $C_{investment}$ is the initial investment cost; NPV_{OM} is the present value of future exploitation and maintenance costs during the life of the HRESs; NPV_r represents the present value of future replacement costs to replace components during the lifetime of the system.

5.2.5 Levelized Cost of Energy (LCOE)

LCOE corresponds, for a given HRESs installation, to the sum of the updated costs of energy production divided by the amount of energy produced by this system, which is also updated. It is typically expressed in \$/kWh (or other currency).

$$LCOE = \frac{\sum_{t=1}^{n} \frac{C_t}{(1+r)^t}}{\sum_{t=1}^{n} \frac{E_t}{(1+r)^t}} \quad (18)$$

where n is the life of the HRESs, C_t represents all the costs, E_t is the annual clean energy production by the HRESs and r the annual discount rate.

5.2.6 Life Cycle Cost (LCC)

The LCC is determined as the addition of the NPVs for the total cost of the HRESs expenses such as investment expenses, operating and maintenance expenses, replacement expenses, etc., minus the net present value of the revenue value S_{NPV} [116, 117].

$$LCC = C + OM_{NPV} + R_{NPV} - S_{NPV} \quad (20)$$

where C is the total cost of HRESs; OM_{NPV} represents the net present value of maintenance and operation; R_{NPV} means the replacement net present value.

5.3 Environmental Index

The environmental index is an indicator for assessing the state of the environment. However, conventional energies generate pollution through the rejection of greenhouse gas emissions, mainly CO_2. As a result, the development of Renewable Energies appease the global energy disturbance, reduce environmental pollution, and enhance sustainable development. Therefore, environmental indices should occupy a significant place in the concept of optimization of HRESs [30].

In this context, the approach to assess the mitigation of the amount of CO_2 avoided admits that each kWh of electricity generated by HRESs replaces each kWh of electricity generated by conventional systems. Thus, the CO_2 emission rates deviated by the capacity of the installed system can be estimated.

6 Conclusion

This chapter aims to shed light on standalone PV-based hybrid renewable energy systems for power generation in rural areas, villages, and remote islands by reviewing various HRESs architectures, formulating basic mathematical background for modeling multiple energy source systems and proposing key performance indicators for the techno-economic assessment of such systems. The use of renewable resources to supply electricity where the grid connection is not technically or economically viable remains a challenge, especially for remote areas. Various configurations of PV hybridization with other renewable resources such as wind, biomass, hydroelectricity, and CSP in off-grid areas have been discussed and the physical modeling of each system is presented. Finally, the indexed performances were expressed including the reliability indexes, the economic indexes, and the environmental indexes. Based on the analysis of various HRES configurations, the main concluding remarks can be outlined as follows;

To successfully satisfy the energy demand, the system configuration is key. It should be carefully tailored by considering available energy sources, the capacity of the system as well as the specifications of the storage. In this way, the cost of energy will be reduced, CO_2 emissions will be minimized and energy requirements will be met.

In PV/Wind configuration, electricity production is obtained mainly from PV panels during the day and from windmill during the night. Hence, this combination is massively adopted given the complementary energy production, its reliability, and adaptability in any weather condition. Moreover, the energy storage system will store excess energy production from hybrid PV–WT combination and meet the energy demand when electricity supply through the system is insufficient.

PV/Biomass configuration is a promising alternative for electricity generation, especially in remote areas where there is an abundant presence of an animal, agricultural, municipal waste, etc. The advantage is that biomass is controllable (present when needed). On the other hand, when the fuel is not available, it must be somewhat purchased which is a standing drawback.

In PV/Hydro configuration, as PV panels are highly affected by ambient conditions, hydroelectricity is, on the other hand, a more flexible, stable, and adjustable sort of energy source. This is another complementary system as there is less solar irradiation but a high level of water during the winter and there is a risk of dry dams but abundant solar irradiation during the summer.

In PV/CSP configuration, there are two different technologies fed by the same energy source. This combination has several advantages as a whole, especially in terms of the cost of electricity and also the large capacities can be provided through these two technologies.

In HRES with three renewable energy sources, and a storage system, PV–Wind–Hydro stands out as a reliable configuration. Although solar and wind power are naturally intermittent due to dependence on the weather condition, hydroelectric, however, is a controllable and stable source. Therefore, this configuration offers certain advantages including higher electric power supply, high profitability, and efficiency comparing the mono source systems.

PV–Wind–Biomass is also popular in remote areas. Even for large capacities, biomass is known for its profitability. The combination of these three technologies with storage system is robust in terms of high power output.

In order to increase the reliability of these systems and create a better balance between supply and demand, such systems require controlling approaches, and storage systems should be adapted to geographical requirements for a continuous supply. Therefore, the integration of storage systems may be of great interest in the design of hybrid renewable energy systems.

References

1. Khare V, Nema S, Baredar P (2013) Status of solar wind renewable energy in India. Renew Sustain Energy Rev 27:1–10
2. Li K, Liu C, Jiang S, Chen Y (2019) Review on hybrid geothermal and solar power systems. J Clean Prod, 119481
3. Vivas FJ, De las Heras A, Segura F, Andújar JM (2018) A review of energy management strategies for renewable hybrid energy systems with hydrogen backup. Renew Sustain Energy Rev 82:126–155
4. Chandel SS, Shrivastva R, Sharma V, Ramasamy P (2016) Overview of the initiatives in renewable energy sector under the national action plan on climate change in India. Renew Sustain Energy Rev 54:866–873
5. Goel S, Sharma R (2017) Performance evaluation of stand alone, grid connected and hybrid renewable energy systems for rural application: a comparative review. Renew Sustain Energy Rev 78:1378–1389
6. Sun X, Lian Z, Wang B, Li X (2009) A Hybrid renewable DC microgrid voltage control. In: 2009 IEEE 6th international power electronics and motion control conference, 2009, pp 725–729
7. Muh E, Tabet F (2019) Comparative analysis of hybrid renewable energy systems for off-grid applications in Southern Cameroons. Renew Energy 135:41–54
8. Haruni AMO, Negnevitsky M, Haque ME, Gargoom A (2012) A novel operation and control strategy for a standalone hybrid renewable power system. IEEE Trans Sustain Energy 4(2):402–413
9. Li X, Jiao X, Wang L (2013) Coordinated power control of wind-PV-fuel cell for hybrid distributed generation systems. In: The SICE annual conference 2013, pp 150–155
10. Robitaille M, Agbossou K, Doumbia ML (2005) Modeling of an islanding protection method for a hybrid renewable distributed generator. In: Canadian conference on electrical and computer engineering, 2005, pp 1477–1481
11. Semaoui S, Arab AH, Bacha S, Azoui B (2013) The new strategy of energy management for a photovoltaic system without extra intended for remote-housing. Sol Energy 94:71–85
12. El-houari H et al (2019) Design, simulation, and economic optimization of an off-grid photovoltaic system for rural electrification. Energies 12(24). https://doi.org/10.3390/en1224 4735
13. Eltamaly Ali M, Mohamed, A.M (2014) A novel design and optimization software for autonomous PV/wind/battery hybrid power systems. Math Probl Eng
14. Liberato MLR, Santos JA, Fernandes C, Pinto JG (2015) Future projections of offshore wind energy potentials in Portugal simulated by the COSMO-CLM regional model, www.icee2015.com, vol 18, p 207

15. Ghafoor A, Munir A (2015) Design and economics analysis of an off-grid PV system for household electrification. Renew Sustain Energy Rev 42:496–502
16. Palzer A, Henning H-M (2014) A comprehensive model for the German electricity and heat sector in a future energy system with a dominant contribution from renewable energy technologies–Part II: Results. Renew Sustain Energy Rev 30:1019–1034
17. Radomes AA Jr, Arango S (2015) Renewable energy technology diffusion: an analysis of photovoltaic-system support schemes in Medellín, Colombia. J Clean Prod 92:152–161
18. de Cerio Mendaza ID, Bhattarai BP, Kouzelis K, Pillai JR, Bak-Jensen B, Jensen A (2015) Optimal sizing and placement of power-to-gas systems in future active distribution networks. In: 2015 IEEE innovative smart grid technologies-Asia (ISGT ASIA), 2015, pp 1–6
19. Erdinc O, Uzunoglu M (2012) Optimum design of hybrid renewable energy systems: overview of different approaches. Renew Sustain Energy Rev 16(3):1412–1425
20. Bahramara S, Moghaddam MP, Haghifam MR (2016) Optimal planning of hybrid renewable energy systems using HOMER: a review. Renew Sustain Energy Rev 62:609–620
21. Moner-Girona M (2009) A new tailored scheme for the support of renewable energies in developing countries. Energy Policy 37(5):2037–2041
22. Moner-Girona M et al (2016) Adaptation of feed-in tariff for remote mini-grids: Tanzania as an illustrative case. Renew Sustain Energy Rev 53:306–318
23. Bhattacharyya SC (2013) Financing energy access and off-grid electrification: a review of status, options and challenges. Renew Sustain Energy Rev 20:462–472
24. Hasan KN, Saha TK, Eghbal M, Chattopadhyay D (2013) Review of transmission schemes and case studies for renewable power integration into the remote grid. Renew Sustain Energy Rev 18:568–582
25. Gómez MF, Silveira S (2012) Delivering off-grid electricity systems in the Brazilian Amazon. Energy Sustain Dev 16(2):155–167
26. Thiam DR (2011) An energy pricing scheme for the diffusion of decentralized renewable technology investment in developing countries. Energy Policy 39(7):4284–4297
27. Izadyar N, Ong HC, Chong WT, Leong KY (2016) Resource assessment of the renewable energy potential for a remote area: a review. Renew Sustain Energy Rev 62:908–923
28. Fathima AH, Palanisamy K (2015) Optimization in microgrids with hybrid energy systems–a review. Renew Sustain Energy Rev 45:431–446
29. Mahesh A, Sandhu KS (2015) Hybrid wind/photovoltaic energy system developments: critical review and findings. Renew Sustain Energy Rev 52:1135–1147
30. Lian J, Zhang Y, Ma C, Yang Y, Chaima E (2019) A review on recent sizing methodologies of hybrid renewable energy systems. Energy Convers Manag 199:112027
31. Twidell J, Weir T (2015) Renewable energy resources. Routledge
32. Khatib T, Mohamed A, Sopian K, Mahmoud M (2015) Optimal sizing of hybrid pv/wind systems for Malaysia using loss of load probability. Energy Sources Part A Recover Util Environ Eff 37(7):687–695
33. IEA 2017 (2020). https://webstore.iea.org/co2-emissions-from-fuel-combustion. Accessed 15 May 2020
34. Tracking Power–Analysis-IEA. https://www.iea.org/reports/tracking-power-2019. Accessed 28 Apr 2020
35. Allouhi A, Kousksou T, Jamil A, El Rhafiki T, Mourad Y, Zeraouli Y (2015) Economic and environmental assessment of solar air-conditioning systems in Morocco. Renew Sustain Energy Rev 50:770–781
36. Eltamaly Ali M, Addoweesh Khaled E, Bawa Umar, Mohamed Mohamed A (2014) Economic modeling of hybrid renewable energy system: a case study in Saudi Arabia. Arabian J Sci Eng 39(5):3827–3839
37. Ismail MS, Moghavvemi M, Mahlia TMI (2013) Techno-economic analysis of an optimized photovoltaic and diesel generator hybrid power system for remote houses in a tropical climate. Energy Convers Manag 69:163–173
38. Soshinskaya M, Crijns-Graus WHJ, van der Meer J, Guerrero JM (2014) Application of a microgrid with renewables for a water treatment plant. Appl Energy 134:20–34

39. Thomas D, Deblecker O, Ioakimidis CS (2016) Optimal design and techno-economic analysis of an autonomous small isolated microgrid aiming at high RES penetration. Energy 116:364–379
40. Fathima H, Palanisamy K (2015) Optimized sizing, selection, and economic analysis of battery energy storage for grid-connected wind-PV hybrid system. Model Simul Eng 2015
41. Das BK, Hoque N, Mandal S, Pal TK, Raihan MA (2017) A techno-economic feasibility of a stand-alone hybrid power generation for remote area application in Bangladesh. Energy 134:775–788
42. Lilienthal P, Lambert T, Gilman P (2004) Computer modeling of renewable power systems
43. Cuéllar AD, Webber ME (2008) Cow power: the energy and emissions benefits of converting manure to biogas. Environ Res Lett 3(3):34002
44. Edwin M, Sekhar SJ (2014) Techno-economic studies on hybrid energy based cooling system for milk preservation in isolated regions. Energy Convers Manag 86:1023–1030
45. Heydari A, Askarzadeh A (2016) Optimization of a biomass-based photovoltaic power plant for an off-grid application subject to loss of power supply probability concept. Appl Energy 165:601–611
46. Kibaara S, Chowdhury S, Chowdhury SP (2012) A thermal analysis of parabolic trough CSP and biomass hybrid power system. PES T&D 2012:1–6
47. Connolly D, Lund H, Mathiesen BV, Leahy M (2010) A review of computer tools for analysing the integration of renewable energy into various energy systems. Appl Energy 87(4):1059–1082
48. Muhsen DH, Khatib T, Haider HT (2017) A feasibility and load sensitivity analysis of photovoltaic water pumping system with battery and diesel generator. Energy Convers Manag 148:287–304
49. Roy A, Kedare SB, Bandyopadhyay S (2010) Optimum sizing of wind-battery systems incorporating resource uncertainty. Appl Energy 87(8):2712–2727
50. Najafi-Shad S, Barakati SM, Yazdani A (2020) An effective hybrid wind-photovoltaic system including battery energy storage with reducing control loops and omitting PV converter. J. Energy Storage 27:101088
51. Kumar P, Deokar S (2018) Optimal design configuration using HOMER. In: Advances in systems, control and automation. Springer, pp 101–108
52. Eltamaly Ali M, Mohamed Mohamed A (2016) A novel software for design and optimization of hybrid power systems. J Braz Soc Mech Sci Eng 38(4):1299–1315
53. Wai R-J, Cheng S, Lin Y-F, Chen Y-C (2013) Installed capacity selection of hybrid energy generation system via improved particle-swarm-optimisation. IET Gener Transm Distrib 8(4):742–752
54. Askarzadeh A, dos Santos Coelho L (2015) A novel framework for optimization of a grid independent hybrid renewable energy system: a case study of Iran. Sol Energy 112:383–396
55. Sanajaoba S, Fernandez E (2016) Maiden application of Cuckoo Search algorithm for optimal sizing of a remote hybrid renewable energy System. Renew Energy 96:1–10
56. Tito SR, Lie TT, Anderson TN (2016) Optimal sizing of a wind-photovoltaic-battery hybrid renewable energy system considering socio-demographic factors. Sol Energy 136:525–532
57. Ghorbani N, Kasaeian A, Toopshekan A, Bahrami L, Maghami A (2018) Optimizing a hybrid wind-PV-battery system using GA-PSO and MOPSO for reducing cost and increasing reliability. Energy 154:581–591
58. Stroe D-I, Zaharof A, Iov F (2018) Power and energy management with battery storage for a hybrid residential PV-wind system–a case study for Denmark. Energy Procedia 155:464–477
59. Papadopoulos V, Desmet J, Knockaert J, Develder C (2018) Improving the utilization factor of a PEM electrolyzer powered by a 15 MW PV park by combining wind power and battery storage–Feasibility study. Int J Hydrogen Energy 43(34):16468–16478
60. Belouda M, Hajjaji M, Sliti H, Mami A (2018) Bi-objective optimization of a standalone hybrid PV–Wind–battery system generation in a remote area in Tunisia. Sustain Energy, Grids Netw 16:315–326

61. Zhang D et al (2019) Research on the configuration and operation effect of the hybrid solar-wind-battery power generation system based on NSGA-II. Energy 189:116121
62. Jung W, Jeong J, Kim J, Chang D (2020) Optimization of hybrid off-grid system consisting of renewables and Li-ion batteries. J Power Sources 451:227754
63. Samy MM, Barakat S, Ramadan HS (2019) Techno-economic analysis for rustic electrification in Egypt using multi-source renewable energy based on PV/wind/FC. Int J Hydrog Energy
64. Hemeida AM, et al. (2019) Optimum design of hybrid wind/PV energy system for remote area. Ain Shams Eng J
65. Benavente F, Lundblad A, Campana PE, Zhang Y, Cabrera S, Lindbergh G (2019) Photovoltaic/battery system sizing for rural electrification in Bolivia: Considering the suppressed demand effect. Appl Energy 235:519–528
66. Huda ASN, Mekhilef S, Ahsan A (2014) Biomass energy in Bangladesh: current status and prospects. Renew Sustain Energy Rev 30:504–517
67. Heydari A, Askarzadeh A (2016) Techno-economic analysis of a PV/biomass/fuel cell energy system considering different fuel cell system initial capital costs. Sol Energy 133:409–420
68. Balaman ŞY, Selim H (2014) A network design model for biomass to energy supply chains with anaerobic digestion systems. Appl Energy 130:289–304
69. Liao X, Li H (2015) Biogas production from low-organic-content sludge using a high-solids anaerobic digester with improved agitation. Appl Energy 148:252–259
70. Monlau F, Sambusiti C, Antoniou N, Barakat A, Zabaniotou A (2015) A new concept for enhancing energy recovery from agricultural residues by coupling anaerobic digestion and pyrolysis process. Appl Energy 148:32–38
71. Scholz M, Melin T, Wessling M (2013) Transforming biogas into biomethane using membrane technology. Renew Sustain Energy Rev 17:199–212
72. Shahzad MK, Zahid A, ur Rashid T, Rehan MA, Ali M, Ahmad M (2017) Techno-economic feasibility analysis of a solar-biomass off grid system for the electrification of remote rural areas in Pakistan using HOMER software. Renew Energy 106:264–273
73. Ganthia BP, Sasmita S, Rout K, Pradhan A, Nayak J (2018) An economic rural electrification study using combined hybrid solar and biomass-biogas system. Mater Today Proc 5(1):220–225
74. Kohsri S, Meechai A, Prapainainar C, Narataruksa P, Hunpinyo P, Sin G (2018) Design and preliminary operation of a hybrid syngas/solar PV/battery power system for off-grid applications: a case study in Thailand. Chem Eng Res Des 131:346–361
75. Singh A, Baredar P (2016) Techno-economic assessment of a solar PV, fuel cell, and biomass gasifier hybrid energy system. Energy Reports 2:254–260
76. Eteiba MB, Barakat S, Samy MM, Wahba WI (2018) Optimization of an off-grid PV/Biomass hybrid system with different battery technologies. Sustain Cities Soc 40:713–727
77. Ghenai C, Janajreh I (2016) Design of solar-biomass hybrid microgrid system in Sharjah. Energy Procedia 103:357–362
78. Pradhan SR, Bhuyan PP, Sahoo SK, Prasad GS (2013) Design of standalone hybrid biomass & PV system of an off-grid house in a remote area. Smruti Ranjan Pradhan al Int J Eng Res Appl 3(6):433–437
79. Chowdhury N, Akram Hossain C, Longo M, Yaïci W (2020) Feasibility and cost analysis of photovoltaic-biomass hybrid energy system in off-grid areas of Bangladesh. Sustainability 12(4):1568
80. Liu L, Wang Q, Lin H, Li H, Sun Q (2017) Power generation efficiency and prospects of floating photovoltaic systems. Energy Procedia 105:1136–1142
81. Shabani M, Mahmoudimehr J (2018) Techno-economic role of PV tracking technology in a hybrid PV-hydroelectric standalone power system. Appl Energy 212:84–108
82. Yibo W, Honghua X (2013) Research and practice of designing hydro/photovoltaic hybrid power system in microgrid. In: 2013 IEEE 39th photovoltaic specialists conference (PVSC), pp 1509–1514
83. Das S, Akella AK (2018) A control strategy for power management of an isolated micro hydro-PV-battery hybrid energy system. In: 2018 4th international conference on electrical energy systems (ICEES), pp 397–401

84. Singh B (2018) PV-hydro-battery based standalone microgrid for rural electrification. In: 2018 5th IEEE Uttar Pradesh section international conference on electrical, electronics and computer engineering (UPCON), pp 1–6
85. Kougias I, Szabo S, Monforti-Ferrario F, Huld T, Bodis K (2016) A methodology for optimization of the complementarity between small-hydropower plants and solar PV systems. Renew Energy 87:1023–1030
86. Shan R, Sasthav C, Wang X, Lima LMM (2020) Complementary relationship between small-hydropower and increasing penetration of solar photovoltaics: evidence from CAISO. Renew Energy
87. Silvério NM, Barros RM, Tiago Filho GL, Redón-Santafé M, dos Santos IFS, de Mello Valério VE (2018) Use of floating PV plants for coordinated operation with hydropower plants: case study of the hydroelectric plants of the São Francisco River Basin. Energy Convers Manag 171:339–349
88. Zurita A, Mata-Torres C, Cardemil JM, Escobar RA (2020) Assessment of time resolution impact on the modeling of a hybrid CSP-PV plant: a case of study in Chile. Sol Energy 202:553–570
89. Costa SCS, Diniz ASAC, Kazmerski LL (2016) Dust and soiling issues and impacts relating to solar energy systems: Literature review update for 2012–2015. Renew Sustain Energy Rev 63:33–61
90. Figgis B, Ennaoui A, Ahzi S, Rémond Y (2017) Review of PV soiling particle mechanics in desert environments. Renew Sustain Energy Rev 76:872–881
91. REN21-Building the sustainable energy future with renewable energy. https://www.ren21.net/. Accessed 30 Apr 2020
92. Renewable Power Generation Costs in 2018. https://www.irena.org/publications/2019/May/Renewable-power-generation-costs-in-2018. Accessed 30 Apr 2020
93. Pan CA, Dinter F (2017) Combination of PV and central receiver CSP plants for base load power generation in South Africa. Sol Energy 146:379–388
94. Bravo R, Friedrich D (2018) Two-stage optimisation of hybrid solar power plants. Sol Energy 164:187–199
95. Green A, Diep C, Dunn R, Dent J (2015) High capacity factor CSP-PV hybrid systems. Energy Procedia 69:2049–2059
96. Starke AR, Cardemil JM, Escobar R, Colle S (2018) Multi-objective optimization of hybrid CSP+PV system using genetic algorithm. Energy 147:490–503
97. Cocco D, Migliari L, Petrollese M (2016) A hybrid CSP–CPV system for improving the dispatchability of solar power plants. Energy Convers Manag 114:312–323
98. Zurita A et al (2018) Techno-economic evaluation of a hybrid CSP+PV plant integrated with thermal energy storage and a large-scale battery energy storage system for base generation. Sol Energy 173:1262–1277
99. Zhai R, Chen Y, Liu H, Wu H, Yang Y (2018) Optimal design method of a hybrid CSP-PV plant based on genetic algorithm considering the operation strategy. Int J Photoenergy 2018
100. Hernández-Moro J, Martinez-Duart JM (2013) Analytical model for solar PV and CSP electricity costs: Present LCOE values and their future evolution. Renew Sustain Energy Rev 20:119–132
101. Aguilar-Jiménez JA et al (2018) Techno-economic analysis of a hybrid PV-CSP system with thermal energy storage applied to isolated microgrids. Sol Energy 174:55–65
102. Awan AB, Zubair M, Praveen RP, Bhatti AR (2019) Design and comparative analysis of photovoltaic and parabolic trough based CSP plants. Sol Energy 183:551–565
103. Ma T, Javed MS (2019) Integrated sizing of hybrid PV-wind-battery system for remote island considering the saturation of each renewable energy resource. Energy Convers Manag 182:178–190
104. Duman AC, Güler Ö (2018) Techno-economic analysis of off-grid PV/wind/fuel cell hybrid system combinations with a comparison of regularly and seasonally occupied households. Sustain Cities Soc 42:107–126

105. Al-Ghussain L, Ahmed H, Haneef F (2018) Optimization of hybrid PV-wind system: case study Al-Tafilah cement factory, Jordan. Sustain Energy Technol Assess 30:24–36
106. Roberts JJ, Cassula AM, Silveira JL, da Costa Bortoni E, Mendiburu AZ (2018) Robust multi-objective optimization of a renewable based hybrid power system. Appl. Energy 223:52–68
107. Bekele G, Tadesse G (2012) Feasibility study of small Hydro/PV/Wind hybrid system for off-grid rural electrification in Ethiopia. Appl Energy 97:5–15
108. Zhang Z et al (2020) Short-term optimal operation of wind-solar-hydro hybrid system considering uncertainties. Energy Convers Manag 205:112405
109. Wei H, Hongxuan Z, Yu D, Yiting W, Ling D, Ming X (2019) Short-term optimal operation of hydro-wind-solar hybrid system with improved generative adversarial networks. Appl Energy 250:389–403
110. Ye L, Sun HB, Song XR, Li LC (2012) Dynamic modeling of a hybrid wind/solar/hydro microgrid in EMTP/ATP. Renew Energy 39(1):96–106
111. Singh S, Singh M, Kaushik SC (2016) Feasibility study of an islanded microgrid in rural area consisting of PV, wind, biomass and battery energy storage system. Energy Convers Manag 128:178–190
112. Balamurugan P, Ashok S, Jose TL (2009) Optimal operation of biomass/wind/PV hybrid energy system for rural areas. Int J Green Energy 6(1):104–116
113. Dhass AD, Santhanam H (2013) Cost effective hybrid energy system employing solar-wind-biomass resources for rural electrification. Int J Renew Energy Res 3(1):222–229
114. Rahman MW, Hossain MS, Aziz A, Mohammedy FM (2014) Prospect of decentralized hybrid power generation in Bangladesh using biomass, solar PV & wind. In: 2014 3rd international conference on the developments in renewable energy technology (ICDRET), pp 1–6
115. Al Busaidi AS, Kazem HA, Khan MF (2014) A review of optimum sizing techniques for off-grid hybrid PV-wind renewable energy systems. Int J Stud Res Technol Manag 2(03):93–102
116. Khatib T, Ibrahim IA, Mohamed A (2016) A review on sizing methodologies of photovoltaic array and storage battery in a standalone photovoltaic system. Energy Convers Manag 120:430–448
117. Anoune K, Bouya M, Astito A, Ben Abdellah A (2018) Sizing methods and optimization techniques for PV-wind based hybrid renewable energy system: a review. Renew Sustain Energy Rev 93:652–673
118. Yang H, Wei Z, Chengzhi L (2009) Optimal design and techno-economic analysis of a hybrid solar–wind power generation system. Appl Energy 86(2):163–169
119. Chauhan A, Saini RP (2014) A review on Integrated Renewable Energy System based power generation for stand-alone applications: configurations, storage options, sizing methodologies and control. Renew Sustain Energy Rev 38:99–120
120. Mahmoudimehr J, Shabani M (2018) Optimal design of hybrid photovoltaic-hydroelectric standalone energy system for north and south of Iran. Renew Energy 115:238–251
121. Tezer T, Yaman R, Yaman G (2017) Evaluation of approaches used for optimization of stand-alone hybrid renewable energy systems. Renew Sustain Energy Rev 73:840–853
122. Kaabeche A, Belhamel M, Ibtiouen R (2011) Sizing optimization of grid-independent hybrid photovoltaic/wind power generation system. Energy 36(2):1214–1222
123. Kanase-Patil AB, Saini RP, Sharma MP (2011) Sizing of integrated renewable energy system based on load profiles and reliability index for the state of Uttarakhand in India. Renew Energy 36(11):2809–2821
124. Maleki A, Khajeh MG, Ameri M (2016) Optimal sizing of a grid independent hybrid renewable energy system incorporating resource uncertainty, and load uncertainty. Int J Electr Power Energy Syst 83:514–524
125. Maleki A, Askarzadeh A (2014) Artificial bee swarm optimization for optimum sizing of a stand-alone PV/WT/FC hybrid system considering LPSP concept. Sol Energy 107:227–235
126. Luna-Rubio R, Trejo-Perea M, Vargas-Vázquez D, Ríos-Moreno GJ (2012) Optimal sizing of renewable hybrids energy systems: a review of methodologies. Sol Energy 86(4):1077–1088
127. Upadhyay S, Sharma MP (2014) A review on configurations, control and sizing methodologies of hybrid energy systems. Renew Sustain Energy Rev 38:47–63

Implementation of Blockchain-Based Security and Privacy in Energy Management

A. D. Dhass, S. Raj Anand, and Ram Krishna

Abstract Digital infrastructure can be used to help communities to meet their energy needs by exploiting the community-based power generation network. The present scenario of power generation can be seen in industries, educational establishments, and other organizations, where these installations are using industrial solar panels in their buildings. The ability to control access of energy with two or more resources has not been convinced by authentication. Commonly, the implementation of an individual solar panel system in each resource is very sensitive and expensive. Hence, it is very important to provide accuracy of the grid environment for social communities, so that they can be benefited with minimal cost for accessing the shared distributed energy and authenticating the access control. Each energy resource is authenticated by the consensus mechanism using the proof of work once connected to the first block as a genesis block. Such energy resource supports web chained database architectures and links to encrypt the large volumes of energy-driven data with blockchain-based distributed cloud storage design. This chapter aims to examine the significance of blockchain technology in accessing its role to date in enhancing energy protection and performance based on its technological advantages. The blockchain technology will be implemented in energy management to secure and understand the consumption and utilization of energy levels to reduce the overburden and waste. In addition, a case study based on energy resource management in educational institutions is provided to demonstrate how well implementation of blockchain technology can be offered alternatives for community-oriented energy security to increase the quality of service.

A. D. Dhass (✉)
Department of Mechanical Engineering, Indus Institute of Technology and Engineering, Indus University, Ahmedabad, Gujarat 382115, India
e-mail: dasaradhan.ad@gmail.com

S. Raj Anand
Department of CSE, VEMU Institute of Technology, Pakala, Chittoor, Andhra Pradesh 517112, India

R. Krishna
Madanapalle Institute of Technology and Science, Madanapalle, Andhra Pradesh, India
e-mail: krishnamme@gmail.com

S. Motahhir and A. M. Eltamaly (eds.), *Advanced Technologies for Solar Photovoltaics Energy Systems*, Green Energy and Technology,
https://doi.org/10.1007/978-3-030-64565-6_18

Keywords Blockchain · Consensus · Smart contract · Proof of work (POW) · Quality of service (QOS)

1 Introduction

Sustainable energy sources have undergone monstrous improvement since the inception, which has been empowered by privatization, the unbundling of the energy division, and motivating the activities through budgetary encouragement and energy strategy [1]. The energy area has a characteristic—power is hard to store for a huge scope as a physical product. The principle issue is that power should be generated on a scale that all individuals need. In any case, this is perfect and it is important to know in advance how much power individuals can accomplish it [2]. The producers of sustainable energy source systems have begun to bring down the expenses for generating power with proficient systems by expanding the financial intensity of sustainable power source assets. Most of the decentralized energy reserves are sustainable power sources which are discontinuous, rendering it hard to predict their capacity yield [3–6]. Throughout the customary power industry, there is a proximity of fundamental lattice to which both the distribution lines and small matrices are associated. The energy stream occurs in one way and the exchange takes place in the other way of the energy stream, demonstrating that the system is brought together. The latest measure for the unified power system is the idea of transactive energy, where the disseminated vitality is regulated directly by its creator. The Distributed Energy Exchange Model is a part of this transactive vitality component [7]. In addition, in this paradigm, residents and partners are progressively becoming a partner in the so-called sustainable power source networks and are partaking in the energy change by investing resources through, creating, selling, and conveying a sustainable power source [8].

Blockchain's invention continues with Bitcoin, standing in as the basis for the cash issuance and flow. The idea of Bitcoin is a disseminated and decentralized accounting innovation dependent on cryptography, with the plan to take care of the decentralization accord issue in the computerized world. Numerous ventures are dealing with how to adapt blockchain technology to their own organizations [9]. Blockchain technology additionally has an essential role in managing energy supply, transportation, storage, usage, and power, including validation of carbon outflow rights, the security of the digital-physical framework, sharing of virtual power assets, and preparation of a multi-energy system network. In any case, the set up focal administrator can be expected to guarantee confidence in the direct prosumer-to-prosumer exchange models. The blockchain stores information obtained from savvy meters and exchanges, while increments are ultimately performed consequently utilizing keen agreements [10]. Moreover, it will likewise encourage the capacity of vitality buyers to monetize their abundance vitality which may have originated from either age or vitality investment funds [11].

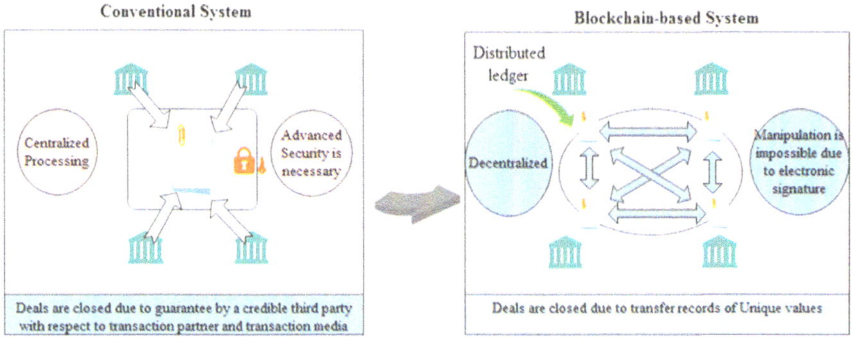

Fig. 1 Conventional and blockchain-based applied systems [12]

Blockchain innovation likewise called conveyed record innovation, is an administration system for the common check, endorsement, control, and sharing of the substance of exchange records among hubs dispersed at numerous center points on a PC arrange, as shown in Fig. 1 [12, 13]. As an independent and solid system innovation, blockchain has significant hypothetical and reasonable worth. It is a safe and self-governing shared system to work and guarantee the successful transmission of energy internet esteems. A settlement strategy is put in place to help with the last reviews and to improve the believability of exchanges, such as the power installment. The framework guarantees the security and validity of Internet of Things (IoT) gadgets and information from source to information sharing, underpins the improvement of vitality internet account, and recognizes the capacity of information on network exchanges dependent on blockchain [14].

Blockchain-based vitality involves exchanging the market stage for private networks with the goal of diminishing by and large network top interest and family power bills. Keen homes inside the network place vitality offer for its accessible, appropriated vitality assets (DERs) for each discrete exchanging period during day, and a twofold closeout instrument is utilized to clear the market and process the market-clearing value (MCP) [15]. Because of blockchain innovation, unalterable information age is conceivable and is for all intents and purposes illustrated. This safe and solid information as a keen agreement, P2P exchange, clean vitality endorsements, service bills without a dispersion system can bring a factor of trust among prosumers and customers [16]. The energy area is an undeniable case of an industry with the possibility to incorporate blockchain innovation and related technology. Power is led at the speed of light and is difficult to follow between two focuses in a power arrangement. Therefore, power markets can be brought together on exchanging stages like stock trades [17].

Decentralized vitality frameworks have been broadly examined in the scholarly community. It is also essential to talk about further progress in the dissemination of frameworks that blockchain innovation may encourage. Figure 2 shows the simplified approach to centralized and decentralized move in power and data streams are emerging. It brought together frameworks where the cause of a partner intended

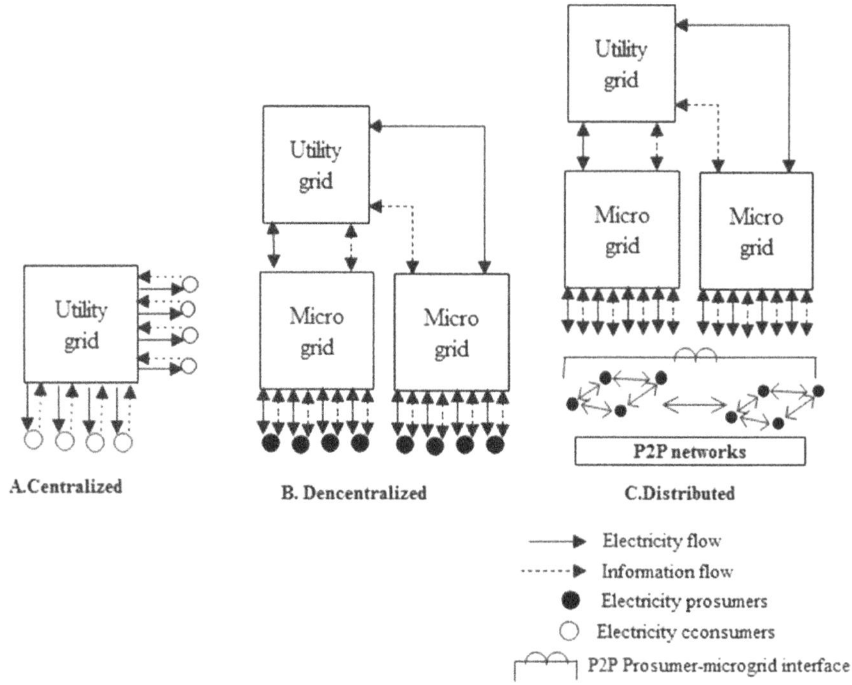

Fig. 2 Simplified model of **a** Centralized, **b** Decentralized, **c** Distributed energy systems with microgrids and Peer-to-Peer (P2P) networks

for the progression of power from high voltage to low voltage downstream loads. Microgrids have progressively added to the decentralized vitality frameworks, with the development of a private, small, and medium-scale sustainable power source age. Power makers, prosumers, and customers can easily purchase and sell power with each other in P2P systems, for which blockchain plays a key role in the development and empowerment of the power distribution innovation depicted in Fig. 2b, c. A few creators likewise suggest power-sharing by the microgrids in multi-microgrid network systems [18].

In fact, the diverse agreement systems are additionally utilized for the approval of squares and after accord, the square turns into the piece of the chain. The characteristics of various blockchain types are listed in Table 1.

Table 1 Types of blockchain and its characteristics [19]

Category	Directness	Decentralization	Write	Read
Open	Any person	Absolute	Any person	Everybody
Personal	Individuals	In part	Specific	Specific
Group	Specific	In part	Specific	Any person

The base of every other sort of agreement systems is Proof of Work (POW). Today's blockchain technology is utilized in various ideal models such as well-being checking, information sharing, customer criticism, decentralized exchange. To limit the odds of single point disappointment, unapproved people are not part of the system to forestall pernicious assaults [19]. Although wireless local area networks (WLAN) and Zigbee technologies are sufficient to manage home area network (HAN) and home automation, large power line access must be used for charging electric vehicles, grid operation, and automatic meter reading.

In an Internet of Energy (IoE) network, electricity delivery among nodes (electric cars, microgrid, smart grid, and smart homes) includes electricity consumption payment bills that are further used for services such as demand estimation, dynamic price estimation, and optimum energy use scheduling [20]. The routing, which satisfies the permission verification criteria and decreases the service center load; when storing the data in the blockchain data system, the hash pointer is used to maintain data access validity and complexity, and database protection is achieved [21]. The blockchain technology was operated by the P2P network, and each node is defined using a public key (PK). Node transfers are authenticated using PKs, and then sent to the network. That node is able to validate the source of the transaction through its digital signature [22]. The central smart contract is responsible for naming the participants and maintaining all the required data relevant to all purchases, P2P smart contract is responsible for handling local market exchange and the P2 G smart contract handles grid energy purchases for the prosumer [23]. Blockchain has been a competitive and quick increasing field of study as a groundbreaking technology. Now, this technology became a lucrative and fast-growing area since it incorporates centralized data management, peer-to-peer sharing, consensus processes, and encryption algorithms to allow verifiable documentation of data and transactions [24].

Key findings from the aforementioned studies have shown that blockchain technology provides secured information of energy data stored and it is expected to be used not only in cryptocurrency and fintech but also in the power and energy system fields. This is revealed that it examines the significance of blockchain technology in accessing its role to date in enhancing energy protection and performance in energy management based on its technological advantages.

2 Blockchain Technology

Blockchain technology is a decentralized process of a distributed ledger. It produces a set of rules for making mutual understanding between the producer and consumer called a smart contract. A smart contract can be applied in many industries, educational institutions, and other organizations for marketing the products and reflecting the information irrespective of buying and selling the procedures. This is an automatic process until it is stored in the blockchain. Each process will be verified by the techniques called a consensus mechanism. When a number of resources are added to the current chain, the POW mechanism has confirmed the authentication. No resources can alter and tamper, once it has been updated in the shared ledger.

Blockchain creates trust because every resource confidently makes a transaction to authorized resources. Hence, it increases confidentiality, authentication, and integrity to improve efficiency and transparency. Proof of Stake (POS) and Proof of Authority (POA) in the consensus technique could be authorized without tampering any transaction data. In the recent scenario, blockchain and photovoltaic are both popular cutting-edge sectors for industries, agriculture, and other organizations. It is optimized the existing security mechanism and produce the combination of a new energy revolution. A decentralized sharing peer-to-peer blockchain-based network to share the energy and particularly it is eliminating the middleman that is the centralized distribution system. Hence, blockchain technology is a very emerging technology to store the accuracy of data and efficiency.

3 Blockchain Architecture

The proposed blockchain technology model used with the photovoltaic system consists of three categories: blockchain smart contracts, consensus, and energy storage in the cloud using a peer-to-peer network. The decentralization of the overall architecture of the blockchain is shown in Fig. 3. In the primary building of educational institutions, the photovoltaic panel should become a genesis block. The smart contracts should be executed automatically for making the mutual agreement between prosumer and consumer. Here, the first block, the genesis block is acted as a prosumer when the new resources are connected with the primary building. Each resource acts as a consumer to consume energy levels locally.

The verification will be approached to each consumer when connecting to a new building with the existing energy level transaction by a consensus mechanism. Each individual consumer is verified the authentication process is called proof of work (POW). In Fig. 3, the photovoltaic panel resets the energy transmission to the maximum value for other buildings. Then the energy process verified in blockchain consensus mechanism to consumer 1 whether it is authenticated or not by POW. This sequence is processed for each consumer (i.e., consumer 2, consumer 3…consumer n). Then the subsystem of the peer-to-peer network has consumed the energy from the respective consumers either in a centralized or decentralized network. Consumption of power in terms of voltage is stored in the cloud. The data is stored rapidly to the cloud for the utilization of total threshold values. Overall efficiency will be monitored in mobile devices through a connected network. It also controls the access of energy between prosumer and consumer which is described in Fig. 3.

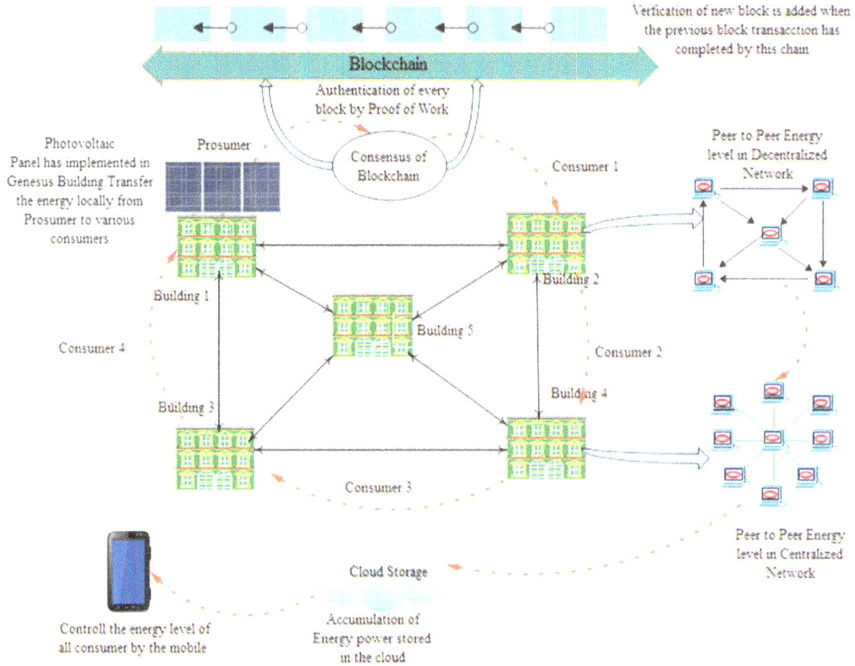

Fig. 3 Performance of blockchain using smart contracts in the photovoltaic panels

4 Energy Management and Blockchain Technology for Educational Institutions

In recent trends, the solar energy level is very popular and common to implement in day-to-day life. It helps the community to implement in the rural sectors. Many educational institutions are facing difficulty in utilizing the electricity in particular transmission time. Moreover, power failures are causing the education of all the students' communities to deny the concentration in order to improve their knowledge.

Therefore, in this portfolio, the entire educational institution will be switched over to the photovoltaic panel to implement the acquisition of more energy consumed throughout the buildings. In order to focus on the solar energy system, the advent of blockchain technology has dramatically increased the capacity of power utilization when the number of buildings in the institutions is increased. This technology can improve the security and authentication of decentralized network connections established in all buildings. This process attempts to clarify the Quality of Service (QoS) to all classrooms without the intervention of any learning methodologies. The solar system will drastically change the power process of all the educational institutions that have emerged in rural sectors. The consumption of energy is much lower in cost and reduces the power discrepancies.

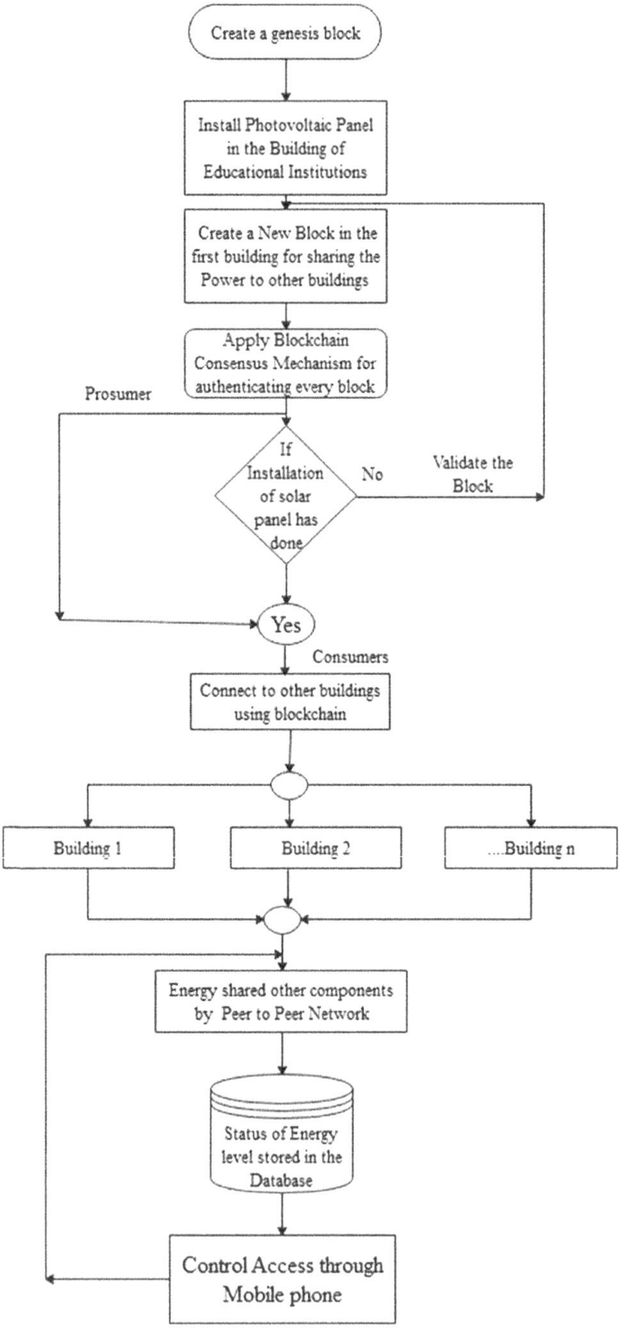

Fig. 4 Block diagram of prosumer and consumer agreement for smart contracts and a consensus mechanism

Figure 4 shows the simple block diagram of blockchain technology that could be adopted in a photovoltaic panel using smart contracts and a consensus mechanism. It shows that the genesis block will be built in the main building of the institution. The photovoltaic panel will then be installed in the main building of the institution for sharing the energy with other buildings. If the number of buildings is connected to the primary block, it should first be authenticated by a consensus mechanism. The POW only verifies all buildings for the accumulation of energy by those blocks. Using the smart contract system of blockchain, the prosumer shares energy to all consumers. Each consumer has to store the energy level in the cloud database.

This mechanism is very sustainable for the collection of a large number of energy levels. Other accessories utilize the power to communicate with peer-to-peer networks. The rural community people could implement these techniques by using blockchain technology to improve electrical energy efficiency and reduce the electricity utilization cost. However, peer-to-peer network sharing is controlled by mobile phones to monitor how power is to be used between the buildings.

India has a milestone in achieving the 20GW power capacity trading target for another two years in the year 2022. The survey of solar power capacity has been shown in Fig. 5, to be established at 6 MW in the year 2009. Then the solar power installed capacity has been increased every year. In Fig. 5, it is described 20 MW capacity of energy power is identified in India in the year 2018. In view of this, the Government of India has planned to implement the solar panel capacity to be implemented with 20GW instead of MW in 2022. Security is also a major factor in the implementation of the solar panel in all industries and other organizations.

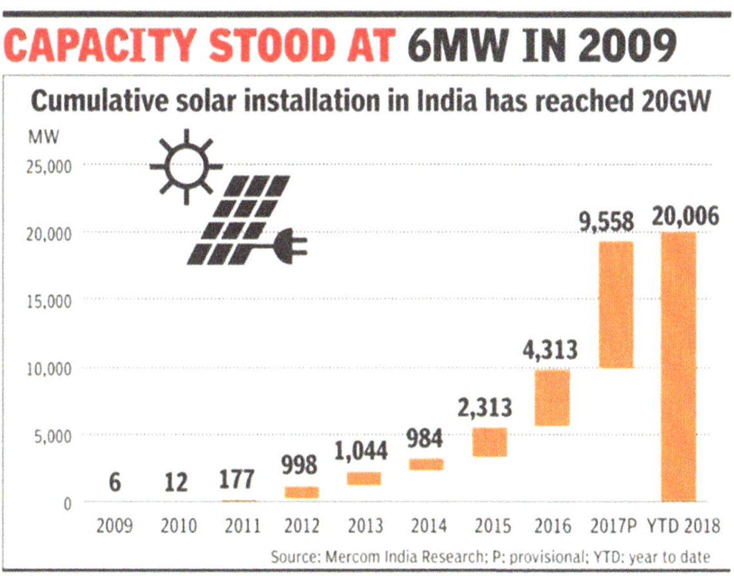

Fig. 5 Survey of solar installation in India [25]

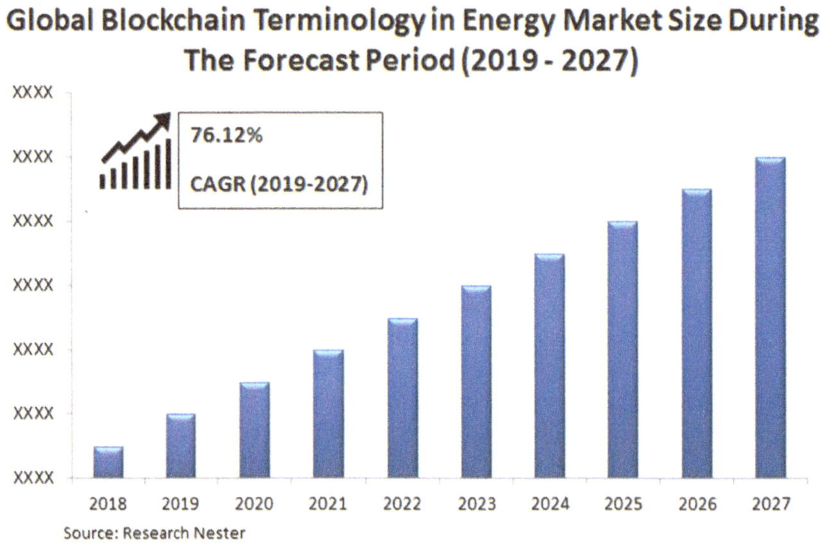

Fig. 6 Energy market size during the Period (2019–2027) using blockchain technology [26]

Figure 6 has depicted how blockchain technology has improved every year for producing security and privacy in the energy market. The survey mentioned in Fig. 6 revealed the energy level using the blockchain at a rate of 76%. There will be a greater manufacturing rate of up to 10 years (2018–2027). Hence, from both the Figs. 5 and 6, the market survey of blockchain technology using solar energy would be more scalable.

In worldwide business, blockchain technology has been implemented smoothly for decentralized mechanisms. The various activities carried out by the businesses, industries, and educational institutions are registered. Figure 7 illustrates the various aspects of security level research behind the blockchain technology. For automation and asset management, peer-to-peer activities will be validated using the energy level consensus algorithms through peer-to-peer transformation, wherever information has been made publicly available such as Proof of Authority (POA), Proof of Capacity (POC), and so on. In Fig. 7, the proof of work (POW) is defined as a major concentration for the verification of all the peer-to-peer network connections with security level up to 55% for the accumulation of solar energy to be connected to other buildings or locations. It is used to verify when a block has been created. At the same time, POA and POC have been establishing the transactions with 13% and 2%, respectively. In order to view Fig. 7, the scalability of every transaction would be increased by using the blockchain technology.

Figure 8 has demonstrated that the plan of action taken to consider every transaction in the business activities. The measurement of the percentage in IoT, smart devices, automation, and asset management by 11% and 19% of the total use of cryptocurrency cases for making the transactions, but 33% will be generated in the

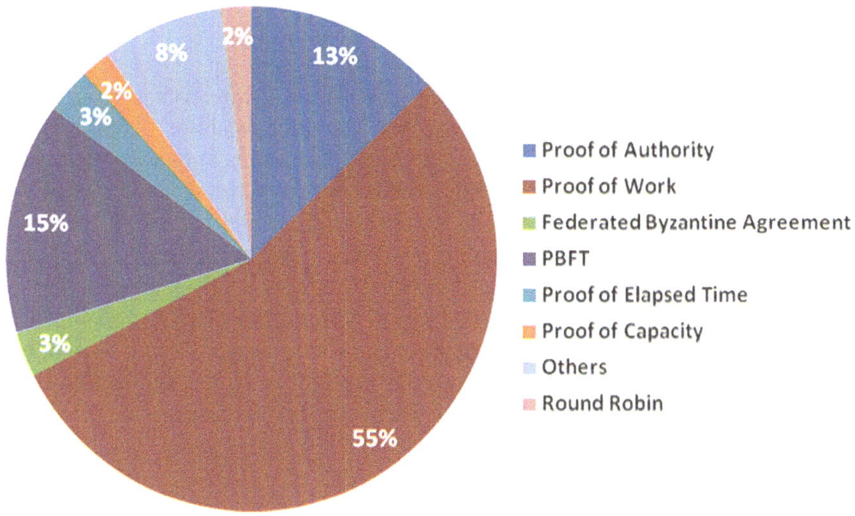

Fig. 7 Security of blockchain using POW in Solar Energy

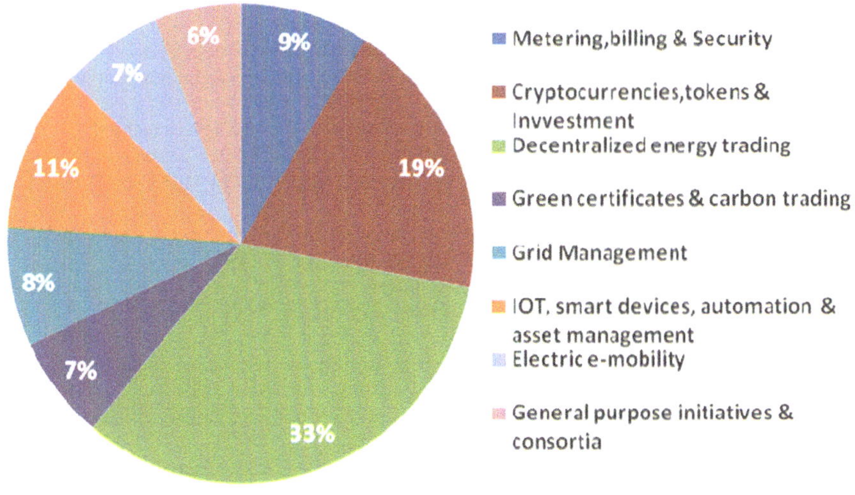

Fig. 8 Percentage level of decentralized energy trading using smart contracts [27]

case of decentralized energy trading. The purpose of this mechanism has been to provide a more secure transaction for various asset management activities. Hence, blockchain is more powerful than a centralized system for producing more accuracy in a decentralized network.

5 Conclusion and Future Outlook

Blockchain technology allows for more efficient and secure data storage while addressing the energy needs of emerging cities. Thus, the scalability of photovoltaic energy levels can be utilized efficiently with blockchain technology using consensus and smart contract mechanisms. Based on the above survey, this technology benefits the rural communities from a peer-to-peer network system with decentralized access to all the energy levels. Such communities will use electricity without any mind to raise electricity costs. The significance of blockchain technology in accessing its role to date in enhancing energy protection and performance in educational institutions on the basis of its technological advancement. It ensures that this technology is more powerful in order to protect energy consumption for the production of quality of service. Moreover, energy production using a photovoltaic panel in the future from all educational institutions, industries, and other organizations would be very scalable to reduce the cost of electricity usage.

References

1. Andoni M, Robu V, Flynn D, Abram S, Geach D, Jenkins D, Peacock A (2019) Blockchain technology in the energy sector: a systematic review of challenges and opportunities. Renew Sustain Energy Rev 100:143-174
2. Golosova J, Romanovs A, Kunicina N (2019) Review of the blockchain technology in the energy sector. In: IEEE 7th IEEE workshop on advances in information, electronic and electrical engineering (AIEEE). Liepaja, Latvia, pp 1–7. https://doi.org/10.1109/AIEEE48629.2019.8977128
3. Cali U, Fifield A (2019) Towards the decentralized revolution in energy systems using blockchain technology. Int J Smart Grid Clean Energy 8(3):245–256
4. Silvente J, Kopanos GM, Pistikopoulos EN, Espuña A (2015) A rolling horizon optimization framework for the simultaneous energy supply and demand planning in microgrids. Appl Energy 155:485–501
5. Eltawil MA, Zhao Z (2010) Grid-connected photovoltaic power systems: technical and potential problems-a review. Renew Sustain Energy Rev 14:112–129
6. Eltamaly AM, Mohamed MA (2014) A novel design and optimization software for autonomous PV/wind/battery hybrid power systems. Math Prob Eng
7. Karthik PK (2020) Energy trading in microgrids using blockchain technology. In: 2020 4th international conference on intelligent computing and control systems (ICICCS), Madurai, India, pp 884–888, https://doi.org/10.1109/ICICCS48265.2020.9121050
8. Svetec E, Nađ L, Pašičko R, Pavlin B (2019) Blockchain application in renewable energy microgrids: an overview of existing technology towards creating climate-resilient and energy independent communities. In: 2019 16th international conference on the European Energy Market (EEM), pp 1–7. IEEE
9. Cao Y (2019) Energy Internet blockchain technology. In: The energy Internet, pp 45–64. Woodhead Publishing
10. Brilliantova V, Thurner TW (2019) Blockchain and the future of energy. Technol Soc 57:38–45
11. Khatoon A, Verma P, Southernwood J, Massey B, Corcoran P (2019) Blockchain in energy efficiency: potential applications and benefits. Energies 12(17):3317
12. Hitachi Ltd. Blockchain. https://www.hitachi.co.jp/products/it/finance/innovation/blockchain/ (in Japanese)

13. Sawa T (2019) Blockchain technology outline and its application to field of power and energy system. Electric Eng Jpn 206(2):11–15
14. Zhu X (2019) Research on key technologies and applications of energy Internet blockchain. In: E3S web of conferences, vol 118, p 01003. EDP Sciences.
15. Saxena S, Farag H, Brookson A, Turesson H, Kim H (2019) Design and field implementation of blockchain based renewable energy trading in residential communities. In: 2019 2nd international conference on smart grid and renewable energy (SGRE), pp 1–6. IEEE.
16. Khan MSA (2019) Scope of blockchain technology in energy sector.
17. Arslan-Ayaydin Ö, Shrestha P, Thewissen J (2020) Blockchain as a technology backbone for an open energy market. In: Regulations in the energy industry, pp 65–84. Springer, Cham
18. Ahl A, Yarime M, Tanaka K, Sagawa D (2019) Review of blockchain-based distributed energy: Implications for institutional development. Renew Sustain Energy Rev 107:200–211
19. Zahid M, Ali I, Khan RJUH, Noshad Z, Javaid A, Javaid N (2019) Blockchain based balancing of electricity demand and supply. In: International conference on broadband and wireless computing, communication and applications, pp 185–198. Springer, Cham
20. Miglani A, Kumar N, Chamola V, Zeadally S (2020) Blockchain for Internet of energy management: review, solutions, and challenges. Comput Commun 151:395–418
21. Rui H, Huan L, Yang H, YunHao Z (2020) Research on secure transmission and storage of energy IoT information based on Blockchain. Peer-To-Peer Netw Appl
22. Li Y, Hu B (2020) A consortium blockchain-enabled secure and privacy-preserving optimized charging and discharging trading scheme for electric vehicles. IEEE Trans Ind Inf
23. Khalid R, Javaid N, Javaid S, Imran M, Naseer N (2020) A blockchain-based decentralized energy management in a P2P trading system. In: ICC 2020–2020 IEEE international conference on communications (ICC), pp 1–6. IEEE.
24. Guan Z, Lu X, Wang N, Wu J, Du X, Guizani M (2020) Towards secure and efficient energy trading in IIoT-enabled energy Internet: a blockchain approach. Future Generat Comput Syst 110:686–695
25. https://www.greenomicsworld.com/india-achieves-20gw-solar-capacity-milestone/. Access on 10 July 2020
26. https://www.researchnester.com/reports/blockchain-terminology-in-energy-market/1402. Access on 10 July 2020
27. Andonia M, Robua V, Flynna D (2019) Blockchain technology in the energy sector: a systematic review of challenges and opportunities. Renew Sustain Energy Rev 100:143–174

Big Data and Deep Learning Analytics for Robust PV Power Forecast in Smart Grids

Yunhui Zhang, Shiyuan Wang, and Payman Dehghanian

Abstract Photovoltaic (PV) power generation is intermittent and volatile in nature, rendering its large-scale deployment a challenge for the smart electricity grid's operation safety, stability, and economic efficiency. Ultra-fast and accurate prediction of PV power helps effectively adjusting the dispatch schedules during different operating states the power grid may undergo. This chapter proposes a deep learning-based PV power forecasting approach, the so-called Chaotic-LSTM, which ensembles the principles of the long short-term memory (LSTM) neural network and chaos theory. The LSTM neural network is used to construct a nonlinear mapping between input and output variables, while the phase space reconstruction technology in chaos theory is used to analyze the nonlinear time series of PV power generation, and extract the intrinsic dynamic characteristics of the PV arrays. Finally, a correlation analysis is applied to extract the external factors influencing the PV arrays. The effectiveness of the Chaotic-LSTM technology is demonstrated by comparing with three state-of-the-art neural network models: back-propagation, radial basis function, and simple recursive Elman neural networks. The accuracy of the proposed method was assessed using four different forecasting time horizons (i.e., one-hour, four-hour, one-day, and four-day-ahead) and three evaluation metrics. Additional tests are conducted with seven levels of signal-noise ratio to provide a measure of model robustness and effectiveness. Numerical results will demonstrate that the proposed Chaotic-LSTM method can significantly improve the prediction accuracy of the short-term PV power generation.

Y. Zhang
Department of Mechanical and Electrical Engineering, Shenzhen Polytechnic, Shenzhen 518055, China
e-mail: zhangyunhui1106@szpt.edu.cn

S. Wang · P. Dehghanian (✉)
Department of Electrical and Computer Engineering, George Washington University, Washington, DC 20052, USA
e-mail: payman@gwu.edu

S. Wang
e-mail: shiyuan1225@gwu.edu

S. Motahhir and A. M. Eltamaly (eds.), *Advanced Technologies for Solar Photovoltaics Energy Systems*, Green Energy and Technology,
https://doi.org/10.1007/978-3-030-64565-6_19

Keywords Renewable energy · Photovoltaic (PV) power generation · Short-term forecast · Chaos theory · Deep learning · Long short-term memory (LSTM) · Chaotic-LSTM method

1 Introduction

Renewable energy developments and technology deployments have recently gained increasing attention. With the rapid growth of photovoltaic (PV) power generation globally, solar energy plays an important role in the generation portfolio of modern smart power systems. However, solar irradiance and PV power generation are characterized by deep intermittency and volatility in short time intervals, making it a challenge to coordinate the operation and control of the large-scale PV power plants connected to the electric power grid [1]. Accurate forecast of the PV power is, therefore, of significant importance in ensuring an appropriate dispatch of the PV power generation in smart grids, thereby resulting in system economic efficiency, enhanced safety, and improved stability in day-to-day operations [2].

1.1 Background

Numerous PV power forecasting techniques in different forecast horizons have been extensively studied in the literature. Even if there is not any widely agreed-upon classification criterion, such techniques can generally fall into one of the following four categories [3–6], which are summarized in Fig. 1.

1. **Very Short-Term Forecasting (0–4-h-ahead)**: The output of such forecasts can be used for PV and energy storage control, real-time dispatch and control, and power quality assessment.
2. **Short-Term Forecasting (4-hour–one-week ahead)**: The output of such forecasts is generally used for power balance and day-ahead economic dispatch, unit commitment, power market transactions, transient stability assessment, etc.

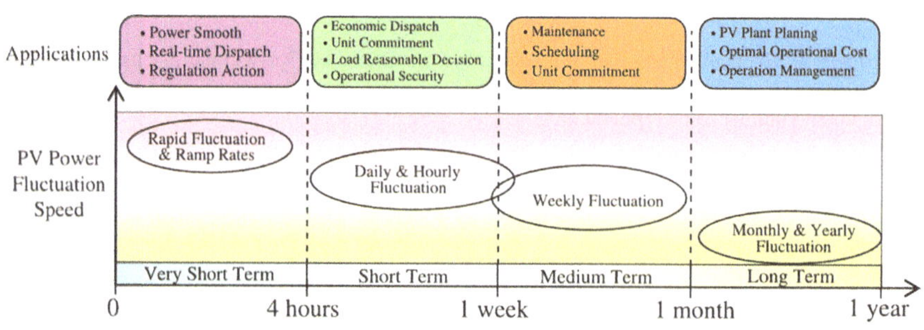

Fig. 1 Forecasting horizons and the corresponding applications in smart grids

3. **Medium-Term Forecasting (one-week–one-month ahead)**: It is commonly used for maintenance scheduling of PV plants, unit commitment, and mid-term dispatch of the power grid.
4. **Long-Term Forecasting (one-month–one-year ahead)**: The output of such forecasts is commonly used for PV plant planning, long-term solar energy assessments, and operation management.

From the perspective of smart grid energy management and operation, short-term prediction of solar power is particularly of importance and urgency for operational security and economic dispatch decisions. Therefore, many studies [5, 6] have focused on developing advanced short-term PV power forecast models. Meanwhile, the state-of-the-art techniques for PV power forecasting are typically classified into the following three categories [7]: (i) physical methods, (ii) statistical methods, and (iii) hybrid methods. Physical methods concern with analyses of the physical photoelectric conversion processes using physical equations for prediction [8, 9]. In physical methods, a large amount of historical data are not required, but accurate weather data and detailed power plant geographic information are requisite. Additionally, the physical formulas may not be accurate, altogether resulting in poor anti-interference ability and low robustness of such models. Statistical methods analyze the historical data to find out the underlying relationships for accurate solar prediction, with no consideration of the complex photoelectric conversion processes. Therefore, while offering simple modeling advantages compared to the physical methods, they often require the collection and processing of large amounts of historical data, making the required data acquisition and processing computationally intensive. There are two types of statistical methods: traditional statistical methods and artificial intelligence-aided algorithms. Traditional statistical methods include Auto-Regression Moving Average [10, 11], Gaussian Process Regression [12–16], Seasonal Auto-Regressive Integrate Moving Average Model [17–19], etc. Traditional statistical methods are usually characterized by poor self-learning characteristics and are only suitable for stationary time series with small data variations. For time series with large, and often sudden, changes in PV power, the traditional statistical methods reveal a large prediction error, thereby, a compromised prediction accuracy. Artificial intelligence (AI) methods, however, do not need to characterize the specific expressions between the inputs and the forecast outputs. By training the AI models using the historical dataset, the nonlinear mapping between the inputs and the output is established effectively, which is then used to predict the PV output power. Among AI algorithms, traditional neural networks [20–27], deep learning [28–30], support vector machine (SVM) [31–34], and fuzzy inference models [35, 36] are commonly used in PV power forecasting. A large amount of historical data are required in these methods, and they feature levels of self-learning and self-adaptive capabilities. However, SVM cannot easily choose suitable kernel functions and can hardly cope with large-scale training samples; traditional neural network methods suffer from the over-learning challenges and can easily fall into the local minimum; single long short-term memory (LSTM) methods can capture the dependencies in the time series, while the choice of input variables can hardly reveal the intrinsic dynamic characteristics of PV arrays. Convolutional neural network (CNN) reduces the complexity of the network because of its local sensing and weight sharing characteristics. However, the pooling layer will lose a lot of valuable information and ignore the correlation between the partial and the entire set.

To improve the prediction accuracy, multiple methods can be integrated in order to synergistically take advantage of their competing potentials. Compared with the single methods, the prediction improvements provided by a hybrid of these methods mainly include: (i) capturing the underlying knowledge in the field data to create features that can unlock the full potential of the forecast model; (ii) optimizing the model training parameters and structure; and (iii) improving the model learning speed and accelerating the algorithm convergence. Among the state-of-the-art techniques, a hybrid model for day-ahead hourly solar forecast using weather data is proposed in [37], where the self-organizing map and learning vector quantization networks are employed to classify the historical PV power dataset, support vector regression is applied to train pairs of input and output data, and the fuzzy inference method is applied to select a trained model for an accurate forecast. References [38–40] focused on the short-term solar power forecasting with the time horizon from five minutes to several days. Their suggested approach includes (i) utilizing a back-propagation neural network (BPNN) as the forecasting model and (ii) implementing a genetic algorithm (GA) to optimize the thresholds and weights of the BPNN. The integration of the two methods could reduce the calculation burden and improve the prediction accuracy of PV power generation. In [41], a combined model is proposed, which implements the global fast search function of particle swarm optimization (PSO) and the local search ability of the BPNN. Application results show that the prediction error of the forecast model is less than 20%. Reference [42] proposed a hybrid forecasting model for PV power generation. In this model, a random forest method is used to select the feature sets, improved grey ideal value approximation is used to select similar days, complementary ensemble empirical mode decomposition is used to decompose the original power time series, and particle swarm optimization algorithm is used to optimize the BPNN parameters. The effectiveness of this method is verified on a real-world photovoltaic power station. In [43], a hybrid method is proposed to predict one-day-ahead hourly PV power generation. The key points of the method are as follows: (i) The historical data of daily PV power generation is classified into various weather types by the fuzzy K-means algorithm; (ii) Training models for different weather types are constructed using radial basis function neural networks (RBFNN); (iii) Fuzzy inference is used to select an appropriate prediction model from the trained models. The results demonstrated that this method provides promising forecasting results than state-of-the-art techniques. Reference [44] uses the fuzzy c-means (FCM) clustering method to divide historical daily samples into several categories and uses historical data with weather conditions similar to the forecast day to build an RBFNN prediction model. The input of the neural network is the most similar two-day data and the current daily average radiation, and the output is the hourly output power prediction. The results proved the acceptable prediction accuracy of the model in practical applications. In [45], historical PV power and external variables are used as the input of the Adaptive Network-based Fuzzy Inference System (ANFIS) model to obtain the initial prediction, and GA is used to optimize the parameters of the ANFIS prediction model; the output of the model together with other variables are used as the input of the FFNN model to obtain the final prediction output. In [46], an ANFIS model is proposed for short-term PV power prediction. This model takes advantage of the neural networks and fuzzy inference system. Through the forecast of one-week PV power generation in the Indian electricity market, the ANFIS model shows significantly improved accuracy. Reference [47] proposed a GA-PSO-ANFIS hybrid approach to forecasting a day-ahead hourly PV power generation. In order to enhance the efficiency of the PV power forecaster, a binary GA is utilized to determine a suitable set of prediction variables. An integrated optimization framework that combines GA and PSO algorithms is finally employed to optimize the ANFIS-based forecast model. This method benefits from the simplicity and effectiveness of the PSO algorithm and the strong global searching capability of the GA to optimize the relatively complex ANFIS structure. A hybrid deep learning model is proposed for PV power generation prediction in [48]. The model uses a wavelet packet decomposition method to decompose the PV power generation time series into four sub-sequences. Then, the prediction models of four sub-sequences are established by LSTM. Finally, the linear weighting method is used to reconstruct the output of the four LSTM models to obtain the final prediction. Reference [49] proposes a combined algorithm, which uses grey relational analysis to select similar hours from historical datasets, and uses LSTM neural network to map the nonlinear relationship between the input and the output. The results show that the algorithm offers promising robustness. A hybrid model based on LSTM and attention mechanism for short-term PV power generation prediction is proposed in [50]. It uses LSTM to extract features from the time series of PV power generation data. The trained attention mechanism is applied to LSTM neural network to process the extracted features, so as to improve the original prediction ability of the LSTM neural network. Reference [51] presents a day-ahead prediction model for PV power. The model uses grey theory to pre-process the data and uses a deep belief network (DBN) to learn the high-level abstraction of historical power data. DBN is composed of multiple RBMs, and a feedforward network (FFN) is used in the last layer. The training process of the RBM network can be regarded as the initialization of the weight parameters of FFN; hence, DBN overcomes the shortcomings of the FNN network as the latter easily falls into the local optima and is featured with a long training time due to the random initialization of the weight parameters. Taking different weather types into account, the CNN regression model is established in [52], and each model parameter is optimized via the SALP group algorithm. The model is verified on a 500kwp photovoltaic power generation system revealing a good performance. In [53], a combined forecasting model for PV power generation is proposed. The input of the model is determined by the correlation coefficient method, and the parameters of the extreme learning machine are optimized by the improved chicken swarm optimization algorithm, resulting in satisfactory prediction performance. In [54], a hybrid deep neural network model is suggested to forecast the PV power output. This model takes weather parameters and the historical PV power as inputs and combines the advantages of CNN for high-dimensional data and LSTM for long-term dependence. It demonstrates superior performance over commonly used forecast models. In [55], a hybrid method based on a deep CNN is proposed for hour-ahead PV power forecasting. First, variational mode decomposition (VMD) is applied to decompose different frequency components from the historical time series of PV power. All components can be constructed into a two-dimensional form for the deep training of a CNN. From simulations, it can be seen that this method can improve the accuracy of short-term PV power forecasting.

1.2 Challenges and the Proposed Solution

The aforementioned artificial intelligence methods and combined forecasting techniques commonly adopt "black-box" forecasting methodologies. However, these models are known with some drawbacks such as the need for a large amount of historical data, certain assumptions on the data distribution, and the difficulty to explain the relationship between the input features and the output prediction. To overcome these shortcomings, a hybrid deep learning-aided approach that adopts and integrates the promising features and complementary benefits of the chaos theory and LSTM neural network is proposed. The introduced deep learning-aided approach, hereafter called the Chaotic-LSTM method, offers improved precision and computational performance in short-term PV power forecasting. In particular, this chapter highlights the following key features and contributions:

1. The phase space reconstruction in chaos theory is introduced for nonlinear time series analysis of PV power, which reveals the inherent dynamic characteristics of PV arrays, and lays the foundation for establishing an effective prediction model.
2. With a unique design structure, the LSTM neural network is suitable for time series prediction. The LSTM is approached to map the nonlinear relationship of the phase point evolution in the chaotic phase space. The chaotic characteristics are used to process the data samples and the input nodes of the neural network. Combining the chaos theory with LSTM neural network, the PV output power forecast model is established.
3. The proposed framework features not only the learning ability of the internal characteristics but also the adaptability to external meteorological conditions. With the prevailing randomness in the meteorological conditions and the resulting intermittency in the PV output power, the proposed Chaotic-LSTM approach applies the correlation analysis to capture the external random variations; it takes the meteorological conditions (obtained from the numerical weather prediction data) as the input and feeds the extracted variables into the LSTM for accurate PV power prediction.

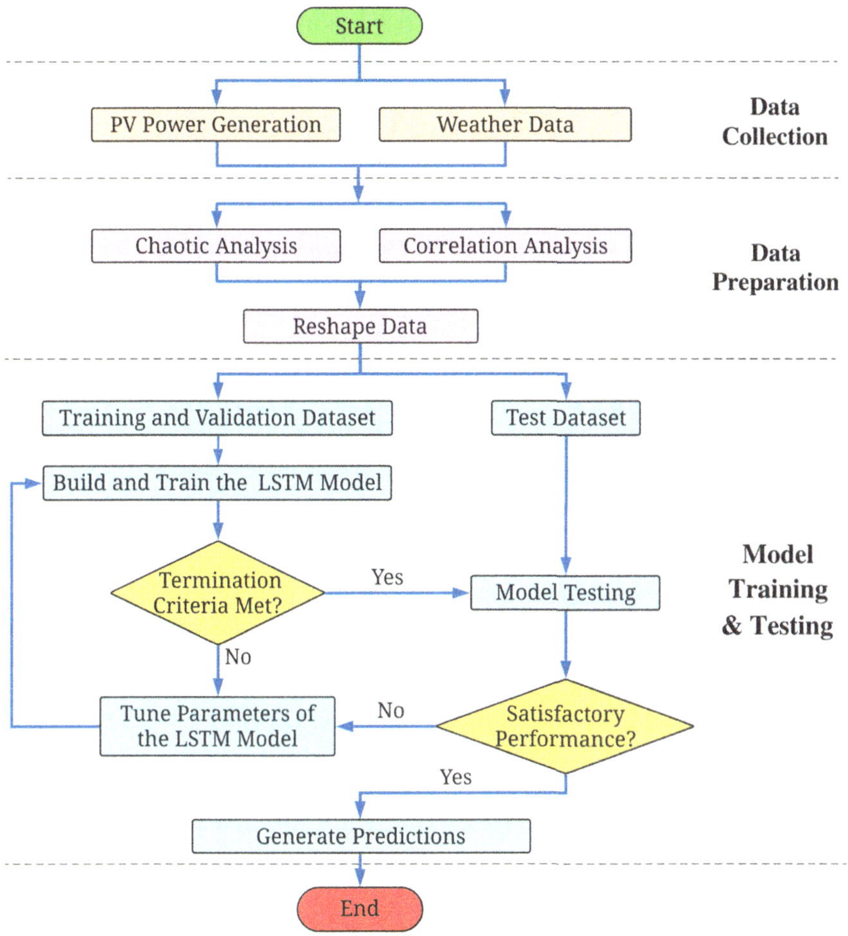

Fig. 2 Big picture architecture of the proposed deep learning-aided PV power prediction model

The remainder of the chapter is organized as follows. A detailed description of the proposed analytics is presented in Sect. 2. Section 3 presents case studies and discusses the numerical results. Finally, Sect. 4 sets forth a summary of the findings and concludes the chapter.

2 Proposed Deep Learning-Aided PV Power Prediction Model

Figure 2 demonstrates the proposed deep learning-aided model implementation procedure for the PV power forecast. The model implementation procedure can be summarized as follows:

1. **Data Collection**: Historical data of PV power generation and the corresponding weather conditions are collected.
2. **Data Preparation**: The collected data must be prepossessed before building the prediction model to ensure acceptable reliability and accuracy of the algorithm. Here, the phase space reconstruction technology in chaos theory is used to analyze the nonlinear time series of PV power generation data, so as to extract the inherent dynamic characteristics of the PV arrays. Correlation analysis is applied to extract the external factors influencing the PV arrays.
3. **Training and Testing**: The original dataset is divided into testing, training, and validation datasets. The training and validation datasets are used to build and train the proposed deep learning-aided PV power prediction model. By checking if the termination criteria are satisfied, one can judge whether additional model training efforts are required, or the model is ready for verification through the testing dataset. If the model performance is satisfactory during the testing process, the mapping relationship between the input variables and the output variable can be obtained by the proposed model.

2.1 Nonlinear Reconstruction Technique Applied to Chaotic Time Series

According to the chaos theory, the inaccurate prediction of a dynamic system in a long run is not driven by the influence of random factors; it is, instead, determined by the inherent dynamic characteristics of the system. The sensitivity of the system to the initial conditions makes the long-term prediction impossible or inaccurate. However, the system motion orbital divergence would be small in shorter time periods, making it feasible to harness the observed data for short-term forecasting. The nonlinear reconstruction of the chaotic time series focuses on excavating the laws hidden in the system, thereby helps in predicting the future trends and behavior of the dynamic system. It is very suitable for complex systems that show deterministic behavior globally, but with some degree of randomness locally.

Phase Space Reconstruction method [56, 57], proposed by Farmer. D. J and laid a solid mathematical foundation by Packard, provides a new mechanism for time series prediction. For a complex dynamic system with many influencing factors, its dynamic behavior can be described by its trajectory in phase space. In [56], it is proved that the phase space reconstruction of a dynamic system can be carried out according to a single-variable time series since the time series itself contains valuable information about all variables involved in the operation of the system. As long as the time delay and the embedding dimension are selected appropriately, a phase space that is topologically equivalent to the original dynamic system can be reconstructed from the one-dimensional chaotic time series.

2.1.1 Phase Space Reconstruction

The phase space reconstruction for the time series $\{a(n),\ n = 1, 2, \ldots, N\}$ is as follows:

$$A(i) = [a(i), a(i+\tau), \cdots, a(i+(m-1)\tau)],\quad i = 1, 2, \cdots, M;\ M = N-(m-1)\tau, \tag{1}$$

where the elements of $A(i)$ include one-time series element $a(i)$ itself and its time-shifted elements. Phase points in the phase space are given as follows:

$$\begin{bmatrix} a(1) & a(1+\tau) & \cdots & a(1+(m-1)\tau) \\ a(2) & a(2+\tau) & \cdots & a(2+(m-1)\tau) \\ \vdots & \vdots & \ddots & \vdots \\ a(N-(m-1)\tau) & a(N-(m-2)\tau) & \cdots & a(N) \end{bmatrix} \tag{2}$$

The embedding dimension m and the optimal time delay τ are determined by the reconstruction of the phase space. Each row of (2), which is represented by vector $A(i)$, stands for a phase point in an m-dimension phase space and the connection between the phase points describes the system evolutionary trajectory in this phase space.

2.1.2 Time Delay and Embedding Dimension Selection

The time delay τ and the embedding dimension m are the key factors to an effective reconstruction of the phase space. Once these two parameters are selected properly, the reconstructed phase space has the same geometric property and information as the actual dynamic system, where it retains all the critical features of the real phase space [56, 57]. There are two common paradigms on how to select time delay τ and the embedding dimension m. One paradigm assumes that τ and m are independent and can be independently selected. There are many methods to estimate τ, such as Auto-correlation [58], information entropy [59], and mutual information [60]. Saturated correlation dimension method [61] and the false nearest neighbor method [62] are effective methods to estimate m. Another paradigm primarily assumes that τ and m are not independent of each other and that they can be assessed simultaneously. A large number of experiments [63–65] show that embedding window width τ_w derives the correlation between τ and m characterized as $\tau_w = (m-1)\tau$. For a specific time series, τ_w is relatively fixed, and improper pairing of τ and m will directly affect the equivalent relationship between the reconstructed phase space and the original dynamic system; hence, joint algorithms for τ and m are generated accordingly, such as C-C method suggested in [66] and time window method in [65],. The C-C method is relatively simple and easy to implement; it has been widely used owing to its promising performance in practical applications. Therefore, in this chapter, the C-C method is chosen to simultaneously estimate the time delay τ and the embedding window width τ_w.

As mentioned earlier in Eq. (2), the phase space of the time series $\{a(n), n = 1, 2, \ldots, N\}$ is reconstructed using the time delay τ and the embedding dimension m. The C-C method is accordingly described as follows [66]: The correlation integral $\varepsilon(m, N, r, \tau)$ of the time series is first described, which represents the probability that the distance between any two points in the phase space is less than r:

$$\begin{cases} \varepsilon(m, N, r, \tau) = \dfrac{2}{M(M-1)} \sum\limits_{i,j=1}^{M} H(r - d_{ij}), \\ r > 0,\ d_{ij} = \|A(i) - A(j)\|, \\ H(\omega) = \begin{cases} 0, & \omega < 0 \\ 1, & \omega > 0 \end{cases}, \end{cases} \tag{3}$$

where $\frac{1}{2}M(M-1)$ denotes the number of all phase point pairs in the phase space. Then the time series $\{a(n), n = 1, 2, \ldots, N\}$ is utilized to build several new matrices, each of which has λ rows, where each row represents a disjoint sub-sequence with the length $int(N/\lambda)$; here, $int(\cdot)$ means taking an integer. The size of the matrices changes depending on the values of λ. For a given λ, the matrix would be expressed as in the following:

$$\begin{bmatrix} a(1) & a(1+\lambda) & a(1+2\lambda) & a(1+3\lambda) & \cdots \\ a(2) & a(2+\lambda) & a(2+2\lambda) & a(2+3\lambda) & \cdots \\ \vdots & \vdots & \vdots & \vdots & \vdots \\ a(n) & a(n+\lambda) & a(n+2\lambda) & a(n+3\lambda) & \cdots \\ \vdots & \vdots & \vdots & \vdots & \vdots \\ a(\lambda) & a(\lambda+\lambda) & a(\lambda+2\lambda) & a(\lambda+3\lambda) & \cdots \end{bmatrix} \tag{4}$$

The statistical measures are then evaluated for each sub-sequence as follows:

$$S(m, N, r, \lambda) = \frac{1}{\lambda} \sum_{\eta=1}^{\lambda} \left(\varepsilon_\eta\left(m, \frac{N}{\lambda}, r, \lambda\right) - \varepsilon_\eta\left(1, \frac{N}{\lambda}, r, \lambda\right)^m \right) \tag{5}$$

For $\lambda = 1$, there is only one-time sequence $\{a(n), n = 1, 2, \ldots, N\}$ at this time; then the following holds

$$S(m, N, r, 1) = \varepsilon_1(m, N, r, 1) - \varepsilon_1(1, N, r, 1)^m. \tag{6}$$

For $\lambda = 2$, there are two disjoint time series with the length $N/2$, i.e., $\{a(1), a(3), a(5), \ldots, a(N-1)\}$and $\{a(2), a(4), a(6), \ldots, a(N)\}$, we then have

$$\begin{aligned} &S(m, N, r, 2) \\ &= \frac{1}{2}\left[\varepsilon_1(m, N/2, r, 2) - \varepsilon_1(1, N/2, r, 2)^m + \varepsilon_2(m, N/2, r, 2) - \varepsilon_2(1, N/2, r, 2)^m\right]. \end{aligned} \tag{7}$$

We define $\Delta S(m, N, \lambda)$ as:

$$\Delta S(m, N, \lambda) = \max(S(m, N, r_i, \lambda)) - \min\big(S\big(m, N, r_j, \lambda\big)\big),\ \ \mathrm{i} \neq \mathrm{j}, \tag{8}$$

where r_i and r_j are different neighborhood radii. $\Delta S(m, N, \lambda)$ indicates the maximum deviation of $S(m, N, r, \lambda)$ for all radii. The values of m and the optimal time delay τ can be estimated properly according to the Brock-Dechert-Scheinkman (BDS) statistics [67] for $N = 3000$, $m = 2, 3, 4, 5$, and $r = \frac{\sigma_s}{2} \sim 2\sigma_s$. σ_s is standard deviation of the time series. The average values of the evaluated quantities are as follows:

$$\begin{cases} \bar{S}(\lambda) = \dfrac{1}{16}\displaystyle\sum_{m=2}^{5}\sum_{j=2}^{4} S\big(m, N, r_j, \lambda\big) \\ \Delta\bar{S}(\lambda) = \dfrac{1}{4}\displaystyle\sum_{m=2}^{5} S(m, N, \lambda) \\ S_{cor}(\lambda) = \Delta\bar{S}(\lambda) + \left|\bar{S}(\lambda)\right| \end{cases} \tag{9}$$

□

The first local minimum of $\Delta\overline{S}(\lambda)$ is taken as the optimal time delay g of the time series. Using the global minimum value of $S_{cor}(\lambda)$ as the embedding window width $(m-1)\tau$ of the time series, the embedding dimension m can be then obtained.

2.1.3 Calculation of the Largest Lyapunov Exponent

Lyapunov exponents reflect the average exponential rate of separation or convergence of the nearby orbits in phase space. If the largest Lyapunov exponent is greater than 0, it means that the adjacent points will eventually be separated corresponding to the local instability of the orbit, which indicates that the system has chaotic behavior, and the chaotic sequence prediction method can be applied. The largest Lyapunov exponent can be calculated by applying the small data sets algorithm [68].

Given the time series $\{a(n), n = 1, 2, \ldots, N\}$, the phase space is reconstructed according to (2). Next, the small data sets algorithm is used to determine the nearest neighbor of each point on the trajectory. By searching for the point with the smallest distance from a specific point $A(i)$, we can get its nearest neighbor $A(i)$, which can be expressed as follows:

$$d_i(0) = \min_{\hat{i}} \left\| A(i) - A\left(\hat{i}\right) \right\|. \tag{10}$$

After μ discrete time steps, the distance between the i-th pair of the nearest neighbors is

$$d_i(\mu) = \left\| A(i+\mu) - A\left(\hat{i}+\mu\right) \right\|, \ \mu = 1, 2, \cdots, \min\left(M - i, M - \hat{i}\right). \tag{11}$$

For each μ discrete time step, the average value of $\ln d_i(\mu)$ is evaluated for i:

$$y(\mu) = \frac{1}{q\Delta t} \sum_{i=1}^{q} \ln d_i(\mu). \tag{12}$$

The largest Lyapunov exponent can be approximately obtained using a least square fit to the curve $y(\mu) \sim \mu$, corresponding to the slope of the regression line.

2.2 Correlation Coefficient Analysis

Pearson correlation coefficient [69] is here employed for correlation analysis. Given the paired data $\{(\alpha_1, \beta_1), \ldots, (\alpha_K, \beta_K)\}$, consisting of K number of pairs, we have

$$r_{\alpha\beta} = \frac{\sum_{i=1}^{K} (\alpha_i - \bar{\alpha})(\beta_i - \bar{\beta})}{\sqrt{\sum_{i=1}^{K} (\alpha_i - \bar{\alpha})^2} \sqrt{\sum_{i=1}^{K} (\beta_i - \bar{\beta})^2}}. \tag{13}$$

$\overline{\alpha} = \frac{1}{K} \sum_{i=1}^{K} \alpha_i$ is the sample mean for variable α and analogously for variable β. The Pearson correlation coefficient reflects the strength of the linear correlation between variable α and variable β. The value of $r_{\alpha\beta}$ falls within the range of $[-1, 1]$. The greater the $r_{\alpha\beta}$, the greater the correlation between the two variables α and β. When $r_{\alpha\beta} > 0$, it indicates a positive correlation between these two variables. That is, when the value of one variable increases, the value of the other variable increases. When $r_{\alpha\beta} < 0$, it indicates a negative correlation between these two variables. That is, when the value of one variable increases, the value of the other variable decreases. When $r_{\alpha\beta} = 0$, this shows that there is no linear correlation between these two variables. $r_{\alpha\beta} = 1$ and -1 indicate that for any positive (negative) increase in one variable, there is a positive (negative) increase of a fixed proportion in the other. When $0 < |r_{\alpha\beta}| < 1$, it indicates that there are different degrees of linear correlations: $|r_{\alpha\beta}| \leq 0.3$ indicates that there is no linear correlation between two variables; $0.3 < |r_{\alpha\beta}| \leq 0.5$ indicates that there is a small linear correlation between two variables; $0.5 < |r_{\alpha\beta}| \leq 0.8$ indicates a significant linear correlation between two variables; and $r_{\alpha\beta} \geq 0.8$ indicates the two variables are highly linearly correlated.

2.3 Long Short-Term Memory (LSTM) Network

Long short-term memory (LSTM) network—a class of recurrent neural network (RNN) models within the general theme of deep learning analytics [70]—features learning long-term dependencies, which is instrumental for certain types of prediction missions that require the network to retain information over long time periods. This is, however, challenging to be achieved in traditional RNNs [71]. The vanishing gradient problem—i.e., how to decide when the neural network stops learning due to the fact that the update rates to the various weights within a given neural network keep decreasing—restricts the memory capabilities of the traditional RNNs by adding too many time steps. LSTM provides a solution to the vanishing and exploding gradient problems and works extremely well in various practical applications [72–74].

LSTM solves the problem of traditional RNNs with only short-term memory. The general architecture of the LSTM neural network is shown in Fig. 3, which can be expressed as a time-expansion chain structure. In Fig. 3, *S* represents an LSTM unit, and each unit is associated with a time instance.

A common LSTM unit is composed of a cell, and three gates—forget gate, output gate, and input gate [75]. Each gate behaves like a switch that regulates the information flowing into and out of the cell. The forget gate selectively forgets the information in the memory cell state. The output gate selects and outputs the necessary information and decides whether the memory cell state affects other neurons. The input gate selectively controls new information into the cell state. At the bottom of Fig. 3, a detailed structure of an LSTM unit is illustrated. Variables (C_{t-1}, h_{t-1}, x_t) are the inputs to each LSTM cell. On the output side of each LSTM cell at time t, there are two states (C_t, h_t) that are transferred to the next cell. The forward calculation formula of the LSTM network is as follows:

$$f_t = \sigma\left(W_{fx}x_t + W_{fh}h_{t-1} + b_f\right) \tag{14}$$

$$i_t = \sigma(W_{ix}x_t + W_{ih}h_{t-1} + b_i) \tag{15}$$

$$o_t = \sigma(W_{ox}x_t + W_{oh}h_{t-1} + b_o) \tag{16}$$

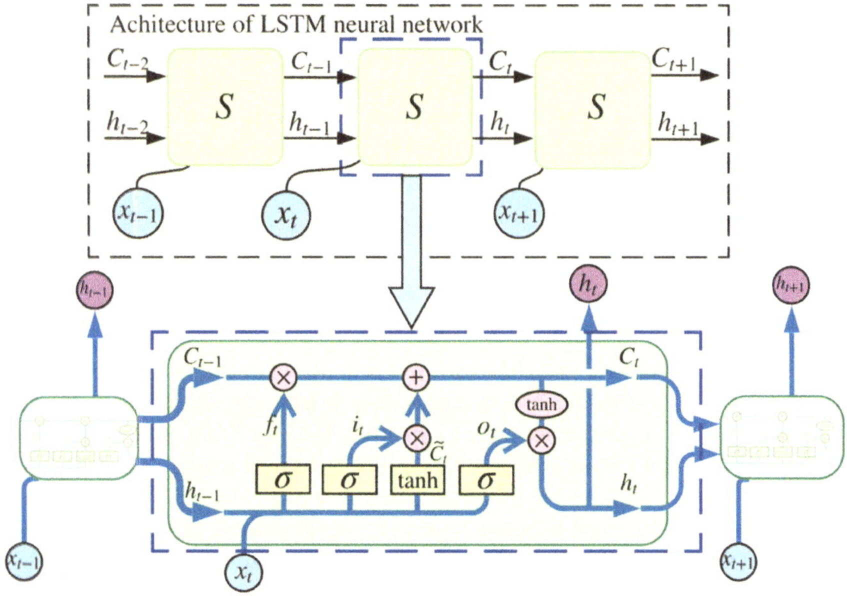

Fig. 3 A generic LSTM neural network architecture

$$\tilde{c}_t = \sigma(W_{cx}x_t + W_{ch}h_{t-1} + b_c) \tag{17}$$

$$c_t = f_t \circ c_{t-1} + i_t \circ \tilde{c}_t \tag{18}$$

$$h_t = o_t \circ \tanh(c_t). \tag{19}$$

3 Numerical Case Studies

PV power forecasting is here described as a regression forecast modeling problem, as expressed in the following formula:

$$y_p\left(t + \hat{h}\right) = G(u(t)). \tag{20}$$

In (20), $u(t)$ denotes input variables, and $y_p\left(t + \hat{h}\right)$ is the output variable, which represents the output of the proposed PV power forecasting model at the time $\left(t + \hat{h}\right)$. PV power is affected by many factors such as the weather, meteorological, as well as geographical environment, which in turn results in random fluctuations and intermittency. These factors can be classified into two categories: (i) the internal factors related to the PV arrays (e.g., installation position, installation angle, type of PV modules and the temperature of PV modules, and aging); (ii) the external factors affecting PV arrays (e.g., season, atmospheric temperature, irradiation intensity, wind speed, irradiance angle, and day type). In the case studies, the phase reconstruction technique of the chaotic time series and the correlation analysis are implemented to pre-process the raw data and obtain the input $u(t)$ in (20) for the forecasting model. The nonlinear reconstruction technique is used to extract the internal dynamics of the PV arrays, which is achieved by considering the PV power variation driven by the internal factors. When dealing with external random factors (the second category), we extract the main influencing factors through correlation analysis. Once the data is prepared, the deep learning forecasting model is developed to learn the nonlinear relationship between the input and output variables. In each case study, a back-propagation (BP) neural network, radial basis function (RBF) neural network, and simple recurrent Elman neural network are used as benchmarks for performance comparison of the proposed Chaotic-LSTM methodology. Four different forecasting time horizons (i.e., one-hour, four-hour, one-day, and four-day-ahead) and three evaluation metrics—Normalized Root Mean Square Error (NRMSE), Mean Absolute Error (MAE), and Mean Absolute Percentage Error (MAPE)—are utilized to evaluate the accuracy of the proposed approach. The simulations are carried out in MATLAB 2019 a programming environment on a PC system with Intel® Core™ i7-7500U CPU @2.70 GHz, and 8 GB RAM.

3.1 Data Collection

Two cases are studied in this Section. The data used in Case I are obtained from the Yulara Solar Power System [76]. The system is located near the iconic Uluru (Ayers Rock) in Australia, with a total capacity of 1.82 MW, distributed in five locations—see Fig. 4. The data covers a time period from January 2018 to the end of 2018.

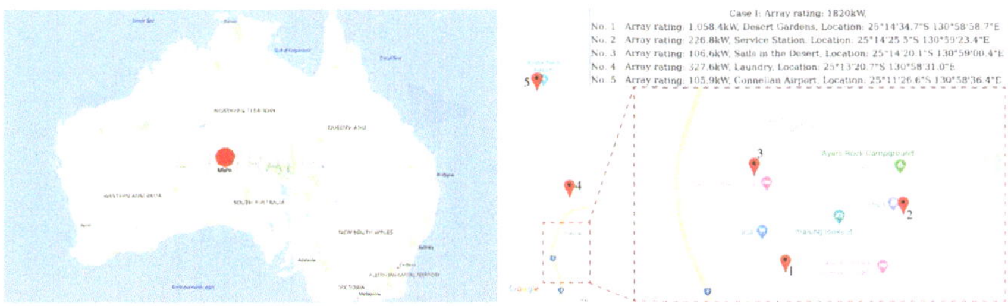

(a) (b)

Fig. 4 Location of the Yulara solar system in Uluru, Australia: Case I

(a) (b)

Fig. 5 Location of the desert knowledge Australia solar centre in Alice Springs, Australia: Case II

The data used in Case II are obtained from the desert knowledge Australia solar centre (DKASC) [77]. DKASC is a small town located in Alice Springs, as shown in Fig. 5. The total output power of all sites is 263 kW. The Case II data covers a time period from January 2019 to the end of 2019. In addition to the PV power generation, meteorological data (weather temperature, wind speed, wind direction, weather relative humidity, diffuse horizontal radiation, and global horizontal radiation) are also provided. The data resolution is 15 min.

Figure 6a, b, respectively, presents the daily trends of the generated PV power throughout the year for Case I and Case II. It can be observed that the PV power generation reaches its peak at noon and drops to zero at night. Figure 7a, b, respectively, represents the monthly distribution of the generated PV power from 7:00 am-7:00 pm for Case I and Case II. From Fig. 7a, it can be seen in Case I that the PV power generation is relatively low from May to October, and relatively high from December to March. In Case II, the annual fluctuation of the PV power generation is relatively small, which can be seen in Fig. 7b.

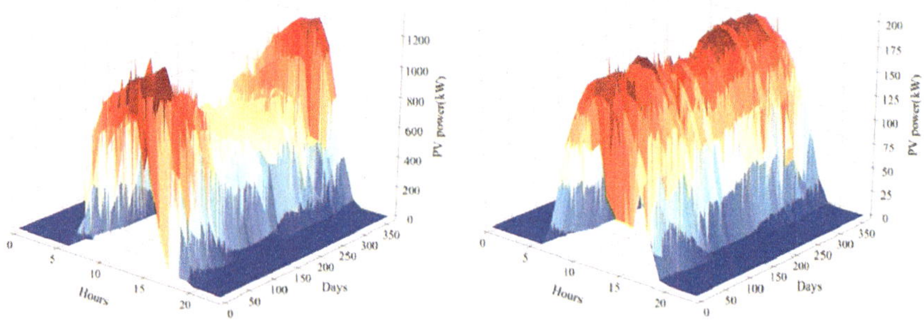

(a) Case I (b) Case II

Fig. 6 PV power data for the studied Case I [76] and Case II [77]

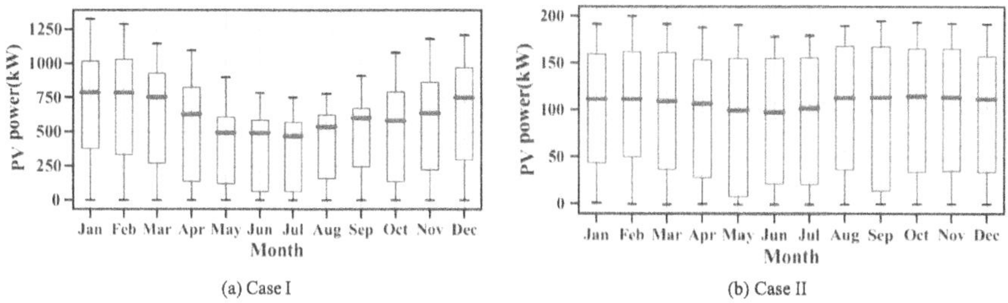

Fig. 7 Monthly distribution of PV power generation in the studied Case I and Case II

3.2 Data Preparation

Once the data is collected, the next step is to consider how to prepare and organize the data in a form ready to be harnessed by the developed forecast engine. As high-quality data can generate better models, data pre-processing becomes more and more important and has become a basic step of the machine learning algorithms. In this chapter, the pre-processing includes data cleaning, the phase space reconstruction, correlation analysis of the input data, and the data reshape processes.

3.2.1 Data Cleaning

Datacleaningmainly includes removing irrelevant and/or duplicate data in the original dataset, as well as dealing with missing and/or abnormal values.

1. Missing Values: In order to improve the data quality, two methods are generally used to process the missing entries:
   - Data Deletion: When there is a large amount of data or multiple fields in a record are missing and it is not convenient to fill in the missing entries, one common approach is to delete the missing values.
   - Data Imputation: It is necessary to estimate the missing values if discarding them is not desired. Generally, the following methods are used to fill the missing values: (i) replacing the missing value by the value (or average value) of the adjacent records; (ii) filling in the missing value with the value on another similar record with the missing value; (iii) building a statistical model for predicting the missing values by estimating the model parameters.
2. Outlier Values: For the treatment of the outliers, one common approach is to use the single-variable scatter plot or box plot, where the points far away from the normal range are regarded as outliers. The processing of outliers includes removing the observations with outliers, treating them as missing values, correcting the average values, and discarding them from the processes. When processing the outliers, one should first analyze the possible causes creating such outliers, and then judge whether the outliers should be discarded.

3.2.2 Phase Space Reconstruction of the Chaotic Time Series

As mentioned earlier in Sect. 2.1.2, since the single-variable time series itself contains valuable information on all variables in the dynamic system, the phase space of the dynamic system can be reconstructed according to the single-variable time series. Accordingly, phase reconstruction technology is here applied to the PV power time series $\{a_{PV}(t), t = 1, 2, .., N\}$ to excavate the intrinsic dynamic characteristics of PV arrays, serving as the foundation for establishing a forecasting model.

The choice of time delay and embedding dimension is critical for phase space reconstruction. In this chapter, the phase space reconstruction is performed through the C-C method. As mentioned in Sect. 2.1.2, three statistical measures $\overline{S}(\lambda)$, $\Delta\overline{S}(\lambda)$, and $S_{cor}(\lambda)$ are evaluated to obtain the optimal values for the time delay τ and the embedding dimension m. The relationships between the time delay and three statistical metrics $\overline{S}(\lambda)$, $\Delta\overline{S}(\lambda)$, and $S_{cor}(\lambda)$ are, respectively, illustrated in Fig. 8a, b for Case I and Case II. The X-axis represents λ, i.e., different time delays, while the Y-axis reflects the above three statistical metrics. For Case I (Fig. 8a), it can be observed that $\Delta\overline{S}(\lambda)$ achieves the first local minimum value when $\lambda = 5$, which is corresponding to the optimal time delay τ. The optimal embedding window width is $(m-1)\tau = 35$, corresponding to the global minimum value of $S_{cor}(\lambda)$. The value of the embedding dimension m is then obtained, which is equal to 8. Similarly, in Case II (Fig. 8b), it can be observed that $\Delta\overline{S}(\lambda)$ achieves the first local minimum value when $\lambda = 5$, which is corresponding to the optimal time delay τ. The optimal embedding window width is $(m-1)\tau = 37$, corresponding to the global minimum value of $S_{cor}(\lambda)$. Therefore, the embedding dimension is obtained as $m = 9$.

To detect whether the PV data time series has chaotic behavior, the small data sets algorithm is applied to calculate the largest Lyapunov exponent in this chapter. The largest Lyapunov exponent is positive, revealing that the time series of PV power generation has chaotic characteristics, which provides the possibility for short-term prediction of PV power generation. In Sect. 2.1.3, Eqs. (10)–(12) described the detailed evaluation method. With the evaluations performed in Case I, the largest Lyapunov exponent is found to be 1.25×10^{-3}, while that of Case II is assessed to be 5×10^{-4}.

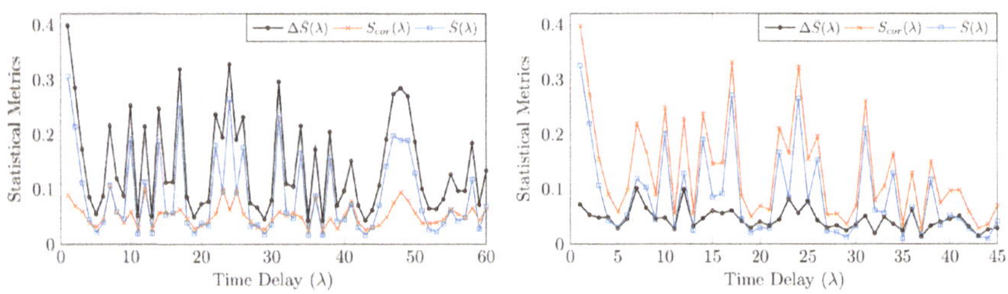

(a) Case I

(b) Case II

Fig. 8 C-C method applied to derive the optimal values for time delay and embedding dimension in the studied Case I and Case II

3.2.3 Correlation Analysis of the Input Data

Various external factors influencing the PV arrays are mutually interdependent. The factors with higher correlation are selected as the characteristic information of the PV power, which is an important reference for sample selection and input node characterization of the PV power forecast model. The correlation between the PV output power and the influencing factors will be extensively analyzed in this Section.

The dataset parameters include the PV output power, weather temperature, weather relative humidity, wind speed, diffuse horizontal radiation, global horizontal radiation, and wind direction. The matrix $r_{\alpha\beta}$ is then evaluated according to Eq. (13), which represents the relationships between each pair of variables in the dataset. The scatter plot matrices for Case I and Case II analyses are shown in Fig. 9 and Fig. 10, respectively, and demonstrated the correlations between multiple variables. The distribution of individual variables can be seen from the histogram on the matrix diagonal entries, while the scatter plot on the upper and lower triangles shows the relationship between every two variables. One can see that the PV power is positively correlated with weather temperature, diffuse horizontal radiation, global horizontal radiation, and wind speed, which indicates that as these variables increase, the PV output power increases. Similarly, PV power is negatively correlated with wind direction and weather relative humidity, which indicates that the increase in these variables will decrease the PV power output.

The heat-maps of the correlation matrices in Case I and II are demonstrated in Fig. 11a, b, respectively, to illustrate the correlation coefficient matrix between variables. From the result in Case I (Fig. 11a), it can be observed that the variations in weather temperature, diffuse horizontal radiation, and global horizontal radiation can significantly influence the PV power. Hence, these three variables are introduced to the proposed Chaotic-LSTM neural network model as the external input variables.

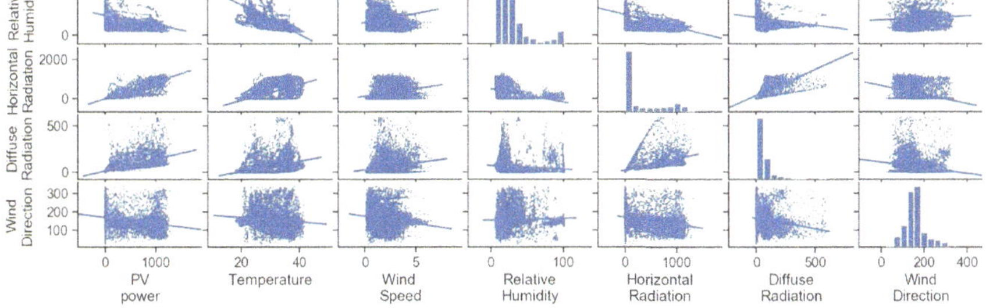

Fig. 9 Scatter plot of the correlation matrix (Case I)

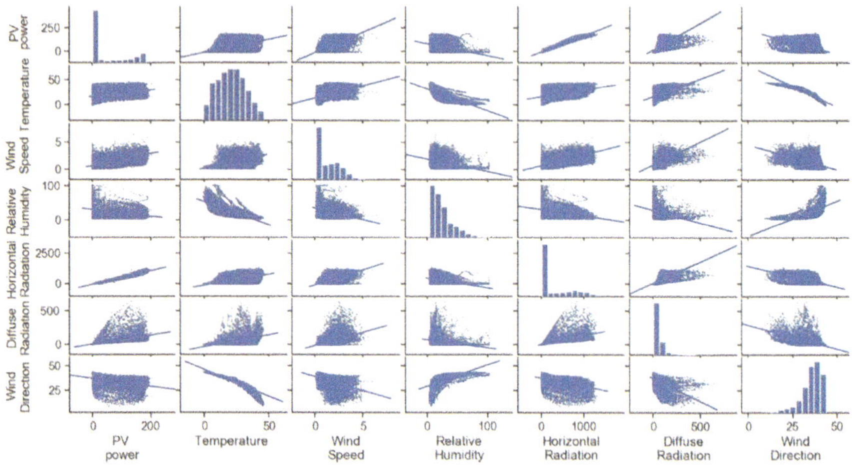

Fig. 10 Scatter plot of the correlation matrix (Case II)

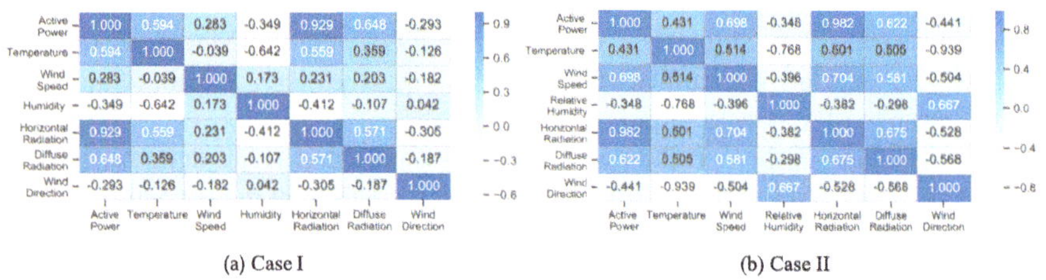

Fig. 11 Heat-map of the correlation matrix for PV power data in Case I and case II

From the result presented in Case II (Fig. 11b), it can be observed that the variations in wind speed, diffuse horizontal radiation, and global horizontal radiation can significantly influence the PV output power. Hence, these variables are introduced to the proposed Chaotic-LSTM neural network model as the external input variables.

3.2.4 Data Reshape

Based on the phase space reconstruction and correlation analyses, two parts of the input variables for $u(t)$ in (20) are achieved, i.e., $u_1(t)$ and $u_2(t)$. $u_1(t)$ is a discrete PV power time series $u_1(t)$ obtained from the phase space reconstruction, and $u_2(t)$ is weather data array obtained from the correlation analyses. The input $u(t)$ to the forecast engine can be then formed as follows:

$$u(t) = [u_1(t), u_2(t)]. \tag{21}$$

Table 1 Input variables of the proposed Chaotic-LSTM network (case I)

PV power data $u_1(t)$	Weather data $u_2(t)$
$a_{PV}(t-(m-1)\tau)$, $a_{PV}(t-(m-2)\tau)$,	Diffuse horizontal radiation at time $(t+\hat{h})$
$\cdots$, $a_{PV}(t-\tau)$, $a_{PV}(t)$	Global horizontal radiation at time $(t+\hat{h})$
with $\tau=5$, $m=8$	Weather temperature at time $(t+\hat{h})$

Table 2 Input variables of the proposed Chaotic-LSTM network (Case II)

PV power data $u_1(t)$	Weather data $u_2(t)$
$a_{PV}(t-(m-1)\tau)$, $a_{PV}(t-(m-2)\tau)$,	Diffuse horizontal radiation at the time $(t+\hat{h})$
$\cdots$, $a_{PV}(t-\tau)$, $a_{PV}(t)$	Global horizontal radiation at the time $(t+\hat{h})$
with $\tau=5$, $m=9$	Wind speed at the time $(t+\hat{h})$

In Case I, $u_1(t) = \left[a_{PV}(t-(m-1)\tau), a_{PV}(t-(m-2)\tau), \ldots, a_{PV}(t-\tau), a_{PV}(t)\right]u_2(t)$; includes diffuse horizontal radiation, global horizontal radiation, and weather temperature at the time $\left(t+\hat{h}\right)$. Here, the embedding dimension of PV power time series m equals to 8, and the time delay τ equals to 5; $a_{PV}(t)$ represents the PV output power at time t; $a_{PV}(t-(m-1)\tau)$ corresponds to the PV output power at the time $(t-35)$. As the sampling resolution is 15 min/sample, $(t-\tau)$ corresponds to the PV power data at 75 min before time t, and $(t-35)$ corresponds to the PV power data at 525 min before time t. The input variables of the proposed Chaotic-LSTM network are listed in Table 1. In Case II,$u_1(t) = [a_{PV}(t-(m-1)\tau), a_{PV}(t-(m-2)\tau), \ldots, a_{PV}(t-\tau), a_{PV}(t)]$; $u_2(t)$ includes diffuse horizontal radiation, global horizontal radiation, and wind speed at the time $\left(t+\hat{h}\right)$. The embedding dimension is set as $m=9$ and the time delay is $\tau=5$. The input variables of the proposed Chaotic-LSTM network are shown in Table 2; a_{PV}. thus corresponds to the PV power at time $(t-40)$. As the sampling interval is 15 min/sample, $(t-40)$ corresponds to the PV power data at 600 min before time t.

The pre-processed data may contain attributes with a mixture of scales for various quantities such as PV power, temperature, and radiation among others. In order to eliminate the impacts of various influencing factors on PV output power due to different scales, data normalization is performed, which is achieved through re-scaling the data so that all values range within 0 and 1. The normalization process is described as follows:

$$u_{norm} = \frac{u - u_{\min}}{u_{\max} - u_{\min}} \tag{22}$$

3.3 Forecast Model Training

The training process of the LSTM model typically includes splitting the dataset, determining the model parameters, and finally deriving a mapping relationship $G(\cdot)$ between the input and the output variables by checking if the termination criteria are satisfied. To derive G, the data is pre-processed as described in Sect. 3.2.3, and then divided into training, testing, and validation datasets. The training samples include 50% of the total number of samples, while the other datasets (for testing and validation) each share 25% of the samples. These three datasets are used to first train the forecast engine parameters, then validate the model's state and convergence during the training process, and finally evaluate the generalization ability of the model. The time windows for datasets are shown in Table 3. The training dataset is employed to train the proposed forecast model parameters, and the validation dataset is employed to validate the model parameters. In the training process of Case I and Case II, the network configuration and the training parameters are given in Table 4. In Case I, the input dimension is $(m + 3)$, which is equal to 11. In Case II, the input

Table 3 Dataset for the proposed Chaotic-LSTM network in Case I and Case II

Dataset	Time window	
	Case I	Case II
Training Dataset	2018/2/20 00:00-2018/8/20 00:00	2019/1/1 00:00-2019/7/1 00:00
Testing Dataset	2018/1/1 00:00-2018/2/20 00:00 2018/11/20 00:00-2018/12/31 00:00	2019/7/1 00:00-2019/10/1 00:00
Validation Dataset	2018/8/20 00:00-2018/11/20 00:00	2019/10/1 00:00-2019/12/31 00:00:00

Table 4 Training parameters for the proposed Chaotic-LSTM network Case I and Case II

Parameters	Values	
	Case I	Case II
Input Dimension	11	12
Output Dimension	1	1
LSTM Layers	1	1
Learning Rate	0.0625	0.0550
Hidden Units	38	30
Number of Training Epochs	1200	1000
Optimization Method	Adam	Adam

Table 5 Comparison of the prediction accuracy for different forecast models (Case I)

Forecast Time Horizon	Method	MAE(kW)	MAPE($\times$100%)	NRMSE
One-hour-ahead	Chaotic-LSTM	64.9432	0.0361	0.0740
	BP	134.3621	0.0746	0.1184
	ELMAN	77.7157	0.0432	0.0771
	RBF	110.5939	0.0614	0.1050
Four-hour-ahead	Chaotic-LSTM	81.0317	0.0450	0.0852
	BP	146.9321	0.0816	0.1376
	ELMAN	134.4136	0.0747	0.1259
	RBF	151.8208	0.0843	0.1237
One-day-ahead	Chaotic-LSTM	84.7732	0.0471	0.0871
	BP	161.8418	0.0899	0.1537
	ELMAN	87.2336	0.0485	0.0893
	RBF	115.8590	0.0644	0.1005
Four-day-ahead	Chaotic-LSTM	100.2353	0.0557	0.1010
	BP	219.5905	0.1220	0.2067
	ELMAN	108.6843	0.0604	0.1049
	RBF	131.4470	0.0730	0.1120

dimension is $(m + 3)$, which is equal to 12. In tuning the neural network, manual adjustment of the hyperparameters is very time-consuming and impractical. Two common methods in searching for optimal hyperparameters are commonly used:

1. Grid Search: This is an exhaustive approach to list all different parameter combinations and to determine the structure with the best performance.
2. Random Search: This is to extract a certain number of candidate combinations from a parameter space with a specific distribution. Commonly used random search algorithms include simulated annealing algorithm, genetic algorithm, evolution strategy, etc.

In this chapter, since the number of network hyperparameters is small, the grid search approach is used to determine the combination of hyperparameters. When the number of hyperparameters increases, the computational complexity of the grid search will increase exponentially, which might lead to a recommendation on the use of the random search instead. Through grid search, it can be observed that for Case I, when the number of hidden layers is 1, the number of hidden layer units is 38, the number of training iterations is 1200, and the initial learning rate is 0.0625; hence, the proposed Chaotic-LSTM model can achieve desirable prediction accuracy. For Case II, it can be seen that the prediction accuracy of the proposed Chaotic-LSTM model is better when the number of hidden layers is 1, the number of hidden layer units is 30, the number of training iterations is 1000, and the initial learning rate is 0.055. Three evaluation metrics are used to estimate the accuracy of the trained

Chaotic-LSTM model compared with the state-of-the-art models of BP, RBF, and ELMAN Neural Networks.

The three evaluation metrics are the normalized root mean squared error (NRMSE), the mean absolute error (MAE), and mean absolute percentage error (MAPE), expressed in the following:

$$\begin{cases} NRMSE = \sqrt{\dfrac{1}{K}\displaystyle\sum_{t-1}^{K}\left(\dfrac{y_{act}(t+h) - y_p(t+h)}{y_{act}(t+h)}\right)^2} \\ MAE = \dfrac{1}{K}\displaystyle\sum_{t-1}^{K}\left|y_{act}(t+h) - y_p(t+h)\right| \\ MAPE = \dfrac{1}{K}\displaystyle\sum_{t-1}^{K}\dfrac{\left|y_{act}(t+h) - y_p(t+h)\right|}{y_{act}(t+h)} \times 100\% \end{cases} \tag{23}$$

where NRMSE measures the deviation between the actual and the forecasted values, while MAE determines the model accuracy. MAPE is the average percentage error (0%–100%), which is used to evaluate whether the forecast value is higher or lower than the actual value.

3.4 Simulations and Analysis: Case I

Following the training process of the proposed PV power forecast model based on Chaotic-LSTM, different forecast time horizons are studied on the testing dataset. In the following analyses, the sample interval for testing the data is selected as 15 min, and the test dataset covers 8 days of PV power time series (from 4:30:00 on Jan 22, 2018 to 4:30:00 on Jan 30, 2018). The PV power forecast results from the four studied methods are illustrated in Figs. 12, 13, 14 and 15, where the forecast time horizon is, respectively, one-hour, one-day, four-hour, and four-day-ahead. In each figure, (a)–(d), respectively, represented the forecast outcome of the proposed Chaotic-LSTM method, BP network, Elman network, and RBF network. Table 5 reports the three evaluation metrics of NRMSE, MAE, MAPE utilized to assess the performance of the proposed Chaotic-LSTM method compared to the other three methods.

From Figs. 12, 13, 14 and 15, the following observations are made:

1. For comparison, test samples cover four days in each figure. One-hour, one-day, four-hour, and four-day-ahead PV output power predictions for the tested period are shown in Figs. 12, 13, 14 and 15, respectively. The corresponding realized PV output power curves used for comparison with the prediction results are also different due to the different forecast time horizons. For instance, the observed curve covers the PV power data from 5:30:00 on Jan 22, 2018 to 5:30:00 on Jan 26, 2018 in Fig. 12, and covers from 8:30:00 on Jan 22, 2018 to 8:30:00 on Jan

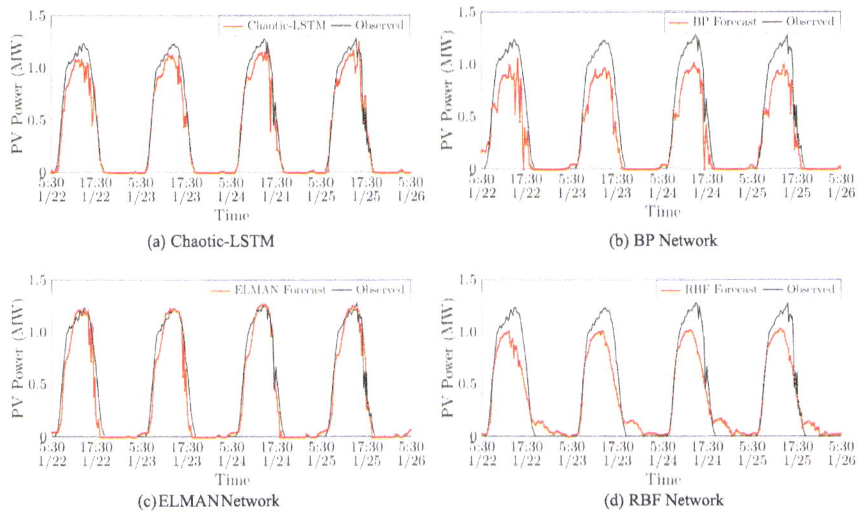

Fig. 12 One-hour-ahead PV power forecast results of different forecast models (Case I)

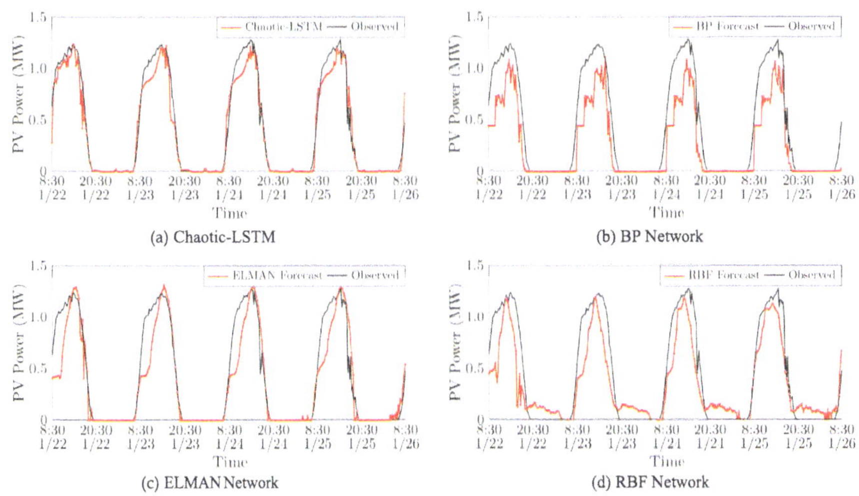

Fig. 13 Four-hour-ahead PV power forecast results of different forecast models (Case I)

26, 2018 in Fig. 13. In Fig. 15, the observed and forecast curve covers the PV power data from 4:30:00 on Jan 26, 2018 to 4:30:00 on Jan 30, 2018.

2. In Figs. 12, 13, 14 and 15, the output of the PV power forecast model versus the realized PV output power are, respectively, highlighted with red and black lines.
3. From Fig. 12, when the proposed Chaotic-LSTM model is applied, the prediction follows the black line almost perfectly, highlighting the fact that the

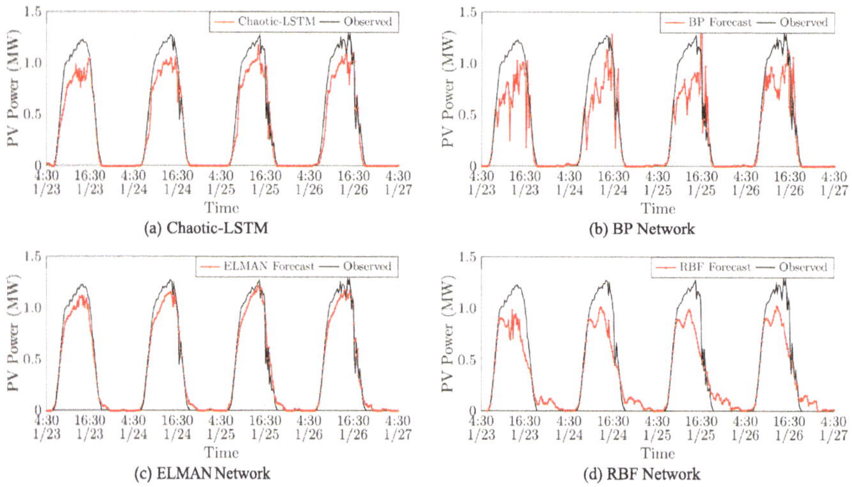

Fig. 14 One-day-ahead PV power forecast results of different forecast models (Case I)

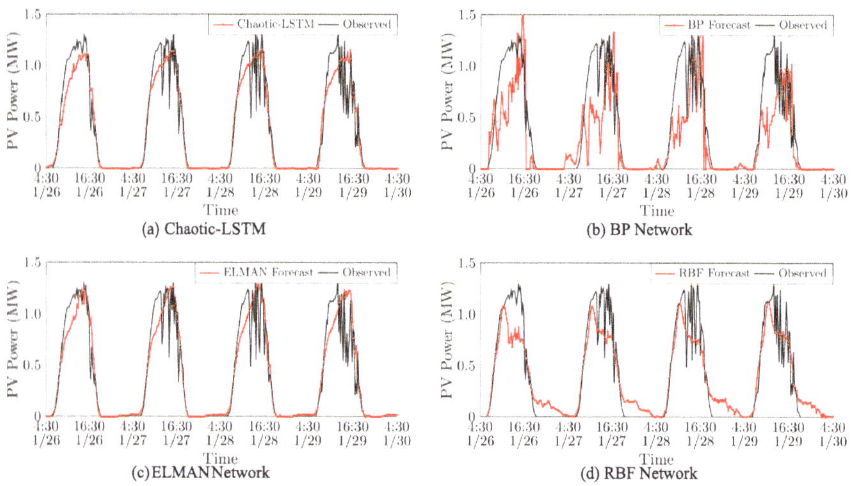

Fig. 15 Four-day-ahead PV power forecast results of different forecast models (Case I)

proposed model outperforms the other three models for one-hour-ahead PV output power forecast. The Elman network can also achieve a promising performance, while the BP network reveals the worst performance on tracking the peaks in the actual observation curve. When the actual observation value slightly changes, the RBF prediction curve reveals a large fluctuation.

4. From Fig. 13, the proposed Chaotic-LSTM model still offers a promising prediction, while its accuracy is a bit lower than the one observed for one-hour-ahead prediction. The other methods can track the overall trend of the observed curve

while presenting poor tracking ability when the observed PV power drastically changes. The RBF network, in particular, has a poor tracking ability when the curve slightly changes where its prediction reflects a large deviation compared to the observed data.

5. From Fig. 14 to 15, the PV output power of the proposed Chaotic-LSTM forecast model is closer to the observed PV power data curve. The BP network model has the worst performance mainly due to its inability to track large and drastic changes in the curve.

From Table 5, the following observations are made:

1. Table 5 presents and compares the NRMSE, MAE, and MAPE metrics corresponding to the four forecast models over different time horizons. When the forecast time horizon is set to one-hour-ahead, the proposed Chaotic-LSTM model shows an obvious advantage over the other models with the best NRMSE of 0.0740, MAE of 64.9432 kW, and MAPE of 3.61%. The worst performance is attributed to the BP method, whose NRMSE is found to be 0.1184 with MAE of 134.3621 kW and MAPE of 7.46%.
2. When the forecast time horizon is set to four-hour-ahead, the RBF method reveals the largest NRMSE of 0.1237, MAE of 151.8208 kW, and MAPE of 8.43%. Compared with the other three models, the proposed Chaotic-LSTM model has the lowest NRMSE, MAE, and MAPE values, with the quantified values of 0.0852, 81.0317 kW, and 4.50% for NRMSE, MAE, and MAPE, respectively.
3. With the expansion of the forecast time horizon, the NRMSE, MAE, and MAPE metrics for each forecast model are found larger than those for the one-hour-ahead forecast. The suggested Chaotic-LSTM approach could achieve the best results of the four forecast models, and the corresponding NRMSE, MAE, and MAPE metrics are the lowest. Therefore, the forecast results provided by the proposed Chaotic-LSTM model are the closest to the actual realized PV output power values.

Based on the conducted tests and numerical analyses, one can conclude the following:

- Under the same forecast time horizon, the proposed Chaotic-LSTM model offers the best performance of the four forecast methods at generalizing the data. The Chaotic-LSTM method can successfully follow the observed PV output power, revealing the minimum values of the NRMSE, MAE, and MAPE metrics.
- With the expansion of the forecast time horizon, the forecast curves corresponding to each model gradually deviate from the actual curve, and the NRMSE, MAE, and MAPE metrics increase consequently. The proposed Chaotic-LSTM model relatively outperforms the other three models, especially when the observed data significantly fluctuates, while the BP and RBF neural network models demonstrated poor tracking performances.

3.5 Simulations and Analysis: Case II

To further demonstrate the effectiveness and robustness of the proposed Chaotic-LSTM forecast engine, a second case study is reported herein in Case II. Similar to Case I, four different forecast time horizons are considered on the testing dataset. The performance of the proposed Chaotic-LSTM model is compared with those of the BP, RBF, and Elman neural networks. With the PV power forecast from 7:00:00 on Sep 16, 2019 to 7:00:00 on Sep 25, 2019 implemented, the results on the four forecast time horizons for the above four methods are illustrated in Figs. 16, 17, 18 and 19, where the forecast time horizon is, respectively, one-hour, one-day, four-hour, and four-day-ahead. In each figure, (a)–(d), respectively, represented the forecast outcome of the proposed Chaotic-LSTM method, BP network, Elman network, and RBF network. The realized PV output power versus the output of the PV power forecast model are, respectively, highlighted with black and red lines. Table 6 reports the three evaluation metrics of NRMSE, MAE, and MAPE utilized to assess the performance of the proposed Chaotic-LSTM method compared to the other three methods.

For comparison, test samples cover nine days in each figure. One-hour, one-day, four-hour, and four-day-ahead PV power predictions for the tested period are, respectively, shown in Figs. 16, 17, 18 and 19. The following observations are made from the presented results:

1. From Figs. 16 to 17, when the proposed Chaotic-LSTM model is applied, the prediction follows the black line almost perfectly. Other methods can also achieve promising performance, but it can be seen that the prediction curves reveal some fluctuations for other methods when the actual output power changes rapidly (from 9/20 morning to 9/22 morning and 9/23 morning to 9/24 morning), indicating that the proposed model is superior to the others.

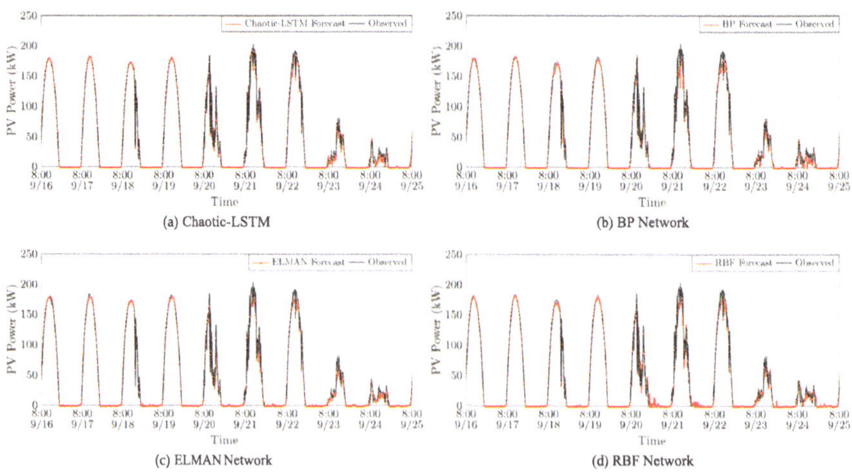

(a) Chaotic-LSTM (b) BP Network (c) ELMAN Network (d) RBF Network

Fig. 16 One-hour-ahead PV output power forecast results of different forecast models (Case II)

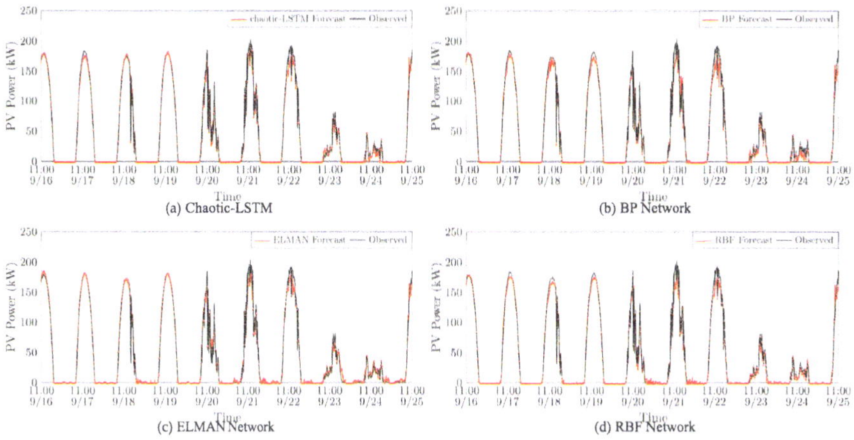

Fig. 17 Four-hour-ahead PV output power forecast results of different forecast models (Case II)

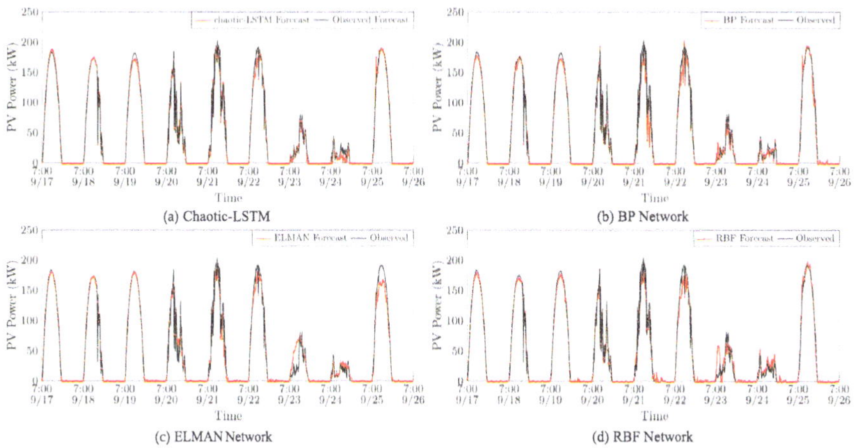

Fig. 18 One-day-ahead PV output power forecast results of different forecast models (Case II)

2. From Fig. 18 to 19, the proposed Chaotic-LSTM model offers a promising prediction. The other methods can track the overall trend of the observed curve while presenting poor tracking ability when the observed PV power drastically changes for the ELMAN network and RBF network. The BP network also reflects a slight deviation compared to the observed data, such as in Fig. 17b at 9/18, 9/21, 9/22 noon and Fig. 18b on 9/25 afternoon.
3. Focusing on the ELMAN and RBF Networks, the forecasted PV power is always found to be nonzero, while the actual PV power is zero at times. Similarly, for the BP network under a four-day-ahead forecast horizon (Fig. 19b), the nonzero forecast can be observed in 9/27. However, the proposed Chaotic-LSTM model could follow the observed curve accurately.

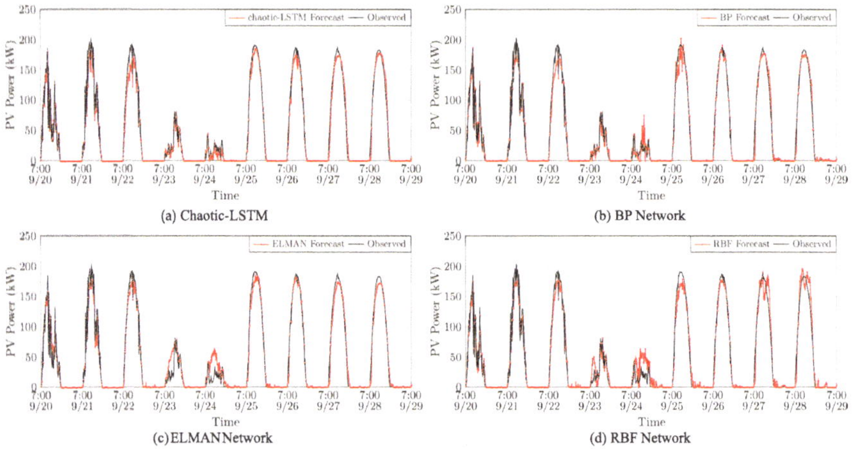

Fig. 19 Four-day-ahead PV output power forecast results of different forecast models (Case II)

Table 6 Comparison of the prediction accuracy for different forecast models (Case II)

Forecast time horizon method		MAE (kW)	MAPE (×100%)	NRMSE
One-hour-ahead	Chaotic-LSTM	1.3070	0.0242	0.0743
	BP	2.3218	0.0515	0.1343
	ELMAN	2.5250	0.0917	0.1613
	RBF	2.0155	0.1013	0.2789
Four-hour-ahead	Chaotic-LSTM	1.7946	0.0292	0.0803
	BP	3.0513	0.0560	0.1271
	ELMAN	2.9385	0.1420	0.3020
	RBF	3.0346	0.0887	0.2014
One-hour-ahead	Chaotic-LSTM	2.4450	0.0462	0.1111
	BP	2.6616	0.0540	0.1530
	ELMAN	3.2117	0.0945	0.2440
	RBF	3.2319	0.1457	0.3565
Four-hour-ahead	Chaotic-LSTM	2.8726	0.0486	0.1169
	BP	3.3549	0.0811	0.1955
	ELMAN	3.5924	0.1335	0.3590
	RBF	4.1966	0.1443	0.4120

From Table 6, the following observations are made:

1. Table 6 presents and compares the NRMSE, MAE, and MAPE metrics corresponding to the four forecast models over different time horizons. For the same forecast time horizon, the proposed Chaotic-LSTM model shows an obvious

advantage over the other models. When the forecast time horizon is set to one-hour-ahead, the best NRMSE of 0.0743, MAE of 1.307 kW, and MAPE of 2.42% are reported for the proposed Chaotic-LSTM model.

2. With the expansion of the forecast time horizon, the NRMSE, MAE, and MAPE metrics for the proposed Chaotic-LSTM model gradually increase compared to those for the one-hour-ahead forecast. The suggested Chaotic-LSTM approach can still achieve the best results of the four forecast models, and the corresponding NRMSE, MAE, and MAPE metrics are the lowest. For example, when the forecast time horizon is set to four-day-ahead, the RBF method reveals the largest NRMSE of 0.4120, MAE of 4.1966 kW, and MAPE of 14.43%; on the other hand, the NRMSE, MAE, and MAPE values corresponding to the proposed Chaotic-LSTM model are the lowest among the four models, with the quantified values of 0.1169, 2.8726 kW, 4.86% for NRMSE, MAE, and MAPE, respectively.

Based on the conducted tests and numerical analysis, one can conclude the following:

- Under the same forecast time horizon, the proposed Chaotic-LSTM model offers the best performance of the four forecast methods at generalizing the data. The Chaotic-LSTM method can successfully follow the observed PV output power, revealing the minimum values of the NRMSE, MAE, and MAPE metrics.
- With the expansion of the forecast time horizon, the NRMSE, MAE, and MAPE metrics increase, for all four forecasting methods.
- The proposed Chaotic-LSTM model relatively maintains the highest forecasting accuracy and outperforms the other three models, especially when the observed data significantly fluctuates.

3.6 Robustness Test

To assess the effectiveness and predictability of the proposed approach, a robustness test is performed. The test is to assess whether the proposed model and the other three can provide forecast results that can still well fit the actual PV power generation data, under the conditions that the input dataset are polluted with undesired interferences. In this Section, white Gaussian noises are added to the input dataset. The signal-noise ratio (SNR) of the polluted input dataset ranges from 10 dB to 60 dB. The robustness test period covers an interval from 8:00 on Sep 17, 2019 to 18:00 on Sep, 26, 2019. For instance, the robustness test results—in terms of the error plots representing the forecast deviation from the observed actual PV power output—under a 10 dB SNR condition are illustrated in Fig. 20. The noise is added from 8:00 on Sep, 18, 2019 to 18:00 on Sep 19, 2019, which falls within the area enclosed by the red dotted line in each figure. For comparison, one-hour, one-day, four-hour, and four-day-ahead PV power deviations from the observed value are, respectively, shown in Fig. 20a–d.

The following observations can be made:

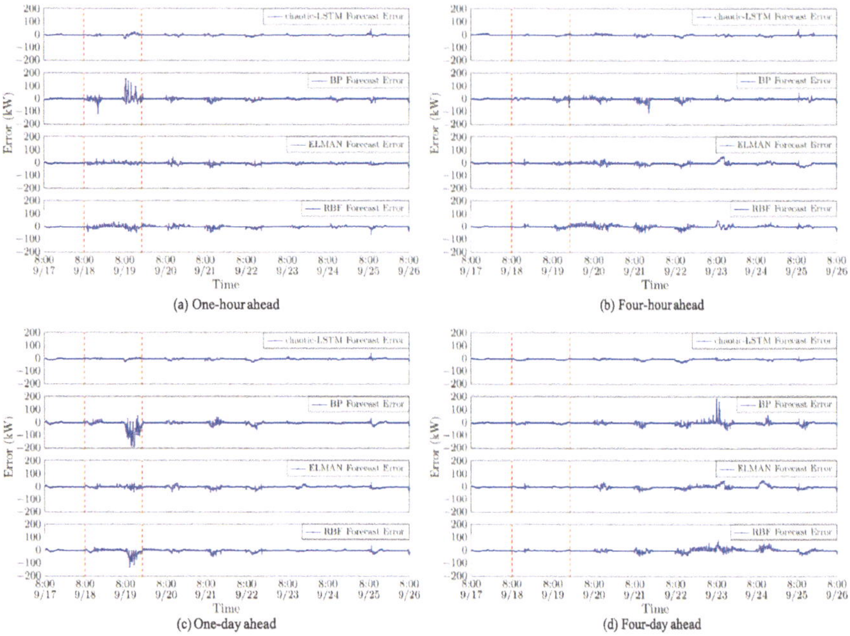

Fig. 20 Forecast error: robustness test under a 10 dB interference scenario

1. From Fig. 20a, all curves show the one-hour-ahead PV power deviations from the observed values during the interfered period. When the 10 dB noise is added to the raw PV power data and one hour after 8:00 on 9/18, it can be seen that there are obvious deviations in the prediction error curves of PV power generation by BP, Elman, and RBF methods. In particular, from 8:00 on 9/18 to 19:00 on 9/19, the deviations for the BP method and the RBF method reveal significant distortions, where the maximum deviation reaches about 200 kW for the BP method. This indicates that the BP method has the worst anti-interference ability. The maximum deviation reaches about 200 kW for the BP method. On the other hand, compared to other methods, the proposed Chaotic-LSTM model offers the least deviation around the observed values during the entire test period, indicating that it has the best forecast accuracy. Also, once the interfered interval is passed, the deviations for other methods are found still larger than those of the proposed Chaotic-LSTM.
2. From Fig. 20b, all curves show the four-hour-ahead PV power output deviations from the observed values when the 10 dB noise is added from 8:00 on 9/18 to 18:00 on 9/19. It can be observed that the PV power forecast error of the proposed Chaotic-LSTM model has the least value than other models, reflecting its high accuracy. Meanwhile, it can be seen that the prediction curves reveal some fluctuations for other methods (for instance from 8:00 on 9/19 to 18:00 on 9/19). Once the noise is removed, one can see that from 18:00 on 9/19 to

8:00 on 9/26, there are larger deviations reported for other methods, driven by the effects of their short-term memory characteristics. However, the suggested Chaotic-LSTM method still reveals the best performance as it can memorize long and short-term information on PV power data and effectively processes the intrinsic dynamic characteristics of PV arrays.

3. From Fig. 20c, all curves show the one-day-ahead PV power output deviations from the observed values when the noise is added from 8:00 on 9/18 to 18:00 on 9/19. One day following the 8:00 on 9/18, 2019, it can be seen that there are large deviations between the observed and the forecasted values for the BP and RBF methods. The maximum deviation reaches about 200 kW for the BP method, and above 100 kW for the RBF method. Once the interference disappears, there are still larger deviations for other methods, while the proposed Chaotic-LSTM model has the least deviation around the observed values.
4. Fig. 20d shows that in a four-day-ahead forecast, all PV power forecast outputs deviate from the actual observed values about 4 days (around 10:00 on 9/22) following the interference impacts on 9/18. Also, it can be seen that there are larger deviations between the forecasted and observed values for the other three methods. The maximum deviation reaches about 200 kW for BP, about 100 kW for RBF, and about 50 kW for ELMAN methods. During the entire testing period, however, the forecast results provided by the proposed Chaotic-LSTM model are still the closest to the actual PV output power values.

Figure 21 and Table 7 present 7 SNR scenarios and correspondingly report the three evaluation metrics of NRMSE, MAE, and MAPE. Figure 21a–c presents the three metrics of MAE, MAPE, and RMSE, respectively, for four forecast models under four different forecast horizons and different SNR scenarios. From the results presented in Fig. 21, the following observations are made:

1. Under the same forecast horizon (one-hour-ahead) in Fig. 21a and SNR = 60 dB, the proposed Chaotic-LSTM method offers the lowest NRMSE, MAE, and MAPE. As the SNR decreases, the three evaluation metrics degrade for the BP, ELMAN, and RBF networks, similar to the suggested Chaotic-LSTM method. The proposed method, however, outperforms the others. The RBF model has the worst performance mainly due to its inability to resist SNR.
2. When the forecast time horizon increases to four-day-ahead, the proposed Chaotic-LSTM model still offers the lowest NRMSE, MAE, and MAPE, while the corresponding values for the RBF network reach significantly higher values compared to the other methods.
3. Under the same forecast time horizon and when SNR = 0 dB across all the forecast models, the NRMSE, MAE, and MAPE values are significantly higher than those of the other SNR values; that is, lower SNR can compromise the forecast accuracy.

From Table 7, the following observations are made:

1. With the expansion of the forecast horizon, the evaluation metrics vary under the same SNR scenario for each forecasting method.

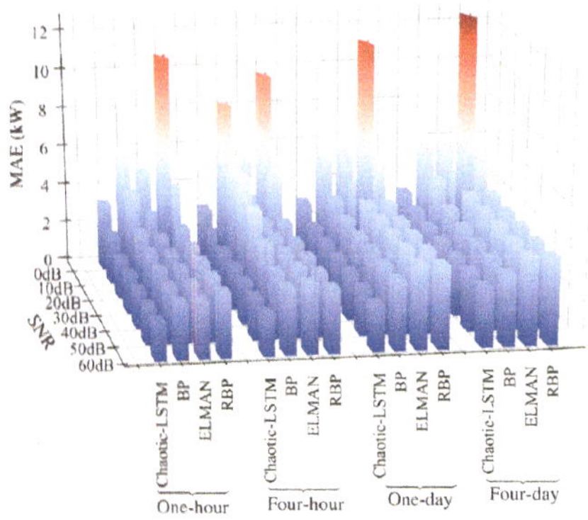

(b) MAE

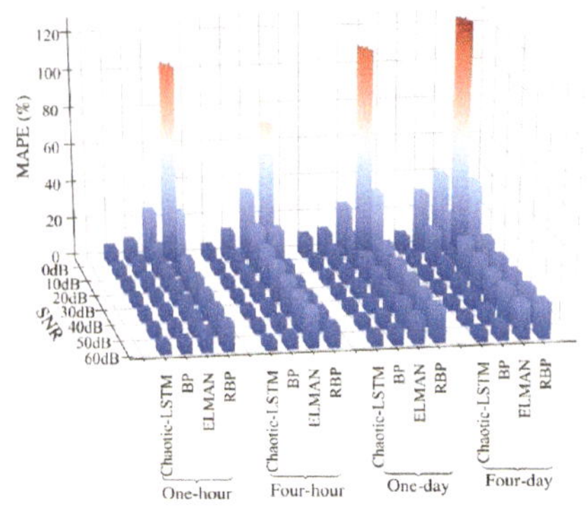

(b) MAPE

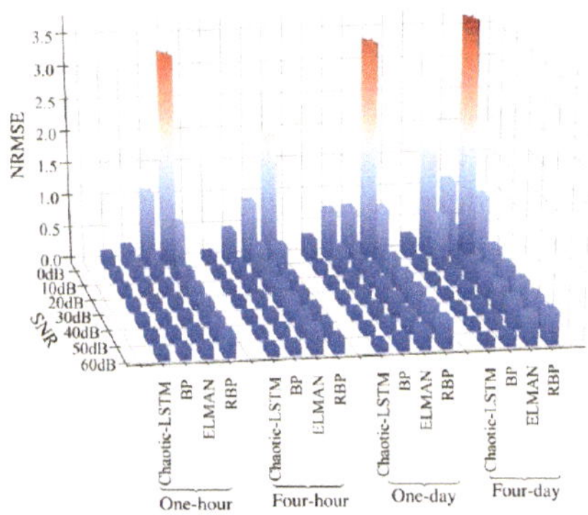

(c) NRMSE

Fig. 21 Evaluation metrics comparison: robustness tests under different SNR scenarios

2. Under the same forecast time horizon, as the SNR increases, the evaluation metrics of NRMSE, MAE, and MAPE decrease consistently across all studied models. For example, the results for the one-hour-ahead forecast time horizon is further elaborated here. When a noise of SNR = 0 dB is added to the raw PV power data, the evaluation metrics of NRMSE, MAE, and MAPE on the Chaotic-LSTM method are, respectively, recorded to be 0.204, 3.405 kW, and 8.37%. Analogously, when a noise of SNR = 60 dB is added to the raw PV power data, the three metrics of NRMSE, MAE, and MAPE on the Chaotic-LSTM method are observed to be 0.1486, 1.9758 kW, and 5.53%, respectively. For the RBF method, three metrics of NRMSE, MAE, and MAPE are, respectively, 3.2797, 10.8993 kW, and 105.31% when SNR = 0 dB; and the three metrics of NRMSE, MAE, and MAPE are found to be 0.3657, 3.2983 kW, and 14.56%, respectively, when SNR = 60 dB.
3. Under the same forecast time horizon and the same SNR scenario, the proposed Chaotic-LSTM method demonstrates a performance with the lowest NRMSE,

Table 7 Prediction accuracy comparison: robustness tests under different SNR scenarios

Forecast time horizon	Method	Metrics	SNR						
			0db	10db	20db	30db	40db	50db	60db
One-hour-ahead	Chaotic-LSTM	MAE	3.4050	2.4673	2.0222	2.0055	1.9790	1.9770	1.9758
		MAPE	0.0837	0.0648	0.0567	0.0556	0.0554	0.0553	0.0553
		NRMSE	0.2040	0.1600	0.1500	0.1488	0.1487	0.1486	0.1486
	BP	MAE	6.1056	4.6572	3.3046	3.0744	2.9228	2.8868	2.9013
		MAPE	0.1154	0.0960	0.0787	0.0719	0.0694	0.0690	0.0692
		NRMSE	0.3046	0.2982	0.2139	0.1772	0.1725	0.1722	0.1729
	ELMAN	MAE	5.0232	3.4529	3.0376	2.9704	2.9677	2.9653	2.9650
		MAPE	0.2792	0.1574	0.1177	0.0.1722	0.1121	0.11170	0.11170
		NRMSE	1.1729	0.3875	0.2001	0.1916	0.1918	0.1913	0.1913
	RBF	MAE	10.8933	4.8698	3.4713	3.3051	3.3014	3.3003	3.2983
		MAPE	1.0531	0.3400	0.1602	0.1461	0.1460	0.1456	0.1456
		NRMSE	3.2797	0.8839	0.3772	0.3667	0.3659	0.3656	0.3657
Four-hour-ahead	Chaotic-LSTM	MAE	2.9463	2.3914	2.2018	2.1772	2.1686	2.1635	2.1612
		MAPE	0.0690	0.0550	0.0524	0.0515	0.0513	0.0512	0.0511
		NRMSE	0.1567	0.1345	0.1311	0.1291	0.129	0.1288	0.1288
	BP	MAE	8.3589	6.6062	5.9708	5.0164	4.0117	3.4225	3.3942
		MAPE	0.1491	0.1055	0.0982	0.0843	0.0763	0.072	0.0719
		NRMSE	0.5186	0.2318	0.2171	0.183	0.1651	0.1594	0.1594
	ELMAN	MAE	5.4853	4.1435	3.6616	3.595	3.5883	3.588	3.5878
		MAPE	0.3661	0.2459	0.2033	0.1979	0.1978	0.1976	0.1977

(continued)

Table 7 (continued)

Forecast time horizon	Method	Metrics	SNR						
			0db	10db	20db	30db	40db	50db	60db
		NRMSE	0.9657	0.4847	0.3957	0.3908	0.3911	0.3909	0.391
	RBF	MAE	9.7831	5.2341	3.5591	3.3376	3.3275	3.3133	3.314
		MAPE	0.7265	0.2302	0.1206	0.1106	0.1113	0.1112	0.1113
		NRMSE	2.0053	0.4849	0.2496	0.2361	0.2365	0.2364	0.2364
One-day ahead	Chaotic-LSTM	MAE	3.1177	2.5319	2.4774	2.4255	2.4277	2.426	2.4255
		MAPE	0.1112	0.0696	0.0598	0.059	0.0587	0.0587	0.0587
		NRMSE	0.3555	0.1632	0.1354	0.1346	0.1341	0.1341	0.1342
	BP	MAE	5.2901	3.931	3.6005	3.5402	3.5477	3.5674	3.5644
		MAPE	0.1332	0.1155	0.0855	0.081	0.0787	0.0786	0.0785
		NRMSE	0.7905	0.3989	0.231	0.2059	0.2026	0.2028	0.2026
	ELMAN	MAE	5.7935	4.6946	4.3807	4.3661	4.3574	4.3549	4.355
		MAPE	0.2719	0.1845	0.1496	0.1437	0.1424	0.142	0.142
		NRMSE	0.8133	0.434	0.335	0.3285	0.3283	0.328	0.3281
	RBF	MAE	11.2828	5.8277	4.3685	4.2249	4.2547	4.2568	4.2546
		MAPE	1.1093	0.4137	0.2024	0.186	0.1879	0.1875	0.1876
		NRMSE	3.3791	1.0113	0.47	0.4565	0.4575	0.4574	0.4574
Four-day-ahead	Chaotic-LSTM	MAE	3.4758	3.2360	3.0994	3.0847	3.0702	3.0601	3.0696
		MAPE	0.0971	0.0731	0.0584	0.0577	0.0572	0.0572	0.0572
		NRMSE	0.3238	0.1736	0.1292	0.1320	0.1303	0.1299	0.1299

(continued)

Table 7 (continued)

Forecast time horizon	Method	Metrics	SNR						
			0db	10db	20db	30db	40db	50db	60db
	BP	MAE	6.3661	4.8156	4.0000	3.7765	3.6869	3.6798	3.6793
		MAPE	0.3298	0.1938	0.1073	0.0922	0.0864	0.0862	0.0863
		NRMSE	1.6792	0.8486	0.2785	0.2268	0.2158	0.2163	0.2164
	ELMAN	MAE	6.3661	5.0676	4.6315	4.5714	4.5517	4.5486	4.5495
		MAPE	0.4264	0.2761	0.2102	0.1980	0.1944	0.1938	0.1939
		NRMSE	1.2073	0.6457	0.4785	0.4651	0.4635	0.4635	0.4636
	RBF	MAE	12.4848	6.3429	4.6825	4.5267	4.5183	4.5101	4.5108
		MAPE	1.2463	0.4524	0.2194	0.1949	0.1930	0.1922	0.1923
		NRMSE	3.6892	1.1644	0.5709	0.5165	0.5169	0.5168	0.5170

MAE, and MAPE metrics, outperforming the other methods. The worst performance is attributed to the RBF method with the largest NRMSE, MAE, and MAPE metrics. For instance, when a noise of SNR = 0 dB is added to the raw PV power data, the RBF method applied to a four-day-ahead forecast scenario has resulted in the evaluation metrics of NRMSE, MAE, and MAPE equal to 3.6892, 12.4848 kW, and 124.63%, respectively.

Based on the conducted tests and numerical analyses, one can conclude the following:

- Overall, the proposed Chaotic-LSTM model can forecast the PV power output with promising performance and accuracy when compared to the observed values.
- Under the same data quality but with different forecast time horizons, the proposed Chaotic-LSTM model has the least deviation around the observed values compared to the other three models.
- Under the same forecast time horizon and a lower SNR scenario, the forecast performance would be compromised consistently across all the forecast models, but the proposed Chaotic-LSTM method was able to provide the forecast outcome that could follow the observed values. When the large disturbance disappears, the proposed approach provides a fairly smooth transition to quickly provide outstanding forecast results.
- With the expansion of the forecast time horizon, the deviation between the forecasted and observed values varies, so do the three evaluation metrics of NRMSE, MAE, and MAPE. The proposed Chaotic-LSTM model significantly outperforms the other three models, especially when the SNR is low.

4 Conclusion

This chapter proposed a deep learning-aided PV power forecast model, which ensembles the chaos theory and the LSTM network. The phase space reconstruction technology from the chaos theory reveals the intrinsic dynamic characteristics of the PV power variations; meanwhile, the correlation analysis was applied to extract the external factors influencing the PV arrays. With the phase space reconstruction and correlation analysis applied, the most relevant features for the PV output power forecast were extracted from the dataset and selected as the input variables of the proposed forecast model. The LSTM neural network is used to achieve a nonlinear mapping between the input variables and the output variable (PV power). The proposed Chaotic-LSTM algorithm revealed a promising learning ability of the internal characteristics, while was simultaneously adaptive to the influences engendered from the external meteorological conditions. Real-world data was utilized to test and verify the performance of the proposed approach. With extensive numerical analyses and comparisons with the Elman, BP, and RBF network models through evaluation metrics (MAE, MAPE, and NRMSE), the proposed model consistently demonstrated an outstanding performance with improved prediction accuracy under four different forecast time horizons. Furthermore, a robustness test was taken as an evaluation approach to demonstrate the effectiveness and predictability of the proposed method. Under different forecast horizons, the proposed Chaotic-LSTM satisfactorily revealed excellent robustness.

References

1. Eftekharnejad S, Heydt GT, Vittal V (2015) Optimal generation dispatch with high penetration of photovoltaic generation. IEEE Transactions on Sustainable Energy 6(3):1013–1020
2. Asrari A, Wu TX, Ramos B (2017) A hybrid algorithm for short-term solar power prediction—sunshine state case study. IEEE Trans Sustain Energy 8(2):582–591
3. Akhter MN, Mekhilef S, Mokhlis H, Mohamed Shah N (2019) Review on forecasting of photovoltaic power generation based on machine learning and metaheuristic techniques. IET Renew Power Generat 13(7):1009–1023
4. Nespoli A, Ogliari E, Leva S, Massi Pavan A, Mellit A, Lughi V, Dolara A (2019) Day-ahead photovoltaic forecasting: a comparison of the most effective techniques. Energies 12(9):1621
5. Yadav HK, Pal Y, Tripathi MM (2015) Photovoltaic power forecasting methods in smart power grid. In: 2015 annual IEEE India conference (INDICON), pp 1–6. IEEE
6. Wan C, Zhao J, Song Y, Xu Z, Lin J, Hu Z (2015) Photovoltaic and solar power forecasting for smart grid energy management. CSEE J Power Energy Syst 1(4):38–46
7. Ceci M, Corizzo R, Fumarola F, Malerba D, Rashkovska A (2017) Predictive modeling of PV energy production: how to set up the learning task for a better prediction? IEEE Trans Industr Inf 13(3):956–966
8. Mondol JD, Yohanis YG, Norton B (2007) Comparison of measured and predicted long term performance of grid a connected photovoltaic system. Energy Convers Manage 48(4):1065–1080
9. Eltamaly AM (2018) Performance of MPPT techniques of photovoltaic systems under normal and partial shading conditions. In: Advances in renewable energies and power technologies. Elsevier, pp 115–161
10. Singh B, Pozo D (2019) A guide to solar power forecasting using ARMA models. In: 2019 IEEE PES innovative smart grid technologies Europe (ISGT-Europe), pp 1–4
11. Chu Y, Urquhart B, Gohari SMI, Pedro HTC, Kleissl J, Coimbra CFM (2015) Short-term reforecasting of power output from a 48 MWe solar PV plant. Solar Energy 112:68–77
12. Payman D, Bei Z, Tatjana D, Mladen K (2018) Predictive risk analytics for weather-resilient operation of electric power systems. IEEE Trans Sustain Energy 10(1):3–15
13. Zhang B, Dehghanian P, Kezunovic M (2016) Spatial-temporal solar power forecast through use of Gaussian conditional random fields. In: 2016 IEEE power and energy society general meeting (PESGM), pp 1–5. IEEE
14. Sheng H, Xiao J, Cheng Y, Ni Q, Wang S (2018) Short-term solar power forecasting based on weighted gaussian process regression. IEEE Trans Indus Electron 65(1):300–308
15. Chong L, Rong J, Wenqiang D, Weicheng S, Xiping M (2016) Short-term PV generation forecasting based on weather type clustering and improved GPR model. In: 2016 China international conference on electricity distribution (CICED), pp 1–5, 2016
16. Kanwal S, Khan B, Ali SM, Mehmood CA, Rauf MQ (2018) Support vector machine and gaussian process regression based modeling for photovoltaic power prediction. In: 2018 international conference on frontiers of information and technology, pp 117–122
17. Zhou L, Wu H, Xu T, Mei F, Li Y, Yuan X, Liu H (2017) Ultra-short term hybrid power forecasting model for photovoltaic power station with meteorological monitoring data. In: 2017 32nd youth academic annual conference of chinese association of automation (YAC), pp 452–456

18. Vagropoulos SI, Chouliaras GI, Kardakos EG, Simoglou CK, Bakirtzis AG (2016) Comparison of SARIMAX, SARIMA, modified SARIMA and ANN-based models for short-term PV generation forecasting. In: 2016 IEEE international energy conference (ENERGYCON), pp 1–6
19. Kushwaha V, Pindoriya NM (2017) Very short-term solar PV generation forecast using SARIMA model: a case study. In: 2017 7th international conference on power systems (ICPS), pp 430–435
20. Laopaiboon T, Ongsakul W, Panyainkaew P, Sasidharan N (2018) Hour-ahead solar forecasting program using back propagation artificial neural network. In: 2018 international conference and utility exhibition on green energy for sustainable development (ICUE), pp 1–7
21. Khan I, Zhu H, Khan D, Panjwani MK (2017) Photovoltaic power prediction by cascade forward artificial neural network. In: 2017 international conference on information and communication technologies (ICICT), pp 145–149
22. Kou J, Liu J, Li Q, Fang W, Chen Z, Liu L, Guan T (2013) Photovoltaic power forecasting based on artificial neural network and meteorological data. In: 2013 IEEE international conference of IEEE region 10 (TENCON 2013), pp 1–4
23. Lu HJ, Chen HT, Chang GW (2018) Multi-day-ahead PV power output for improved radial basis function neural network. In: 2018 IEEE innovative smart grid technologies—Asia (ISGT Asia), pp 957–961
24. Park N, Ahn HK (2019) Multi-layer rnn-based short-term photovoltaic power forecasting using IoT dataset. In: 2019 AEIT international annual conference (AEIT), pp 1–5
25. Wilms H, Cupelli M, Monti A (2018) On the necessity of exogenous variables for load, PV and wind day-ahead forecasts using recurrent neural networks. In: 2018 IEEE electrical power and energy conference (EPEC), pp 1–6
26. Khan I, Zhu H, Yao J, Khan D (2017) Photovoltaic power forecasting based on Elman neural network software engineering method. In: 2017 8th IEEE international conference on software engineering and service science (ICSESS), pp 747–750
27. Ben Ammar R, Oualha A (2017) Photovoltaic power forecasting using recurrent neural networks. In: 2017 14th international multi-conference on systems, signals devices (SSD), pp 537–544
28. He H, Hu R, Zhang Y, Zhang Y, Jiao R (2018) A power forecasting approach for PV plant based on irradiance index and LSTM. In: 2018 37th Chinese control conference (CCC), pp 9404–9409
29. Mohamed A-N, Karar M (2019) Accurate photovoltaic power forecasting models using deep LSTM-RNN. Neural Comput Appl 31(7):2727–2740
30. Huang C-J, Kuo P-H (2019) Multiple-input deep convolutional neural network model for short-term photovoltaic power forecasting. IEEE Access 7:74822–74834
31. Jang HS, Bae KY, Park H, Sung DK (2016) Solar power prediction based on satellite images and support vector machine. IEEE Trans Sustain Energy 7(3):1255–1263
32. Shi J, Lee W-J, Liu Y, Yang Y, Wang P (2011) Forecasting power output of photovoltaic system based on weather classification and support vector machine. In: 2011 IEEE industry applications society annual meeting, pp 1–6
33. Qijun S, Fen L, Jialin Q, Jinbin Z, Zhenghong C (2016) Photovoltaic power prediction based on principal component analysis and support vector machine. In: 2016 IEEE innovative smart grid technologies—Asia (ISGT-Asia), pp 815–820
34. Sun Y, Wang F, Mi Z, Sun H, Liu C, Wang B, Lu J, Zhen Z, Li K (2015) Short-term prediction model of module temperature for photovoltaic power forecasting based on support vector machine. In: International conference on renewable power generation (RPG 2015), pp 1–6
35. Monfared M, Fazeli M, Lewis R, Searle J (2019) Fuzzy predictor with additive learning for very short-term PV power generation. IEEE Access 7:91183–91192
36. Liu F, Li R, Li Y, Yan R, Saha T (2017) Takagi-Sugeno fuzzy model-based approach considering multiple weather factors for the photovoltaic power short-term forecasting. IET Renew Power Gener 11(10):1281–1287

37. Yang H, Huang C, Huang Y, Pai Y (2014) A weather-based hybrid method for 1-day ahead hourly forecasting of PV power output. IEEE Trans Sustain Energy 5(3):917–926
38. Yang Z, Cao Y, Xiu J (2014) Power generation forecasting model for photovoltaic array based on generic algorithm and BP neural network. In: 2014 IEEE 3rd international conference on cloud computing and intelligence systems, pp 380–383
39. Caputo D, Grimaccia F, Mussetta M, Zich RE (2010) Photovoltaic plants predictive model by means of ANN trained by a hybrid evolutionary algorithm. In: The 2010 international joint conference on neural networks (IJCNN), pp 1–6
40. Meng X, Xu A, Zhao W, Wang H, Li C, Wang H (2018) A new PV generation power prediction model based on GA-BP neural network with artificial classification of history day. In: 2018 international conference on power system technology (POWERCON), pp 1012–1017
41. Zhong Z, Tan J, Zhang T, Zhu L (2014) PV power short-term forecasting model based on the data gathered from monitoring network. China Commun 11(14):61–69
42. Niu D, Wang K, Sun L, Wu J, Xu X (2020) Short-term photovoltaic power generation forecasting based on random forest feature selection and CEEMD: a case study. Appl Soft Comput 106389
43. Huang C, Chen S, Yang S, Kuo C (2015) One-day-ahead hourly forecasting for photovoltaic power generation using an intelligent method with weather-based forecasting models. IET Gener Trans Dist 9(14):1874–1882
44. Yan C, Xiu J, Liu C, Yang Z (2014) A high concentrated photovoltaic output power predictive model based on fuzzy clustering and RBF neural network. In: 2014 IEEE 3rd international conference on cloud computing and intelligence systems, pp 384–388
45. Panapakidis IP, Christoforidis GC (2017) A hybrid ANN/GA/ANFIS model for very short-term PV power forecasting. In: 2017 11th IEEE international conference on compatibility, power electronics and power engineering, pp 412–417
46. Yadav HK, Pal Y, Tripathi MM (2018) Short-term PV power forecasting using adaptive neuro-fuzzy inference system. In: 2018 IEEE 8th power India international conference (PIICON), pp 1–6
47. Semero YK, Zhang J, Zheng D (2018) PV power forecasting using an integrated GA-PSO-ANFIS approach and gaussian process regression based feature selection strategy. CSEE J Power Energy Syst 4(2):210–218
48. Pengtao L, Kaile Z, Lu X, Shanlin Y (2020) A hybrid deep learning model for short-term PV power forecasting. Appl Energy 259:114216
49. Biaowei C, Peijie L, Yaohai L, Yunfeng L, Shuying C, Zhicong C, Wu L (2020) Hour-ahead photovoltaic power forecast using a hybrid GRA-LSTM model based on multivariate meteorological factors and historical power datasets. E&ES 431(1):012059
50. Hangxia Z, Yujin Z, Lingfan Y, Qian L, Ke Y, Du Y (2019) Short-term photovoltaic power forecasting based on long short term memory neural network and attention mechanism. IEEE Access 7:78063–78074
51. Chang GW, Lu H-J (2018) Integrating gray data preprocessor and deep belief network for day-ahead PV power output forecast. IEEE Trans Sustain Energy 11(1):185–194
52. Happy A, Hong-Tzer Y, Chao-Ming H (2020) Short-term photovoltaic power forecasting using a convolutional neural network–salp swarm algorithm. Energies 13(8):1879
53. Liu Z-F, Li L-L, Tseng M-L, Lim MK (2020) Prediction short-term photovoltaic power using improved chicken swarm optimizer-extreme learning machine model. J Cleaner Product 248:119272
54. de Jesús DAR, Mandal P, Chakraborty S, Senjyu T (2019) Solar PV power prediction using a new approach based on hybrid deep neural network. In: 2019 IEEE power energy society general meeting (PESGM), pp 1–5
55. Zang H, Cheng L, Ding T, Cheung KW, Liang Z, Wei Z, Sun G (2018) Hybrid method for short-term photovoltaic power forecasting based on deep convolutional neural network. IET Gener Trans Dist 12(20):4557–4567
56. Takens F (1981) Detecting strange attractors in turbulence. In: Dynamical systems and turbulence, Warwick 1980. Springer, pp 366–381

57. Froehling H, Crutchfield JP, Farmer D, Packard NH, Shaw R (1981) On determining the dimension of chaotic flows. Physica D Nonlinear Phenomena 3(3):605–617
58. James T (1990) Estimating fractal dimension. JOSA A 7(6):1055–1073
59. Fraser AM (1989) Information and entropy in strange attractors. IEEE Trans Inf Theory 35(2):245–262
60. Fraser AM, Swinney HL (1986) Independent coordinates for strange attractors from mutual information. Phys Rev A 33(2):1134
61. Grassberger P, Procaccia I (2004) Measuring the strangeness of strange attractors. In *The Theory of Chaotic Attractors*, pages 170–189. Springer, 2004
62. Abarbanel HDI, Kennel MB (1993) Local false nearest neighbors and dynamical dimensions from observed chaotic data. Physical Review E 47(5):3057
63. Tang L, Liang J (2011) CC method to phase space reconstruction based on multivariate time series. In: 2011 2nd international conference on intelligent control and information processing, vol 1, pp 438–441. IEEE
64. Kim HS, Eykholt R, Salas JD (1998) Delay time window and plateau onset of the correlation dimension for small data sets. Phys Rev E 58(5):5676
65. Dimitris K (1996) State space reconstruction parameters in the analysis of chaotic time series—the role of the time window length. Physica D 95(1):13–28
66. Kim HS, Eykholt R, Salas JD (1999) Nonlinear dynamics, delay times, and embedding windows. Physica D Nonlinear Phenomena 127(1–2):48–60
67. Broock WA, Alexandre Scheinkman J, Davis Dechert W, LeBaron B (1996) A test for independence based on the correlation dimension. Econometr Rev 15(3):197–235
68. Rosenstein MT, Collins JJ, De Luca CJ (1993) A practical method for calculating largest Lyapunov exponents from small data sets. Physica D Nonlinear Phenomena 65(1–2):117–134
69. Yona A, Senjyu T, Funabashi T (2007) Application of recurrent neural network to short-term-ahead generating power forecasting for photovoltaic system. In: 2007 IEEE power engineering society general meeting, pp 1–6
70. Yu Y, Cao J, Zhu J (2019) An LSTM short-term solar irradiance forecasting under complicated weather conditions. IEEE Access 7:145651–145666
71. Alex G, Marcus L, Santiago F, Roman B, Horst B, Jürgen S (2008) A novel connectionist system for unconstrained handwriting recognition. IEEE Trans Pattern Anal Mach Intell 31(5):855–868
72. Shinde BT (2020) Real-time stability surveillance in power systems: a deep learning approach. Master's thesis, George Washington University
73. Zhang S, Wang Y, Liu M, Bao Z (2017) Data-based line trip fault prediction in power systems using LSTM networks and SVM. IEEE Access 6:7675–7686
74. Malhotra P, Vig L, Shroff G, Agarwal P (2015) Long short term memory networks for anomaly detection in time series. In: Proceedings, vol 89. Presses universitaires de Louvain
75. Abayomi-Alli A, Odusami MO, Abayomi-Alli O, Misra S, Ibeh GF (2019) Long short-term memory model for time series prediction and forecast of solar radiation and other weather parameters. In: 2019 19th international conference on computational science and its applications (ICCSA), pp 82–92
76. Yulara Solar System data download http://dkasolarcentre.com.au/download?location=yulara. Accessed 14 Feb 2020
77. Alice-springs Solar System data download http://dkasolarcentre.com.au/download?location=alice-springs. Accessed 14 Feb 2020

A MicroGrid System Infrastructure Implementing IoT/Big-Data Technologies for Efficient Energy Management in Buildings

Abdellatif Elmouatamid, Youssef Naitmalek, Radouane Ouladsine, Mohamed Bakhouya, Najib El kamoun, Mohammed Khaidar, and Khalid Zine-Dine

Abstract Recent studies showed that energy consumption in buildings could be efficiently reduced by including recent IoT (Internet of Things) and Big-Data technologies into microgrid systems. In fact, three major aspects could be further considered for reducing energy consumption while maintaining a suitable occupants' comfort, (i) integrating renewable energy sources and storage devices, (ii) integrating programmable and less-energy-consuming equipment, and (iii) deploying innovative information and communication technologies. These aspects might contribute substantially to the improvement of winning and saving energy toward smart and energy-efficient buildings. In this chapter, a microgrid system infrastructure is developed together with a platform for data gathering, monitoring, and processing. We put

A. Elmouatamid (✉) · Y. Naitmalek · R. Ouladsine · M. Bakhouya
LERMA Lab, College of Engineering and Architecture, International University of Rabat, Sala Eljadida 11100, Morocco
e-mail: abdellatif.elmouatamid@uir.ac.ma

Y. Naitmalek
e-mail: youssef.naitmalek@uir.ac.ma

R. Ouladsine
e-mail: radouane.ouladsine@uir.ac.ma

M. Bakhouya
e-mail: mohamed.bakhouya@uir.ac.ma

A. Elmouatamid · N. El kamoun · M. Khaidar
Faculty of Sciences, STIC Laboratory, Chouaïb Doukkali University, CUR"EnR&SIE, El Jadida 24000, Morocco
e-mail: elkamoun.n@ucd.ac.ma

M. Khaidar
e-mail: khaidar.m@ucd.ac.ma

K. Zine-Dine
Faculty of Science, Mohammed V University, Rabat 10000, Morocco
e-mail: khalid.zinedine@um5.ac.ma

Y. Naitmalek
ENSIAS, Mohamed V University, Rabat 10000, Morocco

S. Motahhir and A. M. Eltamaly (eds.), *Advanced Technologies for Solar Photovoltaics Energy Systems*, Green Energy and Technology,
https://doi.org/10.1007/978-3-030-64565-6_20

more emphasis on microgrid systems as crucial infrastructures for leveraging energy-efficient and smart buildings by developing and deploying a holistic IoT/Big-Data platform in which sensing and actuation tasks are performed according to the actual contextual changes. Scenarios are presented in order to show the usefulness of this holistic platform for monitoring, data processing, and control in energy-efficient buildings.

Keywords Energy-efficient buildings · Microgrid system · Energy management · Renewable energy sources and storage devices · IoT and Big-data technologies · Predictive and context-driven control

Acronyms

AC	Alternating Current
D/R	Demand/Response
DC	Direct Current
EEBLab	Energy Efficient Building laboratory
EM	Energy Management
HVAC	Heating Ventilation and Air-Conditioning
ICT	Information and Communication Technologies
IoE	Internet of Energy
IoS	Internet of Service
IoT	Internet of Things
MG	Micro-Grid
PV	Photovoltaic
RES	Renewable Energy Source
SG	Smart Grid
SoC	State-of-Charge
TEG	Traditional Electric Grid

1 Introduction

Buildings are responsible for about 40% of energy consumption and more than 40% of greenhouse gas emissions [1]. Reducing energy consumption and subsequently, CO_2 emissions are highly required since buildings frequently use more energy than anticipated or desired. This need for energy requires the integration of clean energy sources in order to reduce the consumption from TEG, which is generally based on polluted sources (e.g., coal plant, a nuclear plant). Usually, three major aspects could be considered for reducing energy consumption from TEG, (i) integrating RES with efficient energy control and management, (ii) reducing energy consumption by integrating programmable and less-energy-consuming equipment while keeping a good occupants' comfort (e.g., HVAC, lighting), and (iii) reducing energy consumption by integrating innovative ICT concepts for efficient EM of buildings services. These aspects might contribute substantially to the improvement of winning and energy-saving toward smart and energy-efficient buildings [2].

However, buildings have become a producer of electricity due to the RES integration together with the possibility to store and consume locally the electricity without expansion needs for electricity transport and distribution. This integration of distributed generators requires efficient management of energy in order to bring additional benefits for reducing energy consumption and, consequently, CO_2 emissions. In addition, buildings could be capable to control its own energy, from the sources to the end-services, by managing the installed RESs and energy storage systems together with the deployed active/passive equipment (e.g., HVAC, lighting) [3, 4]. Consequently, a platform for data gathering, monitoring, and processing should be installed together with the electrical system making the building "Smart." This new smart building structure presents a main factor for smart grid development, as depicted in Fig. 1. In fact, the controls, automation, and ICTs combined together with the bidirectional communication way with the TEG could be able to make the building components capable to adapt and balance digitally the continuous D/R changes. Additionally, consumers should have the opportunity to anticipate the electricity market and control their electricity consumption accordingly [5, 6]. However, the decentralization of energy production makes the electrical system more complex and more difficult to control in order to keep a suitable electricity balance (i.e., D/R). Consequently, the transition from unidirectional to bidirectional interconnection and from centralized to decentralized energy production requires the use of smart equipment (e.g., smart metering, smart inverter, smart transformer) [7]. This equipment should be able to interact with different building's services taking into consideration

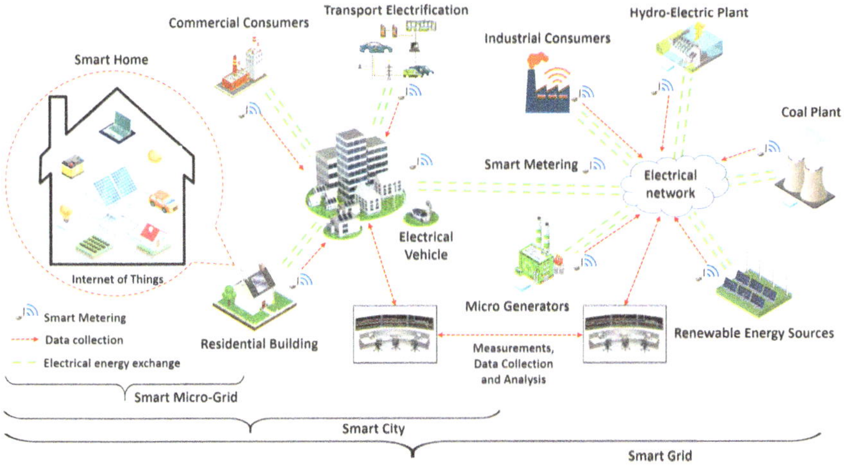

Fig. 1 Global architecture: From smart Grid to smart microgrid

its surrounding environment. The aim is to minimize the usage of electricity while keeping suitable occupants' comfort as well. In this context, an MG system is defined as an "intelligent building" that can produce, consume, and store locally the electrical energy. The MG, via a well-established ICT-based infrastructure, can interact with consumers, with neighboring MGs, and with the TEG.

The main objective of such systems is to connect efficiently the producers and consumers of electricity with a high level of security, stability, and continuity of energy supply (the increase of services quality). As a result, the MG can smoothen the electrical peaks demand in the electrical network, which represents a major challenge for the TEG. It also allows for managing the electricity flows by considering economic and environmental constraints. Accordingly, the electricity bill can be minimized by avoiding peak demands and, therefore, the consumption can be maximized from RESs while minimizing subsequently the carbon impact. Therefore, as state above, the interaction of different buildings' components needs to integrate ICT-based infrastructures for data collection, analysis, and processing. This integration of ICT together with RES and storage has enabled the emergence of "Micro-Grid" (MG) systems [8]. As depicted in Fig. 1, MG systems remain important and necessary building blocks for the development of smart grid systems as well as smart city applications and services [9].

In this chapter, a new holistic architecture of smart buildings is presented by improving the main layers of MG systems. This architecture is proposed in order to integrate all buildings' aspects with the main trade-off is to minimize energy consumption while maintaining a suitable occupants' comfort. In fact, an MG system is structured into three layers following the proposed holistic architecture. More precisely, we shed more light on the MG system's components by putting more emphasis on the integration of recent IoT/Big-Data technologies for data gathering, processing, and control. Several scenarios are presented to show the usefulness of this holistic architecture and its direct relationship with smart microgrids.

The remainder of this chapter is structured as follows. In Sect. 2, the MG system is presented as a part of the "Smart Grid." The operational MG modes together with international standards are detailed in Sect. 3. In Sect. 4, the concept of smart buildings and its relationship with MG systems is introduced by focusing on EM, automation, and control systems. Moreover, an experimental MG system is presented in Sect. 5 by highlighting the main MG components (e.g., ICTs layer, energy layer) and presenting a set of deployed scenarios. In Sect. 6, conclusions and perspectives are presented.

2 Smart Microgrid Systems

The integration of RESs for large-scale production of electrical energy has recently accelerated because of evident climate change, insufficiency of fossil resources, and greenhouse gas emissions. RES are clean and eco-friendly sources and their abundance and renewable nature are among the most important factors for their integration into smart grid networks. However, these green energy sources come with new challenges, mainly their seamless integration with existing electrical networks. In addition, another important challenge for this new electricity infrastructure is real-time monitoring and data processing, which requires new ICT-based infrastructures. The main aim is to ensure sustainable and reliable renewable energy generation systems [10, 11]. Therefore, this integration of ICTs, energy distribution systems, as well as distributed energy generation systems (e.g., RESs), creates what is commonly named "Smart Grid" (SG). In fact, SG represents the new smart electrical network, since it brings the flexibility to integrate new electrical services, such as electrical vehicles, and enables consumers to be energy producers by integrating RESs using bidirectional communication network [12]. This depends mainly on the fast advances in ICT-based infrastructures covering then all aspects of the electricity grid and its associated services. In fact, due to the development of IoT infrastructures (Internet of Things) and their related intelligent services, the electricity grid has new capabilities to monitor, manage, and control its components and then takes advantage of sophisticated bidirectional interactions. Moreover, the ICTs integration enables various smart and automatic services, such as smart metering infrastructure, smart control, and management for D/R balance, advanced electricity marketing, and intelligent energy storage for electrical vehicles integration.

However, despite this progress, some research work stated that the SG is experiencing new issues. Mainly, it is able to manage only electrical energy neglecting other existing types of energy (e.g., thermal, chemical, and electrochemical). In addition, the SG is based on the actual infrastructures of power distribution grids, which are limited by the unidirectional exchange of the electricity [11, 13]. Therefore, face to these challenges, other concepts have been developed together with the revolution of SG, such as the internet of energy (IoE), the internet of things (IoT), and the internet of services (IoS), as mentioned in Fig. 2 [14]. Especially, the development and the emergence of smart MG (microgrid) systems could resolve some of the abovementioned SG challenges. MG could simplify the management of electrical

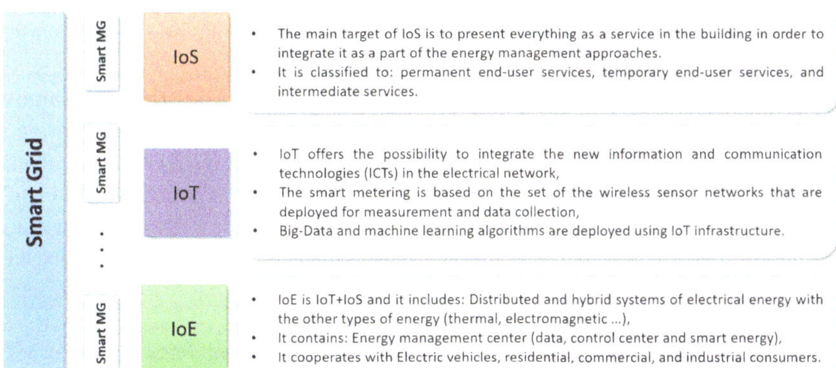

Fig. 2 Smart Grid and smart MG presented as a combination of IoE, IoT, and IoS

energy, from centralized to distributed EM. In addition, in MG systems, different types of energy can be managed locally with the possibility to interconnect different MGs in a distributed manner.

Depending on the scale of the system, numerous definitions for MG systems have been proposed. For two European projects, named "Microgrids" and "More Microgrids," the MG system is considered as a basic feature of future active distribution networks and it is composed of more than one building [15]. For instance, in Greece, the "Kythnos Island Microgrid" is composed of 12 houses connected to solar PV plant and battery bank. The PV plant comprises10 kW of PVs for energy generation, a nominal 53-kWh for the battery bank, and a 5-kW diesel generation. A second PV plant of about 2 kW, mounted on the roof of the control system building, is connected to 32-kWh of the battery bank in order to provide power for monitoring and communication [16]. Another system in Germany for "More Microgrids" project, named "MVV Residential Demonstration," is installed at Mannheim-Wallstadt. The project prepares about 20 families for a continuous long-term field test site that is considered as one MG. In fact, the first goal of the experiment is to involve customers in load management. For that, a total of 30 kW of PV are already installed. Based on PV power availability information from their neighborhood, the families shifted their consumption when it is possible to use directly solar energy. As a result, participating families shifted their consumption significantly from the typical residential evening peak toward hours with the higher solar insolation, and from cloudy days toward sunny days [16, 17].

In the United States, there are many projects in universities and military bases already developed with an estimated global market rise from about 3.2 GW in 2019 to 15.8 GW by 2027 (including all types of MG systems, as it is depicted in Fig. 3), where only the United States accounted for almost 35% of this market in 2018 [18]. The most well-known researches and development project, named "Consortium for Electric Reliability Technology Solutions" (CERTS), is developed for the power system reliability of emerging technology in MG systems. The project is provided for relatively small sites (~<2 MW at the peak) and it is delivered for a research platform, which is considered as an MG installed in a laboratory at the University of Wisconsin, Madison [19]. Another interesting international standard is Japan, which sets ambitious targets for increasing the contribution of RESs in MG systems. In fact, the research funding and management agencies of the Ministry of Economy, Trade, and Industry have started different MG projects. Mainly, a recent project named "Integrating renewables into the Japanese power grid by 2030" is involved. In this project, Japan's Renewable Energy Institute (REI) and "Agora Energiewende" attempt to integrate renewables energy into Japan's power grids without endangering grid stability, the study also promotes data transparency. International experience has shown, however, that several technical measures, not yet widespread in Japan, can be safely implemented to improve the grid stability [20].

All these research projects consider large-scale buildings and RESs plants as MG systems. For instance, according to the MG operation mode, different types of MG systems are classified as depicted in Fig. 3.a. Similarly, by considering the applications and the objectives [21], another classification is presented in Fig. 3.b. Other

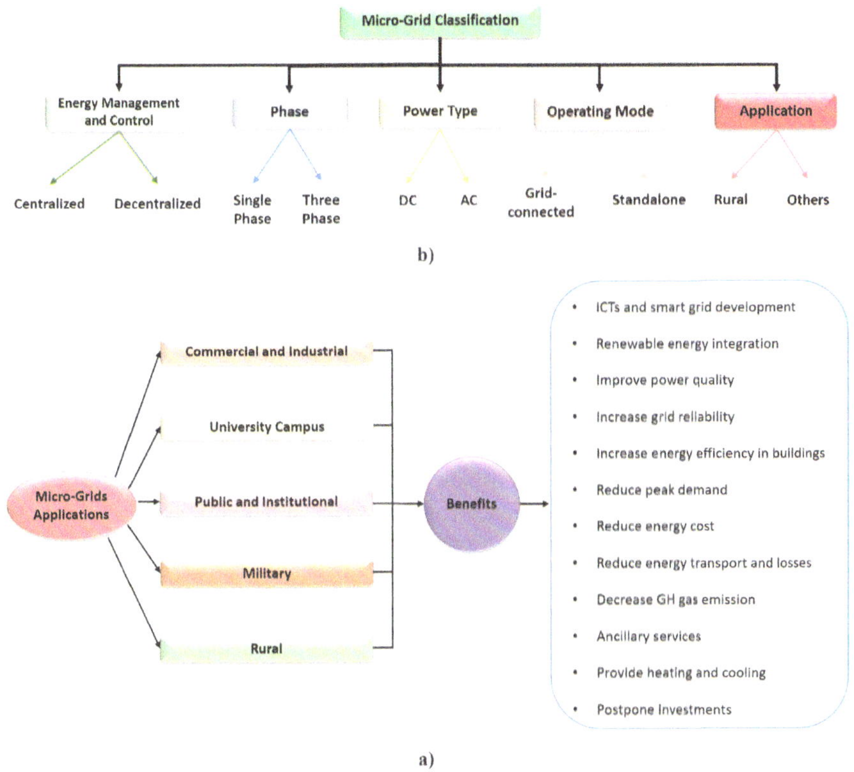

Fig. 3 MG system types and benefits

academic researches present the MG system as a single building, which integrates ICTs infrastructure, RESs, and energy storage with the electrical power grids. For instance, in [22, 23], the MG systems are defined as smart power systems that are grouped within a limited geographic area. They include loads, distributed generation units, and energy storage systems (batteries, electric vehicles, hydraulic storage, etc.). The main advantage of MG is to enable customers to have both a bidirectional communication platform and control devices to manage their energy needs and excesses. In addition, with an adequate communication structure, it is possible to shape the users' load demand curves by means of D/R strategies.

Other works present the MG as a systematic and efficient approach for managing the power system by integrating all the distributed generating sources into a micro-power system [24, 25]. For example, in [24] authors defined the MG system as a low-voltage power network with distributed energy generation (e.g., PV arrays, micro-wind turbines, fuel cell, energy storage), which offers better control capability over network operation. It is considered as a solution to meet the local energy demand by connecting distributed power generation to distribution networks, such as local substations without further expansion of costly centralized utility grids. In addition, the United States Department of Energy (DOE) defines an MG as follows: "An MG is a group of interconnected loads and distributed energy sources within clearly defined electrical boundaries that acts as a single controllable entity with respect to the grid" [26].

3 MG System Architectures

3.1 Operational Modes

MG systems are designed to operate efficiently and resiliently since they are not only dedicated to a high penetration level of RESs and storage systems but also due to their capability to operate in isolated mode when RESs can satisfy the demand or during the faults, which occur in the main electrical network. Therefore, MG systems offer greater reliability and efficiency for the electrical network system, especially by locally controlling the generated power while improving the energy quality, as well as smoothing the power curve by the deployed storage devices. In addition, the losses of energy, which are caused by the transport and distribution system to the end-consumer, are reduced and, consequently, the blackouts of electricity, created by the peak demand, can be avoided.

However, MG systems are operated, as shown in Fig. 4, into two distinctive modes: grid-connected and islanded modes [21, 26]. Other literature works consider another mode, named self-consumption mode, by controlling buildings' services, i.e., identifying those that can be connected to the main grid [27]. For isolated mode, named standalone mode, the RESs production and storage devices are dimensioned in order to satisfy totally the demand. Generally, another source is integrated, such as diesel motors, to satisfy the demand during the low or the absence of RESs generation. This mode is useful in critical applications, such as the isolated site that requires a high cost for electricity transport and distribution. For isolated sites, it is necessary and practical that the hybrid system has total autonomy requiring the use of storage systems not only to smooth the variable nature of RESs but also to ensure the power

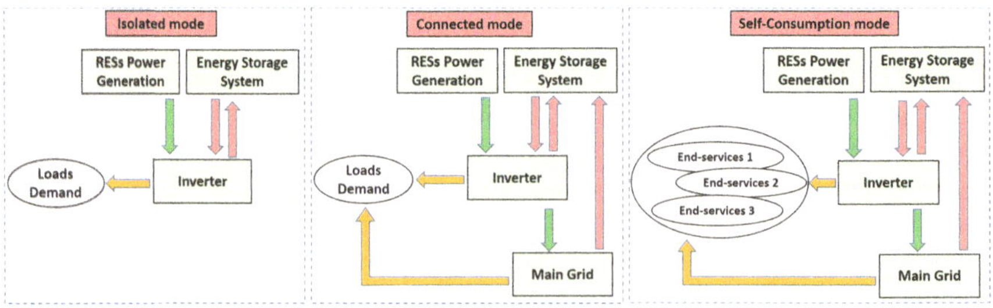

Fig. 4 The MG system operating modes

availability and the continuous supply of energy. However, energy storage represents a very significant part of the cost and maintenance of the MG installation, and the lifetime of storage devices is much lower than PV panels, the wind turbine, and the converters. For that, good strategies are required for sizing the storage devices and the RESs generation. Several works are presented in the literature to dimension the optimal configuration for isolated sites by studying the weather conditions including technical and economic characteristics of all specific MG components [28–31]. In addition, the island mode is studied in various research projects of MG systems [32, 33]. The specific standards IEEE Std 1547.4-2011 is the only international reference for MG systems, which are operating in island mode [34].

Unlike isolated mode, the grid-connected mode is considered as the major key for RESs integration in buildings in order to develop the concept of "Smart Grid" and consequently the MG systems. A real MG system is connected to the electrical network to increase the reliability of the production system and to realize the main objective of such systems. This mode offers a high benefit for both energy and financial cost by reducing the size of the installation (e.g., battery capacity, number of PV panels) on the one hand, and by integrating the cost of energy in the management strategies on the other hand. In fact, the majority of hybrid systems, connected to the electrical grid, have a limited capacity of energy storage systems that are used to reinforce the power quality and to smooth the RESs generation. In this case, they are dimensioned to ensure the power during the failure of TEG or during the perturbed RESs production periods, and as results in minimizing the size of storage devices. However, the architecture of MG connected mode necessitates certainly the inclusion of inverters, both to convert, when necessary, from direct to alternating currents and to provide some level of frequency and voltage control as well. Principally, the inverter is the interface that provides the interconnection to the electrical network by respecting the norms of power quality (e.g., frequency, voltage) and by deploying the EM strategy. Moreover, the inverter supplies the power to the loads offering then the possibility to charge the storage systems, to extract or inject the electricity from/to the electrical network, and to serve potentially heterogeneous sources without loss of synchronization, propagation of harmonics, or loss of system stability. It is worth noting that we have considered the connected mode of MG systems as the main architecture of our deployed MG system. This mode is more adaptable for the actual structure of buildings by offering the possibility to develop the actual building as an MG system. This issue is studied by a set of research projects and several test sites are deployed by considering the connected mode of MG systems [18, 35–37].

Another mode, named self-consumption, depends strongly on the concept of "internet of services" in buildings by coupling the two other operating modes. It requires a high integration of ICTs and IoT/Big-Data technologies to predict and control efficiently the different components and services of the system. In fact, by deploying machine-learning algorithms, the internal and external parameters can be forecasted to control and manage powerfully the power system (i.e., production, consumption) while keeping a high comfort for building's occupants. Mainly, in MG systems, the services can be decomposed into three main categories: (i) permanent end-user services, its energetic assignment plan covers the whole time range;

generally, these services produce directly comfort to occupants; (ii) temporary end-user services, the time range of this services can be modified by the EM system deployed in MG system (e.g., modification of the starting time of a washing machine service, cooking service); (iii) intermediate services, which produce electrical power to the whole end-users by managing the previous end-user services with the energy production services (e.g., RES, storage devices, grid).

Therefore, the self-consumption mode modifies the starting time of temporary end-user services (e.g., electrical vehicle charge/discharge, washing machine service, cooking service) and the buildings can be supplied by electricity from both RESs and electrical network at the same time. In addition, in a given situation, the control strategy can switch some services that consume a high level of energy to the electrical network while keeping the RESs connected only to defined building's loads. This operating mode is more useful to ensure a continuous supply of electricity to some principal services that are not designed to support the cut of electricity (e.g., data-center, networking equipment, IoT/Big-Data platforms). Data centers are considered to be one of the best examples of an industry with relatively established plans for the blackouts of electricity. For example, the Great East Japan Earthquake on March 11, 2011 killed more than 15 000 people, destroyed 4 nuclear generation plants, and left several million people without electricity and no critical damage to data centers was reported [38]. Furthermore, by coupling the IoT, IoE, and IoS concepts, the household equipment can interact with EM strategies in order to minimize the cost of energy while avoiding the cut of electricity during the failure of the energy sources. For example, by considering the electrical vehicle as a service in the building, the control strategy can use its battery as a source of energy during the night by considering that the electrical vehicle is a smart service, which can communicate its SoC and its targets to the communicated system [39]. Different literature works are realized to develop this new concept of service control in buildings [10, 14, 40, 41].

3.2 International MG Standards

The MG concept is relatively new and the regulatory framework is still under development. It should be standardized for being integrated into the existing electrical grid network. In this way, several research groups within the International Electro-technical Commission (CEI) are working on the question of standardizing systems that use renewable energies. The standards consider the power quality (e.g., frequency, voltage, harmonic noise), the components (e.g., inverters, converters), the architecture and design of MG, and the size of the integrated renewable sources (e.g., generated power, low-voltage, medium-voltage). In addition, the MG systems should respect the existing electrical norms and their deployments. Especially, for MG that are connected to the electrical network, the inverter should ensure the electricity quality avoiding then the injection of noise in the utility grid.

Table 1 Voltage-level standards in DC and AC bus of MG

MicroGrid buses	Normalized voltage levels (V)	Micro-grid applications	Principal standards
DC	48	Standalone systems	IEC 60038 and IEEE 2030.10
	380–400	Grid-connected systems Commercial and industrial buildings	IEC 60038
	1500 V	Commercial and industrial buildings	IEEE Std 1709
AC	230 and 400	MG connected to the traditional electrical grid	IEEE 1547

However, realizing specific technical standards is difficult and the standards concerning RESs integration have some differences in different worldwide locations due to the different operational methods, the EM strategies, and the different penetration levels and types of RESs and storage devices. For instance, in the United States, the IEEE 1547 series of standards covers all aspects related to the interconnection of distributed energy resources with the electrical grid. These standards impose requirements on the quality of the energy produced in terms of voltage, frequency, and harmonics. It provides requirements relevant to the interconnection and interoperability performance, operation, testing, safety, maintenance, and security considerations. The first revision of 1547 assembles several participants whose investor affiliations, manufacturers and integrators, test labs, research groups, and academia. The Full revision of 1547 issues, concerns, and updates are being coordinated with corresponding standards and codes, such as the Nippon Denki (NEC) and Underwriters Laboratories (UL) safety standards. This full revision included participants from various states, covering all United States regions and some other regions, such as the United Kingdom, Canada, and Japan [42–44]. Therefore, depending on the MG topologies, buses, and electricity architectures, different standards are considered, as presented in [45, 46] (Table 1).

Alike United States, several works in European Renewable Energy Council (EREC) are urged to improve new integration standards of distributed energy resources. The standardization of the system helps the power system operators to share experiences with manufacturers and developers in order to internationalize their items and consequently normalize the system for future deployment while avoiding the alteration between electricity participants. The main European standards applicable to MG systems are EN 50160 and IEC 61000.

These standards describe and specify the main characteristics of the voltage supplied by a low-voltage, medium-voltage, and high-voltage AC public network under normal operating conditions. They describe the limits and levels of the voltage characteristics that can be expected at each delivery point of the public network [47–49]. Table 2 summarizes the United States and European standards that are appropriate to the deployment of MG systems.

Table 2 International standards for distributed energy integration in MG systems

Standards	Description	Standards specifications
IEEE-1547 (US)	Requirements on power quality and distributed energy sources integration in the electrical grid	● Integration, protection system design and operation of distributed system ● Control/monitoring and application guide ● Interconnection requirement for distributed system higher than 10 MVA ● Testing and measurement techniques ● Rules and guidelines regarding the connection with secondary distribution networks [48] ● Studies on the impact of DES interconnection ● Recommended practice for establishing procedures and methods ● Ideal grid–consumer connection configurations ● Supply methods and loads considerations ● Time tags and synchronization applications ● Verification methods of standards compatibility with measurements ● Phasor Synchronised definition and measurement unit methods ● Specify the main voltage characteristics at the PCC in low, medium and high voltages during steady-state operation ● Determine the power frequency, harmonics, voltage unbalance, voltage variation and flicker limits at PCC ● Describe the indicative values for some power quality events ● Electromagnetic compatibility levels ● Integrity requirements and safety functions ● Requirements for safety and protection ● Short interruptions, voltage sags and voltage variation protection tests ● Mitigation methods and installation guidelines ● Progress on constructing high-performance buildings (near-zero energy buildings) ● Regulations to define the concept of (NZEB) Net Zero Energy Building
IEEE-1547.4 (US)	It includes the planning and operation of the MG systems (IEEE Standards Coordinating Committee 21) [50]. The SCC21 develops a guide to help the operators, the specialists, and the manufacturers to use the technical aspects of the MG operation and implementation	
IEC-61850-7-420 IEC-61968-9 EN-13757-4 & 5	This series of standards concerns: Communications for distributed energy resources, meter reading and control, radio mesh meter bus, wireless meter bus	
IEC 60364-1	Recommendations for human safety, guaranteeing the safety of persons against life dangers, verification of electrical installation of Nominal-Voltages	
IEC 61851	Electrical vehicle integration in MG, charging station regulations for single-phase (levels up to 250 V) and three-phase (levels up to 480 V)	

(continued)

Table 2 (continued)

Standards	Description	Standards specifications
IEEE-C37.95 (US)	It is a guide for grid-consumer interconnection with a number of different protective information. It covers applications involving service to a consumer that normally requires a transformation between the utility's supply voltage and the consumer's utilization voltage	
IEEE-C37.118 (US)	Standard for Phasor Synchronization with power system and data transformation for the grid system operating and interconnection	
IEEE 2030.10	DC energy providers for off-grid system, communication protocols, recommendation for low DC voltage designated to standalone systems	
IEEE 2030.7	EM system, control level associated to the proper operation, configuration, and regardless topology	
IEEE Std 1709	Power quality recommendation and voltage tolerances for Medium-Voltage DC bus	
IEEE Std 115	Electromagnetic compatibility and regulations about power quality limitations for AC and DC buses	
EN-50160 (Europe)	It describes and specifies the main characteristics of the voltage supplied by AC public network under normal operating conditions of distribution systems	
IEC-61000 (Europe)	It contains specifications for Electromagnetic compatibility (emission standards, immunity, installation, testing and measurement techniques), it is required to keep interference between electronic devices under control to reduce disturbance and improve immunity in residential, industrial, and commercial environments	
ISO 52000-1 ISO 52003-1 ISO 52010-1 ISO 52018-1	Standard for energy performance of buildings, which establishes a systematic and comprehensive structure for assessing building energy performance	
ISO 52016-1	Efficient thermal energy in MG, important response time for HVAC to respect building thermal-zone standards, such as the estimation of energy needed for heating and cooling	

However, despite this progress in deploying MG systems and advancing standards, still their integration into existing and smart buildings requires efficient and holistic management platforms. Especially, the integration of recent IoT/Big-data technologies for real-time monitoring and data processing in order to develop new predictive control approaches, which allow ensuring the sustainability and the reliability of these new energy generation systems.

4 Smart Buildings as MicroGrid Systems

MG systems for smart buildings can be seen as socio-technical systems that integrate different heterogeneous entities (e.g., sensors, actuators, lighting, HVAC, occupants, RES, storages), which could interact dynamically and in a collective manner to balance between energy efficiency, occupants' comfort, sustainability, and the adaptability. More precisely, making buildings more energy-efficient while ensuring occupants' comfort require incorporating mechanisms and techniques, which allow entities interacting in order to perform suitable actions (e.g., turning On/Off HVAC and lighting, balancing the fluctuation between power production and consumption) as shown in Fig. 5. As stated in [51–53], systems operating in dynamic environments with these capabilities are qualified as socio-technical Collective Adaptive Systems (CAS). These systems should learn and evolve by performing distributed decisions at different temporal and spatial scales while self-organize when entities join or leave the collective (e.g., occupants' number and presence). For instance, platforms for buildings' EM could react to the dynamic changes (e.g., buildings

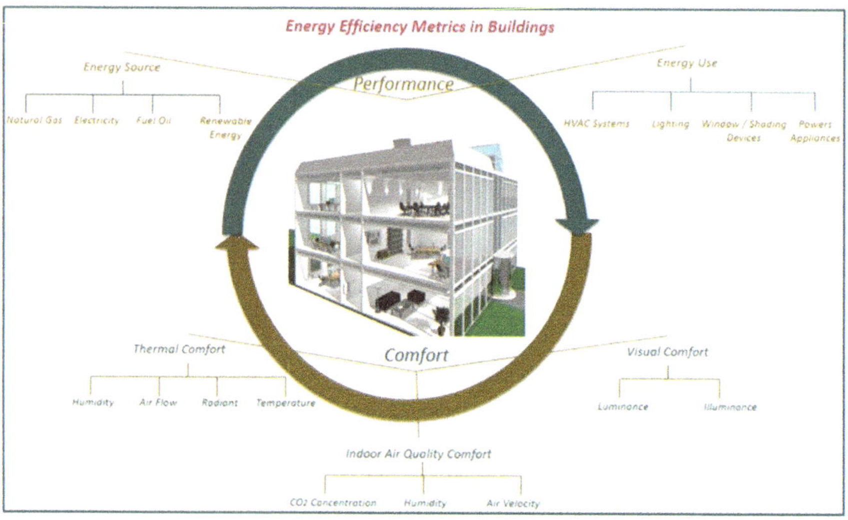

Fig. 5 Energy efficiency and occupants' comfort metrics [51]

occupants' preference, number, presence) for lowering energy consumption while making occupants' life more comfortable and consequently, increasing the energy efficiency in buildings.

Mainly, one of the most important factors that define the "Smart Buildings" is the adaptability. It is defined as the characteristic of buildings to use information gathered from a range of sources to prepare the building for a particular event before that event has happened (e.g., predictive control, occupants forecasting) [54]. The adaptability allows the differentiation between previous generations of buildings and Smart Buildings. In fact, using IoT/Big-Data technologies, the buildings gather data externally (e.g., weather conditions, RES production) and internally (e.g., occupancy, loads consumption) to adapt its operations depending on the context-awareness principles. The collected data is used to develop machine-learning algorithms that are used to forecast the actions, which are required to perform and operate different buildings' services. For example, the forecast of weather conditions can be used to predict the RESs production, which allows flexible management of energy D/R. In addition, by measuring the energy production/consumption and by forecasting the occupant's activities, the adaptive buildings modify the starting time of temporary end-user services (e.g., washing machine service, cooking service).

However, these abovementioned aspects represent the main factors to develop the concept of "Micro-Grid" systems. It is due to the capabilities of recent ICTs techniques (e.g., machine learning) to forecast future events, which are required to develop efficient EM approaches. The next section introduces our MG system's architecture. In this way, we have designed and deployed an MG system for conducting experiments in real-sitting scenarios. In particular, we highlighted the necessity of integrating recent IoT/Big-Data technologies for gathering external and internal data, which have been used to generate predictive actions (e.g., regulating the room temperature by forecasting building's occupancy, ventilation speed variation according to the forecasted CO_2, intelligent and predictive control of energy flows management using forecasted power production, consumption and battery state of charge).

5 The Experimental Platform of MG Systems

As shown in Fig. 6, our MG system is structured into three horizontal layers: passive building layer (e.g., building envelope and insulation, architecture design), active building systems layer (e.g., HVAC system, Lighting), and RESs system layer (e.g., PV, wind, storage). These layers are monitored by one vertical layer for communication and ICTs integration. This layer integrates mainly an IoT/Big-Data platform in order to measure, analyze, predict, and forecast actions depending on the actual and predicted context.

In particular, our MG system is a smart and active building that combines ICTs/Big-data infrastructure, RESs/storage systems, EM/control strategies, and electrical power grids. This new concept of a building is more interactive for both consumers and energy producers. In fact, consumers will reduce the cost of their

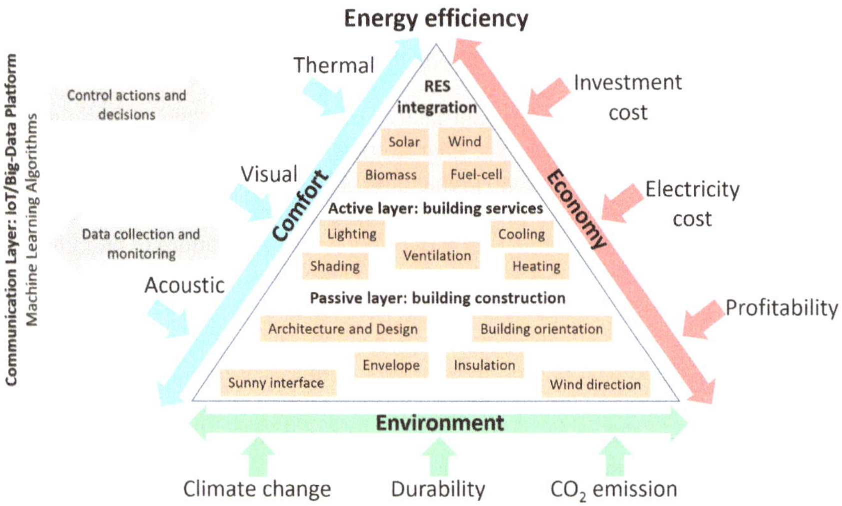

Fig. 6 The main MG system's layers for smart and energy-efficient buildings

energy consumption based on the used control approaches, which take into account the real-time cost of the power and the predictive power generation, for efficient D/R management [55]. In addition, the household equipment (e.g., refrigerators, washing machines, microwaves, lighting) are becoming intelligent devices, which may be actively controlled using IoT devices, as well as adjusted and controlled by interacting with the other systems (e.g., power generation, EM system). Moreover, this MG structure offers the possibility to integrate new buildings' services, such as electrical vehicles, which can be used as a storage device to compensate for the energy in the building by integrating the "Grid-to-Vehicle & Vehicle-to-Grid" techniques.

The rest of this section is dedicated to the description of the deployed MG systems together with the deployed scenarios. The aim was to develop a research test site integrating the different components of an MG system, which is used to test and integrate control strategies for predicting, estimating, and controlling the interaction between power production, storage, and building's demands. As shown in Fig. 7, the system integrates PV panels, wind turbines, batteries, and the TEG connected together in order to supply electricity to the building's services according to actual contexts. The system is monitored by an IoT/Big-Data platform, which is used to collect, analyze, and store the data for EM and control strategies development. Moreover, several scenarios are deployed in order to develop a research platform that considers the concept of MG systems with the different components of the different layers.

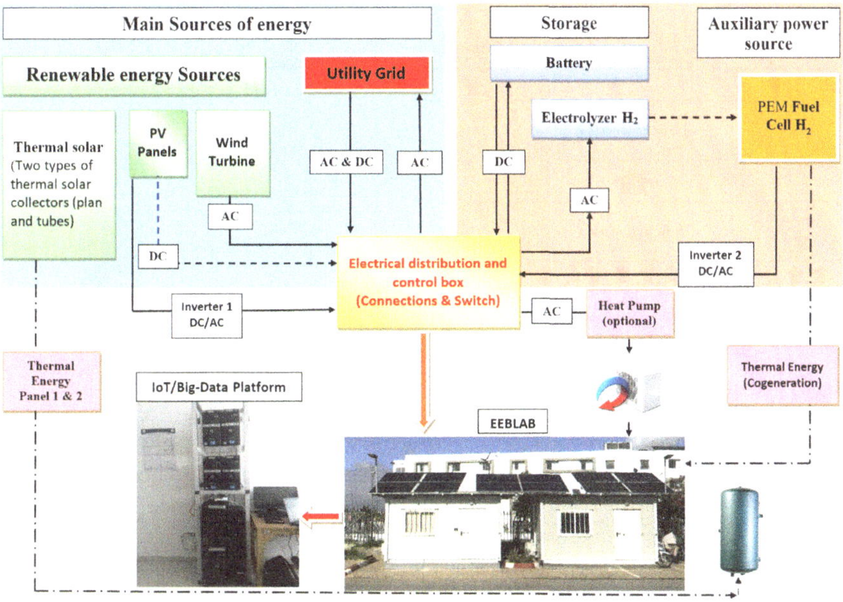

Fig. 7 The holistic model of the deployed MG system

5.1 IoT and Big-Data Platform for Data Monitoring/Processing

Real-time and context-awareness information could be exploited for developing predictive and adaptive context-driven control approaches using recent IoT and Big-data technologies together with real-time and machine-learning algorithms [56, 57]. A platform that uses context-driven technologies, as well as complex-event processing technologies, is deployed for data monitoring and processing in order to develop intelligent and predictive control strategies for EM in MG systems. The platform is composed of four main layers, sensors/actuators layer, data acquisition, data processing, and data visualization/storage together with further services and applications for context-driven control (Fig. 8).

The MG is mainly equipped with a component for measuring the different necessary parameters (e.g., current, voltage, temperature, wind speed), for interacting with the passive and the active equipment, for regulating the comfort for the occupancy, and for managing the power production and consumption. In fact, a set of sensors is installed depending on the desired scenarios. In addition, the first layer includes the actuators that are used to receive and to execute different commands, which are generated by the control strategies for EM or equipment and services control. Regarding the data acquisition layer, a Kaa application is developed (i.e., IoT technique) [59], which is used to receive data from deployed sensors. We have also used MQTT (Message Queue Telemetry Transport), which is a publish-subscribe-based

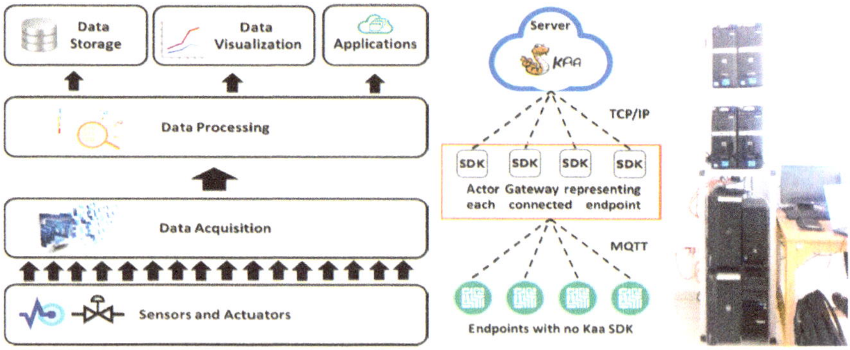

Fig. 8 IoT/Big-Data platform architecture [58]

protocol for IoT applications. For data processing and storage, Storm [60] services are used. Mainly a topology composed of Spouts and Bolts was designed and developed to allow receiving and processing streaming data from sensors. The spouts receive the data from the Kaa application, and then transmit it to the Bolts for processing and storage into the database (e.g., MongoDB) for further in-depth analysis. The services layer includes real-time visualization and storage together with the control of active equipment and RESs power production and consumption monitoring and management.

The platform was used for data gathering and processing of internal and external building's context. For instance, it was used to build occupant information (e.g., number, presence, behavior, activities), since is a major input for control approaches in energy-efficient buildings (e.g., active systems control). In fact, comprehensive fine-grained occupancy information could be integrated to improve the performance of occupancy-driven control of HVAC, lighting, and ventilation systems. A platform for real-time detection of occupants' is deployed (Fig. 9). The platform was adopted by including real-time machine-learning component with the main aim is to analyze, explore, and predict the occupancy information in buildings [54]. However, these predicted values are then used for efficient control of active equipment and for predicting the electricity consumption behavior, which is used for EM.

Regarding the external context, we have deployed a weather station. In fact, for several scenarios, we need to gather internal and external context data. We have built a weather station near to the wind turbine and PVs in order to have as precisely as possible the data concerning wind speed, direction, irradiation, temperature, and humidity. The weather station was deployed and used to collect the data for real-time visualization and processing for further usage by other building's services and applications. All these data are gathered and processed in real-time using our IoT and Big-data platform, as depicted in Fig. 10.

Weather data are collected in order to validate the results obtained from simulations and experimentations by using the same input parameters. For example, radiance and temperature are measured together with the PV power during the same

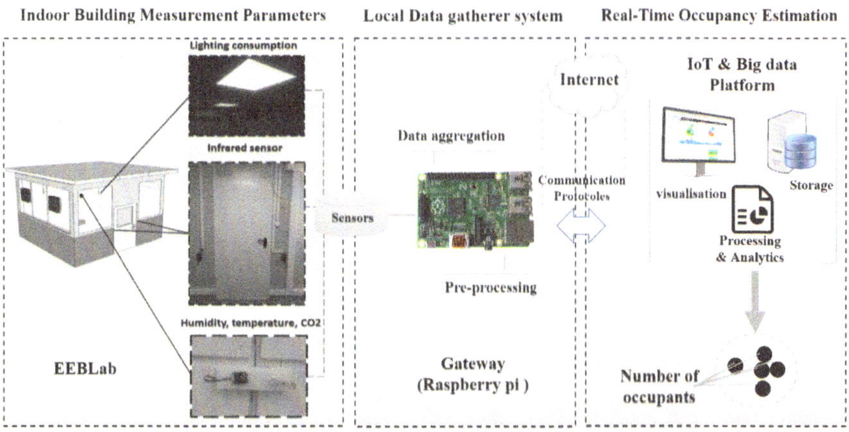

Fig. 9 The architecture for occupants' presence detection/prediction and experimental results [56]

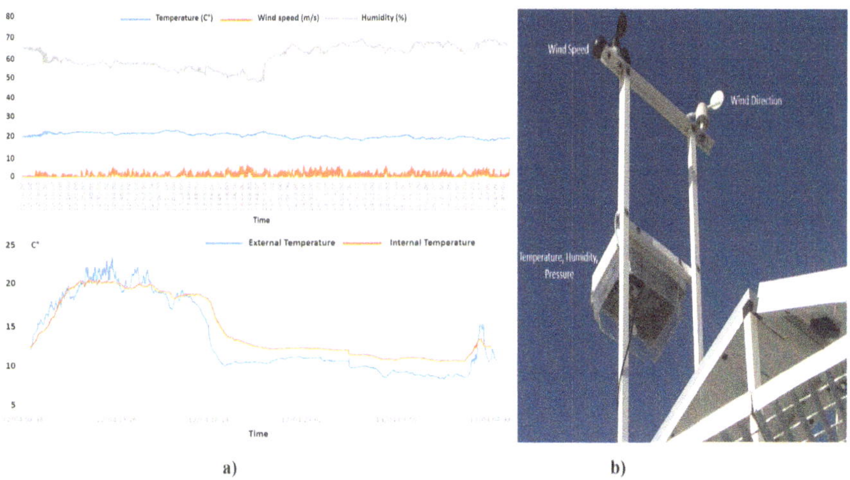

Fig. 10 **a** Internal/external temperature, **b** Weather monitoring

day. Radiance and temperature are used as input parameters to the mathematical PV model, which is developed for conducting simulations and validate experimentations' results using similar contextual data.

5.2 *Building Envelope*

This part concerns the passive layer (Fig. 6), which allows reducing energy consumption by developing less-energy-consuming equipment and materials in buildings.

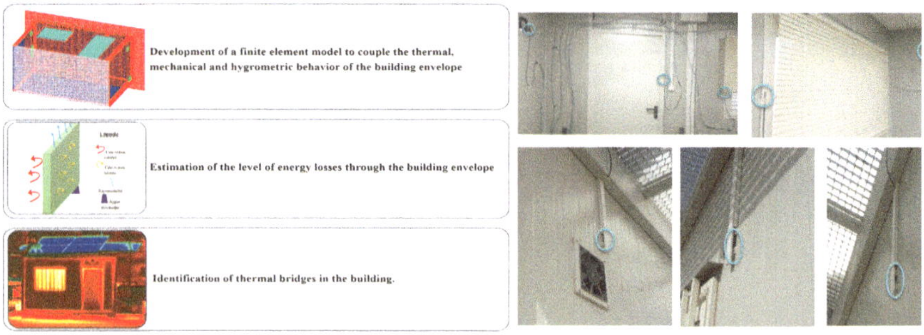

Fig. 11 Thermal treatment of the front wall and the temperature sensors positioning [61]

Emerging devices, which use natural forces without using electricity, such as natural lighting, room relocation, natural ventilation, could be used to increase insulation (Fig. 11). In fact, the architecture design, buildings envelope, and orientations can influence energy reduction. Therefore, the passive design must be considered in the phase of construction in order to reduce the final energy use of the building.

As part of our studies, a work is realized by focusing on the thermo-mechanical characterization of our EEBLab, which mainly consists of galvanized steel, of which expanded polyurethane is injected into the walls and the roof. As well as two types of internal insulation are adapted, namely chipboard for the floor and polyurethane for the roof. The main aim is to thermally study the behavior of the EEBLab, in order to propose good materiel for the insulation and consequently minimizing the use of the HVAC system for heating and cooling [61].

5.3 *Active/Passive Equipment Control*

This part concerns the MG system active layer (Fig. 6) that allows the deployment of context-driven control approaches in order to improve energy consumption. In this layer, the electrical energy can be minimized by optimizing the operation times of the active equipment (e.g., HVAC, ventilation systems), while maintaining occupants' comfort within a good air quality and suitable thermal comfort [62, 63]. In fact, the ventilation systems are normally installed in buildings to improve the air quality by injecting fresher air from outside into inside buildings. These systems automatically act on behalf of occupants by ensuring good indoor air quality, especially in cold or hot periods, or when there are no windows. In fact, the ventilation controller performs this task by adjusting fresh air as much as needed based on actual indoor CO_2 concentration. The aim is to improve the optimal balance between energy efficiency and indoor air quality. For that, a ventilation control system was deployed, as presented in Fig. 12, which maintains the indoor CO_2 concentration at the comfort set-point with an efficient and minimal ventilation rate and energy consumption [64, 65].

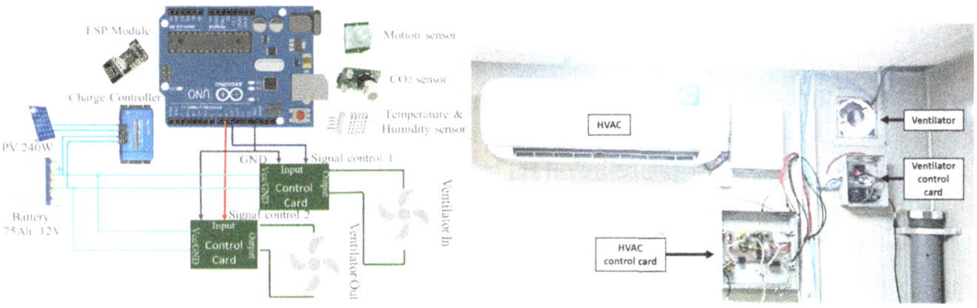

Fig. 12 Context-driven control card and active equipment control [65]

Another study puts more emphasis on developing efficient control approaches in order to deliver acceptable occupants' comfort while maintaining optimal energy consumption. Control approaches are investigated for controlling the deployed HVAC system in our EEBLab [66]. A control card is deployed as illustrated in Fig. 13 in order to interface between all HVAC components and the control device. It allows the regulation of temperature and ON/OFF control of the HVAC system by adjusting the inside ventilator and the compressor based on the desired schedules (heating, conditioning, or only ventilation). The deployed IoT/Big-Data platform is used to measure the hourly electricity consumption of the HVAC system, which is used in our study as a load to test the EM control strategies.

Generally, the HVAC is the most used system for thermal comfort regulation in buildings and is considered as the highest electricity consumer. For that, renewable sources of thermal energy are required to minimize the electricity mainly used for heating, cooling, and air conditioning. In this perspective, we have deployed a geothermal platform [67], an earth-to-air heat exchanger system that could be used for building cooling and heating. In fact, this clean and sustainable source can be deployed and used to minimize the usage of HVAC systems. As illustrated in Fig. 13,

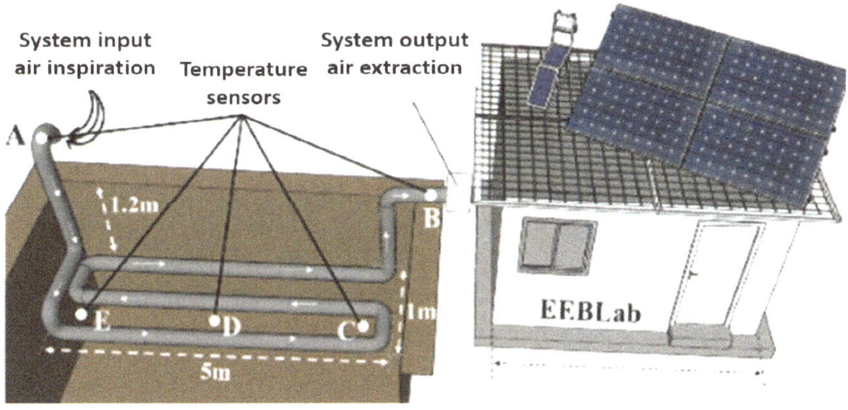

Fig. 13 Geothermal installation in the EEBLab [67]

the system is installed to extract heat from the ground for either cooling or heating purposes. It is basically a buried pipe, deployed at a certain depth in the ground, where air exchanges heat with soil. This system is deployed in the side area of our EEBLab and the pipelines are installed inside a trench of 5 m in length, 2 m width, and 1.5 depth. Different sensors are installed as well in order to measure and control the temperature exchange in order to investigate the performance and the effectiveness of the system in terms of power consumption and comfort.

5.4 RES Integration and Storage Devices

The work focused on this layer (Fig. 6) concerns the deployment of control strategies for EM. After the deployment of the whole components of the MG system, this simple hybrid system, however, needs to be automatically controlled accordingly. In fact, D/R control approaches are therefore required for balancing the intermittent RES generation and the delay might occur between the power production and the actual building's consumption. The main aim is to develop a control card to test the different studied control strategies for EM. Unlike existing systems, which are used as a black-box to collect and manage the energy in MG, the deployed control card allows us to measure, monitor, manage and deploy our algorithms [68, 69]. In fact, the developed card can be seen as an embedded EM system for optimal energy usage according to the actual context (Fig. 14). Therefore, different objective functions can be taken into account when optimizing and designing a control strategy, like the smoothing of the production, the continuity of the power generated to the consumer, the energy cost, and the charge/discharge cycle of the batteries [70]. For that, a control strategy should be deployed to satisfy the constraints designed by the optimization functions. The main communication infrastructure is employed for total energy measurement and management purposes. This infrastructure provides the autonomous operation with the required measurements, decisions, and controls by collecting data through the sensors and producing the commands for the Hw/Sw card, which is connected to the control switches used in the hybrid system [71].

However, a set of current and voltage sensors is installed for power measurement, as shown in Fig. 14. The system contains actuators controlled by an Arduino, which allows collecting the data from different sensors. Furthermore, the system contains a micro-computer (Raspberry pi) for collecting data from different sensors. The sensors transmit analog signals to the microcontroller, which converts them into numerical data. For example, a voltage sensor is used to measure the output PV voltage with an accurate range, which varies from zero to 140 V. For that, a tension divider bridge is used to convert the values from zero to 5 V, which is the Arduino accurate range. In fact, the Arduino program converts obtained values to tension data. Moreover, the Arduino transmits these data to the Raspberry for activating the right action according to the deployed control algorithm. Data are then transmitted to the IoT/Big-data platform for visualization, storage, and further data analytics.

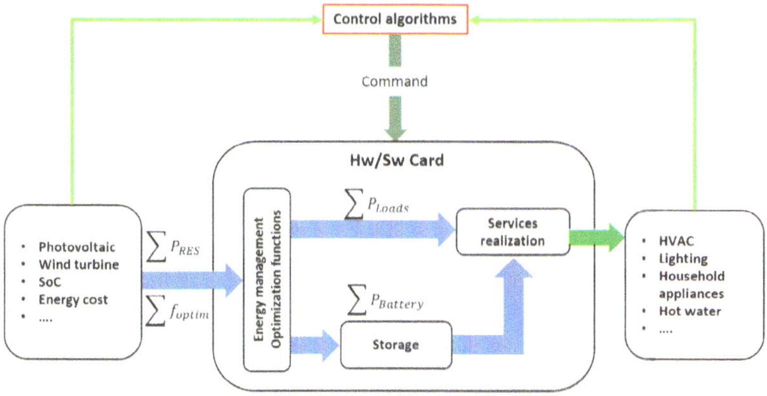

a)

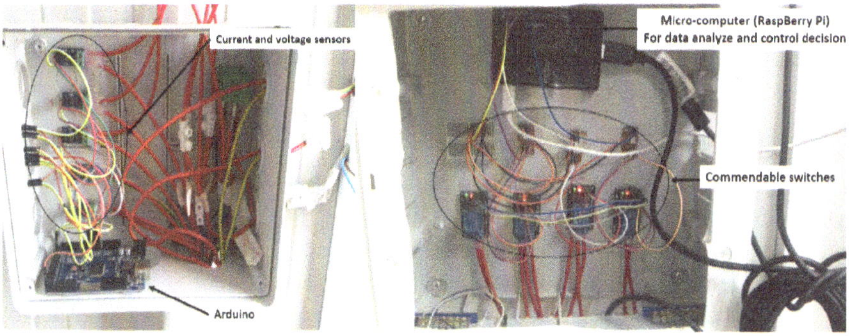

b)

Fig. 14 **a** Schematic view of the control card, **b** the deployed Hw/Sw control card [71]

A case study is presented in which the developed control card and the IoT/Big-Data platform are used to measure and store the data collected by the deployed current and voltage sensors. As shown in Fig. 15, the green curve presents the power generated from a PV panel for 24 h. This power is calculated by measuring the PV current and voltage variability during the day, which depends on the weather conditions (e.g., temperature, irradiance) changeability. At the same time, the battery SoC is calculated using our battery characterization system installed in the MG system. Moreover, the power consumption is measured and stored for the same period. These parameters are the main input for the EM strategy.

Furthermore, as described above, the use of storage devices in the MG system is motivated by the intermittent nature of RESs and the need to regulate the power quality (e.g., frequency, voltage) generated by these generators. The main aim is to store the surplus of the produced power during the peak production for possible usage when there is no production and keeping, at the same time, a maximum state of the health for the storage devices. Therefore, a set of batteries are installed in

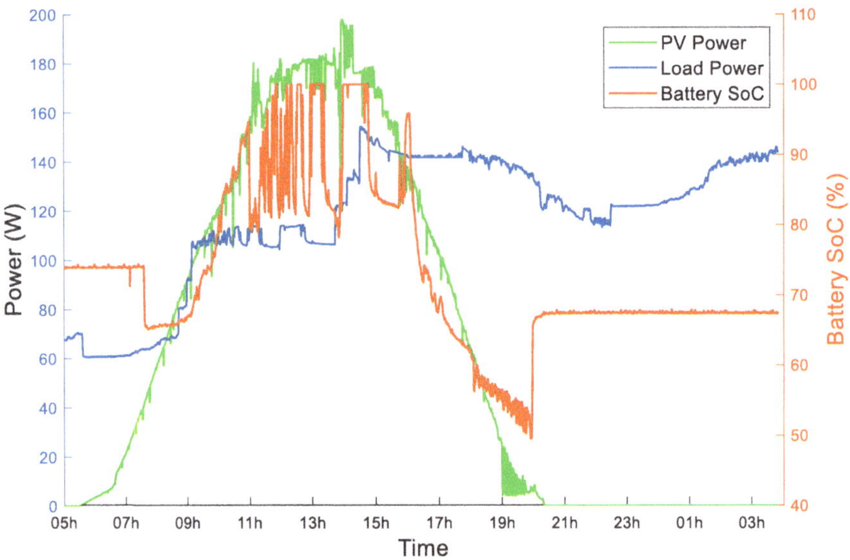

Fig. 15 Power measurement scenario in our deployed MG system platform

our deployed MG system due to their benefits (e.g., fast response, modularity, and good energy efficiency). For that, a platform to study, model, and experiment the batteries is required. This platform offers the possibility to monitor the used battery for better SoC estimation. These parameters are required to develop and deploy control strategies for EM in MG systems. In fact, a battery model is designed to estimate and predict the batteries' performance and behavior because the SoC is used as a critical parameter for our control strategy.

Electrical-circuit models (e.g., the first-order RC model, the second-order RC model) are commonly used for batteries' behavior estimation. These models are composed of a voltage source, resistors, and capacitors, which can simulate its dynamic behavior. They become more and more accurate when the model's order increases (i.e., RC networks). Moreover, for the accurate SoC estimation of the battery, several methods and algorithms are reported in the literature, such as the direct measurement methods, the artificial intelligence methods, and the model-based methods. The direct measurement methods (e.g., Coulomb counting method, Electrochemical method, open-circuit voltage method) use the dynamic measurement of the battery characteristics in order to estimate the battery's SoC. The artificial intelligence algorithms, such as the Neural Network and the Fuzzy logic, can also estimate the battery's SoC with more precision but they are more complex and difficult to deploy for embedded and real-time MG control. Mainly, in our deployed MG system, the model-based methods (e.g., Coulomb counting method, Sliding mode observer, Kalman filter) have been used to estimate the SoC of the battery with more precision and accuracy (Fig. 16) [72, 73].

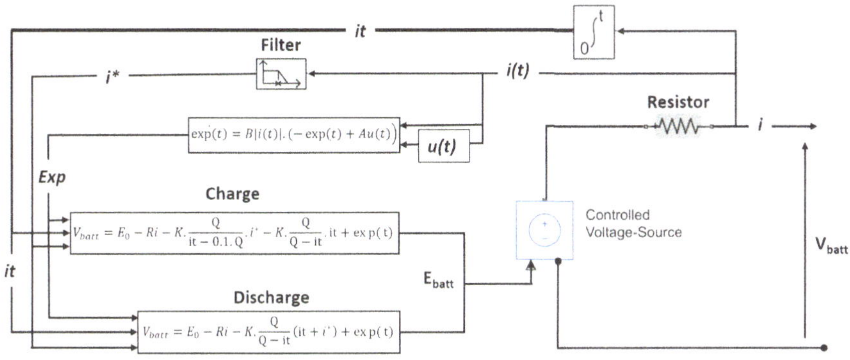

Fig. 16 The deployed battery modeling

In order to determine the battery characteristics, an instrumentation platform is first developed using recent sensing/actuating equipment for gathering important battery's parameters, which are then used for building a model for the battery deployed in our EEBLab (Fig. 17). It is composed of a Lead-acid battery and a set of sensors to extract the battery's voltage and current. The sensors are connected to an acquisition board (e.g., Arduino) used to collect the data, and then send them to a cluster for processing and storage. The developed platform provides other information about the estimation of the battery's SoC by the Coulomb Counting method.

After validating the battery model, it is integrated into our MG for simulations and experiments. As shown in Fig. 18, the blue curve presents the battery SoC variability estimated using the measured battery voltage (orange curve). During this scenario, the battery charge/discharge current and voltage are measured and collected using our deployed IoT/Big-Data platform. The measured parameters are used to estimate

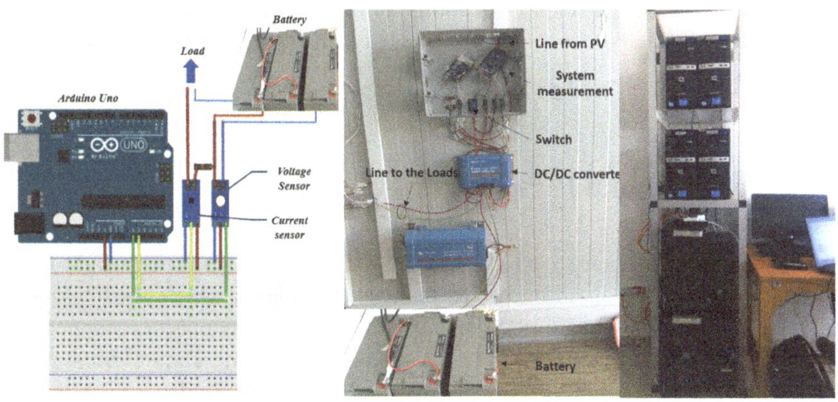

Fig. 17 Battery characterization system [72]

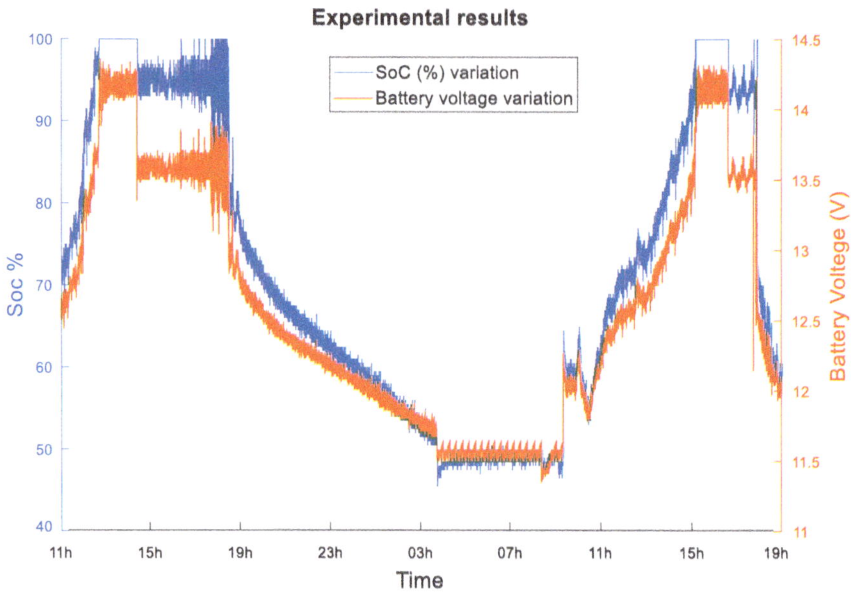

Fig. 18 The battery SoC estimation using the measured voltage by our deployed platform

the SoC variability. Therefore, the SoC is a key parameter for the EM strategies in the MG systems. In fact, as depicted in Fig. 18, from 11:00 AM to around 07:00 PM, the battery is charged by the surplus generated from the RESs. During the night, from 07:00 PM to around 04:00 AM, the battery generates the power to the load because the PV generation is unavailable. However, the battery is at rest from 04:00 AM to around 08:00 AM because the SoC reaches the regulated threshold value, which is fixed by the EM strategy, in order to avoid a deep-discharge of the battery.

Therefore, the proposed IoT/Big-Data platform could be used to measure different parameters in the MG system. Depending on the studied scenario, suitable sensors are selected and can be connected to this platform for data collection, monitoring, and processing.

6 Conclusions and Perspectives

The main aim of the work presented in this chapter is to shed more light on the usefulness of developing an integrated platform in order to enable the deployment of smart MG systems in energy-efficient buildings. The MG platform connects the building's components using sensing/actuating, IoT, and Big Data technologies in order to leverage real-time gathering, data processing, and predictive control. The platform was deployed and several scenarios have been tested and evaluated and preliminary results showed the usefulness of the platform for efficient management of buildings components. The platform will be further enhanced by developing other ongoing scenarios. It will be used for validating the proposed models and results mainly by investigating, (i) the efficient connection, integration, and the management of different RES and storage devices, (ii) the suitable dimensions for energy production and storage devises, (iii) different possible demands/responses and predictive algorithms, (iv) charged and discharged operations on the state-of-health of deployed batteries as well as PV corrosions fault diagnosis, (v) context-aware driven control of deployed equipment, e.g., lighting and HVAC systems. Methods that allow smart management with predictive analytics are still need to be integrated into the platform prototype to handle this type of complex systems. This paves the way to approaches in which an antifragile platform learns and adapts which strategy/action to enact. We envision that future ambient control systems (ACS) will require more and more intelligence as well as the ability to monitor and learn from the experiences, thus realizing an antifragile ACS. Future work shall investigate how to practically realize such an ACS in energy-efficient buildings [51].

Acknowledgements This work is supported by MIGRID project (grant 5-398, 2017-2020), which is funded by USAID under the PEER program. It is also partially supported by HOLSYS project, which is funded by IRESEN (2020-2022).

References

1. International Energy Agency (IEA) (2019) Available at https://www.iea.org/reports/world-energy-outlook-2019/electricity#abstract. Accessed 10 Jan 2020
2. Lee S, Karava P (2020) Towards smart buildings with self-tuned indoor thermal environments—a critical review. Energy Build 110172
3. Moroşan PD, Bourdais R, Dumur D, Buisson J (2010) Building temperature regulation using a distributed model predictive control. Energy Build 42(9):1445–1452
4. Godina R, Rodrigues EM, Pouresmaeil E, Matias JC, Catalão JP (2018) Model predictive control home EM and optimization strategy with demand response. Appl Sci 8(3):408
5. Georgakarakos AD, Mayfield M, Hathway EA (2018) Battery storage systems in smart grid optimised buildings. Energy Proc 151:23–30
6. Buckman AH, Mayfield M, Beck SB (2014) What is a smart building? Smart Sustain Built Environ
7. Llorente IM (2012) Key challenges in cloud computing to enable future internet of things. In: The 4th EU-Japan symposium on new generation networks and future internet
8. Ali AS (ed) (2013) Smart grids: opportunities, developments, and trends. Springer
9. Hatziargyriou N (2014) Microgrids: architectures and control. Wiley-IEEE Press
10. Eltamaly AM, Mohamed MA, Alolah AI (2016) A novel smart grid theory for optimal sizing of hybrid renewable energy systems. Sol Energy 124:26–38
11. Wang K, Hu X, Li H, Li P, Zeng D, Guo S (2017) A survey on energy internet communications for sustainability. IEEE Trans Sustain Comput 2(3):231–254
12. Asaad M, Ahmad F, Alam MS, Sarfraz M (2019) Smart grid and Indian experience: a review. Resour Policy, 101499
13. Wang K, Yu J, Yu Y, Qian Y, Zeng D, Guo S, Wu J (2017) A survey on energy internet: Architecture, approach, and emerging technologies. IEEE Syst J 12(3):2403–2416

14. Tsiatsis V, Karnouskos S, Holler J, Boyle D, Mulligan C (2018) Internet of things: technologies and applications for a new age of intelligence. Academic Press
15. Kanchev H (2014) Gestion des flux énergétiques dans un système hybride de sources d'énergie renouvelable: Optimisation de la planification opérationnelle et ajustement d'un micro réseau électrique urbain. Doctoral dissertation, Ecole centrale de Lille
16. Hatziargyriou N, Asano H, Iravani R, Marnay C (2007) An overview of ongoing research, development, and demonstration projects. IEEE Power Energy Mag 5(4):79–94
17. European Commission (2020b) Available at https://microgrid-symposiums.org/microgrid-examples-and-demonstrations/mvv-mannheim-wallstadt-microgrid/. Accessed 25 Apr 2020
18. Microgrid Knowledge (2020) Microgrids "shining light" for US in World energy markets: report. Available at https://microgridknowledge.com/microgrid-market-aee-report/. Accessed 12 May 2020
19. Consortium for Electric Reliability Technology Solutions (2020) Available at http://certs.lbl.gov. Accessed 1 May 2020
20. Renewable Energy Institute, Agora Energiewende (2018) Integrating renewables into the Japanese power grid by 2030. Study on behalf of Renewable Energy Institute and Agora Energiewende
21. Guimaraes L (ed) (2020) The regulation and policy of Latin American energy transitions. Elsevier Science
22. Zhang Y, Gatsis N, Giannakis GB (2013) Robust EM for microgrids with high-penetration renewables. IEEE Trans Sustain Energy 4(4):944–953
23. Dagdougui H, Ouammi A, Sacile R (2017) Towards a concept of cooperating power network for EM and control of microgrids. Microgrid, pp 231–262. Butterworth-Heinemann
24. Kumar D, Zare F, Ghosh A (2017) DC microgrid technology: system architectures, AC grid interfaces, grounding schemes, power quality, communication networks, applications, and standardizations aspects. IEEE Access 5:12230–12256
25. Khaled U, Eltamaly AM, Beroual A (2017) Optimal power flow using particle swarm optimization of renewable hybrid distributed generation. Energies 10(7):1013
26. Chauhan RK, Chauhan K (eds) (2019) Distributed energy resources in microgrids: integration, challenges and optimization. Academic Press, London
27. Bridier L (2016) Modélisation et optimisation d'un système de stockage couplé à une production électrique renouvelable intermittente. Doctoral dissertation
28. Abbes D (2012) Contribution au dimensionnement et à l'optimisation des systèmes hybrides éoliens-photovoltaïques avec batteries pour l'habitat résidentiel autonome. Ecole Nationale Supérieure d'Ingénieurs-Poitiers
29. Mekontso C, Abubakar A, Madugu S, Ibrahim O, Adediran YA (2019) Review of optimization techniques for sizing renewable energy systems. Comput Eng Appl J 8(1):13–30
30. Aristizábal AJ, Habib A, Ospina D, Castaneda M, Zapata S, Banguero E (2019) RenPower: software for sizing renewable energy microgrids for academic teaching. In: AIP conference proceedings, vol 2123, no 1, p 020011. AIP Publishing LLC
31. Al-Ghussain L, Samu R, Taylan O, Fahrioglu M (2020) Sizing renewable energy systems with energy storage systems in microgrids for maximum cost-efficient utilization of renewable energy resources. Sustain Cities Soc 55:102059
32. Nichols JS, Lasseter RH, Eto JH, Vollkommer HT (2006) Validation of the CERTS microgrid concept the CEC/CERTS microgrid testbed. In: 2006 IEEE Power Engineering Society General Meeting, pp 1–3
33. Bellido M, Rosa L, Pereira M, Falcao D, Ribeiro S (2018) Barriers, challenges and opportunities for microgrid implementation: the case of Federal University of Rio de Janeiro. J Cleaner Prod 188:203–216
34. IEEE (2011) IEEE Std 1547.4-2011: guide for design, operation, and integration of distributed resource Island systems with electric power systems
35. Hirsch A, Paraga Y, Guerrero J (2018) Microgrids: a review of technologies, key drivers, and outstanding issues. Renew Sustain Energy Rev J 90:402–411

36. Castro MAL (2020) Urban microgrids: benefits, challenges, and business models. In: The regulation and policy of Latin American energy transitions, pp 153–172. Elsevier
37. Energy Networks Australia (2020) Behind the News: network reliability. Available at https://www.energynetworks.com.au/news/energy-insider/behind-the-news-network-reliability/. Accessed 10 May 2020
38. International Electro-technical Commission (2014) Micro grids for disaster preparedness and recovery: with electricity continuity plans and systems. IEC, Geneva Switzerland, pp 41–45
39. Veneri O (ed) (2017) Technologies and applications for smart charging of electric and plug-in hybrid vehicles. Springer
40. Basu K, Guillame-Bert M, Joumaa H, Ploix S, Crowley J (2011) Predicting home service demands from appliance usage data. In: International conference on information and communication technologies and applications ICTA
41. Birleanu FG, Bizon N (2020) Control and protection of the smart microgrids using internet of things: technologies, architecture and applications. In: Microgrid architectures, control and protection methods, pp 749–770. Springer, Cham
42. Basso T (2014) IEEE 1547 and 2030 standards for distributed energy resources interconnection and interoperability with the electricity grid (No. NREL/TP-5D00-63157). National Renewable Energy Lab.(NREL), Golden, CO
43. IEEE Standards Association (2020) Working Group Site & Liaison Index. Available at http://grouper.ieee.org/groups/scc21/index.html. Accessed 10 May 2020
44. Gaiceanu M, Arama IN, Ghenea I (2020) DC microgrid control. In: Microgrid architectures, control and protection methods, pp 357–380. Springer, Cham
45. Moussa S, Ghorbal MJB, Slama-Belkhodja I (2019) Bus voltage level choice for standalone residential DC nanogrid. Sustain Cities Soc 46:101431
46. Yamashita DY, Vechiu I, Gaubert JP (2020) A review of hierarchical control for building microgrids. Renew Sustain Energy Rev 118:109523
47. BUILD UP (2020) The European portal for energy efficiency in buildings. Available at https://www.buildup.eu/en/practices/publications/re-thinking-2050-100-renewable-energy-vision-european-union. Accessed 10 May 2020
48. Hannan MA, Tan SY, Al-Shetwi AQ, Jern KP, Begum RA (2020) Optimised controller for renewable energy sources integration into microgrid: functions, constraints and suggestions. J Cleaner Product, p 120419
49. European renewable energy council (EREC) (2020) Intelligent energy Europe. Available at https://ec.europa.eu/energy/intelligent/projects/en/partners/european-renewable-energy-council-1. Accessed 10 May 2020
50. IEEE Standards Coordinating Committee 21 (SCC21) (2020) Fuel cells, photovoltaics, dispersed generation, and energy storage. Available at https://site.ieee.org/sagroups-scc21/standards/. Accessed 10 May 2020
51. Bakhouya M, NaitMalek Y, Elmouatamid A, Lachhab F, Berouine A, Boulmrharj S, Elkamoune N (2017) Towards a context-driven platform using IoT and big data technologies for energy efficient buildings. In: 2017 3rd international conference of cloud computing technologies and applications (CloudTech), pp 1–5. IEEE
52. Lachhab F, Bakhouya M, Ouladsine R, Essaaidi M (2017) Energy-efficient buildings as complex socio-technical systems: approaches and challenges. In: Advances in complex societal, environmental and engineered systems, pp 247–265. Springer, Cham
53. De Florio V, Bakhouya M, Coronato A, Di Marzo G (2013) Models and concepts for socio-technical complex systems: towards fractal social organizations. Syst Res Behav Sci 30(6):750–772
54. Elkhoukhi H, NaitMalek Y, Berouine A, Bakhouya M, Elouadghiri D, Essaaidi M (2018) Towards a real-time occupancy detection approach for smart buildings. Proc Comput Sci 134:114–120
55. ElMouatamid A (2020) MAPCAST: an adaptive control approach using predictive analytics for energy balance in micro-grid systems. Int J Renew Energy Res (IJRER) 10(2):945–954

56. Elkhoukhi H, NaitMalek Y, Bakhouya M, Berouine A, Kharbouch A, Lachhab F, Essaaidi M (2019) A platform architecture for occupancy detection using stream processing and machine learning approaches. Concurren Comput Pract Exp, e5651.
57. Hadri S, Naitmalek Y, Najib M, Bakhouya M, Fakhri Y, Elaroussi M (2019) A comparative study of predictive approaches for load forecasting in smart buildings. Proc Comput Sci 160:173–180
58. Kharbouch A, Bakhouya M, Maakoul AE, Ouadghiri DE (2019) A holistic approach for heating and ventilation control in EEBs. In: Proceedings of the 17th international conference on advances in mobile computing & multimedia, pp 236–241
59. Kaa, IoT technique (2020) Available at https://www.kaaproject.org/. Accessed 23 May 2020
60. Storm, Apache Storm (2020) Available at http://storm.apache.org/. Accessed 23 May 2020
61. Berrabah S, Moussa MO, Bakhouya M (2020) Towards a thermo-mechanical characterization approach of buildings' envelope. Energy Reports 6:240–245
62. Lachhab F, Bakhouya M, Ouladsine R, Essaaidi M (2017) Monitoring and controlling buildings indoor air quality using WSN-based technologies. In: 2017 4th international conference on control, decision and information technologies (CoDIT), pp 0696–0701. IEEE
63. Berouine A, Akssas E, Naitmalek Y, Lachhab F, Bakhouya M, Ouladsine R, Essaaidi M (2019) A fuzzy logic-based approach for HVAC systems control. In: 2019 6th international conference on control, decision and information technologies (CoDIT), pp 1510–1515. IEEE
64. Lachhab F, Bakhouya M, Ouladsine R, Essaaidi M (2018) Towards an intelligent approach for ventilation systems control using IoT and big data technologies. Proc Comput Sci 130:926–931
65. Berouine RO, Bakhouya M, Essaaidi M (2020) Towards a real-time predictive management approach of indoor air quality in energy efficient buildings. Energies J
66. Lachhab F, Ouladsine R, Bakhouya M, Essaaidi M (2017) An energy-efficient approach for controlling heating and air-conditioning systems. In: 2017 international renewable and sustainable energy conference (IRSEC), pp 1–7. IEEE
67. Kharbouch A, El Maakoul A, Bakhouya M, El Ouadghiri D (2018) Modeling and performance evaluation of an air-soil exchange system in energy efficient buildings. In: 2018 6th international renewable and sustainable energy conference (IRSEC), pp 1–6. IEEE
68. El Mouatamid A, Ouladsine R, Bakhouya M, Felix V, Elkamoun N, Zine-Dine K, Abid R (2017) Modeling and performance evaluation of photovoltaic systems. In: 2017 International renewable and sustainable energy conference (IRSEC), pp 1–7. IEEE
69. Elmouatamid A, Bakhouya M, Ouladsine R, Zine-Dine K, Khaidar M, Abid R (2018) Deployment and experimental evaluation of micro-grid systems. In: 2018 6th international renewable and sustainable energy conference (IRSEC), pp 1–6. IEEE
70. Elmouatamid A, Ouladsine R, Bakhouya M, El Kamoun N, Zine-Dine K, Khaidar M (2019) A model predictive control approach for EM in micro-grid systems. In: 2019 international conference on smart energy systems and technologies (SEST), pp 1–6. IEEE
71. Elmouatamid A, NaitMalek Y, Bakhouya M, Ouladsine R, Elkamoun N, Zine-Dine K, Khaidar M (2019) An EM platform for micro-grid systems using Internet of Things and Big-data technologies. Proc Instit Mech Eng Part I J Syst Control Eng 233(7):904–917
72. Naitmalek Y, Najib M, Bakhouya M, Essaaidi M (2019) Forecasting the state-of-charge of batteries in micro-grid systems. In: 2019 4th world conference on complex systems (WCCS), pp 1–6. IEEE
73. NaitMalek Y, Najib M, Bakhouya M, Essaaidi M (2019) On the use of machine learning for state-of-charge forecasting in electric vehicles. In: 2019 IEEE international smart cities conference (ISC2), pp 408–413. IEEE

The manufacturer's authorised representative in the EU is Springer Nature Customer Service Centre GmbH, Europaplatz 3, 69115 Heidelberg, Germany. If you have any concerns regarding our products, please contact ProductSafety@springernature.com

Printed and bound by CPI Group (UK) Ltd, Croydon, CR0 4YY
30/09/2026
02225295-0018